HANDBOOK OF MODERN PHARMACEUTICAL ANALYSIS
2nd Edition

Volume 10 of

SEPARATION SCIENCE AND TECHNOLOGY

A reference series edited by Satinder Ahuja

HANDBOOK OF MODERN PHARMACEUTICAL ANALYSIS 2nd Edition

Edited by

Satinder Ahuja

Ahuja Consulting
Calabash, NC, USA

Stephen Scypinski

Analytical Research & Development, Bristol-Myers Squibb Company
New Brunswick, NJ, USA

ELSEVIER

AMSTERDAM • BOSTON • HEIDELBERG • LONDON
NEW YORK • OXFORD • PARIS • SAN DIEGO
SAN FRANCISCO • SINGAPORE • SYDNEY • TOKYO
Academic Press is an imprint of Elsevier

Academic Press is an imprint of Elsevier
30 Corporate Drive, Suite 400, Burlington, MA 01803, USA
525 B Street, Suite 1900, San Diego, CA 92101-4495, USA
84 Theobald's Road, London WC1X 8RR, UK
Radarweg 29, PO Box 211, 1000 AE Amsterdam, The Netherlands

First edition 2001
Second edition 2011

Library of Congress Cataloging-in-Publication Data

A catalog record for this book is available from the Library of Congress

British Library Cataloguing in Publication Data

A catalogue record for this book is available from the British Library

ISBN: 978-0-12-375680-0

ISSN: 0149-6395

For information on all Academic Press publications visit our website at www.books.elsevier.com

Printed and bound in The USA

11 12 10 9 8 7 6 5 4 3 2 1

CONTENTS

6. Novel Dosage Form Analysis

JOSEPH ETSE

7. Method Development for Pharmaceutical Chiral Chromatography

GREGORY K. WEBSTER AND LAILA KOTT

8. Characterization and Analysis of Biopharmaceutical Proteins

YIN LUO, TATJANA MATEJIC, CHEE-KENG NG, BRIAN NUNNALLY, THOMAS PORTER, STEPHEN RASO, JASON ROUSE, TANYA SHANG, AND JOHN STECKERT

9. Preformulation and Early Phase Method Development

KEVIN COLEMAN BYNUM

10. In-Process Control Testing

CHRISTOPHER WOOD

11. Validation of Analytical Test Methods

ALISON E. BRETNALL AND GRAHAM S. CLARKE

12. Stability Studies

JESSICA CHA, TIMOTHY GILMOR, PHILIP LANE, AND JOSEPH S. RANWEILER

13. Analytical Methodology Transfer

STEPHEN SCYPINSKI AND JOEL YOUNG

14. Pharmaceutical Analysis Documentation

HITESH CHOKSHI, KATHLEEN SCHOSTACK, AND RICHARD STEINBACH

PREFACE

The first edition of the *Handbook of Modern Pharmaceutical Analysis* addressed a need for a text that highlighted the importance of analytical chemistry during the entire life cycle of a pharmaceutical drug product, from discovery of an active molecule, through the various phases of clinical development, product registration, and technical transfer from R&D to the quality and manufacturing arena. We are highly gratified that the text has been widely embraced over the past nine years, and because of the process-oriented arrangement of the subject material, it has remained fairly current in that regard.

The pharmaceutical industry has continued to evolve. New regulatory guidelines have had their effect on all facets of drug development and commercialization. The socioeconomic trends in the emerging markets of the world have also changed the development paradigm, as companies have more carefully considered the demand for new medicines to treat unmet medical needs in these markets. The routine development of biopharmaceutical proteins has increased dramatically over the past several years. Finally, technology has played a major role in altering the business of drug development. More sophisticated instrumentation with higher sensitivity has allowed the routine quantification of trace and ultratrace level impurities in both active pharmaceutical ingredients and dosage forms. Spectroscopic techniques with the capability of monitoring synthetic reactions and drug product processing steps have led to increased understanding of such operations. And the trend toward electronic and paperless systems in every area of the industry, from laboratory notebooks to regulatory filings, has caused a reexamination of processes based on the usage of paper.

Careful considerations of the changes in the industry have influenced the structure and content of this edition. The chapters repeated in this text from the first edition are still relevant today and have been updated in terms of literature references; and they cite any pertinent changes to regulatory authority guidance documents. The chapter covering dissolution adds a focus on modeling, while the stability studies section has been enlarged in scope to consider requirements of countries other than the United States. The chapter on analysis of novel drug delivery systems has also been expanded to reflect the variety of new approaches being considered to increase bioequivalence, especially in the case of low-solubility active ingredients. Given the increased scrutiny by the authorities and quality of in-process control methods, a separate chapter devoted to in-process testing has been added. The chapter on technical transfer pays more attention to the requirements of "rest of the world" countries.

Notable additions to the text by way of new chapters include the important subjects of quality by design (QbD) and process analytical technology (PAT), genotoxic impurities (GTIs), method development of chiral compounds, and characterization and analysis of biopharmaceutical proteins. QbD has been the subject of numerous meetings, seminars, and publications over the past several years. Embracement of the QbD paradigm relies heavily on the movement away from empirical thinking to one that puts an increased emphasis on modeling and PAT. A greater understanding of the relevant guidance documents is necessary to allow an analyst (working with process development) to implement a sound control strategy that can be defended with the regulatory authorities. It is expected that QbD and PAT will play an ever-increasing role in pharmaceutical development and testing in the years to come.

The discussion of GTIs incorporates the regulatory requirements as well as analytical approaches for the determination of these analytes. It is felt that a better background in the reality of the expectations of the authorities is required for a pharmaceutical analyst to make the best choices with regard to techniques. Greater knowledge of the guidance documents will encourage informed decisions. The importance of chiral compounds both in terms of pharmacological and toxicological activity cannot be overstated. Chiral methods present unique challenges in that the molecules have the same physical and chemical properties, except their optical activity. Readers will find this chapter fascinating. Given the increased importance of biomolecules, a separate chapter has been devoted to the analytical challenges associated with the characterization and analysis of biopharmaceutical proteins.

We believe that the additions and changes made in the second edition of *The Handbook of Modern Pharmaceutical Analysis* will enhance its stance as a useful and informative text for those engaged in pharmaceutical drug discovery, development, manufacturing, and quality control and assurance.

Satinder Ahuja
Stephen Scypinski

CONTRIBUTORS

Numbers in parentheses indicate the page on which the authors' contributions begin.

Satinder Ahuja (1) Ahuja Consulting, Calabash, North Carolina, USA

Karen M. Alsante (59) Pfizer Global Research & Development Division, Research Analytical Department, Groton, Connecticut, USA

Steven W. Baertschi (59) Eli Lilly and Company, Lilly Research Laboratories, Indianapolis, Indiana, USA

Robert W. Bondi, Jr. (195) Mylan School of Pharmacy, Duquesne University, Pittsburgh, Pennsylvania, USA

Alison E. Bretnall (429) Bristol-Myers Squibb, Moreton, UK

Harry G. Brittain (11) Center for Pharmaceutical Physics, Milford, New Jersey, USA

Kevin Coleman Bynum (361) Boehringer Ingelheim Pharmaceuticals, Inc., Ridgefield, Connecticut, USA

Jessica Cha (459) Johnson & Johnson Pharmaceutical Research and Development, Raritan, New Jersey, USA

Hitesh Chokshi (527) Hoffmann-La Roche, Inc., Nutley, New Jersey, USA

Graham S. Clarke (429) Bristol-Myers Squibb, Moreton, UK

Michael Coutant (59) Pfizer Global Research & Development Division, Research Analytical Department, Groton, Connecticut, USA

James K. Drennen, III (195) Mylan School of Pharmacy, Duquesne University, Pittsburgh, Pennsylvania, USA

Joseph Etse (225) Pharmaceutical and Analytical Development, Novartis Pharmaceuticals Corporation, East Hanover, New Jersey, USA

Timothy Gilmor (459) Johnson & Johnson Pharmaceutical Research and Development, Raritan, New Jersey, USA

Laila Kott (251) Analytical Development Small Molecules, Millennium Pharmaceuticals, Cambridge, Massachusetts, USA

Philip Lane (459) Johnson & Johnson Pharmaceutical Research and Development, Raritan, New Jersey, USA

Yin Luo (283) Analytical Research & Development, BioTherapeutics Pharmaceutical Sciences, Pfizer, Andover, Massachusetts, USA

Brian L. Marquez (59) Pfizer Global Research & Development Division, Analytical Research and Development Department, Groton, Connecticut, USA

Tatjana Matejic (283) Analytical Research & Development, BioTherapeutics Pharmaceutical Sciences, Pfizer, Andover, Massachusetts, USA

Scott A. Miller (171) Bristol-Myers Squibb Co., Analytical Research and Development, New Brunswick, New Jersey

Chee-Keng Ng (283) Analytical Research & Development, BioTherapeutics Pharmaceutical Sciences, Pfizer, Andover, Massachusetts, USA

Brian Nunnally (283) Site Technical Services, Pfizer Global Manufacturing, Pfizer, Sanford, North Carolina, USA

Rodney L. Parsons (171) Bristol-Myers Squibb Co., Process Research and Development, New Brunswick, New Jersey

Thomas Porter (283) Analytical Research & Development, BioTherapeutics Pharmaceutical Sciences, Pfizer, Andover, Massachusetts, USA

Joseph S. Ranweiler (459) Johnson & Johnson Process Research and Development, Raritan, New Jersey, USA

Stephen Raso (283) Analytical Research & Development, BioTherapeutics Pharmaceutical Sciences, Pfizer, Andover, Massachusetts, USA

Jason Rouse (283) Analytical Research & Development, BioTherapeutics Pharmaceutical Sciences, Pfizer, Andover, Massachusetts, USA

Kathleen Schostack (527) Hoffmann-La Roche, Inc., Nutley, New Jersey, USA

Stephen Scypinski (507) Analytical Research & Development, Bristol-Myers Squibb Company, New Brunswick, New Jersey, USA

Tanya Shang (283) Analytical Research & Development, BioTherapeutics Pharmaceutical Sciences, Pfizer, Andover, Massachusetts, USA

Thomas R. Sharp (59) Pfizer Global Research & Development Division, Research Analytical Department, Groton, Connecticut, USA

John Steckert (283) Analytical Research & Development, BioTherapeutics Pharmaceutical Sciences, Pfizer, Andover, Massachusetts, USA

Richard Steinbach (527) Hoffmann-La Roche, Inc., Nutley, New Jersey, USA

Gregory K. Webster (251) Global Analytical Research and Development, Abbott Laboratories, Abbott Park, Illinois, USA

Christopher Wood (397) Analytical Chemistry Research & Development, Bristol-Myers Squibb Pharmaceutical Development, New Brunswick, New Jersey, USA

Joel Young (507) Analytical Research & Development, Bristol-Myers Squibb Company, New Brunswick, New Jersey, USA

Todd C. Zelesky (59) Pfizer Global Research & Development Division, Analytical Research and Development Department, Groton, Connecticut, USA

1

OVERVIEW OF MODERN PHARMACEUTICAL ANALYSIS

SATINDER AHUJA

Ahuja Consulting, Calabash, North Carolina, USA

I. INTRODUCTION

Modern pharmaceutical analysis encompasses much more than analysis of active pharmaceutical ingredients (APIs), inert ingredients (excipients), or formulated drug product (DP). The primary goal of pharmaceutical analysis is to assure drug quality. It is well known that quality cannot be tested into a product; however, well-planned testing with suitable methodology and instrumentation can help build quality into a DP.[1] A thorough understanding of interactions of drug substances with excipients is necessary, especially when residual solvents (including moisture) are present. It is also essential to understand potential degradation reactions that may occur in the formulated product under various stress conditions that might be encountered during storage and in shipment of the final package. And dissolution tests should correlate well with the bioavailability. In short, the aim of pharmaceutical analysis is to help build and assure quality of DPs. Recognition of these principles has encouraged the Food and Drug Administration (FDA) to issue the new initiative quality by design (QbD) outlined in the May 2007 FDA report "Pharmaceutical Quality for the 21st Century: A Risk-Based Approach." The focus of this concept is that quality should be built into a product by means of a thorough understanding of the product and the process by which it is developed and manufactured, along with a knowledge of the risks involved in manufacturing the product and how best to mitigate those risks through continuous product improvement.

This second edition of *Handbook of Modern Pharmaceutical Analysis* addresses QbD and other new developments such as the best ways to monitor genotoxic impurities (GTIs), separations of chiral compounds, and characterization and analysis of biopharmaceutical proteins. The first edition published in 2001 has been a popular text among individuals in the pharmaceutical industry who are involved in drug development and the associated analytical function area. That book provided a journey through the drug development process, with emphasis on the role of analytical chemistry. It covered various separation methods such as thin layer chromatography (TLC), gas chromatography (GC), high-pressure or high-performance liquid chromatography (HPLC), and capillary electrophoresis (CE) that are commonly used in the pharmaceutical industry for the evaluation of a large variety of samples. These methods still remain the methods of choice for checking purity of new drug candidates, monitoring changes or scaleup of synthetic procedures, evaluating new formulations, and implementing quality control/assurance of final DPs. Hyphenated methods that combine a chromatographic method with spectroscopic methods such as mass spectrometry (MS) or nuclear magnetic resonance (NMR) have been found useful for characterizing impurities; these methods are fully covered in that text. The aim of this second edition is to supplement and complement the first edition and other complementary volumes on pharmaceutical analysis[2–4] and to provide a comprehensive text of modern pharmaceutical analysis that incorporates various relevant requirements and new initiatives to address DP quality. The key features of the second edition are listed below:

- Provides the latest analytical technology that allows greater sensitivity and selectivity.
- Includes methods for monitoring GTIs.
- Highlights implications for pharmaceutical analysis based on the new paradigm, "Quality by Design."
- Covers characterization and analysis of biopharmaceutical proteins, as these molecules are gaining significance.
- Provides up-to-date information on documentation in light of regulatory guidance, technology, and industry trends.

This book has been designed to be particularly useful for both novice and experienced method development chemists in the pharmaceutical industry who are seeking to update their knowledge. It should continue to serve as a definitive reference source on pharmaceutical analysis for researchers, analysts, managers, and pharmaceutical industry regulators.

II. SOLID-STATE STUDIES

A systematic approach to the physical characterization of pharmaceutical solids outlined in Chapter 2 serves as a useful pedagogical device for the classification of the many available methods of physical characterization. Within this system, physical properties are classified as being associated with the molecular level (those associated with individual molecules), the particulate level (those pertaining to individual solid particles), or the bulk level (those associated with an assembly of particulate species). One of the areas where the physical characterization of solids has become extremely important is in the study of polymorphs and solvatomorphs. The nature of the crystal structure adopted by a given compound upon crystallization exerts a profound effect on the solid-state properties of that system, and these variations can translate into significant differences in properties of pharmaceutical importance. It is now accepted that an evaluation of the polymorphism available to a drug substance must be thoroughly investigated in the early stages of development. The results of these studies must be included in the chemistry, manufacturing, and control section of a new drug application (NDA), and this information is required to demonstrate control over the manufacturing process.

III. DEGRADATION AND IMPURITIES STUDIES

Forced degradation studies are used to develop stability-indicating analytical methodology, to gain a better understanding of API and DP stability, and to provide information about degradation products and pathways. To fulfill development and regulatory needs, Chapter 3 provides a road map for performing these studies, helpful tools in designing rugged scientific studies, and guidance on how to record and communicate results. The primary goal of stress testing is efficiently producing degradation samples realistic of those formed during manufacture, handling, and normal storage conditions (as specified by the International Conference on Harmonization—ICH) of the API and DP. Overstressing can destroy degradation products (degradants) that are relevant or generates unrealistic degradants, whereas understressing may fail to generate important degradation products. From a regulatory perspective, forced degradation studies should provide data to support the following:

- Identification of possible degradants
- Degradation pathways and intrinsic stability of the drug molecule
- Validation of stability-indicating analytical procedures

Specific issues addressed in regulatory guidances related to forced degradation are provided by the US FDA and from private industry on regulatory requirements for investigational new drug (IND) and NDA filings. It should be noted that regulatory guidance does not specifically address issues such as exact experimental conditions for forced degradation studies (temperatures, duration, extent of degradation, etc.) and experimental design.

IV. ANALYTICAL CONSIDERATIONS FOR GENOTOXIC AND OTHER IMPURITIES

The assessment and control of GTIs in pharmaceutical products is of great importance (Chapter 4). Shortly after the ICH formally adopted the drug substance quality guidance, ICH Q3A(R), a number of health authorities raised the concern that this guidance document did not adequately address the issue of highly toxic impurities in APIs and did not provide guidance for safe levels. To address this gap, the European Medicines Agency (EMEA) Safety Working Party (SWP) of the Committee for Medical Products for Human Use (CHMP) issued a draft position paper in late 2003 to provide guidance on the limits of GTIs. This position paper was the genesis for the eventual EMEA CHMP GTI guidance document that was issued in 2006. The EMEA guidance utilized an assessment procedure that is based on a concept of threshold of toxicological concern (TTC) to assign an acceptable exposure limit for GTIs in APIs. The successful implementation and execution of any GTI control strategy relies on the ability to adequately demonstrate that these impurities are controlled to a safe level in the final API. To accomplish this goal, the analytical methodologies used for the determination of GTIs at low levels should be very reliable.

V. QUALITY BY DESIGN AND PROCESS ANALYTICAL TECHNOLOGY

To achieve risk-based DP development and manufacturing, it is imperative to apply modern science-based principles throughout the entire life cycle of a DP (Chapter 5). Such modern approaches include the utilization of quality risk management principles, process analytical technology, and quality systems principles. In its entirety, the collective implementation of modern science-based principles will direct pharmaceutical development in a new direction that will build quality into products, as opposed to merely testing quality into products. This forms the core of QbD in that risks associated with pharmaceutical development and manufacturing will be addressed and mitigated throughout the life cycle of the product. After the release of the final draft of "Pharmaceutical CGMPs for the 21st Century—A Risk-Based Approach," the FDA and the ICH have released various documents outlining the necessary components for the application of QbD. Such efforts to unify regulatory bodies and industry's direction in improving pharmaceutical quality provide the driving force and framework for modernizing pharmaceutical development and manufacturing to a state that ensures delivery of high-quality DPs to consumers.

VI. NOVEL DOSAGE FORMS

Novel pharmaceutical dosage forms cover a broad range of formulation delivery platforms such as tablets, capsules, cachets, sustained-release dosage forms, parenteral dosage forms, transdermal dosage forms, metered dose inhalants, solutions, emulsions, and suspensions that are designed to improve patients' acceptance and/or to assure maximum absorption following administration (Chapter 6). Excipients in the pharmaceutical processing operations may originate from a wide variety of sources including plants (e.g., starches, sugars, celluloses), animals (e.g., gelatin, lactose, shellac), minerals (e.g., dicalcium phosphate dehydrate, magnesium stearate), and from synthetic origins (e.g., polyvinylpyrolidone, polysorbates, polyethylene glycols, EUDRAGIT®). Excipients exhibit a wide range of solubilities that influence the type of sample preparation technique that may be employed. Since the goal of any sample preparation technique is to get the drug substance into solution to achieve quantitative recovery (>98%) of the API, the number of sample treatment steps and choice of solvents used become a critical consideration. Sample treatment steps, whether manual or automated, involve laboratory sampling, weighing, sample extraction/dilution, and subsequent injection of an aliquot of the sample for analysis generally by HPLC. Unlike excipients, the API or drug substance (DS) is typically well characterized and may be organic acids or bases that exhibit different solubility in water. Solubility of the dosage form is a key factor when it comes to analysis. In view of the importance of API solubility in the HPLC analysis of pharmaceutical dosage forms, this chapter examines some of the strategies that may be employed to assure optimum quantitative recovery of the API during the sample preparation steps prior to HPLC analysis. In addition, emphasis on sample preparation strategies used for tablets and capsules has been exemplified, as they are the most commonly used delivery platforms for novel pharmaceutical dosage forms.

VII. METHOD DEVELOPMENT FOR CHIRAL COMPOUNDS

Separation chemistry has progressed to a point where it can be used to resolve isomers that are identical except for their ability to rotate the plane of polarized light. In today's pharmaceutical industry, the vast majority of this work is now done by HPLC. Regardless of whether a traditional normal phase HPLC, reversed phase HPLC, or SFC (supercritical fluid chromatography) technique is used, routine analytical chiral method development screening is a very effective and efficient means for developing chiral selective methods in the pharmaceutical industry (Chapter 7). Of course, for preparative chromatography, normal and supercritical fluid chromatographic systems provide significant advantages in recovering enantiomers from the mobile phase. However, for analytical operations, analysts overwhelmingly prefer not to have to switch to normal phase modes for chiral analysis. Thus, chiral method development screening has become quite important in all these chromatographic formats. As a more diverse stream of chemical entities is being developed for market, the analytical chemist must continually look for ways to optimize the screening system. This is required to ensure the best opportunity for success in yielding a robust analytical method for the new chemical entities in development.

VIII. CHARACTERIZATION AND ANALYSIS OF BIOPHARMACEUTICAL PROTEINS

An analytical development program for biomolecules such as proteins comprises in-depth product characterization, assay development, and routine testing activities (Chapter 8). The goals of characterization efforts are to define the structure of the major product substances and product-related impurities, to relate structure to function, and to identify and quantify protein/DNA impurities derived from the host cell.[3,5–7] In addition to fundamental product knowledge, a comprehensive characterization effort allows the future implementation of meaningful comparability protocols that support manufacturing changes. The goals of assay development are to develop and implement relevant, routine assays that address biological activity, safety, efficacy, and quality. These assays support process development efforts, manufacturing investigations, drug substance and DP release, and stability programs. Many assays evolve from the characterization activities. This chapter provides a brief overview of analytical methods commonly used to assess the structure, purity, safety, stability, and potency of recombinant protein pharmaceuticals. These methods are technically challenging to develop because the protein products are large in size, complex in structure and function, and marginally stable in aqueous solution. The basic biochemical, biophysical, and cell biological principles that support these methods are described in the chapter.

IX. PREFORMULATION AND EARLY PHASE METHOD DEVELOPMENT

The transition from discovering a new therapeutic agent to developing a new pharmaceutical product begins with an investigation of the physical and chemical properties of the API, a process commonly referred to as "preformulation" (Chapter 9). This is a key milestone in the drug development process, as the focus shifts from looking at the therapeutic agent as a molecular entity to how it can be effectively made into a product. Preformulation studies, when effectively conducted, can be accurate predictors of the challenges that will be encountered in combining the API with a suitable system that will deliver this new therapeutic agent in a safe and effective manner to a patient. Following preformulation, formulation development activities begin with the eventual goal of developing a robust bioavailable, efficacious, and safe dosage form that can be manufactured at a suitable scale for commercialization. Analytical chemists play an important role in the preformulation process by developing and validating the optimal methods to support these studies.

X. IN-PROCESS CONTROL TESTING

Most pharmaceutical analysts are very familiar with developing methods and analyzing the intermediates, APIs, and DPs of chemical, biological, or pharmaceutical processes. They are also very familiar with the challenges that working with these samples can provide. For the analyst dealing with in-process testing, there are some key additional challenges that include the many different sample matrices, sample or matrix instabilities, sample reactivities, and, in early development, the uncertainty of the sample matrix from batch to batch. The challenges can lead to having to modify the way we qualify and validate the methods and how the technical transfer performed. Chapter 10 deals with analyzing samples taken during a unit operation within a processing step and discusses a typical processing step and how one can determine what, when, and how to test the samples, and the technical transfer of these test methods to a production facility.

XI. METHOD VALIDATION

Because of the current accuracy and precision in analytical instrumentation, reagents, and capabilities of modern data processing systems, even poor methods may be acceptably validated; however, this does not necessarily certify a method as being "good," robust, or suitable for a control environment (Chapter 11). Ideally, the method validation should be integral to the development process and should be utilized for the method optimization. This chapter reviews a systematic and scientific approach to the method development and validation of analytical methods, which should demonstrate to regulatory agencies that methods used in the testing of pharmaceutical products at each stage of development, and ultimately at commercialization, are fit for the purpose. To ensure that the data are accurate and reliable, qualified and trained laboratory analysts should always perform methods on qualified equipment, using suitable standards. It should always be recognized that the validation of such methods is not required solely to comply with good laboratory practices or to satisfy regulatory guidance. In reality, it is done to ensure that data used to approve medicinal products are accurate and that patients can receive life-enhancing medicines without compromise to their welfare or safety.

XII. STABILITY STUDIES

Conducting a dynamic, compliant stability program requires more than merely knowing and adhering to regulatory requirements. It also requires management of the stability samples, the environmental chambers, and all the associated documentation. Standard operating procedures, processes for protocol amendments/deviations, and out-of-trend or out-of-specifications investigations are all key elements of a compliant stability program. Chapter 12 describes how to manage both the operational elements and the regulatory compliance issues to ensure a successful application and inspection by any regulatory agency. The ICH guidelines and the WHO guidelines related to stability are also discussed in detail.

XIII. METHOD TRANSFER

Presently, there is no technology transfer guidance available from either the FDA or from the ICH. Good analytical method transfer requires that a method or test procedure work in an equivalent fashion at two or more different sites or laboratories. This is evidenced by the results of the transfer meeting all predefined acceptance criteria (Chapter 13). The process is driven by compliance and governed by a statistical treatment of the resulting data. Method transfer can be defined as "the introduction of a validated method into a designated laboratory so that it can be used in the same capacity for which it was originally developed." The second portion of the technology transfer process concerns the transfer of technical ownership from one laboratory to another. This latter type of transfer is usually associated with the movement of drug development projects from research and development (R&D) to the commercial release environment; however, it can also be associated with transfer to and from Contract Research Organizations (CROs) or Contract Manufacturing Organizations (CMOs). In many case studies involving technical transfer, the process of ensuring that the receiving laboratory is sufficiently familiar with all scientific aspects of the project is not carried through to fruition. In such cases, problems with the operation of the methods can manifest themselves later during testing of commercial products. Indeed, rapid and complete transfers are crucial to the success of process validation experiments for pharmaceutical dosage forms. The importance of analytical transfers is underscored by the incidence of 483 observations that laboratories have received from the FDA.

XIV. DOCUMENTATION

Pharmaceutical analytical documentation should accomplish the key mission of analytical R&D: to monitor and ensure the identity, purity, stability, and consistency of drug substances/APIs and dosage forms used during preclinical, clinical, and marketing phases in accord with the governing regulatory guidance and policies (Chapter 14). Analytical data are the foundation and backbone for pharmaceutical development leading to approval and production of new drugs for market. Analytical documentation provides the critical links during the evolution and life cycle of a new pharmaceutical product—beginning from earliest studies, enabling entry into humans, through product launch and postapproval changes. Prior to marketing approval, analytical R&D personnel support API synthesis and process development, as well as dosage form design, process development, and optimization activities. Following approval, quality control personnel provide the data to assure consistent quality and stability for the marketed product and to support the inevitable postapproval quality and stability for the marketed product and to support the necessary postapproval changes that occur in every product's life cycle. Analytical scientists have welcomed the concepts of ICH M4 Q—Common Technical Document (CTD) as a means of providing globally accepted information and scientific data in a common format. Analytical data generated with a better understanding of QbD can play a pivotal role in describing the API and DP, as well as at the interface. To concisely and effectively communicate the increased level of drug development phase-dependent information, scientifically sound, consistent, and compliant documents become essential elements of the development process. Good documentation of the product development also helps to assure that the continued quality of the marketed product is achieved.

REFERENCES

1. Ahuja, S. Drug quality, Plenary Lecture presented at Indian Pharmaceutical Congress, Madras, India, December 29, 1974.
2. Ahuja, S. and Scypinski, S. *Handbook of Modern Pharmaceutical Analysis*, 1st ed., Academic, San Diego, 2001.
3. Ahuja, S. and Rasmussen, H. *HPLC Method Development for Pharmaceuticals*, Academic, San Diego, 2007.
4. Ahuja, S. and Dong, M. *Handbook of Pharmaceutical Analysis by HPLC*, Elsevier, Amsterdam, 2005.
5. Ahuja, S. *Handbook of Bioseparations*, Academic, San Diego, 2000.
6. Garfin, D. and Ahuja, S. *Handbook of Isoelectrofocusing and Proteomics*, Elsevier, Amsterdam, 2005.
7. Ahuja, S. and Jimidar, I. *Capillary Electrophoresis Methods for Pharmaceutical Analysis*, Elsevier, Amsterdam, 2008.

2 CHARACTERIZATION OF PHARMACEUTICAL COMPOUNDS IN THE SOLID STATE

HARRY G. BRITTAIN

Center for Pharmaceutical Physics, Milford, New Jersey, USA

I. INTRODUCTION

The essential component of a solid dose form is, of course, the active pharmaceutical ingredient, and since most drug substances are administered via some type of solid dosage form,[1] the characterization of pharmaceutical compounds in the solid state remains of utmost importance. Systematic approaches to the physical characterization of pharmaceutical substances are typically considered within the context of preformulation programs,[2–7] where investigators pay particular attention to issues associated with drug stability,[8] analytical testing,[9] and the physical property aspects of the drug substance.[10,11] Equally important are studies of the crystallographic state of the drug substance, and how these effects translate into properties of pharmaceutical utility.[12–14]

The acquisition of a sufficiently detailed body of physical information can allow a formulator to transcend an ability to cope with unanticipated crises. For a well-understood system, it is theoretically possible to design an automated or semiautomated manufacturing scheme for which the processing variables would be appropriately controlled and so minimize the possibility of batch failure. Materials passing the hurdles of physical test specifications would be predictable in their performance and could therefore be blended, granulated, dried, compressed, and delivered into containers without operator intervention. It is presently recognized that to avoid problems during drug development, the physical characterization of bulk drugs, excipients, and blends of these should become part of the normal process. The degree of physical testing would necessarily vary with the particular formulation, but would include any and all test methods deemed appropriate.

A systematic approach to the physical characterization of pharmaceutical solids has been outlined,[10,15] which serves as a useful pedagogical device for the classification of the many methods of physical characterization available. Within this system, physical properties are classified as being associated with the *molecular* level (those associated with individual molecules), the *particulate* level (those pertaining to individual solid particles), or the *bulk* level (those associated with an assembly of particulate species).

One of the areas where the physical characterization of solids has become of extreme importance is in the study of polymorphs and solvatomorphs.[12–14] The nature of the crystal structure adopted by a given compound upon crystallization exerts a profound effect on the solid-state properties of that system, and these variations can translate into significant differences in properties of pharmaceutical importance. It is now accepted that an evaluation of the polymorphism available to a drug substance must be thoroughly investigated early during the stages of development. The results of these studies must be included in the chemistry, manufacturing, and control section of a new drug application, and such information is required to demonstrate control over the manufacturing process.

A number of additional factors provide an impetus and requirement for the inclusion of detailed solid-state characterization at essentially all stages of drug development. One of the most important of these concerns the intellectual property associated with crystal forms of drug substances. Given the current state of polymorph prediction, it still remains true that one cannot predict *a priori* how many polymorphs or solvatomorphs would be associated with a given drug substance, or even which ones would be isolated from a given set of crystallization conditions. Consequently, many patents have been filed and issued covering crystal forms of active pharmaceutical ingredients, and many of these have been the source of contentious litigations between brand companies and their generic competition. Complete characterization of drug substances is perhaps the only way a company can obtain total control over this aspect of the patent space of their compounds.

On a more practical note, it is generally understood that alternate crystal forms will exhibit sufficiently different solid-state properties so that one could anticipate encountering measurable differences in properties of pharmaceutical importance. For example, it has been reported that polymorphs of some drug substances exhibit different equilibrium solubilities and dissolution rates, and these differences have sometimes led to the existence of nonequivalent bioavailabilities for the different forms. As long as no legal obstacles lie in the way, one would clearly prefer to develop the most thermodynamically stable form of a drug substance, as that would be most amenable to large-scale production and secondary processing. For instances when that particular crystal form is not available and development of a metastable form is required, then one must fully understand the conditions under which the forms could interconvert if disaster is to be avoided.

In this chapter, a comprehensive program of solid-state characterization of substances will be outlined, whose implementation would go a long way in the amelioration of the problems just discussed.

II. PROPERTIES ASSOCIATED WITH THE MOLECULAR LEVEL

Molecular properties may be defined as those material characteristics that theoretically can be measured for a small ensemble of individual molecules. Due to the minimal sample requirements, molecular properties are often determined at the earliest stages of drug development, and are generally derived from the spectroscopy of the molecule. The value of molecular level techniques for the characteristics is based on the ability of the physical state of the substance to influence a spectroscopy property of interest.

A. UV/Vis Diffuse Reflectance Spectroscopy

When the differing crystal structures of solid substances lead to perturbations of molecular orbital energies, techniques associated with ultraviolet or visible spectroscopy (i.e., UV/vis reflectance or fluorescence spectroscopy) can be used to study the solids. With the exception of single-crystal transmission work, most solids are too opaque to permit the conventional use of UV/vis electronic spectroscopy. As a result, such work must be performed using diffuse reflection techniques.[16–18] Studies have been conducted where UV/vis spectroscopy was used to study the reaction pathways of various solid-state reactions. Other applications have been made in the fields of color measurement and color matching, areas that can be of considerable importance when applied to the coloring agents used in formulations.

It was recognized some time ago that diffuse reflectance spectroscopy would be a very useful tool for the study of interactions among various formulation components, and the technique has been successfully used in the characterization of many solid-state reactions.[19] Investigations conducted under appropriately designed stress conditions have been useful in the study of drug–excipient interactions, drug degradation pathways, and alterations in bioavailability owing to chemisorption of the drug onto other components in the formulation.

For example, Connors and Jozwiakowski have used diffuse reflectance spectroscopy to study the adsorption of spiropyrans onto pharmaceutically relevant solids.[20] The particular adsorbants studied were interesting in that the spectral characteristics of the binary system depended strongly on the amount of material bound. At low concentrations, the pyran sorbant exhibited its main absorption band around 550 nm, but as the degree of coverage was increased, the 550 nm band was still observed and a much more intense absorption band at 470 nm became prominent. This secondary effect was attributed to the presence of pyran–pyran interactions, which became more important as the concentration of sorbant was increased.

The perception of color is subjectively developed in the mind of an individual, and consequently different people can perceive a given color in various ways. Such variability in interpretation causes great difficulty in the evaluation of color-related phenomena, leading to problems in making objective judgments. The development of quantitative methods for color determination was undertaken to eliminate the subjectivity associated with visual interpretative measurements.

The most successful quantitative expression of color is that known as the CIE (Commission Internationale de l'Éclairage) system.[21] This methodology assumes that color may be expressed as the summation of selected spectral components (blue, green, and red hues) in a three-dimensional manner. The CIE system is based on the fact that human sight is trichromatic in its color perception, and that two stimuli will produce the same color if each of the three tristimulus values (X, Y, and Z) are equal for the two. Detailed summaries of the CIE and other quantitative systems for color measurement are available.[22,23]

In a recent application, the appearance testing of tablets through measurement of color changes has been automated using fiber optic probes and factor analysis of the data.[24] Good correlation between measured chromaticity parameters and visual subjective judgment was demonstrated, with samples of differing degrees of whiteness being used to develop the correlation. The methodology was complicated since surface defects on the analyzed materials could compromise the quality of the correlation.

One of the more interesting uses of reflectance spectroscopy has been its utility in determining the relative acidity or basicity of a solid substance. While the relative degree of acidity or basicity of a substance dissolved in an aqueous medium can be adequately defined in terms of the solution pH, a comparable expression for the acidity or basicity of the surface of a solid is not as easily expressed. The concept of the acid strength of a surface was first discussed in a systematic manner by Walling,[25] who sought to differentiate the acid strength of a surface from the stoichiometry of acidic sites. Walling extended the indicator theory developed by Hammett to explain the color changes associated with certain substances as the acidity of their medium was changed[26–28] to solids. In particular, Walling defined the acidity of a solid surface as the ability of the surface to convert an adsorbed neutral base to its conjugate acid.

Walling defined a quantity, H_0, as the acid strength for solids, and defined this function analogously to pH, such that decreasing values of H_0 signified increasing acid strength of the surface. Therefore, the acid strength of a surface should be readily measurable by the observation of the colors of absorbed indicators and quantifiable using reflectance spectroscopy. In particular, the appearance of the color characteristic of the acid form of the indicator would be taken as indicating a value of the H_0 function for the surface lower than the pK_a of the indicator used. This phenomenon is illustrated in Figure 1, which shows the pH dependence of the absorption spectrum associated with aqueous solutions of bromcresol green.[29]

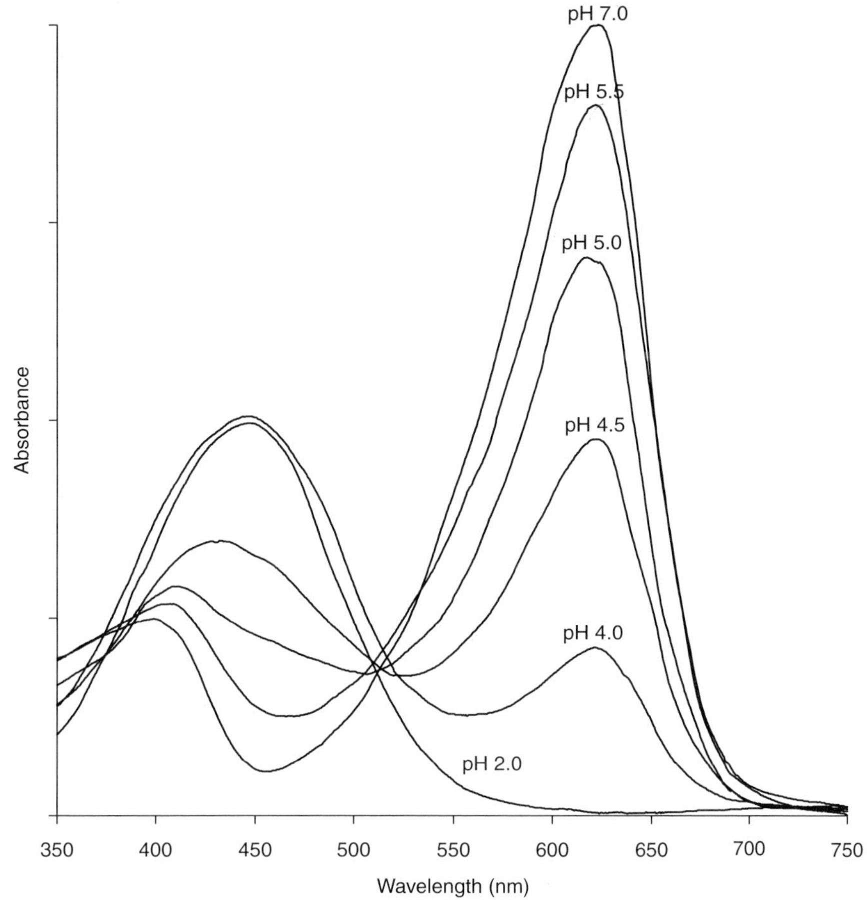

FIGURE 1 Absorption spectra of aqueous solutions of bromcresol green as a function of pH (H. G. Brittain, unpublished results).

Walling[25] pointed out a number of requirements as essential for the successful use of basic indicators for the determination of the acid strength of surfaces:

1. The indicators used are restricted to those for which the basic form is neutral, and which are converted to their conjugate acids by the addition of a single proton.
2. The indicators must be capable of adsorbing onto the surface under investigation.
3. The solvent used to interact with the indicator with the solid surface must not perturb the properties of the surface itself.
4. The method requires substrate surfaces whose inherent color does not interfere with the color determination of the adsorbed indicator.

Walling then continued to define the indicator types that would eventually be termed the Hammett indicators. Any substance that is neutral or slightly basic in water, that has a different color in concentrated sulfuric acid, and that is a 1:1 electrolyte can act as a basic indicator at acidity strengths between those of water and sulfuric acid. Furthermore, the substance must owe its color change to protonation by a single proton, and not by the addition of water or some other molecule. In the original work, 15 such indicators were indicated as having appropriate properties.[25]

Although the Hammett indicators could be used to provide visual determinations of the acidity or basicity of surfaces, the use of reflectance spectroscopy facilitated an extension of these studies. In one series of studies, Leermakers and coworkers used UV spectroscopy to study the electronic spectra and structure of a number of molecules absorbed onto silica gel.[30–33] Since the n,π^* excited state is less polar than that of the ground-state molecule, after sorption onto the support, the $n \rightarrow \pi^*$ transition would be observed at higher energy and therefore shifted to shorter wavelengths. Conversely, since the π,π^* excited is more polar than the ground state, the $\pi \rightarrow \pi^*$ transition would be shifted to lower energies and longer wavelengths.

Schmidt and coworkers have carried out determinations of the surface acidity of pharmaceutical excipients and used a variety of absorption indicators to span the ranges of observed the acid strengths. In one study, the surface acidity of microcrystalline cellulose and dicalcium phosphate anhydrate was evaluated using thymol blue, bromcresol green, bromcresol purple, and phenol red as the absorption indicators.[34] Using the pH dependence of dye absorbance published in the paper, it was concluded that H_0 was approximately 1.0 for dicalcium phosphate anhydrate, and approximately 4.0 for microcrystalline cellulose.

The surface acidities of a wide variety of excipients used in solid dosage forms have been determined using absorption indicators.[35] A wide number of fillers, binders, disintegrants, and lubricants were found to exhibit H_0 values in the range of 3–4, and only a few excipients exhibited H_0 values larger than this. In fact, potassium bicarbonate ($H_0 = 8.05$), sodium bicarbonate ($H_0 = 8.27$), and sodium carbonate ($H_0 = 8.58$) were found to be the only excipients studied, whose surfaces exhibited basic properties. As shown in Figure 2, bromcresol green exists primarily in its acid form when adsorbed onto anhydrous dicalcium phosphate, primarily in its basic form when adsorbed onto sodium bicarbonate, but is distributed between its acidic and basic form when adsorbed onto sorbitol.[29]

B. Vibrational Spectroscopy

When the differing crystal structures of solid substances lead to perturbations of the pattern of intramolecular motion, techniques associated with vibrational spectroscopy (i.e., infrared absorption or Raman scattering) can be used to study the solids. The patterns of intramolecular motion existing in molecules are characterized by repetitious oscillations about the center of gravity, and these correlated motions are termed the vibrational modes of the molecule. Energies associated with the lowest energy vibrational modes of a chemical compound will lie within the range of 400–4000 cm^{-1}, a spectral region of the electromagnetic spectrum denoted as the mid-infrared. Extensive discussions of the fundamental principles of vibrational spectroscopy[36–45] are available, and the utility

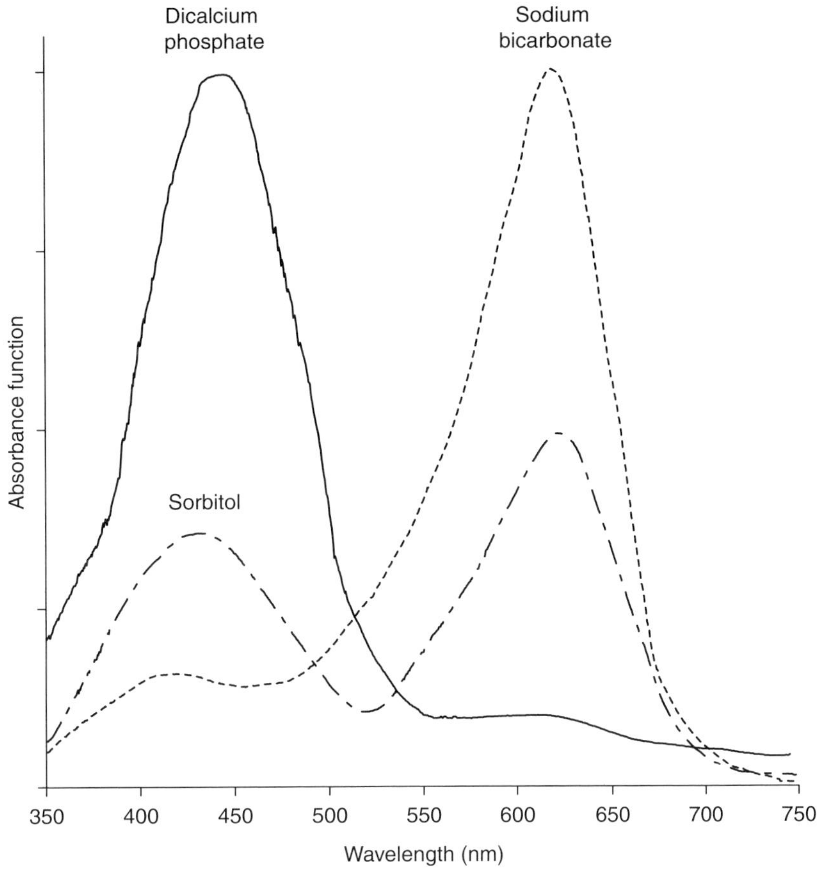

FIGURE 2 **Diffuse reflectance spectra, transformed to an absorbance scale, of bromcresol green adsorbed onto dicalcium phosphate anhydrate (solid trace), sorbitol (broken trace), and sodium bicarbonate (dashed trace) (H. G. Brittain, unpublished results).**

of vibrational spectroscopy in the physical characterization of pharmaceutical compounds is equally well reviewed.[46–58]

Transitions among states associated with vibrational modes can be observed directly through their absorbance of energy in the mid-infrared region of the spectrum. Infrared absorption spectroscopy is an enormously useful technique for the physical characterization of solids having pharmaceutical interest. For example, being especially sensitive to changes in molecular conformation between different solid-state forms, infrared absorption spectra are frequently used to evaluate the polymorphic space of a drug substance system. In addition, the technology can be very useful in studies of solvent molecules contained within a solvatomorph.

Alternatively, one may also study transitions among states associated with vibrational modes through an inelastic scattering of incident energy via the Raman effect. Since most compounds of pharmaceutical interest are characterized by little or no molecular symmetry, the same bands observed in the infrared absorption spectrum will necessarily be observed in the Raman spectrum. However, the fundamentally different nature of the selection rules associated with the Raman effect leads to the existence of significant differences in intensity among peaks as measured by the two methods. In general, symmetric vibrations and nonpolar groups yield the most intense Raman scattering bands, while antisymmetric vibrations and polar groups yield the most intense infrared absorption bands.

Transitions involving the simultaneous excitation of multiple vibrational transitions are observed as overtone bands or combination bands, since these are derived from the same fundamental bands detected in infrared absorption or Raman spectroscopies. Overtone or combination bands are observed in the near-infrared (NIR) region of the electromagnetic spectrum (4000–13,350 cm^{-1}). Some of the spectral features that have found the greatest utility are those functional groups that contain unique hydrogen atoms. For example, studies of water in solids can be easily performed through systematic characterization of the characteristic –OH band, usually observed around 5170 cm^{-1}. The determination of hydrate species in an anhydrous matrix can also be easily performed using NIR analysis.

1. Infrared Absorption Spectroscopy

The acquisition of high-quality infrared spectra on solid materials has been made possible using spectrometers that employ Fourier transform infrared (FTIR) technology, since the use of this methodology minimizes transmission and beam attenuation problems. Detailed descriptions of the components of FTIR spectrometers are available.[59,60] While many sampling techniques may be used to obtain good FTIR spectra, only a few of these are widely used for the study of solids having pharmaceutical interest. The classic alkali halide pellet method is probably the least useful, as the sample may undergo a solid-state transformation due to the pressure used to form the pellet, or one may encounter halide exchange between the KBr or KCl matrix and the sample.[61] The use of mineral oil mulls avoids the problems noted for the pellet technique, but the mineral oil itself exhibits a number of intense absorption bands (2952, 2923, 2853, 1458, and 1376 cm^{-1}) that may overlap important absorption bands associated with the sample and obscure important spectral regions. Being noninvasive, diffuse reflectance sampling avoids most problems,[62,63] but usually requires dilution of the analyte with either KBr or KCl at a level of 1–5% (w/w).

Probably, the most useful sampling method for FTIR spectroscopy is that of attenuated total reflectance.[64,65] In the ATR technique, infrared radiation is passed through a crystal at an angle less than the critical angle, which causes the light to undergo total internal reflection. At each such reflection, the radiation penetrates a small distance beyond the crystal surface so that the internally reflected energy will be attenuated at those frequencies corresponding to changes in molecular vibrational states of analyte in physical contact with the crystal. The advantage of the ATR technique is that it requires effectively no sample preparation, since one simply clamps the analyte onto the surface of the crystal with moderate pressure to ensure a sufficient degree of optical contact. Since the internal reflectance process does not permit the infrared beam to pass very deeply into the sample, it is typical to determine the composition of an analyte up to a sampling depth in the range of 5–10 μm.

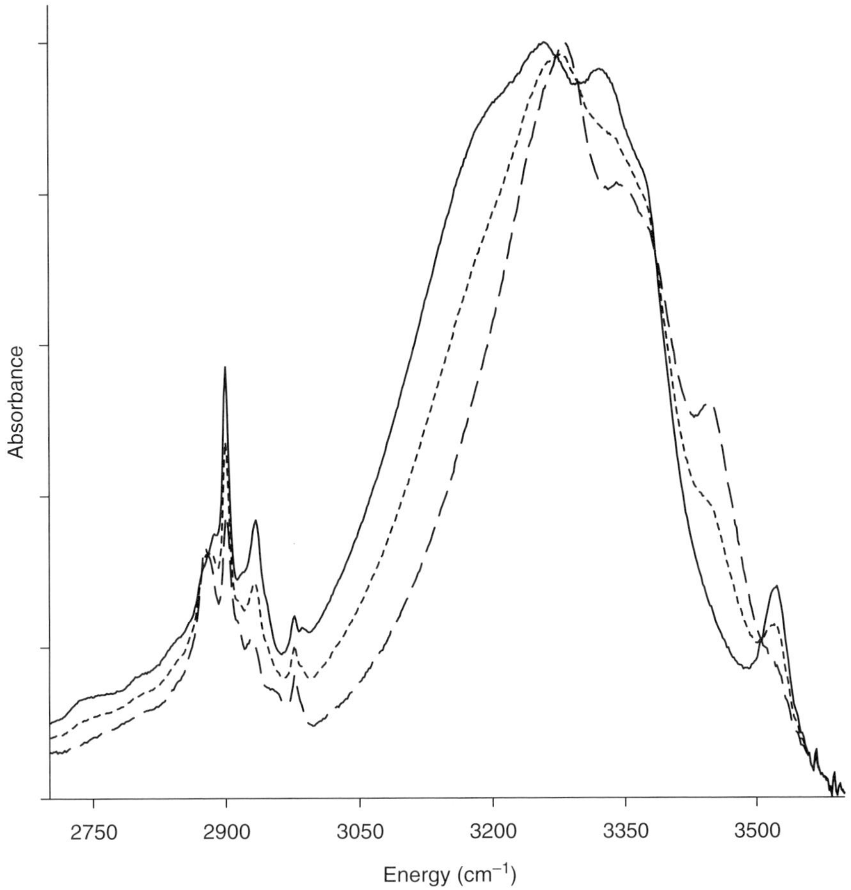

FIGURE 3 **Infrared absorption spectra (obtained using attenuated total reflectance sampling) of lactose monohydrate (solid trace), lactose anhydrate (broken trace), and a 1:1 blend of the monohydrate and anhydrate (dashed trace) (H. G. Brittain, unpublished results).**

One of the most important applications for infrared absorption spectroscopy is in the qualitative identification of materials having pharmaceutical interest. For example, as shown in Figure 3, the anhydrous and monohydrate forms of lactose can be readily differentiated on the basis of their characteristic absorption spectra in the high-frequency region (i.e., the peak at 3523 cm^{-1} being most useful), and the technique is even able to distinguish lactose materials consisting of mixtures of the two.[66]

The sensitivity of infrared absorption spectroscopy toward subtle changes in crystal structure has led to its widespread use in the investigations of pharmaceutical compounds that exhibit polymorphism. For example, three concomitant polymorphs of 1,3-bis(*m*-nitro-phenyl)urea had been reported in 1899 as yellow prisms, white needles, and yellow tabular crystals, and FTIR spectroscopy was used during a more detailed investigation of the system.[67] In this work, complete assignments for the absorption bands associated with hydroxyl, amide, nitro, and benzene-ring functional groups were developed in order to obtain a deeper understanding of the conformational differences in the molecules constituting the various crystal forms. In another example, it was reported that the carbonyl frequency of niclosamide anhydrate and two of its monohydrates was particularly sensitive to the crystal form of the drug substance, enabling infrared spectroscopy to be used as a means to study the conformational state of the molecule in the solvatomorphs.[68]

Since the pattern of vibrational modes is usually sufficiently perturbed by crystallographic differences, infrared absorption spectroscopy can be profitably used to study the phase transformation process between polymorphic forms of a substance. Infrared spectroscopy has been used to follow the isothermal transformation of mefenamic acid form-I into form-II at a sufficient number of temperatures so as to calculate the activation energy for the process.[69] The value of 71.6 kcal/mol, obtained using the infrared spectroscopic method, was significantly smaller than the previously reported value of 86.4 kcal/mol that had been determined using of differential scanning calorimetry (DSC). Conventional FTIR spectroscopy in conjunction with FTIR microspectroscopy has been used to study the polymorphic interconversion of famotidine metastable form-B into stable form-A during grinding in a ceramic mortar.[70] The particle size reduction resulting from the grinding process was also found to lead a decrease in the transition temperature for the form-B to form-A phase transformation.

2. Raman Spectroscopy

When substances are irradiated with intense beams of monochromatic radiation, the majority of the photons are scattered at the same frequency as that of the incident beam. It has been found that a small amount of the incident light will also be scattered at nonresonant frequencies, with the differences in frequency between the incident radiation and the shifted frequencies corresponding to the frequencies of molecular vibrations. Although Raman spectra can be obtained using dispersive technology, the best spectra are acquired through the use of Fourier transform technology since FT-Raman systems can provide superior wavelength accuracy and minimization of any background fluorescence.[42–44] Consequently, the use of Raman spectroscopy as a tool to solve problems having pharmaceutical interest has become widespread.[49–58]

Unlike infrared absorption spectroscopy (which operates via an electric dipole mechanism), only molecular transitions characterized by a change in polarizability will exhibit Raman transitions of appreciable intensity. Since the transition moment of a Raman transition depends on the magnitude of polarizability, it follows that the vibrational modes exhibiting the strongest intensities in a Raman spectrum will be those that are associated with functional groups that are characterized by high degrees of polarizability. Examples of such functionalities typically exhibiting strong Raman bands include the C–S, S–S, C–C, N=N, and C=C groups.

As in the case of infrared absorption spectroscopy, the bands in a Raman spectrum can be assigned through the use of group frequency correlation tables. Significant insight can be obtained from the compilations of functional group vibrational frequencies associated with infrared absorption spectroscopy,[71] but of greater utility is the compilation of group frequencies specific to Raman spectroscopy.[72] Although excitation of a given vibrational mode gives rise to infrared absorption bands and Raman scattering bands having identical frequencies, the fact that the infrared absorption process is an electric dipole transition and the Raman process is based on changes in polarizability leads to the observation of differences in relative spectral intensities for a given material.

As with infrared absorption spectroscopy, Raman spectroscopy can be used to study polymorphic solids when the differing crystal structures results in a perturbation of the pattern of molecular vibrations. The question of how significant difference in the energies of Raman bands would need to be in order to demonstrate the existence of polymorphism was addressed through the use of statistics, where it was proposed that the shift in the energy of a Raman peak would need to exceed 1.6 cm^{-1} to indicate the existence of polymorphism.[73] More often than not, the shifts commonly observed among polymorphic substances are frequently sufficiently large to permit the use of Raman spectroscopy as a triage method in screening protocols. For example, polymorphic form-A of famotidine exhibits a unique scattering band at 651.6 cm^{-1} that easily distinguishes it from polymorphic form-B[74] (Figure 4).

A combination of infrared and Raman spectroscopic investigations was used to characterize two polymorphs of olanzapine, with spectral assignments being deduced for all observed bands in the two solid-state forms in order to obtain insight into the crystalline structures.[75] The spectrum of form-1 in the high-frequency region was found to be characteristic of a hydrogen-bonded molecule, while the spectrum of form-2 contained information consistent with atoms involved directly or indirectly to hydrogen bonds associated primarily with the N–H group. Raman spectroscopy has been found to be useful for the quantitative analysis of phase composition. For example, two polymorphs of buspirone hydrochloride were characterized by a full range of techniques, with unique peaks in the Raman spectrum being used in conjunction with chemometric analysis to develop quantitative methods for phase composition in mixtures of the two forms.[76]

Owing its ready adaptability as an *in situ* method of characterization, Raman spectroscopy has found applications in the study of solution-mediated phase transformation reactions. For example, a number of methods (including Raman spectroscopy) were used to follow the phase change of the metastable α-form of (*L*)-glutamic acid into its stable β-form, facilitating the development of a population balance model to understand the first-order kinetics for nucleation and growth kinetics of both polymorphs.[77] In another work, flufenamic acid was used as the model system in a study of *in situ* Raman spectroscopy as a method for the determination of transformation kinetics and transition temperatures, with good agreement being reported between results obtained by the spectroscopic method and those obtained using the more conventional van't Hoff approach.[78]

Raman spectroscopy has been shown to be useful in the study of the kinetics associated with the solution-mediated aqueous transformation of anhydrous carbamazepine to its dihydrate phase. Using a Raman immersion probe to establish the phase composition, the thermodynamics associated with the system were studied in ethanol–water mixtures through measurements of the solubility of both forms over the temperature range of 0–60 °C.[79] Raman spectroscopy was also

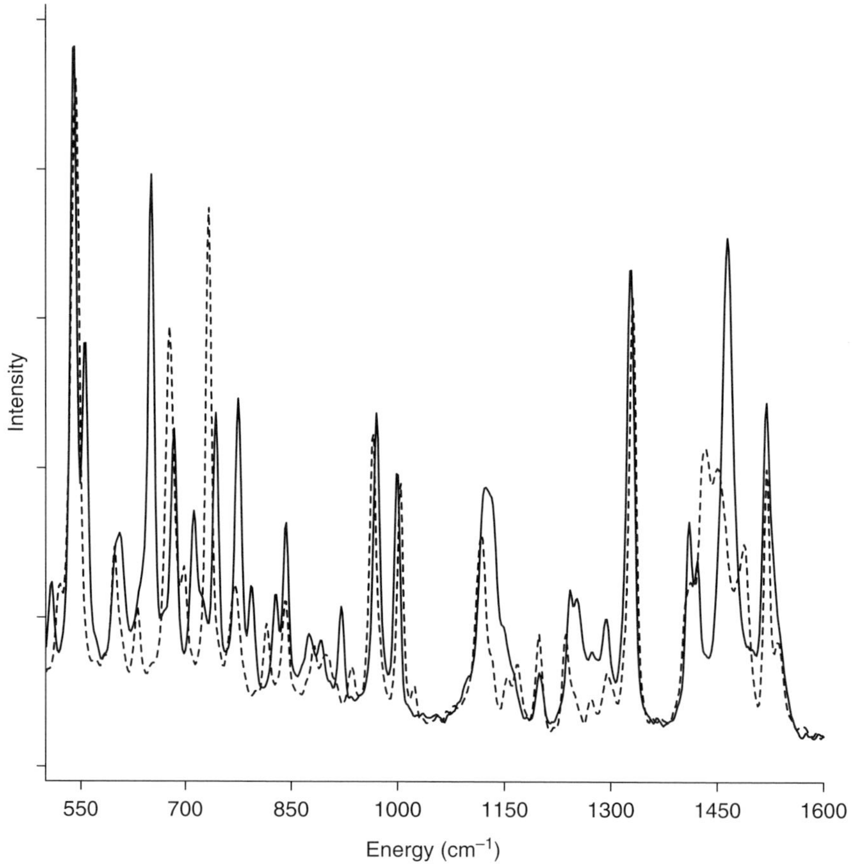

FIGURE 4 **Raman spectra in the fingerprint region of famotidine form-A (solid trace) and form-B (dashed trace) (H. G. Brittain, unpublished results).**

used to follow the kinetics associated with the conversion of the form I, II, and II anhydrates to the dihydrate, where it was found that the morphology of the starting material seemed to determine the rates of reaction.[80] The Raman technique, in combination with partial least-squares regression analysis, was found to be the most robust method for the detection and quantification of mixtures of the carbamazepine solvatomorphs.[81]

3. NIR Spectroscopy

The NIR region of the electromagnetic spectrum is generally considered to span 750 nm (13,350 cm^{-1}) to 2500 nm (4000 cm^{-1}). NIR spectra are due to the excitation of multiple fundamental vibrational modes of molecules, with these being classified as being either overtone transition (multiple excitation of a single vibrational mode) or combination transitions (simultaneous excitation of different vibrational modes).[82–86] In general, the spectral features of greatest utility entail overtone transitions associated with functional groups that contain unique hydrogen atoms.

The spectral features of greatest utility in NIR spectroscopy are those that involve overtone and combination bands of hydrogen atoms. For example, studies of water in solids can be easily performed through systematic characterization of the characteristic –OH band, typically observed around 5170 cm^{-1}. This characteristic has been exploited in the development of NIR methods for moisture determination.[87] Through the use of derivative spectra combined with multicomponent analysis, NIR spectroscopy has been successfully used to develop methods to study the course of powder blending.[88–90]

The real power of NIR spectroscopy is demonstrated by the many methods that have been developed for quantitative analysis. For example, the technique was used to monitor the degree of conversion between stable form-A of a new chemical entity and its metastable form-B that was produced at the elevated temperature and humidity conditions of its wet granulation.[91] Since a reference method was not available for quantitation of form-B in bulk quantities of form-A, a calibration set was developed from the NIR spectra of the drug substance, the premix blend, and wet granulated samples, since narrow spectral regions unique to form-B were found that were insensitive to differences in physical properties between the premix blend and wet granulation.

In-line NIR spectroscopy was used to demonstrate that the process-induced transformation of erythromycin A dihydrate to its dehydrated form did not take place during pellet manufacture via extrusion/spheronization and drying, although partial phase transformation was noted for pellets that had been dried at 60 °C.[92] NIR spectroscopy was one of several techniques used to obtain quantitative determinations of phase compositions of a developmental compound in its bulk drug substance and in compressed tablets, with a partial least-squares regression algorithm being used to obtain good multivariate calibration.[93]

Distinctions between the solvated and nonsolvated forms of a drug substance are often easy to detect using NIR spectroscopy, making the technique a valuable part of multidisciplinary studies. The factors associated with the quantitative analysis of anhydrate/hydrate powder mixtures have been discussed,[94] and these issues illustrated in work establishing the phase boundaries in the anhydrate/hydrate system of caffeine.[95] The interconversion between the anhydrate, monohydrate, and dihydrate forms of azithromycin have been studied, with the spectral region associated with the first overtone of water having been found to be the most useful.[96]

C. Solid-State Nuclear Magnetic Resonance Spectrometry

The ultimate molecular level characterization of a pharmaceutical material is performed on the level of individual chemical environments of each atom in - the compound, and this information is best obtained using solid-state nuclear magnetic resonance (SS-NMR) spectroscopy.[97] Although effectively any nucleus that can be studied in the solution phase can also be studied in the solid state, most of the published work has involved on studies involving the ^{13}C nucleus.

The local magnetic field (B_{loc}) at a ^{13}C nucleus in an organic solid is given by:

$$B_{loc} = \pm \frac{h\gamma_H}{4\pi} \frac{3\cos^2\theta - 1}{r^3}$$

where γ_H is the magnetogyric ratio of the proton, r is the internuclear C–H distance to the bonded proton, and θ is the angle between the C–H bond and the external applied field (B_0). The $\pm$ sign results from the fact that the local field may add to or subtract from the applied field depending on whether the neighboring proton dipole is aligned with or against the direction of B_0. In a microcrystalline organic solid, there is a summation over many values of θ and r, resulting in a proton dipolar broadening of many kilohertz. A rapid reorientation of the C–H internuclear vectors (such as those associated with the random molecular motions which take place in the liquid phase) would result in reduction of the dipolar broadening. In solids, such rapid isotropic tumbling is not possible, but since the term ($3\cos^2\theta - 1$) equals zero if θ equals approximately 54°44′, spinning the sample at this "magic angle" with respect to direction of the applied magnetic field results in an averaging of the chemical shift anisotropy. In a solid sample, the anisotropy reflects the chemical shift dependence of chemically identical nuclei on their spatial arrangement with respect to the applied field. Since it is this anisotropy that is primarily responsible for the spectral broadening associated with ^{13}C samples, spinning at the magic angle makes it possible to obtain high-resolution ^{13}C NMR spectra of solid materials.

An additional method for the removal of ^{13}C–^{1}H dipolar broadening is to use a high-power proton decoupling field, often referred to as dipolar decoupling. One irradiates the sample using high power at an appropriate frequency, which results in the complete collapse of all ^{13}C–^{1}H couplings. With proton dipolar coupling alone, the resonances in a typical solid-state ^{13}C spectrum will remain very broad (on the order of 10–200 ppm). This broadening arises from the fact that the chemical shift of a particular carbon is directional, depending on the orientation of the molecule with respect to the magnetic field.

Even though high-resolution spectra can be obtained on solids using the MAS technique, the data acquisition time is lengthy due to the low sensitivity of the nuclei and the long relaxation times exhibited by the nuclei. This problem is circumvented using cross polarization (CP), where spin polarization is transferred from the high-abundance, high-frequency nucleus (^{1}H) to the rare, low-frequency nucleus (^{13}C). This process results in up to a fourfold enhancement of the normal ^{13}C magnetization, and permits a shortening of the waiting periods between pulses. The CP experiment also allows the measurement of several relaxation parameters that can be used to study the dynamic properties of the solid under investigation.

When the crystallography of compounds related by polymorphism is such that nuclei in the two structures are magnetically nonequivalent, it will follow that the resonances of these nuclei will not be equivalent. Since it is normally not difficult to assign organic functional groups to observed resonances, solid-state NMR spectra can be used to deduce the nature of polymorphic variations, especially when the polymorphism is conformational in nature. Such information has proved to be extremely valuable during various stages in the development of numerous pharmaceutical substances.[98–102]

Ampicillin has been known for a long time to be capable of existing in anhydrate and trihydrate solvatomorphic forms. As shown in Figure 5, the various resonance bands observed in the solid-state ^{13}C NMR spectra of the trihydrate and one of the anhydrate crystal forms are grouped according to the functional groups of the differing carbons in the compounds.[103] The effects of hydrogen bonding are seen most clearly in the bands associated with the three carbonyl groups, with these being fully resolved for the anhydrate and only partially resolved for the trihydrate. On the other hand, the trihydrate exhibits a well-resolved sequence of peaks for the carbons of the aromatic ring, while the degree of resolution is much worse for the aromatic carbons in the anhydrate structure.

During the development of fosinopril sodium, a crystal structure was solved for the most stable phase, but no such structure could be obtained for its metastable phase.[104] The compound contains three carbonyl groups, and the solid-state ^{13}C NMR spectra of two of these are effectively equivalent. The third carbonyl, located on the acetal side chain, was found to resonate at different chemical shifts in the two structures. When combined with the observations obtained using

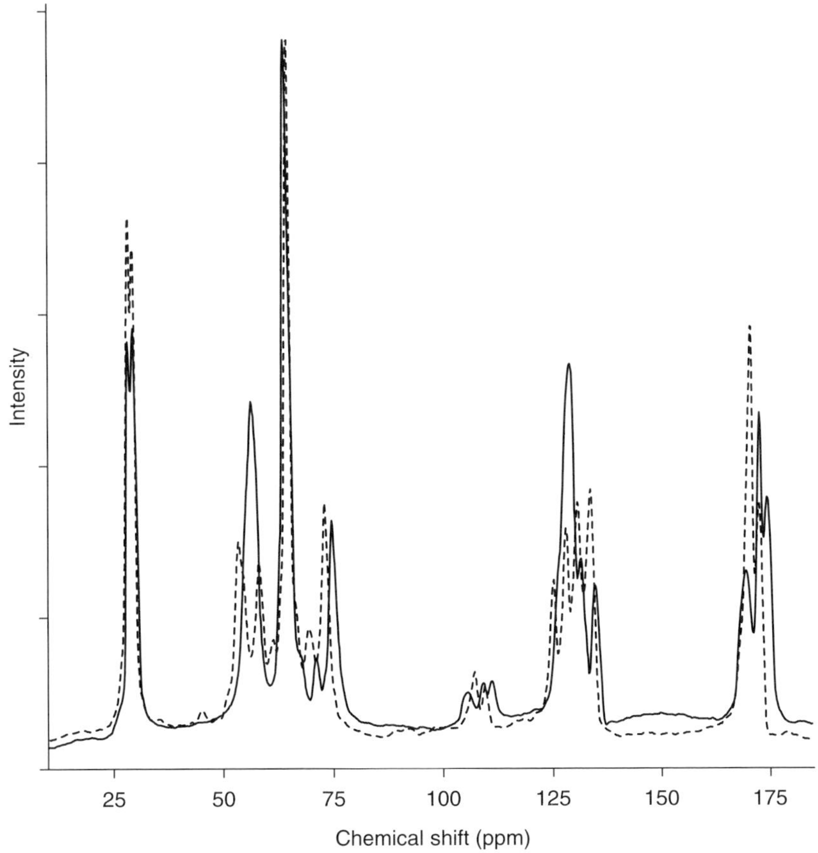

FIGURE 5 **Solid-state ^{13}C nuclear magnetic resonance spectra of the anhydrate form of ampicillin (solid trace) and its trihydrate phase (dashed trace) (H. G. Brittain, unpublished results).**

vibrational spectroscopy, these results permitted the deduction that the solid-state polymorphism was associated with different conformations of this side chain. The NMR data also suggested that additional conformational differences between the two polymorphs were associated with *cis–trans* isomerization along the peptide bond, which in turn results in the presence of nonequivalent molecules existing in the unit cell. In the absence of solved crystal structures for the two polymorphs, this information would not have been otherwise obtainable.

Owing to its ability to report on the chemical environment of nuclei within functional groups, SS-NMR spectroscopy can play an extremely useful role in the evaluation of interactions between drug substances and excipients in a formulation. For example, the interactions between ibuprofen free acid and its sodium salt with Eudragit RL-100 in solid dispersions have been evaluated using both ^{1}H and ^{13}C NMR spectroscopies.[105] After full assignment of the resonances for both forms of ibuprofen that permitted an understanding as to what was taking place in formulations, it was determined that the degree of drug–polymer interaction was much strong for ibuprofen free acid than it was for the sodium salt. These results correlated with those obtained from *in vitro* dissolution testing, where the drug release of ibuprofen free acid was slowed and modulated by the Eudragit matrix.

^{13}C SS-NMR spectroscopy was used to investigate the effect of encapsulation on conformation of molecularly dispersed acylated and esterified derivates of salicylic acid.[106] It was found that alteration of the microenvironment of the incorporated solutes had accompanied the formulation process. For example, increasing the aliphatic character of the acyl side chain resulted in an increase in the upfield shift of the acyl-bearing aromatic ring carbon of an incorporated substance relative to the resonance of the corresponding free compound. In addition, a downfield shift of the resonance of the free acid-bearing aromatic ring carbon was also observed. The electrostatic shielding within the microenvironment in the proximity of the ester carbonyl was attributed to an increase in the association of the probe molecule with the polymer subunits, suggesting that the incorporated compounds are essentially shielded from hydrolytic attack being liberated into an aqueous environment.

III. PROPERTIES ASSOCIATED WITH THE PARTICULATE LEVEL

Particulate properties are defined as those material characteristics that can effectively be determined by the analysis of a relatively small ensemble of particles. Since the sample requirements for these assay methods are not extravagant, one often finds that these properties are also studied during early development once the drug substance is available in at least milligram quantities.

A. Microscopy

Evaluation of the morphology of a pharmaceutical solid is of extreme importance, since this property exerts a significant influence over the micromeritic and bulk powder properties of the material.[107,108] Microscopy is also useful as a means to obtain estimations of the particle size distribution in a powdered sample. A determination can be easily made regarding the relative crystallinity of the material, and skilled workers can deduce crystallographic information as well. Unknown particulates can often be identified solely based on their microscopic characteristics, although it is useful to obtain confirmatory support for these conclusions with the aid of microscopically assisted techniques.

Both optical and electron microscopies have found widespread use for the characterization of pharmaceutical solids. Optical microscopy is limited in the range of magnification suitable for routine work, with an upper limit of approximately 600× is associated with problems in maintaining depth of focus. However, this magnification limit does not preclude the investigation of most pharmaceutical materials, and the use of polarizing optics introduces a power into the technique not available with other methods. Electron microscopy work can be performed at extraordinarily high-magnification levels (up to 90,000× on most units), and the images that can be obtained contain a considerable degree of three-dimensional information.

The two methods are complimentary in that each can provide information inaccessible to the other. When these techniques are used in conjunction, substantial characterization of a solid material becomes possible. This information can be extremely useful during the early stages of drug development, since only a limited amount of the drug candidate is normally available at that time. Electron microscopy yields excellent topographic and shape information, and is most useful in forensic situations involving trace evidence characterization and identification. Light microscopy is most useful to obtain information on the internal properties of small particles, fibers, and films. When polarizing optics are used in light microscopy, the optical properties of the crystals under investigation can also be determined.[107] In that case, molecular (rather than elemental) information can be obtained on the analyte. A general procedure for the conduct of microscopic analysis and standards for reporting on the morphology of crystals can be found in the United States Pharmacopoeia.[109]

In the study of solids, probably the most useful light microscopy technique involves the use of polarizing optics.[107] The polarizing microscope is essentially a light microscope equipped with a linear polarizer located below the condenser, and an additional polarizer mounted on top of the eyepiece. A rotating stage is also found to be very useful, as are the ability to add other optical accessories (such as phase contrast). Polarization optical analysis is based on the action of the analyte crystal on the properties of the transmitted light. This method can yield several directly measured parameters, such as the sign and magnitude of any observed birefringence, knowledge of the refractive indices associated with each crystal direction, what the axis angles are, and what are the relations among the optical axes.

The refractive index of light passing through an isotropic crystal will be identical along each of the crystal axes, and such crystals therefore possess *single refraction*. Anisotropic substances will exhibit different refractive indices for light polarized with respect to the crystal axes, thus exhibiting *double refraction*. Crystals within the hexagonal and tetragonal systems possess one isotropic direction, and are termed *uniaxial*. Anisotropic crystals possessing two isotropic axes are termed *biaxial*, and include all crystals belonging to the orthorhombic, monoclinic, or triclinic systems. Biaxial crystals will exhibit different indices of refraction along each of the crystal axes.

Isotropic samples are characterized by the existence of equivalent crystal axes, therefore exhibit *isotropic extinction*, and have no effect on the polarized light no matter how the crystal is oriented. When a sample is capable of exhibiting double refraction, the specimen will then appear bright against a dark background. For example, when a uniaxial crystal is placed with the unique *c*-axis horizontal on the stage, it will be alternately dark and bright as the stage is rotated. Furthermore, the crystal will be completely dark when the *c*-axis is parallel to the transmission plane of the polarizer or analyzer. If the crystal has edges or faces parallel to the *c*-axis, then it will be extinguished when such an edge or face is parallel to one of the polarizer directions, a condition is known as *parallel extinction*. At all intermediate positions, the crystal will appear light and usually colored. A rhombohedral or pyramidal crystal will be extinguished when the bisector of a silhouette angle is parallel to a polarization direction, and this type of extinction is termed *symmetrical extinction*. For biaxial crystals, similar results are obtained as with uniaxial crystals. The exception to this rule is that in monoclinic and triclinic systems, the polarization directions need not be parallel to faces or to the bisectors of face angles. If the prominent faces or edges of an extinguished crystal are not parallel to the axes of the initial polarizer, the extinction is said to be *oblique*. Knowledge of the type of extinction therefore permits one to make a determination of the system to which a given crystal belongs.

The ability to observe optical properties of crystals during heating and cooling processes is termed thermal microscopy, and this can be a profoundly useful technique during the study of polymorphs and solvatomorphs.[110,111] Crystal polymorphs ordinarily exhibit different melting points, with the order of melting points being indicative as the order of stability at the elevated temperature condition. The interconversion of such crystal forms is classified as being either *enantiotropic* or *monotropic*, according to whether the transformation of one modification into the other is reversible or not. Enantiotropic modifications interchange reversibly at the ordinary transition point, and each form is characterized by having its own stability range of temperature. Monotropic substances are characterized by the existence of a purely hypothetical transition point, since this point is predicted to be higher in temperature than the melting point of one of the polymorphic forms. Monotropic polymorphs are characterized by the fact that one form is stable at all temperatures below its melting point, while the second form is metastable at all temperatures.[108]

Scanning electron microscopy (SEM) is the technique of choice when one seeks to obtain information at high-magnification levels, or when a three-dimensional view of a particle surface is required. A conventional SEM is similar to an inverted light microscopy, in that the source lies above the specimen, the interrogating electron beam focused by a series of lenses, and the image constructed on the basis of scattered electromagnetic radiation. Samples are usually coated with conductive materials to reduce the deleterious effects of surface charging, although newer systems are not necessarily limited in this regard. SEM analysis is often combined with X-ray analysis, whereby one may obtain maps of the elemental distribution within a heterogeneous sample.

The morphology of excipients plays an important role is their physical properties, which in turn affects their application as formulation ingredients. One rather comprehensive study linked morphology and functionality of 14 direct compression excipients, and SEM analysis was the technique employed throughout.[112] For instance, croscarmellose is a polymeric substance that is commonly used in solid dose forms as a disintegrant. When the substance is produced as an ensemble of short fibers, one obtains improved flowability characteristics and good blending ability without adversely affecting the disintegrant properties.

In another particularly interesting application, SEM analysis was used to study the growth of carbamazepine crystals on the surface of tablets that had been stored at elevated temperatures.[113] This crystal growth was found only to take place when stearic acid was used as the tablet lubricant, and it was shown in this work that the carbamazepine drug substance could dissolve in the stearic acid. The lubricant provided a mechanism to transport the drug to the tablet surface, where it could crystallize.

B. X-ray Diffraction

The technique of X-ray diffraction is exceedingly important to pharmaceutics since it represents the primary method whereby one can obtain fundamental structural information on the structure of a crystalline substance. For example, it is only by pure coincidence that two compounds might form crystals in which the three-dimensional spacing of planes was identical in all directions. One such example is provided by the trihydrate phases of ampicillin and amoxicillin,[114] but such instances are uncommon. Typical applications of X-ray diffraction methodology include the determination of crystal structures, the evaluation of polymorphism and solvate structures, the evaluation of degrees of crystallinity, and the study of phase transitions.

Bragg[115] explained the diffraction of X-rays by crystals using a model where the atoms of a crystal are regularly arranged in space, and that they can be regarded as lying in parallel sheets separated by a definite and defined distance. Then he showed that scattering centers arranged in a plane act like a mirror to X-rays incident on them, so that constructive interference would occurs for the direction of specular reflection. Within a given family of planes, defined by a Miller index of (hkl) and each plane being separated by the distance d, each plane produces a specular reflectance of the incident beam. If the incident X-rays are monochromatic (having wavelength equal to λ), then for an arbitrary glancing angle of θ, the reflections from successive planes are out of phase with one another. This yields destructive interference in the scattered beams. However, by varying θ, a set of values for θ can be found so that the path difference between X-rays reflected by successive planes will be an integral number (n) of wavelengths, and then constructive interference will occurs. One ultimately obtains then expression known as Bragg's law that explains the phenomenon:

$$2d \sin\theta = n\lambda$$

Unlike the case of diffraction of light by a ruled grating, the diffraction of X-rays by a crystalline solid leads to observation of constructive interference (i.e., reflection) occurs only the critical Bragg angles. When reflection does occur, it is stated that the plane in question is reflecting in the *n*th order, or that one observes *n*th order diffraction for that particular crystal plane. Therefore, one will observe an X-ray scattering response for every plane defined by a unique Miller index of (*hkl*).

The analysis of single-crystal X-ray diffraction data is divided into three parts.[116–120] The first of these is the geometrical analysis, where one measures the exact spatial distribution of X-ray reflections and uses these to compute the size and shape of a unit cell. The second phase entails a study of the intensities of the various reflections, using this information to determine the atomic distribution within the unit cell. Finally, one looks at the X-ray diagram to deduce qualitative information about the quality of the crystal or the degree of order within the solid. This latter analysis may permit the adoption of certain assumptions that may aid in the solving of the crystalline structure.

While single-crystal X-ray diffraction is the most powerful technique for the study of crystalline solids, the necessity of obtaining perfect crystals and the complicated nature of the process does not lend it to the routine evaluation of polycrystalline solids. For such work, X-ray powder diffraction (XRPD) is much more useful.[121–125] Since a powdered sample will present all possible crystal faces at a given interface, the diffraction off this powdered surface will therefore provide information on all possible atomic spacings (i.e., defined by the crystal lattice). The powder pattern consists of peaks having varying intensities that are detected at various scattering angles. These angles, and their relative intensities, are correlated with computed *d*-spacings to provide a full crystallographic characterization of the powdered sample.

To measure a powder pattern, a randomly oriented powdered sample is prepared so as to expose all the planes of a sample. The scattering angle is determined by slowly rotating the sample and measuring the angle of diffracted X-rays (typically using a scintillation detector) with respect to the angle of the incident beam. Alternatively, the angle between sample and source can be kept fixed while moving the detector to determine the angles of the scattered radiation. Knowing the wavelength of the incident beam, the spacing between the planes (identified as the *d*-spacings) is calculated using Bragg's law.

Perhaps the most typical application of XRPD is in the qualitative identification of the crystallographic nature of a given sample, and a general guidance is available in the United States Pharmacopeia.[126] This technique is especially useful in establishing the phase identity of a polycrystalline solid that is capable of existing in multiple polymorphic forms. For example, as illustrated in Figure 6, the two polymorphs of famotidine can be readily distinguished on the basis of their characteristic XRPD patterns.[127] The relevance of XRPD methodology for the characterization of compounds having pharmaceutical interest has been reviewed.[128–130]

If one defines polymorphism as the situation where a given compound crystallizes in more than one structure and crystals of the two structures yield exactly the same elemental analysis, then two possibilities are available for polymorphic solids.[131–134] Should the molecule be constrained to exist as a rigid

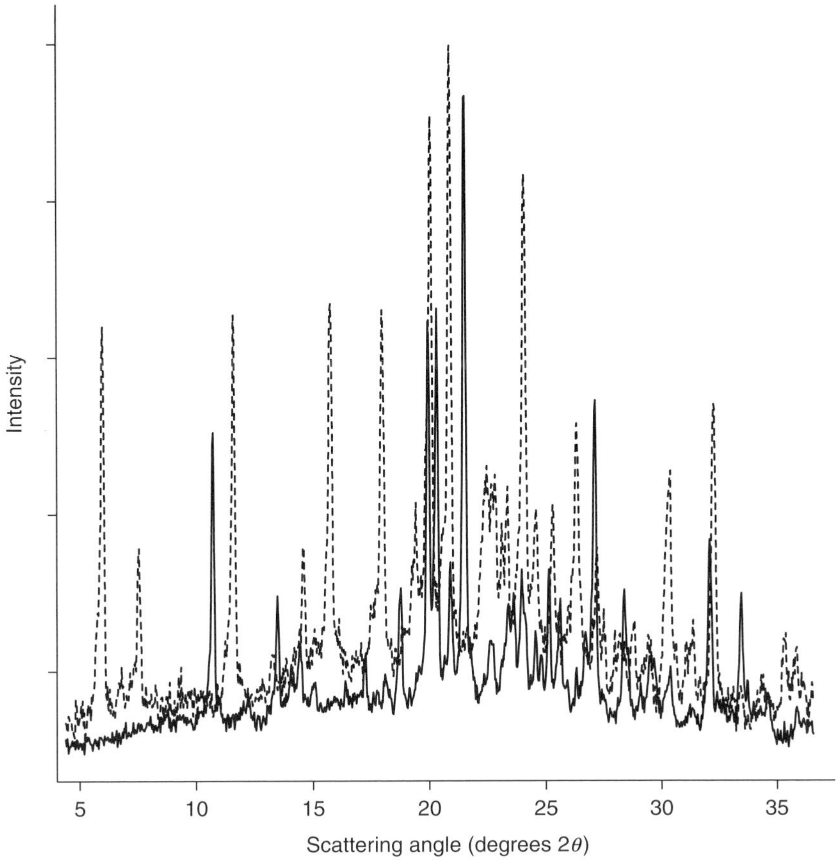

FIGURE 6 X-ray powder diffraction patterns for famotidine polymorphic form-A (solid trace) and polymorphic form-B (dashed trace) (H. G. Brittain, unpublished results).

grouping of atoms, these may be stacked in different motifs to occupy the points of different lattices. This type of polymorphism is then attributable to packing phenomena, and is termed packing polymorphism. On the other hand, if the molecule in question is not rigidly constructed and can exist in distinct conformational states, then it can happen that each of these conformationally distinct modifications may crystallize in its own lattice structure. This latter behavior has been termed conformational polymorphism.[135]

One of the earliest studies in this area was concerned with a case of packing polymorphism, where Robertson initially reported the structure of resorcinol (1,3-dihydroxy-benzene).[136] The crystal studied in this work corresponded to the form obtained at room temperature, and which was later termed the α-form. However, it was subsequently learned that this α-form was in fact a metastable form, and that it could undergo a phase transformation into a denser crystalline modification (denoted as the β-form) when heated at about 74 °C. The structure of this newer form was found to be completely different, and the crystallography permitted an explanation of the anomalous density change.[137]

Probucol (4,4′-[(1-methylethylidene)bis(thio)]-bis-[2,6-bis(1,1-dimethylethyl) phenol]) is a cholesterol-lowering drug that has been reported to exist in two forms.[138] Form-II has been found to exhibit a lower melting point onset relative to form-I, and samples of form-II spontaneously transform to form-I upon long-term storage. The conformations of the probucol molecule in the two forms were found to be quite different. In form-II, the C–S–C–S–C chain is extended, but this molecular symmetry is lost in the structure of form-I, where now the torsional angles around the two C–S bonds deviate significantly from 180°. Steric crowding of the phenolic groups by the *t*-butyl groups was evident from deviations from trigonal geometry at two phenolic carbons in both forms. Using a computational model, the authors found that the energy of form-II was 26.4 kJ/mol higher than the energy of form-I, indicating the less symmetrical conformer to be more stable.

A very useful supplement to XRPD is to use a variable temperature hot stage as the sample holder, enabling one to heat a sample to any desired temperature and thus study the crystallographic consequences of thermally induced phenomena. Such a VT-XRPD system equipped with a heatable sample holder has been described which permitted accurately controlled heating up to 250 °C.[139] The system was used to study the phase transformation of phenanthrene and the dehydration of caffeine hydrate.

Studies of the desolvation of solvatomorphs can be conducted using VT-XRPD. For instance, after the dehydration of a hydrate phase, one may obtain either a crystalline anhydrate phase or an amorphous phase and the XRPD pattern will clearly indicate the difference. In addition, should one encounter an equivalence in powder patterns between the hydrate phase and its dehydrated form, this would indicate the existence of channel-type water (as opposed to lattice bound water). In one study, the solid-state properties of the isomorphic desolvates of cephalexin, cefaclor, erythromycin A, and spirapril hydrochloride were investigated, with the hygroscopicity of the compounds being evaluated using a vacuum moisture balance and the structural relaxation measured using a combination of VT-XRPD and isothermal microcalorimetry.[140]

The commercially available form of aspartame is hemihydrate form-II, which has been shown to transform into hemihydrate form-I when milled.[141,142] VT-XRPD has been used to study the desolvation and ultimate decomposition of the various hydrates. When heated to 150 °C, both hemihydrate forms dehydrate into the same anhydrous phase, which then cyclizes to 3-(carboxymethyl)-6-benzyl-2,5-dioxopiperazine if heated to 200 °C. A 2.5-hydrate solvatomorph was shown to dehydrate to hemihydrate form-II when heated to 70 °C, and this product was then shown to undergo the same decomposition sequence as directly crystallized hemihydrate form-II.

It is now possible to determine the structure of a polycrystalline sample using XRPD data of sufficient quality, typically obtained using synchrotron radiation as the source.[142–146] To solve a structure from powder data, one first indexes the powder pattern to determine the dimensions of the unit cell by assigning each of the observed diffraction peaks to a particular (*hkl*) diffraction plane. The number of atoms per unit cell is calculated from the size of the unit cell, the chemical composition, and the density of the specimens. After this, a structure is assumed and tested by calculation of the XRPD pattern expected for this structure. Different possible structures are thus evaluated, leading to an initial approximate structure model. Finally, one refines the structure by using the Rietveld profile refinement technique[147,148] to identify the most promising structure.

High-resolution XRPD data have been used to determine the structures of three crystalline modifications of telmisartan,[149] the elusive form-II of enalapril maleate,[150] and bupivacaine free base.[151]

C. Thermal Methods of Analysis

Thermal methods of analysis can be defined as those techniques in which a property of the substance under study is determined as a function of an externally applied and programmed temperature. Dollimore has defined the practice of thermal analysis as consisting of technology where the physical property and the sample temperature are measured continuously, both the property and the temperature are recorded automatically, and the temperature of the sample is changed at a predetermined rate.[152] Thermal analysis methods are used to monitor both endothermic processes (i.e., melting, boiling, sublimation, vaporization, desolvation, solid–solid phase transitions, and chemical degradation) and exothermic processes (i.e., crystallization and oxidative decomposition). The technology of thermal analysis, and its application to pharmaceutical compounds, has been the subject of several books.[153–157]

Significant insight into the practice of thermal analysis can be gleaned from a consideration of the determination of ordinary melting points through study of fusion curves, for which a typical example is provided in Figure 7. To determine this type of curve, one places the sample within a suitable container, immerses it in a bath whose temperature is increased at a fixed rate, and the temperature of the sample monitored. As long as the substance is a solid, it has a fixed and finite heat capacity, so its temperature linearly increases at a rate governed by the rate of heat flow. Once the substance begins to melt, the heat capacity of the system goes to infinity as all absorbed heat is used to transform the solid phase into the liquid phase. The temperature of the sample cannot change during this process owing to the infinite value of the heat capacity and the fact that any heat entering the system acts only to redistribute the relative amounts of the two phases. Once the entire sample has melted, however, the heat capacity again becomes fixed and finite since the substance is now a simple liquid. Further heating results in a linear rate of temperature increase up to the next phase transition.

Differential thermal analysis (DTA) is an improvement to the melting point determination in that one monitors the difference in temperature between the sample and a reference as a function of temperature.[158,159] As long as no thermal transitions take place, the temperature of the sample and reference will be the same since the heat capacities of the two will be roughly equivalent. However, differences in temperature between the sample and reference will be manifest when changes occur that require a finite heat of reaction. If ΔH for the transition is positive (endothermic reaction), the temperature of the sample will lag behind that of the reference (since more heat will be absorbed by the sample than by the reference), and this event will be recorded as a negative-going peak in the thermogram. If the ΔH is negative (exothermic reaction), the temperature of the

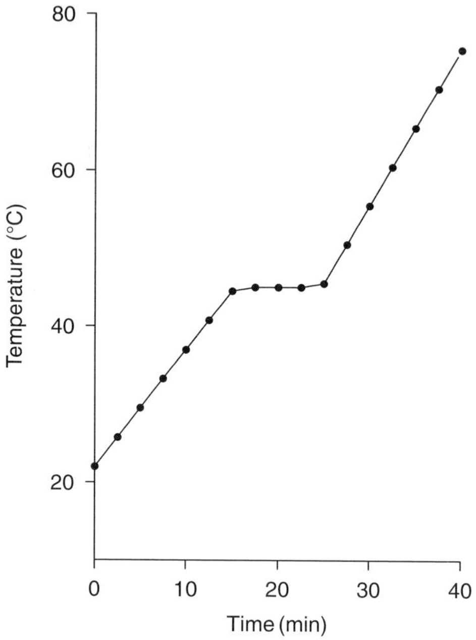

FIGURE 7 **Fusion curve for the melting of a hypothetical compound.**

sample will exceed that of the reference (since the sample itself will be a source of additional heat), and the event will be recorded as a positive-going peak in the thermogram. Thus, DTA analysis is an excellent qualitative technique useful to deduce the temperature ranges associated with a variety of thermal events, as well as assigning the endothermic or exothermic nature of these reactions.

DSC represents the next stage in the growth of thermal analysis, and can be used to obtain analogous information about the same types of thermally induced reactions as DTA. However, the instrumentation used to measure DSC thermograms is such that quantitative analyses are easy to perform, and consequently, DSC has become the most widely used method of thermal analysis.

In the DSC method, the sample and reference are maintained at the same temperature and the heat flow required to keep the equality in temperature is measured. DSC plots are therefore obtained as the differential rate of heating (in units of W/s, cal/s, or J/s) against temperature. The area under a DSC peak is directly proportional to the heat absorbed or evolved by the thermal event, and the integration of these peak areas yields the heat of reaction (in units of cal/s/g or J/s/g). Owing to its ability to facilitate quantitative data interpretation, the use of DSC analysis has virtually supplanted the use of DTA analysis. The relevance of the DSC technique as a tool for pharmaceutical scientists has been amply documented in numerous reviews.[156,157,160–165]

When a compound is observed to melt without decomposition, DSC analysis can be used to determine the absolute purity.[166] If the impurities are soluble in the melt of the major component, the van't Hoff equation applies:

$$T_s = T_0 - \frac{R(T_0)^2 X_i}{F \Delta H_f}$$

In the preceding equation, T_s is the sample temperature, T_0 is the melting point of the pure major component, X_i is the mole fraction of the impurity, F is the fraction of solid melted, and ΔH_f is the enthalpy of fusion of the pure component. A plot of T_s against $1/F$ should yield a straight line, whose slope is proportional to X_i. This method can therefore be used to evaluate the absolute purity of a given compound without reference to a standard, with purities being obtained in terms of mole percent. The method is limited to reasonably pure compounds which melt without decomposition. The assumptions justifying the van't Hoff equation fail when the compound purity is below approximately 97 mol%, and the method cannot be used in such instances. The DSC purity method has been critically reviewed, with the advantages and limitations of the technique being carefully explored.[167]

The DSC analysis has proved to be of great use in the study of the relative phase stability of polymorphic solids. One of the more useful empirically established relations is the heat of fusion rule, which states that if the higher melting form has a lower heat of fusion relative to the lower melting form, then the two forms are related by enantiotropy. The DSC of a metastable polymorph that exists as part of an enantiotropic pair is shown in Figure 8, which is seen to consist of a melting endotherm of the metastable phase, followed by crystallization and melting transitions of the more stable phase.[168] The nature of the enantiotropy was established by systematically changing the sample heating rates, where it was found that the heat of fusion of the lower melting form was greater than that of the higher melting form.

One other commonly used thermoanalytical technique is that of thermogravimetry (TG), where one measures the thermally induced weight loss of a material as a function of the applied temperature.[169,170] TG analysis is restricted to studies that involve either a mass gain or loss (usually loss), and is most commonly used to study desolvation processes and compound decomposition. The major use of TG analysis is in the quantitative determination of the total volatile content of a solid. When a solid can decompose by means of several discrete, sequential reactions, the magnitude of each step can be separately evaluated. TG analysis of compound decomposition can also be used to compare the stability of similar compounds. The higher the decomposition temperature of a given compound, the more negative would be the ΔG value and therefore the greater would be the stability.

An example will be cited to illustrate the synergistic relationship existing between DSC and TG analysis. The anhydrate, dihydrate, and trihydrate phases of chemical pure magnesium stearate and magnesium palmitate have been prepared and characterized as to their structural characteristics.[171] The crystal structures of the various materials were found to be very similar to each other, differing primarily in the magnitude of the long (001 face) crystal spacing. The full thermal characterization of magnesium stearate dihydrate is shown in Figure 9. The nature of the lowest temperature endothermic transition seen in the DSC thermogram is established as a dehydration transition, since a substantial weight loss is observed to take place over the same temperature interval. On the other hand,

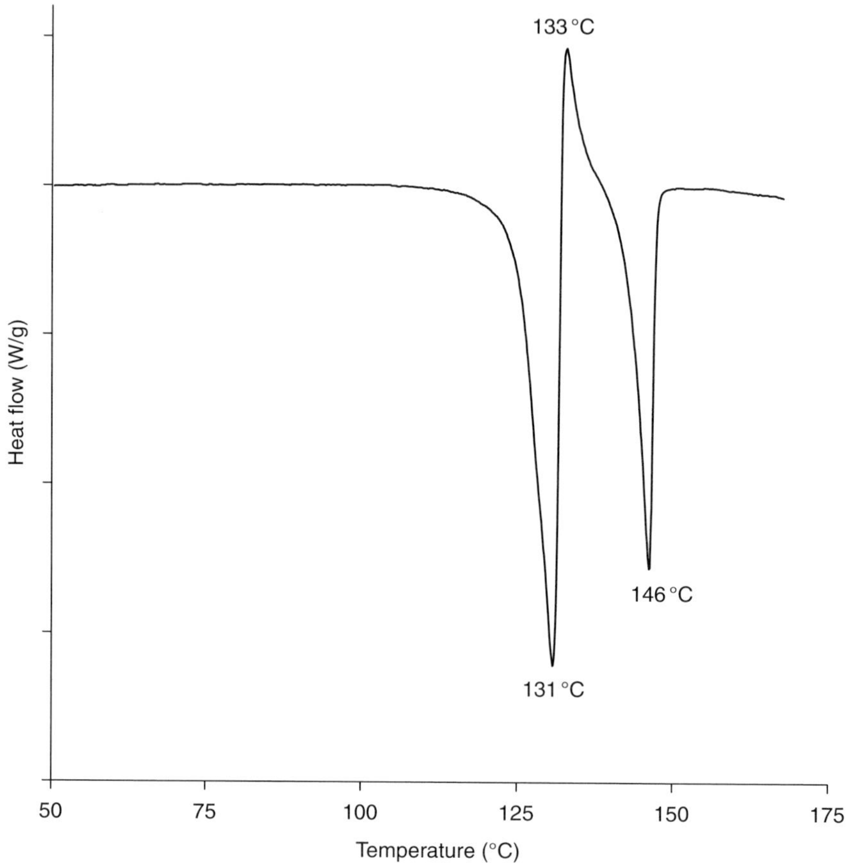

FIGURE 8 **DSC thermogram obtained for the metastable phase of a new chemical entity that undergoes a melting transition (131 °C peak), subsequent crystallization as the stable phase (133 °C peak), and melting of the stable phase (146 °C peak) (H. G. Brittain, unpublished results).**

the higher temperature endothermic transition in the DSC thermogram is confirmed as a melting transition since this event takes place without any accompanying loss in mass.

IV. PROPERTIES ASSOCIATED WITH THE BULK LEVEL

Bulk material properties may be conveniently defined as those characteristics of a solid that can only be measured for a large ensemble of particles. Given the sheer magnitude of the materials issues involved, it is safe to state that once a solid formulation has reached the bulk manufacturing stage, bulk physical properties become of the highest degree of importance. The ability to totally control a large-scale formulation process will be governed by the degree of knowledge amassed for the system in question, and the most pertinent body of knowledge concerns the properties that directly relate to those of the bulk powders involved.

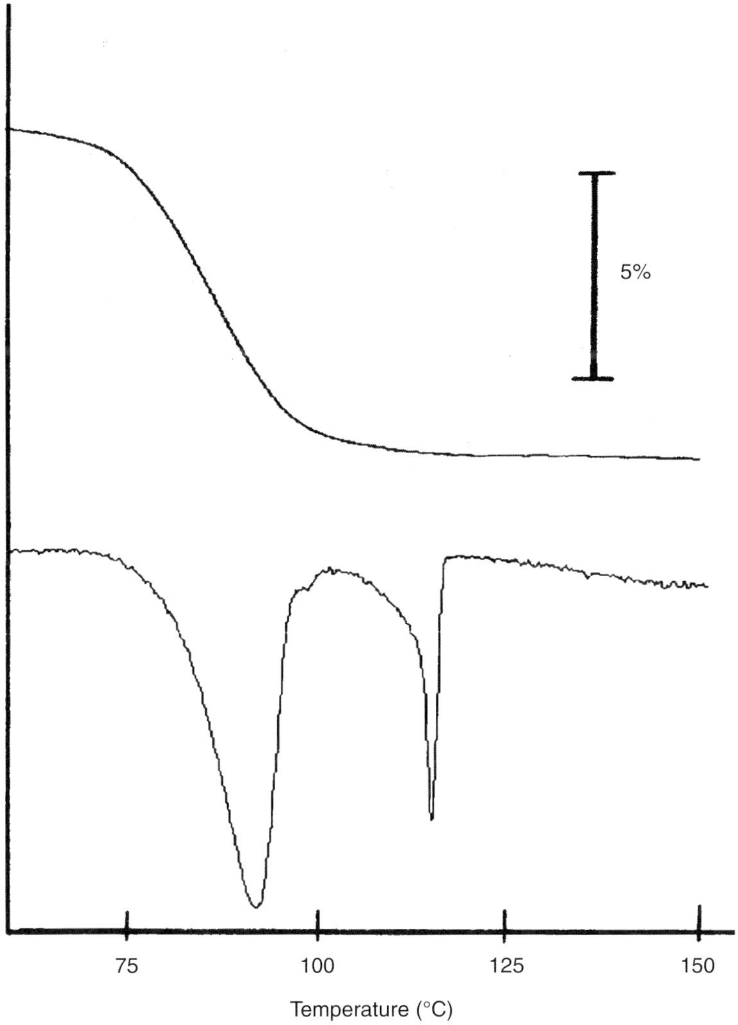

FIGURE 9 **Differential scanning calorimetry (lower trace) and thermogravimetry (upper trace) thermograms obtained for the dihydrate phase of magnesium stearate. The data have been adapted from Ref. 171.**

It is also true that reproducibility in all aspects of lot-to-lot behavior can be significantly improved through the implementation of properly designed specifications. The testing of raw materials is particularly important at this stage, since only those species passing appropriate challenges should be used for manufacturing processes.

Once a given drug candidate has reached the latter stages of development, the physical characteristics associated with the other formulation ingredients assumes a high level of importance. Excipients are included in formulations to execute a desirable physical effect, such as enhancing powder compaction, promoting dosage form dissolution, modifying drug release rates, improving powder flowability, etc. Since there is really no excipient that can be termed an "inactive component," it is almost self-evident that assays suitable for the evaluation of functionality properties of excipients would be essential to the development of robust pharmaceutical formulations. This requires the establishment of comprehensive programs for the physical characterization of excipient materials, especially with respect to properties related to use and functionality.

It is beyond the scope of the present coverage to discuss sampling effects, but it cannot be stressed too often that the ability to obtain relevant information on the bulk properties of a material depends critically on the sampling plan used. It is fairly easy to unintentionally introduce bias during the act of taking samples, and no amount of method development and validation in subsequent steps can alleviate this problem. When milligram to gram samples are removed from a multikilogram bulk, those samples must be representative of the bulk properties or else the analytical work will be absolutely valueless. Issues pertaining to the sampling of powdered samples have been discussed in detail,[172–177] and such issues can only be ignored at one's peril.

A. Particle Size Distribution

It is not possible to discuss the size of a particle, or any distribution of the sizes of an ensemble of particles, without first considering the three-dimensional characteristics of the particle itself. This is because the size of a particle is expressed in terms of linear dimension characteristics derived from its shape, or in terms of its projected surface or volume. As will be shown, some methods of expressing particle size discard any concept of particle shape, and instead express the size in terms of a spherical equivalent to a chosen geometric shape.

An appropriate starting place for a discussion of particle shape can be found in general test <776> of the United States Pharmacopoeia, which categorizes particles as being acicular (slender, needle-like particle of similar width and thickness), columnar (long, thin particle with a width and thickness that are greater than those of an acicular particle), flakes (thin, flat particle of similar length and width), plates (flat particles of similar length and width but with greater thickness than flakes), laths (long, thin, and blade-like particle), or equant (particles of similar length, width, and thickness; both cubical and spherical particles are included). In addition, the USP provides other terms for aggregated or agglomerated structures, such as lamellar (stacked plates), aggregate (mass of adhered particles), agglomerate (fused or cemented particles), conglomerate (mixture of two or more types of particles), spherulite (radial cluster), and drusy (particle covered with tiny particles).

The most commonly used definitions of particle sizes are derived from particle images, where one may define a particle length (the longest dimension from edge to edge of a particle oriented parallel to the ocular scale) and its width (the longest dimension of the particle measured at right angles to the length). Closely related to these properties are two other descriptors of particle size, namely Feret's diameter (the distance between imaginary parallel lines tangent to a randomly oriented particle and perpendicular to the ocular scale), and Martin's diameter (the diameter of the particle at the point that divides a randomly oriented particle into two equal projected areas).

Defining spatial descriptors for freely tumbling particles is considerably more difficult, and consequently, the use of a number of derivative particle descriptors has arisen. These derived descriptors for particle size all begin with a homogenization of the length and width descriptors into either a circular or spherical equivalent, and make use of the ordinary geometrical equations associated with the derived equivalent. For instance, the perimeter diameter is defined as the diameter of a circle having the same perimeter as the projected outline of the particle. The surface diameter is taken as the diameter of a sphere having the same surface area as the particle, and the volume diameter is defined as the diameter of a sphere having the same volume as the particle. One of the most widely used derived descriptors is the projected area diameter, which is the diameter of a circle having the same area as the projected area of the particle resting in a stable position.

Several other derived descriptors of particle diameter have been used for different applications. For instance, the sieve diameter is the width of the minimum square aperture through which the particle will pass. Other descriptors that have been used are the drag diameter (the diameter of a sphere having the same resistance to motion as the particle in a fluid of the same viscosity and at the same velocity), the free-falling diameter (the diameter of a sphere having the same density and the same free-falling speed as the particle in a fluid of the same density and viscosity), and the Stokes' diameter (the free-falling diameter of a particle in the laminar flow region).

The particles making up samples of powdered substances will not consist not of mono-sized particles, but instead will exhibit a range of shapes and sizes. Particle size determinations are therefore undertaken to obtain information about the size characteristics of an ensemble of particles. Furthermore, since the particles under study are not of the same exact size, information is required regarding the average particle size and the distribution of sizes about this average. The simplest representation of a particle size distribution would be that of a bell-shaped curve, and this type of system is known as a normal distribution. Samples that conform to the characteristics of a normal distribution are fully described by a mean particle size and the standard deviation.

The arithmetic *mean* of the ensemble of particle diameters is calculated using the relation:

$$d_{\mathrm{av}} = \frac{\sum n d_{\mathrm{i}}}{\sum n}$$

where n is the number of particles having a diameter equal to d_{i}. The standard deviation in the distribution is then calculated using:

$$\sigma = \left[\frac{(d_{\mathrm{av}} - d_i)^2}{n} \right]^{1/2}$$

The most commonly occurring value in the distribution is the mode, and is the value at which the frequency representation is a maximum. The median divides the frequency curve into two equal parts, and equals the particle size at which the cumulative representation equals 50%. In a rigorous normal distribution, the mean, mode, and median have the same value.

It would be highly advantageous if powder distributions could be described by the normal function, since all of the statistical procedures developed for Gaussian distributions could be used to describe the properties of the sample. However, it turns out that unless the range of particle sizes is extremely narrow, the distributions of most powder samples cannot be adequately described by the normal distribution function. The size distribution of the majority of real powder samples is usually skewed toward the larger end of the particle size scale, and such powders are better described using the log-normal distribution type. This terminology has arisen since when the particle distribution is plotted using the logarithm of the particle size, the skewed curve is transformed into one closely resembling a normal distribution.

The distribution in a log-normal representation can be completely specified by two parameters, the *geometric median* particle size (d_g) and the standard deviation in the geometric mean (σ_g). The geometric median is the particle size pertaining to the 50% value in the cumulative distribution, and is calculated using:

$$d_g = \text{anti} - \log \frac{\sum n \log(d_i)}{\sum n}$$

where n is the mass of particles having particle size equal to d_i. Two samples having identical d_g and σ_g values can be said to have been drawn from the same total population and exhibit properties of characteristics of the total population.

Although the distribution in the log-normal representation is completely specified by the geometric median particle size and the geometric mean standard deviation, a number of other average values have been derived to define useful properties. These are especially useful when the physical significance of the geometric median particle size is not clear. The *arithmetic mean* (d_{av}) particle size is defined as the sum of all particle diameters divided by the total number of particles, and is calculated using:

$$d_{av} = \frac{\sum n d_i}{\sum n}$$

The *surface mean* (d_s) particle size is defined as the diameter of a hypothetical particle having an average surface area, and is calculated using:

$$d_s = \left[\frac{\sum n d_i^2}{\sum n}\right]^{1/2}$$

The *volume mean* (d_v) particle size is the diameter of a hypothetical particle having an average volume, and is obtained from:

$$d_v = \left[\frac{\sum n d_i^3}{\sum n}\right]^{1/3}$$

The volume–surface mean (d_{vs}) particle size is the average size based on the specific surface per unit volume, and is calculated using:

$$d_{vs} = \frac{\sum n d_i^3}{\sum n d_i^2}$$

A variety of methods are available for the determination of the particle size distribution of powdered solids,[178–181] with the most important of these being optical microscopy (usually combined with image analysis), sieve analysis, and laser light scattering.

The most straight-forward method of particle size determination is that of microscopy, especially when combined with image analysis, since here one performs direct observation of the particles.[178] Calibration of an observed image is easily effected through the use of stage micrometers, and once calibrated, a given set of optics will not require recalibration. In the automated methodology, microscope parameters are adjusted so as to optimize the contrast between the background and particles to be sized. A video image of the powder is transmitted to a computer system, which then counts the number of pixels making up a particle. The size of each pixel is easily converted to micrometers, and the data analyzed as desired by the analyst. Average particle sizes, full weight distributions, or shape information can be generated. The advantage of the optical microscope method is that it provides direct and absolute information on the particles under characterization. Its chief disadvantage is that it can only provide data on the particles on the slide, and can therefore be biased by the method preparing the slide.

In principle, sieve analysis is the simplest method for the determination of particle sizes, and is certainly one of the most widely used methods.[182] In sieving analysis, particles are allowed to distribute among a series of screens, and the amount of material retained on each screen determined. The smaller particles passing through a screen are termed the fines, while the larger particles remaining on the screen are the coarse particles. When using multiple screens, the intermediate sized particles that pass through one or more screens (but which are retained on a subsequent screen) are called the medium fractions. A variety of facilitation methods can be provided during the sieving process, with vibration, ultrasound, or air suspension being used to assist the particles with passage through the various screens. A proper size determination requires the use of five to six sieves, whose sizes are selected to obtain approximately equal amounts of powder on each screen and past the smallest sieve. The data are most commonly displayed as the percent of material retained on each sieve, the cumulative percent of sample retained, and the percent of sample passing each sieve.

In many cases, the particle size distribution of a real sample turns out to be adequately represented by a log-normal distribution. In that case, the distribution can be specified by the geometric median particle size (d_g) and the geometric mean standard deviation (σ_g). Two samples having identical d_g and σ_g values can be said to have been drawn from the same total population. The value of d_g is equal to the 50% value of the cumulative distribution, and consequently, the geometric median particle size is often referred to as the d_{50} value. The value of σ_g is obtained by dividing the 84.1% value of the distribution by the 50% value. Sieving results are most commonly plotted on three-cycle log paper to compare the particle size with the cumulative percentage of undersize particles. If the plot is linear over the entire range, then the material is characterized by a log-normal distribution. If the line is curved, or consists of two or more linear segments, the distribution is polymodal.

The use of laser light scattering for the determination of particle size distribution has become widespread. Owing to the nature of light scattering by particles, the technology is generally divided into two general methods. One of these uses high-angle light scattering, and is suited for determining the particle size of ultrafine particles. The other technique makes use of low-angle scattering, and is appropriate for the characterization of larger particles. It must be understood that neither technique actually directly measures the size of particles, but instead interpretations of light scattering phenomena are used to deduce particle size information.[183]

Other than nanoparticles, most pharmaceutically interesting particles have sizes in the typical 1–200 μm size range. Samples may be dispersed in a laser beam either in the form of suspensions in an inert fluid or in the form of dry powders aspirated directly into the analyzing beam. Here, the validation of the sample dispersion methodology is crucial, since it must be verified that the method of sample handling and processing does not alter the particle size distribution of the substance. Particle sizing by laser light scattering entails an interpretation of the scattering of collimated laser light (at low angles of angles of incidence) as caused by suspended particles, and particle size distributions are then deduced from the scattering intensities. After passing through the sample, the laser light is passed through a Fourier lens to produce a diffraction pattern that is allowed to fall on a detector that consists of a sequence of concentric rings. The light intensity impinging on each ring is measured, yielding a pattern or radial intensity values.

To deduce a particle size distribution, the detector response must be deconvoluted by means of a simulation calculation. The scattering particles are assumed to be spherical in shape, and the data subjected to one of three different computational methods. One system uses the unimodal model-dependent method, which begins with the assumption of a model (such as log-normal) for the size distribution. The detector response expected for this distribution is simulated, and then the model parameters are optimized by minimizing the sum of squared deviations from the measured and the simulated detector responses. The model parameters are finally used to modify the originally chosen size distribution, and it is this modified distribution that is presented to the analyst as the final result.

A second approach uses the unimodal model-independent method. Here, one begins with the assumption that the size distribution consists of a finite number of fixed size classes. The detector response expected for this distribution is simulated, and then the weight fractions in each size class are optimized through a minimization of the sum of squared deviations from the measured and simulated detector responses. The third system uses the multimodal model-independent method. For this, one simulates diffraction patterns for known size distributions, superimposes random noise on the patterns, and then calculates the expected element responses for the detector configuration. The patterns are inverted by the same minimization algorithm, and these inverted patterns compared with known distributions to check for qualitative correctness.

To illustrate some basic principles of particle size representation and distribution, an experimentally derived size distribution of a sample of sodium starch glycolate suspended in water is shown in Figure 10 along with its cumulative distribution.[184] Where the size axis transformed to a logarithmic scale, one would observe that the sample is characterized by an effective log-normal distribution. For this particular sample, the median particle size (i.e., the d_{50} value) was calculated to be 78.7 μm, the mode was found to be 90.1 μm, and the arithmetic mean was calculated to be 89.4 μm.

B. Micromeritics

When applied to powders, micromeritics is taken to include the fields that relate to the nature of the surfaces making up the solid. Of all the properties that could be measured, the surface area, its porosity, and the density of a material are generally considered to the most pharmaceutically relevant parameters.

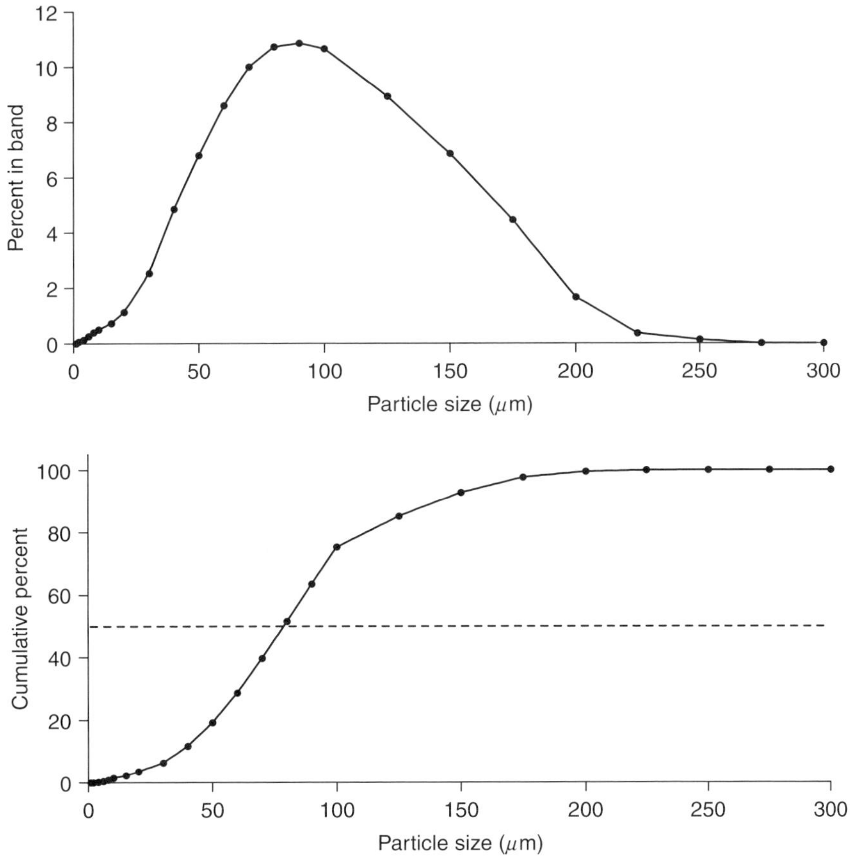

FIGURE 10 **Particle size distribution (upper curve) and the associated cumulative distribution (lower curve) obtained for a sample of sodium starch glycolate dispersed in water (H. G. Brittain, unpublished results).**

The surface area of a solid material is important in that it provides information on the available void spaces on the surfaces of a powdered solid.[185] In addition, the dissolution rate of a solid is partially determined by its surface area. The most reproducible measurements of the surface area of a solid are obtained by adsorbing a monolayer of inert gas onto the solid surface at reduced temperature, and subsequently, desorbing this gas at room temperature. The sorption isotherms obtained in this technique are interpreted using the equations developed by Brunauer, Emmett, and Teller (BET), and therefore, the technique is referred to as the BET method.[186] The surface area is obtained in units of square meters of surface per gram of material.

Any condensable, inert gas can be used for BET measurements, but the preferred gases are nitrogen and krypton. Nitrogen is used for most samples exhibiting surface areas of 2 m^2/g or greater, but materials with smaller surface areas should be measured using krypton. The gas to be adsorbed (the adsorbate) is mixed with an inert, noncondensable, carrier gas (usually helium). A range of 5–30% adsorbate in carrier gas is commonly used, and the use of multiple adsorbate gas levels in a BET determination is recommended.

Relations between the internal surface area and tablet properties have been drawn through the characterization of a variety of lactose compacts.[187] For example, a given bulk sample of anhydrous α-lactose was sieved into selected size fractions, and compacts of these prepared through compression at 37.5 mPa. Nitrogen gas adsorption was used to evaluate the surface area of the compacts, and the tablet crushing strengths of each were determined. The crushing strength of the compacts was found to be directly proportional (and almost linearly dependent) on the surface area of the compacted material. It was also found that compacts prepared from the coarser size fractions contained significantly lower surface areas than compacts prepared from the fine fractions.

Although a variety of methods are available to characterize the interstitial voids of a solid, the most useful of these is that of mercury intrusion porosimetry.[188] This method is widely used to determine the pore size distribution of porous materials, and the void size of tablets and compacts. The method is based on the capillary rise phenomenon in which excess pressure is required to force a nonwetting liquid into a narrow volume.

Mercury, with its contact angle on glass of approximately 140°, is most commonly used as the intrusion fluid. The mercury is forced into the pores of the sample using an externally applied pressure, with the smallest pores requiring the highest pressures to effect the filling. The Washburn equation, as applied to circular pore openings, is used to relate the applied pressure and the pore size opening.

Measurements of particle porosity are a valuable supplement to studies of specific surface area, and such data are particularly useful in the evaluation of materials used in direct compression processes. For example, both micromeritic properties were measured for several different types of cellulose-type excipients.[189] BET surface areas were used to evaluate the all types of pore structures, while the method of mercury intrusion porosimetry used could not detect pores smaller than 10 nm. The data permitted a ready differentiation between the intraparticle pore structure of microcrystalline and agglomerated cellulose powders.

Another extremely important micromeritic parameter is that of powder density, which is defined as the ratio of mass to volume. Three types of density are normally differentiated, which differ in their determination of volume occupied by the powder. *Bulk* density is obtained by measuring the volume of a known mass of powder sample (that has been passed through a mesh screen) into a suitable volume-measuring apparatus.[190] When a graduated cylinder is used for the measurement, the volume is determined to the nearest milliliter. The bulk density is then obtained by dividing the mass of solid by the unsettled apparent volume. A measurement of *tapped* density is normally obtained at the same time, with the volume of the solid being measured after subjecting the system to a number of controlled shocks. The repeated mechanical stress causes the powder bed to pack into a smaller volume, so it follows that the tapped density will always be higher than the bulk density.

The *true* density of a solid is the average mass per unit volume, exclusive of all voids that are not a fundamental part of the molecular packing arrangement.[191] This density parameter is normally measured by helium pycnometry, where the volume occupied by a known mass of powder is determined by measuring the volume of gas displaced by the powder. The true density of a solid is an intrinsic property characteristic of the analyte, and is determined by the composition of the unit cell.

Inverse gas chromatography (IGC) is a relative new method that has been used to evaluate the surface energy of powdered materials. IGC differs from conventional gas chromatography in that the stationary phase is the powdered solid under study, and one measures the interaction of selected probe molecules with the surface.[192] The technique was developed as a superior technique to contact angle measurements, as IGC does not require compaction of the powder for a measurement to be made.[193] In one work, it was reported that the surface energy for mixtures of amorphous and crystalline lactose became equivalent to that of the amorphous form once the amorphous content exceeded 15% w/w, but even so, it was concluded that the IGC results did not simply measure the high-energy sites in the sample, but instead reflected properties that were not weighted averages of the surface energy of the two components.[194] However, guidelines are available that indicate the appropriate use of the technique,[195] which has been used to study the surface properties of racemic propranolol hydrochloride,[196] racemic and resolved mannitol,[197] and various physical forms of lactose.[198,199]

C. Mechanical Properties of Powders

Evaluation of the mechanical characteristics of powdered solids is vitally important to the processing of these materials. Information can be obtained on bulk powders prior to their processing, during the compaction process through the instrumentation of tablet presses, or on tablet compacts after these have been compressed. In the first instance, work is generally centered on determinations of the degree of flowability associated with a given powder. Data obtained during the compaction of powdered solids can be an invaluable source of information to optimize the consolidation process. Measurements conducted on consolidated materials are also used during process optimization, and can also be employed as part of the quality control testing. It should be recognized, however, that particle–particle interactions are at the center of these investigations, and consequently, the appropriate methods respond to such interactions.[200–203]

When the developmental stage for a drug product reaches large-scale manufacture, one of the more important parameters of interest is the ability of powdered solids to flow.[204–206] Processability of these materials is greatly affected by flowability concerns, since the materials invariably need to be moved from place to place. For example, when tablets are to be compressed at high speeds, the efficiency of the machine will only be suitable if the powder feed can be delivered at a sufficiently high rate. Since many pharmaceutical compounds are cohesive in nature, their flow characteristics tend to be undesirable. One of the aims of granulation is reduce the cohesive nature of the individual components, producing a uniformly blended material whose physical properties are more suitable for processing.

An extremely useful system was described by Carr, who devised a system that can be used for the evaluation of the flowability of powdered solids.[207,208] In his approach, Carr defined a number of parameters related to flow, which are scored after their measurement according to a weighting system. Powder flowability is evaluated using the angle of repose (defined as the angle formed when a cone of powder is poured onto a flat surface), the angle of spatula (defined as the angle formed when material is raised on a flat surface out of a bulk pile), compressibility (obtained from measurement of the bulk and tapped material densities), and cohesion (relating to the attractive forces which exist on particle surfaces). The overall summation of these permits deductions regarding the degree of powder flowability.

When powders flow, they do so either in a steady controlled fashion (as in the case of dry sand), or in an uncontrolled gushing manner (as would damp sand, for which the entire bulk tries to move in a solid mass). This latter condition is termed floodable flow, and is most characteristic of the flow of cohesive, sticky powders. The floodability of a powder is determined by its overall flowability (the determination of which has just been described), the angle of fall (obtained as the new repose angle when the powder cone is mechanically shocked), the dispersibility (ability of a given powder to become fluidized), and the angle of difference (obtained as the numerical difference between the angle of fall and angle of repose).

To illustrate the utility of Carr's method in the evaluation of powder flow, full characterization of 12 lots of microcrystalline cellulose have been carried out.[209] Measurements of the actual mass flow rates were obtained, as well as each of the parameters specified by Carr. As evident in Figure 11, the overall flowability index proved to be a reliable predictor of the relative degree of bulk powder flow for granular material grades, but the correlation with actual flow rates did not hold for fine material grades. Detailed comparisons of the possible relation of each parameter with the mass flow rates were conducted with the aim of determining which parameter exerted the largest influence over the mass flow rate. For the microcrystalline cellulose samples evaluated in this study, it was learned that the powder flow rates were largely determined by the sample compressibility and degrees of cohesion. Fairly well defined, but separate, inverse linear relationships were deduced between these properties and the mass flow rates. Interestingly, only

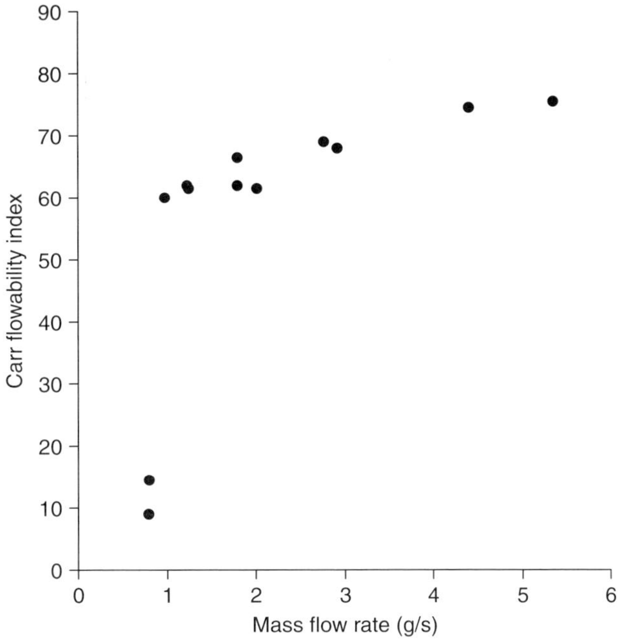

FIGURE 11 **Correlation between actual mass flow rates and the Carr flowability indices of 12 samples of microcrystalline cellulose, each differing in their particle size distribution (H. G. Brittain, unpublished results).**

a poor correlation between actual powder flow rates and the angle of repose was detected, in spite of the generally held notion that the repose angle is a reliable indicator of powder flow.

More recently, avalanching powder flow has been used to evaluate the flowability of powdered solids.[210,211] In this approach, one uses measurements of relative frequency as a function of time between successive powder cascades to deduce powder flowability under simulated working conditions. In one study, a systematic study of the powder avalanches of a number of different materials was used to deduce two indices describing fluidity and cohesion.[212] The flowability index (not to be confused with Carr's index of the same name) illustrates a resistance characteristic of a powder to initiate movement during an avalanche, while the cohesion information provides information on the capacity of the powder to be agglomerated.

Since ultimately the majority of pharmaceutical formulations are administered as solid dose forms, it is no surprise that a great deal of research work has been conducted on tablets and other solid dose forms.[213–215] The compaction of powdered solids has been reviewed at length,[216] and detailed studies have been conducted concerning fundamentals of powder compression[217,218] and powder densification.[219,220] In addition, a detailed series of studies have been conducted on the consolidation and compaction of powder mixtures, using various particle size fractions and solvatomorphic forms of lactose.[221–223]

Studies involving instrumented compaction equipment can be extremely useful in the development of dosage forms, especially when the amount of drug substance is limited in quantity. Dynamic studies of powder compaction[224] and the compaction simulator[225] have been found to be useful during all stages of the development process to acquire information on the compactibility of formulations. The use of the compaction simulator is particularly important since it is possible to subject the formulation to high-speed tableting cycles in order to evaluate strain forces in the compacts. This work can be of great importance, since strain rates can exert strong influences on a variety of tablet properties.[226] The information gathered in this way can be used to aid in the transfer of the manufacturing process among different types of instrumentation, and aid in the technology transfer process.

In one study, the formulation efficiency of several direct compression materials was evaluated using instrumented press methodology.[226] It was found that subtle changes in the structure of the component particles could lead to the observation of significantly different behaviors upon compression. The tablet hardness and compressibility of differently sourced sucrose materials, obtained at comparable compressional forces, was found to vary significantly with the source of the compound. To quantitative the results, an elastic recovery index was defined, and it was found that the indices of direct compressible materials were lower than those of poorly compressible powders.

The use of compression indices dates back to Hiestand and coworkers, who developed several indices of tableting performance that are indicative of the performance of powdered solids during their compaction.[227] The bonding index is an estimation of the survival of tablet strength following the decompression that takes place after the tablet is ejected from the press. The brittle fracture index is a measure of the brittleness of a material, and provides a measure of the ability of a compact to relieve stress through plastic deformation. Finally, the strain index is an indicator of the relative strain that forms in a compact following its decompression. These indices and subsequent compaction studies were eventually used by Hiestand to develop a theoretical model for tablet bonding.[228–230]

The theory of percolation has been applied to powdered solids,[231] and tableting performance indices have been used to study its application to powder compaction.[232] It was found that when the relative density of excipients characterized by plastic deformation behavior was increased, the bonding index underwent a concomitant increase. In another work, a new index was reported that was used to quantitatively predict and estimate the degree of capping that took place during the tableting process.[233] The theory predicted ways in which one could minimize the occurrence of capping, and these suggestions were found to be in accord with previously recorded experimental trends.[234]

Brittle fracture indices were obtained for a series of compacted direct compression excipients which had been exposed to various degrees of ambient relative humidity.[235] Compacts of essentially nonhygroscopic microcrystalline cellulose exhibited essentially the same degree of brittleness regardless of its exposure to relative humidity. On the other hand, compacts of more hygroscopic processed sugar materials exhibit a substantial decrease in tablet brittleness as the tablets are exposed to higher degrees of relative humidity. These phenomena were ascribed to the plasticizing effects of the adsorbed water.

V. SUMMARY AND CONCLUSIONS

Physical characterization of drug substances and drug products represents the process whereby one obtains the information needed to support the development of a formulation from its initial concept to its commercial embodiment. Physical characterization will remain an essential part of drug development for the foreseeable future.

REFERENCES

1. Byrn, S. R., Pfeiffer, R. R. and Stowell, J. G. *Solid State Chemistry of Drugs.* 2nd ed., SSCI Inc., West Lafayette, IN, 1999.
2. Fiese, E. F. and Hagen, T. A. Preformulation. In *The Theory and Practice of Industrial Pharmacy.* (L. Lachman, H. A. Lieberman and J. L. Kanig, eds.), 3rd ed., Lea and Febiger, Philadelphia, pp. 171–196, 1986. Chapter 8.
3. Wells, J. I. *Pharmaceutical Preformulation: The Physicochemical Properties of Drug Substances.* Halsted Press, New York, 1988.
4. Wadke, D. A., Serajuddin, A. T. M. and Jacobson, H. Preformulation testing. In *Pharmaceutical Dosage Forms: Tablets.* (H. A. Lieberman, L. Lachman and J. B. Schwartz, eds.), 2nd ed., Marcel Dekker, New York, pp. 1–73, 1989. Chapter 1.
5. Carstensen, J. T. *Pharmaceutical Preformulation.* CRC Press, Boca Raton, 1998.
6. Gibson, M. *Pharmaceutical Preformulation and Formulation.* CRC Press, Boca Raton, 2001.
7. Adeyeye, M. C. and Brittain, H. G. *Preformulation in Solid Dosage Form Development.* Informa Healthcare Press, New York, 2008.
8. Carstensen, J. T. and Rhodes, C. T. *Drug Stability: Principles and Practices.* 3rd ed., Marcel Dekker, New York, 2000.
9. Ohannesian, L. and Streeter, A. J. *Handbook of Pharmaceutical Analysis.* Marcel Dekker, New York, 2002.
10. Brittain, H. G. *Physical Characterization of Pharmaceutical Solids.* Marcel Dekker, New York, 1994.
11. Carstensen, J. T. *Advanced Pharmaceutical Solids.* Marcel Dekker, New York, 2001.
12. Brittain, H. G. *Polymorphism in Pharmaceutical Solids.* Marcel Dekker, New York, 1999.
13. Bernstein, J. *Polymorphism in Molecular Crystals.* Clarendon Press, Oxford, 2002.
14. Hilfiker, R. *Polymorphism in the Pharmaceutical Industry.* Wiley-VCH, Weinheim, 2006.
15. Brittain, H. G., Bogdanowich, S. J., Bugay, D. E., DeVincentis, J., Lewen, G. and Newman, A. W. Physical characterization of pharmaceutical solids. *Pharm. Res.* 8:963, 1991.
16. Kortüm, G. *Reflectance Spectroscopy.* Springer-Verlag, New York, 1969.
17. Wendlandt, W. W. and Hecht, H. G. *Reflectance Spectroscopy.* Interscience, New York, 1966.
18. Frei, R. W. and MacNeil, J. D. *Diffuse Reflectance Spectroscopy in Environmental Problem Solving.* CRC Press, Cleveland, OH, 1973.
19. Pope, D. G. and Lach, J. L. Some aspects of solid-state stability and diffuse reflectance spectroscopy. *Pharm. Acta Helv.* 50:165, 1975.
20. Jozwiakowski, M. J. and Connors, K. A. Studies on adsorptiochromism II: Diffuse reflectance spectroscopy of adsorptiochromic spiropyrans adsorbed to some pharmaceutically useful solids. *J. Pharm. Sci.* 77:241, 1988.
21. CIE Publication 15.2. *Colorimetry,* 2nd ed, Central Bureau of CIE, Vienna, Austria, 1986.
22. Judd, D. B. and Wyszecki, G. *Color in Business, Science, and Industry.* 2nd ed., John Wiley & Sons, New York, 1963. pp. 264–361.
23. Berns, R. S. *Principles of Color Technology.* 3rd ed., Wiley-Interscience, New York, 2000.
24. Wirth, M. Instrumental color measurement: A method for judging the appearance of tablets. *J. Pharm. Sci.* 80:1177, 1991.
25. Walling, C. The Acid Strength of Surfaces. *J. Am. Chem. Soc.* 72:1164, 1950.

26. Hammett, L. P. and Deyrup, A. J. A series of simple basic indicators. I. The acidity functions of mixtures of sulfuric and perchloric acids with water. *J. Am. Chem. Soc.* 54:2721, 1932.
27. Hammett, L. P. and Deyrup, A. J. A series of simple basic indicators. II. Some applications to solutions in formic acid. *J. Am. Chem. Soc.* 54:4239, 1932.
28. Hammett, L. P. and Paul, M. A. A Series of Simple Basic Indicators. III. The Zero Point of the Acidity Function Scale. *J. Am. Chem. Soc.* 56:827, 1934.
29. H.G. Brittain, unpublished results obtained in the visible region of the spectrum using a Varian DMS-200 UV/VIS spectrometer, with the samples being contained in 1-cm square quartz cuvettes mounted at one of the focal points of a diffuse reflectance sampling accessory. A sample of optical grade magnesium oxide was mounted at the other focal point of the accessory to serve as the sample reference. After digitization, the spectra were baseline corrected so as to yield diffuse reflectance visible absorption spectra by the removal of spurious effects due to specular reflectance.
30. Leermakers, P. A. and Thomas, H. T. Electronic Spectra and Photochemistry of Adsorbed Organic Molecules. I. Spectra of Ketones on Silica Gel. *J. Am. Chem. Soc.* 87:1620, 1965.
31. Leermakers, P. A., Thomas, H. T., Weis, L. D. and James, C. Spectra and Photochemistry of Molecules Adsorbed on Silica Gel. IV. *J. Am. Chem. Soc.* 88:5075, 1966.
32. Evans, T. R., Toth, A. F. and Leermakers, P. A. Medium effects of photochromism. Reversible photobleaching of a spiropyran on silica gel. *J. Am. Chem. Soc.* 89:5060, 1967.
33. Weis, L. D., Evans, T. R. and Leermakers, P. A. Electronic spectra and photochemistry of adsorbed organic molecules. VI. Binding effects of silica as a mechanistic probe in systems of photochemical interest. *J. Am. Chem. Soc.* 90:6109, 1968.
34. Glombitza, B. W., Oelkrug, D. and Schmidt, P. C. Surface acidity of solid pharmaceutical excipients. I: Determination of the surface acidity. *Eur. J. Pharm. Biopharm.* 40:289, 1994.
35. Scheef, C.-A., Oelkrug, D. and Schmidt, P. C. Surface acidity of solid pharmaceutical excipients III. Excipients for solid dosage forms. *Eur. J. Pharm. Biopharm.* 46:209, 1998.
36. Wilson, E. B., Decius, J. C. and Cross, P. C. *Molecular Vibrations: The Theory of Infrared and Raman Vibrational Spectra.* McGraw-Hill Book Co., New York, 1955.
37. Rao, C. N. R. *Chemical Applications of Infrared Spectroscopy.* Academic Press, New York, 1963.
38. Conley, R. T. *Infrared Spectroscopy.* Allyn and Bacon, Boston, MA, 1966.
39. Colthup, N. B., Daly, L. H. and Wiberley, S. E. *Introduction to Infrared and Raman Spectroscopy.* 2nd ed., Academic Press, London, 1975.
40. Nakanishi, K. and Solomon, P. H. *Infrared Absorption Spectroscopy.* Holden-Day, San Francisco, CA, 1977.
41. Grasselli, J. G., Snavely, M. K. and Bulkin, B. J. *Chemical Applications of Raman Spectroscopy.* Wiley-Interscience, New York, 1981.
42. Grasselli, J. G. and Bulkin, B. J. *Analytical Raman Spectroscopy.* John Wiley & Sons, New York, 1991.
43. Ferraro, J. R. and Nakamoto, K. *Introductory Raman Spectroscopy.* Academic Press, New York, 1994.
44. Lewis, I. R. and Edwards, H. G. M. *Handbook of Raman Spectroscopy.* Marcel Dekker, New York, 2001.
45. Brittain, H. G. Molecular motion and vibrational spectroscopy. In *Spectroscopy of Pharmaceutical Solids*, (H. G. Brittain, ed.), Taylor and Francis, New York, pp. 205–233, 2006. Chapter 7.
46. Markovich, R. J. and Pidgeon, C. Introduction to Fourier transform infrared spectroscopy and applications in the pharmaceutical sciences. *Pharm. Res.* 8:663–675, 1991.
47. Threlfall, T. L. Analysis of organic polymorphs. *Analyst* 120:2435–2460, 1995.
48. Bugay, D. E. and Williams, A. C. Vibrational spectroscopy. In *Physical Characterization of Pharmaceutical Solids*, (H. G. Brittain, ed.), Marcel Dekker, New York, pp. 59–91, 1995. Chapter 3.
49. Brittain, H. G. Spectral methods for the characterization of polymorphs and solvates. *J. Pharm. Sci.* 86:405–412, 1997.
50. Findlay, W. P. and Bugay, D. E. Utilization of Fourier transform Raman spectroscopy for the study of pharmaceutical crystal forms. *J. Pharm. Biomed. Anal.* 16:921–930, 1998.
51. Bugay, D. E. and Findlay, W. P. *Pharmaceutical Excipients: Characterization by IR, Raman, and NMR Spectroscopy.* Marcel Dekker, New York, 1999.
52. Frank, C. J. Pharmaceutical applications of Raman spectroscopy. In *Analytical Applications of Raman Spectroscopy*, (M. J. Pelletier, ed.), Blackwell Science, Oxford, pp. 224–275, 1999. Chapter 6.
53. Bugay, D. E. Characterization of the solid-state: Spectroscopic techniques. *Adv. Drug Del. Rev.* 48:43–65, 2001.

54. Brittain, H. G. and Medek, A. Polymorphic and solvatomorphic impurities. In *Handbook of Isolation and Characterization of Impurities in Pharmaceuticals*, (S. Ahuja and K. M. Alsante, eds.), Academic Press, pp. 39–73, 2003. Chapter 3.
55. Bugay, D. E. and Brittain, H. G. Infrared absorption spectroscopy. In *Spectroscopy of Pharmaceutical Solids*, (H. G. Brittain, ed.), Taylor and Francis, New York, pp. 235–269, 2006. Chapter 8.
56. Bugay, D. E. and Brittain, H. G. Raman spectroscopy. In *Spectroscopy of Pharmaceutical Solids*, (H. G. Brittain, ed.), Taylor and Francis, New York, pp. 271–312, 2006. Chapter 9.
57. Brittain, H. G. Spectroscopic methods for the characterization of drug substances. In *Preformulation in Solid Dosage Form Development*, (M. C. Adeyeye and H. G. Brittain, eds.), Informa Healthcare Press, New York, pp. 253–277, 2008. Chapter 3.5.
58. Brittain, H. G. Vibrational spectroscopy. In *Polymorphism in Pharmaceutical Solids*, (H.G. Brittain, ed.), Informa Press, New York, pp. 345–378, 2009. Chapter 10.
59. Griffiths, P. R. and de Haseth, J. A. *Fourier Transform Infrared Spectrometry.* John Wiley & Sons, New York, 1986.
60. Fredericks, P., Rintoul, L. and Coates, J. Vibrational spectroscopy: Instrumentation for infrared and Raman spectroscopy. In *Ewing's Analytical Instrumentation Handbook*. (J. Cazes, ed.), 3rd ed., Marcel Dekker, New York, pp. 163–238, 2005. Chapter 7.
61. Mutha, S. C. and Ludemann, W. B. Solid-state anomalies in IR spectra of compounds of pharmaceutical interest. *J. Pharm. Sci.* 65:1400–1403, 1976.
62. Bugay, D. E., Newman, A. W. and Findlay, W. P. Quantitation of cefepime dihydrochloride in cefepime dihydrochloride monohydrate by diffuse reflectance IR and powder X-ray diffraction techniques. *J. Pharm. Biomed. Anal.* 15:49–61, 1996.
63. Culler, S. R. Diffuse reflectance spectroscopy: Sampling techniques for qualitative/quantitative analysis of solids. In *Practical Sampling Techniques for Infrared Analysis*, (P. B. Coleman, ed.), CRC Press, Boca Raton, FL, pp. 93–105, 1993.
64. Compton, S. V. and Compton, D. A. C. Optimization of data by internal reflectance spectroscopy. In *Practical Sampling Techniques for Infrared Analysis*, (P. B. Coleman, ed.), CRC Press, Boca Raton, FL, pp. 55–92, 1993.
65. Urban, M. W. *Attenuated Total Reflectance Spectroscopy of Polymers.* American Chemical society, Washington, DC, 1996.
66. H.G. Brittain, unpublished results obtained at a resolution of 4 cm^{-1} using a Shimadzu model 8400 S Fourier-transform infrared spectrometer, with each spectrum being obtained as the average of 40 individual spectra. The data were acquired using the attenuated total reflectance sampling mode, where the samples were clamped against the ZnSe crystal of a Pike MIRacleTM single reflection horizontal ATR sampling accessory.
67. Rafilovich, M., Bernstein, J., Harris, R. K., Apperly, D. C., Karamertzanis, P. G. and Price, S. L. Groth's original concomitant polymorphs revisited. *Cryst. Growth Design* 5:2197–2209, 2005.
68. van Tonder, E. C., Maleka, T. S. P., Liebenberg, W., Song, M., Wurster, D. E. and de Villiers, M. M. Preparation and physicochemical properties of niclosamide anhydrate and two monohydrates. *Int. J. Pharm.* 269:417–432, 2004.
69. Gilpin, R. K. and Zhou, W. Infrared studies of the thermal conversion of mefenamic acid between polymorphic states. *Vibrat. Spectrosc.* 37:53–59, 2005.
70. Lin, S. Y., Cheng, W. T. and Wang, S. L. Thermodynamic and kinetic characterization of polymorphic transformation of famotidine during grinding. *Int. J. Pharm.* 318:86–91, 2006.
71. Socrates, G. *Infrared and Raman Characteristic Group Frequencies—Tables and Charts.* 3rd ed., John Wiley & Sons, Chichester, 2001.
72. Dollish, F. R., Fateley, W. G. and Bentley, F. F. *Characteristic Raman Frequencies of Organic Compounds.* John Wiley & Sons, New York, 1974.
73. Mehrens, S. M., Kale, U. J. and Qu, X. Statistical analysis of differences in the Raman spectra of polymorphs. *J. Pharm. Sci.* 94:1354–1367, 2005.
74. H.G. Brittain, unpublished results obtained in the fingerprint region using a Raman Systems model R-3000HR spectrometer, operated at a resolution of 5 cm^{-1} and using a laser wavelength of 785 nm. The data were acquired using front-face scattering from a thick powder bed contained in an aluminum sample holder.

75. Ayala, A. P., Siesler, H. W., Boese, R., Hoffmann, G. G., Polla, G. I. and Vega, D. R. Solid-state characterization of olanzapine polymorphs using vibrational spectroscopy. *Int. J. Pharm.* 326:69–79, 2006.
76. Sheikhzadeh, M., Rohani, S., Jutan, A., Manifar, T., Murthy, K. and Horne, S. Solid-state characterization of buspirone hydrochloride polymorphs. *Pharm. Res.* 23:1043–1050, 2006.
77. Schöll, J., Bonalumi, D., Vicum, L. and Mazzotti, M. In-situ monitoring and modeling of the solvent-mediated polymorphic transformation of L-glutamic acid. *Cryst. Growth Design* 6:881–891, 2006.
78. Hu, Y., Wikström, H., Byrn, S. R. and Taylor, L. S. Estimation of the transition temperature for an enantiotropic polymorphic system from the transformation kinetics monitored using Raman spectroscopy. *J. Pharm. Biomed. Anal.* 45:546–551, 2007.
79. Qu, H., Louhi-Kultanen, M. and Kallas, J. Solubility and stability of anhydrate/hydrate in solvent mixtures. *Int. J. Pharm.* 321:101–107, 2006.
80. Tian, F., Zeitler, J. A., Strachan, C. J., Saville, D. J., Gordon, K. C. and Rades, T. Characterizing the conversion kinetics of carbamazepine polymorphs to the dihydrate in aqueous suspension using Raman spectroscopy. *J. Pharm. Biomed. Anal.* 40:271–280, 2006.
81. Tian, F., Zhang, F., Sandler, N., Gordon, K. C., McGoverin, C. M., Strachan, C. J., Saville, D. J. and Rades, T. Influence of sample characteristics on quantification of carbamazepine hydrate formation by X-ray powder diffraction and Raman spectroscopy. *Eur. J. Pharm. Biopharm.* 66:466–474, 2007.
82. Burns, D. A. and Ciurczak, E. W. *Handbook of Near-Infrared Analysis.* 2nd ed., Marcel Dekker, New York, 2001.
83. Ciurczak, E. W. and Drennen, J. K. *Pharmaceutical and Medical Applications of Near-Infrared Spectroscopy.* Marcel Dekker, New York, 2002.
84. Cogdill, R. P. and Drennen III, J. K. Near-infrared spectroscopy. In *Spectroscopy of Pharmaceutical Solids*, (H. G. Brittain, ed.), Taylor & Francis, New York, pp. 313–412, 2006. Chapter 10.
85. Ciurczak, E. W. Uses of near-infrared spectroscopy in pharmaceutical analysis. *Appl. Spectrosc. Rev.* 23:147–163, 1987.
86. MacDonald, B. F. and Prebble, K. A. Some applications of near-infrared reflectance analysis in the pharmaceutical industry. *J. Pharm. Biomed. Anal.* 11:1077–1085, 1993.
87. Roggo, Y., Chalus, P., Maurer, L., Lema-Martinez, C., Edmond, A. and Jent, N. A review of near infrared spectroscopy and chemometrics in pharmaceutical technologies. *J. Pharm. Biomed. Anal.* 44:683–700, 2007.
88. El-Hagrasy, A., D'Amico, F. and Drennen III, J. K. A process analytical technology approach to near-infrared process control of pharmaceutical powder blending. Part I: D-optimal design for characterization of powder mixing and preliminary spectral data evaluation. *J. Pharm. Sci.* 95:392–406, 2006.
89. El-Hagrasy, A., Delgado-Lopez, M. and Drennen III, J. K. A process analytical technology approach to near-infrared process control of pharmaceutical powder blending. Part II: Qualitative near-infrared models for prediction of blend homogeneity. *J. Pharm. Sci.* 95:407–421, 2006.
90. El-Hagrasy, A. and Drennen III, J. K. A process analytical technology approach to near-infrared process control of pharmaceutical powder blending. Part III: Quantitative near-infrared calibration for prediction of blend homogeneity and characterization of powder mixing kinetics. *J. Pharm. Sci.* 95:422–434, 2006.
91. Li, W., Worosila, G. D., Wang, W. and Mascaro, T. Determination of polymorph conversion of an active pharmaceutical ingredient in wet granulation using NIR calibration models generated from the premix blends. *J. Pharm. Sci.* 94:2800–2806, 2005.
92. Römer, M., Heinämäki, J., Miroshnyk, I., Sandler, N., Rantanen, J. and Yliruusi, J. Phase transformations of erythromycin A dihydrate during pelletization and drying. *Eur. J. Pharm. Biopharm.* 67:246–252, 2007.
93. Xie, Y., Tao, W., Morrison, H., Chiu, R., Jona, J., Fang, J. and Cauchon, N. Quantitative determination of solid-state forms of a pharmaceutical development compound in drug substance and tablets. *Int. J. Pharm.* 362:29–36, 2008.
94. Rantanen, J., Wikström, H., Rhea, F. E. and Taylor, L. S. Improved understanding of factors contributing to quantification of anhydrate/hydrate powder mixtures. *Appl. Spectrosc.* 59:942–951, 2005.

95. Krzyaniak, J. F., Williams, G. R. and Ni, N. Identification of phase boundaries in anhydrate/hydrate systems. *J. Pharm. Sci.* 96:1270–1281, 2007.
96. Blanco, M., Valdés, D., Llorente, I. and Bayod, M. Application of NIR spectroscopy in polymorphic analysis: Study of pseudo-polymorphs stability. *J. Pharm. Sci.* 94:1336–1342, 2005.
97. Fyfe, C. A. *Solid State NMR for Chemists.* CFC Press, Guelph, 1983.
98. Bugay, D. E. Solid-state nuclear magnetic resonance spectroscopy: Theory and pharmaceutical applications. *Pharm. Res.* 10:317–327, 1993.
99. Bugay, D. E. Characterization of the solid state: Spectroscopic techniques. *Adv. Drug Del. Rev.* 48:43–65, 2001.
100. Stephenson, G. A., Forbes, R. A. and Reutzel-Edens, S. M. Characterization of the solid state: Quantitative issues. *Adv. Drug Del. Rev.* 48:67–90, 2001.
101. Tishmack, P. A., Bugay, D. E. and Byrn, S. R. Solid-state nuclear magnetic resonance spectroscopy: Pharmaceutical applications. *J. Pharm. Sci.* 92:441–474, 2003.
102. Medek, A. Solid-state nuclear magnetic resonance spectrometry. In *Spectroscopy of Pharmaceutical Solids*, (H. G. Brittain, ed.), Taylor & Francis, New York, 2006. Chapter 11.
103. H.G. Brittain, unpublished results for solid-state ^{13}C nuclear magnetic resonance spectra obtained at a frequency of 270 MHz, using a combination of magic-angle spinning and cross-polarization. The spectra were obtained using a contact time of 1 millisecond and a 3 second recycle time.
104. Brittain, H. G., Morris, K. R., Bugay, D. E., Thakur, A. B. and Serajuddin, A. T. M. Solid state characterization of fosinopril sodium polymorphs. *J. Pharm. Biomed. Anal.* 11:1063, 1993.
105. Geppi, M., Guccione, S., Mollica, G., Pignatello, R. and Veracini, C. A. Molecular properties of ibuprofen and its solid dispersions with Eudragit RL100 studied by solid-state nuclear magnetic resonance. *Pharm. Res.* 22:1544–1555, 2005.
106. Vachon, M. G. and Nairn, J. G. The use of ^{13}C solid state NMR to elucidate physicochemical association in Eudragit RS100 microencapsulated acyl esters of salicylic acid. *Eur. J. Pharm. Biopharm.* 45:9–21, 1998.
107. McCrone, W. C., McCrone, L. B. and Delly, J. G. *Polarized Light Microscopy.* Ann Arbor Science Publishers, Ann Arbor, MI, 1978.
108. Rochow, T. G. and Rochow, E. G. *An Introduction to Microscopy by Means of Light, Electrons, X-rays, or Ultrasound.* Plenum Press, New York, 1978.
109. "Optical Microscopy", General Test <776> in *United States Pharmacopeia*, Vol. 1, United States Pharmacopeial Convention, Rockville, Maryland, 2009, pp. 302-304
110. McCrone, W. C. *Fusion Methods in Chemical Microscopy.* Interscience Publishers, New York, 1957.
111. Kuhnert-Brandstätter, M. *Thermomicroscopy in the Analysis of Pharmaceuticals.* Pergamon Press, Oxford, 1971.
112. Shangraw, R. F., Wallace, J. W. and Bowers, F. M. Morphology and functionality of tablet excipients for direct compression. *Pharm. Tech.* 5(10):44–60, 1981.
113. Matthews, G. P., Lowther, N. and Shott, M. J. Crystallization of carbamazepine in tablets stored at elevated temperatures. *Int. J. Pharm.* 50:111, 1989.
114. Boles, M. O., Girven, R. J. and Gane, P. A. C. The structure of amoxicillin trihydrate and a comparison with the structures of ampicillin. *Acta Cryst.* B34:461, 1978.
115. Bragg, W. H. and Bragg, W. L. *X-Rays and Crystal Structure.* G. Bell & Sons, London, 1918.
116. Woolfson, M. M. *X-Ray Crystallography.* Cambridge University Press, Cambridge, 1970.
117. Stout, G. H. and Jensen, L. H. *X-Ray Structure Determination: A Practical Guide.* Macmillan Co., New York, 1968.
118. Rousseau, J. J. *Basic Crystallography.* John Wiley & Sons, Chichester, 1998.
119. Hammond, C. *The Basics of Crystallography and Diffraction.* 2nd ed., Oxford University Press, Oxford, UK, 2001.
120. Cullity, B. D. and Stock, S. R. *Elements of X-Ray Diffraction*, 3rd ed., Prentice-Hall, New York, 2001.
121. Klug, H. P. and Alexander, L. E. *X-Ray Diffraction Procedures for Polycrystalline and Amorphous Materials.* 2nd ed., Wiley-Interscience, New York, 1974.
122. Jenkins, R. and Snyder, R. *Introduction to X-Ray Powder Diffractometry.* John Wiley & Sons, New York, 1996.
123. Pecharsky, V. and Zavalij, P. *Fundamentals of Powder Diffraction and Structural Characterization of Materials.* Springer, New York, 2005.

124. Dinnebier, R. E. and Billinge, S. J. L. *Powder Diffraction: Theory and Practice.* Springer, New York, 2008.
125. Clearfield, A., Reibenspies, J. and Bhuvanesh, N. *Principles and Applications of Powder Diffraction.* Wiley-Blackwell, New York, 2008.
126. "X-ray Diffraction", General test <941>, in *United States Pharmacopeia*, Vol. 1, United States Pharmacopeial Convention, Rockville, Maryland, 2009, pp. 390–392.
127. H.G. Brittain, unpublished results obtained using a Rigaku MiniFlex powder diffraction system, equipped with a horizontal goniometer operating in the θ/2θ mode. The X-ray source was nickel-filtered Kα emission of copper (1.54184 Å). Samples were packed into the sample holder using a back-fill procedure, and were scanned over the range of 3.5 to 40 degrees 2θ at a scan rate of 0.5 degrees 2θ/min. Using a data acquisition rate of 1 point per second, the scanning parameters equate to a step size of 0.0084 degrees 2θ.
128. DeRanter, C. J. Applications of X-ray diffractometric techniques in the analysis of drugs. *J. Pharm. Biomed. Anal.* 4:747–754, 1986.
129. Suryanarayanan, R. X-ray powder diffractometry. In *Physical Characterization of Pharmaceutical Solids*, (H. G. Brittain, ed.), Marcel Dekker, New York, pp. 187–221, 1995. Chapter 7.
130. Brittain, H. G. X-ray diffraction of pharmaceutical materials. In *Profiles of Drug Substances, Excipients, and Related Methodology*, (H. G. Brittain, ed.), Elsevier Academic Press, Amsterdam, pp. 273–319, 2003. Chapter 7.
131. Byrn, S. R., Pfeiffer, R. R. and Stowell, J. G. *Solid-State Chemistry of Drugs.* 2nd ed., SSCI Inc., West Lafayette, IN, 1999.
132. Bernstein, J. *Polymorphism in Molecular Crystals.* Clarendon Press, Oxford, UK, 2002.
133. Hilfiker, R. *Polymorphism in the Pharmaceutical Industry.* Wiley-VCH, Weinheim, 2006.
134. Brittain, H. G. *Polymorphism in Pharmaceutical Solids.* 2nd ed., Informa Press, New York, 2009.
135. Bernstein, J. Conformational polymorphism. In *Organic Solid State Chemistry*, (G. R. Desiraju, ed.), Elsevier, Amsterdam, pp. 471–518, 1987. Chapter 13.
136. Robertson, J. M. The structure of resorcinol: A quantitative X-ray investigation. *Proc. R. Soc. Lond.* A157:79–99, 1936.
137. Robertson, J. M. and Ubbelohde, A. R. A new form of resorcinol: Structure determination by X-rays. *Proc. R. Soc. Lond.* A167:122–135, 1938.
138. Gerber, J. J., Caira, M. R. and Lötter, A. P. Structures of two conformational polymorphs of the cholesterol-lowering drug probucol. *J. Cryst. Spect. Res.* 23:863–869, 1993.
139. Epple, M. and Cammenga, H. K. Investigation of solid state reactions and solid-solid phase transformations with time- and temperature-resolved X-ray powder diffractometry. *Ber. Bunsenges. Phys. Chem.* 96:1774–1778, 1992.
140. Stephenson, G. A., Groleau, E. G., Kleemann, R. L., Xu, W. and Rigsbee, D. R. Formation of isomorphic desolvates: Creating a molecular vacuum. *J. Pharm. Sci.* 87:536–542, 1999.
141. Leung, S. S. and Grant, D. J. W. Solid state stability of model dipeptides: Aspartame and aspartylphenylalanine. *J. Pharm. Sci.* 86:64–71, 1997.
142. Rastogi, S., Zakrzewski, M. and Suryanarayanan, R. Investigation of solid-state reactions using variable temperature X-ray powder diffractometry. I. Aspartame hemihydrate. *Pharm. Res.* 18:267–273, 2001.
143. Poojary, D. M. and Clearfield, A. Applications of X-ray powder diffraction techniques to the solution of unknown crystal structures. *Acc. Chem. Res.* 30:414–422, 1997.
144. Stephenson, G. A. Structure determination from conventional powder diffraction data: Application to hydrates, hydrochloride salts, and metastable polymorphs. *J. Pharm. Sci.* 89:958–966, 2000.
145. Harris, K. D. M. New opportunities for of molecular materials directly from powder diffraction data. *Crys. Growth Design* 3:887–895, 2003.
146. David, W. I. F., Shankland, K., McCusker, L. B. and Baerlocher, C. *Structure Determination from Powder Diffraction Data.* Oxford University Press, Oxford, UK, 2006.
147. Rietveld, H. M. A profile refinement method for nuclear and magnetic structures. *J. Appl. Crystallogr.* 2:65–71, 1969.
148. Will, G. *Powder Diffraction: The Rietveld Method and the Two Stage Method to Determine and Refine Crystal Structures from Powder Diffraction Data.* Birkhäuser, Basel, Switzerland, 2006.

149. Dinnebier, R. E., Sieger, P., Nar, H., Shankland, K. and David, W. I. F. Structural characterization of three crystalline modifications of telmisartan by single crystal and high-resolution X-ray powder diffraction. *J. Pharm. Sci.* 89:1465–1479, 2000.
150. Kiang, Y.-H., Huq, A., Stephens, P. W. and Xu, W. Structure determination of enalapril maleate form II from high-resolution X-ray powder diffraction data. *J. Pharm. Sci.* 92:1844–1853, 2003.
151. Cheung, E. Y., Harris, K. D. M., Johnston, R. L., Kitchin, S. J., Hadden, K. L. and Zakrzewski, M. Rationalizing the structural properties of bupivacaine base—a local anesthetic—directly from powder X-ray diffraction data. *J. Pharm. Sci.* 93:667–674, 2004.
152. Dollimore, D. Thermoanalytical instrumentation and applications. In *Analytical Instrumentation Handbook*. (G. W. Ewing, ed.), 2nd ed., Marcel Dekker, New York, 1997. Chapter 17.
153. Schwenker, R. F. and Garn, P. D. *Thermal Analysis: Instrumentation, Organic Materials, and Polymers*. Academic Press, New York, 1969.
154. Wendlandt, W. W. *Thermal Methods of Analysis*. 2nd ed., Wiley-Interscience, New York, 1974.
155. Wendlandt, W. W. and Collins, L. W. *Thermal Methods of Analysis*. Dowden, Hutchinson & Ross, Stroudsburg, PA, 1976.
156. Ford, J. L. and Timmins, P. *Pharmaceutical Thermal Analysis*. Ellis Horwood, Ltd., Chichester, 1989.
157. Craig, D. Q. M. and Reading, M. *Thermal Analysis of Pharmaceuticals*. CRC Press, Boca Raton, FL, 2007.
158. Smykatz-Kloss, W. *Differential Thermal Analysis*. Springer-Verlag, Berlin, 1974.
159. Pope, M. I. and Judd, M. D. *Differential Thermal Analysis*. Heyden, London, 1977.
160. Ghiron, D. Applications of thermal analysis in the pharmaceutical industry. *J. Pharm. Biomed. Anal.* 4:755–770, 1986.
161. Ghiron-Forest, D., Goldbronn, C. and Piechon, P. Thermal analysis methods for pharmacopoeial materials. *J. Pharm. Biomed. Anal.* 7:1421–1433, 1989.
162. Townsend, I. A basic strategy for the thermal stability assessment of pharmaceutical synthetic intermediates and products. *J. Therm. Anal.* 37:2031–2065, 1991.
163. Barnes, A. F., Hardy, M. J. and Lever, T. J. A review of the applications of thermal methods within the pharmaceutical industry. *J. Therm. Anal.* 40:499–509, 1993.
164. McCauley, J. A. and Brittain, H. G. Thermal methods of analysis. In *Physical Characterization of Pharmaceutical Solids*, (H. G. Brittain, ed.), Marcel Dekker, New York, pp. 223–251, 1995. Chapter 8.
165. Murphy, D. K. and Rabel, S. Thermal analysis and calorimetric methods for the characterization of new crystal forms. In *Preformulation in Solid Dosage Form Development*, (M. C. Adeyeye and H. G. Brittain, eds.), Informa Healthcare Press, New York, pp. 279–321, 2008. chapter 3.6.
166. Blaine, R. L. and Schoff, C. K. *Purity Determinations by Thermal Methods*. ASTM Press, Philadelphia, 1984.
167. van Dooren, A. A. and Muller, B. W. Purity determinations of drugs with differential scanning calorimetry: A critical review. *Int. J. Pharm.* 20:217–233, 1984.
168. H.G. Brittain, unpublished DSC results obtained using a TA Instruments 2910 thermal analysis system. Samples of approximately 1–2 mg were accurately weighed into an aluminum DSC pan, and then covered with an aluminum lid that was inverted and pressed down so as to tightly contain the powder between the top and bottom aluminum faces of the lid and pan. The samples were then heated over the temperature range of 20–175 °C, at a variety of heating rates.
169. Duval, C. *Inorganic Thermogravimetric Analysis*. 2nd ed., Elsevier, Amsterdam, 1963.
170. Keattch, C. J. and Dollimore, D. *Introduction to Thermogravimetry*. 2rd ed., Heyden, London, 1975.
171. Sharpe, S. A., Celik, M., Newman, A. W. and Brittain, H. G. Physical characterization of the polymorphic variations of magnesium stearate and magnesium palmitate hydrate species. *Struct. Chem.* 8:73, 1997.
172. Muzzio, F. J., Robinson, P., Wightman, C. and Brone, D. Sampling practices in powder blending. *Int. J. Pharm.* 155:153–178, 1997.
173. Kraemer, J., Svensson, J. R. and Melgaard, H. Sampling bias in blending validation and a different approach to homogeneity assessment. *Drug Dev. Indust. Pharm.* 25:217–222, 1999.
174. Saranadasa, H. RSD requirement for different sample size for blend sampling. *Drug Dev. Indust. Pharm.* 26:1213–1216, 2000.

175. Garcia, T. P., Wilkinson, S. J. and Scott, J. F. The development of a blend-sampling technique to assess the uniformity of a powder mixture. *Drug Dev. Indust. Pharm.* 27:297–307, 2001.
176. Brittain, H. G. Particle size distribution. 2. The problem of sampling powdered solids. *Pharm. Tech.* 26(7):67–73, 2002.
177. Venables, H. J. and Wells, J. I. Powder sampling. *Drug Dev. Indust. Pharm.* 28:107–117, 2002.
178. Kaye, B. H. *Direct Characterization of Fine Particles.* John Wiley & Sons, New York, 1981.
179. Barth, H. G. *Modern Methods of Particle Size Analysis.* John Wiley & Sons, New York, 1984.
180. Allen, T. *Particle Size Measurement*, 5th ed., Chapman and Hall, London, 1997. Vols. 1 and 2.
181. Brittain, H. G. Evaluation of the particle size distribution of pharmaceutical solids. In *Profiles of Drug Substances, Excipients, and Related Methodology*, (H. G. Brittain, ed.), Elsevier Academic Press, Amsterdam, Vol. 31, pp. 379–419, 2004.
182. Fayad, M. E. and Otten, L. *Handbook of Powder Science and Technology.* Van Nostrand Reinhold, New York, 1984.
183. Chu, B. *Laser Light Scattering.* Academic Press, New York, 1974.
184. H.G. Brittain, unpublished particle size results obtained using a Malvern Mastersizer 200 laser light scattering. Increasing amounts of sample were dispersed in a recirculating water system until an obscuration of approximately 20% was attained.
185. Lowell, S. and Shields, J. E. *Powder Surface Area and Porosity.* 2rd ed., Chapman and Hall, London, 1984.
186. Brunauer, S., Emmett, P. H. and Teller, E. Adsorption of gases in multimolecular layers. *J. Am. Chem. Soc.* 60:309, 1938.
187. Leuenberger, H., Bonney, J. D., Lerk, C. F. and Vromans, H. Relation between crushing strength and internal specific surface area of lactose compacts. *Int. J. Pharm.* 52:91, 1989.
188. Hayes, J. M. and Rossi-Doria, P. *Principles and Applications of Pore Structural Characterization.* J.W. Arrowsmith, Bristol, 1985.
189. Pesonen, T. and Paronen, P. The effect of particle and powder properties on the mechanical properties of directly compressed cellulose tblets. *Drug Dev. Indust. Pharm.* 16:31, 1990.
190. "Bulk and Tapped Density", General Test <616> in *United States Pharmacopeia*, Vol. 1, United States Pharmacopeial Convention, Rockville, Maryland, 2009, pp. 226–227.
191. "Density of Solids", General Test <699> in *United States Pharmacopeia*, Vol. 1, United States Pharmacopeial Convention, Rockville, Maryland, 2009, pp. 261–262.
192. Ticehurst, M. D., Rowe, R. C. and York, P. Determination of the surface properties of two batches of salbutamol sulfate by inverse gas chromatography. *Int. J. Pharm.* 111:241–249, 1994.
193. Dove, J. W., Buckton, G. and Doherty, C. A comparison of two contact angle measurement methods and inverse gas chromatography to assess the surface energies of theophylline and caffeine. *Int. J. Pharm.* 138:199–206, 1996.
194. Newell, H. E. and Buckton, G. Inverse gas chromatography: Investigating whether the technique preferentially probes high energy sites for mixtures of crystalline and amorphous lactose. *Pharm. Res.* 21:1440–1444, 2004.
195. Planinsek, O. and Buckton, G. Inverse gas chromatography: Considerations about appropriate use for amorphous and crystalline powders. *J. Pharm. Sci.* 92:1286–1294, 2003.
196. York, P., Ticehurst, M. D., Osborn, J. C., Roberts, R. J. and Rowe, R. C. Characterization of the surface energetics of milled dl-propranolol hydrochloride using inverse gas chromatography and molecular modeling. *Int. J. Pharm.* 174:179–186, 1998.
197. Grimsey, M., Sunkersett, M., Osborn, J. C., York, P. and Rowe, R. C. Interpretation of differences in the surface energies of two optical forms of mannitol by inverse gas chromatography and molecular modeling. *Int. J. Pharm.* 191:43–50, 1999.
198. Newell, H. E., Buckton, G., Butler, D. A., Thielmann, F. and Williams, D. R. The use of inverse phase gas chromatography to measure the surface energy of crystalline, amorphous and recently milled lactose. *Pharm. Res.* 18:662–666, 2001.
199. Newell, H. E., Buckton, G., Butler, D. A., Thielmann, F. and Williams, D. R. The use of inverse phase gas chromatography to study the change of surface energy of amorphous lactose as a function of relative humidity and the processes of collapse and crystallization. *Int. J. Pharm.* 217:45–56, 2001.
200. Hiestand, E. N. Powders: Particle-particle interactions. *J. Pharm. Sci.* 55:1325–1344, 1966.
201. Pilpel, N. Cohesive pharmaceutical powders. *Adv. Pharm. Sci.* 3:173–219, 1971.

202. Luckham, P. F. The measurement of interparticle forces. *Powder Tech.* 58:75–91, 1989.
203. Millili, G. P., Wigent, R. J. and Schwartz, J. B. Autohesion in pharmaceutical solids. *Drug Dev. Indust. Pharm.* 16:2383–2407, 1990.
204. Neumann, B. S. The flow properties of powders. *Adv. Pharm. Sci.* 2:181–221, 1967.
205. Amidon, G. E. and Houghton, M. E. Powder flow testing in preformulation and formulation development. *Pharm. Manuf.* 2(7):20–30, 1985.
206. Tan, S. B. and Newton, J. M. Powder flowability as an indication of capsule filling performance. *Int. J. Pharm.* 61:145–155, 1990.
207. Carr, R. L. Evaluating flow properties of solids. *Chemical Engineering*163–168, 1965. January 18 issue.
208. Carr, R. L. Classifying flow properties of solids. *Chemical Engineering*69–72, 1965. February 1 issue.
209. H.G. Brittain, unpublished flowability results obtained using a model PT-D Powder Characteristics Tester manufactured by the Hosokawa Micron Company. The Carr flowability index was determined from the individual indices derived from measurements of the angle of repose, compressibility, angle of spatula, and cohesion.
210. Rastogi, S. and Klinzing, G. E. Characterizing the rheology of powders by studying dynamic avalanching of the powder. *Particle Syst. Char.* 11:453–456, 1994.
211. Lee, Y. S., Poynter, R., Podczeck, F. and Newton, M. Development of a dual approach to assess powder flow from avalanching behavior. *AAPS PharmSciTech* 1:, 2000. Article 21.
212. Lavoie, F., Cartilier, L. and Thibert, R. New methods characterizing avalanche behavior to determine powder flow. *Pharm. Res.* 19:887–893, 2002.
213. Augsburger, L. L. and Hoag, S. W. *Pharmaceutical Dosage Forms: Tablets*, 3rd ed., Unit Operations and Mechanical Properties. Vol. 1. Informa Healthcare Press, New York, 2008.
214. Augsburger, L. L. and Hoag, S. W. *Pharmaceutical Dosage Forms: Tablets*, 3rd ed., Rational Design and Formulation. Vol. 2. Informa Healthcare Press, New York, 2008.
215. Augsburger, L. L. and Hoag, S. W. *Pharmaceutical Dosage Forms: Tablets*, 3rd ed., Manufacture and Process Control. Vol. 3. Informa Healthcare Press, New York, 2008.
216. Alderborn, G. and Nyström, C. *Pharmaceutical Powder Compaction Technology.* Marcel Dekeker, New York, 1996.
217. Leuenberger, H. and Rohera, B. D. Fundamentals of powder compression. I. The compactibility and compressibility of pharmaceutical powders. *Pharm. Res.* 3:12–22, 1986.
218. Leuenberger, H. and Rohera, B. D. Fundamentals of powder compression. II. The compression of binary powder mixtures. *Pharm. Res.* 3:65–74, 1986.
219. Lum, S. K. and Duncan-Hewitt, W. C. Powder densification. 1. Particle-particle basis for incorporation of viscoelastic material properties. *J. Pharm. Sci.* 88:261–276, 1999.
220. Lum, S. K., Hoag, S. W. and Duncan-Hewitt, W. C. Powder densification. 2. Viscoelastic material properties in modeling the uniaxial compaction of powders. *J. Pharm. Sci.* 87:909–916, 1998.
221. Riepma, K. A., Lerk, C. F., de Boer, A. H., Bolhuis, G. K. and Kussendrager, K. D. Consolidation and compaction of powder mixtures. I. Binary mixtures of same particle size fractions of different types of crystalline lactose. *Int. J. Pharm.* 66:47–52, 1990.
222. Riepma, K. A., Veenstra, J., de Boer, A. H., Bolhuis, G. K., Zuurman, K., Lerk, C. F. and Vromans, H. Consolidation and compaction of powder mixtures. II. Binary mixtures of different particle size fractions types of α-lactose monohydrate. *Int. J. Pharm.* 76:9–15, 1991.
223. Riepma, K. A., Zuurman, K., Bolhuis, G. K., de Boer, A. H. and Lerk, C. F. Consolidation and compaction of powder mixtures. III. Binary mixtures of different particle size fractions of different types of crystalline lactose. *Int. J. Pharm.* 85:121–128, 1992.
224. Marshall, K. Monitoring punch forces and punch movements as an aid to developing robust tablet formulations. *Drug Dev. Ind. Pharm.* 15:2153–2176, 1989.
225. Celik, M. and Marshall, K. Use of a compaction simulator system in tableting research. *Drug Dev. Ind. Pharm.* 15:759–800, 1989.
226. Celik, M. and Travers, D. N. The use of an elastic recovery index as a criterion of compactional behavior of some direct compression bases. *Drug Dev. Ind. Pharm.* 11:299–314, 1985.
227. Hiestand, E. N. and Smith, D. P. Indices of tableting performance. *Powder Tech.* 38:145–159, 1984.
228. Hiestand, E. N. Tablet bond. I. Theoretical model. *Int. J. Pharm.* 67:217–229, 1991.

229. Hiestand, E. N. Tablet bond. II. Experimental check of model. *Int. J. Pharm.* 67:231–246, 1991.
230. Hiestand, E. N. Principles, tenets, and notions of tablet bonding and measurements of strength. *Eur. J. Pharm. Biopharm.* 44:229–242, 1997.
231. Leuenberger, H. and Ineichen, L. Percolation theory and physics of compression. *Eur. J. Pharm. Biopharm.* 44:269–272, 1997.
232. Imbert, C., Tchorelogg, P., Leclerc, B. and Couarraze, G. Indices of tabletuing performance and application of percolation theory to powder compaction. *Eur. J. Pharm. Biopharm.* 44:229–242, 1997.
233. Sugimori, K.-I. and Kawashima, Y. A new practical index to predict capping occurring during the tableting process. *Eur. J. Pharm. Biopharm.* 44:323–326, 1997.
234. Sugimori, K.-I., Mori, S. and Kawashima, Y. Application of a newly defined capping index in evaluation of the compressibility of pharmaceutical powders. *Adv. Powder Tech.* 1:25–37, 1990.
235. Malamataris, S., Goidas, P. and Dimitriou, A. Moisture sorption and tensile strength of some tableted directed compression excipients. *Int. J. Pharm.* 68:51–60, 1991.

3

DEGRADATION AND IMPURITY ANALYSIS FOR PHARMACEUTICAL DRUG CANDIDATES

KAREN M. ALSANTE,* STEVEN W. BAERTSCHI,‡ MICHAEL COUTANT,* BRIAN L. MARQUEZ,† THOMAS R. SHARP,* AND TODD C. ZELESKY†

**Pfizer Global Research & Development Division, Research Analytical Department, Groton, Connecticut, USA*

†Pfizer Global Research & Development Division, Analytical Research and Development Department, Groton, Connecticut, USA

‡Eli Lilly and Company, Lilly Research Laboratories, Indianapolis, Indiana, USA

I. FORCED DEGRADATION STUDIES

A. How Far Is Too Far?

Forced degradation (also referred to as stress testing[1]) is a form of experimentation that is used to determine the stability of a given system or entity. It involves testing beyond normal operational capacity, often to a breaking point, in order to observe the results.

Forced degradation studies facilitate the development of stability-indicating analytical methodology, to gain a better understanding of active pharmaceutical ingredient (API) and drug product (DP) stability, and to provide information about degradation pathways and degradation products. To fulfill development and regulatory needs, this chapter provides a roadmap for when and how to perform studies, helpful tools in designing rugged scientific studies, and guidance on how to record and communicate results.

The primary goal of stress testing focuses on efficiently producing degradation samples realistic of those formed during manufacturing, handling, and normal storage conditions (as specified by the International Conference on Harmonization (ICH)) of the API and DP. Overstressing can destroy degradants that are relevant or generate unrealistic degradants whereas understressing may fail to generate important degradation products. Limits to how much stress should be applied will be discussed in the following sections.

Forced degradation samples are used in the process of developing stability-indicating methods early in development prior to initiating formal stability studies. The samples are analyzed for API peak purity (coeluting impurities), specificity, and mass balance. Efforts are also focused on determination degradation pathways/mechanisms for API and DP.

B. Regulatory Requirements

According to ICH Guidelines on Impurities in New Drug Products, a degradation product is defined as chemical change in the drug molecule brought about over time and/or by action of, for example, light, temperature, pH, or water, or by reaction with an excipient and/or the immediate container/closure system (also called decomposition product).[1]

From a regulatory perspective, forced degradation studies provide data to support the following:

- Identification of possible degradants.
- Degradation pathways and intrinsic stability of the drug molecule.
- Validation of stability-indicating analytical procedures.

Specific issues addressed in regulatory guidances related to forced degradation include:

- Forced degradation studies are typically carried out using one batch of material.
- Forced degradation conditions are more severe than accelerated stability testing including:

  the effect of temperatures (in 10 °C increments, e.g., 50 °C, 60 °C, etc., above that for accelerated testing), humidity (e.g., 75% RH or greater), where appropriate, oxidation, and photolysis on the drug substance. The testing should also evaluate the susceptibility of the drug substance to hydrolysis across a wide range of pH values when in solution or suspension. Photostability should be an integral part of stress testing.[1]

- Degradation products that do not form in accelerated or long-term stability may not have to be isolated or have their structure determined.
- Mass balance should be considered.

Issues not specifically addressed in regulatory guidance:

- Exact experimental conditions for forced degradation studies (temperatures, duration, extent of degradation, etc.) are not specified.
- Experimental design is left to the applicant's discretion.

Guidance is available from the US Food and Drug Administration (FDA) and from private industry on regulatory requirements for investigational new drug (IND) and new drug application (NDA) filings.[2,3] This section gives a global perspective on regulatory requirements (e.g., USA, Europe, and Japan) based on current regulations and guidances.

1. Requirements at the IND Stage

The reporting of forced degradation study conditions or results is not required in phase 1 or phase 2 IND applications. However, preliminary studies are encouraged to facilitate the development of stability-indicating methodology. It is recommended that forced degradation testing conducted as early in the development of API and DP as possible. Studies can be conducted on the API and developmental formulations to examine for degradation by thermolysis, hydrolysis, oxidation, and photolysis to evaluate the potential chemical behavior of the active. A guidance document suggests that results of one-time forced degradation studies should be included in phase 3 INDs.[4]

2. Requirements for Marketing Application

Completed studies of the degradation of the API and DP are required at the NDA stage, including isolation and/or characterization of significant degradation products and a full written account of the degradation studies performed.[5]

Requirements at the time of registration include[6]:

- Forced degradation products should be accurately characterized and the reaction kinetics established.
- Structural elucidation of degradation products should be attempted, even if not successful, should be referenced in the NDA.
- Mass balance should be determined or at least attempted.
- Main band peak purity should be confirmed.
- Any degradants present in ICH stability samples that are present at greater than the identification threshold should be isolated and identified.

Information from these studies should be referenced in the filing and should provide:

- Degradation pathways of the API, alone and in DP.
- Discussion of any possible polymorphic or enantiomeric substances.
- Differentiation between drug-related degradation and excipient interferences.

That outlined in the experimentation section detail the general protocol of tests and conditions recommended for regulatory submissions.

C. Forced Degradation Timing and Strategy

The requirements for forced degradation testing depend on project needs and the stage of development of the compound. For example, preclinical through phase 2 project needs dictate intense method development, and the rate of compound attrition is high.[7] Therefore, when developing a rational study design, forced degradation deliverables should be focused on method development activities, and not isolation and identification of degradants. As a compound progresses into later phase 2 through registration, method development activities center on optimization. The focus of stress testing should be directed to characterization and elucidation of degradants to ensure a complete understanding of the degradation pathways. Table 1 outlines the timing and strategy for carrying out forced degradation experiments.

1. Degradation Discussion

Degradation background discussion is a critical first step in the process. In these initial discussions, degradation prediction, background knowledge, and historical degradation knowledge gained can be shared. The purpose of the discussion is to review potential or known stability issues and possible degradation mechanisms for the API and DP in a team-based environment to be used as a resource tool to aid analysts in performing forced degradation studies. Degradation discussions are held to facilitate meeting milestone deliverables, such as stability-indicating methodology. Participants can include analytical chemists, process chemists, formulators, and discovery representatives. This discussion should be revisited for API process or salt changes, DP formulation changes as well as line extension efforts.

TABLE I Forced Degradation Timing and Strategy

	Development timing	Actions	Recommendations/rational study design
Early studies	Prephase 1	Predict API degradants	Focus on experiments resulting in approximately 5–20% degradation
		Design experimental protocol	If degradation by a certain pathway is not predicted and/or unlikely, minimal effort should be exerted on that condition
		Perform experiments	
IND studies	Prephase 1 + 6–12 weeks (or after lab experiments are complete)	Assessment of API forced degradation data	Forced degradation data used for rugged method development
		Selection of key predictive sample set (KPSS) for analytical method development	Mass balance not required
		Challenge analytical methodology developed with KPSS	Track only significant degradants Track and/or ID significant peaks by RRT = relative retention time and MW (LC-MS) only Isolation and structure elucidation at this stage not required
	Milestone: Initial IND	• Provide analytical methodology with data to support stability-indicating confidence • Analytical methods delivered with expectation of future method development with change of process and/or salt form	
Phases I and II	Formulation development (phase I–II)	Comprehensive forced degradation experiments for DP and API	Design experiments to highlight process, salt form and/or formulation change
		Review excipient compatibility	Peak purity assessment as necessary

(*Continued*)

TABLE 1 *(Continued)*

	Development timing	Actions	Recommendations/rational study design
		Update KPSS for DP and API	ID significant peaks by RRT and MW (LC-MS) only, unless more work is necessary for RRF (relative response factor) determination or project needs
		Challenge current analytical methodology with updated KPSS	
	Milestone	Rugged analytical method with high confidence in stability-indicating ability; no further method development activities expected (unless API/ formulation process changes warrant further work)	
Phase III	Phase III: ICH stability start	Attempt full characterization of significant degradants (including NMR) when standards are not available	Mass balance and peak purity assessment completed Isolation, mechanistic understanding and structure elucidation as required Significant degradants that are fully characterized should include those seen on real-time stability
	Milestone: Registration	Full characterization of significant degradation products completed	
Key	KPSS = key predictive sample set		

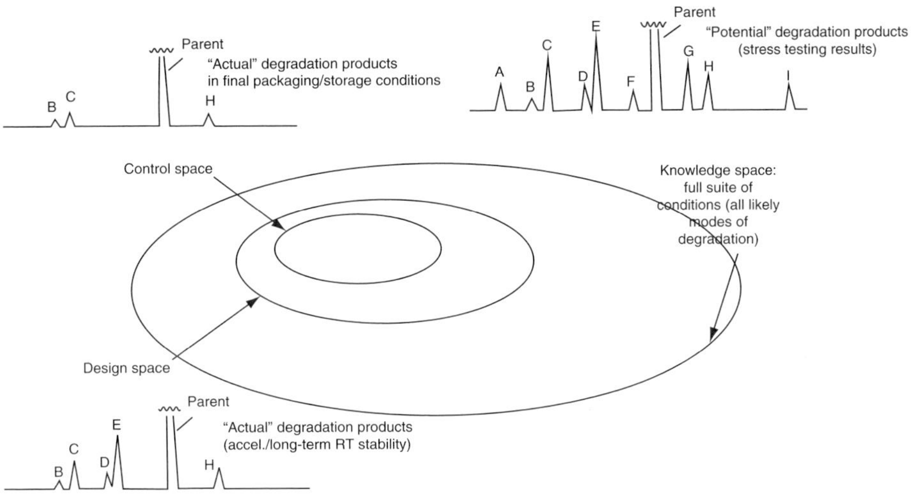

FIGURE 1 Quality by design model.[9]

2. Quality by Design Principles

In an effort to establish consensus and develop consistency, industry and regulators have frequently described quality by design (QbD) by dividing it into three fundamental, interrelated concepts: control strategy, design space, and knowledge space.[8] Figure 1 describes a QbD approach for developing design space, establishing control strategy, and delineating knowledge space for forced degradation studies utilizing a map to illustrate how these conceptual elements were fit for the forced degradation process flow. A chemistry-guided approach to experimental design aids in targeting most likely reactive functional groups and experimental conditions to focus on enabling scientists to narrow in on the relevant knowledge space (streamline efforts).

A preliminary assessment of the QbD strategy for forced degradation typically begins early in development when the analyst conducts a predictive chemistry analysis. This exercise, if done well, can outline potential degradation reactions of the molecule, providing insight into the knowledge space of possible degradation chemistry and helping in the design of stress testing studies. The knowledge space is further clarified by carrying out actual stress testing studies under a full suite of stress conditions, as outlined by ICH to generate all potential degradation products. The model can be applied for API as well as DP. To establish design space, a forward-looking risk assessment is executed in accordance with ICH Q9.[10] The design space is the acceptable operating space where outside these boundaries resides the failure to operate space. As shown in the chromatogram above depicting design space, this circle in Figure 1 refers to the actual degradation products generated upon accelerated stress testing conditions or long-term RT stability. Finally, control space narrows in on the actual degradation products observed under real-time room temperature storage in the actual packaging over the shelf-life of the product.

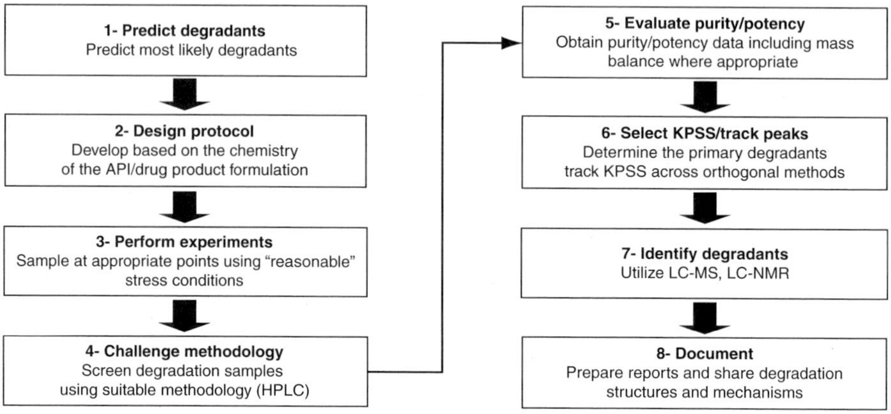

NOTE: KPSS=key predictive sample set used to develop the analytical methodology with appropriate resolution and detection limits, See H.2 for a more detailed explanation.

FIGURE 2 **Forced degradation process flow map.**

3. Degradation Process Flow Map

Figure 2 depicts the complete forced degradation process flow. The remaining sections of the forced degradation section of the chapter will outline the key elements of each step of the process.

D. Degradation Prediction Tools

1. CAMEO

CAMEO[11] is a computer program that predicts the products of organic reactions given starting materials, reagents, and conditions (see Figure 2, Step 1: Predict degradants). The analyses cover the following key degradation conditions: basic/nucleophilic, acidic/electrophilic, radical, oxidative/reductive, and photochemical as well as mechanistic interpretations of these reactions. In general, the CAMEO algorithms have been designed to give product mixtures that err on predicting more degradation products than actually observed. This is preferable to rules that are too restrictive and reject a key product observed in actual degradation or ICH stability studies. It is also likely that certain products predicted can undergo further decomposition. Because of these limitations with this prediction program, tracking historical degradation data in terms of functional groups along with CAMEO prediction data provides a more thorough approach to degradation prediction exercises. In terms of CAMEO limitations, predictions often stop at the primary degradation product(s) and can overlook secondary or ternary degradants, which may be the degradants that are actually seen in experiments. CAMEO is particularly successful at predicting certain types of chemistry—for example, amide hydrolysis, oxidation at benzylic sites, sulfoxide and sulfone formation, imine formation, and subsequent cleavage; however, CAMEO is extremely limited at others such as photochemistry (Figure 3).

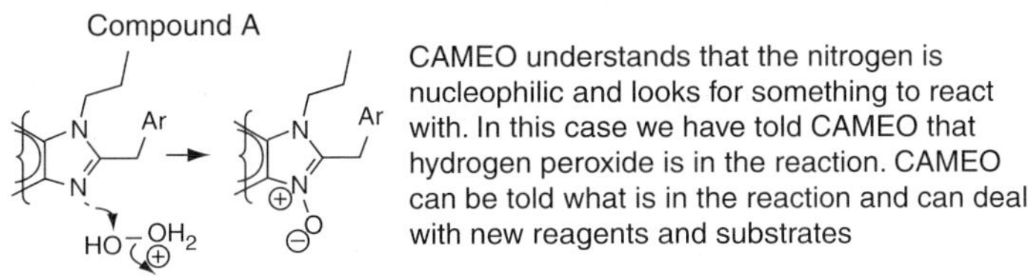

FIGURE 3 CAMEO example.

2. Degradation Expert Leading to Pharmaceutical Insight

DELPHI (Degradation Expert Leading to PHarmaceutical Insight)[12] is a Pfizer-proprietary expert system, developed by Pfizer colleagues. DELPHI is also a forward-looking expert system. DELPHI perceives and assesses an input molecular structure for substructures, reactive sites, and functional groups, then predicts reaction products under given conditions based upon known general chemical principles and specialized example rules. In contrast to CAMEO, DELPHI was specifically designed to predict reactivity and degradation of molecules. Also, in contrast to CAMEO, DELPHI proceeds beyond a primary reactive degradant, to subsequent degradants of degradants.

3. Zeneth

Improved degradation predictions software resulting in streamlined degradation protocols and targeted experiments is critically needed. This will enable more efficient method development and provide scientific rationale for minimized prototype stability. The infrastructure for building such an expert system already exists, having been developed by LHASA Ltd. for the METEOR drug metabolism predictive system. LHASA Ltd., a nonprofit organization that promotes data and knowledge sharing for chemistry and life sciences, also developed and maintains the DEREK genotoxicity expert system.[13,14] Substantial portions of programming used in DEREK and METEOR are transferable. LHASA Ltd. has made significant progress in developing Zeneth, the next generation degradation prediction software. A dictionary of degradation transformations has been created. Reaction conditions are taken into account through reasoning rules. Predictions are made by "rules" developed by chemists, experts, etc. (and not via computational chemistry results). The transformations, in conjunction with reasoning rules, are used to make predictions about the possibility of reactions and pathways and will include drug/excipient interaction chemistry lacking in DELPHI/CAMEO. Kinetics are indirectly addressed (e.g., "very likely, likely, neutral, unlikely, very unlikely"). This software development is well under way.

E. Experimental Design

Forced degradation studies of API and DP include appropriate solid-state and solution-state stress conditions (e.g., acid/base hydrolysis, heat/humidity, oxidation, and light exposure) in accordance with ICH guidelines (Figure 2, Steps 2 and 3).[15,16] Forced degradation studies should be conducted whenever a stability-indicating

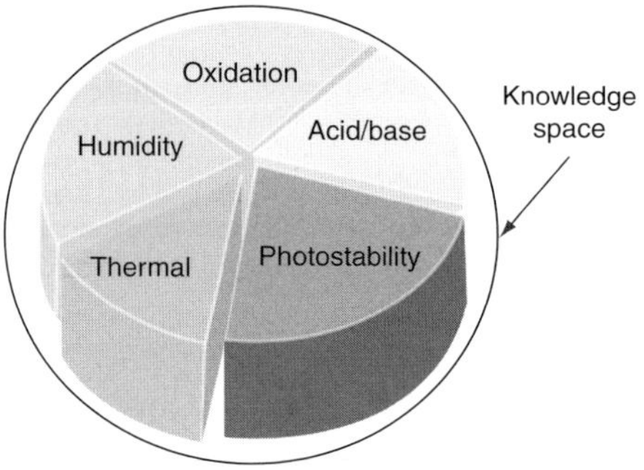

FIGURE 4 Conditions recommended for forced degradation studies to develop a complete QbD knowledge space.

method is required (Figure 4). Studies may need to be repeated as the methods, processes, or formulations change. The tables that follow in this section outline general protocol of tests and conditions that may be used to generate data for regulatory submissions.

1. Active Pharmaceutical Ingredient

The specified stress conditions (intensity and duration) should result in approximately 5–20% degradation of the API or represent a reasonable maximum stress condition achievable for the API. The specific conditions used will depend on the chemical stability characteristics of the API (i.e., how fast does it degrade under a given stress condition). The stressed sample should be compared to the unstressed sample (control) and the appropriate blank. A compound may not necessarily degrade under a given stress condition. No further stressing is advised in these cases.[2] The following experimental guidance represents pharmaceutical industry best practices with suggested starting and maximum stress conditions.

a. Acid

Studies should be carried out in the solution state.[17] To study acid/base hydrolysis, the recommended pH range is 1–2 using 0.1–1.0 M HCl or H_2SO_4 (Table 2). For certain APIs that are partially soluble or insoluble in the described acidic solution, addition of an appropriate cosolvent, or adjustment of solution pH in the acidic range may be required to achieve dissolution; alternatively, the APIs can be stressed as suspensions (Table 3). Special attention to the API structure should be paid when choosing the appropriate cosolvent (i.e., do not use alcohols for acidic conditions because of their reactivity, Table 4). Dimethylsulfoxide, acetic acid, and propionic acid are useful under acidic conditions as solvents. Additionally, the sample may be heated for a defined time/temperature to accelerate degradation, depending on the API sensitivity to heat. It is recommended not to stress higher than 70 °C because of the evaporation of water and deviation from mechanisms observed on real-time stability. It is also not recommended to stress longer than 3 weeks time duration or to greater than 5–20% degradation.

TABLE 2 Acid and Base Degradation Conditions (API)

API concentration	0.1–1.0 mg/mL
pH range	1–2, As Is, 12–13, neutral if it is not neutral
pH adjustment	HCl/NaOH (0.1–1.0 M) for low and high ranges, phosphate buffer for mid-range
Cosolvents (if needed for poorly soluble compounds)	Acetonitrile (methanol or DMSO as an alternate) *Note*: Select inert solvent to avoid solvent adduct artifacts
Temperature	Room temperature to 70 °C (avoid >70 °C because of the change in mechanism—nonpredictive)
Duration	5–20% degradation—no longer than 1 week at 70 °C 2 weeks at RT
Neutralization	Not recommended: Benefit does not outweigh the risk with artifacts (secondary reaction/artifacts in chromatography)
Containers	Flint glass vials with rubber septa closures; airtight so no solvent evaporation when using cosolvents

TABLE 3 Acid and Base Degradation Conditions (for Solution Drug Product)

API concentration	Formulation dependent
pH range	±2 pH units around the target pH
pH adjustment	HCl/NaOH for low/high ranges, phosphate buffer for mid-range
Temperature	Up to 70 °C
Duration	5–20% degradation or 3 weeks maximum

TABLE 4 Cosolvent Selection Guide

Cosolvent	Pros	Cons
Acetonitrile	Completely miscible in water. UV transparent. Cosolvent of choice for photochemistry	Too volatile at 80 °C, hydrolyzes in acid/base, can form reactive per-acid like compounds under oxygen
Methanol	Majority of APIs soluble in methanol. UV transparent	Reactive toward many functional groups, especially in acid/base. May contain trace formaldehyde; reactive with amines
Dioxane	Completely miscible in water, inert to base. UV transparent	Rapidly oxidizes under neutral and acidic conditions
THF	Completely miscible in water, inert to base. UV transparent	Too volatile, rapidly oxidizes under neutral and acid conditions. Contains stabilizers
DMSO	Good universal solvent, inert toward APIs, completely miscible in water, can be removed by lyophilization but otherwise nonvolatile. Useful for acid/base studies	Peak in UV. Complexity if assaying at wavelengths where DMSO absorbs

b. Base

Base studies should also be carried out in the solution state at a pH range of 12–13. Example bases include NaOH, LiOH, or KOH (0.1–1 mol/L solution). For certain APIs which are partially soluble or insoluble in the described basic solution, addition of an appropriate cosolvent, or adjustment of solution pH may be required to achieve dissolution; or the APIs can be run as suspensions. Additionally, the sample may be heated for a defined time/temperature to accelerate degradation, depending on the API sensitivity to heat, with the same maximum time/temperature duration (up to 70 °C and 5–10% degradation/3 week maximum). Inert organic cosolvents may be needed for poorly soluble APIs but the potential to react under given stress conditions must be recognized (Table 4[18]). Glyme, diglyme, triglyme, 1,4-dioxane, and DMSO are useful for obtaining dissolution under basic conditions while remaining relatively inert.[19]

c. Oxidation

Oxidation chemistry can be very complex and presents the most challenges for the forced degradation experimental strategy. The conditions that follow probe for the following oxidative reactions: autoxidation, free radical, peroxides and other oxidizing agents, dilute hydrogen peroxide peracids, electron transfer metals, photochemically induced oxidation (singlet oxygen mediated and radical mediated).[20,21]

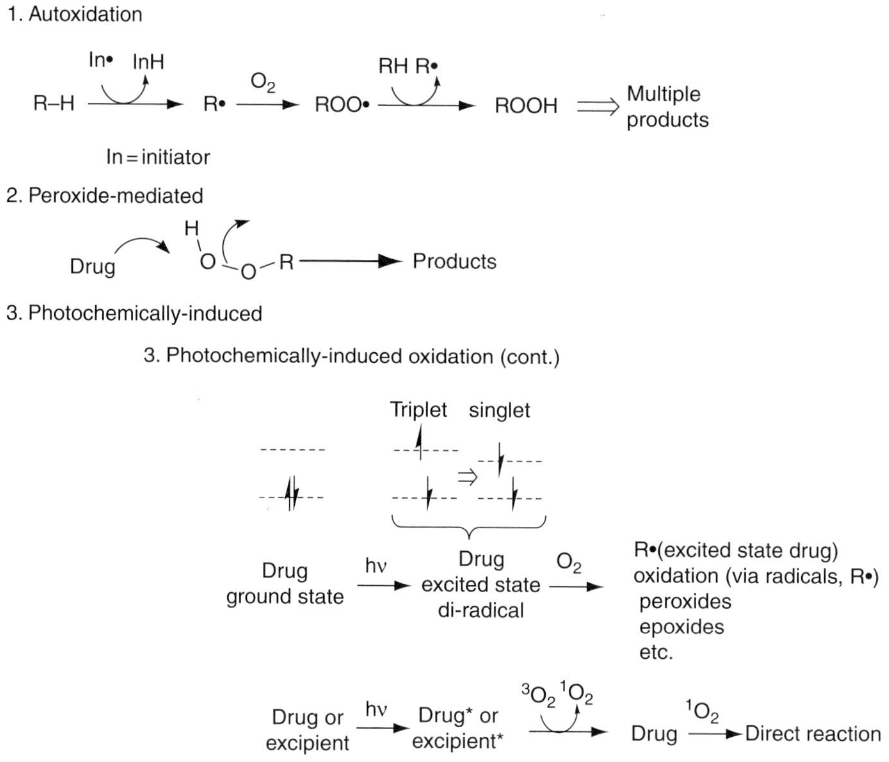

Oxidation can be carried out under an oxygen atmosphere or in the presence of peroxides (Table 5). The use of molecular oxygen in the presence of free-radical initiators is a more realistic model. Generally, a free-radical initiator and peroxide will produce all primary oxidation degradation products observed on real-time

TABLE 5 Oxidative Screening Conditions

Radical chain initiators	Initiator	AIBN (organic soluble) ACVA (water soluble)
	API concentration	0.1–1 mg/mL
	Initiator concentration	5–20 mol% of API concentration
	Solvent	Acetonitrile/water
	Temperature	Ambient-60 °C
	Duration	5–20% degradation or 14 days maximum
Hydrogen peroxide	Reagent concentration	Starting concentration: 0.3% (0.02–3% if needed)
	API concentration	0.1–1 mg/mL
	Temperature	Ambient-30 °C
	Duration	5–20% degradation or 7 days maximum

stability. Therefore, free-radical and/or hydrogen peroxide conditions are strongly recommended at all stages of development.

For solution-state free-radical stress conditions, dissolve the API utilizing an appropriate solvent, add 5–20 mol% of a free-radical initiator at atmospheric pressure. To increase the solubility of oxygen in the solution, the reaction can be performed in a reaction vessel pressurized at 50–300 psi with molecular oxygen; however, most reactions have adequate oxygen solubility under atmospheric pressure at 60 °C.[22] Additionally, the system can be heated to accelerate degradation. The temperature depends on the free-radical initiator selected. For water soluble APIs, 4,4′-azo-bis(4-cyanovaleric acid) (ACVA), AIBN (azo-bis-isobutyrl-nitrile) is a good selection for a free-radical initiator with a half-life of 10 h at 60 °C. For organic soluble APIs, AIBN is a good selection with a similar half-life to ACVA.[23]

For peroxide conditions, hydrogen peroxide reagent (recommended starting conditions 0.3%, v/v) can be used. As previously indicated, the addition of an appropriate cosolvent may be necessary, depending on API solubility. Hydrogen peroxide stress testing can be useful in DP studies where hydroperoxides are common impurities in excipients.[24,25] Solid-state stress conditions may be similarly investigated by placing the API (as is) in suitable closed containers filled with an oxygen headspace versus an argon or nitrogen control headspace. Additionally, the sample may be heated for a defined time/temperature to accelerate degradation, depending on the API sensitivity to heat.

For later stage development compounds when more time and effort can be focused on mechanistic understanding, the following investigative oxidation conditions can be applied (Table 6). The addition of metal ions to solutions of API can indicate whether there is a tendency for the API to be catalytically oxidized. Iron and copper ions are routinely found in APIs and formulation excipients.[26]

Transition metal ions can also reduce peroxide to generate hydroxyl radicals in a Fenton-type reaction. In addition, light can also effect oxidation reactions. Light absorbed by a photosensitizer can react with molecular oxygen to form the more reactive singlet oxygen species.

d. Thermal/Humidity

Solid-state stability can be evaluated utilizing accelerated storage temperatures in general greater than 50 °C and 75% relative humidity (Table 7).[27,28] The duration of exposure is dependent on the API sensitivity. If the forced degradation thermal/

TABLE 6 Oxidative Investigation Conditions

Transition metals	Metal	$Cu(II)Cl_2$ or $Fe(III)Cl_3$ (separate solutions)
	Concentration	1.0 mM
Note: Samples should be prepared so oxygen is available	API concentration	0.1–20 mg/mL
	Temperature	30–40 °C
	Duration	5–20% degradation or 7 days maximum
Fenton conditions (for investigations related to oxidative degradation)	Metal	$Fe(II)SO_4$ or $Fe(II)Cl_2$
	Metal concentration	1–5 mM
	Peroxide concentration	0.3–3% (choose one)
	API concentration	0.1–20 mg/mL
	Temperature	0 °C-RT
	Duration	5–20% degradation or 1 day maximum
Singlet oxygen (for investigations related to oxidative degradation)	Photosensitizer	Rose bengal
	Photosensitizer concentration	0.1 mM
	API concentration	0.1–20 mg/mL
	Temperature	30–40 °C
	Light source	Xenon or fluorescent lamp (1000–10,000 lux)
	Duration	5–60 min

For investigations related to oxidative: peracetic acid, MCPBA, oxone, formic acid, formaldehyde.
Note: Dependent on potential impurities in the formulation.

TABLE 7 Thermal/Humidity Best Practice Conditions

Container	Open
Sample	API—Use representative synthetic route and physical form DP—Use high and low potencies of definitive product
Temperature	50–70 °C (typical start: 70 °C)
Relative humidity	25–75% (optional based on instrumentation and resource capabilities at the early development stage; typical start: 20–30% RH and 75% RH) *Note*: Relative humidity can be controlled using desiccators containing saturated salt solutions *Reference*: Connors, K. *Chemical Stability of Pharmaceuticals* (2nd ed., Wiley, 1986), p. 19
Maximum duration	3 weeks (add longer duration for late-stage compounds) (see relative rate factors of degradation)

humidity conditions produce a phase change, it is recommended to also run thermal/ humidity conditions below the critical thermal/humidity that produces the phase change.

TABLE 8 Relative Rate Factors of Degradation[a]

Temperature (°C)	Relative rate (E_a = 12 kcal/mol) conservative	Relative rate (E_a = 15 kcal/mol) recommended	Relative rate (E_a = 20 kcal/mol) aggressive
25	1	1	1
40	2.62	3.37	5.05
50	4.78	7.10	13.65
70	14.20	27.7	83.98

Number of days of storage to simulate 18 months' storage at 25 °C

Temperature (°C)	E_a = 12 kcal/mol (days)	E_a = 15 kcal/mol (days)	E_a = 20 kcal/mol (days)
40	209	162	108
50	115	77	40
70	39	20	7

[a]Ref. 31.
Calculation: Number of days storage at T (°C) = (simulated number of days at 25 °C)/(relative rate factor). Assume E_a = 15 kcal/mol: number of days' storage at 70 °C = (547.5 days)/(27.7 days) = 20 days.

Arrhenius kinetics may be used to establish an appropriate temperature and maximum duration of thermal degradation studies. Using an appropriate assumption of activation energy, the duration of controlled room temperature storage that is simulated by the study can be estimated. Table 8 provides a guide to that conversion. In general, a relatively conservative activation energy assumption of 15 kcal/mol is recommended. Energies of activation for a number of drugs have been studied and documented previously. The average E_a was shown to be 19.8 kcal/mol.[29] Where particular concerns exist about the stability and degradation rate of a particular compound, the energy of activation should be established to allow a more accurate prediction of degradation rate at lower temperatures using the Arrhenius relationship. Deviation from Arrhenius kinetics is increasingly expected above 70–80 °C, and the risk of such deviation should be considered during experimental design. A more detailed discussion on accelerated aging of DP and prediction of product shelf-life is provided as a reference.[30]

e. Photostability

Studies should be performed in accordance with ICH photostability guidelines (Table 9).[32–36] The goal of the photostability studies is to force the degradation of drug substances via ultraviolet (UV) and fluorescent conditions over time in order to determine the primary degradation products. UV and visible light are the most energetic electromagnetic radiation to which pharmaceutical drug substances and DPs are typically exposed. A molecule absorbs light when an absorption band exists which overlaps to some extent with the incident light energy and a valence electron in the relevant chromophore is raised to an excited state.[37] Light stress conditions can also induce photooxidation by free-radical mechanisms. The photoexcited molecule undergoes dissociation and the free-radical formed is

TABLE 9 Photostability Conditions

Visible light exposure	At least 2× the ICH recommendation of 1.2 million lux hours
UV light exposure	At least 2× the ICH recommendation of 200 $W\text{-}h/m^2$
API in solution	Optional—recommended for IV, suspensions, and other liquid dosage forms
ICH option	1 or 2 (if using option 1 and base the exposure on 1.2 million lux hours; exposure in the UV region will be exceeded)
Dark control	Cover container completely with aluminum foil
Container	Quartz or borosilicate glass container
Duration	Instrument dependent

scavenged by molecular oxygen. The resultant peroxy free radical may then undergo reactions yielding a mixture of oxidized products. Functional groups likely to introduce drug photoreactivity are as follows: carbonyl, nitroaromatic, N-oxide, alkenes, aryl chlorides, weak C–H and O–H bonds, sulfides, and polyenes.[38]

There are two types of studies used in pharmaceutical photostability testing: stress testing and confirmatory testing. Purposeful degradation is used to evaluate the overall photosensitivity of the material (unprotected drug substance and DP) for method development purposes and/or degradation pathway elucidation. Confirmatory tests are used to determine if there is a need to protect the final product from light. The ICH Harmonized Tripartite Guideline covering the stability testing of new drug substances and products notes that photostability should be an integral part of stress testing.[39] The guideline recommends a systematic approach to photostability testing and lists light sources to be used, which are described below. The only requirements are that the source(s) be continuous over the near UV and visible spectral regions.[40]

Option 1. Any light source that is designed to produce an output similar to the D65/ID65 emission standard such as an artificial daylight fluorescent lamp, combining visible and UV outputs, xenon lamp, or metal halide lamp. D65 is the internationally recognized standard for outdoor daylight as defined in ISO 10977 (1993). ID65 is the equivalent indoor indirect daylight standard. For a light source emitting significant radiation below 320 nm, an appropriate filter (s) may be fitted to eliminate such radiation.

Option 2. For Option 2, the same sample should be exposed to both the cool white fluorescent and near UV lamp. A cool white fluorescent lamp designed to produce an output similar to that specified in ISO 10977(1993); and a near UV fluorescent lamp having a spectral distribution from 320 to 400 nm with a maximum energy emission between 350 and 370 nm; a significant portion of the UV should be in both bands of 320–360 nm and 360–400 nm. The sample must be exposed to both sources; however, the exposure can be in a sequential or simultaneous set-up.

(*Note*: The ISO 10977 standard contains standards spectral power distributions for these light sources as well as transmission for standard window glass).[41]

The guideline does not recommend one option over the other. Currently, our approach focuses on Option 1 to perform forced degradation. The goal of these forced degradation studies is to determine the primary photodegradation products and to generate key predictive samples for HPLC method development and validation. The ultimate goal is to develop photostability-indicating HPLC methods as soon as possible.

> For confirmatory studies, samples should be exposed to light providing an overall illumination of not less than 1.2 million lux hours and an integrated near ultraviolet energy of not less than 200 watt hours/square meter to allow direct comparisons to be made between the drug substance and drug product.

According to the ICH guidelines, "the design of the forced degradation experiments is left to the applicant's discretion although the exposure levels should be justified."

The following considerations should be taken into account when planning photostability experiments and recommended practice is outlined in Table 9. One key concern is competing thermal reactions that may complicate photostability purposeful degradation studies. The samples should be irradiated under temperature controlled conditions to minimize the thermal variable. In stress testing, the samples are placed close to the light source to take advantage of the inverse square dependence of light intensity.

Light measurements should be taken with an actinometer (visible light measured as illuminance and UV light measured as irradiance). Actinometry, the calculation of the number of molecules reacted per photon absorbed, is performed with either a physical device or a chemical system. For physical actinometers, it is critical that the actinometer is calibrated for the wavelength region of interest. For example, separate actinometers are required for Option 2 (a UV radiometer and fluorescence photometer). The ICH Harmonized Tripartite Guideline on Stability Testing provides details on a quinine chemical actinometer for monitoring exposure to a near UV fluorescent lamp.[39] It is particularly useful to bench mark the measurements of the quinine system for Option 2 UV experiments in comparison to the radiometer measurements. This can give one an understanding as to how well the radiometer has been positioned (a radiometer calibration check). However, this quinine chemical actinometer is not accurate for the fluorescence region for Option 2 or for the Option 1 Xenon light source. A potassium ferrioxalate actinometer solution absorbs photons in the range 250–577 nm and this actinometer can be applied to fluorescence systems.[42]

Option 1 and/or Option 2 conditions can be used. According to the ICH guideline, "the design of the forced degradation experiments is left to the applicant's discretion although the exposure levels should be justified." The recommended exposures for confirmatory stability studies are an overall illumination of not less than 1.2 million lux hours and an integrated near UV energy of not less than 200 W h/m^2. For forced degradation studies, the samples should be exposed to at least 2× the ICH exposure length to ensure adequate exposure of the sample. For solution studies, acetonitrile is the recommended cosolvent. Methanol can produce more artifact degradation products from methoxy radicals produced from light exposure.[43]

For Option 1, stress testing can be performed using an Atlas Suntest Tabletop Sunlight Exposure System.™ For the region of interest 300–800 nm, the UV and infrared radiation have been removed by filters. Also, the output is similar to sunlight (D65/ID65) which is considered to be the worst-case scenario radiation, an ideal photostability stress system. The system meets requirements of ID65 "indoor standard daylight—behind window glass." Lastly, the spectral output for the Suntest™ Xenon source also contains the 400 nm gap region left out in Option 2.[44]

One other factor of particular concern is sample depth. Photodegradation typically occurs on the surface in the solid state and a thin layer of material works best for forced degradation analysis. As per the ICH Guidelines on Photostability Testing, "solid drug substances should be spread across the container to give a thickness of typically not more than 3 millimeters." It is recommended to prepare a thin layer for API stress testing (<1 mm depth). Recommendation for DP is represented in Figure 5 to enable a representative sample of tablets exposed to light. Exposure to an entire bottle of tablets and sampling for analysis would not provide an accurate representation of photostability issues.[44]

2. Drug Product

DP degradation cannot be predicted solely from the stability studies of the API in the solid state or solution. The non-APIs can also react with the API or catalyze degradation reactions. Impurities in the excipients can also lead to degradation in the DP not originally observed in the API. For solid oral DP formulations, heat, light, and humidity are often used as appropriate stress conditions. As indicated for API, DP stress conditions should result in approximately 5–20% degradation of the API or represent a reasonable maximum condition for a given formulation. The specific conditions used will depend on the chemical characteristics of the DP. For a solid DP, key experiments are thermal, humidity, and photostability. For solution formulations, additional experiments recommended in addition to thermal and photostability include acid/base hydrolysis and oxidation.

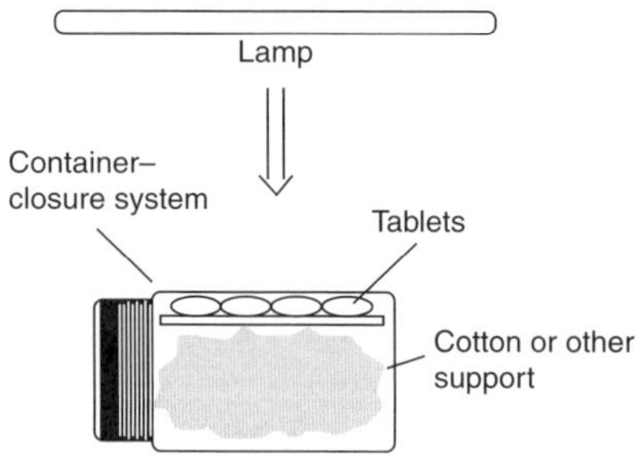

FIGURE 5 Recommended sample presentation for drug product tablet exposure.

For aqueous formulations, obviously no humidity studies would be required for thermal conditions. It is recommended to compare stressed samples with unstressed samples and an appropriate blank. For DP studies, the blank sample is an appropriate placebo. The stressed placebo sample will provide information about whether an observed degradation product is related to the API or is simply related to excipient degradation. It is advised to take kinetic time points along the reaction pathway for API and DP degradation studies to determine primary degradants and a better understanding of the degradation pathway.

F. Stability-Indicating Method Development

A stability-indicating method is defined as an analytical method that accurately quantitates the active ingredients without interference from degradation products, process impurities, excipients, or other potential impurities. A method that accurately quantitates significant degradants may also be considered stability indicating. A proactive approach to developing a stability-indicating HPLC method involves forced degradation as the first step in development. If forced degradation studies are performed early, method development and identification of primary degradation products and unknown impurities can be run in parallel. Using this process, a validated HPLC analytical assay, mechanisms of degradation, and the impurity/degradant information for filing can all be generated without delays in the project timeline.

1. Mass Balance

Mass balance is defined in the ICH Impurity Guidelines[1] as "adding together the assay value and levels of degradation products to see how closely these add up to 100 percent of the initial value, with due consideration of the margin of analytical error." Assessment of mass balance may be informative in assuring that the chosen analytical strategy controls all significant degradants (Figure 2, Step 5). The Guidelines recognize that it can be difficult to determine mass balance because of unknown analytical precision and differences in response factor. Additional guidance on helping the analyst obtain or approximate mass balance is given by Nussbaum *et al.*[45]

2. Key Predictive Sample Set

The key predictive sample set (KPSS) (Figure 2, Step 6) for a given compound is equal to the significant degradants plus process impurities that can include intermediates, starting materials, and by products.[46] Process-related impurities and known degradation products might be available as reference standards for use in method development. Unknown degradation products can also be critical in the development of a stability specific method. Forced degradation studies may generate complex mixtures of degradants, but method development should consider only significant degradants. Although project-specific factors may influence judgments of degradant significance, the examples that follow describe guidelines that may generally be applied, considering the stage of development. We propose that a degradant in a stress testing sample may be judged not significant, and hence discounted, if it fails to exceed either threshold described.

Example 1. A degraded sample from an exploratory API forced degradation study shows 25% main band loss. Analysis shows a major degradant (16%), a second degradant (3%), and several further degradants at levels NMT 2%. The 16% peak is the largest degradant, representing 60% of total degradation and is judged significant. Although the 3% component comprises more than 10% of total degradation, its level is less than 25% that of the largest degradant and is thus judged not-significant. The components at or below 2% are less than 10% of total degradation, as well as being less than 25% the level of the largest degradant. All are thus judged not significant.

Example 2. A forced photodegradation experiment results in 10% main band loss. Many individual degradants are observed, of which the largest is 0.8%. Since no individual impurity exceeds 10% total degradation, none are judged significant.

Figure 2 details the process involved in the key degradation impurity sample generation and selection. Work should then be performed to identify the unknown significant degradants, as appropriate. The integrity of these analyte peaks can be verified by liquid chromatography (LC)/mass spectrometry (MS) selective ion monitoring. The purity of the peaks can also be verified by collecting the peak of interest and chromatographing it by another method (thin-layer chromatography (TLC), GC, or a second HPLC system). Usually, key predictive samples can be generated on a large-scale basis for preparative isolation. Small-scale synthesis is also an option when a likely structure is identified and the chemistry can be accomplished in a few short steps.

3. Stereochemical Stability

Chiral APIs should be assessed for their stereochemical stability during forced degradation studies on a case-by-case basis. CAMEO analysis can be a useful tool to evaluate concerns with racemization of chiral centers in the API. If the degradation prediction suggests racemization to be likely by any condition, stereochemical stability should be explored. APIs with one or two chiral centers should be analyzed with a chiral method. Based on predictive data and chemical knowledge, choose degradation conditions that are most likely to cause racemization. If the chosen conditions do not invert or racemize the API, then chiral analysis does not need to be part of further forced degradation protocols. APIs with three or more chiral centers most likely convert to diastereoisomers and could be analyzed with an achiral method. Stereoisomers should be treated like any other API-related impurity with respect to quantitation, identification, and qualification thresholds.[47] A recommended procedure is to choose either the last timepoint or a timepoint where 5–20% degradation for each condition, and look for racemization of parent. If there is no racemization of parent under any stress conditions, then that information is used to justify why a chiral method does not need to be part of the control strategy.

4. Physicochemical Stability

A polymorph appearing in the late stage of drug development may require reformulation, redevelopment of analytical method, and change of manufacturing procedures. In addition to this, some physical form change only occurs in the solid state.[48,49] For these reasons, the solid form change should be monitored during forced degradation studies including stressing solvates and hydrates in closed and open containers. If a solid form change is detected, degradation rates could change for the new form; therefore, chemical stability should be evaluated in addition to physical stability.

5. Identify Degradants

Degradation product structure elucidation requires a multidimensional approach, typically involving MS and NMR analysis (and if needed, other spectroscopic techniques such as UV and IR and chromatographic techniques such as 2D-LC). Such multidimensional approaches can be greatly facilitated by collaborative efforts involving analytical, process, formulation, and spectroscopic scientists in a team setting. (Figure 2, Step 7)[50]

Typically, the focus will be on collecting LC/MS data only through the phase 1 clinical stage. At the phase 2 clinical stage and beyond, more time is invested in isolation, synthesis and structural identity using NMR characterization of forced degradation products of concern.

G. Pharma D3 Degradation Database

A structure-searchable degradation profile database has been created to compile API/DP degradation data and share degradation knowledge globally at Pfizer (Figure 2, Step 8).[51] The database is a web-based novel "hybrid" publishing/database paradigm where both chemical structures and associated reports are stored in a structure and word-searchable database called the "Pharmaceutical Drug Development Database," otherwise known as "Pharma D3." The use of Pharma D3 enables a proactive approach to drug stability by establishing trends with functional groups to allow enhanced prediction of degradation results. It was developed to address certain key goals including: improve communication, minimize duplication of effort, and establish a proactive approach to drug stability, predict problems and identify trends in data. With the degradation profile database, an analyst can search by structure and produce tabulated results of all the identified degradants observed under each degradation condition. The database also contains a substructure search function that allows the analyst to retrieve all records on compounds with similar structure. This function is extremely powerful in that it enables better prediction of degradation reactions and better use of our library of historical degradation studies. The following data is tracked in the database: compound number, notebook reference, salt form, parent structures, degradant structures, molecular weights/change in molecular weights of degradants, degradation experimental conditions, HPLC procedure references, conditions for performing degradant HPLC screening analysis, functional group involved in the degradation pathway, structure elucidation data, report references, links to electronic report, and proposed mechanisms for degradants (Figure 6).

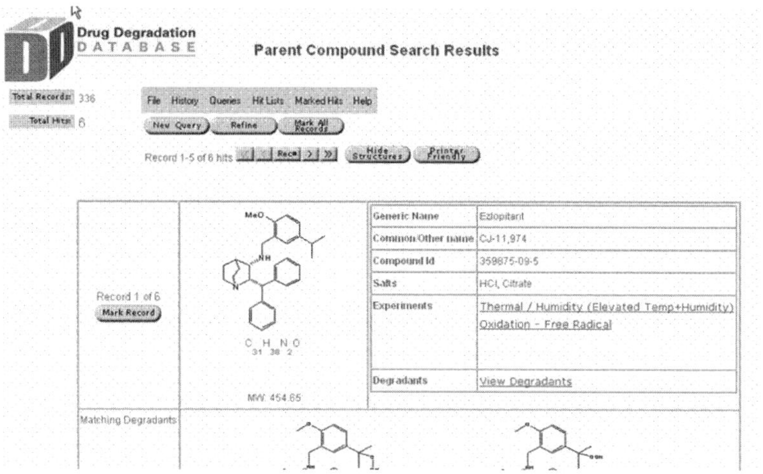

Benzylic funtionality: *Oxidation will be emphasized in our degradation experiments*

FIGURE 6 Degradation database screen shot example.

II. ISOLATION OF IMPURITIES/DEGRADANTS

A. Introduction

The objective of this section is to provide guidance for the isolation step in identifying both process related as well as degradation product-related impurities from pharmaceutical drug candidates. Impurity structure elucidation supports a drug candidate's progress through preclinical studies, clinical trials as well as through launch and provides structural support to commercially available products when necessary. The impurity structure elucidation process is a multidisciplinary and collaborative effort between a structure elucidation group (SEG) and the process chemist, analytical chemist, formulator, and degradation experts.[52–57] A structure elucidation group includes a team of isolation chemists, mass spectrometrists, and NMR spectroscopists who work together to effectively streamline the impurity identification process. The SEG works toward the common goal of providing critical structural information. Isolation is the first step in a multistep process. The process described in this section uses a designed approach for the impurity and/or degradant identification, which focuses on efficiency so that the success of data collection is maximized. There are a number of activities other than collecting experimental data, even though the experiments are central to the process. The activities associated with this work are captured in the process flow chart provided in Figures 7 and 8.

As a candidate moves successfully through the drug development process, it is critical to identify drug-related impurities. Once a compound is handed off from Discovery, the role of the structure elucidation chemists takes shape as the compound traverses through the development lifecycle. When the structure elucidation

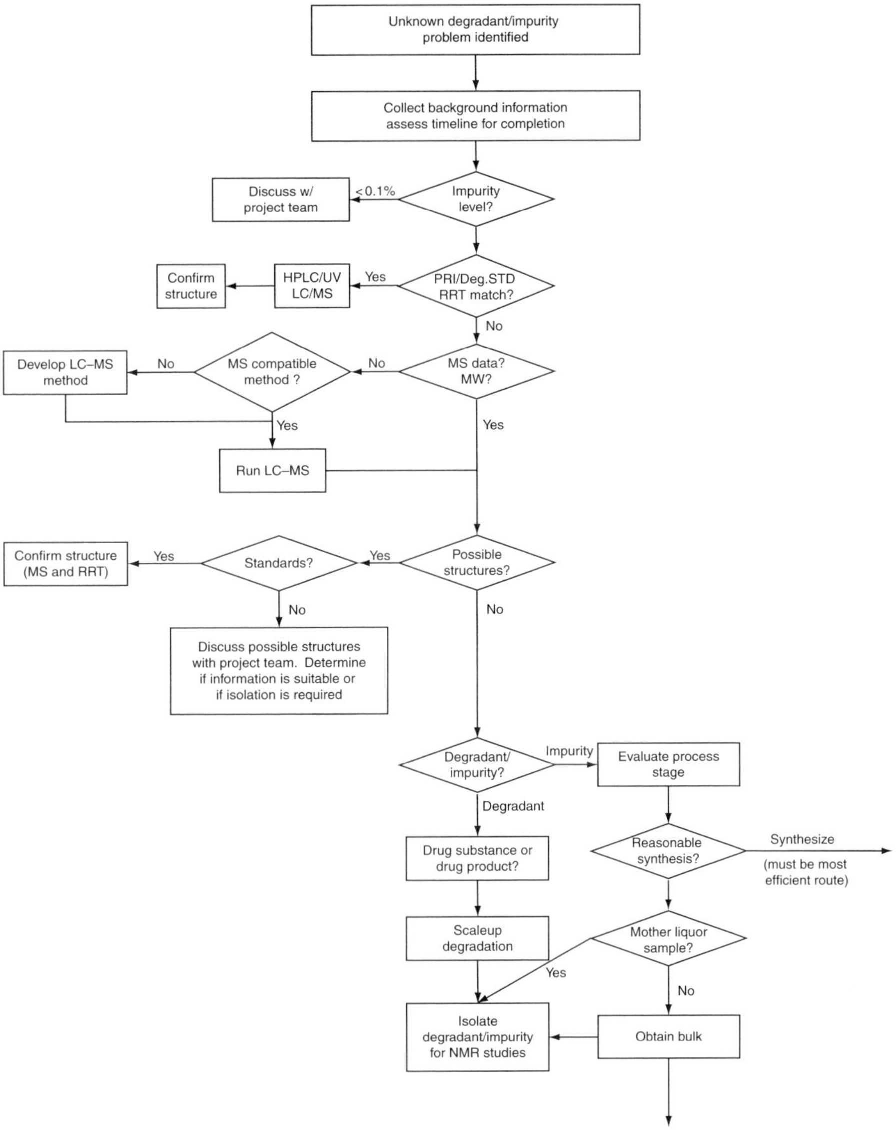

FIGURE 7 **Impurity/degradant isolation and identification process flow chart.**

process is performed at an early stage of drug development, there is adequate time for improvements in the drug substance process and DP formulation to prevent these impurities and degradants long before the filing stage. Degradation product structural information can provide an understanding of impurity formation and define degradation mechanisms. Representative samples submitted to a Structure Elucidation Group for impurity identification include (1) low-level (0.1–0.5%) process-related impurities[58] that are present in intermediate step reaction products or final API lots; (2) low-level degradants[59] of API lots; (3) process chemist reaction by-products generated from reaction optimization decisions (~3%

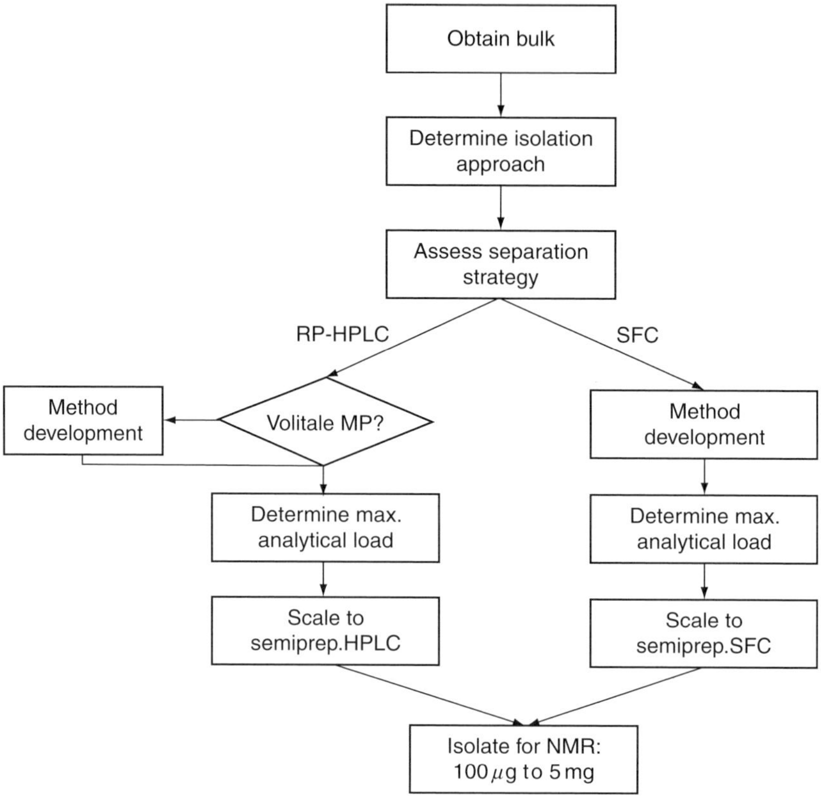

FIGURE 8 **Impurity/degradant isolation and identification process flow chart.**

level—major components); and (4) low-level drug formulation related impurities (process or stability related). Degradation product and process-related impurity (PRI) structural information allows the analytical chemist to set appropriate impurity specifications in API and DP lots needed to support packaging of materials used in clinical trials and commercial products. Process chemists benefit when problematic by-product structures are known. Structural characterization of by-product impurities can be instrumental to the process chemists in moving forward with key reaction optimization decisions[60] giving the chemist insight into ways of increasing yields or determining how to purge these impurities from reaction products and final API lots. Characterization of these impurities allows efficiencies to be built into these chemists workflows.

The focus of this section will be on the process, techniques and tools by which low-level impurities (i.e., $\geq 0.1\%$ level) are isolated for structure elucidation purposes. The collection of background information and a clear understanding of the customers needs are central to the success of the impurity isolation step. These factors will determine the isolation approach and strategy one chooses. The isolation step is a first step from which many downstream experiments depend.

Semipreparative supercritical fluid chromatography (SFC) and RP-HPLC are effective tools and the preferred techniques to isolate low-level drug-related impurities. Semipreparative SFC has proved itself as a first-line approach in the application of impurity isolation. The physical properties of a supercritical carbon dioxide mobile phase allow for chromatographic gains to be realized. Advantages of SFC over HPLC for impurity isolation include speed and solvent cost savings. Each of these techniques has advantages and disadvantages. These advantages and disadvantages, along with impurity project specifics and customer needs, determine which technique will be utilized. This section will discuss the benefits of using SFC for impurity isolation work and present case studies to highlight these advantages. Isolation turnaround times and the importance of monitoring the isolation process, with regards to process control, will also be discussed.

B. The Isolation Process

1. Collecting Background Information

The identification process of impurities and/or degradants begins with the gathering of background information specific to each isolation and structure elucidation project request. Collecting background information specific to each isolation/structure elucidation project request is a very important first step of the overall isolation and identification process. Collection of background information includes: (1) obtaining all pertinent sample information and (2) obtaining a clear and concise definition of the problem to be solved including timelines for completion. The first question to be asked is: What is the level of the unknown impurity? Determining the level of the unknown is a very important piece of information to be obtained.

According to the ICH Guidelines on Impurities in New Drug Substances[61]:

> The studies conducted to characterize the structure of actual impurities present in a new drug substance at a level greater than (>) the identification threshold (0.1% for a maximum daily dose of less than or equal to 2 g/day or 0.05% for a maximum daily dose of greater than 2 g/day) (e.g., calculated using the response factor of the drug substance) should be described. Note that any impurity at a level greater than (>) the identification threshold in any batch manufactured by the proposed commercial process should be identified. In addition, any degradation product observed in stability studies at recommended storage conditions at a level greater than (>) the identification threshold should be identified.

According to the ICH Guidelines on Impurities in New Drug Products[62]:

> Any degradation product observed in stability studies conducted at the recommended storage condition should be identified when present at a level greater than (>) the identification thresholds (1% for a maximum daily dose of <1 mg to 0.1% for a maximum daily dose of >2 g).

Identification of impurities below the 0.1% level is generally not considered to be necessary unless the potential impurities are expected to be unusually potent or toxic.[61,63] Therefore, it is imperative to determine the level of the unknown impurity and/or degradant early in the process. If the unknown is below the 0.1% threshold, then a discussion will need to take place among the project team members in order to determine if isolation and identification are necessary. However, if the unknown is at or above the 0.1% limit, then effort should be put forth for identification.

It is imperative that discussions at this stage include scientists intimately involved and familiar with the project and the sample being submitted. The analytical chemist, process chemist, formulator and the degradation chemists are all valuable resources in collecting this information. "A few minutes consulting about the complete background of the sample can save hours of time. Depending upon the sample history, different types of information should be secured."[60] Additional sample related questions that need to be answered at this stage are: Is the request for an API or DP impurity? Is this a PRI or degradant problem? If it is a PRI, at what step of the process is it formed? Is the synthetic route available? Are starting materials available? If it is a degradant, under what specific degradation condition is it formed? If it is a formulated DP is a list of all excipients available? What is the analytical HPLC test method conditions used to generate chromatography from which the target impurity is observed? Is a representative HPLC chromatogram available for the sample submitted from which the target impurity of interest can be identified and tracked? Are enriched samples with the unknown impurity/degradant available? How much material is available? Is the analytical method using an MS friendly mobile phase? Is MS data (i.e., molecular ion) available for the target impurity? Are there proposed structures for the target impurity? What are solubility and solution stability characteristics and of sample source?

2. Understanding Exactly What Is Requested—Identify Isolation Goal

"A clear and concise definition of the problem to be solved is the second most important piece of data"[60] an SEG team can obtain. It is important to have a clear understanding around exactly what level of structural information the requesters (customers) actually needs. The amount of structural information and the level of confidence required are dictated by the customers and their project needs. The customer's needs are generally shaped by the stage at which the drug candidate is in the development process and the specific reason by which the structural information has been requested. As a drug candidate progresses through the development process it becomes increasingly important to elucidate impurity structures (degradants, PRI's, reaction by-products, etc.) with higher degrees of confidence. Understanding the level of structural information the customer needs facilitates decisions around the level of resource devoted to an isolation request. It is equally important for the structure elucidation team (isolation chemists, MS spectrometrists, and NMR spectroscopists) to explain to the customer the relative effort and resources that will be necessary to provide the customer with the structural information they are requesting. It is instructive for the structure elucidation team to explain to the customer the types and varying levels of structural support the SEG can provide. The structure elucidation team and customer reach a decision in aligning a level of effort appropriate to support the drug candidate project needs. This helps the SEG evaluate and determine an isolation approach that will best match the customer's needs. By gathering all relevant information, the first step in successfully isolating an impurity is accomplished before one ever enters the lab.

Once a decision has been made to identify an unknown and impurity background information has been obtained the next logical step is to evaluate all known process-related impurities, precursors, intermediates, and degradation products to determine whether or not the impurity of interest is truly unknown. By observing the HPLC relative retention time (RRT) of the unknown impurity with that of available standards: (1) a known PRI, (2) a precursor and (3) an intermediate, then the unknown can be identified using HPLC (with UV photodiode array (PDA) detection) and LC-MS. The identity is confirmed by correlating the retention time, UV spectra and mass spectra of the unknown impurity with that of the standard. A spiking experiment of the standard into the bulk lot of material generating a single peak consistent with the retention time of the target impurity can be used to further increase confidence.

3. Determine the Isolation Approach

When an impurity is determined to be truly unknown, an SEG team approach is implemented. Understanding the needs of the customer dictates the level of effort and amount of resources that will be necessary to meet the customer needs. The isolation chemist efforts toward supporting an MS only characterized impurity structure is very different than the level of effort and resources needed to support an MS and NMR characterized impurity. An MS only request may or may not require the assistance of the isolation chemist. Molecular mass, fragmentation, and accurate mass data are collected via HPLC-MS from bulk materials. It is then determined if the MS data evaluation yields sufficient structural information satisfying the customers needs. If MS data is unobtainable from bulk materials, an enriched sample source will be needed by the mass spectrometrist. The isolation chemist can use bulk materials for enrichment purposes. The isolation step is one on which many downstream MS experiments can depend and all tube NMR experiments depend. It is not uncommon for the MS chemist to have a bulk sample with chromatographic conditions that do not produce the amount of ionization of sufficient signal strength needed to collect quality data. As a result confidence in proposed structures can be of low confidence. In addition, at times the target impurity coelutes or is not fully resolved from other sample related impurities. The role of the isolation chemist is to simplify the sample matrix by providing an enriched or isolated target impurity sample source at concentrations that allow MS spectrometrist to obtain a higher quality, more comprehensive data set. It may be possible to collect an enriched impurity sample source for MS analysis from one or two fraction collections on an overloaded analytical or a standard load semipreparative method. The amount of isolated material that results may be appropriately suited for the needs of the MS chemist. Isolation support for an MS proposed structure does not require a pure target impurity sample source, when MS friendly chromatographic conditions exist. However, simplification of the matrix at the retention time of the target impurity of interest in very helpful. In addition, the amount of target impurity material needed to perform a full suite of MS experiments is significantly less than the amount needed for NMR because of the sensitivity of this technique.

The isolation chemist efforts toward supporting a full MS and NMR structure characterization are far greater than the level of effort needed to support an MS only characterized impurity request. A typical workflow sequence of events for a request of this nature is as follows: (1) the isolation of the target impurity work is performed in parallel with MS data collection from bulk or enriched materials, (2) MS and NMR data collection on the isolated impurity material is performed, (3) followed by spectral impurity characterization. The job of the isolation chemist is threefold when working with an NMR spectroscopist: (1) to deliver an appropriate amount of isolated impurity material consistent with NMR instrumentation sensitivity requirements, (2) deliver high purity isolated impurity material, and (3) deliver a high quality isolated impurity material free of excess salts, residual solvents, and nonsample related impurities. NMR instrumentation and component specifics dictate the minimum amount of isolated target materials needed for analysis. NMR instrumentation available today can collect quality data across a full suite of experiments with less than one milligram of material, however, regardless of the NMR instrumentation specifics; the amount of material isolated is inversely proportional to the time necessary to run a full suite of NMR experiments. Refer to the NMR section of this chapter for more information.

4. Isolation Strategy

Two strategies can be implemented as one enters the laboratory faced with isolating a low level or major component impurities from an isolation sample source. The strategy will help identify analytical method from which to scale. The first approach employs analytical screens as a starting point for identifying a method for scale. The ultimate goal of an analytical screen is to identify a method that best resolves all drug-related components as well as excipient-related sample components from the target impurity of interest. All other components can coelute with each as long as the target impurity is well resolved from them. The second approach is to critically evaluate the information one has acquired through the background information collection process when developing a suitable analytical method from which to scale. This evaluation process serves as a jump off point from which successful analytical method development can begin. The goal is not to identify final analytical conditions, but to give a starting point from which analytical method optimization can occur. Regardless of the isolation strategy chosen it is preferred that all method development be performed using a PDA and single quad MS detector when possible. Obtaining molecular mass data from the mass spectrometrist enables the isolation chemist to track the correct peak using LC-MS. A unique UV spectrum and a molecular ion of the target impurity of interest is an effective way to peak track and identify your target during method development.

a. Analytical Screens

Analytical screens are performed with both reverse-phase RP-HPLC and SFC isolation techniques. Analytical SFC should be screened first unless instrumentation availability or project background specifics dictate otherwise. Screening achiral column bonded phases varying in polarity and functionality against different mobile-phase solvent choices is effective for identifying analytical methods for the purpose of impurity isolation. There are currently many unique achiral SFC bonded phase column choices commercially available (2-ethyl pyridine, diethyl amino, dinitrophenyl, pyridine urea, diol, cyano, etc.). SFC column choice provides the most impact in manipulation of relative selectivity for individual components of a sample in a chromatogram. Alcohol modifiers are typically used in SFC method development with methanol most commonly used. Generic gradients are used to screen columns and modifiers. Run times are on the order of 10 minutes. Volatile acidic and basic additives added directly to the modifier and used at times to maintain peak shape and prevent excessive peak tailing. RP HPLC analytical screens are performed when SFC is determined to not be a viable option. RP column bonded phase choice and mobile pH changes provide the most impact in manipulation of relative selectivity for individual components of a sample in a chromatogram. Screening column bonded phases varying in polarity and functionality combined with screening low, neutral, and high pH mobiles has proved very useful for identifying analytical methods for the purpose of impurity isolation. Acetonitrile is the preferred organic mobile component. Volatile additives, like trifluoracetic acid (TFA), formic acid, ammonium acetate, and ammonium hydroxide are preferred pH modifiers.

b. Critically Evaluate Background Information

The second approach is to critically evaluate the information one has acquired through the background information collection process. It is important to note that every isolation project request is unique with regards to all the specifics that define the particular request. Before one ever enters the laboratory evaluation of these specifics can lead to more direct paths in developing a suitable analytical method from which to scale. Project details and specifics combined with scientific reasoning are at times the quickest way to develop an analytical method for scale. Identifying ways to generate enriched sample sources for low-level impurity requests as well as identifying ways to increase sample solubility are part of this process. It is important to keep in mind the chromatographic task at hand when developing a method that targets an impurity for isolation and determines a logical line of approach.

c. Enriched Sample Sources

One of the most important factors to consider when approaching an isolation experiment is to determine whether the unknown is an impurity and/or degradant, and the level at which it exists in the sample source submitted for isolation. Isolating low-level impurities can prove to be very cumbersome and time consuming. A request for full structure elucidation on a low-level degradant impurity in a low-dose formulation with a limited bulk supply will necessitate the need to identify an enriched sample source. Low-level impurity isolations with ample bulk supply can alleviate the necessity to obtain an enriched sample source, but this is not ideal. Therefore, the ultimate goal is to find a sample that contains an enriched quantity of the unknown. Upfront time invested in identifying enriched sample sources can decrease the overall isolation turnaround time for a targeted amount of isolated impurity material. Enriched sample sources can also assist in generating larger quantities of isolated material for NMR analysis. This will, in turn, decrease the NMR data collection time. Two great resources of enriched samples are retained mother liquor[64] samples and purposeful degradation/stability samples. If the unknown is an API degradant, then the degradation reaction can be scaled up to generate a large quantity of the unknown. If it is a DP degradant, then effort should be put forth to form the degradant in the API so that extraction from the excipients is not required. Whenever enriched samples are not available, the unknown must be isolated from the bulk API supply or DP sample source.

d. Supportive Isolation Tools

There are other useful techniques available to the isolation chemist that can be incorporated into the isolation process to support the overall isolation goal. Solid-phase extraction (SPE) and liquid–liquid extraction are commonly used techniques that are part of the isolation chemist's tool chest that can be used in clever ways. SPE also offers great potential in purifying the isolated product because of its universality.[65–68] These supportive techniques are at times used as a means to manipulate an isolation sample source prior to injection onto SFC or HPLC semipreparative isolation methods. The sample source must be adequately soluble in an organic solvent prior to injection onto SFC. SPE and liquid–liquid extraction are effective in converting less than ideal samples for SFC injection into workable samples. For example, API salts that have limited solubility in organic solvents or have a counter ion that may absorb in the UV spectrum and complicate the chromatography. Parenteral DP sample sources in aqueous media cannot be injected directly onto an SFC as water impacts chromatography negatively. Free basing drug-related components using a basic aqueous media (aq.) and then extracting into organic solvents can result in a sample source that is free of counter ion and has favorable SFC solubility characteristics. These supportive techniques are also quite useful post-fraction collection as a means to decrease overall isolation turnaround times. RP semipreparative HPLC fraction collection yields fractions containing large amounts water. A time effective way of removing water from these fractions involves both SPE and liquid–liquid extraction. The organic solvent component, present in collected fractions, is removed under vacuum. Resulting aqueous isolate mixture can be passed through a RP SPE sorbant to effectively dewater and desalt the isolate in one step. This aqueous isolate mixture can also be extracted with an organic solvent to effectively dewater and desalt the isolate.

e. Maintaining Process Control

Regardless of the isolation technique chosen, isolation process control and understanding are critical components to the overall success of an isolation. Having control over the isolation process is essential. Process control and understanding is accomplished using analytical HPLC methods to monitor the progress along each isolation step. These analytical methods used are typically obtained from the customer. This is the same method from which the target impurity was observed and submitted along with the sample for isolation and identification. The steps along the isolation process that require monitoring in order to maintain control over the process include: (1) confirming the correct target has indeed been isolated, (2) relative purity of collected fractions, and (3) indications of potential fractionated impurity solution stability issues. Monitoring isolation progress is the first step in devising an approach to resolve roadblocks when they occur. It is not uncommon to have loss of purity result from of solution stability issues of the target impurity post-fraction collection. Monitoring progress and having control over the isolation process allows the chemist to identify and critically evaluate areas of concern and assist in identifying ways to overcome these challenges. There are a number of actions that can be taken to help navigate through difficult portions of challenging isolation projects. There are many different ways to successfully navigate through an impurity isolation. However, there are also many points along the process in which decisions can lead to the failure in impurity isolation.

f. NMR Solvent Choice—Isolation Chemist's Role

It is the job of the isolation chemist to make available isolated materials solubility and solution stability learnings to the NMR spectroscopist. Identifying which solvents the isolated impurity material is soluble in will assist in solvent choice for the NMR analysis. Identifying solvents that could be cause for concern with regards to solution degradation is very important. It is not uncommon to have an impurity that is pH sensitive or labile in protic solvents. NMR sample preparation will avoid DMSO for impurities sensitive to high pH as DMSO is basic in nature. NMR sample preparation will avoid chloroform for impurities sensitive to low pH as chloroform is acidic in nature.

C. Preparative Chromatography

1. Introduction

The impact of preparative chromatography support a drug candidate's lifecycle is felt from Discovery's candidate seeking stage through the development and manufacture of materials for clinical supply, through the launch of a drug candidate into the marketplace. Preparative chromatography is the process of using liquid or supercritical fluids to isolate a sufficient amount of material for other experimental or functional purposes.[69] This section describes the use of preparative chromatography in isolating up to milligram quantities of unknown compound(s) for the purpose of structure elucidation by spectroscopic techniques, which is often referred to as semipreparative HPLC and semipreparative SFC.[70–72] This section will focus primarily on preparative methods with the following parameters:

Column length	10–25 cm
Column width	10–30 mm
Particle size	5 μm
Column load	<1 g
Flow rate	5–100 mL/min

The scale of preparative chromatography is larger than that of conventional analytical chromatography.[73] Therefore, a practical starting point is to develop an analytical separation that optimizes the isolation conditions. Optimization of the analytical method implies seeking conditions which combine maximum resolution of the peak of interest and minimum elution time, under the restriction of a limited pressure drop.[74] The optimized conditions determine the column, mobile phase, flow rate, and sample loading capacity for the particular column. The conditions may be either normal phase (NP) or RP.

2. Scaling up a Method

Scaleup from an analytical method to a preparative method for structure elucidation purposes is a four-step process. First, optimize an MS friendly analytical scale method at analytical load conditions. Second, determine the maximum analytical load of the analytical column and method being used. This is limited by diluent strength, sample solubility, buffering capacity of the mobile phase, and injection volume. Maximum analytical load tends to decrease for difficult separations and samples with complicated matrices, however, may increase with simpler matrices with well-resolved components. Sample load is optimized such that the target impurity resolution is maintained from all other components in the matrix. This is an important step in the scaleup procedure and used to determine the loading capacity of the preparative column that will be used. Third, calculate preparative HPLC or SFC flow rate and column load using the scaleup factor in Equations (1–3). Finally, perform preparative HPLC or SFC injections at calculated flow rate and load amounts and then optimize by fine-tuning conditions. For ease of transition it is vitally important that the preparative column used be of the same manufacturer, column brand and particle size, as the analytical column, for the scaleup to occur in any predictable fashion. Column sets containing an analytical methods development column and a preparative isolation column should ideally be packed from the same lot of packing material. Columns should be purchased with this in mind. This provides assurance that any separation developed on the analytical column will scaleup directly on the matched preparative column with minimal modifications necessary. The use of these column sets eliminates potential selectivity differences caused by different types of silica and different particle and pore size packings by providing matched columns. The analytical isolation can be scaled up using Equation (1):

Scaleup equation[75]

$$\text{Scaleup factor} = \frac{(D_\mathrm{p})^2 \times L_\mathrm{p}}{(D_\mathrm{a})^2 \times L_\mathrm{a}} \qquad (1)$$

where D_p is the diameter of preparative column; D_a, diameter of analytical column; L_p, length of preparative column; L_a is the length of analytical column.

The column is the most important component of the process. The scaleup factor is used to predict how much material can be loaded onto the preparative column and what the flow rate should be. For example, if a separation optimized on a 4.6 mm × 100 mm column was scaled to a 20 mm × 150 mm column, the scaleup factor for the sample load would be 28. Thus, the scaleup factor multiplied by the maximum analytical load estimates how much material can be loaded on the preparative column (see Equation (2)).

Determination of maximum preparative load[75]

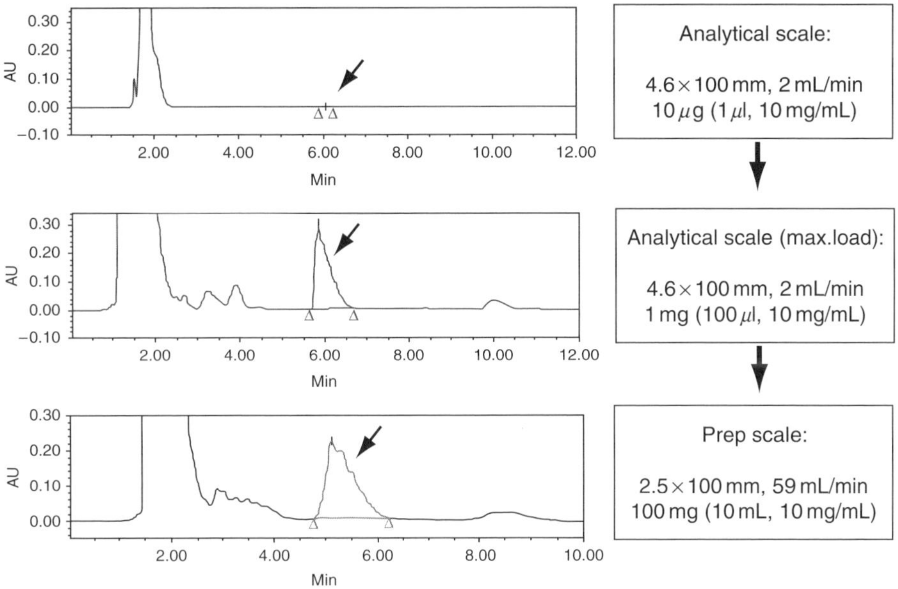

FIGURE 9 Case study of analytical HPLC flow rate and sample load scaled to semi-preparative HPLC.

$$M_p = M_a \times \frac{(D_p)^2 \times L_p}{(D_a)^2 \times L_a} \quad (2)$$

where M_p is the maximum mass load on the preparative column and M_a is the maximum mass load on the analytical column.

To maintain the same resolution when scaling-up a method, the flow rate also needs to be scaled proportionally. The preparative flow rate can be estimated by using the scaleup factor. For this estimation, the scaleup factor is multiplied by the analytical flow rate to estimate the preparative flow rate (see Equation (3)).

Determination of preparative flow rate[75]

$$F_p = F_a \times \frac{(D_p)^2 \times L_p}{(D_a)^2 \times L_a} \quad (3)$$

where F_p is the preparative flow rate and F_a is the analytical flow rate.

Figure 9 (top panel) shows an analytical scale RP-HPLC method optimized under analytical load conditions. Maximum analytical load conditions were then determined (second panel). Note that in this example, the maximum analytical load was limited by the solubility of the sample (10 mg/mL) and the analytical HPLC 100 μl injection loop size. The scaleup factor is then used to calculate preparative flow and sample load conditions. Preparative load (third panel) was optimized and significantly increased above the calculated load in this example while maintaining resolution of target impurity in order to maximize throughput and decrease overall collection time. It is important to note that with any semipreparative column dimension choice the flow rate and load increases (1) linearly when column length increases and column diameter remains fixed and (2) exponentially when column diameter increases and column length remains fixed.

When the preparative method has been optimized, injections are made and the compound of interest is collected using a fraction collector. Fractionation of target peaks is triggered by time, UV threshold or by mass.[76–78] The fractions are pooled together in a collection vessel. The stability of the isolated product should be assessed off a small number of pooled fractions prior to consuming your entire sample source for isolation. This will enable decision as to if special collection conditions are required or if alternative chromatographic phases or condition are necessary. In addition, be sure to use clean glassware. Contamination can occur if the glassware has not been thoroughly cleaned. Also, carefully select the tubing used for collection of the unknown (i.e., avoid using Tygon tubing because it contains phthalates). The isolated product is concentrated using conventional sample concentration techniques (i.e., distillation, precipitation, freeze-drying, and solvent extraction). Rotary evaporation and centrifugal type vacuum evaporation systems are the most commonly used to recover isolated products from the mobile phase.[66] After the product has been recovered, it should be dried under high vacuum to remove all residual solvents. Washing the isolated sample with deuterated solvent several times also helps to prepare the sample for NMR. Once the sample has been purified, it is submitted for LC-MS and NMR analysis.

As was mentioned earlier, isolation of low-level impurities and/or degradants can be cumbersome and time consuming. Consider a 0.1% level impurity present in an API bulk lot. Based on traditional NMR experiments (see Section IV), 1 mg of the impurity would be needed to obtain structural confirmation. Fraction collection time is determined by sample solubility (i.e., 3 vs. 40 mg/mL), loadability (i.e., 6 vs. 60 mg), method run time (i.e., 7 vs. 18 min), % level of impurity (i.e., 0.1% vs. 25%), and amount of impurity targeted for recovery (i.e., 0.5 vs. 10 mg). Figure 10 illustrates that each of these factors has an additive affect impacting the overall fraction collection time. This time frame does not include the time needed to develop an analytical method for scale, scaleup to preparative

Case 1:
Solubility = 3 mg/mL
Loadability = 2 mL (6 mg)
Run time = 18 min
% Impurity = 0.1%
Target isolate amount = 1 mg
→ 50 h on prep-LC (167 inj.)

Case 2:
Solubility = 40 mg/mL
Loadability = 2 mL (80 mg)
Run time = 7 min
% Impurity = 0.5%
Target isolate amount = 1 mg
→ 21 min on prep-LC (3 inj.)

FIGURE 10 **Factors that impact preparative HPLC and SFC fraction collection time.**

chromatography systems and fraction solvent removal, all necessary to complete the isolation. It is important to keep in mind too that the MS and NMR impurity data collection and interpretation time are additive to the total time it takes to turn a structure around to the customer. Enriched sample sources can significantly impact fraction collection time. A 5% level enriched sample source would drastically reduce Case 1 collection time to four injections over 72 min to reach a calculated 1.2 mg of isolated material. A 5% level enriched sample source would reduce Case 2 collection time to one injection over 21 min to reach a calculated 4 mg of isolated material. Note that Figure 10 assumes 100% recovery, which is not a practical expectation. Actual recoveries are generally closer to 50% for low-level (0.1% level) isolations, however, do increase as the impurity level increases. Actual fraction collection times and the number of injections performed would likely be closer to 2× that which is reported in Figure 10 in order to recover the targeted 1 mg amount. This particular isolation would likely require approximately 2 g of bulk.

3. Analytical HPLC as an Isolation Tool

Using analytical scale HPLC systems as a means to isolate impurities is a useful tool but should be applied only when the project specifics dictate or when semipreparative scale instrumentation is not available for use. It is possible to successfully fractionate the proper amount of material from low-level impurities (0.2% level) in low-dose tablets or even in API bulk sample sources for the purpose of MS data collection. However, it is not realistic to fractionate the proper amount of material from such a sample in this manner for NMR purposes. Analytical HPLC does have the capacity of being exploited as an isolation tool and can be accomplished as follows. Most analytical HPLC systems can flow up to 5–10 mL/min. It is possible to collect material for NMR when target impurity levels are sufficiently high enough to collect a target amount of impurity material in a reasonable amount of time. First, develop an analytical method for scale at 1 mL/min on a 4.6 mm ID analytical column with a column length from 100 to 250 mm. Loadability is increased fivefold by scaling the analytical method to a 10 mm ID column of the same length. The scaleup factor is five, hence, flow scales to 5 mL/min and loadability increases 5× the predetermined maximum analytical load amount from the 4.6 mm ID column. Figure 11 case study illustrates the turnaround time differences between analytical versus semipreparative HPLC. In this example, a 0.3% level impurity was collected from an analytical HPLC system. The analytical project laboratory targeted 50 μg of impurity material for MS analysis over 17 h. The method was then modified for preparative HPLC when identification of the 0.3% level impurity via NMR analysis was determined to be necessary. The power of preparative chromatography shows that six preparative injections were performed over 1.5 h to obtain sufficient material for NMR analysis. Had this been performed on an analytical system, fraction collection would have taken 13 days! In addition, sample material for MS was accessible from one 15.5 min preparative injection versus a 17 h analytical collection time (Figure 11).

Prep-HPLC:
Solubility = 6 mg/mL
Loadability = 10 mL (60 mg)
Run time = 15.5 min
% Impurity = 0.3%
Targeted isolate amount = 1 mg
→ 1.5 h on prep-LC

vs.

Analytical fraction collection:
Solubility = 3.6 mg/mL
Loadability = 100 μl (0.36 mg)
Run time = 20 min
% Impurity = 0.3%
Targeted isolate amount = 0.05 mg
→ 17 h on analytical

Targeted isolate amount = 1 mg → 13 days

FIGURE 11 Case study illustrating the power of semipreparative HPLC. Targeted isolate amounts listed above are calculated values that typically deviate from experimentally determined recoveries.

D. Isolation Techniques

1. TLC and Flash Chromatography

Over the years a number of techniques and approaches have proved to be useful tools to successfully isolate low-level impurities and degradants. TLC is most useful when an impurity or degradant is identifiable by LC-MS and above the 1% level. In cases, where NMR analysis is essential for identification, semipreparative SFC, semipreparative HPLC, and flash chromatography are more suitable techniques. Please refer to the *Handbook of Pharmaceutical Analysis*, 1st ed.[15] for a more in-depth explanation of TLC[79] and flash chromatography's[80,81] use for impurity isolation. As mentioned earlier, SPE and liquid–liquid extractions are at times incorporated into the process. These tools can quickly convert bulk supply materials into a more suitable form for SFC or HPLC injection. SPE is also a useful tool in dewatering and desalting final RP-HPLC isolated materials obtained from solvents containing mobile-phase additives.

2. Practical Aspects of Preparative HPLC

As of recent the method of choice to isolate impurities structure elucidation purposes was RP or NP semipreparative HPLC. Turnaround time for an isolation would be on the order of 3 days to weeks, depending on the project specifics. There are many reasons why preparative RP-HPLC is often not first choice for the isolation of impurities today. These reasons include, but are not limited to, stability of the target isolate post-fraction collection and turnaround time necessary to generate a quality sample for NMR experiments. Stability issues can often be overcome in a RP approach but consume an enormous amount of time and solvent to troubleshoot and work through. In addition to long turnaround times, solvent costs are high relative to SFC. An average RP-HPLC isolation of a low-level impurity can consume 10 L of solvents or more. This volume is dependent on the level of the impurity, the amount of material targeted for isolation, loadability and injection cycle time.

Fraction volumes are on the order of 200 mL to 1 L or more. A low 0.1% level impurity, with poor sample loadability and a complicated sample matrix can aggravate the situation with regards to project turnaround times and total solvent costs. Impurity fraction solution stability issues exacerbate the situation even more. For example, substituted benzylic alcohols can dehydrate under acidic HPLC conditions, or carboxylic esters can hydrolyze in aqueous mobile phase. A RP-HPLC isolation can yield solvent costs on the order of $50–200 with a turnaround time of approximately 1 week to yield the quality sample necessary for NMR analysis.

3. Isolation Techniques of Today: SFC and RP Preparative HPLC

Two of the most utilized techniques currently used today for the isolation of low-level drug-related impurities are semipreparative SFC, which is a NP technique, and semipreparative RP-HPLC. SFC has proved itself as a first-line approach in the application of impurity isolation for structure elucidation purposes.[82] SFC is an effective isolation tool that results in enormous time savings as well as significant solvent cost savings. The impact as a result of the implementation of SFC is nothing less than impressive. Semipreparative SFC often delivers a targeted amount of isolated impurity faster and results in equal or higher quality and purity when compared to semipreparative RP-HPLC. Projects can be turned around with impurities ready for NMR analysis in 1–2 days for most projects. This includes SFC method development, scaleup, identifying and confirming the target peak of interest via spiking experiments, fraction collection and dry downtime. Isolated materials can be turned around and ready for NMR in as few as 3 h and it is not unusual to have an isolation complete in less than a day.

SFC can be successfully used for the vast majority of drug-related impurity identification requests requiring isolation. Semipreparative SFC is an effective tool in generating high-purity and high-quality impurity isolates. Organic solvents such as alcohols are typically used as a strong solvent (the modifier) to assist in elution of sample components. Target impurities are recovered from the fractionated organic modifier component of flow. A diverse and large selection in SFC achiral bonded phase column choices gives an advantageous selectivity impact over RP-HPLC. Tunable mobile-phase composition, temperature and outlet pressure (P_o) give additional levels of selectivity manipulation. Typical run times are on the order of 6–8 minutes compared to semipreparative RP-HPLC run times of 10–15 minutes. Shorter overall fraction collection time is observed when targeting the same amount of material for recovery. Stacking SFC injections has the advantage of reducing injection cycle times further. Total volumes of solvent resulting from SFC collected fractions are on the order of 3–40 mL. Minimal time is necessary for organic solvent removal.

a. Isolation Turnaround Time

Isolation timelines can vary significantly from project to project when the end goal is to generate material for NMR analysis. This is directly linked, in part, to the amount of material one wishes to recover from the sample source used. Turnaround times can vary from less than a day to weeks. There are many factors that impact an impurity isolation turnaround time including project background specifics, customer needs and isolation process specifics. However, the choice between using SFC or RP-HPLC is single most important choice that will impact turnaround times. SFC can save significant amounts of time. In addition, more complicated projects may require sample preparation prior to developing an analytical method for scale. At times a second purification pass using the same method or a different preparative method is necessary to increase purity values from the first preparative collection to levels that are acceptable for NMR analysis.

b. Use of SFC as a Purification Tool in Discovery

Discovery purification laboratories have embraced SFC technology[83–86] as a method of choice for chiral purifications over recent years but it has yet to gain traction as an impurity isolation tool with structure elucidation as the end goal. Few literature references report on this impurity application.[69,87–89] Discovery purification laboratories are in the business of achiral and chiral purifications on a milligram to multigram scale. SFC has gained explosive popularity, primarily focusing on chiral applications.[90–92] Implementation of SFC has resulted in increased productivity and throughput for these purification laboratories which, in turn, help expedite candidate decision-making processes. In addition to increased productivity, SFC delivers high-purity fractions with acceptable recoveries at significant solvent cost saving, resulting in substantial long-term operational cost savings. Preparative SFC has become so successful that Discovery purification laboratories have made a shift from preparative HPLC to preparative SFC as their primary method of choice for purification.

c. CO_2 as a Mobile Phase

There are many benefits to using CO_2 as a mobile phase. SFC is NP chromatography that replaces the conventional liquid mobile phase with CO_2. Commercial grade carbon dioxide is a low-cost solvent alternative with purity at $\geq$99.5%. A typical lab-sized cylinder of commercial grade CO_2 containing 50 lbs of liquid CO_2 costs approximately \$10-15 per cylinder. Carbon dioxide is captured and recycled for use from large-scale industrial processes that emit CO_2 as a waste by-product. In addition to low purchase costs, it has the added benefit of being free of monetary costs to dispose of. As the CO_2 is pumped through and exits the SFC system, it is vented through laboratory exhaust ventilation lines. Cylinders are a practical option for semipreparative systems with flow rates up to 10 mL/min. Additional benefits of working with CO_2 include its inert nature, nonflammable properties and its role in reducing a scientist's exposure to more toxic HPLC solvents like acetonitrile.

d. Supercritical Fluids and SFC

A key feature of SFC is the supercritical nature of the CO_2 and the advantageous physical properties that result. Supercritical fluids have intermediate physical properties of both gas and liquid phases and exhibit the positive chromatographic qualities of both. Low viscosity and high solute diffusivity, relative to HPLC, result in faster analysis times, shorter reequilibration times, and higher throughput without loss in efficiency. A supercritical fluid is a defined state, not an additional state of matter. A fluid is supercritical when it remains above its critical temperature (T_c) and critical pressure (P_c). For CO_2, T_c is 31 °C and P_c is 74 bar. Transition from a gas to a supercritical state occurs without a phase transition and without significant changes in viscosity or density. The same is true when a liquid to a supercritical state transition occurs. A back pressure regulator is incorporated at the back end of the instrument so that upstream chromatography is conducted at or near supercritical conditions. SFC operates efficiently as long as phase separation does not occur.[93]

e. SFC Considerations

SFC does have additional capital investment costs as well as chromatographic limitations. A semipreparative SFC system suitable for isolation of impurities will have a capital cost approximately twice that of a comparable semipreparative HPLC system. In addition, there are sample considerations when evaluating SFC as an isolation tool. The sample injected must be dissolved in a diluent that is miscible with CO_2 and typically is the modifier. Care must also be taken when components of a sample have a reactive functionality (e.g., carboxylic acid) when alcohol modifiers are used. Ester formation of sample injected or impurity collected can complicate matters.

f. Impurity Isolation SFC Case Studies: Speed and Cost-effectiveness

Numerous case study examples illustrate the operational cost savings SFC has on the workflow and customer base for the isolation of drug-related impurities.[82] These include significant project cycle time savings as well as a significant reduction in organic solvent purchase and disposal costs. A greater than 50% time reduction has been observed in the time needed to turnaround the isolation portion of a structure elucidation project when compared to semipreparative RP-HPLC. Typical solvent cost savings are on the order of 70–95%. HPLC grade organic solvent purchase costs can vary from approximately \$2/L to over \$9/L. Solvent disposal costs are approximately \$5/L. Note that all solvent costs cited in the below case studies include CO_2 and methanol purchase *and* disposal costs. Case studies will illustrate the significant time savings as well as solvent cost savings when using SFC as an impurity isolation tool.

Case study #1. A request was made for the isolation and structure elucidation of a 3.8% level PRI present in a non-GMP range finding toxicology API lot. SFC can significantly reduce injection cycle times and fraction collection time when injections are stacked (Figure 12). A complex impurity sample matrix also lends SFC as a method of choice because of high peak efficiencies that typically result. Optimized SFC method conditions were determined to be 10 mL/min on a 2-ethylpyridine 1 cm × 25 cm column. Excellent separation was observed in isocratic methanol cosolvent conditions at 23/77, methanol (0.2% diethylamine, 0.2% formic acid)/CO_2. Stacking injections reduced a 5.5 min run time to a 3.5 min injection cycle time. The arrow in Figure 12 identifies the target impurity of interest.

Case study #2. A process chemist generated a problematic by-product in an intermediate reaction step. The chemist was able to purify this impurity to an 80% level from chromatography. SFC is an

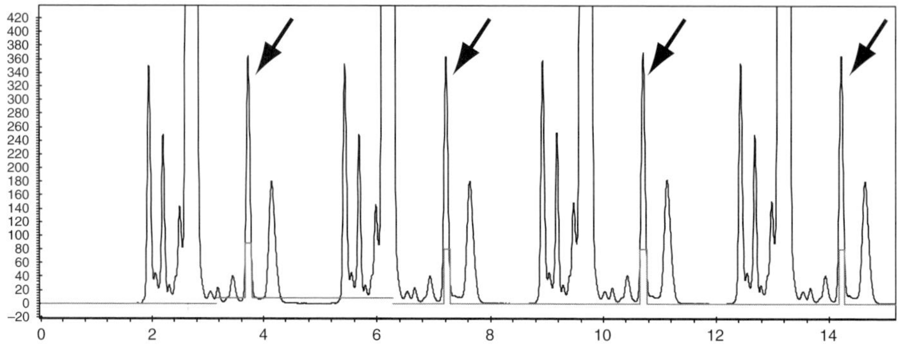

FIGURE 12 **Semipreparative SFC chromatogram illustrating stacked injections for impurity isolation.**

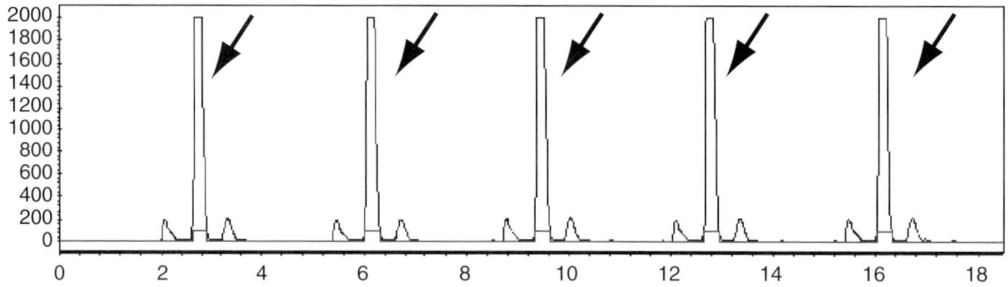

FIGURE 13 **Semipreparative SFC chromatogram of a major component impurity purification.**

effective purification tool when major sample components need cleaning up. Increased sample purity and quality typically provide a much better NMR sample resulting in higher quality NMR data collection. Optimized SFC method conditions were determined to be 10 mL/min on a cyano 1 cm × 25 cm column using isocratic methanol cosolvent 15/85, methanol/CO_2 conditions. Forty-six milligrams of the process chemists 80% pure reaction material were injected over 10 stacked injections, as shown in Figure 13. A 5 mL fraction volume was taken to dryness yielding 25 mg of material for NMR. Project turnaround time was only 2 h with less than $2 in solvent costs.

Case study #3. A process chemist received a vendor prepared 10 g naphthylpiperazine sample representative of a reagent to be used in a synthetic route. Vendor claimed 99% purity, however, the customer's analytical chemists observed a 10% level impurity. The material is expensive and purchased from a preferred vendor. Simple two component matrix impurity isolations can be quickly turned around using SFC. Optimized SFC method conditions were determined to be 10 mL/min on a cyano

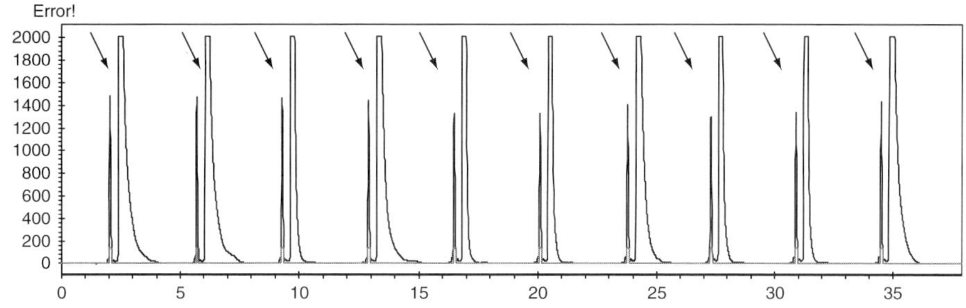

FIGURE 14 **Semipreparative SFC chromatogram of a two component sample containing a 10% target impurity.**

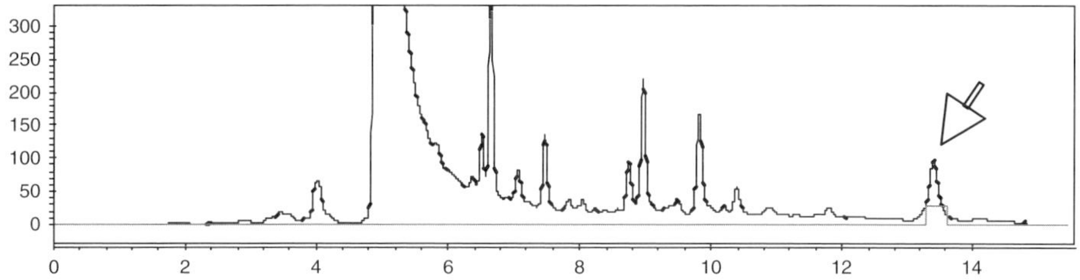

FIGURE 15 **Semipreparative SFC chromatogram of a complicated sample matrix with the target peak identified.**

1 cm × 25 cm column using isocratic methanol cosolvent 35/65, methanol/CO_2 conditions. Fifty injections were stacked over 2.7 h, as shown in Figure 14. Eleven milligrams of the target impurity was recovered for NMR analysis in 5 h with $5 in solvent costs.

Case study #4. A request was made for the isolation and structure elucidation of a 0.5% level PRI present in a starting material used for a step 1 reaction in an ICH synthetic route. This material was purchased from an outside vendor where lack of control for this particular impurity was an issue. Full structure elucidation on the target impurity provided critical information allowing the vendor to gain control for future lots. The project lab's RP-HPLC analytical method showed this particular impurity eluting in the middle of the impurity matrix and was not well resolved from nearby impurities. The orthogonal nature of SFC and multiple available column choices can assist in fully resolving the target impurity. Optimized SFC method conditions were determined to be 10 mL/min on a DEAP 1 cm × 25 cm column using a 10–60% methanol cosolvent gradient. Three hundred and fifty milligrams of material were loaded over 4.5 h in 18–1.5 mL injections, as shown in Figure 15. Approximately 2 mg of target impurity was recovered for NMR analysis. Project turnaround time was 8 h with $5 in solvent costs.

In addition to targeting specific impurities of interest, the isolation chemist uses SFC for a process Pfizer has termed "main band elimination" (MBE).[82] This experiment proves useful as a means to enrich low-level impurities when LC-NMR and other structure elucidation experiments are desired. LC-NMR is a technique that proves challenging when impurities are at the less than 1% level. SFC is a very efficient and fast tool for the removal of the major sample component. This MBE process effectively removes a reaction product from the sample matrix. The resulting sample significantly increases impurity relative area percents allowing for increased loadability and signal intensity for the targeted experiment.

The complementary nature of SFC and RP-HPLC has advantages in pulling low-level impurities out of API bulk supply or tablet formulations. Often, based on the specifics of the request and the nature of the sample, a two-pass isolation is the quickest, most effective way to generate a high-purity, high-quality impurity sample for MS and NMR. If possible a first pass SFC enrichment followed by a second pass SFC isolation occurs. However, preparative RP-HPLC is at times used as a first pass means to generate an enriched impurity sample. For example, taking the impurity from a 0.2% level to a 70% level can at times require minimal effort and be performed in a timely manner. This enriched sample, with a simpler matrix, is then used as a sample source for a quick SFC clean up. The orthogonality between techniques is exploited.

III. MASS SPECTROMETRY IN THE IDENTIFICATION OF IMPURITIES

A. Introduction

MS, by itself and in various combinations with other analytical instrumentation, is the first logical technique to use to probe unknown structures. It is a structurally sensitive technique, giving the potential of molecular mass and structure-indicating pieces of information in one observation. Coupled with chromatography, it provides orthogonal chromatographic retention time information to support any assumptions of purity, and provides clues to identity from retention time behavior. It is a quantitatively sensitive technique. The amplitude of the signals directly represents the amount of material introduced into the instrument over a relatively wide dynamic range. MS requires small amounts of sample to obtain significant amounts of structural information on a target compound.

MS is, however, a destructive technique—it consumes sample, albeit very small amounts. It is not a universal detector. There are entire classes of compounds that respond differentially in the mass spectrometer under varied conditions, poorly in general, or not at all. The inference of structure from mass spectra is also highly empirical. Regardless of past significant effort, the theoretical prediction of the mass spectrum from a given structure has not been perfected. One is obliged to compare carefully the behavior of accepted known structures with unknowns. The structure that is inferred is often not complete. Although MS is a powerful structure elucidation tool, it cannot differentiate some chemical structural features. Its use is not without risk, and should always be used in conjunction with data from other techniques and with good chemical sense.

The most important concept to remember when evaluating the value of MS—or of any analytical technique—is what the technique cannot tell one about the problem. This is just as important as what the technique can tell one about the problem. It is vital not to mislead oneself by thinking that the technique under consideration can address the problem when it really cannot.

B. The Process

How does one logically proceed after deciding that MS is an appropriate technique to use? History of a problem is collected, the most appropriate samples are identified, and the necessary supporting methods assembled—for example, for an LC-MS experiment, an LC-MS compatible assay that separates the components of interest is identified. Any decision regarding "confident answers" implies that the experiments have provided clear answers, such as clear molecular mass and fragmentation observations that are reproducible. It is essential to convince oneself that the facts are correct before expending effort trying to interpret those facts.

Comparison of the observed behavior of the unknown(s) with that of the parent drug substance is often helpful. (It is essential to have prior characterization of the parent drug substance so one knows how the structure class behaves.) Simple differences between the behavior of the parent molecule and of the impurity will suggest obvious structures. For example, a difference of 14 Da suggests a change in hydrocarbon chain length, presence of a methyl ester or hydroxylation followed by oxidation to a carbonyl. An increase of 16 Da suggests oxidation. A loss of 18 Da suggests dehydration. These simple "magic number" differences will often give immediate suggestions. In our environment, such simple variants are often available for comparison, or can be easily synthesized. This can lead rapidly to a confirmed identification.

C. The Tools

The technique of MS has been evolving for approximately 100 years. Most practitioners target J. J. Thomson's 1910–1912 reports[94] on his work with the parabola mass spectrograph as the origin. The field has evolved into a number of techniques that have found significant initial utility, then either proved their worth over the long term as general purpose tools, settled into a particular niche application area, or faded entirely from view.

To perform MS, one must make ions from neutral molecules. Ionization methods have advanced from the classic electron ionization (EI), through chemical ionization (CI),[95] field desorption (FD),[96] fast atom bombardment (FAB)[97,98] and thermospray[99] to the atmospheric pressure ionization (API) techniques currently favored. EI is classic, but is restricted in its applicability to thermally stable, volatilizable compounds. FD was always a specialized niche technique applicable to some larger compounds. FAB enjoyed a meteoric rise and fall in use; first reported in 1981, but has all but disappeared now, being replaced by the API techniques: atmospheric pressure CI (APCI)[100] and electrospray ionization (ESI).[101] Matrix-assisted laser desorption ionization (MALDI)[102] has shown significant utility for characterizing larger proteins, approximately 100 kDa and larger.

The magnetic sector mass analyzer is the prototype instrument. A number of double focusing sector geometries have been developed, but the sector instrument has largely become obsolete in recent years. Its expense, complexity and difficulty of operation have caused its demise in favor of quadrupole instruments, time-of-flight (TOF) instruments, and ion traps. Improvements in the reflectron TOF (rTOF) mass analyzer have even challenged the one remaining niche that the sector instruments, and the equally technologically intense Fourier-transform mass spectrometers (FT-MS) and orbitrap instruments[103] occupy, namely the high resolution accurate mass measurement arena.

Applying two or more mass analyzers in series produces a tandem mass spectrometer (an MS-MS),[104] the best known of which is the triple quadrupole. A number of tandem sector geometries have been built, some of which have been made commercially available. Hybrid geometries, combining a magnetic sector instrument with a quadrupole or an ion trap, have also been built and characterized. The ion traps, triple quadrupoles and, more recently, the tandem quadrupole-TOF instruments are the more abundant and user-friendly instruments. The quadrupole and the sector geometries provide the ability to perform tandem-in-space MS-MS experiments. The ion traps and FT-MS instruments permit performing tandem-in-time MS-MS, permitting MS^n (mass spectrometry raised to the *n*th degree—sequential collisional activation and decomposition) experiments.

The evolution of the ionization techniques, discussed briefly above, has progressed from the hard EI technique (hardness and softness of ionization refer to the energetics of forming an ion from a neutral molecule, and the extend of spontaneous fragmentation resulting from the ionization process. Hard ionization techniques induce extensive fragmentation. Soft ionization techniques induce minimal or no fragmentation), which produces extensive fragmentation, to the soft ionization API techniques, which produce in most cases only molecular ions. In achieving the goal of being able to ionize larger, more polar, less thermally stable molecules, and in optimizing for production of molecular ions, much of the structure-indicating fragmentation information has been lost. MS-MS experiments restore structural information content. They select the molecular ion produced by the soft ionization technique in the first mass analyzer of a tandem instrument and activate it by collision with a target gas. The product ion mass spectrum of fragments resulting from collision-induced decomposition (CID) of the parent ion in a later mass analyzer is recorded. Surface-induced decomposition (colliding an ion against a solid surface and detecting the fragments rebounding from that surface) is another collisional activation technique receiving significant attention in studying the structure of proteins.[105,106]

The effect of inducing fragmentation by collisional activation can also be achieved through collision with solvent molecules in the API ion source. Various implementations of this procedure are called up-front CID, cone voltage fragmentation and "poor man's MS-MS." Results obtained this way are comparable to true MS-MS product ion spectra. The strict interpretation of the results in terms of precursor–product ion relationships is lost, but this limitation does not negate the utility of the method when a true tandem instrument is not available.

Inlet systems—the mechanisms through which one introduces samples into the mass spectrometer—have also evolved substantially. Heated expansion

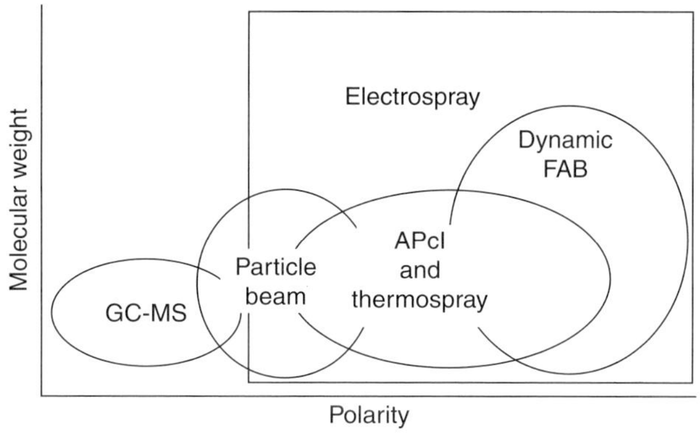

FIGURE 16 **Applicability of various chromatography–mass spectrometry techniques to the molecular domain.**

volumes and direct insertion probes have given way to interfaces which couple with chromatographic devices—gas, liquid, supercritical fluid, capillary electrophoresis, and thin layer—to the inlet of the mass spectrometer. This has given rise to a variety of combined chromatography-MS techniques. Applicability of these various examples depends on the properties of the target compounds being analyzed. Figure 16 maps two common ways to classify organic compounds against each other to help visualize which techniques are appropriate for which types of molecules.

By examining this brief history of the development of instrumentation and methods, it should be clear what parameters must be considered to identify the instrument of choice for analysis of pharmaceutically important molecules. Such an instrument is one capable of performing LC-MS, typically using a RP-HPLC separation and one of the API ionization techniques. MS-MS capability is desirable—using an ion trap, a triple quadrupole or other tandem mass analyzer instrument. Accurate mass measurement capability is also desirable. These attributes make up a prioritized list of capabilities for the "ideal" full-purpose instrument (or instruments) to be used in a pharmaceutical research environment. As one progresses through this list, the expense of the instrument increases, the sophistication of the instrument increases, and the intellectual and technological commitment required to do these experiments increases.

D. The MS Literature and Other Resources

Now that we know what some of the tools are, the following will illustrate the utility of MS in the arena of impurity/degradant identification using these tools. Recognize that we have left out a large variety of very valuable and interesting areas and aspects of MS, in order to focus on the analysis of pharmaceutically important compounds. Resources and information on other aspects of MS can be found in the following.

Recent reasonably comprehensive textbooks on the general topic of MS have been published by Gross,[107] deHoffmann and Stroobant,[108] Ekman *et al.*,[109] and in particular by Watson and Sparkman.[110] A treatise on the general subject of MS, slightly more up to date than the textbooks, has been published by Burinsky.[111] A description of the state of the art on combined chromatography-MS (including a discussion on LC-NMR) has been published by Sharp and Marquez.[112] Volume 101, issue 2 (2001), of *Chemical Reviews* covers many of the current active areas of development in MS.

The MS research literature is scattered through many journals and publications. *The Journal of the American Chemical Society* and *Analytical Chemistry* have been consistent contributors. In recent years, several journals specializing in fundamentals and applications of MS have come and gone or consolidated: *The International Journal of Mass Spectrometry*,[113] and *The Journal of Mass Spectrometry*.[114] *Mass Spectrometry Reviews* started in 1982. *Rapid Communications in Mass Spectrometry* started in 1987 as a means of providing (as the title implies) rapid dissemination of new results. *The Journal of the American Society for Mass Spectrometry* started in 1990 as the official archive of the American Society for Mass Spectrometry. *European Journal of Mass Spectrometry*, published since 1995, contains fundamental and applied work. Specific applications are also located in the specialty journals for those particular areas.

Two primary MS conferences occur on a regular basis. The International Mass Spectrometry Conference is held every 3 years, and is typically held in a European city.[115,116] The proceedings of this conference, in the form of extended abstracts of presented papers, are often published in book form as the series, titled *Advances in Mass Spectrometry*. Proceedings have been published occasionally as a special volume of the *International Journal of Mass Spectrometry and Ion Processes*. The annual conference of the American Society for Mass Spectrometry rotates among numerous cities on the North American continent.[116] The Society originated as the American Society for Testing and Materials Committee E-14 on MS, starting with an annual meeting in 1953. The Society was formally organized in 1969. Extended abstracts of papers presented at this meeting are also collected and distributed to members of the Society and to conference attendees. A growing archive of presented papers is becoming available to ASMS members.

E. How to Use the Tools

1. The Universal Detector (Not Really)

The mass spectrometer has been touted as the universal chromatographic detector. This claim is not true! Like any other detector, the mass spectrometer exercises some degree of preference for materials being presented to it. Consistent with this thesis, we have accumulated a growing list of molecules that are mass spectrometrically silent. Small molecules—in particular small aromatic molecules—exhibit a combination of volatility comparable with the mobile-phase solvents and a reduced ionizability under the ESP soft ionizing conditions to show minimal or no response in the mass spectrometer. Not every compound has a significant UV–vis chromophore, but many UV-silent compounds respond very well in the mass spectrometer. In particular, a number of commonly used excipient molecules do not have useful chromophores. Even though the UV–vis detector on the HPLC does not observe significant interferences from excipients, such as lactose or cellulose, the mass spectrometer suffers mightily from these UV-silent small molecules and polymers.

HPLC chromatographic methods should be developed off-line in the client project laboratories, and not on an LC-MS system. It is, however, still essential to have a UV–vis detector in-line with the mass spectrometer. The mass spectrometer detector has different molar responses to compounds than a UV–vis detector. When the total ion current chromatogram does not resemble the UV–vis chromatogram generated on another instrument, a simultaneously recorded UV–vis chromatogram of the separation presented to the mass spectrometer is an invaluable piece of information. This chromatogram can help resolve disputes over whether the desired chromatography is being reproduced by the LC-MS instrument. Does a PDA detector provide any significant advantage over a single wavelength UV–vis detector? The primary advantage of a PDA detector is during method development when wavelength maxima are potentially unknown. Assuming the separation method is developed offline and a detection wavelength has been chosen, we propose that there is little advantage to be gained over a single-wavelength detector. The added complexity, expense, computational sophistication, and manipulation and handling of very large data files produced by the PDA detector diminishes the return on the effort investment.

2. LC-MS Friendly Mobile Phases

With LC-MS experimentation as the goal, we assume the availability of a chromatographic method that adequately separates the components to be identified. The HPLC mobile phase must also be compatible with the mass spectrometric instrumentation. All components of the mobile phase should be volatile. Buffers need to be composed of a salt, both components of which are volatile compounds. Unfortunately, phosphate buffers and most ion pairing agents are incompatible. Trifluoroacetic (to access very low pH), formic, and acetic acids are the preferred acids. TFA has been claimed in the literature to compromise sensitivity,[117] but it is one of only a few volatile acids that can provide very low pH. The ammonium counterion is preferred for pH adjustment of any of these three acids. Unfortunately, the necessity to use these carboxylic acids eliminates the UV-invisibility advantage that the phosphate buffers have, compromising the use of low UV detection wavelengths. Orthogonal spray devices in the LC-MS interface introduce a degree of immunity toward nonvolatile components in the sample and the mobile phase. At least two instrument manufacturers have produced commercially available orthogonal spray interfaces. Both show an increase in immunity toward nonvolatiles. One manufacturer in particular has come close to eliminating the interferences caused by phosphate mobile phases.

Trifluoroacetic acid shows some reasonable ion pairing behavior. We have also been able to use low concentrations of heptafluorobutyric acid as a reasonable substitute for the conventional octanesulfonate ion pairing agents. Methanesulfonic acid is another ion pairing agent coming into wider use. HBF_4, in low concentrations, also provides ion pairing behavior.[118]

Volatile buffer components alone are not sufficient to determine LC-MS compatibility. Triethylamine (TEA) is often included in a mobile phase to inhibit chromatographic peak tailing. Although it is a volatile component, it tends to quench the ionization process occurring in the API ion source of the mass spectrometer. TEA is also very latent, in that TEA signals can be seen for days in non-TEA-containing systems, following a session with a TEA mobile phase.

3. Ionization Methods

a. Electrospray Ionization

Malcolm Dole[119] described his group's attempts to measure the relative molecular masses of oligomers of synthetic polymers based on ESI in 1968. John Fenn's group elaborated on the Dole observations, and built the electrospray ion source. In essence, liquid sample containing molecules of interest is flowed into the source through a small tube, upon which a high electrostatic voltage is placed. The liquid is sprayed from the tube tip, forming a Taylor cone and subsequently a stream of droplets, each electrostatically charged. A countercurrent flow of dry nitrogen gas opposes the spray, causing evaporation of solvent. As the droplets become smaller because of solvent loss, they reach the Rayleigh instability limit, where Coulombic charge repulsion overcomes solvent surface tensions and the droplets fragment further. Repetition of this process ultimately produces solvent-free solute ions, which are subsequently swept into the sampling orifice of the mass spectrometer and detected.

The detailed mechanism of electrospray is still an active and lively topic of discussion at current scientific conferences. Volume 35, issue 7 (2000), of the *Journal of Mass Spectrometry* published a series of five papers[120–124] from research groups working in this area, followed by a lengthy printed discussion[125] in the subsequent issue.

Two models have been discussed repeatedly in the literature—ion evaporation, attributed to Iribarne and Thomson,[126] and coulombic repulsion, attributed to Dole. Both models rely on the Rayleigh instability of charged droplets to become smaller and smaller. They differ in that coulombic repulsion relies on the continuous repeating of the process until one is left with a single isolated desolvated ion. Ion evaporation invokes the idea that, before a droplet becomes small enough to contain a single solute molecule, the electrostatic field strength at the droplet surface becomes sufficiently intense to eject ions from the droplet into the ambient gas. Kebarle reviewed the data available in 1991,[127] and updates and interprets[125] the summed results to propose that the data indicate that the mechanisms constitute a continuum. Elements of both interplay, depending upon physical properties such as hydrophobicity–hydrophilicity, the size of the molecule under consideration—small ions are produced by ion evaporation, large ions are produced by coulombic repulsion—and other environmental factors such as ion concentration in solution. Vestal[128] gives an additional concise summary of the current thoughts on how ESI works, but more importantly summarizes a number of design variations of ESI ion sources extant on modern instruments.

b. Atmospheric Pressure Chemical Ionization

APCI is similar to electrospray in that ionization occurs at atmospheric pressure. The effluent of an HPLC column is passed through a heated nebulizer and sprayed into the source chamber. Ionization occurs in a corona discharge, struck in the plume of effluent. The solvent vapors provide a source of protonating reagent for proton transfer CI reactions in the vapor. The corona discharge region of the spray plume is sampled through an orifice by the mass spectrometer. When this experiment was first reported, the discharge was struck by proximity with the ionizing radiation of a radioactive nickel needle.[101] A needle with an applied high voltage later replaced the radioactive needle. APCI ionization is soft, as the majority of compounds produce abundant protonated molecular ions. It is, however, somewhat harsher than electrospray, and is often used for compounds which do not have readily protonatable sites in the molecule, and require higher energies to make ions. Regarding the "quality" of the spectra, some comparisons have been made to FAB, using extent of fragmentation as a measure of quality.

c. APCI Versus ESI

Which ionization method should one choose for a particular compound? Both APCI and ESI are considered soft ionization techniques, although ESI is the softer method. They are often implemented in the same ion source, with very simple conversion procedures to go from one to the other. APCI will also occasionally produce fragments that are not mass spectral fragments, but rather ones caused by the high operating temperatures typical of the APCI probe. ESI, being the softer method, is more prone to cluster formation. Both situations can produce misleading results.

The quinolone and naphthyridinone antibiotics, of which trovafloxacin (Structure 1) is an example, produce a fragmentation anomaly under APCI conditions.[129] With a carboxylic acid moiety, it might be reasonable to expect to see a decarboxylation fragmentation (loss of 44 Da) in the mass spectrum of this compound. The APCI mass spectrum (Figure 17, top panel) shows a prominent $[M+H-44]^+$ fragment at *m/z* 372. An ESI spectrum of trovafloxacin, or an MS-MS product ion spectrum of the $[M+H]^+$ generated by either APCI or ESI, does not contain this fragment (Figure 17, lower panel). This observation is consistent with a number of literature observations[130,131] on compounds of this structural class, which indicate that this decarboxylation is a fragmentation thermally induced by the high APCI probe

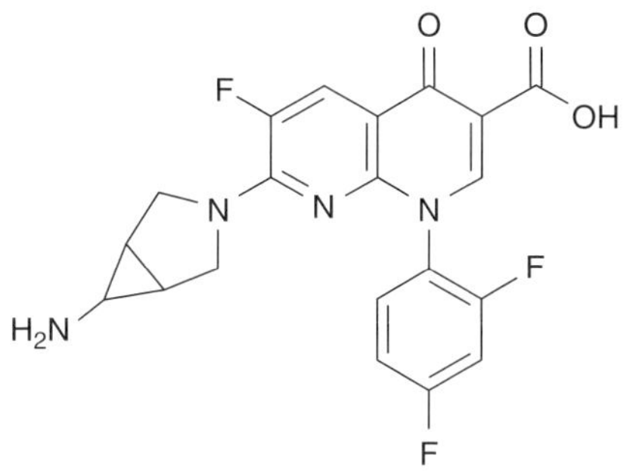

STRUCTURE 1 **Trovafloxacin**

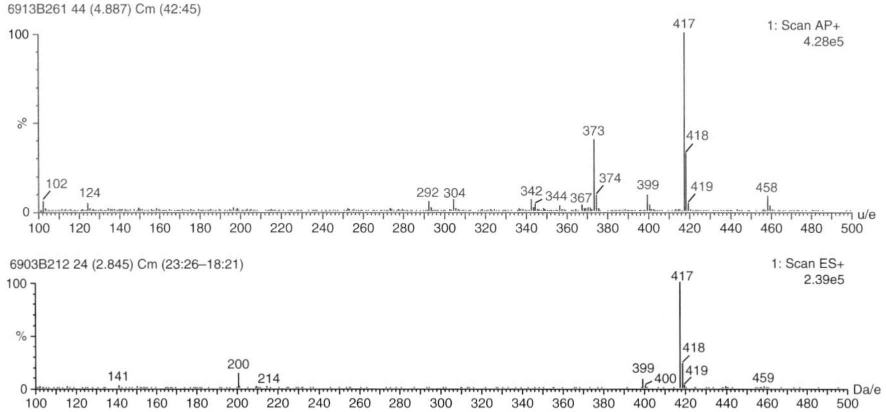

FIGURE 17 **APCI (top panel) and ESP (bottom panel) mass spectra of trovafloxacin.**

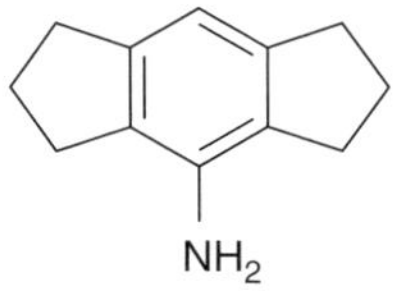

STRUCTURE 2 **Hexahydroindacene**

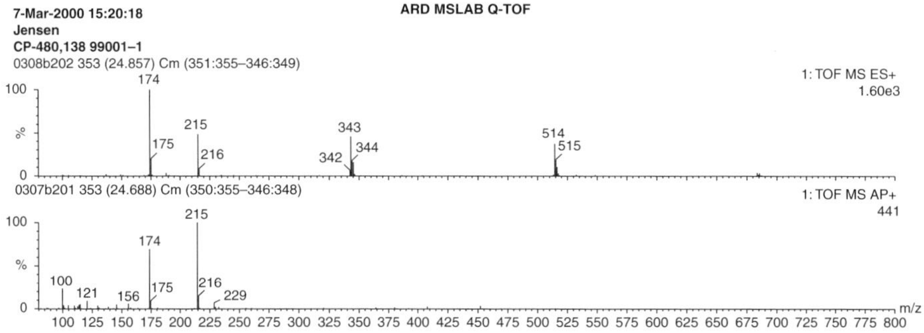

FIGURE 18 **APCI (top panel) and ESP (bottom panel) mass spectra of 4-aminohexahydroindacene.**

temperature—upwards of 500 °C. The more characteristic mass spectral fragmentation is dehydration from the carboxylate functionality (*m/z* 399 in Figure 17, lower panel).

An instance where the APCI behavior is more directly interpretable, involves 4-aminohexahydroindacene (Structure 2). This compound shows extensive "clustering" in the ESI ion source,[132] producing species in the mass spectrum of much higher mass than the expected *m/z* 174 $[M+H]^+$ (Figure 18, upper panel). The spectrum is misleading when attempting to assign an $[M+H]^+$ to an unknown compound. APCI (Figure 18, lower panel) is required to obtain a clearer, less ambiguous indication of the true $[M+H]^+$ of this compound. Even so, one must remain cognizant of the fact that the experiment is being conducted using an acetonitrile-containing solvent, and that the *m/z* 215 species is an acetonitrile adduct of the target molecule.

d. Open Air Ionization Methods

Two new ionization methods to appear recently are desorption ESI (DESI) and direct analysis in real time (DART). They are the first in a new area of ionization mechanisms that are collectively referred to as open-air ionization. Both DART and DESI form ions in an "open atmosphere" and sample the resulting plume through an entrance cone into the mass spectrometer.

DESI has been described in detail.[133,134] Figure 19 shows the configuration of the source and sample. A spray of solvent, generated by an electrospray spraying device, is impinged on a surface. This beam of impinging particles consists of gas-phase solvent ions, ionic clusters, and charged microdroplets. These charged projectiles act similar to the bombarding particles (ions, atoms, photons) in desorption ionization methods, chemically sputtering target analyte molecules off the surface. The nearby mass spectrometer entrance orifice samples this collection of sputtered chemical species. The process has been likened to a "power-washing" process.[135] In a very short time, variations on the open air ionization concept have been reported,[136] numerous applications have been described, from analyzing thin-layer chromatographic separations, analyzing explosive residues on human skin,[137] and doing two-dimensional (2D) imaging of surfaces and of thin sections of biological tissues.[138]

The applicability of DESI to the 2D mapping of a surface resurrects significant interest in the combination of MS and TLC. Reports have periodically appeared in the literature on developing devices and techniques for accomplishing this coupling, such as enclosing an entire *x–y*-translatable stage (such as implied by Figure 19, and actually used by Takáts *et al.*[133]) within a vacuum chamber and adding a layer of matrix for FAB ionization.[139] A relatively recent review[140] summarizes those efforts. The innovation comes here in that the entire *x–y*-translatable stage apparatus no

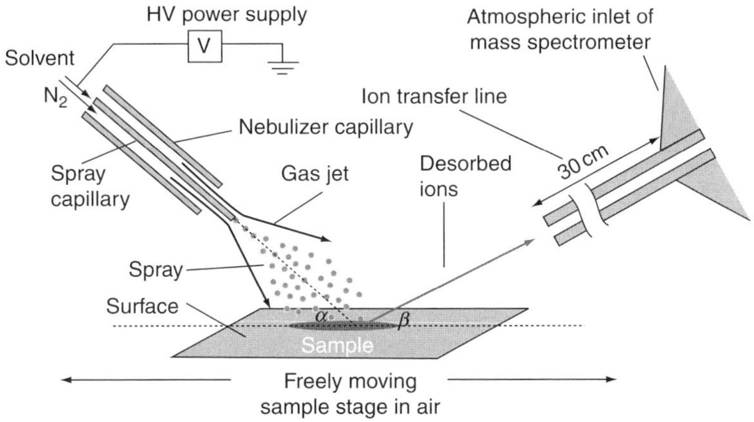

FIGURE 19 Desorption electrospray ionization source. Reproduced from Ref. 133.

longer needs to be enclosed within a vacuum chamber, thus eliminating all of the mechanical and logistical problems encountered with previous attempts.

The DART ionization source[141] uses an electric discharge to produce a plasma of electrons, ions, and excited states of a gas, typically nitrogen, which is then formed into a high-velocity gas stream. This gas stream can be directed toward a liquid or solid sample, or it can interact with vapor-phase samples. As with the DESI source, the angular and distance orientation of the ionizing gas stream toward the sample surface and the mass spectrometer entrance influences the quantitative aspects of DART sampling. However, regarding efficient operation of the technique, and for qualitative applications, these orientation parameters are not critical. This noncritical nature makes sample preparation and presentation for study by this technique very easy. Therein lies the attraction, especially when identity of substances is the primary goal of the experiment and absolute amounts or percentage composition of mixtures are less important to the analysis.

e. Atmospheric Pressure Photoionization

Atmospheric pressure photoionization (APPI) is a variation on the electrospray and APCI techniques. Instead of a corona discharge or a high-voltage potential to effect ionization, a high-intensity UV lamp is used to effect photoionization of analytes in a chromatographic effluent stream.[142] Photoionization detection combined with gas chromatography has been studied and available since the 1970s. Photoionization, coupled with addition of dopants such as toluene or methanol into the chromatographic effluent to control the photochemistry occurring in the sprayed effluent, has been demonstrated to present advantages of sensitivity and extending the range of applicable types of compounds. Developers claim to be able to address a group of nonpolar analytes that are not accessible by either electrospray or APCI. Study of the mechanism of APPI,[143] where the range of applicability is discussed, has been published.[144]

f. Matrix-Assisted Laser Desorption Ionization

Hillenkamp and Karas first described this technique in 1985,[145] and review early progress.[103] Laser desorption of target analytes from surfaces had been exploited extensively, but was restricted to certain classes of compounds, in particular those of no more than approximately 1000 Da. Matching wavelength of irradiating light from a laser with the absorption maxima of target analytes imposed further technical difficulties. By contrast, dispersing a target analyte (or mixture of analytes) in a matrix of a light-absorbing small molecule matrix, and irradiating with a wavelength at which the matrix absorbed, effected absorption and energy transfer into the target analytes without extensive fragmentation of the analytes and without the necessity of tuning the irradiating wavelength to the absorption maximum of the target. A number of other advantages also presented. The molecular mass boundary of laser desorption could be extended to intact proteins. In the process of examining this application, a number of small aromatic molecules that absorbed in the 270 nm region were examined for utility as matrices.[146] Attempts to find the magic matrix, choosing or tailoring the matrix to the application, resulted in a significant amount of research activity, similar to searching for that special FAB matrix. Today, MALDI competes effectively with electrospray as an ionization method of choice, in particular for large proteins and other macromolecules. It is somewhat less useful for small molecules.

4. Monoisotopic Versus Chemical (Average) Molecular Mass

The relative molecular mass[147] that is determined by the mass spectrometer is numerically different from the chemical (average) molecular mass used to calculate molar concentrations or reaction stoichiometries. For the elements commonly found in organic molecules, the numbers are very close and in many contexts are interchangeable. However, when one discusses an accurate mass measurement made by a mass spectrometer, a molecule which contains significantly polyisotopic elements such as chlorine or bromine, or when the mass of the molecule gets large, the numbers become significantly different. One must exercise care and be explicit about which kind of molecular mass number one is discussing.

Further complicating the situation, mass spectroscopists often discuss nominal (integral) masses and exact (or accurate) masses. The term nominal mass derives from the fact that the relative atomic masses of all the isotopes are very close to integral values—for example, carbon is 12, hydrogen is 1, and oxygen is 16. The nominal molecular mass of a molecule is calculated using these nominal atomic mass values. Nominal masses are the values discussed in most instances. Accurate (exact) masses derive from increasing the precision with which the relative masses of atoms and molecules are measured. This increased precision reveals that the masses of the isotopes are not exactly integral. They all—with the exception of ^{12}C, which is arbitrarily assigned to be 12.0000—exhibit a mass defect amounting from a few thousandths to a few hundredths of a Dalton above or below the nominal integral mass. Mass deficient isotopes (oxygen, sulfur, phosphorus, and the halogens) have an exact mass less than the integer. Mass-sufficient isotopes (^{13}C, hydrogen, nitrogen) have an exact mass greater than the integer. A list of the accurate masses of the isotopes and their natural abundances can be found in most MS textbooks and is published in several places throughout the general chemical literature.[148]

Even mass spectroscopists use the terms exact mass and accurate mass interchangeably when they probably should not. An exact molecular mass is the mass calculated from the accepted exact atomic masses of the isotopes for a specified empirical formula. It is the value that one would expect to observe if one could measure a molecular mass exactly![149] An accurate mass is a value measured carefully, with high precision, on an instrument capable of making such precise measurements, typically expressed to at least four decimal places—the nearest 0.1 mDa. An accurate mass measurement is compared to the exact masses of empirical formulae being considered. Sufficiently accurate measurements can be used to "assign" empirical formulae to peaks in a mass spectrum. Knowing that these two terms are commonly used interchangeably, but that there may be subtle differences in the way practitioners use them, is sufficient for our discussions here. We will comment more on the utility of accurate mass measurements later.

5. When is a Molecular Ion Not an $[M+H]^+$?

A molecular ion species should not always be automatically assumed to be a protonated species, one mass unit heavier than the neutral molecular species. Cationic species, such as quaternary ammonium compounds, will appear in the mass

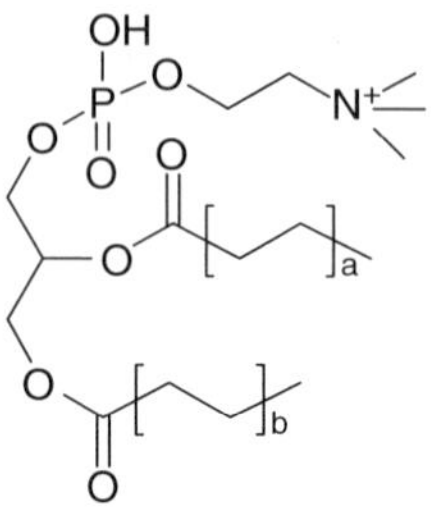

Lecithin general structure
(phosphatidylcholine)
for a, b = 7, dipalmitoyl phosphatidylcholine
$C_{40}H_{81}NO_8P$
r.m.m. 734.5700

STRUCTURE 3 **Lecithin general structure**

spectrum at the mass of the cationic species. A biologically relevant example is the family of phosphatidylcholines (also known as the lecithins), represented by Structure 3. Mass spectra of these compounds, recorded under appropriate conditions, show M^+ species, not $[M+H]^+$. The molecular ion species for the dipalmitoyl phosphatidylcholine species, whose empirical formula and relative molecular mass are given with Structure 3, would appear at *m/z* 734.6. Note also the 0.6 Da of excess mass above the integral value because of the large number of mass-sufficient hydrogen atoms in this molecule.

6. Cluster Ions and Adduct Ions

The quest for an ionization method that guarantees a molecular ion for all compounds has driven the development of softer and softer methods. While approaching this goal, a wealth of new exceptions to the rule has been observed. Cluster ions and adduct ions form in the matrix (for FAB ionization) or in the spray of the API methods depending on the nature of the compound, the matrix, solvent, or extraneous salts present in the sample or in the instrument. While complicating the interpretation of such results and requiring the appropriate care in interpretation, the presence of these species can be useful.

The ubiquitous nature of the sodium cation often leads to Na^+ cationization and the presence of $[M+Na]^+$ in the mass spectrum in addition to, or in place of, the $[M+H]^+$. Efforts to either remove or add Na^+ from the sample result in changes in the relative abundances of the $[M+H]^+$ and $[M+Na]^+$. Addition will often cause replacement of other labile protons with additional sodium atoms. Addition of lithium or potassium can replace the Na, with the appropriate mass changes, to confirm the identity of the Na adducts. Ammonium cationization will often occur from LC-MS mobile phases containing an ammonium buffer salt. Acetonitrile solvent adducts will also often form.

An example of a molecule that exhibits an extreme dependence on solvent conditions is shown in the partial spectra of Figure 20. The molecule has a 404 Da relative molecular mass. The material is produced as the sodium salt. Samples of the exact same material have been dissolved in and introduced into the electrospray ion

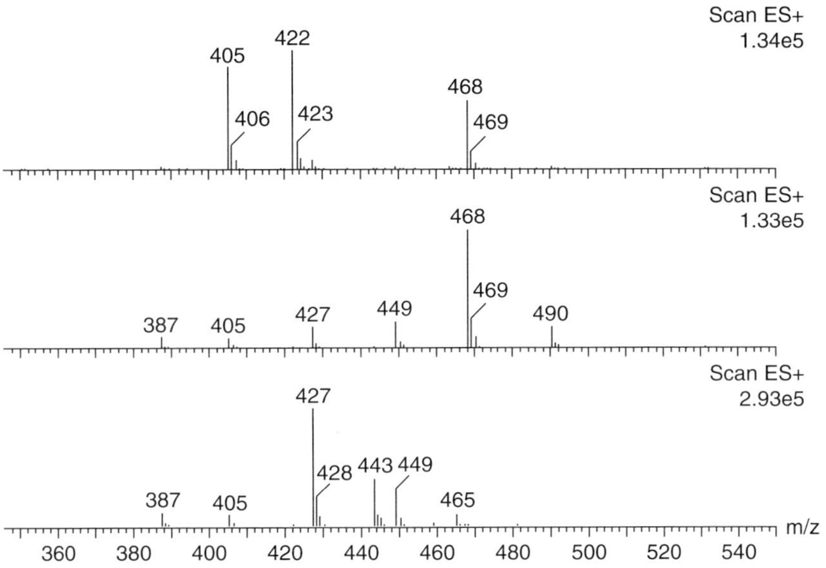

FIGURE 20 **Formation of different adduct ions under different solvent conditions. Samples of a 404 Da molecule dissolved in and introduced in (top panel) 1:1 acetonitrile:20 mM aqueous ammonium acetate, (middle panel) 1:1 acetonitrile:water, and (bottom panel) 1:1 methanol:water.**

source using three different solvents. The *m/z* 405 species is the $[M+H]^+$. The *m/z* 422 species is an $[M+NH_4]^+$. The *m/z* 427 species is [M+Na]. The *m/z* 468 species corresponds to an $[M+Na+\text{actronitrile}]^+$. Examining any of these three spectra in isolation, one cannot easily tell which is the real $[M+H]^+$. However, once one knows that a particular class of molecule—such as this one—behaves in this manner, one can use the presence of the cluster ions as additional supporting evidence for assigning the molecular mass of the compound. This phenomenon adds uncertainty about assuming the behavior of unknowns. If one can safely assume that an unknown impurity or degradant is sufficiently structurally similar to the parent molecule to behave in the same manner, then the adduct formation helps to confirm molecular mass assignments. Looking for 17 Da (NH_4^+), 22 Da (Na^+), 38 Da (K^+), 41 Da (acetonitrile) mass differences, or various combinations of these numbers, will provide some level of comfort in assuming adduction behavior.

7. Polyisotopic Elements and Isotope Patterns

The consideration of molecular mass is further complicated by the existence of a number of significantly polyisotopic elements. Isotope patterns in a mass spectrum are the manifestation of the relative abundances of the naturally occurring isotopes. A number of elements are polyisotopic, some of the more dramatic examples being bromine and chlorine. Tin is the most dramatic, having a total of 10 naturally occurring stable isotopes of significant abundance spread over a 12 Da range, seven of which have 5% relative abundance or greater. Given this complication, determination of the correct molecular mass is controversial. Polyisotopic elements immediately divide practitioners into two camps. One camp says the molecular mass is that calculated using the lowest mass isotope of each element in the molecule. The other camp says to calculate using the mass of the most abundant isotope of each element. In most instances, these two rules give rise to the same number. The most abundant isotope of carbon, hydrogen, oxygen, nitrogen, sulfur, phosphorus, and all of the halogens is indeed the lowest mass isotope. Boron-containing molecules or organometallic compounds containing a polyisotopic transition metal or heavy metal may more appropriately deserve application of the second rule.[150,151] Applying either of these rules, however, becomes even more complicated when a molecule contains, for example, three or more chlorine atoms, two or more bromine atoms, or more than one metal atom.

a. Isotope Patterns in Normal Mass Spectra

In conventional mass spectra, the isotope patterns deriving from the presence of polyisotopic elements are striking and quite noticeable. An excellent example is the ESI spectrum shown in Figure 21 for a compound with a $C_{14}H_{12}NOFCl_2$ empirical formula. The *m/z* 300 $[M+H]^+$ and the *m/z* 269 neutral loss fragment (resulting from neutral loss of methylamine) clearly show Cl_2 isotope patterns which match well with simulations.[152] The *m/z* 234 fragment results from further loss of one of the chlorine atoms, illustrating a clear Cl_1 isotope pattern, and the utility of the isotope patterns in assigning structure.

The close match of simulation (Figure 22, upper panel) and observation (Figure 22, lower panel) of the $[M+H]^+$ of another chlorine-containing compound, along with some chemical considerations of the history of the sample, permitted assigning the structure of the unknown compound based solely upon the isotope pattern.

b. Isotope Patterns in MS-MS Product Ion Spectra

The interpretation of isotope patterns changes when applied to MS-MS product ion spectra. Consider the product ion spectrum, in the top panel of Figure 23, of the same *m/z* 300 $[M+H]^+$ discussed above, in Figure 21. The experiment was

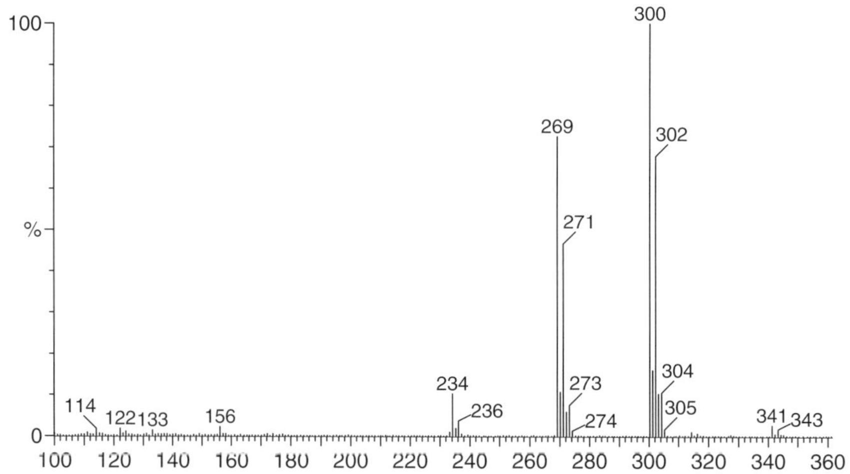

FIGURE 21 ESP mass spectrum of a molecule containing two chlorine atoms.

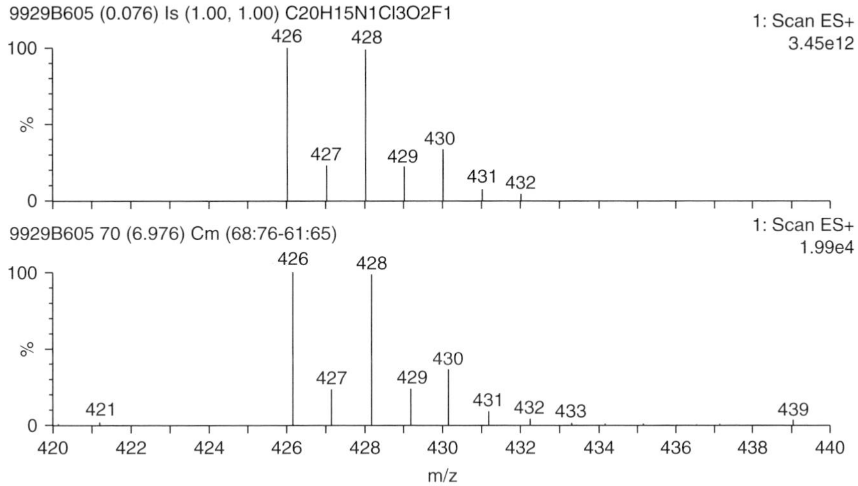

FIGURE 22 **Comparison of the isotope pattern for the observed $[M+H]^+$ (bottom panel) with the predicted model (top panel) for $C_{20}H_{15}NO_2FCl_3$.**

performed on a triple quadrupole instrument by selecting the *m/z* 300 species with the first analyzer under unit resolution conditions. The *m/z* 300 species was collisionally activated in the second quadrupole collision cell. The product ion spectrum resulting from CID was recorded in the third quadrupole mass analyzer. Recall that the empirical formula of this molecular ion is $C_{14}H_{13}NOFCl_2$. Why have the very prominent isotope patterns disappeared? Because one has selected the *m/z* 300 species as the parent, there is no longer any consideration of polyisotopy. This parent is a monoisotopic parent—namely $C_{14}H_{13}NOF^{35}Cl_2$. Methylamine neutral loss produces a single species—*m/z* 269. Further fragmentation, sequentially losing chlorine atoms, produces single species—*m/z* 234 and *m/z* 199—because the only chlorine isotope that can be lost from this selected *m/z* 300 parent is a ^{35}Cl atom.

Why does an isotope pattern reappear in the product ion spectrum of the *m/z* 302 parent (Figure 23, middle panel)? The elemental composition of the parent is now $C_{14}H_{13}NOF^{35}Cl_1{}^{37}Cl_1$. The *m/z* 271 species shows no polyisotopic signature because its formation results from loss of CH_3NH_2 from the parent to give $C_{13}H_8OF^{35}Cl_1{}^{37}Cl_1$—only one possible elemental composition. Loss of the first chlorine atom, however, now permits producing two possible elemental compositions—$C_{13}H_8OF^{35}Cl_1$ and $C_{13}H_8OF^{37}Cl_1$ at *m/z* 234 and *m/z* 236. They are of equal abundance because of the equal probability of losing either a ^{35}Cl or a ^{37}Cl atom in the fragmentation process. The fact that the natural abundance of the chlorine isotopes is approximately three to one is irrelevant here. Selecting the *m/z* 302 species as the parent sets the isotope ratio of ^{35}Cl to ^{37}Cl in this experiment to be one to one. The isotope pattern disappears on losing the second chlorine to give the *m/z* 199 species since the source of polyisotopy is now completely gone.

All evidence of isotope patterns in the product ion spectrum of the *m/z* 304 parent (Figure 23, bottom panel) disappears because the selected parent is monoisotopic—$C_{14}H_{13}NOF^{37}Cl_2$. Justification for the absence of isotope patterns

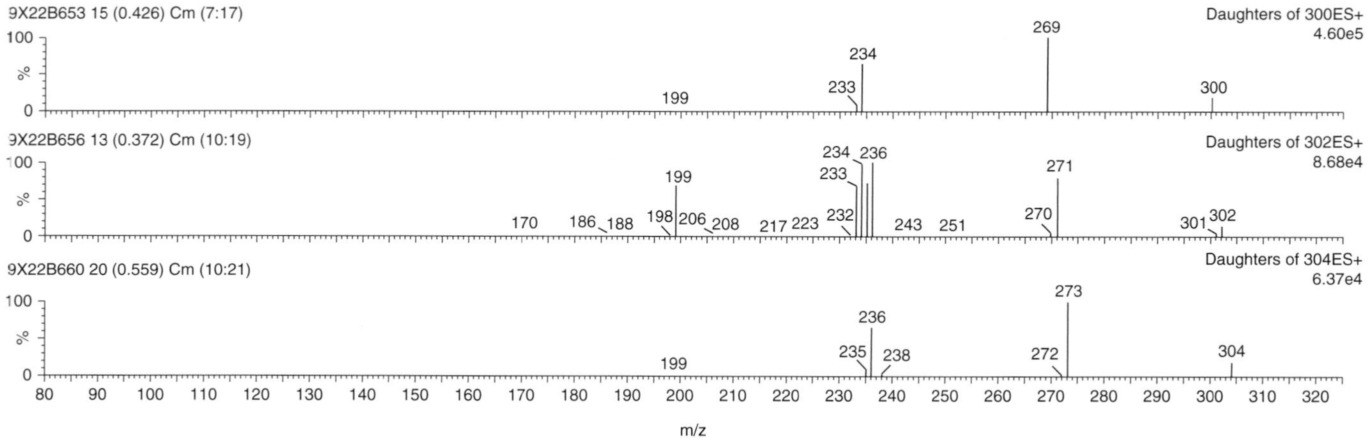

FIGURE 23 MS-MS product ion spectra of the *m/z* 300 (top panel), *m/z* 302 (middle panel), and *m/z* 304 (bottom panel) $[M+H]^+$.

is identical to that of interpretation of the *m/z* 300 product ion spectrum. The detailed explanation of this excellent example serves to illustrate the care one needs to exercise in interpreting isotope patterns (or the lack thereof) in MS-MS spectra. At the same time, it demonstrates the structure-indicating power of careful interpretation.

With the recent development of the combined quadrupole TOF tandem instrument geometry, MS-MS product ion spectra are starting to appear in the literature with isotope patterns reflecting the natural abundance of the involved elements. Considering the discussion above, how does this occur? Note that our selection of parent ions for CID in the experiments described above was done under unit resolution conditions. With a tandem instrument composed of a quadrupole and a TOF mass analyzer, the resolution of the first mass analyzer can be set to substantially less than unit resolution, such that the entire isotope pattern can be selected as the "parent." This approach preserves the isotope pattern information while providing MS-MS product ion information. This new consideration must be added while interpreting isotope patterns in MS-MS spectra—namely to identify on what kind of instrument the experiment is being done, and how the instrument is set up.

8. Deuterium Isotope Exchange

Substantial amounts of work appear in the literature in recent years on the study of backbone amide hydrogen–deuterium exchange studies of protein folding, based upon the pH dependence of the rates of H–D exchange of amide protons.[153] The rate minimum for this exchange is between pH 2 and pH 3 at 0 °C. Half-lives for exchange can be greater than 1 h at these conditions for unprotected amides. This slowness of exchange permits conducting exchange experiments on intact proteins, then cleaving those proteins with acid-active peptidases and determining the regions of the original protein accessible to exchange. These constitute solution state probes of tertiary and higher structure of proteins. Review papers by Englander and colleagues[154,155] describe this important technique.

The utility of deuterium exchange, however, is not limited to large molecules. Identifying and confirming the structures of small molecules can benefit considerably by determining the numbers (and nature) of exchangeable hydrogen atoms in the molecule. Olsen *et al.*[156] report on the general utility of deuterium exchange LC-MS experiments for structure elucidation and impurity identification. However, one should not forget the importance of the chemical nature of the exchanging hydrogen. Exchange is not instantaneous and universal for all exchangeable hydrogen atoms. Such a serendipitous observation figured prominently in the identification of a PRI[52] in a pharmaceutical drug candidate. It exemplifies the benefit of an inventory of exchangeable hydrogen atoms in impurity identification.

An example where deuterium exchange played an important role to imply a keto–enol tautomerization in a pharmaceutical drug substance has been reported.[157] An impurity exhibiting the same relative molecular mass as the drug substance was observed. Conducting a deuterium exchange LC-MS experiment showed that the drug substance contained five exchangeable hydrogen atoms, while the impurity contained six exchangeable hydrogens, the structural explanation being that the two molecular forms were keto and enol tautomers. The

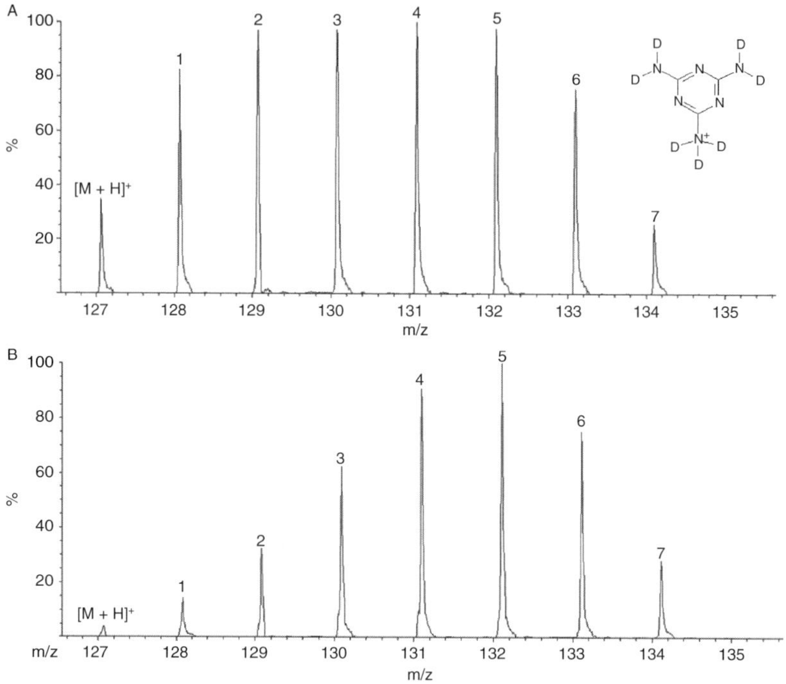

FIGURE 24 **Hydrogen–deuterium exchange of melamine in a DART ion source. Reproduced from Ref. 156.**

chromatographically distinguishable enol tautomer contained an additional hydroxyl group with a readily exchangeable hydrogen.

The DART ionization source permits performing even more fascile hydrogen–deuterium exchange experiments. Investigations into the contamination of pet foods by melamine are reported by Vail *et al.*[158] The hydrogen–deuterium exchange was conducted simply by placing a source of D_2O vapors in proximity with the ion source and the sample. Exchange was effected in the gas phase between the melamine molecules and the D_2O (Figure 24). The six exchangeable hydrogen atoms in the melamine molecule plus the additional deuterium to form the $[M+D]^+$ are accounted for in the spectrum of the exchanged compound.

9. Unimolecular Decompositions Versus Granddaughter [159] Ions

The theoretical concept behind MS-MS involves unimolecular decomposition to establish ion-genetic relationships between fragments and their parent structures. The difficulty of interpreting a conventional mass spectrum consisting of more than one component is determining which fragment belongs to which parent. Another problem occurs when a fragment gives rise to another fragment via a sequence of fragmentations. This dilemma is tremendously improved by having a chromatographic separation on the front of the mass spectrometric experiment, giving some assurance that one is looking at the mass spectrum of a single compound, and that all fragments are ultimately derived from one parent. This is not the case when one still has coeluting components. MS-MS experiments seek to eliminate this uncertainty. These experiments assume an “infinitely thin” collision zone and single collisions of the parent with a collision gas, depositing only sufficient energy to induce one unimolecular decomposition reaction. These assumptions are not sustained when utilizing a physically large quadrupole collision cell. Multiple collisions at high energy are possible. Product ions can also undergo collisional activation before they leave the collision cell. Granddaughter and great-granddaughter ions can be formed and detected in the MS-MS spectrum. How does one tell a unimolecular decomposition from a multistep decomposition? Fragmentations requiring the breaking of multiple bonds at multiple sites in the molecule can indicate multistep decomposition.

Examination of the collision energy dependence of the product ion spectrum is a functional test. To illustrate, the product ion spectra of the *m/z* 300 $[M+H]^+$ of a molecule of empirical formula $C_{14}H_{12}NOFCl_2$, at three different collision energies, are shown in Figure 25. This is the same molecule upon which the isotope pattern discussion in Section III.E.7 is based. The top panel, at low collision energy, shows significant survival of the $[M+H]^+$ and loss of a 31 Da species, corresponding to neutral loss of methylamine—a sensible fragmentation, as the original molecule contains an aliphatic N-methylamine side chain. Only just appearing is an *m/z* 234 fragment, corresponding to further loss of one of the chlorine atoms. That the methylamine and chlorine moieties are distant from each other in the original structure implies that these are two separate events, and that the *m/z* 269 species is losing the chlorine. If chlorine were to be lost directly from the parent ion, an *m/z* 265 species would be present. It is not.

Increasing the collision energy, as illustrated in the two lower panels of Figure 25 further decreases the abundance of the $[M+H]^+$ and induces more extensive fragmentation of the *m/z* 269 species. It increases the relative abundance of the first chlorine loss, and induces a further loss of the second chlorine atom, producing the *m/z* 199 species. The collision energy dependence on the abundance and appearance of the product ions indicates that these species are the result of a sequence of fragmentations, and helps one correlate back to the original structure of the molecule.

10. Multiply Charged Ions

ESI-generated multiply charged ions are most frequently discussed in the examination of peptides and intact proteins. The general observation is that peptides need to be on the order of 1500–2000 Da or larger in order to form abundant multiply charged ions (i.e., doubly charged ions). Although doubly charged molecular ions have been reported in the EI mass spectra of polycyclic aromatic hydrocarbons,[160] mention of multiply charged ions in the context of classic mass spectra of small molecules is rare. In our work on impurities and degradants of small pharmaceutical molecules, however, we are starting to see an increasing frequency of doubly charged molecular ions in the electrospray mass spectra—molecules with relative molecular masses as low as 500–600 Da. The factors that determine whether or not and to what extent a molecule will form doubly charged molecular ions are still under investigation. How does one tell if one has a doubly charged species?

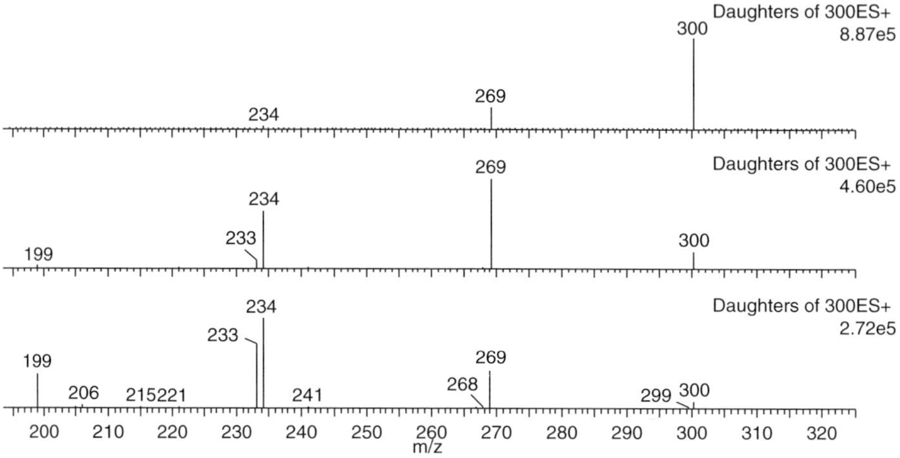

FIGURE 25 MS-MS product ion spectra of the *m/z* 300 $[M+H]^+$ at 10 V (top panel), 20 V (middle panel), and 30 V (bottom panel) interface cone voltages.

ESI mass spectra of compounds often contain van der Waals clusters of the molecule. Numerologically, these species appear at two times the relative molecular mass of the molecule plus one. Consider the example spectrum shown in Figure 26. The *m/z* 403 species could be assigned to be the $[M+H]^+$ of the compound in question, indicating a relative molecular mass of 402 Da. It was indeed the most abundant peak in the spectrum. A proton-bound dimer would correspond to a $[2M+H]^+$ peak at *m/z* 805, consistent with the expanded spectral portion shown in Figure 26. The relative abundance of the *m/z* 805 species is approximately 10% that of the *m/z* 403 species, consistent with typical relative abundance observations of $[M+H]^+$ and $[2M+H]^+$. The convention would be to assign the $[M+H]^+$ to be the most abundant peak in the spectrum. The fact that a 402 Da molecular mass exhibited an illogical mass difference between this and the drug substance molecule under investigation prompted reexamination of the *m/z* 403 region and the putative assignment of this species as the $[M+H]^+$.

Close examination reveals that the peaks comprising the isotope pattern of the *m/z* 403 species in Figure 26 appear at 0.5 Da intervals. The nature of the instrument on which this spectrum was obtained, and the confidence in mass assignments, indicates that these mass assignments were indeed correct, and that this spectrum represents a doubly charged ion. Recall that the mass spectrometer does not measure mass, but rather mass-to-charge ratio. Typically, for small molecules, ions carry only a single electrostatic charge, and the *m/z* value reduces to the mass. The interval between isotope peaks in a pattern such as this should be 1 Da. The 0.5 Da spacing indicates that the *m/z* 403 species is really a $[M+2H]^{2+}$. The *m/z* 805 species is the true $[M+H]^+$, and the molecular mass is really 804 Da. This reassigned molecular mass and a consideration of the synthetic chemistry of the drug substance being studied were sufficient to propose a reasonable identity of the impurity.

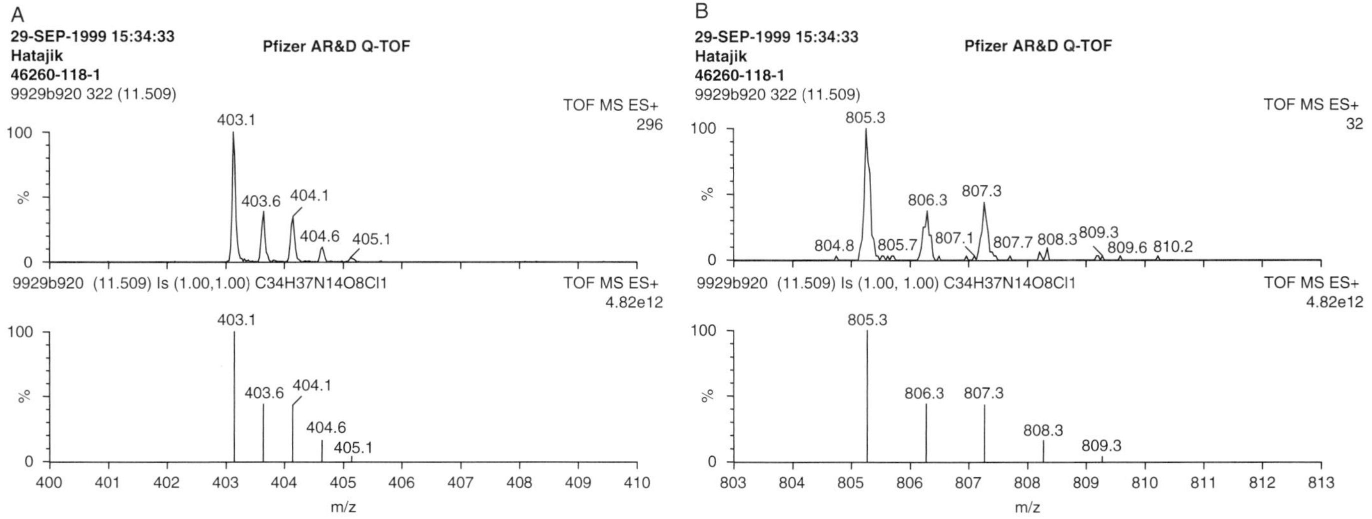

FIGURE 26 **(A) Comparison of the isotope pattern of the doubly charged molecular ion region (top panel) and the isotope pattern modeled (bottom panel) from the $C_{34}H_{37}N_{14}O_8Cl$ elemental composition of the *m/z* 403 doubly charged molecular ion. (B) Comparison of the isotope pattern of the singly charged molecular ion region (top panel) and the isotope pattern modeled (bottom panel) from the $C_{34}H_{37}N_{14}O_8Cl$ elemental composition of the *m/z* 805 singly charged molecular ion.**

11. Polymer Patterns Versus Multiply Charged Ion Series

Another striking feature that can be seen in the overall appearance of a mass spectrum is a repeating series of peaks at a regular interval—a "polymer pattern." It is almost as striking as an isotope pattern. Such a pattern is illustrated by the spectrum shown in Figure 27, an ESI mass spectrum of a mixture of octoxynol ((1,1,3,3-tetramethyl)butylphenyl (poly)ethylene glycol, Triton X), a commonly used surfactant (Structure 4). The spectrum is of a mixture of molecules of commercially available octoxynol with an average molecular mass of 643 Da.[161] The average molecular mass can be calculated directly from the spectral information. The constant interval between the repeating peaks is 44 Da, the mass of the repeating unit. This spectrum illustrates the phenomenon of NH_4^+ cationization by these polymers—the species in the spectrum are ammoniated cations, $[M+NH_4]^+$. Recognizing these features, the complicated appearing spectrum can be easily interpreted. Recognizing polymer patterns has been key in identifying drug–excipient interactions in formulations containing such polymers.

With the increasing application of ESI to the examination of intact proteins (illustrated by the example of Figure 28), one observes a pattern outwardly resembling a polymer pattern. This feature is not in fact a polymer pattern but a multiply charged ion envelope of a single chemical species. The protein used here is bovine serum albumin, a single molecule with a chemical molecular mass calculated from this spectrum to be 66,424 Da. There are two additional higher mass proteins in this sample. The definitive way to differentiate a polymer pattern from a multiply charged ion series of a protein is by evaluating the spacing of the peaks. A polymer pattern will show peaks spaced at a constant mass interval across the entire series. A multiply charged ion series will show a steadily increasing mass spacing as one traverses the pattern from low mass to high mass.

While this monograph is primarily concerned with the characterization of small molecules, it is useful to point out some of the differences in interpretation and thought process needed to interpret mass spectra of proteins. The shift from monoisotopic molecular masses in small molecules to chemical molecular masses in peptides and proteins is necessary when the chemical entities exceed 1500–2000 Da. Mass calibration procedures and philosophies of instrument operation change. The concept and application of chemical molecular mass accuracy changes from 0.1 to 0.2 Da accuracy for nominal monoisotopic measurements to 0.01–0.02% for proteins (acceptable 0.01% accuracy for the BSA example used here is ± 7 Da). In short, the MS of proteins and other intact biopolymers deserves an entire monograph of its own.

12. Accurate Mass Measurements

The accurate mass measurement technique is being actively rediscovered because of recent improvements in instrumentation. John Beynon[162,163] established the validity of using high-precision mass measurements (to at least four decimal places of precision) to determine elemental compositions of ions with high certainty, capitalizing on the nonintegral nature of the precise atomic masses of the isotopes. Accurate mass measurements were until recently the nearly exclusive province of magnetic sector instruments. Making such measurements presented

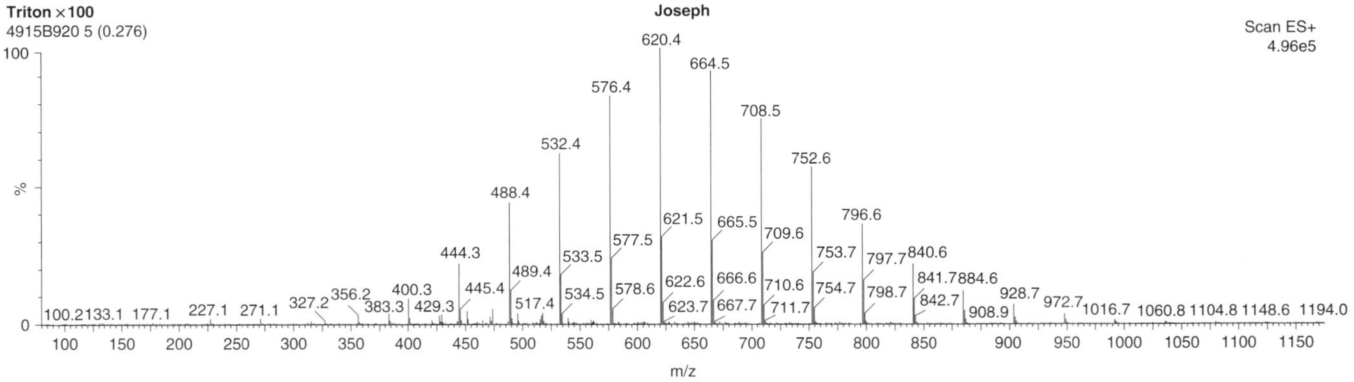

FIGURE 27 Electrospray mass spectrum of octoxynol (triton X).

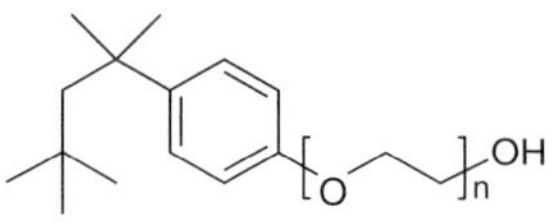

"Triton X"
$C_8H_{17}\ C_6H_4\ [C_2H_4O]_nOH$
r.m.m. 206 + $[44]_n$

STRUCTURE 4 **Octoxynol.**

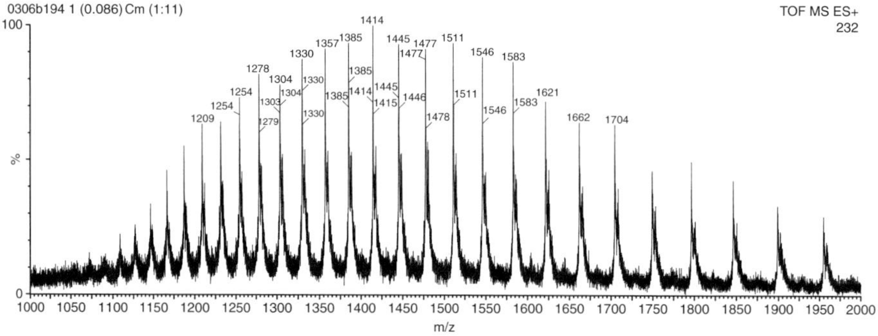

FIGURE 28 **Positive ion ESI mass spectrum of bovine serum albumin.**

difficulties, and required careful attention to detail. TOF instruments and FT-MS instruments, and recently the orbitrap mass analyzer,[164] have challenged magnetic sector dominance of this area and facilitated an easier route to making these measurements. Accurate mass measurements became accepted in the synthetic organic chemistry community as an appropriate measure of the "proof" of a structure. Acceptance criteria for including accurate mass measurements in the suite of data defending structure assignment of a newly synthesized compound are given in the instructions for authors of the synthetic and other journals,[165,166] even though Clayton Heathcock[167] (then editor-in-chief of the *Journal of Organic Chemistry*) acknowledged that combustion analysis, for which accurate mass measurements are often used as a substitute, is really a measure of purity and not of identity. Accurate mass measurements are a measure of identity.

One must also accept the challenge and responsibility of using accurate mass measurements correctly. Kondrat[168] and Gross[169] illustrate some of the prevalent misconceptions about the utility and appropriateness of accurate measurements, and illustrate the confusion that still exists on the differences between definition of the terms "high resolution" and "accurate mass." Busch illustrates the use of and limitations of accurate mass measurements.[170] Biemann's discussion drives the point home in much more detail.[171] Biemann lists three typical uses for accurate mass measurements: elucidation of fragmentation pathways, confirmation of structure of the synthetic products, and elucidation of completely unknown structures. The demands for necessary precision, accuracy and care of measurements increase, in the order given. One MS laboratory in a pharmaceutical company[172,173] routinely did accurate mass measurements on all samples, in support of the synthetic organic laboratories, with a magnetic sector instrument.

The instrument manufacturers often imply that accurate mass measurements (and therefore elemental composition determinations) are a single measurement solution to many structural problems. This is an oversimplification. A recent report, stemming from a metabolomics application, presents data that highly accurate mass measurements, even at sub-1 ppm accuracy, are insufficient to "identify" metabolites.[174] Isotopic abundance pattern analysis is a necessary additional constraint to eliminate false elemental composition candidates.

13. Mass Spectral Fragmentation Compendia

The concept of the organic functional group—that a particular organic functional group will behave in essentially the same way, regardless of other functional groups attached to it—holds true in the mass spectrometer, too. For this reason, compilations of mass spectral fragmentation data for organic compounds are useful sources for help when interpreting mass spectra. Two of the better-known libraries are the NIST/EPA/MSDC mass spectral database, maintained by the United States National Institute of Standards and Technology, and the Wiley-McLafferty database, vended by John Wiley Publishing Company. These two differ in that the NIST collection contains only one spectrum representing a given compound. The Wiley-McLafferty collection contains a larger number of spectra, but includes multiple spectra for some compounds, intending to capture information on variability of spectra from observation to observation and instrument to instrument. There are a number of smaller and more specialized collections. The Thermodynamics Research Center at Texas A&M University still collects compounds and compiles data, which eventually are incorporated into the NIST and Wiley-McLafferty libraries. Several compendial books have been published. They are dated, but nevertheless effectively capture the collective fragmentation information prior to their publication.[175–177] Sparkman[178] captures a concise history of the published mass spectral databases. Unfortunately, all of these information sources discuss EI spectra. EI fragmentation rules can be of limited assistance in interpreting MS-MS product ion spectra. One small collection of CI information has been published.[96]

Compilation of spectral catalogs for the soft ionization techniques—CI, FAB, API—has not occurred, for at least two reasons. First, the exact nature of a soft ionization spectrum is much more dependent upon instrument and sampling conditions. Second, for many compounds analyzed by the soft ionization techniques, the spectrum consists principally of a molecular ion. A compilation of soft ionization spectra, in its ideal form, would be a compilation of molecular masses. Movements have surfaced several times to promote the compilation of MS-MS product ion spectra. As we have seen from some of the discussions above, these too are dependent upon exact operating conditions. An acceptable standard set of conditions has eluded efforts to make such a project happen.

14. Interpreting Spectra for Structure

All of the aspects discussed above for extracting structural information from mass spectral data come in to play in some way when one is presented with a structure elucidation problem. Guidebooks on mass spectral interpretation help.[179–181] However, this is not the complete set of tools and techniques. Finding that one obscure but revealing literature reference, flashes of insight in the middle of the night, snippets from one's collective personal and professional experiences, interactions with others practicing the art, are all ill-definable but nevertheless important contributors to this process. One can apply a degree of systematic practice, but structure elucidation from spectral (and other) data is still largely an empirical method, and still requires significant amounts of time simply staring at the data.

IV. NMR

A. Introduction

NMR spectroscopy is a widely used technology for the study of many different aspects of the development of pharmaceutical entities. Arguably, NMR spectroscopy is used throughout the entire discovery and development process. Examples of this include screening of compound libraries against pharmaceutically relevant targets, metabonomics, metabolite ID, medicinal chemistry support, impurity and degradant identification, in-line reaction monitoring, kinetic analysis, quantitation, and binding studies in formulations to name a few. There are many books available that provide in depth examples of many of the above-mentioned applications of NMR spectroscopy.[182,183] The focus of this section will be on the use of liquids NMR spectroscopy and its information rich data to the structure elucidation of small molecular entities in the pharmaceutical environment.

The ability of NMR to provide information regarding the specific bonding structure and stereochemistry within a molecule has created a broad applicability across physics, chemistry, biology, and medicine.[184–188] NMR provides a powerful analytical tool for structural elucidations. Unfortunately, NMR has traditionally been sensitivity limited compared to other analytical techniques. Historically, sample requirements for NMR were on the order of greater than 25 mg, as compared with mass spectroscopy, for example, which requires much less than 1 mg. For this reason, NMR spectroscopy historically was not typically the first approach for a chemist when identifying an unknown chemical entity.

Recent technology advancements in the field of magnetic resonance have made significant strides in improving sensitivity levels.[189–191] These improvements range across not only hardware elements (extremely stable and fast console components and cryoprobes) of the spectrometer, but also in the design of very creative NMR pulse sequences (DPFGSE NOE) and applications (RxnNMR). These types of advances become particularly important in the structural characterization of natural products, drug impurities, and degradants.

The nondestructive, noninvasive nature of NMR spectroscopy makes it a valuable tool for the characterization of compounds of limited quantity. In addition, NMR can be considered close to a "universal detector" for hydrogen, carbon, fluorine, phosphorous, nitrogen, as well as many other magnetically active nuclei. This is both good and bad, since all signals are detected—those arising from the compound of interest as well as all other components in the sample, such as solvent, counterions, and starting materials. In addition, a valuable and inherent characteristic of NMR spectroscopy is its ability to quantitate materials dissolved in the NMR tube. Quantitation is accurate over a dynamic range of nominally four orders of magnitude, although not as precise as other analytical tools particularly at low levels, accuracies of 1% can be achieved if the right procedures are taken into account.[192] This poses a challenge for studying mixtures containing low-level components. It is therefore preferred, if possible, to isolate a given impurity prior to study by NMR rather than to analyze a mixture.

The following discussion focuses exclusively on what is termed "small molecules," namely compounds with a molecular mass on the order of 2000 Da or less. The study of proteins, polymers, and other such "macromolecules" by NMR warrants an entirely different approach that is beyond the scope of this book. Similarly, we will restrict our discussion to liquid-state NMR spectroscopy, since the quantity of sample available for the characterization of a degradant or impurity is typically far below reasonable amounts needed for most solid-state NMR (SSNMR) techniques. The overall characterization strategy we will take for NMR is shown schematically in Figure 29.

B. Information Gathering

The art of structure elucidation is built upon a foundation of many different techniques. In particular are those that have been previously described in this book. However, there are many that should always remain a part of one's arsenal of tools to determine the structures of small organic molecules. These include, infrared spectroscopy, UV spectroscopy, circular dichroism spectroscopy to name a few. Structure elucidation is not a single technique science. That being said, the focus of this section will be on the application and utility of NMR spectroscopy.

Arguably, one of the most important pieces of non-NMR data used by the NMR spectroscopist is the likely molecular formula determined through accurate mass measurements. Although not explicitly required for proposing potential structural via interpretation of the NMR data, this information greatly facilitates the characterization process by reducing the number of possible structures and allows one to have a better idea if non-NMR-active nuclei are present. As learned in the previous section, certain mass changes are indicative of specific structural fragments. Furthermore, in many cases fragmentation patterns can identify which portion of the molecule differs from the parent species, thus pinpointing the site of degradation. One should be careful however to remember the capabilities and limitations of each technique when identifying possible structures. All available MS information must be balanced with the NMR data when considering plausible structures.

Another fruitful source for identifying possible structures comes from knowing the sample's origin. Was the sample extracted from mother liquor, isolated from a bulk

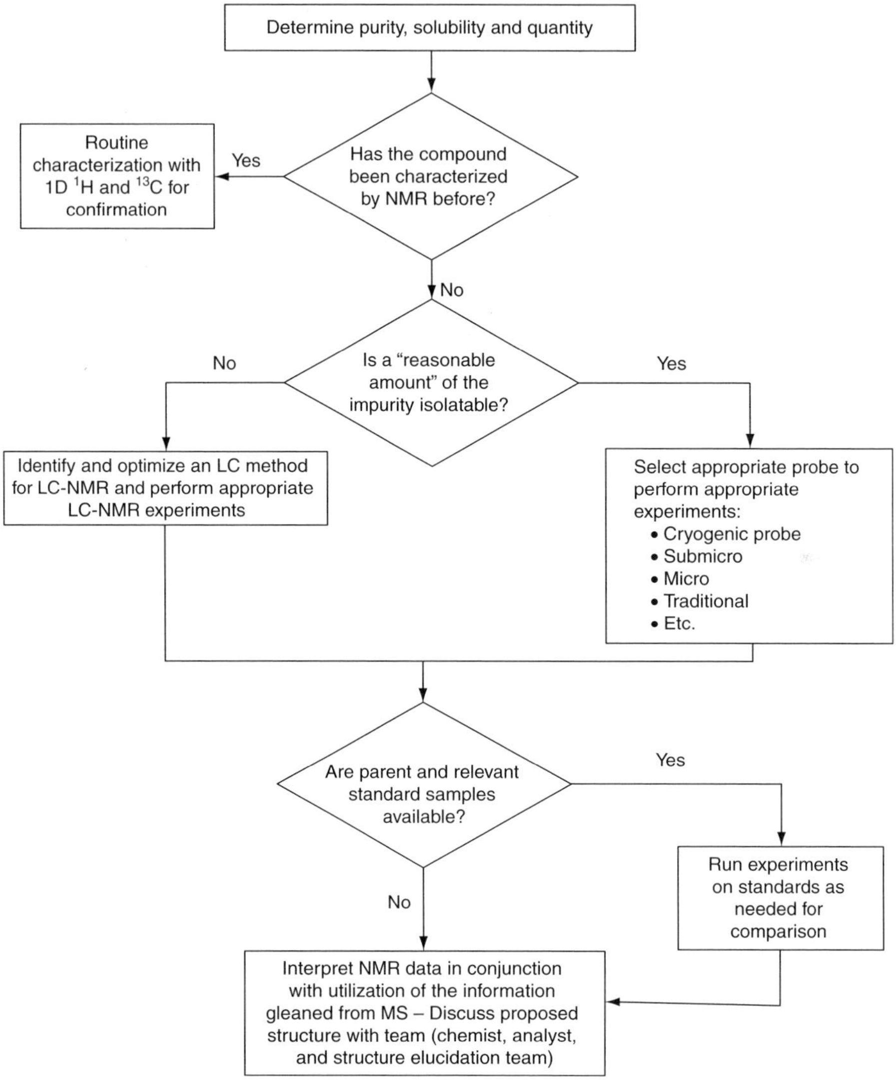

FIGURE 29 The overall characterization strategy for structural characterization of impurities and degradants by NMR is shown schematically above.

lot of synthesized material or formed during the synthesis of a precursor? It is critical to know the synthetic route or isolation pathway followed to prepare the sample. Not only will this provide clues of the possible structure based on predicted chemistry, it will also allow one to look at possible functionality and handles for data analysis. Understanding the chemistry that is behind the compounds being analyzed is paramount. It is highly recommended that if the spectroscopist conducting the experiments and analyzing the data does not have a reasonable understanding of organic chemistry that they partner with a synthetic organic chemist for consultation.

In many cases, little is known about the inherent properties of a degradant or impurity before it has been fully characterized. There are several properties, if the information is available, that are useful to know for proper handling of the NMR sample. First, it is essential to have an approximate idea regarding the quantity and purity of the sample, since this will determine which NMR hardware is most appropriate. We will discuss specific quantity requirements in the NMR instrumentation discussion, but for now, let it suffice that approximately 1 mg of targeted compound is required. Of course, this value is directly related to the sample purity, since 1 mg of sample with a purity level of 50% would only contain 500 μg of targeted compound. For traditional NMR, that is, not LC-NMR, one would prefer a purity level of at least 80%. However, the purity of the sample is key and the better the purity the better. Of course, the quantity of sample observed by NMR relates to the solubility in the selected solvent. Preliminary solubility tests therefore are important to identify an appropriate NMR solvent and hence maximize sensitivity. One important thing to consider when utilizing samples that have been generated through the isolation of small quantities (~0.2%, mol/mol) of desired material from crude material is the potential for high salt content in the resulting isolate. It is not uncommon when utilizing typical RP chromatographic methods to end up with several 100 μg of pure material in a liter or more mobile phase. If this mobile phase contains buffers/modifiers then the sample will contain a great deal of salt (either as counter ions to the isolated molecule or resulting from utilizing nonvolitive salts in the mobile phases. A large amount of salt in the sample, that is several times the amount of degradant, significantly changes the magnetic susceptibility of the sample. This may be beyond the tuning range of the probe and will therefore make it challenging or impossible to tune and match the probe. The effect on the resulting data is poor resolution and coherence selection, which is essential for indirect detect experiments, as well as reduced sensitivity.

Information regarding sample stability can be critical if available. Sample stability (e.g., sensitivity to light, air, protonated solvents, etc.) is useful for appropriate sample handling. Unfortunately, this information is usually not known *a priori*. Therefore, one should minimize risk by taking suitable precautions as described in the sample preparation section. Since this is an unknown research sample, safety hazards are also unknown, so one should proceed with at least the same level of caution that applies for the parent species.

The parent is often a useful starting point to assist in the interpretation of the NMR spectra of the unknown degradant or impurity. A sample of the parent compound ideally should be run in the same solvent used for the impurity for the most straightforward comparison. This is because chemical shifts and resolution may change somewhat in different solvents, particularly in the proton spectrum. Of course, solubility differences between the impurity and parent may preclude using the same solvent. Both the one-dimensional (1D) proton and carbon spectra of the parent should be obtained. 2D proton–carbon correlation spectra may also be necessary to make all resonance assignments.

Additional information may come from a variety of sources. Certainly a proposed structure based on plausible chemistry and the total molecular mass is helpful. However, one must be careful to keep in mind that proposed structures are based on preliminary data only and thus may not be consistent with subsequently collected data. If chromatographic-based hyphenated NMR techniques (LC-NMR, LC-NMR/MS, etc.) are to be performed, then it is essential to obtain details of the chromatographic method to be used. Other information may include color to suggest a highly conjugated system, IR absorption to detect carbonyl stretches, and relative chromatographic retention time to evaluate potential polarity differences compared to the parent.

C. NMR Instrumentation

The three main factors that determine the appropriate selection of NMR equipment are: quantity, solubility, and stability. The quantity of sample determines the detection limits required. Because we are usually sample limited, it is important to identify the NMR probe that can accommodate the entire quantity of sample without unnecessarily diluting the sample in excessive solvent. This then relates to the sample's solubility in the selected solvent. Obviously one should choose a solvent that yields high solubility to maximize the amount of sample in solution. If the solubility is low, using a smaller sample tube size will not help to concentrate the sample. Lastly, stability is important, since we wish to ensure sample integrity over the course of acquiring the data. For unstable compounds, it may be necessary to separate additional degradants from the target species. For this purpose, chromatographic techniques can be combined with traditional NMR techniques.[193]

Regardless of individual sample characteristics, there are several general instrumentation guidelines to perform the experiments described. Beginning with the NMR console itself, it is imperative that the system can perform quadrature detection, since this significantly minimizes instrument artifacts that appear in the spectra.[194,195] Antivibration legs reduce spectral noise that arises from floor vibrations. This is essential at high field strengths, 400 MHz and above. Similarly, temperature control minimizes thermal instabilities, which may produce unwanted ridges in 2D data sets. Temperature control is also critical for reasons, which include but are not limited to, the study of rotamers or *cis/trans* amides and other chemical exchange phenomenon, hydrogen bonding, kinetic analysis, etc. Digital filtering capabilities, while not required, improve spectral quality of 2D data sets by filtering out noise outside the spectral window.[196] A minimum of 500 MHz is recommended to achieve the resolution and sensitivity necessary for reliable low-level impurity characterizations. The experiments described here are for protons, carbons, and nitrogen so three radio frequency (also called "RF") channels optimized for these frequencies are required. Ideally, one channel is dedicated to protons while another (sometimes designated "X") channel covers a broad range of frequencies that include carbon. A single *Z*-axis gradient channel is required to significantly improve the coherence selection and/or rejection and hence the overall spectral quality of 2D experiments.

Regarding general probe specifications, single *Z*-axis gradient is sufficient for most small molecule NMR experiments. The probe must also have the appropriate coil configuration to achieve the experiments that one needs to perform, in addition to the deuterium coil used to lock the signal. Probes are designed such that one coil is concentrically inside the other. The inner coil yields the highest sensitivity. Therefore, for proton-detected experiments, one ideally should choose a probe with the proton coil as the inner coil. Alternatively, for optimal carbon detection, one should designate the X coil to be the inner coil. The field strength of the probe must also match that of the magnet; at least 500 MHz is recommended.

Several probe sizes and configurations are commercially available, and if one intends to do impurity and degradant characterizations routinely, it is valuable to have a representative range of sizes. The probe size is expressed in terms of the diameter of the sample tube for which it is designed. The most commonly used size is a 5 mm probe. This is an appropriate choice if one is not limited by sample quantity. The sample size is ideally 5–20 mg in 0.5–0.75 mL of solvent, although as little as 1 mg of material can be used. Obviously, increasing the sample concentration will reduce the experiment time required to obtain adequate signal-to-noise ratios. It should be noted that increasing the sample concentration too much can have deleterious effects that manifest themselves as signal broadening, chemical shift changes, and unwanted chemical exchange phenomenon.

If the amount of sample is limited, a good alternative is the family of "microprobes." Depending on vendor, microprobes are available in a multitude of sizes ranging from 3 to 1 mm. The sample volume is approximately 150–10 μL. These probes are suitable for samples on the order of hundreds of micrograms to hundreds of nanograms.[186–188] It is important to note that decreasing the sample tube size effectively concentrates the sample by reducing the amount of solvent required. Concentrating the sample in the active coil region increases the number of spins detected and thus increases observed signal-to-noise ratios in the NMR spectra. Going to smaller tube sizes therefore will yield no significant advantage if there are solubility limitations preventing an increase in concentration.

A very important development in the last 10 years is the commercialization of the cryogenically cooled NMR probes. These probes have opened many doors to the practicing NMR spectroscopist, particularly those working in the area of small molecule research. The specifics of the varying vendors' specifications for their respective cryogenically cooled probes are beyond the scope of this paper. Suffice it to say, both work on the same principle and that is to reduce the amount of electronic noise to afford an increase in the signal-to-noise ratio. This of course results in the ability to explore many more types of NMR experiments on an ever decreasing amount of available sample. These types of probes come in varying sizes just as their room temperature counterparts. The recent release of the 1.7 mm CryoProbe has opened the doors for scientists to truly explore sample sizes that have not been easily accessible in the past.

Another probe choice is that of the flow variety. These probes come in several formats; however, they all have one thing in common. This common feature is that of a flow cell. There are many different varieties of flow cells customized to the types of analysis that need to be done. These differences vary from either the thickness of the flow cell (SFC-NMR) to the actual volume the cell holds (e.g., flow NMR, LC-NMR, LC-SPE-NMR, etc.). An additional element that has truly enhanced the world of LC-NMR is the introduction of the flow cell inserts for cryogenically cooled probes. This has really taken this application of NMR to a much more productive level. The NMR community refer to flow cell sizes based on either the total cell volume or the volume of the flow cell that is contained within the active region of the NMR probe, namely within the coil of the probe. In this chapter, the use of total cell volume will be used.

Flow NMR has been around for many years and is used in varying applications ranging from combinatorial library analysis to reaction NMR (RxnNMR).[197] The more heavily utilized of the "flow" NMR techniques is LC-NMR and its many variants.[190] LC-NMR, which combines the mixture separation capabilities of LC with the structural elucidation capabilities of NMR spectroscopy allows the analysis of compounds off column without a great deal of manipulation.[190] To perform LC-NMR experiments, one must have both the LC accessory for the NMR console and an LC-NMR probe. The LC accessory consists of an LC system with a seemingly endless possibility in detection methods. Examples of the types of detectors that can be attached to an LC system include but are not limited to UV (variable and photo-diode array, evaporative light scattering (ELS), MS and of course the NMR instrument itself. It is imperative to remember that when performing LC-NMR-type experiments that the NMR should be treated as a detector and the time a researcher puts into these types of experiments should be on providing the best possible sample to the detector flow cell, whether NMR, UV, etc.

There are also more specialized varieties of flow probes that allow a researcher to analyze data acquired for samples in the submicrogram level. These probes include the many variants of the protasis/MRM microcoil and the inserts that one can utilize in the cryogenically cooled probes mentioned above. Both of these probes offer exceptional capabilities to acquire data on samples of very limited mass as well as very good spectral quality and reproducibility. In addition, the utilization of the smaller cell volumes also requires less effort in the experimental set-up around the suppression of residual solvent inherent in the samples of this size.

It must be noted that LC-NMR is a niche technique in this authors eyes. It is very appropriate for samples where a great deal of information is known about the family of molecules (metabolism for example) or that the sample under analysis is in a quite large percentage (say 5% relative to the main component). In situations where the sample is very limited and much more structural information is needed it is recommended that a traditional approach of compound isolation and tube-based (or hand injected) experiments be performed. An additional important application of LC-NMR is for unstable compounds because it is able to separate multiple components in a sample mixture without the need for sample dry down and reconstitution.

D. Sample Preparation

Deuterated solvents are used in NMR for two reasons. First, deuterium provides a signal on which to lock the spectrometer frequency, thus enhancing long-term stability. Second, it dramatically reduces the proton signal that arises from the solvent, which would otherwise overwhelm the signal of the targeted compound. Solvents should be $99.9^{+}\%$ deuterated. Many vendors provide high-quality deuterated solvents premade in sealed glass ampoules containing differing volumes as needed for ones particular application. This ensures the integrity of the solvent and avoids contamination during sample preparation. Most importantly it prevents absorption of water. Some common deuterated solvents used for NMR studies include dimethyl sulfoxide, chloroform, methanol, water, acetone, and acetonitrile. DMSO is considered by many to be a "universal solvent," since it provides excellent solubility for a wide variety of compounds. In addition, its high boiling point makes it ideal for variable temperature experiments, which are useful for dynamics studies. One point to note is that DMSO absorbs water over time, so careful storage conditions are essential. Protic solvents such as water and methanol provide proton sources for rapid exchange with labile protons. This makes NMR detection of species such as N–H, O–H, or labile carbon-bound protons difficult in these solvents. Also, many compounds, such as peroxides will likely decompose in protic solvents.

As we learned in our discussion of NMR instrumentation, the volume of solvent depends on the specific probe and tube size chosen. In general, one should minimize the total amount of solvent in order to concentrate the sample and hence increase the observed signal-to-noise ratio. Concurrently, adequate solvent should be used so that the sample height is above the top of the coil to ensure proper field homogeneity across the sample. Ideally, the sample should be symmetric about the coil region for maximum field homogeneity. To minimize the degree of shimming required to achieve good resolution, the sample height should be roughly equal to that of the standard NMR sample used to create the shim file. Otherwise, there may be a significant difference in the shim values necessary to optimize the line shape. That being said, there are many instances where one needs to minimize the solvent volume as much as possible and therefore additional shimming will need to be done to compensate. In addition, the use of gradient shimming facilitates drastic recovery from poor shims allows minor touch-ups to be made by hand. A standard proton sample used to measure resolution is 1% chloroform in deuterated acetone. To properly reference the spectra, TMS is often added as an internal reference standard. The chemical shift is set equal to 0.0 ppm for both protons and carbons. One must be extremely careful to add only a miniscule amount of TMS to the sample so as not to overwhelm the target signals. Typically, not more than vapor is added. Alternatively, one can use the solvent signal itself as an internal reference for chemical shift.

Clean glassware is essential for characterization of low-level impurities and degradants. Remember that one is possibly working with the world's supply of the sample, so it is imperative to avoid contamination. For this reason, we recommend not reusing NMR tubes for this purpose, but rather use a new tube for each impurity or degradant characterization. Special NMR tubes with plugs are also available that are susceptibility matched to most deuterated solvents. This will increase sensitivity, assuming appropriate solubility although the cost of these tubes and plugs are significantly higher than for traditional glass tubes. These tubes are always reused. For LC-NMR, disposable syringes are available as an alternative to cleaning glass syringes, although one must be certain that the sample is inert to the plastic material of the syringe. The LC-NMR system, including the probe and any loops/parking/trapping systems to be used, should be washed for several minutes prior to use. The LC pump and degasser should be running for at least 30 min prior to use. Using deuterated solvents enhances detection capabilities by greatly reducing solvent signal contributions. This facilitates data interpretation.

Unless otherwise known, proceed with a level of safety precaution at least as rigorous as for the parent species. To prepare the sample, pipette or syringe a minimal amount of solvent to the sample vessel. Samples should be provided in the smallest vessel possible, preferably with a conical bottom, to reduce loss. The pipette should also be appropriately small to minimize sample loss. Use the pipette to dissolve as much of the sample as possible. Transfer this solution to the NMR tube (or small vessel for LC-NMR). Repeat until the desired solvent volume is used, and the sample is completely transferred to the NMR tube. Cap or seal the tube. Inspect the tube for any residual solids. Use caution not to heat the sample significantly. Alternatively, one can filter the solution, but this may cause significant sample loss. Also, be sure the degradant or impurity is not expected to react with the filter. Lastly, before placing the sample into the magnet be sure to eliminate any air bubbles, since bubbles will make shimming more challenging and will cause adverse effects in the LC-NMR system.

Based on available stability information, keep the sample out of light by covering it completely with aluminum foil when not in the magnet. To avoid thermal instability, do not heat the sample above ambient conditions. Keep the sample capped or sealed to minimize exposure to air. Dissolve the sample in solvent only immediately prior to running NMR experiments to evaluate its stability in solvent. Run a standard proton spectrum immediately after making up the sample as a stability reference. Reacquire a proton spectrum following the final NMR experiment to verify that sample integrity was maintained. Reacquire the proton spectrum if any additional experiments are subsequently required.

E. NMR Experiments

1. Getting Started

Once the sample is prepared, load it into the magnet. Tune and match the probe for both protons and carbons, lock on the deuterium signal, and shim the probe. If the sample does not adequately tune or match, there may be an excessive amount of residual salt present, in which case the sample must be further manipulated. Spinning the sample improves resolution by reducing field inhomogeneities in the *x–y* plane. However, it is important to remember that spinning sideband signals arise in the proton spectrum at multiples of the spinning speed on either side of each observed resonance.[198] The intensity of sideband signals depends on the quality of shimming. Poor shimming gives rise to more intense sideband signals. This may confuse the interpretation of the data, particularly when deciphering low-level contributions to the spectrum. We therefore recommend running all experiments described here without spinning the sample for ease of data interpretation and optimal stability. Throughout the remainder of this NMR section of the chapter ibuprofen will be used as a model compound (Structure 5).

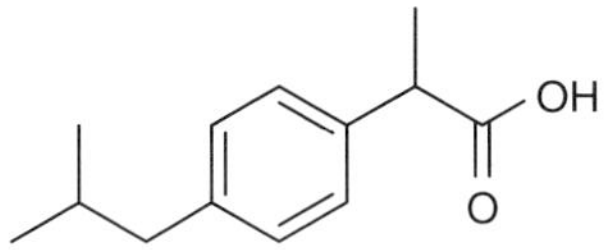

STRUCTURE 5 Ibuprofen

2. 1D Experiments

For all the 1D experiments described, the number of points should be enough to define the peaks in the spectrum. The transformed peaks should each have roughly 15 points defining them for reasonable resolution. The receiver gain can be set using the autogain feature of the NMR software. In general, the gain should be maximized to enhance sensitivity while not saturating the receiver. If the signal is amplified beyond the receiver limits, the signal will be clipped in the time domain. The Fourier transform of the resulting square wave in the time domain is a sinc function in the frequency domain, which produces a detrimental wave pattern in the baseline.

The most basic NMR experiment is the one pulse proton experiment.[193,199,200] It is comprised of a 90° pulse on the proton channel followed by acquisition. The pulse duration should be set equal to the measured 90° proton pulse for the probe. The recycle time, or time between scans, should be greater than five times T_1 of the sample if one wants to achieve full relaxation profiles for applications such as quantitative NMR. In practice, a pulse with a $<90°$ flip angle is used and one second is often used as a default value. The use of reduced flip angles and relaxation delays to achieve full relaxation is done. This is generally referred to as the Ernst angle.[201] Proton chemical shifts typically range from 0 to 10 ppm, so the spectral width should be set at least this large. A good approach is to set the spectral width to a larger value, such as 15 ppm, to identify the actual limits of the resonances observed for a given sample. Then the spectral width can be reset to a smaller value more appropriate to the sample. Typically, 32,768 (32 K) points should be used to cover a 15–18 ppm range. Once the proton spectrum is integrated, one can estimate the sample purity to determine if further purification is necessary. There should be no more than roughly 20% contaminants in the sample, preferable the sample should be as close to 100% pure as possible (Figure 30).

The standard carbon experiment decouples protons from carbons to enhance sensitivity.[202] In the absence of proton decoupling, each carbon resonance is split by one more than the number of protons attached to that carbon. Since, unlike for protons, the natural abundance of carbon-13 is only 1.1%, it is critical to maximize sensitivity in order to detect carbon in a reasonable amount of time.[193,203] The most commonly used proton decoupling scheme is "WALTZ-16."[204] Usually, protons are the only nuclei that are decoupled in a standard carbon experiment. Other spin-1/2 nuclei, such as fluorine, are not decoupled and thus split each carbon resonance by one more than the number of attached fluorines (e.g., see Figure 31). This can be particularly advantageous for identifying fluorinated carbon entities. Since the carbon signals are often too weak to observe in a small number of scans, one needs to make a reasonable approximation for the spectral

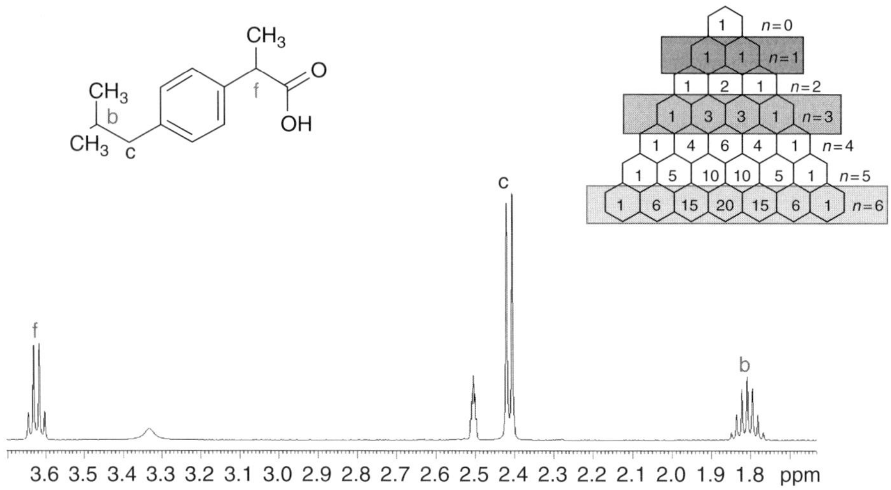

FIGURE 30 **A homonuclear *J*-coupling analysis of the aliphatic portions of ibuprofen.**

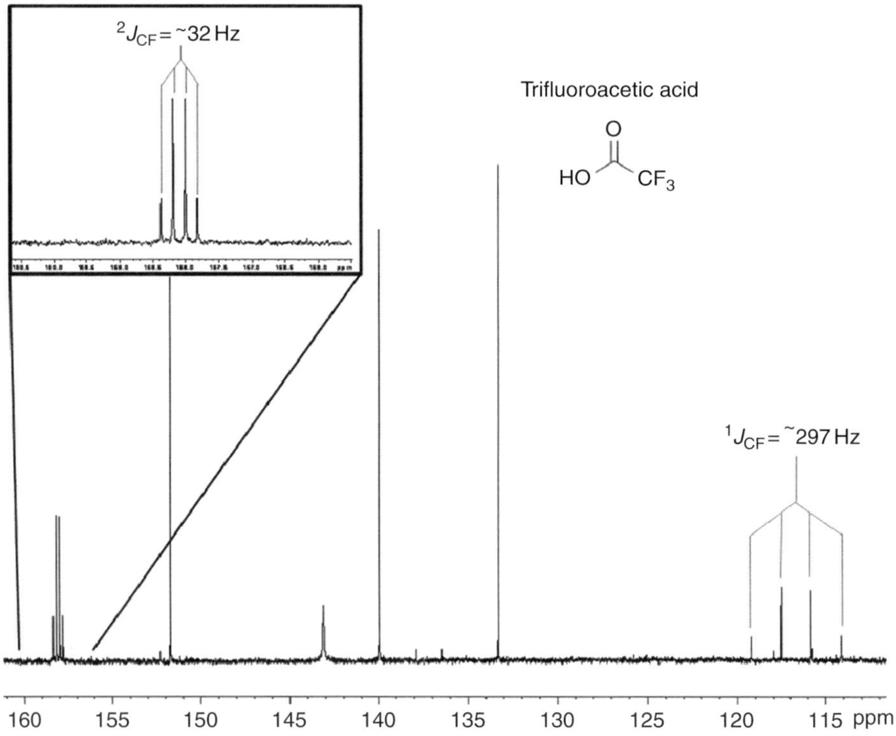

FIGURE 31 **One-dimensional carbon spectrum showing splitting that is due to tri-fluoro group in a mixture of an API and trifluoroacetate. Notice that the affected resonances appear as quartets because of splitting by three equivalent spin-1/2 nuclei. The size of the coupling constant between the proton and fluorine increases dramatically with proximity to the fluorine atoms. This greatly facilitates the assignment of carbon resonances.**

width. A standard carbon spectrum covers the range 0 –220 ppm with 32–64 K data points. Because of the relatively long T_1 relaxation time of carbons, common practice is to use less than a full 90° pulse, such as a 30° pulse, and a recycle time of 0.5 s. A line broadening of 1 Hz improves sensitivity without significantly reducing resolution.

The DEPT experiment, or "distortionless enhanced polarization transfer," is a carbon selectivity experiment.[180,193,205–209] Depending on the pulse length selected, one can selectively observe different types of carbon entities. We recommend setting the DEPT proton pulse length to 135°. In this case, quarternary carbons are suppressed, methylenes are inverted, and methine and methyl carbons appear upright. Methines and methyls are distinguished based on chemical shift and 2D proton correlations. Methines usually appear downfield of methyls. Alternatively, if time permits, the entire series of DEPT experiments can be performed to conclusively distinguish methine from methyl resonances. One second is a reasonable default value for the recycle time. The spectrum should be set to capture the resonances of interest with 32–64 K data points and four dummy scans. While this is an easily interpreted data set, the multiplicity edited HSQC provides much more robust information, both multiplicity as well and the ^{1}H–^{13}C one-bond connectivity ($^{1}J_{CH}$). Please see the next section for greater detail.

3. 2D Scalar Coupling Experiments

There are several general considerations for identifying appropriate 2D experiments to use for structural elucidation. The experiments described within this section all utilize gradients. Gradients are used in these experiments to improve coherence selection that is otherwise performed using elaborate phase cycle schemes.[210,211] It is therefore possible to achieve improved signal selection and reduced spectral artifacts with fewer acquisitions using gradient selection. The primary disadvantage of gradient coherence selection is reduced sensitivity. Therefore, if the low quantity of sample available requires a large number of scans for sensitivity reasons, one may prefer to use nongradient versions of the experiments described and instead rely upon phase cycling for signal selectivity. One should weigh the sensitivity losses with the fact that spectral artifacts can cause significant issues with data interpretation.

On the topic of sensitivity, the 2D experiments described here are all proton-detected experiments. The reason is that protons are an abundant and hence easily detected nucleus while ^{13}C has a low natural abundance and thus is more challenging to detect. The corresponding magnetogyric ratios also make protons more readily detectable than carbons. Experiments that traditionally were based on carbon detection thus often have modern day equivalent versions that transfer the carbon magnetization to protons and detect the corresponding proton signal. This is known as "inverse detection" or "indirect detection," since the carbon signal is indirectly detected via its effect on the observed proton signal.[212–215] Inverse detection experiments are particularly valuable for the characterization of impurities and degradants, since the sample quantity is often quite limited. For the 2D experiments described here, the spectral width of the proton dimension should be equal to that determined for the 1D proton spectrum, and the spectral width of the carbon dimension should be equal to that determined for the 1D carbon spectrum. If the 1D carbon spectral width has not yet been determined, then 0–220 ppm is a reasonable default range. A good compromise of data set size versus adequate resolution is 2048 points in the t_2 dimension and 256 points in the t_1 dimension. This data set size is appropriate for all the 2D experiments described in this section. The use of linear prediction in the indirect dimension (t_1) is typically performed to provide resolution enhancement. One should take care to not "over" linear predict as artifacts can be introduced that will lead to a reduction in the quality in the data as it pertains to interpretation.

The experiment traditionally used to detect proton–proton (homonuclear) scalar couplings is known as a "COSY"[193,216] for *c*orrelation *s*pectroscop*y*. The 1D proton spectrum is shown along each of the two axes, and off-diagonal peaks show correlations between neighboring protons resulting from homonuclear scalar coupling (commonly referred to as *J*-coupling). Each proton also exhibits a correlation to itself, which appears along the diagonal axis of the spectrum. These self-correlation peaks do not provide any additional information, and depending on resolution limitations, may obscure cross peaks near the diagonal. It may therefore be advantageous to utilize experiments such as the COSY45 or to add a "double quantum filter" to the COSY experiment to reduce contributions along the diagonal. The *d*ouble *q*uantum *f*iltered COSY, or "DQFCOSY,"[217–219] provides the same information as a standard COSY with improved resolution near the diagonal because of reduced self-correlation signals. However, depending on the spin system under investigation coupling constants can be measured in the DQFCOSY experiment (other variants exist that allow the facile measurements of homonuclear coupling constants via the directly correlated proton spins through active and passive couplings, namely the ECOSY experiment[193,220]). For these 2D homonuclear experiments, the spectral width is typically set equal to that of the 1D proton spectrum in both dimensions. The recycle time (the time between individual experiments) is typically 1 s.

Regarding heteronuclear correlations, "HETCOR" historically was the standard experiment used to detect proton–carbon scalar couplings.[221–223] The disadvantage of HETCOR is that it is a carbon detection experiment. While the direct detection of carbon is considered a disadvantage relative to the sensitivity if one has the capability to obtain significantly enhanced S/N (e.g., ^{13}C observe cryoprobe) one can take advantage of the high resolution of ^{13}C in the t_2 dimension while still having plenty of digital resolution in t_1, which in this case is ^{1}H. However, in many cases spectroscopists are limited to room temperature probes and are therefore limited in sensitivity as previously discussed. There are two modern inverse detection experiments for the observation of short-range ($^1J_{CH}$) proton–carbon scalar couplings. They are heteronuclear single quantum coherence, or "HSQC,"[193,224,225] and heteronuclear multiple quantum coherence, or "HMQC."[193,211,226] Both of these experiments are available in many different varieties.[193] One of the more useful pulse sequence modifications is the ability to distinguish between protonated carbon types (methine, methylene, and methyl). These are multiplicity edited experiments and should be used as the amount of added information is affectively free relative to the experiment time and setup.[193] This information is complementary to that obtained from the DEPT experiment, since one can also match proton integrals to corresponding carbons to distinguish methine, methylene, and methyl carbons. All inverse heteronuclear correlation experiments show the 1D proton spectrum along t_2 and the 1D carbon spectrum or projection along t_1. Cross peaks correlate protons directly bonded to carbons. A proton resonance showing no correlations is likely bonded to a noncarbon heteroatom, for example, O–H or N–H groups. Similarly, a carbon resonance showing no correlations is likely to be a quarternary carbon. One can set the carbon spectral width of these 2D experiments to only cover the protonated carbon region, nominally 0–180 ppm, to improve resolution in this dimension. The 0–180 ppm recommendation is typically valid unless a moiety (e.g., aldehyde) is suspected from MS information and 1D 1H data that is protonated and falls out of this chemical shift range. In these cases, care should be taken in experiment setup and interpretation. Comparing the HSQC and HMQC experiments, the HSQC tends to yield higher resolution (because of the use of single quantum coherence selection), while the HMQC tends to yield higher sensitivity. Thus, if sample quantity is limited, the HMQC is preferred. However, one should look at the systems under study and consider other HSQC variants, such as the sensitivity improved HSQC depending on the level of complexity of the ^{1}H multiplicities. In most cases, the HSQC is preferred. A coupling constant of $^1J_{C,H}$ of 145 Hz is recommended for either experiment as a compromise to try and ensure observation of both aliphatic and aromatic $^1J_{C,H}$. A 1 s recycle time should be sufficient.

Similar to the short-range correlation experiments described, a long-range scalar coupling experiment is also available. The heteronuclear multiple bond correlation experiment, or "HMBC," detects scalar couplings between protons and carbons that are separated by multiple bonds.[212,227–229] We recommend setting the long-range coupling constant, $^nJ_{C,H}$, to 8 Hz. Most useful variants of the first described heteronuclear multiple bond correlation (HMBC) experiment contain what is called a "low-pass filter."[230] This filter is used to reduce the deleterious effects of the $^1J_{CH}$ couplings and is typically set to 145 Hz. "Filters" have been developed using many different principles to varying levels of success.[180] In HMBC experiments, the most typical couplings observed are residual $^1J_{CH}$ which are observed as correlations split in the direct detection dimension by the $^1J_{CH}$, $^2J_{CH}$ (geminal couplings), and $^3J_{CH}$ (vicinal couplings). The HMBC experiment in its original form (as well as many variants) does not allow the discrimination of the two- and three-bond couplings. There is a general "rule of thumb" that suggests the sign of the coupling of the three-bond is the same as the one-bond. Special experiments are needed to determine this and there are many examples where this rule breaks down and should be used with care. One of the most sought after long-range heteronuclear correlation experiments is to design a pulse program element to distinguish between $^2J_{CH}$ and $^3J_{CH}$. This has been achieved to an extent with using a variety of filters based on homonuclear coupling evolution.[231,232] Because the HMBC experiment detects weak couplings, it is an inherently insensitive technique. It is therefore wise to allow more time for acquisition of the HMBC as compared to the previous experiments described. A 1 s recycle time should be sufficient. HMBC combined with COSY, multiplicity edited HMQC or HSQC, and 1D proton and carbon sequences comprise the majority of the standard set of experiments used as a first pass for structural elucidations of impurities and degradants.

4. Additional Experiments

If more information is required than is available from the standard set of experiments, one need to perform additional experiments. Common examples include information regarding stereochemistry of fairly rigid moieties is necessary or several heteroatoms/multiple quaternary carbon centers preclude through bond homonuclear or heteronuclear experiments. The 2D experiments previously described all relate to through bond correlations and not spatial relationships. One set of experiments that specifically addresses through space interactions is known as the nuclear Overhauser effect (nOe) experiments.[181,233–235] The NOE experiment shows correlations between nuclei that are in close proximity to each other, no greater than approximately 5 Å. Today there are typically two types of experiments that are used for acquiring 1D NOE data. These experiments are the difference NOE experiment (steady-state NOE) and 1D selective excitation experiments (transient NOE).[231] In the difference NOE experiments, a target signal is saturated for a period of time which alters the intensity of signals from neighboring nuclei. The resulting alteration of signal intensity is known as the nOe and represents a through-space interaction (properly referred to as dipolar couplings). Taking the difference between the irradiated spectrum and a standard spectrum leaves only the target signal, which is inverted, and the resulting nOe's. This 1D proton experiment requires very high instrument stability, so if available on the instrument, it is useful to employ temperature control to ensure thermal stability. Undoubtedly, one of the most useful NOE experiments in the double pulsed field gradient spin echo NOE (DPFGSE NOE).[236] This experiment utilizes excitation sculpting providing exquisitely selective excitation (irradiation) of a resonance of interest. This selective excitation followed by a NOE buildup period provides very convenient and easily interpreted data set. A recent version utilizing a zero quantum coherence removal scheme that provides resonances with very clean phase properties and little baseline distortion. There are also 2D versions of the NOE. This experiment is aptly named the nuclear Overhauser effect spectroscopy (NOESY). In most small molecule structure elucidation or characterization in the pharmaceutical industry projects a spectroscopist deals with compounds that have a correlation time that allows one to observe negative NOE correlations which is desired. A molecular correlation time roughly refers to how a molecules motion is in solution. This motion is roughly correlated to molecular weight and the spatial arrangement of the compound in solution. At a particular correlation time the NOE build up approaches zero and ultimately becomes positive. When the NOE enhancement is zero it is commonly referred to as the NOE null.[231] If one looks at correlation time as a function of molecular weight an approximation of 1000 Da is where the "null" occurs. However, since molecular motion of a molecule changes with both the viscosity of the solvent used and the intra- and intermolecular interactions, it is often observed that a compound of 600 Da molecular mass will approach or even pass through the "null." The caveat that should be watched for when experiencing data with a positive NOE is that peaks affiliated with chemical exchange are also observed as positive peaks in NOE-based experiments. When this situation is encountered or the NOEs for a molecule of interest are very small (approaching the null) then a different through space experiment should be used. This experiment is based on what is called the rotating frame Overhauser effect (ROE).[237] This experiment comes in the same varieties as the NOE-based experiments. Because of the design of the experiment ROE-based experiments do not suffer from the same issues as the NOE experiments. That being said, it is not the case that the ROE is superior to the NOE experiment in terms of the data acquired.

Other common experiments include 1D spectra of phosphorous and fluorine and other nuclei of pharmaceutical interest. The other nuclei include atoms such as boron, lithium, nitrogen, oxygen, etc., these active nuclei typically provide more complexity in terms of experiment set up and are in the case of nitrogen and oxygen have very low natural abundance and very rapid relaxation times with respect to oxygen.[200] Nitrogen detection is typically done via 2D methods utilizing coherence selection starting with protons in an indirect detection configuration. This type of experiment, particularly the ^{1}H–^{15}N HMBC is an invaluable tool in the analysis of pharmaceutical compounds. As we discussed earlier, fluorine couplings can be identified by splitting patterns in the 1D carbon spectrum. For example, a trifluorinated carbon will appear as a quartet in the 1D carbon spectrum (see Figure 31 for example). The size of the coupling constant will determine the number of bonds separating the fluorine and carbon, since fewer bonds in between the two increase the strength of interaction and hence the size of the coupling constant.

Oftentimes during the course of analysis of chemical compounds by NMR multiple resonances are observed for the same proton in the compound. This is typically due to two phenomena often observed. These are typically the presence of rotamers and isomerization, collectively referred to as chemical exchange. These end up showing up as two distinct resonances if the rate of interconversion (e.g., hindered rotation, *cis*/*trans* isomerization) is slow relative to the NMR time scale. In times of intermediate exchange the resonance will not be resolved and peak broadening will be observed. When the rate of interconversion is rapid on the NMR time-scale resonances will be observed as a single resonance. The fast exchange case is what is most often observed. However, in cases of restricted rotation (e.g., steric hindrance) or isomerization (e.g., tertiary amide *cis*/*trans* isomerization) on the slow time, elevating the temperature increases the rate of kinetic energy such that, at the coalescence temperature, pairs of peaks coalesce to one, and an average conformation is observed.[238] If temperature does not allow coalescence of the resonances and therefore doubt as to the presence of a chemical exchange issue occurs, the use additional 1D techniques can be performed. Useful experiments to detect chemical exchange are the EXSY and/or the use of NOE experiments with long NOE buildup times ($\sim$1 s).[231]

The experiments described thus far detect proton–carbon and proton–proton interactions and do not specifically address protons bonded to other nuclei. As mentioned when discussing NMR solvents, the presence of protic solvents causes labile protons to exchange rapidly rendering them unobservable by NMR. We can use this to our advantage. A proton spectrum is first acquired using an aprotic solvent. A few drops of deuterium oxide are then added to the sample, and the proton spectrum is reacquired. Labile protons will exchange with the D_2O

and hence disappear from the spectrum. With this arsenal of techniques, we are now prepared to obtain sufficient data to begin elucidating a structure.

F. Data Interpretation

1. Primary Interpretation

An important point to make before continuing with this section is that structure elucidation is a very interpretive science and many approaches can be taken with the same outcome. There is no true prescriptive approach to take when determining the connectivities of atoms and proposing a structure based on all available data. When working on unknown compounds for structure elucidation (e.g., degradation products, process-related impurities, etc.) it is advisable to acquire a proton spectra to get an idea of the compound purity, the behavior of the compound in the solvent you have chosen as well as the concentration depending on solubility and sample volume considerations. If the purity and behavior of the compound is good then one should proceed with setting up the primary set of experiments for structure elucidation. In the author's case a ^{1}H, COSY, multiplicity edited HSQC, HMBC, NOESY, and ^{13}C (if enough sample is on hand) will be set up and acquired. A logical beginning is the 1D proton and carbon spectra. In an age of 2D experiments, the plethora of information contained in the 1D data is often overlooked. From the 1D data it is possible to deduce the class of compound that the spectroscopist is working with. In addition to the chemical shift information if the resonances are resolved the numbers of protons can be deduced as well as the multiplicities of the protons. The couplings constants can be measured (assuming first-order coupling) and deductions can be made as to the contiguous proton spin systems. The multiplicity of the proton resonance will identify the number of neighboring protons, since the multiplicity is equal to one more than the number of neighboring protons.[239]

Once the information is gleaned from the 1D data, the 2D data should be analyzed. A reasonable approach to take is to begin with the multiplicity edited HSQC (see Figure 32) to get the information concerning the numbers of protons attached to a respective carbon. This allows one to get a handle on the types of carbon that are present and get an idea of chemical environment base of the multiplicity as well as the chemical shift. For example, a methyl group with a carbon and proton chemical shift of approximately 70 and 3.5 ppm, respectively would represent a methyl ether moiety.

Once the protonated carbons and nonprotonated carbons are determined (quaternary carbons can be deduced through a combination of the multiplicity edited HSQC and the 1D ^{13}C data) a good approach is to begin the analysis of the COSY data (see Figure 33). This will allow the spectroscopist to look for the presence or absence of contiguous 1H spin systems. While much of this can be deduced from a detailed understanding of the 1D ^{1}H NMR data the use of the COSY data will both confirm previous data as well as enable one to look at spin system that have non-first-order spin systems.

Once the HSQC and COSY data have been analyzed and multiple spin systems and fragments (see Figures 32 and 33) of the molecule have been deduced

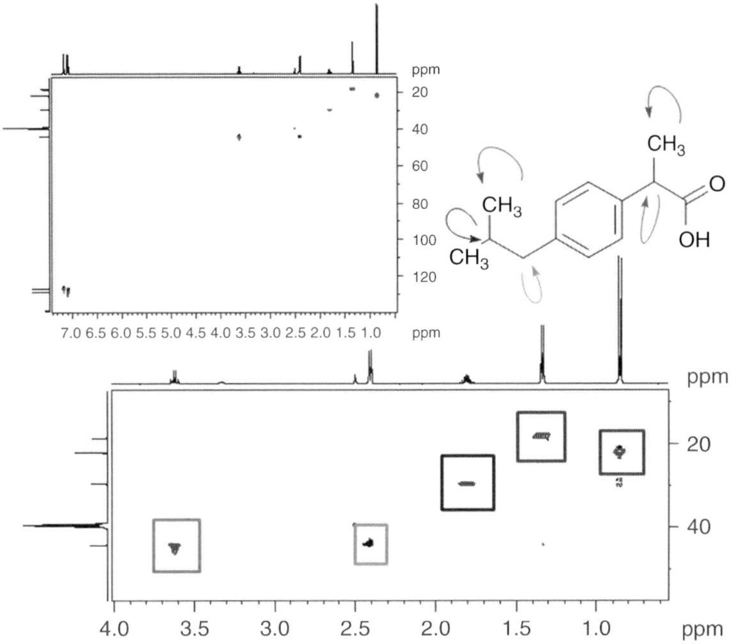

FIGURE 32 **Multiplicity edited HSQC spectrum of ibuprofen. The boxes are color-matched to the arrows indicating the $^{1}J_{CH}$ heteronuclear scalar coupling. (See Color Insert.)**

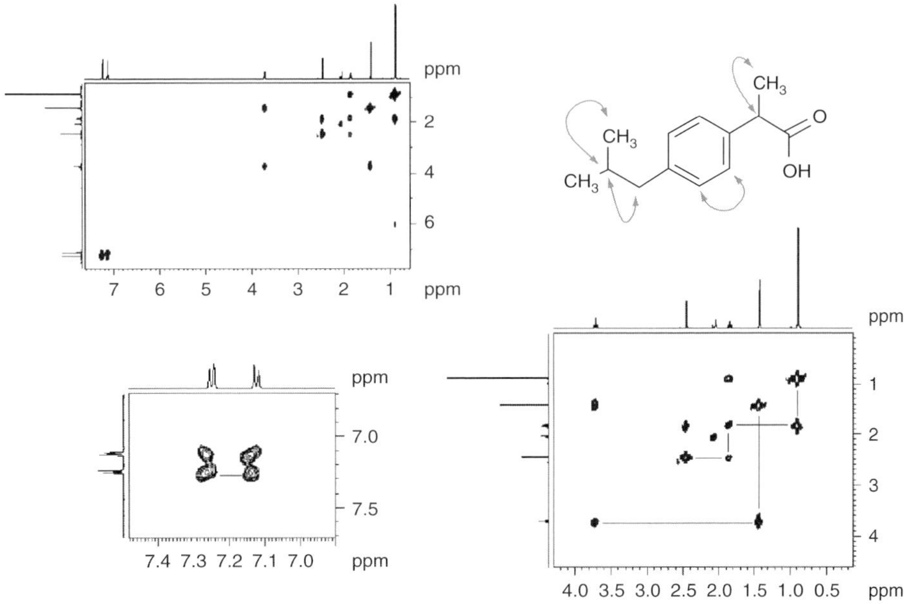

FIGURE 33 **COSY correlations are indicated on the structure of ibuprofen as well as in the two-dimensional expansions taken from the full two-dimensional data matrix in the top left-hand corner of the figure.**

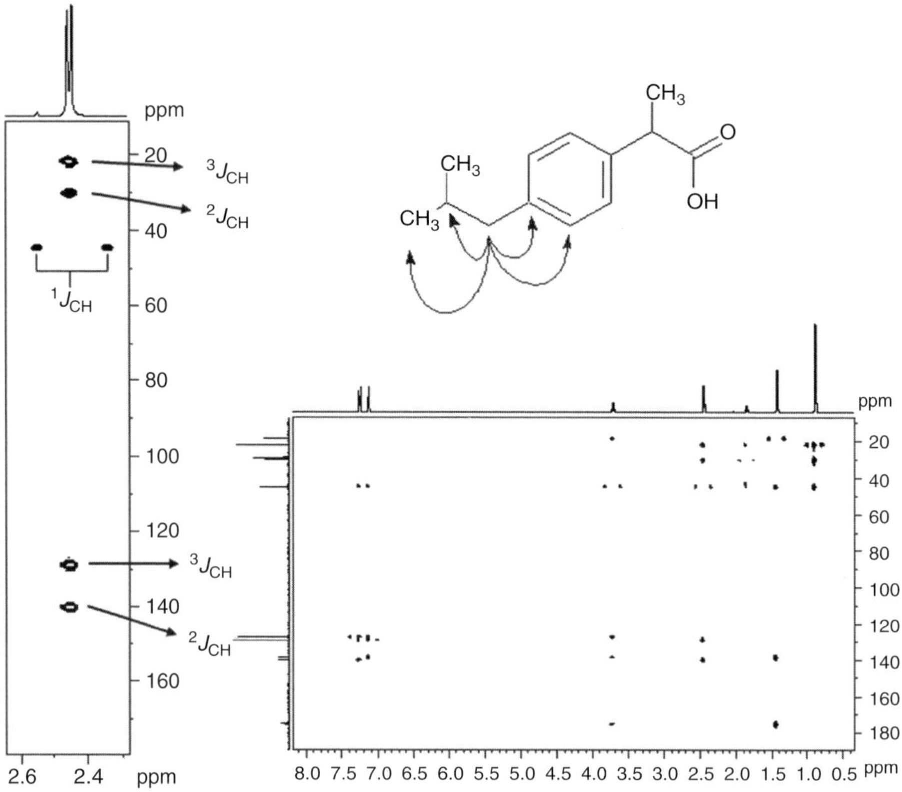

FIGURE 34 HMBC data for ibuprofen. The correlations indicated in curved arrows are shown and labeled in the expansion on the left of the figure.

it is a good idea to begin looking at your long-range heteronuclear correlation data (HMBC in this case) as well as your dipolar coupling information (NOESY in this case). The HMBC data (Figure 34) is arguably one of the most important pieces of information when solving unknown chemical structures. This data allows the analysts to connect fragments through scalar coupling. These correlations are observed through heteroatoms as well as quaternary carbons which coupled with chemical shift information and partial structures from your previous analyses allow one to stitch together a proposed structure.

NOESY data (Figure 35) is also very valuable as it allows one to observe the molecule in its spatial environment. To this end, the data can provide supportive information to the structure that has been assembled by the rest of the 1D and 2D data. In addition, the dipolar coupling data can be used to understand regiochemistry as well is some stereochemical issues. In addition, if properly set up the dipolar coupling experiments can allow one to observe chemical exchange processes.

Once all the proton and carbon assignments have been made, one should have a reasonable picture of the actual structure. Recall that we have not directly investigated any nuclei other than protons and carbons, so be sure to keep in mind

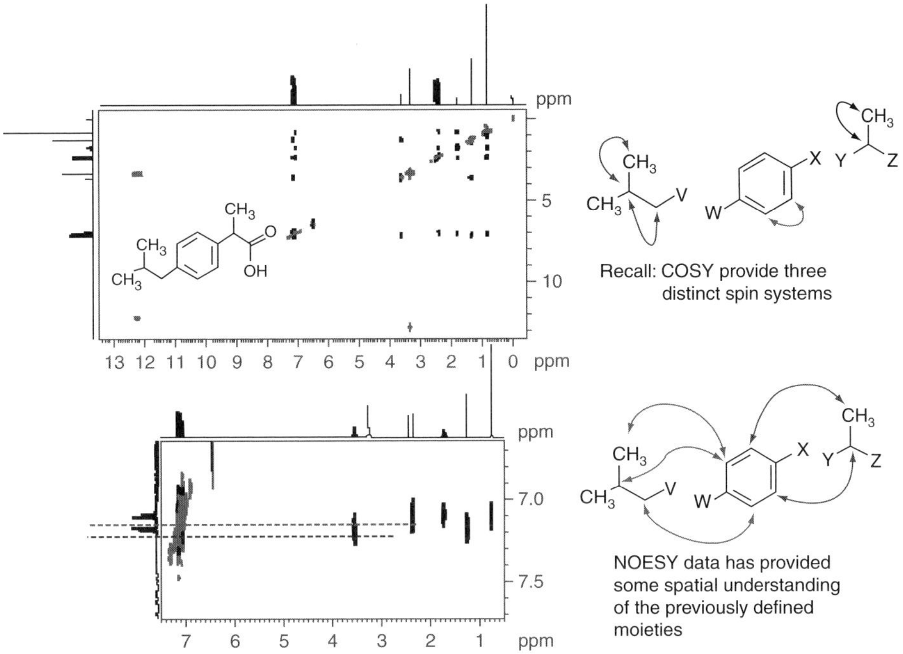

FIGURE 35 **NOESY data for ibuprofen. The partial structures are labeled to indicate the information that can be determined through the COSY data. The arrows show the data that is used to stitch the partial structures together using homonuclear dipolar couplings. Note that this information can also be gleaned through the HMBC data for this particular molecule.**

FIGURE 36 **Overview of the experiments discussed as related to the information that can be gleaned about stitching together atoms to form a molecular structure.**

other possible heteroatoms that may be present. Certainly proton and carbon chemical shifts and correlations (or lack thereof) will provide clues to the presence of other nuclei. Also, as previously discussed, NMR is limited to identifying magnetically inequivalent nuclei.[240] Therefore, the NMR spectra of a particular molecule often will be indistinguishable from the corresponding dimer, trimer, etc. Possible structures will therefore include these equivalent structures as well. We need to rely on molecular mass and other complimentary information to make this distinction. This brings us to the secondary interpretation (Figure 36).

2. Secondary Interpretation

As mentioned, for the structural elucidation of impurities and degradants, it is useful to compare the proton and carbon assignments to those of the parent compound. It will be immediately obvious which peaks have changed or disappeared. This will help to identify the site of degradation or reactivity. The carbon and proton resonances immediately neighboring the oxidation site should shift to reflect the change in the electronic environment. If there is no clear resemblance between the parent and impurity spectra at all, then the impurity may have derived from an alternative source. In this case, one should refer to the list of possible sources previously described to identify a likely candidate. One can then proceed in a similar fashion by comparing the proton and carbon assignments for the impurity to the corresponding chemical shifts for the excipient, precursor, or other identified source. Also bear in mind whether the parent or other source was run as a salt form or a free base, and similarly for the impurity sample, since proton and carbon resonances may arise from the salt. These peaks should be easily identifiable by comparison to proton and carbon chemical shifts for known salts, assuming a protonated salt.

The next step is to compare the data and structural information obtained by NMR to any other information gathered. More specifically, is the structural information obtained by NMR consistent with any proposed structure? If not, identify the particular areas of inconsistency to propose new structures. Even if a structure is identified to be consistent with the NMR data, it is important to consider if any other structures may also be consistent with the data. As mentioned, NMR cannot distinguish magnetically equivalent nuclei, and the characterization strategy outlined here does not explicitly detect nuclei other than carbon and hydrogen. One may also use spectral simulations to lend credibility to a specific structure.[241] However, simulations should not be taken as proof of structure but rather be used to suggest plausible structures. The structure proposed based on the NMR data must be consistent with the total molecular mass, as determined by MS. This is especially important for distinguishing monomers from dimers, trimers, etc.

While the entire structure elucidation team should be communicating throughout the process it is a good idea at this point to convene and discuss the structural information obtained by all of the techniques used in the structural assignment. Specifically, the group should identify areas of consistency or inconsistency of information gathered from other sources compared with the NMR conclusions. For example, the degradation site, or other deviation from known compounds, identified by NMR should be consistent with the location identified by mass spectral fragmentation patterns. It should be noted that there have been some cases in which the fragmentation was inconclusive for determining the reaction site. The proposed structure should also be consistent with any other information available, such as carbonyl stretches observed by IR and relative polarity determined by HPLC. If a structure has passed all the requirements thus far, one needs to determine if it can be derived from a plausible synthetic route. In other words, is it consistent with the reaction pathway and chemical entities present in the reaction vessel? It must be noted that there are plenty of examples where the proposed compound does not have a straight forward synthetic rationale. However, this does not mean that the proposal is not correct, it means that one should be aware of the inconsistency and keep it in mind when compiling the entire data set. If additional structural information is required, it is necessary at this point to identify appropriate follow up experiments to perform. For example, if the stereochemistry at a chiral center is critical to the understanding of the system, additional experiments may be useful to investigate spatial/torsional relationships. Finally, the team should draw conclusions based on the identified structure, including potential toxicity, plausible synthetic routes and ways to prevent its formation.

V. CASE STUDIES

A. A Collaborative Effort

Impurity identification is a highly collaborative effort and every impurity poses its own unique challenges. To quickly and efficiently resolve the identity of an unknown requires a team working closely together. An effective team often includes the isolation, degradation, MS, and NMR labs working together in conjunction with the project chemist and analyst. The case studies discussed in this chapter are meant to highlight the technology, teamwork and thought process that is involved in the identification of an impurity. Many other examples of impurity structure elucidation exist within the literature, some of which are referenced here.[52,54,65,242–251]

B. Solid-State Photodegradation of a Substituted 1,2,4-Triazolo[4,3-a]pyridine

CP-944,629 is a substituted 1,2,4-triazolo[4,3-a]pyridine. This compound was designed as a selective inhibitor of the stress-activated kinase p38α, and was nominated as potential disease-modifying drug for rheumatoid arthritis (RA). Purposeful degradation experiments performed on this compound was analyzed by RP-HPLC with UV detection. Analysis of the solid-state photostability samples indicated a lack of mass balance with an assay value of 25% impurities totaling 18% by total area percent. Closer inspection of the sample revealed undissolved material in the bottom of the volumetric flask. This case study focuses on the identification of the degradation product that resulted in the undissolved material and source of the lack of mass balance.

Analysis of CP-944,629 samples exposed to photodegradation conditions for 1, 3, and 24 h revealed that the discrepancy in mass balance grew with exposure time. CP-944,629 is highly soluble in acetone, which facilitated the isolation of the photodegradation product. Isolation of the degradation product was achieved by stirring the exposed sample in the presence of excess acetone, followed by filtration of the undissolved material. The isolated material displayed no discernable solubility in any of the following solvents: water, acetonitrile, 2-propanol, methanol, acetone, hexane, dichloromethane, DMSO, chloroform, and THF. Many of the traditional impurity identification techniques require that the sample be in solution; therefore, the initial identification of the degradation product was attempted through the use of several solid-state techniques.

IR and Raman spectroscopy revealed that the degradation product retained structural similarities to CP-944,629 but the spectra contained too few significant differences to allow for the proposal of a structure. A powder X-ray diffraction pattern was obtained to determine if the impurity was an observed polymorph of CP-944,629, but the diffraction pattern of the degradation product revealed a unique pattern leaving open the possibility of a new, insoluble solid form of CP-944,629. SSNMR experiments were also performed on both of the know forms of CP-944,629. ^{19}F SSNMR experiments determined that the formation of the degradation product was consistent between forms and also more prevalent and similar, but not identical, to one of the forms. Analysis of the ^{13}C data confirmed the ^{19}F results and revealed the presence of major peaks at 136, 59, and 41 ppm that were not observed in the parent compound (Figure 37). In the absence of additional data, the team was unable to successfully propose a structure using the data obtained from the solid-state techniques.

Eventually, the degradation product was found to have appreciable solubility (ca. $\geq$0.6 mg/mL) in TFA, trifluoroethanol, hydrochloric acid, 1% TFA in acetonitrile, 0.1% $HClO_4$ in acetonitrile, and 0.5% HBF_4 in acetonitrile. Once a suitable dissolving solvent was identified, it was possible to analyze the degradation product by HPLC, HPLC-MS, and NMR. Analysis of the dissolved degradation product by HPLC did not show any new, major peaks in the chromatogram. To this end further method development work was performed and the original assay was modified to achieve separation of the degradation product from the parent compound. This was achieved through the addition of 0.1% $HClO_4$ to the aqueous portion of the mobile phase. The degradation product eluted at a RRT of 1.04. Analysis of the original purposeful degradation 10$\times$ ICH photostability sample showed the presence of the degradation product at a level of 53% by area which corresponded well to the previously observed lack of mass balance.

The chromatographic method was further modified to facilitate the use of MS. This was accomplished by substituting HBF_4 for the perchloric acid in the original assay. To achieve solubility and separation, the concentration of HBF_4 was increased to 0.5% in the aqueous portions of the dissolving solvent and mobile phase. Using this method, both CP-944,629 and it is photodegradation product were analyzed by both MS and MS^n. MS analysis of CP-944,629 showed the expected $[M+H]^+$ of 373.2 with three major daughter ions (208, 290, 373) observed in the MS experiment (Figure 38). Two major ions (745, 373) were observed in the MS of the degradation product. MS/MS of the *m/z* 745 ion resulted in one daughter (373), identical to that of the API. Further fragmentation resulted in three granddaughters (208, 290, 373), identical to the daughters of the parent compound.

The *m/z* of 745 was attributed to a proton-bound dimer formed within the mass spectrometer. The identification of the impurity as a dimer was also discounted because of the RRT of the degradation product with respect to CP-944,629. These peaks initially coeluted, before the addition of perchloric acid or HBF_4 to the mobile phase. This information indicated that the degradation product was not a dimer, which would typically be more nonpolar, resulting in a longer retention time, and have a corresponding *m/z* of $[2M-1+H]^+$ where M is the

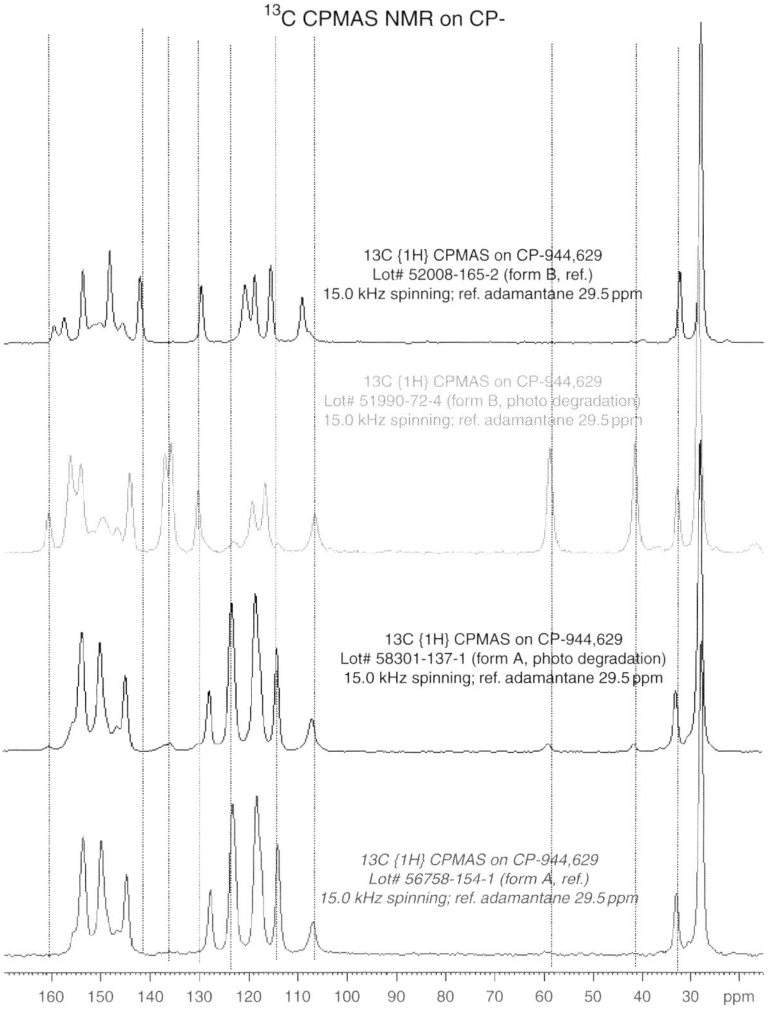

FIGURE 37 Comparison of ^{13}C CPMAS NMR spectra for CP-944,639 and isolated degradation product for both Form A and Form B of the CP-944,629.

molecular weight of CP-944,629. This combination of the chromatographic and mass spectral data led to the conclusion that the degradation product was an isomer of CP-944,629.

1D and 2D solution-state NMR experiments were performed using 99.9% deuterated TFA. Figure 39 shows the proton NMR spectra for both CP-944,629 and the photodegradation product, along with the structure of CP-944,629 with each carbon labeled. Major shifts upfield in the 1D proton and carbon spectra were seen at positions 4, 5, and 10. HSQC (short-range ^{1}H–^{13}C) correlations showed the same number of methines and methyls in both CP-944,629 and the

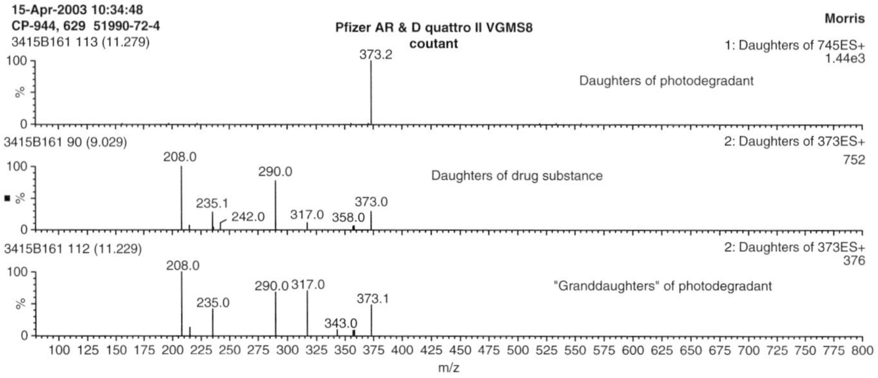

FIGURE 38 **MS and MS/MS of CP-944,629 and its photodegradation product.**

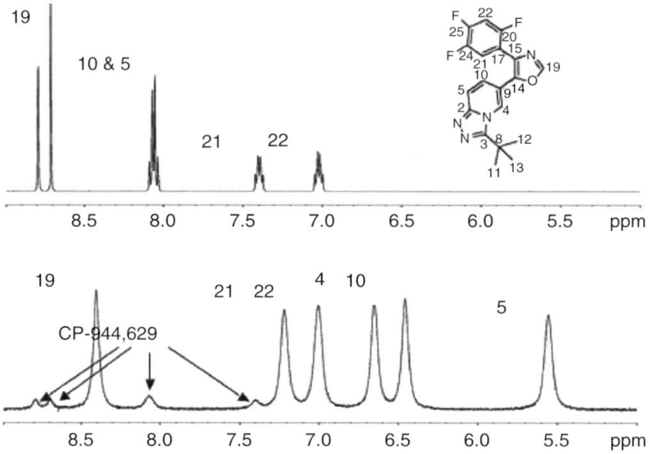

FIGURE 39 **^{1}H NMR spectra for CP-944,629 (top) and the photodegradation product (bottom). Peaks are labeled to correspond to the hydrogens attached to carbons labeled in the CP-944,629 structure (insert).**

degradation product. HMBC (long-range ^{1}H–^{13}C) also showed the same correlations in the parent and the degradation product. The COSY (^{1}H–^{1}H correlations) experiment showed only one difference: correlation of H5–H10 and H4. The MS data and the corresponding assumption that the impurity was an isomer of CP-944,629 was used during the initial evaluation of the NMR data. Using this information several potential structures were proposed, but none of them were reasonable when considering the known degradation pathways.

While the NMR data were being evaluated and structures were being proposed, the preparation of a single crystal from the isolated degradation product was attempted. The growth of a single crystal was eventually achieved using trifluoroacetic acid (TFA) as the crystallization solvent. Using this crystal a structure of the degradation product was generated using X-ray crystallography. Upon

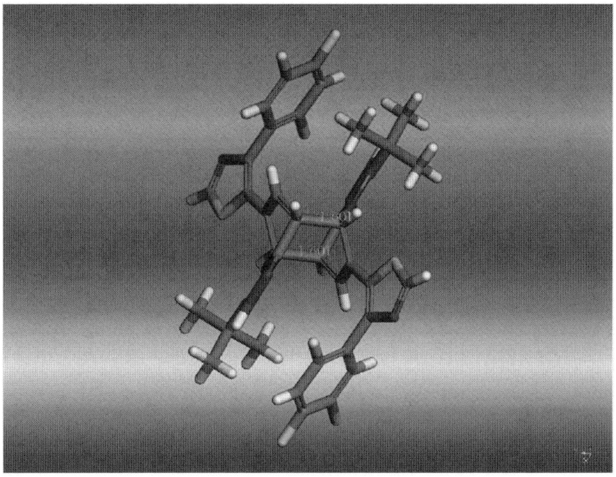

FIGURE 40 Structure of the photodegradation product of CP-944,629—a head-to-tail doubly bound dimer.

analysis of the X-ray data the structure, shown in Figure 40 was determined to be the doubly bound dimer of CP-944,629 with two molecules of TFA incorporated into the lattice. This configuration is consistent with all of the previous analytical data. The mass of this dimer would be exactly twice that of CP-944,629 giving the same molecular ion as a proton-bound dimer. In the NMR data, the COSY data indicated that H5 was correlated to H10 and H4. The correlation to H4 is due to the new face-to-face bond formed between H4 and H5. The structure of the degradation product is also consistent with the results obtained from the analysis of different crystal forms of CP-944,629. The crystal packing of the form that more readily formed the degradation product consists of the parent compound stacked in a head-to-tail orientation that would facilitate the formation of the dimer.

Face-to-face photodimerizations, analogous to those observed with CP-944,629, have been well documented in the literature. A large portion of this work has focused on the study of anthracenes.[252,253] The majority of the anthracene studies were performed in solution, with a much lower yield observed for the solid-state reaction.[254] The resulting dimers displayed poor solubilities with 1 l of benzene necessary to recrystallize ca. 400 mg of dianthracene.[255] The photodimerization of 1,2,4-triazoles was investigated in a series of papers by Potts *et al.*[256–258] The resulting photodimers were described as highly insoluble. Similar behavior was seen in 1,2,4-triazolo[4,3-]quinoline derivatives.

A similar correlation between solid crystal packing and resulting photochemical reactions was documented by Schmidt and coworkers.[259–261] It was demonstrated that the arrangement of molecules within their crystal lattice predicts the photoproduct. This was well illustrated with *trans*-β-2-thienylacrylic acid.[257] This more stable form, referred to as the g-modification, produced a photopolymer upon radiation. The metastable form, known as the b-form, primarily yielded a dimer with the two molecules forming a cyclobutane ring. The differences in the photoproducts were directly correlated with their crystal packing.

The identification of the CP-944,629 photodegradation impurity emphasizes the importance of understanding all of the relevant data and working within a team to elucidate the structure of an unknown impurity. In this case, the initial evaluation of the HPLC and MS data led the team to originally pursue the incorrect assignment of the impurity as an isomer. Combining this information with the NMR data and known degradation pathways pushed the team to continue gathering data and to pursue additional possibilities. In the end a quick and efficient identification of the impurity was achieved with communication and teamwork key to reaching a consensus.

C. Biocatalysis—Identification of Enzyme Reaction By-products

The integrated isolation and identification approach advocated here supported the purification and identification of two major by-products resulting from an aldolase enzymatic process used to catalyze the stereoselective condensation of a functionalized aldehyde with acetaldehyde. SFC and FTICR-MS were two technologies critical to fast turnaround time on this problem and increased confidence on the correctness of the results.

Semipreparative SFC was the isolation tool used for the purification of both impurity A and impurity B as indicated in Figure 41. See Isolation Section of this chapter for an in-depth explanation of SFC and its benefits when using as an isolation tool for structure elucidation purposes. [69,262] The FT-MS is capable of achieving ultra high mass resolution and excellent mass accuracy across a wide mass range.[263–265]

A reaction sample consisting of the desired lactol product and the impurity A by-product was obtained as a sample source for isolation and identification of

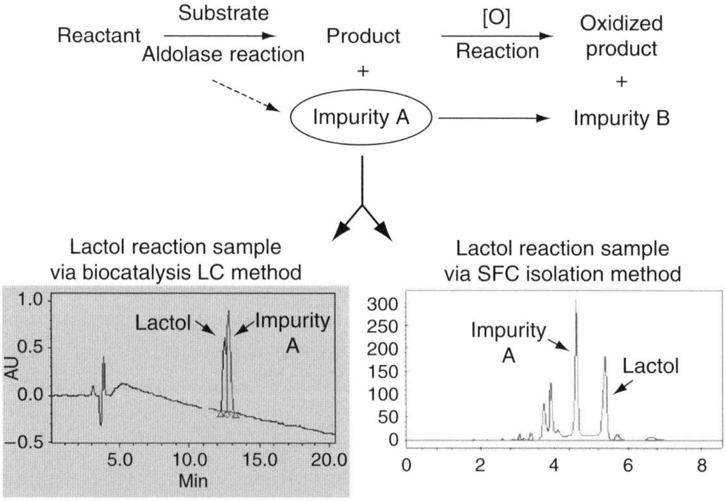

FIGURE 41 **Biosynthetic reaction scheme (top panel) representative reaction sample assayed via the biocatalysis process chemist's analytical RP-LC method (bottom left panel). Representative reaction sample assayed via SFC semipreparative purification method (bottom right panel).**

impurity A. This crude reaction product is carried through an oxidation step *in situ* and as a result a pure lactol standard was not available for MS and NMR spectral comparison studies, therefore, it was isolated from the sample containing impurity A for data reference purposes. However, this oxidation reaction step produces the oxidation product of impurity A, thus named impurity B. Figure 41 shows the reaction scheme and the isolation sample source assayed by both the biocatalysis chemists RP-HPLC analytical method conditions and the SEG semipreparative SFC method conditions used for purification. The RP-HPLC method showed only two peaks in the submitted sample. Of particular note is the complexity of this same sample matrix observed with numerous by-product peaks via semipreparative SFC chromatogram (as shown in Figure 41 bottom right panel) which are not represented in the RP-HPLC chromatogram acquired under the method used. The orthogonal nature of SFC relative to RP-HPLC has been exploited. The SFC chromatogram illustrates speed, efficiency, selectivity impact, and resolving power of this purification method for this sample.

Figure 42 shows stacked injections. In this case, as a result of using stacked injections, the single 8 min injection time observed in Figure 40 is reduced to a 4.5 min injection cycle time allowing for the purification of two targets to be completed in only 47 minutes. An optimized SFC method took approximately 1 h to develop. Optimized SFC method conditions were determined to be 10 mL/min on a 2-ethylpyridine 1 cm × 25 cm column using isocratic methanol cosolvent 40/60, methanol/CO_2 conditions. The outlet pressure was set at 140 bar and the column temperature set at 35 °C. Ten injections were stacked over 47 minutes.

The lactol product target (RT 5.23–5.48 min) peak and impurity A target (RT 4.48–4.70 min) peak were collected via forced time window mode. Fraction dry downtime of MeOH solvent took 1 h with 6 mg of lactol product and 8 mg of impurity A material recovered for NMR analysis. Isolation turnaround time was only 3 h with less than $2 in CO_2 and MeOH solvent costs. Impurity A purity was determined to be approximately 95% and subsequently used as a sample source for FT-MS and NMR. The lactol fraction purity was 74% with a single impurity observed. As a result of the purification process impurity A was observed growing into the lactol sample. Indicating that the compound was chemically labile in the solution conditions used for the isolation. Therefore, a semipreparative NP-LC method using 2-propanol and hexane was developed for the

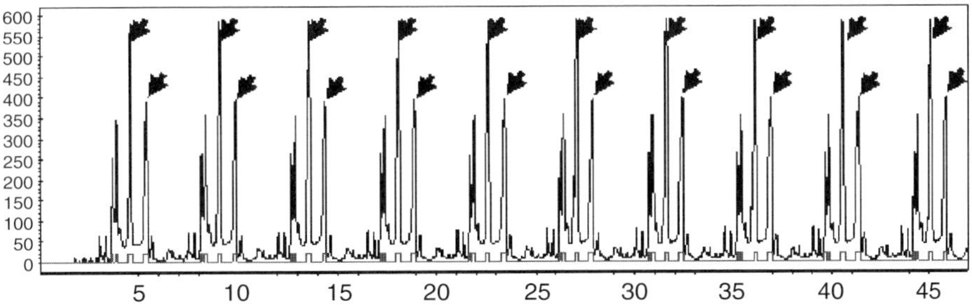

FIGURE 42 Semipreparative SFC stacked chromatogram of biocatalysis reaction sample source. Target impurity A is labeled with topmost arrow and lactol product with lower arrow.

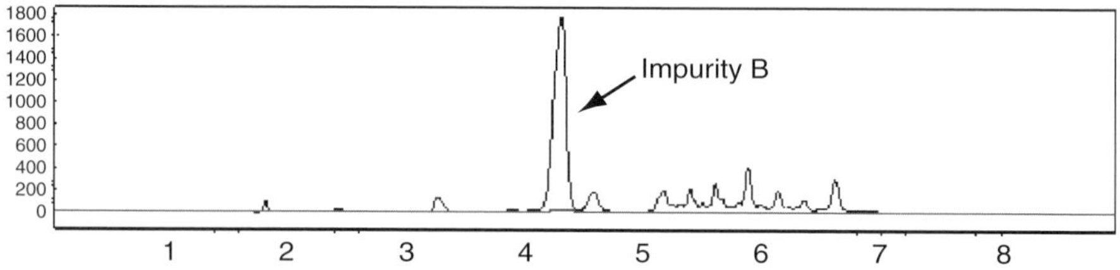

FIGURE 43 **Representative semipreparative SFC chromatogram of biocatalysis-enriched impurity B reaction sample.**

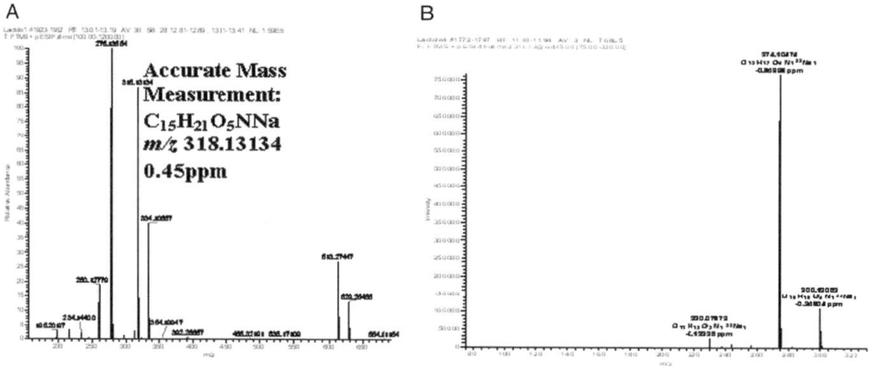

FIGURE 44 **FT-MS spectra of lactol. Panel A shows the full scan data with the $[M+Na]^+$ at *m/z* 318, $[M+K]^+$ at *m/z* 334, $[2M+Na]^+$ at *m/z* 613, and $[2M+K]^+$ at *m/z* 629. Panel B shows the collision-induced dissociation (CID) spectrum for the $[M+Na]^+$ *m/z* 318 ion, producing ions suggesting the loss of water (at *m/z* 300), followed by fragmentation of the lactol ring at *m/z* 274 and *m/z* 230.**

removal of impurity A present in the lactol sample. A 92% pure lactol sample resulted, which was analyzed via NMR.

The biocatalysis group submitted an Impurity B sample source which was purified via a very similar SFC method with only small changes in isocratic CO_2/MeOH mobile percents made. Seven injections were performed over 49 min with 3 mg of purified impurity B recovered. Total time spent on this purification including method development, fractionation collection, and dry downtime was 3 h. Total CO_2 and MeOH solvent costs were approximately $2. See Figure 43 for a representative SFC chromatogram. A purified impurity B sample was submitted for FT-MS and NMR analysis. An important point to make is that the amount of sample that was isolated was far greater than the amount needed to obtain high-quality NMR data. However, the speed and efficiency observed in the isolation allowed the isolation of an abundance of both target impurities. This allows greater flexibility and timely NMR data acquisition.

The mass spectrum for the lactol does not exhibit a characteristic $[M+H]^+$ ion (*m/z* 296, Figure 44). However, the spontaneous dehydration of the protonated lactol is noted at *m/z* 278. The molecular mass may be deduced by the adduct ions present in the spectrum. The most prominent is the $[M+Na]^+$ observed at *m/z* 318.13134, compared to the theoretical $[M+Na]^+$ of 318.13119 (0.45 ppm difference). Further evidence of the molecular mass is

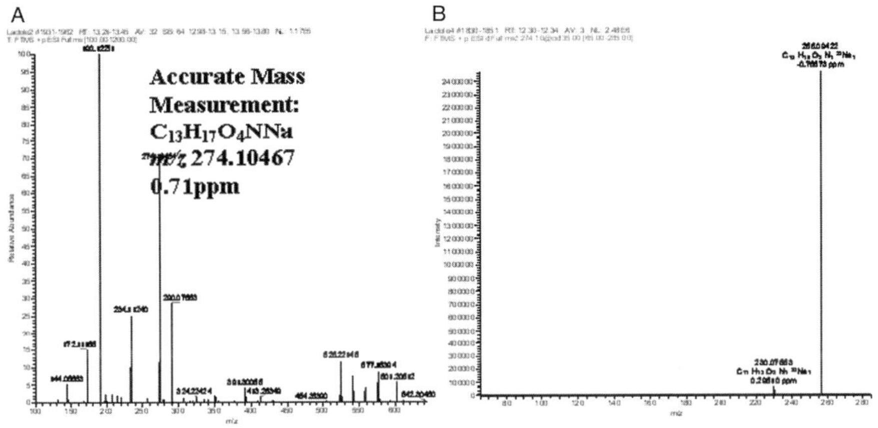

FIGURE 45 FT-MS spectra of impurity A. Panel A shows the full scan data with the $[M+Na]^+$ at *m/z* 274, $[M+K]^+$ at *m/z* 290, $[2M+Na]^+$ at *m/z* 525, and $[2M+K]^+$ at *m/z* 541. Panel B shows the collision-induced dissociation (CID) spectrum for the *m/z* 274 molecular ion produces fragments indicating the loss of water at *m/z* 256 and the fragmentation of the piperidine ring at *m/z* 230.

shown by the $[M+K]^+$ at *m/z* 334, $[2M+Na]^+$ at *m/z* 613, and $[2M+K]^+$ at *m/z* 629. The collision-induced dissociation (CID) spectrum for the $[M+Na]^+$ 318 ion produces ions suggesting loss of water at *m/z* 300, followed by fragmentation of the lactol ring at *m/z* 274 and 230.

Interpretation of the mass spectrum for impurity A can be approached in a fashion analogous to that employed for the lactol. Once again, an $[M+H]^+$ is not observed. However, the dehydrated $[M+H]^+$ is seen at *m/z* 234 (Figure 45). A prominent $[M+Na]^+$ is present at *m/z* 274.11240 which, compared to the theoretical $[M+Na]^+$ exact mass of 274.10498 (0.71 ppm). Further evidence of the molecular mass is provided by the $[M+K]^+$ at *m/z* 290, $[2M+Na]^+$ at *m/z* 525, and $[2M+K]^+$ at *m/z* 541. The CID spectrum for the *m/z* 274 molecular ion produces fragments indicating the loss of water (*m/z* 256) and the fragmentation of the piperidine ring at *m/z* 230.

The mass spectrum for impurity B produces the $[M+Na]^+$ at *m/z* 272.08896, compared with the theoretical $[M+Na]^+$ exact mass of 272.08933 (0.9 ppm, Figure 46). The $[M+K]^+$ at *m/z* 288, $[2M+Na]^+$ at *m/z* 521, and $[2M+K]^+$ at *m/z* 537 are also present. In contrast to the lactol and impurity A, a prominent $[M+H]^+$ is present at *m/z* 250.10713, consistent with a theoretical $[M+H]^+$ of 250.10738 (1.0 ppm). Fragmentation of the *m/z* 250 $[M+H]^+$ (spectrum not shown) produces fragments indicative of the loss of CO_2 from the lactam ring along with the subsequent loss of water.

NMR experiments were performed to allow the connectivities to be determined for all of the isolates. These isolates include the lactol, impurity A, and impurity B. As seen with the MS data and the NMR data in Figure 47, it is clear that all of the compounds are structurally related. Extensive 2D data was acquired to stitch together the atoms into the final structure.

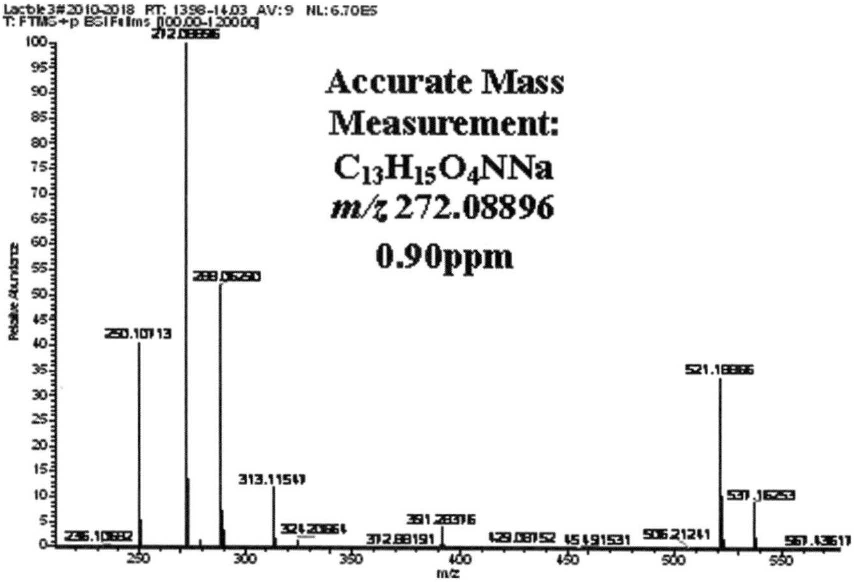

FIGURE 46 FT-MS spectra of impurity **B**. The full scan data showing the $[M+H]^+$ at *m/z* 250, $[M+Na]^+$ at *m/z* 272, $[M+K]^+$ at *m/z* 288, $[2M+Na]^+$ at *m/z* 521, and $[2M+K]^+$ at *m/z* 537.

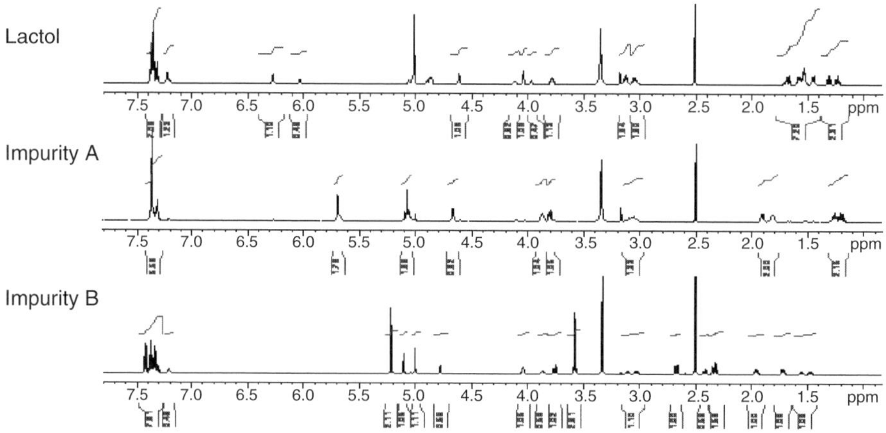

FIGURE 47 Proton NMR spectra of the desired lactol (top trace), impurity **A** (middle panel), and impurity **B** (top panel).

All 1D and 2D data were collected at 298 K using a Bruker-Biospin 5-mm gradient TCI cryoprobe on a Bruker-Biospin AV NMR spectrometer operating at 700 MHz. Because of the large amount of material isolated and the quality of the isolates, data were acquired in approximately 3 h per compound. This allowed extremely rapid turnaround for determination of the structures. The following data were collected: 1D proton, 1D carbon, $^1H–^1H$ COSY, $^1H–^{13}C$ multiplicity

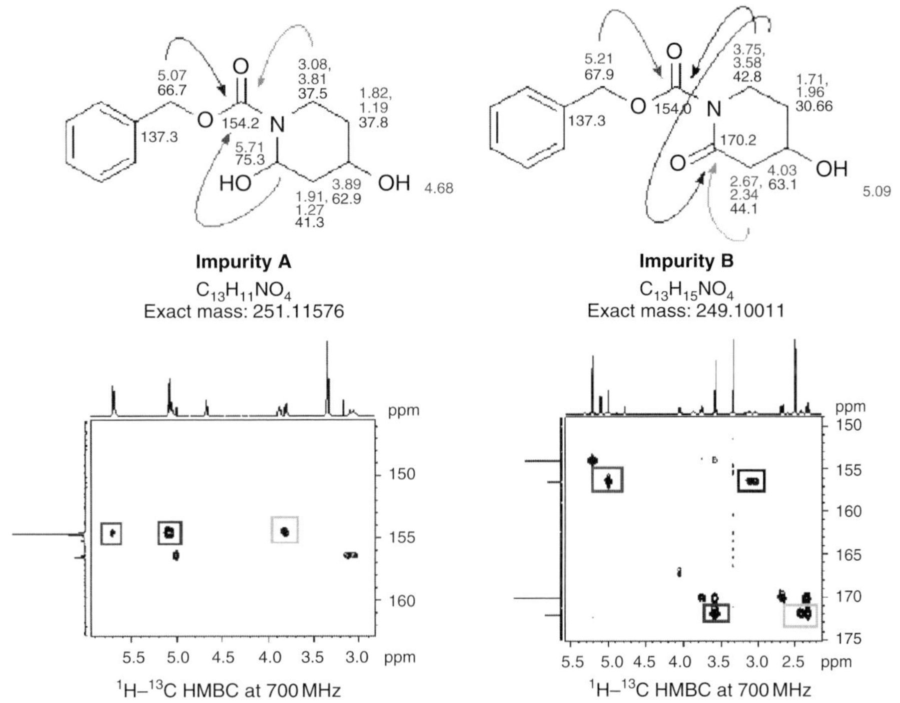

FIGURE 48 **The proposed structures and chemical shift assignments for the two elucidated impurities. The arrows on the structures indicate the key HMBC correlations observed for these molecules. Below are two expansions of the HMBC data for each of the impurities, respectively, from left to right. The colors are meant to quickly indicate the actual correlation in the 2D NMR plot of the HMBC data. (See Color Insert.)**

edited HSQC, ^{1}H–^{1}H NOESY (*n*uclear *O*verhauser and *e*xchange *s*pectroscop*y*), and ^{1}H–^{13}C HMBC.

The key correlations that were used to define the structures are shown in Figure 48. The two spectra in Figure 48 show an expansion of the ^{1}H–^{13}C HMBC data. This experiment allows the detection of protons that are long-range bound to ^{13}C atoms. This experiment is often the key experiment in putting together partial structures of homonuclear spin systems. The colored arrows drawn on the assigned structures in Figure 48 correlate to the same colored boxes placed around the 2D correlation in the spectra shown. These data confirm the connectivities of the structures for impurity A and impurity B.

VI. SUMMARY

The objective of the chapter is to give guidance with degradation and impurity structure elucidation for pharmaceutical drug candidates. Stress testing knowledge is provided for guidance in developing and executing forced degradation experiments. Impurity and degradant structure elucidation is a collaborative effort involving the degradation chemist, analytical chemist, process chemist, and/or formulator as well as the isolation chemist, MS and NMR experts. Updates in this second addition include: (1) QbD model as applied to stress testing practices, expanded oxidative experimental analysis and expanded information on predictive programs. (2) The impact SFC as a tool for impurity isolations along with expanded information on process workflow. (3) Mass spectrometry; additional examples have been provided as well as a section on the use of accurate mass information in the role of structure elucidation. (4) An expansion on the use of NMR as applied to structure characterization and elucidation is included. (5) In depth case studies illustrate collaborative efforts made between disciplines as well as the impact new technologies have on the structure elucidation process.

REFERENCES

1. ICH Stability Testing of New Drug Substances and Products: Q1A(R2), Published in the *Federal Register*, Vol. 68, No. 225, Friday, November 21, 2003.
2. Reynolds, D. W., Facchine, K. L., Mullaney, J. F., Alsante, K. M., Hatajik, T. D. and Motto, M. G. Available guidance and best practices for conducting forced degradation studies. *Pharm. Technol.* 26:48–54, 2002.
3. Baertschi, S. W. and Reynolds, D. W. Chapter 1: Stress Testing: Introduction. In *Pharmaceutical Stress Testing: Predicting Drug Degradation*, (S. W. Baertschi, ed.), Taylor & Francis Group, Boca Raton, FL, pp. 1–12, 2005.
4. FDA Guidance for industry INDs for phase 2 and phase 3 studies: chemistry, manufacturing and controls information. *Federal Register* 68, 27567–27568 (issued May 2003).
5. FDA, Submitting Documentation for the Stability of Human Drugs and Biologics, CDER, February 1987.
6. ICH Stability Testing of New Drug Substances and Products, Published in the *Federal Register*, Vol, 68, No. 225, pages 65717–18, 21, November 2003.
7. Baertschi, S. W., Olsen, B. A. and Alsante, K. M. Chapter 5: Stress Testing: Relation to the Drug Development Timeline. In *Pharmaceutical Stress Testing: Predicting Drug Degradation*, (S. W. Baertschi, ed.), Taylor & Francis Group, Boca Raton, FL, pp. 173–180, 2005.
8. ICH Harmonized Tripartite Guideline Pharmaceutical development Q8(R1) published in the *Federal Register*, Vol. 71, No. 98, May 2006.
9. Baertschi, S. W. Forced degradation and its relation to real time drug product stability. In *Pharmaceutical Stability Testing to Support Global Markets*, (K. Huyn-ba, ed.), Springer, New York, NY, 2009.
10. ICH Q9: Quality Risk Management. Published in the *Federal Register*, Vol. 71, No 106, pages 32105–32106, 2 June 2006.
11. Jorgensen, W. L., Laird, E. R., Gushurst, A. J., Fleischer, J. M., Gothe, S. A., Helson, H. E., Paderes, G. D. and Sinclair, Shenna CAMEO: A program for the logical prediction of the products of organic reactions. *Pure Appl. Chem.* 62:1921–1932, 1990.
12. Pole, D. L., Ando, H. Y. and Murphy, S. T. Prediction of drug degradants using DELPHI: An expert system for focusing knowledge. *Mol. Pharm.* 4:539–549, 2007.
13. Corey, E. J., Long, A. K. and Rubenstein, S. D. Computer-assisted analysis in organic synthesis. *Science* 228:408–418, 1985.
14. Marchant, C. A., Briggs, K. A. and Long, A. In silico tools for sharing data and knowledge on toxicity and metabolism: Derek for windows, meteor, and vitic. *Toxicol. Mech. Methods* 18:177–187, 2008.
15. Alsante, K. M., Friedmann, R. C., Hatajik, T. D., Lohr, L. L., Sharp, T. R., Snyder, K. D. and Szczesny, E. J. Chapter 4: Degradation and impurity analysis for pharmaceutical drug candidates.

In *Handbook of Modern Pharmaceutical Analysis*, (S. Ahuja and S. Scypinski, eds.), Academic Press, Boston, MA, pp. 85–172, 2001.

16. Alsante, K. M., Ando, A., Brown, R., Ensing, J., Hatajik, T. D., Kong, W. and Tsuda, Y. The role of degradant profiling in active pharmaceutical ingredients and drug products. *Adv. Drug Deliv. Rev.* 59:29–37, 2007.
17. Waterman, K. C., Adami, R. C., Alsante, K. M., Antipas, A. S., Arenson, D. R., Carrier, R., Hong, J., Landis, M. S., Lombardo, F., Shah, J. C., Smith, S. W. and Wang, H. Hydrolysis in pharmaceutical formulations. *Pharm. Dev. Technol.* 7:113–146, 2002.
18. Dan W. Reynolds (2007). IIR Forced Degradation Pre-Conference Workshop. An Experimental Approach for Conducting Forced Degradation Studies, 4th Annual Master the Science & Strategy Behind FORCED DEGRADATION STUDIES, January 22, 2007.
19. Reynolds, D. W. Forced degradation of pharmaceuticals. *Am. Pharm. Rev.* 7:56–61, 2004.
20. Boccard, G. Chapter 7: Oxidative susceptibility testing. In *Pharmaceutical Stress Testing: Predicting Drug Degradation*, (S. W. Baertschi, ed.), Taylor & Francis Group, Boca Raton, FL, pp. 205–234, 2005.
21. Baertschi, S. W. and Jansen, P. J. Chapter 2: Stress testing: A predictive tool. In *Pharmaceutical Stress Testing: Predicting Drug Degradation*, (S. W. Baertschi, ed.), Taylor & Francis Group, Boca Raton, FL, pp. 13–50, 2005.
22. Nelson, E. D., Harmon, P. A., Szymanik, R. C., Teresk, M. G., Li, L., Seburg, R. A. and Reed, R. A. Evaluation of solution oxygenation requirements for azonitrile-based oxidative forced degradation studies of pharmaceutical compounds. *J. Pharm. Sci.* 95(7):1527–1539, 2006.
23. Baertschi, S. W., Olsen, B. A. and Alsante, K. M. Chapter 7: Stress testing: Oxidative susceptibility testing. In *Pharmaceutical Stress Testing: Predicting Drug Degradation*, (S. W. Baertschi, ed.), Taylor & Francis Group, Boca Raton, FL, p. 214, 2005.
24. Wasylaschuk, W. R., Harmon, P. A., Wagner, G., Harman, A. B., Templeton, A. C., Xu, H. and Reed, R. A. Evaluation of hydroperoxides in common pharmaceutical excipients. *J. Pharm. Sci.* 96:106–116, 2007.
25. Artauer, K. J., Arbuthnot, G. N., Baertschi, S. W., Johnson, R. A., Luke, W. D., Pearson, N. G., Rickard, E. C., Tingle, C. A., Tsang, P. K. S. and Wiens, R. E. Influence of peroxide impurities in povidone and crospovidone on the stability of raloxifene hydrochloride in tablets: Identification and control of an oxidative degradation product. *Pharm. Dev. Technol.* 5:303–310, 2000.
26. Waterman, K. C., Adami, R. C., Alsante, K. M., Hong, J., Landis, M. S., Lombardo, F. and Roberts, C. J. Stabilization of pharmaceuticals to oxidative degradation. *Pharm. Dev. Technol* 7(1):1–32, 2002.
27. Abendroth, R. P. Selecting light-resistant containers. *ChromConnection.* July:4–6, 1994.
28. Greenspan, L. Humidity fixed points of binary saturated aqueous solutions. *J. Res. National Bureau of Standards—Phys. Chem.* 81A:89–96, 1977.
29. Kennon, L. Use of models in determining chemical pharmaceutical stability. *J. Pharm. Sci.* 53:815–818, 1964.
30. Waterman, K. C. Understanding and predicting pharmaceutical product shelf-life. In *Handbook of Stability Testing in Pharmaceutical Development: Regulations, Methodologies, and Best Practices* (K. Huyn-Ba, ed.), New York, NY, pp. 115–138, 2009.
31. Carstensen, J. T. Drug Stability Principles and Practices. 2nd ed., Marcel Dekker, Inc., New York, 1995.
32. FDA (May 16, 1997). Guideline for the photostability testing of new drug substances and new drug products. *Federal Register* 62(95), 27115–27122.
33. Tonnesen, H. H., ed., *In* Photostability of Drugs and Drug Formulations. 2nd ed., CRC Press, Boca Raton, FL, 2004.
34. Albini, A. and Fasani, E., eds., *In* Drugs: Photochemistry and Photostability. The Royal Society of Chemistry, Cambridge, UK, 1998.
35. Thatcher, S. R., Mansfield, R. K., Miller, R. B., Davis, C. W. and Baertschi, S. W. Pharmaceutical photostability: A technical and practical interpretation of the ICH guideline and its application to pharmaceutical stability: Part II. *Pharm. Technol.* 25(30):98–110, 2001.
36. Thatcher, S. R., Mansfield, R. K., Miller, R. B., Davis, C. W. and Baertschi, S. W. Pharmaceutical photostability: A technical and practical interpretation of the ICH guideline and its application to pharmaceutical stability: Part I. *Pharm. Technol.* 25(4):50–62, 2001.

37. Moore, D. E. Principles and practice of drug photodegradation studies. *J. Pharm. Biomed. Anal.* 5:441–453, 1987.
38. Allen, J. M. *Basic Principles of Drug Photostability Testing*, Washington, DC, 1999.
39. ICH Stability Testing: Photostability Testing of New Drug Substances and Products, Published in the *Federal Register*, Vol. 62, No. 95, pages 27115–27122, 16 May 1997.
40. Tonnesen, H. H. *Pharmazie* 46:263–265, 1991.
41. (ISO), 1993, First Edition.
42. Allen, J. M. *Principles of Actinometry*, Washington, DC, 1999.
43. Santafianos, D. *Identification and Mechanistic Studies of Photodegradants: Case Study 1 & 2*, Pharmaceutical Photostability 2004, London, England.
44. Baertschi, S. W. and Thatcher, S. R. Sample presentation for photostability studies: Problems and solutions. In *Pharmaceutical Photostability and Stabilization Technology*, (J. Piechocki, ed.), Taylor & Francis, New York, 2006.
45. Nussbaum, M. A., Jansen, P. J. and Baertschi, S. W. Chapter 6: Role of "mass balance" in pharmaceutical stress testing. In *Pharmaceutical Stress Testing: Understanding Drug Degradation*, (S. W. Baertschi, ed.), Taylor and Francis Group, New York, pp. 181–204, 2005.
46. Clarke, H. J. and Norris, K. J. Chapter 7: Sample selection for analytical method development. In *Handbook of Isolation and Characterization of Impurities in Pharmaceuticals*, (S. Ahuja and K. M. Alsante, eds.), Academic Press, New York, pp. 145–163, 2003.
47. FDA. Q6A International Conference on Harmonization, Guidance on Q6A Specifications: Test Procedures and Acceptance Criteria for New Drug Substances and New Drug Products: Chemical Substances, December 2000.
48. Byrn, S. J., Pfeiffer, R. R. and Stowell, J. G. Theoretical analysis of solid-state chemical reactions. In *Solid State Chemistry of Drugs*. (S. J. Byrn, R. R. Pfeiffer and J. G. Stowell, eds.), 2nd ed., SSCI, Inc., West Lafayette, Indiana, p. 307, 1999.
49. Byrn, S. J., Pfeiffer, R., Ganey, M., Hoiberg, C. and Poochikian, G. Pharmaceutical solids: a strategic approach to regulatory considerations. *Pharm. Res.* 12:945–954, 1995.
50. Alsante, K. M., Hatajik, T. D., Lohr, L. L., Santafianos, D. P. and Sharp, T. R. Chapter 14: Solving impurity/degradation problems: Case studies. In *Handbook of Isolation and Characterization of Impurities in Pharmaceuticals*, (S. Ahuja and K. M. Alsante, eds.), Academic Press (Elsevier), New York, pp. 361–400, 2003.
51. Alsante, K. M., Snyder, K. D., Swartz, M. and Parks, C. Applications development using out-of-the-box software: A structure searchable degradation/impurity database. *Scientific Computing and Instrumentation*, June, 30–37, 2002.
52. Alsante, K. M., Boutros, P., Couturier, M. A., Friedmann, R. C., Harwood, J. W., Horan, G. J., Jensen, A. J., Liu, Q., Lohr, L. L., Morris, R., Raggon, J. W. and Reid, G. L., *et al.* Pharmaceutical impurity identification: A case study using a multidisciplinary approach. *J. Pharm. Sci.* 93:2296–2309, 2004.
53. Biba, M., Welch, C. J., da Silva, J. O. and Mathre, D. J. Impurity isolation in support of pharmaceutical process research. *Am. Pharm. Rev.* 8:68–84, 2005.
54. Ahuja, Satinder and Alsante, Karen M., eds., *In* Handbook of Isolation and Characterization of Impurities in Pharmaceuticals. Elsevier Publishing, Amsterdam, Netherlands, p. 414, 2003.
55. Alsante, K. M., Hatajik, T. D., Lohr, L. L., Santafianos, D. P. and Sharp, T. R. Abstracts of Papers, 224th ACS National Meeting, Boston, MA, United States, August 18–22, 2002.
56. Alsante, K. M., Hatajik, T. D., Lohr, L. L. and Sharp, T. R. Isolation and identification of process related impurities and degradation products from pharmaceutical drug candidates, Part I. *Am. Pharm. Rev.* 4:70–78, 2001.
57. Lohr, L. L., Sharp, T. R., Alsante, K. M. and Hatajik, T. D. Isolation and identification of process related impurities and degradation products from pharmaceutical drug candidates: Part II. The roles of NMR and mass spectrometry. *Am. Pharm. Rev.* 4:104–113, 2001.
58. Sahasrabuddhey, B., Nautiyal, R., Acharya, H., Khyade, S., Luthra, P. K. and Deshpande, P. B. Isolation and characterization of some potential impurities in ropinirole hydrochloride. *J. Pharm. Biomed. Anal.* 43:1587–1593, 2007.
59. Kumar, D., Tomar, R. S., Deolia, S. K., Mitra, M., Mukherjee, R. and Burman, A. C. Isolation and characterization of degradation impurities in docetaxel drug substance and its formulation. *J. Pharm. Biomed. Anal.* 43:1228–1235, 2007.

60. Cummings, P. G., Offen, P., Olsen, M. A., Kennedy-Gabb, S. and Zuber, G. LC/MS, LC/NMR, FTIR: an integrated approach to impurity identification in pharmaceutical development and formulation. *Am. Pharm. Rev.* 6:88–92, 2003.
61. International Conference on Harmonization Steering Committee. Q3A(R2): Impurities in New Drug Substances (Revised Guideline), *Federal Register*, Vol, 68, No. 68, p. 6924–6925, 2003.
62. International Conference on Harmonization Steering Committee.Q3B(R2): Impurities in New Drug Products (Revised Guideline), Federal Register, Vol, 68, No. 220, p. 64628–64629, 2003.
63. Sigvardson, K. W., Adams, S. P., Bradford Barnes, T., Blom, K. F., Fortunak, J. M., Haas, M. J., Reilly, K. L., Repta, A. J. and Nemeth, G. A. The isolation and identification of a toxic impurity in XP315 drug substance. *J. Pharm. Biomed. Anal.* 27:327–334, 2002.
64. Chen, H., Wang, Y.-F., Yang, Z.-D. and Li, Y.-C. Isolation and identification of novel impurities in spironolactone. *J. Pharm. Biomed. Anal.* 40:1263–1267, 2006.
65. Landis, M. S. The use of mixed-mode ion-exchange solid phase extraction to characterize pharmaceutical drug degradation. *J. Pharm. Biomed. Anal.* 44:1029–1039, 2007.
66. Porsch, B. Some specific problems in the practice of preparative high-performance liquid chromatography. *J. Chromatogr.* 658:179–194, 1994.
67. Seger, C., Godejohann, M., Tseng, L.-H., Spraul, M., Girtler, A., Sturm, S. and Stuppner, H. LC-DAD-MS/SPE-NMR hyphenation. A tool for the analysis of pharmaceutically used plant extracts: Identification of isobaric iridoid glycoside regioisomers from *Harpagophytum procumbens*. *Anal. Chem.* 77:878–885, 2005.
68. Nyberg, N. T., Baumann, H. and Kenne, L. Solid-phase extraction NMR studies of chromatographic fractions of saponins from *Quillaja saponaria*. *Anal. Chem.* 75:268–274, 2003.
69. Terfloth, G. Preparative isolation of impurities. *Analysis of Drug Impurities* 215–234, 2007.
70. Aranyi, A. Isolation of impurities by (semi)preparative HPLC. *Progress in Pharmaceutical and Biomedical Analysis* 240–251, 2000.
71. Ashok, C., Golak, M., Adwait, D., Krishna, V., Himani, A., Umesh, P., Amol, G., Srinivas, S., Sharad, M., Deepali, J. and Atul, C. Isolation and structural elucidation of two impurities from a diacerein bulk drug. *J. Pharm. Biomed. Anal.* 49:525–528, 2009.
72. Kumar, D., Tomar, R. S., Deolia, S. K., Srivastava, R., Mitra, M. and Tyagi, S. Isolation and characterization of degradation impurities in epirubicin hydrochloride injection. *J. Chromatogr. B* 869:45–53, 2008.
73. Xia, F., Mallet, C. and Diehl, D. M. LCGC North America 74–75, 2005.
74. Knox, J. H. and Pyper, H. M. Framework for maximizing throughput in preparative liquid chromatography. *J. Chromatogr.* 363:1–30, 1986.
75. Waters Corporation. Waters Preparative Liquid Chromatography Catalog, 1998
76. Goetzinger, W., Zhang, X., Bi, G., Towle, M., Cherrak, D. and Kyranos, J. N. High throughput HPLC/MS purification in support of drug discovery. *Int. J. Mass Spectrom.* 238:153–162, 2004.
77. McClain, R. T., Dudkina, A., Barrow, J., Hartman, G. and Welch, C. J. Evaluation and implementation of a commercially available mass-guided SFC purification platform in a high throughput purification laboratory in drug discovery. *J. Liq. Chromatogr. Relat. Technol.* 32:483–499, 2009.
78. Kagan, M., Chlenov, M., Melnikov, S., McConnell, O., Bach, A.C., Carter, G., Failli, A., Caggiano, T.J., Shumsky, J. S. and Lubda, D. Normal-phase automated mass-directed HPLC purification of a pyrrolobenzodiazepine library with vasopressin agonist activity. *Journal of Combinatorial Chemistry* 11(4):704–719, 2009.
79. Kochana, J., Wilamowski, J. and Parczewski, A. TLC profiling of impurities of 1-(3, 4-methylenedioxyphenyl)-2-nitropropene an intermediate in MDMA synthesis. Influence of sample preparation methods and conditions. *J. Liq. Chromatogr. Relat. Technol.* 27:2463–2470, 2004.
80. Lui, Jane S. S. L., Dasgupta, Tara P., Maragh, Paul T., Beckford, F. A. Mechanistic studies of reactions of [Fe2(CN)10]4- with sulfite ions in slightly acidic aqueous solutions. Abstracts of Papers, 222nd ACS National Meeting, Chicago, IL, United States, August 26–30, 2001.
81. Clark Still, W., Kahn, M. and Mitra, A. Rapid chromatographic technique for preparative separations with moderate resolution. *J. Org. Chem.* 43:2923, 1978.
82. Zelesky, T. C. Supercritical fluid chromatography (SFC) as an isolation tool for the identification of drug related impurities. *Am. Pharm. Rev.* 11:56–62, 2008.
83. Miller, L. and Potter, M. Preparative supercritical fluid chromatography (SFC) in drug discovery. *Am. Pharm. Rev.* 11:112–117, 2008.

84. Yan, T. Q., Bradow, J., Chang, S.-P., Depianta, R. and Philippe, L. *LCGC North America* 27:340–345, 2009.
85. Kraml, C. Pushing the limits of SFC to resolve chiral molecules. *Am. Pharm. Rev.* 11:80–83, 2008.
86. McClain, R. T., Dudkina, A., Barrow, J., Hartman, G. and Welch, C. J. Evaluation and implementation of a commercially available mass-guided SFC purification platform in a high throughput purification laboratory in drug discovery. *J. Liq. Chromatogr. Relat. Technol.* 32:483–499, 2009.
87. Cretier, G., Neffati, J. and Rocca, J. Preparative LC and preparative SFC: Two complementary techniques in the fractionation of an impurity from a major component. *J. Chromatogr. Sci.* 32:449–454, 1994.
88. Zelesky, T. C., Riley, F. and Marquez, B. L. *234th ACS National Meeting*, Boston, MA, USA, August 19–23, 2007.
89. Riley, F., Zelesky, T. C., Marquez, B. L., and Brunelli, C. *2nd International Conference on SF*, Zurich, Switzerland, September 30th to October 1st, 2008.
90. Miller, L. and Potter, M. Preparative chromatographic resolution of racemates using HPLC and SFC in a pharmaceutical discovery environment. *J. Chromatogr. B* 875:230–236, 2008.
91. Wu, D.-R., Leith, L., Balasubramanian, B., Palcic, T. and Wang-Iverson, D. *Am. Lab.* 38:24–26, 2006.
92. Van Anda, J. The use of supercritical fluid chromatography (SFC) for chiral pharmaceutical analytical and preparative separations. *Am. Pharm. Rev.* 12:48–53, 2009.
93. Berger, T. A. Packed column SFC. In R. Smith (ed.), Cambridge, UK: The Royal Society of Chemistry.
94. Thomson, J. J. XIX. Further experiments on positive rays. Philosophical Magazine 24:209–253, 1912.
95. Harrison, A. G. *Chemical Ionization Mass Spectrometry*, 2nd ed., CRC Press, Boca Raton, FL, 1992. p. 224.
96. Wood, G. W. Field desorption mass spectrometry: Applications. *Mass Spectrom. Rev.* 1:63–102, 1982.
97. Barber, M., Bordoli, R. S., Donald Sedgwick, R. and Tyler, A. N. Fast atom bombardment of solids (F.A.B.): A new ion source for mass spectrometry. *J. Chem. Soc. Chem. Commun.* 325–326, 1981.
98. Surman, D. J. and Vickerman, J. C. Fast atom bombardment quadrupole mass spectrometry. *J. Chem. Soc. Chem. Commun.* 324–325, 1981.
99. Vestal, M. L. Ionization techniques for nonvolatile molecules. *Mass Spectrom. Rev.* 2:447–480, 1983.
100. Carroll, D. I., Dzidic, I., Stillwell, R. N., Haegele, K. D. and Horning, E. C. Atmospheric pressure ionization mass spectrometry: Corona discharge ion source for use in liquid chromatograph-mass spectrometer-computer analytical system. *Anal. Chem.* 47:2369–2373, 1975.
101. Yamashita, M. and Fenn, J. B. Electrospray ion source: Another variation on the free-jet theme. *J. Phys. Chem.* 88:4452–4459, 1984.
102. Hillenkamp, F. and Karas, M. Mass spectrometry of peptides and proteins by matrix-assisted ultraviolet laser desorption ionization. *Meth. Enzymol.* 193:280–294, 1990.
103. Marshall, A. G. and Hendrickson, C. L. High resolution mass spectrometers. *Annu. Rev. Anal. Chem.* 1:579–599, 2008.
104. McLafferty, F. W. Tandem Mass Spectrometry. Wiley Intersciences, New York, 1983.
105. Wysocki, V. H., Jones, C. M., Galhena, A. S. and Blackwell, A. E. Surface-induced dissociation shows potential to be more informative than collision-induced dissociation for structural studies of large stems. *J. Am. Soc. Mass Spectrom.* 19:903–913, 2008.
106. Benesch, J. L. Collisional activation of protein complexes: Picking up the pieces. *J. Am. Soc. Mass Spectrom.* 19: 2008.
107. Gross, J. H. Mass Spectrometry: A Textbook. Springer-Verlag, Berlin, 2004. p. 518.
108. deHoffmann, E. and Stroobant, V. *Mass Spectrometry: Principles and Applications*, 2nd ed., John Wiley & Sons, Ltd, Chichester, UK, 2002. p. 407.
109. Ekman, R., Silberring, J., Westman-Brinkmalm, A. and Kraj, A. Mass Spectrometry: Instrumentation, Interpretation and Applications. John Wiley & Sons, New York, 2009. p. 371.

110. Watson, J. T. and David Sparkman, O. *Introduction to Mass Spectrometry: Instrumentation, Applications and Strategies for Data Interpretation*, 4th ed., John Wiley & Sons, Chichester, UK, 2007. p. 819.
111. Burinsky, D. J. Chapter 11: Mass spectrometry. In *Comprehensive Analytical Chemistry. Volume 47: Modern Instrumental Analysis*, (S. Ahuja and N. Jespersen, eds.), Elsevier, Amsterdam, pp. 319–396, 2006.
112. Sharp, T. R. and Marquez, B. L. Chapter 19: Hyphenated methods. In *Comprehensive Analytical Chemistry. Volume 47: Modern Instrumental Analysis*, (S. Ahuja and N. Jespersen, eds.), Elsevier, Amsterdam, pp. 691–754, 2006.
113. The International Journal of Mass Spectrometry and Ion Physics began publication in 1968. The title was altered to The International Journal of Mass Spectrometry and Ion Processes, then simplified to IJMS.
114. Organic Mass Spectrometry started in 1968, specializing in organic chemical applications of mass spectrometry. Biomedical Mass Spectrometry (later changing to Biomedical and Environmental Mass Spectrometry, then to Biological Mass Spectrometry) started in 1986 as a companion journal specializing in the biomedical and environmental application areas, as the title implies. These two journals combined in 1995 to become JMS.
115. Quayle, A. Recollections of mass spectrometry of the fifties in a UK petroleum laboratory. *Org. Mass Spectrom.* 22:569–585, 1987.
116. Hamming, M. C. and Foster, N. G. *Interpretation of Mass Spectra of Organic Compounds*, Academic Press, New York, 1972. Chapter 1.
117. Kuhlmann, F. E., Apffel, A., Fischer, S. M., Goldberg, G. and Goodley, P. C. Signal enhancement for gradient reverse-phase high-performance liquid chromatography-electrospray ionization mass spectrometry analysis with trifluoroacetic and other strong acid modifiers by postcolumn addition of propionic acid and isopropanol. *J. Am. Soc. Mass Spectrom.* 6:1221–1225, 1995.
118. Roberts, J. M., Diaz, A. R., Fortin, D. T., Friedle, J. M. and Piper, S. D. Influences of the Hofmeister series on the retention of amines in reversed-phase liquid chromatography. *Anal. Chem.* 74:4927–4932, 2002.
119. Dole, M., Mack, L. L., Hines, R. L., Mobley, R. C., Ferguson, L. D. and Alice, M. B. Molecular beams of macroions. *J. Chem. Phys.* 49:2240–2249, 1968.
120. Cole, R. B. Some tenets pertaining to electrospray ionization mass spectrometry. *J. Mass Spectrom.* 35:763–772, 2000.
121. Van Berkel, G. J. Electrolytic deposition of metals on to the high-voltage contact in an electrospray emitter: Implications for gas-phase ion formation. *J. Mass Spectrom.* 35:773–783, 2000.
122. Amad, M. H., Cech, N. B., Jackson, G. S. and Enke, C. G. Importance of gas-phase proton affinities in determining the electrospray ionization response for analytes and solvents. *J. Mass Spectrom.* 35:784–789, 2000.
123. Gamero-Castaño, M. and de la Mora, J. F. Kinetics of small ion evaporation from the charge and mass distribution of multiply charged clusters in electrospray. *J. Mass Spectrom.* 35:790–803, 2000.
124. Kebarle, P. A brief overview of the present status of the mechanisms involved in electrospray mass spectrometry. *J. Mass Spectrom.* 35:804–817, 2000.
125. de la Mora, J. F., Van Berkel, G. J., Enke, C. G., Cole, R. B., Martinez-Sanchez, M. and Fenn, J. B. Electrochemical processes in electrospray ionization mass spectrometry. *J. Mass Spectrom.* 35:939–952, 2000.
126. Iribarne, J. V. and Thomson, B. A. On the evaporation of small ions from charged droplets. *J. Chem. Phys.* 64:2287–2294, 1976.
127. Blades, A. T., Ikonomou, M. G. and Kebarle, P. Mechanism of electrospray mass spectrometry. Electrospray as an electrochemical cell. *Anal. Chem.* 63:2109–2114, 1991.
128. Vestal, M. L. Methods of ion generation. *Chem. Rev.* 101:361–375, 2001.
129. Sharp, T. R. and Horan, G. J. Mass spectrometry of trovafloxacin and related compounds. *45th ASMS Conference on Mass Spectrometry and Allied Topics*, 1–5 June 1997, Palm Springs, CA, 1997.
130. D'Agostino, P. A., Hancock, J. R. and Provost, L. R. Electrospray mass spectrometric characterization of fluoroquinolone antibiotics: Norfloxacin, enoxacin, ciprofloxacin and ofloxacin. *Rapid Communications in Mass Spectrometry* 9:1038–1043, 1995.

131. Colorado, A. and Brodbelt, J. Class-selective collisionally activated dissociation/ion-molecule reactions of 4-Quinolone antibiotics. *Anal. Chem.* 66:2330–2335, 1994.
132. The peaks at m/z 343 and m/z 514 are not simple noncovalent clusters! The masses correspond to multiples of the 173 dalton molecular mass minus two protons per molecule. We do not fully understand this behavior.
133. Takáts, Z., Wiseman, J. M., Gologan, B. and Cooks, R. G. Mass spectrometry sampling under ambient conditions with desorption electrospray ionization. *Science* 306:471–473, 2004.
134. Takáts, Z., Wiseman, J. M., Gologan, B. and Cooks, R. G. Ambient mass spectrometry using desorption electrospray ionization (DESI): Instrumentation, mechanisms and applications in forensics, chemistry, and biology. *J. Mass Spectrom.* 40:1261–1275, 2005.
135. The power-washing analogy is attributed to Dr. Gary van Berkel (2007).
136. Chipuk, J. E. and Brodbelt, J. S. The influence of material and mesh characteristics on transmission mode desorption electrospray ionization. *J. Am. Soc. Mass Spectrom.* 20:584–592, 2009.
137. Chen, H., Hu, B., Hu, Y., Huan, Y., Zhou, Z. and Qiao, Z. Neutral desorption using a sealed enclosure to sample explosives on human skin for rapid detection by EESI-MS. *J. Am. Soc. Mass Spectrom.* 20:719–722, 2009.
138. Seeley, E.H., Oppenheimer, S.R., Deming, M., Chaurand, P., Caprioli, R.M. Enhancement of protein sensitivity for MALDI imaging mass spectrometry after chemical treatment of tissue sections. *J. Am. Soc. Mass. Spectrom.* 19:1069–1077, 2008.
139. Busch, K. L. Mass spectrometric detectors for samples separated by planar electrophoresis. *J. Chromatogr. A* 692:275–290, 1995.
140. Wilson, I. D. The state of the art in thin-layer chromatography-mas spectrometry: A critical appraisal. *J. Chromatogr. A* 856:429–442, 1999.
141. Cody, R. B., Laramée, J. A. and Dupont Durst, H. Versatile new ion source for the analysis of materials in open air under ambient conditions. *Anal. Chem.* 77:2297–2302, 2005.
142. Robb, D. B., Covey, T. R. and Bruins, A. P. Atmospheric pressure photoionization: An ionization method for liquid chromatography-mass spectrometry. *Anal. Chem.* 72:3653–3659, 2000.
143. Syage, J. A. Mechanism of $[M+H]^+$ formation in photoionization mass spectrometry. *J. Am. Soc. Mass Spectrom.* 15:1521–1533, 2004.
144. Syage, J. A., Short, L. C. and Cai, S.-S. Atmospheric pressure photoionization—The second source for LC-MS? *LCGC North America* 26(3):286–296, 2008.
145. Karas, M., Bachmann, D., Bahr, U. and Hillenkamp, F. Matrix-assisted ultraviolet laser desorption of non-volatile compounds. *Int. J. Mass Spectrom. Ion Process.* 78:53–68, 1987.
146. Beavis, R. C. and Chait, B. T. Factors affecting the ultraviolet laser desorption of proteins. *Rapid Commun. Mass Spectrom.* 3:233–237, 1989.
147. Molecular weight is an incorrect scientific term! The numerical value being referenced is a relative molecular mass. It is a mass, measuring a quantity of matter, not the influence on that matter by an external gravitational field. It is a relative mass, not an absolute mass measured in grams, because it is expressed relative to the standard atomic mass of 12.000000 units (daltons), assigned to the 12C isotope of carbon. It is a relative molecular mass because it is a sum of all the relative atomic masses of the elements contained within the molecule.
148. DeBievre, P. and Barnes, I. L. Table of the isotopic composition of the elements as determined by mass spectrometry. *Int. J. Mass Spectrom. Ion Process.* 65:211–230, 1985.
149. The term "exact mass" should perhaps not be used at all in any context!
150. Sharp, T. R., Smith, D. E., Heah, P.-C., Buhro, W. E., Crocco, G. L., Bodner, G. S., Fernandez, J. M. and Gladysz, J. A. EI, CI and FAB mass spectral characterization of neutral and cationic organorhenium complexes. *Adv. Mass Spectrom.* 10B:1359–1360, 1986.
151. Kowalski, M. H., Sharp, T. R. and Stang, P. J. FAB mass spectral observations on new cationic platinum organometallic and related complexes. *Org. Mass Spectrom.* 22:642–643, 1987.
152. The well-done mass spectrometry data systems all have isotope pattern modeling programs built in. A number of other stand-alone simulation programs are also available.
153. Bai, Y., Milne, J. S., Mayne, L. and Englander, S. W. Primary structure effects on peptide group hydrogen exchange. *Proteins Struct. Funct. Genet.* 17:75–86, 1993.
154. Krishna, M. M. G., Hoang, L., Lin, Y. and Englander, S. W. Hydrogen exchange methods to study protein folding. *Methods* 34:51–64, 2004.

155. Englander, S. W. Hydrogen exchange and mass spectrometry: A historical perspective. *J. Am. Soc. Mass Spectrom.* 17:1481–1489, 2006.
156. Olsen, M. A., Cummings, P. G., Kennedy-Gabb, S., Wagner, B. M., Nicol, G. R. and Munson, B. The use of deuterium oxide as a mobile phase for structural elucidation by HPLC/UV/ESI/MS. *Anal. Chem.* 72:5070–5078, 2000.
157. Chen, G., Khusid, A., Daaro, I., Irish, P. and Pramanik, B. N. Structural identification of trace level enol tautomer impurity by on-line hydrogen/deuterium exchange HR-LC/MS in a LTQ-Orbitrap hybrid mass spectrometer. *J. Mass Spectrom.* 42:967–970, 2007.
158. Vail, T. M., Jones, P. R. and David Sparkman, O. Rapid and unambiguous identification of melamine in contaminated pet food based on mass spectrometry with four degrees of confirmation. *J. Anal. Toxicol.* 31:304–312, 2007.
159. Some individuals have taken offense at the usage of the originally coined term "daughter ion" in describing the ion genetic relationships resulting from a collision-induced decomposition experiment. The more palatable term "product ion" has come into use as an alternative. We have attempted to refer to product ions in our discussions. Notwithstanding, the term "granddaughter" conveys a certain meaning about the skipping-of-generations relationship between a product formed by two or more sequential chemical reactions from its original parent. No comparable companion term to "product ion" has yet arisen which conveys this meaning. For this reason, we still use the term "granddaughter ion" here in this limited context.
160. Sharp, T. R., Lee, H., Ferguson, A., Marsh, K. N. and Harvey, R. G. Electron impact mass spectra of polycyclic aromatic hydrocarbons: A reference collection *36th Annual Conference on Mass Spectrometry and Allied Topics*, San Francisco, CA, 1988.
161. Note here that the term “average molecular mass” applied to a polymer implies yet another nuance in the usage of the term molecular mass.
162. Beynon, J. H. Qualitative analysis of organic compounds by mass spectrometry. *Nature (London)* 174:735–737, 1954.
163. Beynon, J. H. Mass Spectrometry and Its Applications to Organic Chemistry. Elsevier, Amsterdam, 1960.
164. Marshall, A. G. and Hendrickson, C. L. High resolution mass spectrometers. *Annu. Rev. Anal. Chem.* 1:579–599, 2008.
165. Journal of Organic Chemistry (1993) 58, 7A–12A and (2000) 65, 14A–19A.
166. Any recent issue of the Journal of the American Society for Mass Spectrometry.
167. Heathcock, C. Editorial. *J. Org. Chem.* 55:8A, 1990.
168. Kondrat, R. High resolution mass spectrometry. *J. Am. Soc. Mass Spectrom.* 10:661, 1999.
169. Gross, M. L. Accurate masses for structure confirmation. *J. Am. Soc. Mass Spectrom.* 5:57, 1994.
170. Busch, K. L. Using exact mass measurements. *Spectroscopy* 9(7):21–22, 1994.
171. Biemann, K. Utility of exact mass measurements. *Meth. Enzymol.* 193:295–304, 1990.
172. Gilliam, J. M., Landis, P. W. and Occolowitz, J. L. Accurate mass measurement in fast atom bombardment mass spectrometry. *Anal. Chem.* 55:1531–1533, 1983.
173. Gilliam, J. M., Landis, P. W. and Occolowitz, J. L. On-line accurate mass measurement in fast atom bombardment mass spectrometry. *Anal. Chem.* 56:2285–2288, 1984.
174. Kind, T. and Fiehn, O. Metabolomic database annotations via query of elemental compositions: Mass accuracy is insufficient even at less than 1 ppm. *BMC Bioinform.* 7:234–246, 2006.
175. Budzikiewicz, H., Djerassi, C. and Williams, D. H. Mass Spectrometry of Organic Compounds. Holden-Day, Inc., San Francisco, CA, 1967.
176. Porter, Q. N. Mass Spectrometry of Heterocyclic Compounds. 2nd ed., John Wiley & Sons, New York, 1985.
177. McLafferty, F. W. and Venkataraghavan, R. *Mass Spectral Correlations*, 2nd ed. Washington, DC: American Chemical Society Advances in Chemistry Series #40, p. 124, 1982.
178. David Sparkman, O. A review of elecronic mass spectral databaes from John Wiley and Sons. *J. Am. Soc. Mass Spectrom.* 20:R22–R27, 2009.
179. McLafferty, F. W. *Interpretation of Mass spectra*, 3rd ed., University Science Books, Mill Valley, CA, 1980. p.303.
180. Lee, T. A. A Beginner’s Guide to Mass Spectral Interpretation. John Wiley & Sons, New York, 1998. p. 186.

181. Martin Smith, R. *Understanding Mass Spectra*, 2nd ed., Wiley-Interscience, New York, 2004. p. 372.
182. Holzgrabe, U., Wawer, I. and Diehl, B. NMR Spectroscopy in Drug Development and Analysis. Wiley-VCH, Verlag, GmbH, 1999. p. 284.
183. Claridge, T. D. W. High-Resolution NMR Techniques in Organic Chemistry. Pergamon Press, Oxford, UK, 1999.
184. Derome, A. E. Modern NMR Techniques for Chemistry Research. Pergamon Press, Oxford, UK, 1987.
185. Ernst, R. R. Principles of Nuclear Magnetic Resonance in One and Two Dimensions. Clarendon Press, Oxford, UK, 1987.
186. Abragam, A. Principles of Nuclear Magnetism. Clarendon Press, Oxford, UK, 1989.
187. Slichter, C. P. Principles of Magnetic Resonance. 3rd ed., Springer-Verlag, Berlin, 1990.
188. Harris, R. K. Nuclear Magnetic Resonance Spectroscopy. Longman Scientific & Technical, New York, 1990.
189. Martin, G. E. Small-volume and high-sensitivity NMR probes. *Annu. Rep. NMR Spectrosc.* 56:1–96, 2005.
190. Martin, G. E. Applications of cryogenic NMR probe technology for the identification of low-level impurities in pharmaceuticals. *Mod. Magn. Reson.* 2:1187–1194, 2006.
191. Godejohann, M., Tseng, L.-H., Braumann, U., Fuchser, J. and Spraul, M. Characterization of a paracetamol metabolite using on-line LC-SPE-NMR-MS and a cryogenic NMR probe. *J. Chromatogr. A* 1058(1–2):191–196, 2004.
192. Pauli, G. F., Jaki, B. U. and Lankin, D. C. Quantitative 1H NMR: Development and potential of a method for natural products analysis. *J. Nat. Prod.* 68:133–149, 2005.
193. Albert, K. On-line LC-NMR and Related Techniques. John Wiley and Sons, West Sussex, England, 2002.
194. Derome, A. E. Modern NMR Techniques for Chemistry Research. Pergamon Press, Oxford, UK, 1987. pp. 77–80.
195. Traficante, D. D. Phase-sensitive detection. Part II: Quadrature phase detection. *Concepts in Magnetic Resonance* 2:181–195, 1990.
196. Braun, S. and Berger, S. 200 and More Basic NMR Experiments. Wiley-VCH, Weinheim, 2004.
197. Maiwald, M., Steinhof, O., Sleigh, C.,Bernstein, M., Hasse, H. Quantitative high-resolution online NMR spectroscopy in pharmaceutical reaction and process monitoring. *NMR Spectroscopy in Pharmaceutical Analysis*, Amsterdam, The Netherlands, Elsevier, pp. 471–491, 2008.
198. Derome, A. E. Modern NMR Techniques for Chemistry Research. Pergamon Press, Oxford, UK, 1987. p. 41.
199. Sanders, K. K. and Hunter, B. K. Modern NMR Chemistry. 2nd ed., Oxford University Press, Oxford, UK, 1993.
200. Friebolin, H. P. Basic One- and Two-Dimensional NMR Spectroscopy. 2nd ed., VCH, Weinheim, 1993.
201. Ernst, R. R. and Anderson, W. A. Application of Fourier transform spectroscopy to magnetic resonance. *Rev. Sci. Inst.* 37:93–102, 1966.
202. Kalinowski, H.-O., Berger, S. and Braun, S. Carbon-13 NMR Spectroscopy. Wiley, Chichester, 1988.
203. Harris, R. K. Nuclear Magnetic Resonance Spectroscopy. Longman Scientific & Technical, New York, 1990. p. 236.
204. Shaka, A. J., Keeler, J. and Freeman, R. An improved sequence for broadband decoupling: WALTZ-16. *J. Magn. Reson.* 52:335–338, 1983.
205. Bendall, M. R., Pegg, D. T., Doddrell, D. M. and Field, J. Inverse DEPT sequence. Polarization transfer from a spin-1/2 nucleus to n spin-1/2 heteronuclei via correlated motion in the doubly rotating reference frame. *J. Magn. Reson.* 52:520–526, 1983.
206. Bendall, M. R. and Pegg, D. T. Identification of isolated methyl groups by detection of cross-correlation effects using polarization transfer. *J. Magn. Reson.* 53:40–48, 1983.
207. Bendall, M. R. and Pegg, D. T. Complete accurate editing of decoupled carbon-13 spectra using DEPT and a quaternary-only sequence. *J. Magn. Reson.* 53:272–296, 1983.
208. Schenker, K. V. and Von Philipsborn, W. Off-resonance effects and their compensation in the multiple-pulse sequences INEPT, DEPT, and INADEQUATE. *J. Magn. Reson.* 66:219–229, 1986.

209. Doddrell, D. M., Pegg, D. T. and Bendall, M. R. Distortionless enhancement of NMR signals by polarization transfer. *J. Magn. Reson.* 48:323–327, 1982.
210. Hurd, R. E. Gradient-enhanced spectroscopy. *J. Magn. Reson.* 87:422–428, 1990.
211. Tolman, J. R., Chung, J. and Prestegard, J. H. Pure-phase heteronuclear multiple-quantum spectroscopy using field gradient selection. *J. Magn. Reson.* 98:462–467, 1992.
212. Tyburn, J. M., Brereton, I. M. and Doddrell, D. M. Coherence selection in gradient-enhanced, heteronuclear correlation spectroscopy. *J. Magn. Reson.* 97:305–312, 1992.
213. Geerten, W., Vuister, R. B., Kaptean, R., Hurd, R. E., John, B. K. and van Zijl, P. C. M. Gradient-enhanced HMQC and HSQC spectroscopy. Applications to ^{15}N-labeld Mnt repressor. *J. Am. Chem. Soc.* 113:9688, 1991.
214. Hurd, Ralph E. and John, B. K. Gradient-enhanced proton-detected heteronuclear multiple-quantum coherence spectroscopy. *J. Magn. Reson.* 91:648, 1991.
215. Willker, W., Leibfritz, D., Kerssebaum, R. and Bermel, W. Gradient selection in inverse heteronuclear correlation spectroscopy. *J. Magn. Reson.* 31:287–292, 1993.
216. Jeener, J. Ampere International Summer School, proposal. Basko Polje, 1971.
217. Piantini, U., Sorensen, O. W. and Ernst, R. R. Multiple quantum filters for elucidating NMR coupling networks. *J. Am. Chem. Soc.* 104:6800–6801, 1982.
218. Brereton, I. M., Crozier, S., Field, J. and Doddrell, D. M. Quadrature detection in F1 induced by pulsed field gradients. *J. Magn. Reson.* 93:54–62, 1991.
219. Braun, S., Kalinowski, H.-O. and Berger, S. 150 and More Basic NMR Experiments. Wiley-VCH, Weinheim, 1998. pp. 168–371.
220. Griesinger, C., Sorensen, O. W. and Ernst, R. R. Practical aspects of the E.COSY technique. Measurement of scalar spin-spin coupling constants in peptides. *J. Magn. Reson.* 75:474–492, 1987.
221. Martin, G. E. and Zektzer, A. S. *Two-Dimensional NMR Methods for Establishing Molecular Connectivity*, VCH, Weinheim, 1988. Chapter 3.
222. Bax, A. and Morris, G. A. An improved method for heteronuclear chemical shift correlation by two-dimensional NMR. *J. Magn. Reson.* 42:501–505, 1981.
223. Freeman, R. and Morris, G. A. Experimental chemical shift correlation maps in nuclear magnetic resonance spectroscopy. *J. Chem. Soc. Chem. Commun.* 684–686, 1978.
224. Bodenhausen, G. and Ruben, D. J. Natural abundance nitrogen-15 NMR by enhanced heteronuclear spectroscopy. *Chem. Phys. Lett.* 69:185–189, 1980.
225. Kay, L. E., Keifer, P. and Saarinen, T. Pure absorption gradient enhanced heteronuclear single quantum correlation spectroscopy with improved sensitivity. *J. Am. Chem. Soc.* 114:10663–10665, 1992.
226. Muller, L. Sensitivity enhanced detection of weak nuclei using heteronuclear multiple quantum coherence. *J. Am. Chem. Soc.* 101:4481–4484, 1979.
227. Bax, A. and Summers, M. F. Proton and carbon-13 assignments from sensitivity-enhanced detection of heteronuclear multiple-bond connectivity by 2D multiple quantum NMR. *J. Am. Chem. Soc.* 108:2093–2094, 1986.
228. Ruiz-Cabello, J., Vuister, G. W., Moonen, C. T., van Gelderen, P., Cohen, J. S. and van Zijl, P. C. M. Gradient-enhanced heteronuclear correlation spectroscopy. Theory and experimental aspects. *J. Magn. Reson.* 100:282–303, 1992.
229. Martin, G. E. and Zektzer, A. S. Two-Dimensional NMR Methods for Establishing Molecular Connectivity. VCH, Weinheim, 1988. pp. 267–273.
230. Kogler, H., Sorensen, O. W., Bodenhausen, G. and Ernst, R. R. Low-pass J filters. Suppression of neighbor peaks in heteronuclear relayed correlation spectra. *J. Magn. Reson.* 55:157–163, 1983.
231. Krishnamurthy, V. V., Russell, D. J., Hadden, C. E. and Martin, G. E. ^{2}J, ^{3}J-HMBC: A new long-range heteronuclear shift correlation technique capable of differentiating ^{2}JCH from ^{3}JCH correlations to protonated carbons. *J. Magn. Reson.* 146:232–239, 2000.
232. Nyberg, N. T., Duus, J. O. and Sorensen, O. W. Heteronuclear two-bond correlation: Suppressing heteronuclear three-bond or higher NMR correlations while enhancing two-bond correlations even for vanishing 2JCH. *J. Am. Chem. Soc.* 127:6154–6155, 2005.
233. Kinns, M. and Sanders, J. K. M. Improved frequency selectivity in nuclear Overhauser effect difference spectroscopy. *J. Magn. Reson.* 56:518–520, 1984.
234. Neuhaus, D. and Williamson, M. P. *The Nuclear Overhauser Effect in Structural and Conformational Analysis*, Wiley-VCH Inc., 2000.

235. Noggle, J. H. and Schirmer, R. E. The Nuclear Overhauser Effect—Chemical Applications. Academic Press, San Diego, 1972.
236. Stott, K., Stonehouse, J., Keeler, J., Hwang, T.-L. and Shaka, A. J. Excitation sculpting in high-resolution nuclear magnetic resonance spectroscopy: Application to selective NOE experiments. *J. Am. Chem. Soc.* 117:4199–4200, 1995.
237. Bothner-By, A. A., Stephens, R. L., Lee, J., Warren, C. D. and Jeanloz, R. W. Structure determination of a tetrasaccharide: Transient nuclear Overhauser effects in the rotating frame. *J. Am. Chem. Soc.* 106:811–813, 1984.
238. Harris, R. K. Nuclear Magnetic Resonance Spectroscopy. Longman Scientific & Technical, New York, 1990. pp. 119–131.
239. Harris, R. K. Nuclear Magnetic Resonance Spectroscopy. Longman Scientific & Technical, New York, 1990. p. 26.
240. Abragam, A. Principles of Nuclear Magnetism. Clarendon Press, Oxford, UK, 1989. pp. 480–481.
241. Williams, T., Blinov, K., Elyashberg, M. and Martisrosian, E. (2000). Recent advances in the use of NMR prediction and automated structure elucidation software for natural product characterization, presented at the *SMASH NMR Conference*, Argonne, IL.
242. Lee, H., Shen, S. and Grinberg, N. Identification and control of impurities for drug substance development using LC/MS and GC/MS. *J. Liq. Chromatogr. Relat. Technol.* 31:2235–2252, 2008.
243. Sharp, T. R. Impurity identification in pharmaceuticals: An integrated approach and examples of how well it works! (Part 1). *Am. Pharm. Rev.* 9:84–86, 2006.
244. Sharp, T. R. Impurity identification in pharmaceuticals: An integrated approach and examples of how well it works! (Part 2). *Am. Pharm. Rev.* 9:88–91, 2006.
245. Qiu, F. and Norwood, D. L. Identification of pharmaceutical impurities. *J. Liq. Chromatogr. Relat. Technol.* 30:877–935, 2007.
246. Jansen, P. J., Oren, P. L., Kemp, C. A., Maple, S. R. and Baertschi, S. W. Characterization of impurities formed by interaction of duloxetine HCl with enteric polymers hydroxypropyl methylcellulose acetate succinate and hydroxypropyl methylcellulose phthalate.3. *J. Pharm. Sci.* 87 (1):81–85, 1998.
247. Franolic, J. D., Lehr, G. J., Barry, T. L. and Petzinger, G. Isolation of a 2:1 hydrochlorothiazide–formaldehyde adduct impurity in hydrochlorothiazide drug substance by preparative chromatography and characterization by electrospray ionization LC-MS. *J. Pharm. Biomed. Anal.* 26:651–663, 2001.
248. Glazkov, I. N., Bochkareva, N. L. and Revel'skii, I. A. Determination of organic impurities in pharmaceutical preparations. *J. Anal. Chem.* 60:107–118, 2005.
249. Martin, G. E. A systematic approach to impurity identification. In *Analysis of Drug Impurities*, (R. J. Smith and M. L. Webb, eds.), Blackwell Publishing Ltd., Oxford, UK, pp. 124–155, 2007.
250. Hartauer, K. J., Arbuthnot, G. N., Baertschi, S. W., Johnson, R. A., Luke, W. D., Pearson, N. G., Rickard, E. C., Tingle, C. A., Tsang, P. K. S. and Wiens, R. E. Influence of peroxide impurities in povidone and crospovidone on the stability of raloxifene hydrochloride in tablets: Identification and control of an oxidative degradation product. *Pharm. Dev. Technol.* 5(3):303–310, 2000.
251. Baertschi, S. W., Brunner, H., Bunnell, C. A., Cooke, G. G., Diseroad, B., Dorman, D. E., Jansen, P. J., Kemp, C. A. J., Maple, S. R., McCune, K. and Speakman, J. L. Isolation, identification, and synthesis of two oxidative degradation products of olanzapine (LY170053) in solid oral formulations. *J. Pharm. Sci.* 97:883–892, 2007.
252. Bouas-Laurent, H., Desvergne, J.-P., Castellan, A. and Lapouyade, R. Photodimerization of anthracenes in fluid solution: Structural aspects. *Chem. Soc. Rev.* 29:43–55, 2000.
253. Bouas-Laurent, H., Desvergne, J.-P., Castellan, A. and Lapouyade, R. Photodimerization of anthracenes in fluid solutions: (part 2) mechanistic aspects of the photocycloaddition and of the photochemical and thermal cleavage. *Chem. Soc. Rev.* 30:248–263, 2001.
254. Wan, J. K. S., McCormick, R. N., Baum, E. J. and Pitts Jr., J. N. The roles of molecular structure and environment in the reactivity of excited states. *J. Am. Chem. Soc.* 87:4409–4414, 1965.
255. Bouas-Laurent, H., Castellan, A. and Desvergne, J.-P. From anthracene photodimerization to jaw photochromic materials and photocrowns. *Pure Appl. Chem.* 52:2633–2648, 1980.

256. Potts, K. T., Brugel, E. G. and Dunlap, W. C. 1, 2, 4-Triazoles—XXXI: Photodimerization of s-triazoleo [4, 3-a]pyridines: Formation of cyclobutane dimmers. *Tetrahedron* 33:1247–1251, 1977.
257. Potts, K. T., Dunlap, W. C. and Brugel, E. G. 1, 2, 4-Triazoles—XXXIII: Photodimerization of some sym-di(3-s-triazolo-[4,3-a]pyridyl)-alkanes and related derivatives. *Tetrahedron* 33:1253–1262, 1977.
258. Potts, K. T., Dunlap, W. C. and Apple, F. S. 1, 2, 4-Triazoles—XXXIV: Photodimerization of some 1, 2, 4-triazo[4, 3–]quinoline and 1, 2, 4-triazolo[3, 4-a]isoquinoline derivatives. *Tetrahedron* 33:1263–1271, 1977.
259. Filippakis, S. E. and Schmidt, G. M. J. Topochemistry. Part XVI. The crystal structure of trans-β-2-furylacrylic acid. *J. Chem. Soc. B* 229–232, 1967.
260. Block, S., Filippakis, S. E. and Schmidt, G. M. J. Topochemistry. Part XVII. The crystal structures of two modifications of trans-β-2-thienylacrylic acid. *J. Chem. Soc. B* 233–238, 1967.
261. Lahav, M. and Schmidt, G. M. J. Topochemistry. Part XVIII. The solid-state photochemistry of some heterocyclic analogues of trans-cinnamic acid. *J. Chem. Soc. B* 239–243, 1967.
262. Berger, T. A. Practical advantages of packed column supercritical fluid chromatography in supporting combinatorial chemistry. *ACS Symp. Ser.* 748:203–233, 2000.
263. Schmidt, D. G., Grosche, P., Bandel, H. and Jung, G. FTICR-MS Spectrometry for high-resolution analysis in combinatorial chemistry. *Biotechnol. Bioeng.* 71:149–161, 2001.
264. Amster, Jonathan. Fourier transform mass spectroscopy. *J. Mass Spectrom.* 31?1325–1337, 1996.
265. Marshall, A. G., Hendrickson, C. L. and Jackson, G. S. Fourier transform ion cyclotron resonance mass spectrometry: A primer. *Mass Spectrom. Rev.* 17:1–35, 1998.

4 ANALYTICAL CONSIDERATIONS FOR GENOTOXIC AND OTHER IMPURITIES

RODNEY L. PARSONS* AND SCOTT A. MILLER†

*Bristol-Myers Squibb Co., Process Research and Development, New Brunswick, New Jersey
†Bristol-Myers Squibb Co., Analytical Research and Development, New Brunswick, New Jersey

I. INTRODUCTION

Synthetic chemistry requires the use of reactive functional groups to forge new chemical bonds and construct molecules. The functional groups that impart reactivity to reagents and intermediates and make them synthetically useful may also incure the liability in that they may be biologically reactive as genotoxins and carcinogens. Use of these reactive chemical agents is required to construct molecules that comprise the active ingredient in medicinal agents (active pharmaceutical ingredient (API)). Because of this, residual levels of these species may be present in the API and may have the potential to increase the risk of undesired toxicities, such as genotoxicity. The assessment and control of genotoxic impurities (GTIs) in pharmaceutical products is a significant challenge for the pharmaceutical industry and has become the focus of considerable attention. This chapter provides an introduction into the concept of GTIs, the evolution of the established exposure limits and current regulatory landscape. In addition, it will also describe some of the issues related to assessment of GTIs in APIs and will provide context on how these issues influence the development of an appropriate risk-based GTI control strategy. The successful implementation and execution of any GTI control strategy relies on the ability to adequately demonstrate that these impurities are controlled to a safe level in the API. To accomplish that, the analytical methodologies used for the determination of GTIs at low levels is of critical importance. The analytical strategy should utilize the same risk-based approach and take into consideration the stage of clinical development and current health authority expectations. The analytical challenges and examples of the methodologies that are currently being utilized for GTI assessment are also a primary focus of this chapter.

II. CONCEPT OF GTI RELATIVE TO STANDARD IMPURITIES AND THE CURRENT REGULATORY THINKING

Shortly after the International Conference on Harmonization (ICH) formally adopted the drug substance quality guidance, ICH Q3A(R), a number of health authorities raised the concern that this guidance document did not adequately address the issue of highly toxic impurities in APIs and did not provide guidance for safe levels.[1,2] To address this gap, the European Medicines Agency (EMEA) Safety Working Party (SWP) of the Committee for Medical Products for Human Use (CHMP) issued a draft position paper in late 2003 to provide guidance on the limits of GTIs.[3] This position paper served as the genesis for the eventual EMEA CHMP GTI guidance document that was issued in 2006.[4] The EMEA guidance utilized an assessment procedure that is based on a concept of threshold of toxicological concern (TTC) to assign an acceptable exposure limit for GTIs in APIs. The TTC concept had originally been developed by Rulis in 1990s and then adopted by the United States FDA for the regulation of trace substances in food contact materials.[5] It is important to note that the EMEA guideline acknowledges that it is impossible to define a zero risk for genotoxic carcinogens that operate without a threshold mechanism and the practical fact is that complete elimination is unachievable. The TTC concept is based on an analysis of the tumorigenic potency in rodents of over 700 genotoxic and nongenotoxic carcinogens from which it was estimated that daily exposure to 0.15 μg/day for most genotoxic carcinogens is not likely to induce a lifetime cancer risk of greater than one in a million.[6] For pharmaceuticals, the EMEA argued that a TTC of 1.5 μg/day equates to a one in 100,000 lifetime cancer risk and this is acceptable based on the expected benefit of medicinal products. It is important to note that the EMEA guidance was issued for marketed products, and thus was intended for extended exposure durations (i.e., greater than 1 year in length), but did not address the shorter (variable) exposure durations encountered during clinical trials. In 2005, the US Pharmaceutical Research and Manufactures Association (PhRMA) technical sub-team drafted a guidance document that provided recommended exposure limits for GTIs based on the duration of clinical exposure.[7] One critical aspect of the PhRMA guidance document was the introduction of a staged TTC concept which allowed for higher exposure limits with

TABLE 1 Staged TTC Exposure Limits from the EMEA Guidance on Genotoxic Impurities

	Duration of Clinical Trial Exposure					
	Single Dose	≤1 month	≤3 months	≤6 months	≤12 months	>12 months
Allowable daily intake (μg)	120	60	20	10	5	1.5

shorter clinical exposures (a concept which had been introduced in 2004 by Bos *et al.*).[8] In addition, Müller also points out that a GTI control strategy is influenced by the state of the chemical process knowledge and analytical controls, and these are often immature in the early phases of clinical development. The PhRMA guidance document clearly establishes an expectation that the GTI control strategy will progress as the manufacturing process and analytical controls develop in line with the clinical phase of the product development. In 2008, the EMEA responded to a number of questions on GTI controls and exposure limits, specifically addressing the expectations for clinical development and the use of a staged approach.[9] Significantly, the EMEA endorsed the stage approach concept outlined in the publication by Müller *et al.*, but with reduced limits (factor of 2; see Table 1).

In 2008, the FDA published a draft guidance that further justified the use of the TTC concept for controlling the risks associated with mutagenic and carcinogenic impurities and provided a similar staged approach for GTI control during clinical development.[10] Prior to 2008, Jacobson-Kram and Jacobs had discussed the FDA's position on GTI control.[11]

Overall, the current thinking expressed by the regulators for GTI control has prompted the publication of numerous articles, many of which challenge critical aspects of GTI assessment and control.[12] Specifically, the arduously low control limits derived from use of the TTC concept have been the focus of a number of articles.[13] Many of the authors referenced strive to conclude that the use of a 1.5 μg/day TTC-derived limit will impede the development of new medicines while not truly providing an appropriate cancer risk-avoidance benefit to patients. The contentious nature of this topic will certainly continue to generate further dialog and debate as both regulators and industry optimize the processes to control GTIs in APIs.

III. VIRACEPT® GTI CONTAMINATION ISSUE

The recent events surrounding the Viracept® GTI contamination issue highlight the impacts that can result from the unexpected discovery of GTI contamination in a commercial product. In the case of Viracept, the presence of ethylsulfonate (EMS) in the drug product at a level above the TTC-derived exposure limits led to a suspension of the marketing authorization, a product recall, an extensive investigation into the source of the GTI contamination and subsequent risk to patients.[14] As a result of this issue, the EMEA also issued letter to all marketing authorization holders for medicinal products, instructing them to conduct a risk assessment for the occurrence of contamination of sulfonate esters in pharmaceuticals.[15] The EMEA letter subsequently stimulated a number of pharmaceutical industry studies to evaluate the propensity for sulfonate esters to form and the risk for API contamination of these esters.[16] In the end, the Viracept sponsor determined that the root cause for the EMS contamination was the result of a reaction of residual cleaning solvent, ethanol, with methane sulfonic acid (MSA) in a storage tank which was used in the API manufacture.[17] By establishing the root cause, the sponsor was able to design a control strategy to avoid the potential for future contamination and regain marketing approval. In addition, the sponsor was required to address the risk to patients who had received the GTI-containing drug, and this was done through a series of elegant toxicological studies conducted by Müller *et al.*[18] These studies clearly demonstrated that EMS exhibits a threshold limit for mutagenicity and exposures below that level do not result in genotoxicity. These findings have provided direct evidence that even established mutagens may operate by a threshold mechanism and further challenge the conservative principles engrained in the TTC-derived exposure limits. This issue clearly pointed out to the pharmaceutical industry that a thorough GTI evaluation including a risk assessment and design of an appropriate control strategy was an established health authority expectation.

IV. GTI RISK ASSESSMENT AND CONTROL STRATEGY DESIGN

The current state of analytical technology makes it possible to measure impurities in APIs down to very low levels and thus has exponentially increased the number of impurities which can be observed in API. Control of all these impurities down to low ppm levels is neither practical nor feasible. Despite this, control of mutagenic impurities is essential to minimize the health risks associated with exposure to genotoxins. To do this effectively, it is imperative to have a risk-based, strategic approach to identification and control of GTIs in APIs. Overall, risk assessment and control strategy development is fairly harmonized among many of the development organizations within the pharmaceutical industry.[19] The stages used in this process are broadly categorized in Table 2.

The subsequent discussion will serve to describe in detail the steps involved in the risk assessment for GTIs and its use in the design of a control strategy based on process and analytical controls to ensure patient safety.

TABLE 2 Stages in GTI Assessment and Control

Stage 1	Toxicological assessment: analysis of synthetic route and chemical structures
Stage 2	Risk assessment and control strategy design
Stage 3	Control strategy: execution and verification
Stage 4	Reevaluation of GTI risk and control strategy

A. Stage I: Toxicological Assessment: Analysis of Synthetic Route and Chemical Structures

Once a chemical route has been selected for use in the manufacture of an API, an assessment to identify potential genotoxins should be conducted by a multidisciplinary team of process chemist, analytical chemist, and toxicologist. The structural assessment should include starting materials, reagents, intermediates, side products, and impurities. For side products and impurities, greater emphasis is warranted when the presence of a genotoxic substance is expected based on the synthetic route or transformation. All of the chemical structures identified are then evaluated using a combination of *in silico* tools, literature databases, and expert opinion. The standard *in silico* tools used for the *in silico* structure activity assessments are DEREK[20] and MultiCASE.[21] These programs identify alerting structural features of compounds based on functional group-mutagenicity algorithms derived from an extensive database of compounds. From this assessment, the structures of concern, those containing alerting functionality, are identified. Dobo *et al.*[15b] have proposed a classification system for the potential GTIs, where they are grouped into one of five classes based on their genotoxicity. It should be noted that these categorizations are consistent with those proposed earlier by Müller *et al.*[7] The classifications are as follows:

Category 1: Known mutagenic carcinogens
Category 2: Mutagens with unknown carcinogenic potential
Category 3: Alerting structure—unique structure, unknown mutagenic potential
Category 4: Alerting structure—but analogous to an established nonmutagen
Category 5: No structural alert

The classification provides guidance for prioritizing compounds for testing in subsequent *in vitro* genotoxicity assays. The bacterial reverse mutation assay (Ames assay)[22] has been established as a definitive test to determine if an impurity is mutagenic and thus should be classified as a genotoxin. Known mutagens, compounds grouped into Categories 1 and 2 (i.e., sulfonate esters, hydrazine, hydroxylamine, etc.), are typically not tested unless they possess unique structural elements that may influence their potential activity as mutagens. Per the EMEA question and answers to the GTI guidance document,[9] compounds that elicit a negative result in the Ames assay are considered nongenotoxic and are therefore controlled following the guidance outlined in ICH Q3(A)R2. However, compounds that give a positive result in the Ames assay are classified as genotoxic and need to be controlled in the API to a level consistent with TTC or to as low as reasonably practical (ALARP). Use of the ALARP principle in the GTI control strategy will be discussed in a subsequent section. It is important to note that the timing for conducting the Ames test is a point of differentiation among the pharmaceutical companies who have published their approaches to GTI control. Some companies use a decision tree matrix that involves a risk assessment before submitting compounds for Ames testing.[15] Conversely, some companies first test all the unique compounds with alerting structures to define the genotoxins and then use these results in the risk assessment and design of the control strategy. This is an interesting divergence of practice since the Ames test is standard, simple, and relatively inexpensive test to conduct and the outcome provides a definitive classification of the compound as a GTI or typical impurity. Having identified the GTIs, a risk assessment should be conducted to determine the probability that the GTIs will be present in the API.

B. Stage 2: Risk Assessment and Control Strategy Design

The goal of the risk assessment is to determine the probability that a GTI will be present in API above the exposure limit based on clinical dose, duration and nature of the synthesis and to provide the basis for design of the eventual GTI control strategy. Table 3 provides a list of the key aspects which are considered when conducting the risk assessment to develop the GTI control strategy.

Each of these factors should be considered when assessing risk and designing the control strategy. The type of mutagen will have a direct impact on the risk. Highly potent genotoxic carcinogens (i.e. Category 1 compounds) pose a greater safety risk to patients and typically require more stringent control levels (in some cases below 1.5 μg/day). The increased risk of these mutagens makes the design of an appropriate control strategy more challenging. The medical benefit from the drug vs. the risk due to potential contamination by a trace level GTI should be assessed against the intended disease indication. Both the EMEA and FDA GTI guidance documents point out that the presence of genotoxic and carcinogenic impurities can be acceptable in some situations depending on clinical circumstances (e.g., cancer chemotherapies), but overall impurities in APIs and drug products generally do not have any beneficial effect. In these situations, use of a control strategy based on an ALARP approach may result in a justifiable control limit that is higher than that derived from TTC. The exposure limits for any GTI is directly linked to the clinical dose and duration, thus these two aspects will impact the risk. High dose drugs or those used for extended durations will require lower levels of the GTI to achieve the TTC derived control limits. The proximity of the GTI to the API and consideration of the subsequent chemical transformation are considered when determining the risk. In general, a GTI that enters the synthesis route greater than four steps prior to the API will have a reduced risk relative to those that are introduced in the final intermediate or API step. The number of subsequent isolations is an important factor to consider when assessing risk and designing the control strategy. Finally, highly reactive mutagens, those with functional groups which are

TABLE 3 Factors Considered in Risk Assessment

- Nature of the mutagen
- Disease indication and patent population
- Clinical dose and exposure duration
- Origin point in the synthetic sequence and proximity to API
- Number of subsequent purifications before API
- Functional group(s) of the GTIs and reactivity
- Chemical tolerance in the subsequent transformations

incompatible with the subsequent chemical stages pose less risk for carry over to the API than those that are unreactive. The chemical reactivity of the GTI and down stream processing steps should be considered when conducting the risk assessment.

C. Stage 3: Control Strategy: Execution and Verification

The ultimate goal of the control strategy is to insure that all GTIs are below the limits defined in the toxicological assessment. A successful control strategy typically requires the combined use of process chemistry elements and analytical testing to accomplish the GTI control. As noted in the previous discussion, the factors which impact the risk of GTI carry over to the API also directly influence the design of the control strategy. In situations where the risk for GTI contamination in the API is significant or the GTI is a member of the Category 1 group, an alteration of the synthetic route should be considered. If a modified synthesis route that avoids use or formation of GTIs is possible then this option should be considered by the sponsor. The following discussion will provide example control strategy scenarios based on different GTI situations.

1. Control of GTIs in Medicines for Life-Threatening Diseases

As noted previously, both the EMEA and FDA guidances indicate that higher levels of GTIs may be acceptable in certain circumstances, like cancer chemotherapies. For drugs in these therapeutic classes, the control strategy should be designed around the ALARP principle. An ALARP-based approach involves the development of process and analytical controls designed to reduce the GTI levels to those achievable by practical nonextreme methods. Demonstration of sound scientific due diligence into the technical feasibility should be used to support the limits derived from an ALARP approach. For example, control through standard chemical processing operations or straightforward reprocessing steps should be evaluated when conducting the technical due diligence. The analytical methods employed in these approaches should also be fit-for-purpose such as limits test or semiquantitative methodologies. For any ALARP-based control strategy, a dialog with the regulatory authorities about the approaches to be employed, conclusions from the due diligence conducted, and proposed limits should be considered.

2. Presence of GTIs in the Penultimate or API Step

GTIs that are introduced in the final intermediate step or API step typically require methods with sensitivity down to the TTC limit (typically to ppm level) to verify that the GTI is controlled. It is often preferable to implement the process control at the final intermediate stage with a verification test when possible, rather than testing at the API. However, if control to below TTC cannot be achieved, then a subsequent API specification test for the GTIs may be required. The use of lab scale GTI fate and tolerance data should also be incorporated into the data set to help craft the scientific basis for the control strategy. Fate and tolerance studies conducted using higher spiked levels of GTIs can often be supported using standard analytical methods and may not require low ppm-level analytical sensitivity although having the low ppm-level methods can often further strengthen the justification of the control strategy. However, confirmatory analytical testing of the API manufactured for use in clinical studies at a larger scale may be required even when supportive laboratory data exists.

3. GTIs Introduced Three to Four Steps Upstream of the API

In these situations, upstream controls with analytical verification directly after the point of introduction should be considered versus API specification testing. For upstream controls which reduce the GTI level to below the API TTC limit, no further evaluation or testing should be required. Knowledge of the GTI chemical reactivity in the subsequent steps with supportive fate and tolerance data may be required to justify the upstream control arguments when the GTI is not reduced to below TTC limits. The analytical approach used in these situations could be a combination of ppm limit tests and fate and tolerance studies using analytical methods based on standard ICH detection limits.

4. GTI Introduced Greater than Four Steps Upstream of the API

For most GTIs (exclusive of Category 1) in this situation, a control strategy based on chemical arguments should be pursued, rather than one based on a low-level GTI testing approach. Knowledge of the GTI chemical reactivity in the subsequent steps with supportive fate and tolerance data should be sufficient to justify the upstream control arguments. The number of purifications employed between the GTI introduction step and the API should be incorporated into the justification arguments.

Due to the toxicity risks associated with Category 1 GTIs, verification testing to TTC limits may be required. For these situations, upstream testing to the TTC-derived limit should be considered.

D. Reevaluation of GTI Risk and Control Strategy

The GTI risk assessment and control strategy should be reviewed on an ongoing basis as a compound progresses in clinical development. It is expected that the chemical process and analytical methodologies will evolve as a compound progresses and the impact of these changes on the risk for GTI contamination in the API needs to be assessed. Certainly, any new intermediates or reagents used will require a toxicological assessment for potential genotoxicity. Furthermore, any changes to the dose or clinical exposure duration will impact the GTI exposure limits and the process/analytical controls required to achieve the modified limits. Finally, generation of comprehensive fate and tolerance data for GTI impurity control should be generated to support the controls defined in the registrational dossier. The next sections describe examples of the analytical controls used in different control strategy situations.

V. ANALYTICAL CHALLENGES

The most significant analytical challenge for the pharmaceutical industry related to GTIs is the ability to develop, validate, and transfer methods to quality control (QC)/manufacturing sites that are sensitive, selective, reproducible (rugged), and robust.[23] GTI methods require sensitivities in the ng/mL range for drug concentrations in the several mg/mL range or ppm (relative to drug). The control level is determined by the allowable daily intake of the GTI based on the staged TTC level or TTC of 1.5 μg/day and the maximum daily dose of the drug. Table 4 illustrates the GTI control level in ppm as a function of maximum daily dose (MDD) of the drug at the TTC of 1.5 μg/day.

While several analytical techniques can commonly provide this level of sensitivity, it is imperative that the sensitive method also be highly selective. The need for selectivity is paramount and often more difficult than obtaining the required sensitivity since there are many impurities and degradants in APIs and drug products at these levels and the chance for interference and potential false positives increases with poor selectivity. These selective and sensitive methods must be able to support the testing of products from early development to late development and postlaunch. The need to conduct these trace-level analyses after the drug is on the market further challenges the analytical chemist as these methods also need to be rugged and robust such that they can be transferred and

TABLE 4 GTI Control Level as a Function of Maximum Daily Dose of the Drug

MDD of drug (mg)	GTI control level (ppm) at TTC (1.5 μg/day)
2000	0.8
1500	1.0
1000	1.5
750	2.0
500	3.0
250	6.0
200	7.5
150	10.0
100	15.0
75	20.0
50	30.0
25	60.0
20	75.0
10	150.0
5	300.0
1	1500.0

successfully executed at marketed product testing sites around the world. This challenge is not only related to method development but also to instrument resources (potentially complex equipment, such as mass spectrometers) and trained scientist capable of troubleshooting, maintaining the instrumentation, and executing the test on a routine basis. These challenges are further discussed in the following sections on analytical strategies for GTI testing, selecting appropriate methodologies and techniques, and method development and validation.

A. Analytical Strategies for GTI Testing

An analytical strategy for determining GTIs in intermediates, APIs, and/or drug products is to develop and ultimately validate sensitive, selective, robust, and rugged methods appropriate for the GTI control strategy and for the stage of development. In addition to developing methods to support the clinical testing and release of a drug, other methods may be required to determine the level of GTIs at multiple steps of a synthetic process to support the GTI control strategy. These methods will allow greater understanding of the process capabilities especially related to purgability (or tolerance) and may provide data that can justify testing upstream in the process at TTC levels or levels higher than the TTC. The use of reaction kinetics data that show the reduction of the GTI in the process stream to levels below a safety concern is also valuable in establishing a control strategy or mitigation of testing.[24] As projects mature and move through development, the length of clinical studies increases which in turn will cause a change in the allowable GTI level as per the staged TTC discussed above.[7–10] This will require methods with greater sensitivity and, potentially, selectivity.

The use of quality by design (QbD) principles in drug development should also be considered when testing for GTIs. The principle of QbD is that quality is built into the product from an in-depth knowledge of the process and appropriately established process controls. As knowledge is gained about the process, studies should be conducted within and at the boundaries of the manufacturing operating design space to understand the impact on the control of GTIs. By providing a thorough understanding of the level of GTIs in a process within the design space and proven acceptable ranges of the process parameters, mitigation of testing or testing up-stream at levels higher than the established TTC in commercial manufacture may be justified. Such a justification can only be accomplished through a well defined control strategy, data to support the strategy, understanding of the process, and direct dialog with health authorities regarding the intended GTI control strategy early in the drug development cycle. An example of utilizing QbD to justify upstream controls at levels above the TTC was recently published by Liu *et al.* for pazopanib hydrochloride.[25] These authors established process knowledge via testing for multiple GTIs at multiple stages of a process to determine the purgability and process tolerance of the GTIs. They also determined which process parameters influenced the purgability to assist in establishing allowable levels in early intermediates at the 0.1% level rather that at the TTC. This approach to GTI testing provides the ability to control the quality of the final product rather than waiting to test in the API, which would be too late.

B. Testing Methodology

As with any analytical determination, the physical and chemical characteristics of the analyte will determine the best technique and methodology. For GTI determinations, some form of a highly selective separation technique is required, and depending on the nature of the analyte, either LC[26–33] or GC[34–37] is most commonly utilized. There are examples of SFC,[38] CZE,[39] and IC[40] to perform GTI separations; however, these separation techniques have received little attention in the literature as GTI methods. For volatile, low-molecular-weight (MW) GTIs, GC may be a preferred technique for analysis while less volatile, API-related substances would LC. The separation technique should be chosen such that it provides a high level of selectivity especially if the detector is not particularly specific to the GTI.

The physical and chemical characteristics of the analyte and separation technique will also influence the best choice of a detector. The type of detector should be highly sensitive and even more advantageous, a detector with high sensitivity and selectivity for the GTI. For example, if an analyte is highly conjugated, then UV may be the preferred detector, or if the GTI contains a fluorophore, then fluorescence may be the best choice. While UV can be very sensitive, it tends not to provide selectivity for GTIs that have similar chromophores as the API and its related substances; thus UV detection is typically dependent on the separation to provide selectivity. GTIs with acidic or basic groups that are readily ionizable by atmospheric pressure ionization techniques may work well with a mass spectrometer as the detector. Mass spectrometry (MS) can provide very high selectivity as the detection is specific to the ionized mass-to-charge ratio (m/z) of the analyte. This high selectivity reduces the requirements for the separation and may allow faster method development and shorter run times. MS can also be very sensitive. The use of tandem MS can provide even greater selectivity and sensitivity as needed.[28,30] Other types of detectors such as charged aerosol detector (CAD), evaporative light scattering detector (ELSD), and electrochemical detection may also be viable for detecting GTIs. While CAD and ELSD can detect analytes that do not contain a chromophore, they tend to suffer from a lack of sensitivity for volatile analytes and limit the choice of mobile phase modifiers and composition.[28,41] The use of electrochemical detection[42] and fluorescence has been limited or nonexistent in the literature for detecting GTIs but may be used in situations where the analyte is amendable to these types of techniques. The most common detectors used with GC for GTI analyses are FID[19a,27] and MS.[34–36] Both types of detectors can provide high sensitivity and increased selectivity when using MS.

Examples of commonly used detectors with LC are discussed below, providing advantages and limitations when using these techniques to determine GTIs.

1. LC/UV

API and its related substances are most often determined using LC/UV, the pharmaceutical "gold standard" for performing impurity/purity assessments of APIs and drug products. The ability to use LC/UV methods developed for standard, nongenotoxic, impurities for GTI analyses is dependent on the dose of the drug and allowable level of the GTI. Table 5 shows the calculated w/w% as a function of the maximum daily dose of the drug and maximum daily intake of a GTI at 120 μg/day (single dose), 60 μg/day (less than 4 week exposure), and 1.5 μg/day (greater than 1 year exposure). From this table, one can determine based on the maximum daily dose of the drug, whether or not the standard LC/UV impurity method used to release the API or drug product for clinical trials has sufficient sensitivity to test for a GTI. For example, if one assumes the LC/UV impurity method has a method quantitation limit (MQL) of 0.03% (as noted with dark line in Table 5), sufficient selectivity for the GTI, and a known relative UV response factor for the GTI relative to the API (to provide w/w% measurement), then the standard LC/UV impurity method would have sufficient sensitivity for the GTI if (1) the maximum daily dose of drug is 400 mg or less and is intended for single day exposure,

TABLE 5 GTI Concentration as w/w% for GTI Allowable Daily Intake (ADI) of 1.5, 60, and 120 μg/day as a Function of Maximum Daily Dose of the Drug

	GTI (w/w%)		
MDD of drug (mg)	ADI of GTI 120 μg	ADI of GTI 60 μg	ADI of GTI 1.5 μg
1000	0.0120	0.0060	0.00015
900	0.0133	0.0067	0.00017
800	0.0150	0.0075	0.00019
700	0.0171	0.0086	0.00021
600	0.0200	0.0100	0.00025
500	0.024	0.0120	0.00030
400	0.030	0.0150	0.00038
300	0.040	0.0200	0.00050
200	0.060	0.030	0.00075
150	0.080	0.040	0.00100
100	0.120	0.060	0.00150
90	0.133	0.067	0.00167
80	0.150	0.075	0.00188
75	0.160	0.080	0.00200
70	0.171	0.086	0.00214
50	0.240	0.120	0.00300
40	0.300	0.150	0.00375
30	0.400	0.200	0.00500
25	0.480	0.240	0.00600
20	0.600	0.300	0.00750
10	1.20	0.600	0.015
5	2.40	1.200	0.030
1	12.00	6.000	0.15
0.5	24.00	12.00	0.30

or (2) the exposure is less than 4 weeks and a maximum daily dose is 200 mg or less, or (3) the exposure is greater than a year and the maximum daily dose of the drug is 5 mg or less. Similar calculations can be made for the other staged TTC limits, drug dose, and MQLs. In general, the likelihood of using the LC/UV release method diminishes as the exposure duration increases unless the drug is very potent (MDD <5 mg). This does not, however, preclude the use of LC/UV as a viable technique to perform GTI analyses at levels much lower than 0.03% (300 ppm).

Much lower levels can be achieved using LC/UV by increasing the amount of material injected on column either by increasing the sample concentration or by increasing the injection volume (or both). Examples of this are shown in Figures 1 and 2 and others can be found in the literature.[32,33] Typical column loading in quantitative non-GTI methods is on the order of 2–3 μg of the main component (i.e., the API). In Figure 1, 4 mg of API was loaded on column (20 mg/mL sample concentration, 200 μL injection) and shows a tightly focused main peak with minimal tailing due to the weak sample diluent (water) and shows sufficient sensitivity

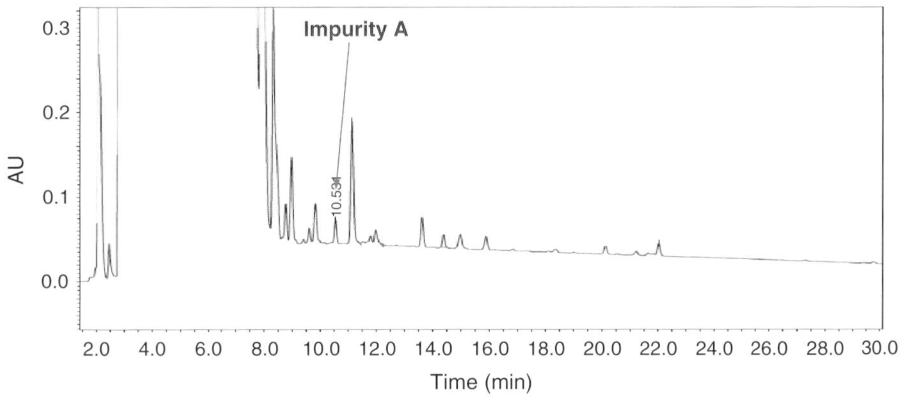

FIGURE 1 **Example of a chromatogram using high column loading to observe a minor impurity (5 ppm) in an acetaminophen derivative.**

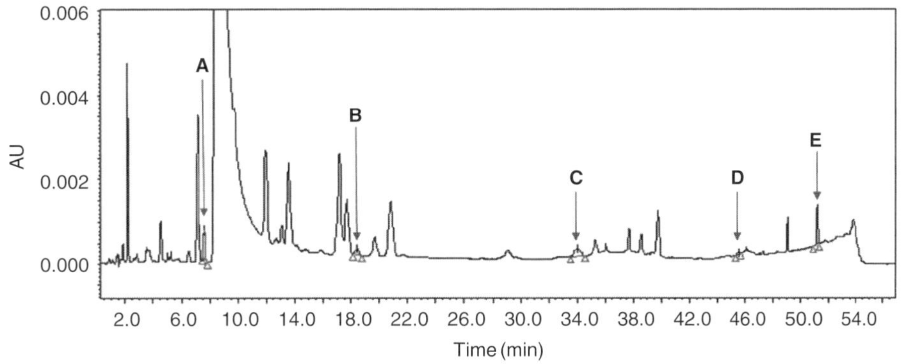

FIGURE 2 **Chromatogram of five GTIs at 50 ppm in a drug penultimate using high column loading.**

and selectivity to detect "Impurity A" at 5 ppm or 100 ng/mL. A more typical high column loading chromatogram is shown in Figure 2 where considerable tailing is observed with 30 μg of a drug penultimate injected on column (3 mg/mL sample concentration, 10 μL injection). Five GTIs at 50 ppm (150 ng/mL) were separated and labeled A–E. In both Figures 1 and 2, there are a considerable number of impurities at similar levels as the GTIs of interest. One should note that given the number and level of impurities, varying impurity profile from batch to batch, as well as the tailing of the main component, there is a considerable risk of interference from the coelution of other impurities or interference from the partial overlap with the main component. LC/UV can often provide the required sensitivity but often lacks the necessary specificity; thus, the success of implementing a routine LC/UV GTI assay may be limited. Chances of success improve if the test article is of higher purity, the GTI elutes prior to the main component, or some additional selectivity can be obtained when the analyte has a UV absorbance that is significantly different than the API or its related substances, for example, nitroaromatics which can exhibit strong absorption in the 300–400 nm range[43] where the drug and its related substance may have limited absorption in this range.

2. LC/MS

The use of LC/MS provides a greater level of specificity compared to LC/UV and often allows for faster method development with fewer interferences and reduced run times. LC method development with MS detection only requires the separation of the analyte from the main component or other high-level impurities that could cause ion suppression or enhancement thus greatly reducing the method development time. This advantage is derived by the inherent selectivity of MS. The measurement of an analyte's *m/z* versus a UV absorbance at an expected retention time is far more specific. Figure 3 shows an example of an LC/MS method using

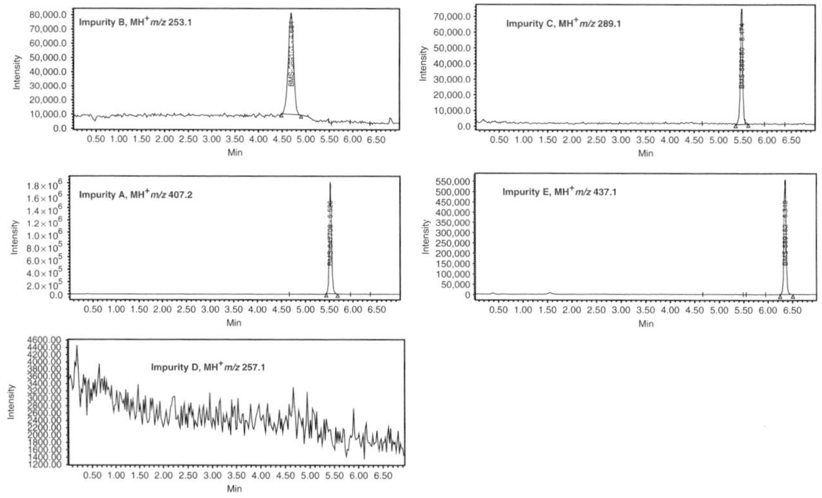

FIGURE 3 **Ion Chromatograms of five GTIs at 50 ppm in a drug penultimate using LC/MS.**

single ion monitoring (SIM) that was developed to determine the same GTI impurities as shown in Figure 2. This method was able to reduce the run time from 60 to 7 min with sufficient sensitivity for four of the five GTIs. It's important to note that atmospheric pressure ionization techniques (electrospray and atmospheric pressure chemical ionization) are not universal and thus, some molecules, such as Impurity D, would not ionize efficiently to detect it at the 50 ppm level.

Pierson *et al.* published a similar example as the one shown above for the determination of hyrdazine.[19a] In this example, hydrazine was derivatized with benzaldehyde to form 1,2-dibenzylidenehydrazine to improve the sensitivity of the LC/UV method. Despite this effort, sufficient selectivity was not obtained due to interference from the sample matrix. The use of LC/MS with SIM at *m/z* 209.1 to detect 1,2-dibenzylidenehydrazine provided sufficient selectivity and sensitivity to determine less than 1 ppm in derivatized samples.

3. LC/MS/MS

LC/MS/MS provides increased selectivity over LC/MS in cases where interference occurs from coeluting matrix interferences that produce a signal at the same *m/z* as the GTI. This can occur from an impurity or matrix component with the same MW as the GTI or may occur from an isotope of an impurity with a similar MW as the GTI. MS/MS can also provide additional sensitivity, especially in cases where high chemical noise may limit sensitivity in an LC/MS assay. For nearly two decades, the use of LC/MS/MS to quantitate drugs and their metabolites in biological matrices has been a successful and powerful analytical tool.[44] In a biological sample, the analyte is structurally very different from the matrix and can be selectively detected. For the determination of a GTI that is a related substance of the API, the matrix is very similar structurally and in large excess 1,000,000× as opposed to a biological matrix that may be 100–10,000× more concentrated than the analyte of interest. For determination of GTIs, there is a greater propensity of interference even when using a highly specific methodology like LC/MS/MS for a GTI that is a related substance of the API compared to a bioanalytical assay. In these situations, the LC separation must still be utilized to provide sufficient specificity from potential interference. Such an example was recently published by Borman *et al.* which detailed the development, validation, and transfer to manufacturing sites of an LC/MS/MS method used to determine (4-(4-fluorophenyl)-1-methyl-1,2,3,6-tetrahydropyridine), FMTP, a highly neurotoxic impurity of paroxetine.[30] The LC/MS/MS method used selective reaction monitoring (SRM) to detect FMTP. The SRM transition used the protonated molecule of FMTP at *m/z* 192 and a product ion at *m/z* 44 produced from a Retro-Diels–Alder rearrangement upon low-energy collision-induced dissociation in the triple quadrupole mass spectrometer as shown in Figure 4.

It was found that thermal degradation or in-source fragmentation of the API (paroxetine) in the ion source of the mass spectrometer could also produce signals in the SRM channel. An example is shown in Figure 5 of the API degrading in the ion source to form a different precursor ion at *m/z* 192 that can fragment in the collision cell to produce the same product ion at *m/z* 44.

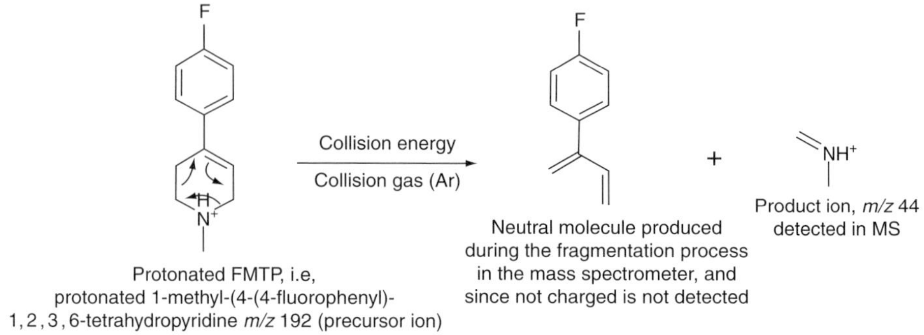

FIGURE 4 **Fragmentation of FMTP (4-(4-fluorophenyl)-1-methyl-1,2,3,6-tetrahydropyridine). Reproduced with permission from *J. Pharm. Biomed. Anal.* 2008, 48, 1082–1089.**

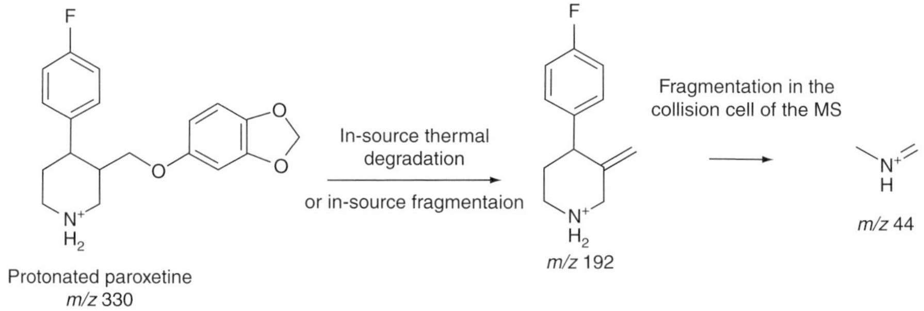

FIGURE 5 **In-source thermal degradation or fragmentation of paroxetine to produce an interference in the SRM channel for FMTP. Reproduced with permission from *J. Pharm. Biomed. Anal. 2008*, 48, 1082–1089.**

Such interferences, as pointed out in the Borman manuscript, require the chromatographic separation to provide the additional specificity for the GTI by insuring that the API and its potentially coeluting related substances that can interfere with the GTI are sufficiently separated.

4. LC/Fluorescence

Fluorescence can provide high sensitivity, and in some cases, selectivity as is the case for the determination of *meta*-anisidine, a known mutagen.[45,46] Multiple approaches to the selective determination of trace *meta*-anisidine in an API are viable. One example is the use of LC/MS to determine *meta*-anisidine, a potential degradant of a drug in a tablet formulation as shown in Figure 6. The LC/MS method utilized here is capable of determining sub-ppm levels of *meta*-anisidine in a tablet using selected ion monitoring (SIM) at *m/z* 124 and electrospray ionization in the positive ion mode.

The LC/MS method is fast with a run time of 1.5 min, selective, and sensitive. However, other approaches are also possible such as LC/fluorescence (Figure 7). For this example, the drug and its related substances as well as the excipients from

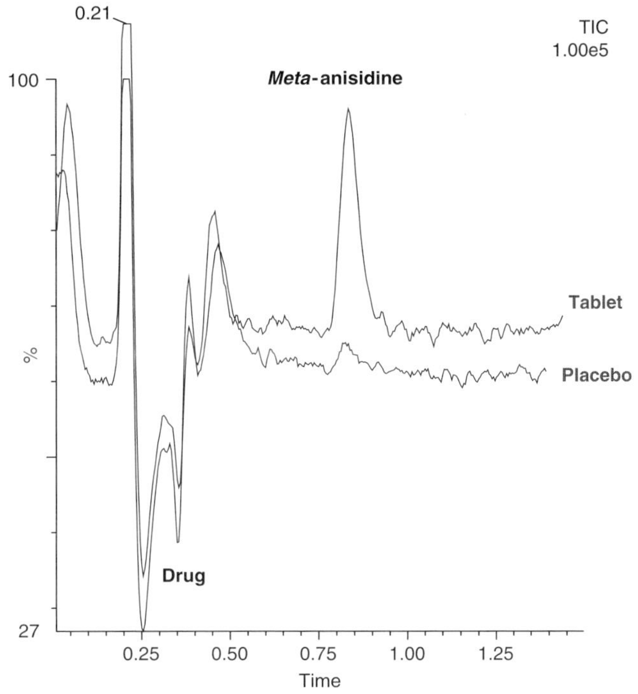

FIGURE 6 Trace-level determination of *meta*-anisidine in a tablet formulation using LC/MS.

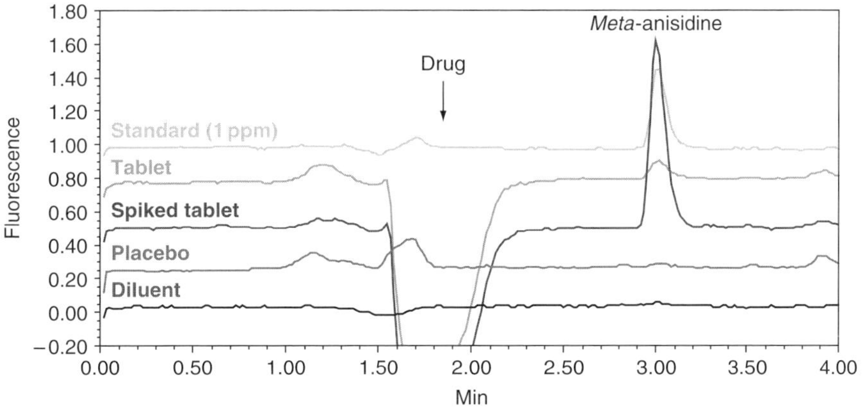

FIGURE 7 Trace-level determination of *meta*-anisidine in a tablet formulation using LC/fluorescence.

the tablet formulation do not fluoresce providing a very selective and sensitive sub-ppm level, alternative to LC/MS. The use of fluorescence as a detector for amendable GTIs may be an attractive alternative to LC/MS especially in cases where the method will be transferred to a QC environment.

The use of LC or GC coupled with MS is the preferred approach given its inherent specificity, sensitivity, fast method development, and broad applicability to many GTIs. LC/MS/MS is a useful technique especially in cases where additional sensitivity or specificity is needed. Fluorescence may also prove to be a highly sensitive and selective detector for some GTIs. Other sensitive detectors may also be useful including UV, electrochemical, or FID for GC. In general, one should use the technique that is most suitable for the GTI of interest that provides the required selectivity, sensitivity, and method ruggedness and robustness.

C. Method Development and Validation

The process of developing and validating trace-level GTI methods is similar to other types of quantitative assays. Methods must be developed to provide sufficient specificity, sensitivity, precision, accuracy, linearity, and range and must be validated for their intended use.[23] Other considerations in developing a GTI method are how the sample preparation will be conducted and whether the method will be a limit test or a quantitative test. The approach to sample preparation and factors that can influence the choice of whether to develop a limit test or quantitative test will be discussed.

1. Sample preparation

A critical feature that is critical to GTI method development is sample preparation. Most GTIs are reactive by nature. Establishing appropriate diluents which are amendable to the separation technique, are capable of solubilizing the sample at high concentration (mg/mL), and provide sufficient analyte (GTI) stability is essential to develop a rugged and robust GTI method. Care needs to be taken as the GTI should be stable throughout the solution preparation process, which may include heating or sonication, etc., and also should not be generated during sample preparation. In addition, the sample preparation can also play a role in overcoming matrix interference when the selectivity of the separation technique or detector is insufficient.

Isolated solid samples (API, intermediates, etc.) are favored for GTI testing as this allows for generating sample solutions at high concentration, typically at several mg/mL levels, to improve the sensitivity of the method. Trying to determine GTIs in a formulated drug product solution, tablet, or suspension rather than an isolated solid is often more challenging, as the concentration of the drug may be far too low to readily conduct a trace-level GTI analysis and the excipients generate a more complex matrix which may cause interference. In situations where sufficient sample concentration cannot be obtained or the sample matrix is interfering with the measurement, enrichment can be obtained using extraction techniques such as solid-phase extraction (SPE), solid-phase microextraction (SPME), liquid–liquid extraction (LLE), or liquid-phase microextraction (LPME).[47] While some extraction techniques can be labor intensive and increase the probability of contamination, interference, or poor recovery (analyte loss, instability), they can also be effective at analyte enrichment. Colon *et al.* assessed the use of SPME, LPME, and SPE for determining seven alkyl sulfonate esters in an API at the 5 ppm level by GC/MS.[37] The authors suggested that the best approach for detection of the sulfonate esters in their APIs was achieved using SPME; however, LPME and SPE also were viable alternatives especially for APIs of limited aqueous solubility. Miller *et al.* developed a GC/MS method to determine trace levels of ethyl methane sulfonate (EMS) which used LLE to overcome *in situ* generation of EMS when analyzing drug product solutions containing an API that was a MSA salt and ethanol.[24] Upon injection of these solutions, direct or diluted, into the heated injection liner of the GC, MSA and ethanol would react to form EMS. This was overcome by utilizing an LLE procedure with dichloromethane to remove the MSA and eliminate the formation of EMS in the injector.

A useful tool in determining trace level, unstable analytes is to use chemical derivatizations.[48] Reacting a GTI with a reagent can provide a more stable product with improved detectability and/or specificity. Reagents can be chosen to provide a useful chromophore, fluorophore, or a particularly basic or acidic group to improve the detectability of the GTI by UV, fluorescence, or MS. They can also provide products with different polarity to improve specificity or extractability to overcome or eliminate matrix interference. Examples of using chemical derivatizations in the GTI literature have focused on a few classes of common GTIs including sulfonate esters,[29,34,35] arylamines and aminopyridines,[31] and hydrazine.[19a,36]

Pentafluorothiophenol (PFTP) is a commonly used reagent for various sulfonate esters, as it improves the volatility, stability, and makes the product less polar and easier to extract by head space GC methods. Alzaga *et al.* developed and validated a general head space GC/MS method to determine ppb levels of various mesylates, besylates, tosylates, and sulfates utilizing sample derivatization with PFTP.[35] Not only is this a sensitive general method for sulfonate esters, but this also provides a means of significantly reducing potential matrix interference by analyzing only the volatile components of the sample, as the nonvolatile components (API and impurities) are not transferred to the GC. This allows much greater sample concentrations to be used without jeopardizing the analysis. An *et al.* developed and validated an LC/MS method for determining sulfonate esters by using triethylamine (TEA) as the reagent.[29] When TEA reacts with a sulfonate ester, it produces quaternary ammonium ions that are readily detected by electrospray LC/MS in the positive ion mode and greatly improves the sensitivity of the method compared to analyzing the GTI directly. The products of TEA and sulfonate esters are also much more stabile compared to the native ester. This does not preclude the direct analysis of sulfonate esters as shown by Taylor *et al.* but may require "prep and inject" procedures due to a lack of analyte stability.[26] Vanhoenacker *et al.* developed and validated an LC/MS method to determine sub-ppm levels of a select set of arylamines and aminopyridines in pharmaceutical products using hexylchloroformate (HCF) as a derivatization reagent.[31] The acyl-derivatives provided flexibility in method selectivity as addition of the acyl chain increased the chromatographic retention compared to their underivatized form. By changing the column, mobile phase composition, or method gradient, it can be tailored to a particular set of arylamines or aminopyridines in various APIs. Finally, Sun *et al.* provided an example of a general head space GC/MS method for the determination of hydrazine in various APIs by using acetone as the derivatization reagent.[36] The reaction of acetone with hydrazine produces an acetone azine with a boiling point amendable to head space GC. As previously mentioned, head space GC reduces or eliminates matrix interference as only the volatile components of the mixture are transferred to the GC column thus allowing for high sample loading. This approach provided a method sensitivity for the acetone azine of 0.1 ppm. For APIs that may have used acetone in the synthetic process, acetone azine may exist in trace amounts. For these situations, Sun *et al.* utilized acetone-d_6 as the derivatization reagent to change the m/z of the product and successfully determine residual levels of hydrazine in the API.

Another facet of sample preparation for GTI methods is the potential for contamination. Given that most GTI methods are detecting analytes with concentrations of ng/mL or pg on column sensitivity, it only takes a few micrograms of the GTI to contaminate a sample and potentially provide a false positive failure of a batch. The probability of contamination increases especially in laboratories that test a particular process intermediate (also a GTI) at standard concentrations for an impurity/degradant method. Trying to analyze for trace levels of the process intermediate with a GTI method can be readily complicated from glassware contamination, contamination of a spatula, balance, or LC vials. It can also be problematic if the same LC or GC is used to test both the impurity/degradant method and trace GTI method. Extra care must be taken to prevent cross contamination.

2. Quantitative Versus Limit Test

A critical aspect to consider when developing a GTI method is whether the method will be a limit test or a quantitative assay. A limit test method will only provide a pass or fail result while a quantitative assay method provides a determined level of the analyte. The use of a limit test is desirable, as it require less validation and less time to execute (fewer injections).[23] However, such considerations must be evaluated on a case-by-case basis and will depend on a number of situations including the stage of development, number of GTIs, certainty of the clinical dose, and purgability or control of the GTI(s) in the process. A risk assessment should be conducted to determine if the choice of method will have a potential, significant impact on the ability to release API or drug product. This is a more common issue for early stage compounds as the clinical dose and process are often still being optimized.

For compounds where a single GTI or multiple GTIs have a unique toxicological pathway (i.e., each GTI class has its own limit) and are well controlled in the process, a limit test may be the best approach. For drugs in early development where the process control or clinical dose is not well defined or multiple GTIs with a common toxicological pathway must be summed, quantitative assay method may be more appropriate. By developing a limit test method for multiple GTIs that must be summed, a single GTI may surpass its individual limit and fail a batch while the sum of the multiple GTIs may pass the summed control limit. While these examples provide extreme opposite situations, those cases where conditions are not so well defined will require further risk assessment as to what type of method to ultimately develop.

VI. SUMMARY AND CONCLUSIONS

The control of GTIs in investigational medical products has become an established requirement for drug development based on the expectations defined in the recently issued health authority guidance documents. This is a rapidly evolving area, and changes are expected as these requirements further mature. Meeting the stringent control levels defined by the TTC requires the use of a multipronged approach that combines risk assessment tools, process control strategies, and specialized analytical techniques. In this chapter, we have described a variety of analytical approaches which can be used to quantitate impurities at the ppm level with an assessment of advantages and limitations. In most cases, selectivity is the most difficult figure of merit to meet. The use of chromatography along with a selective detector amendable to the GTI is critical to developing a reliable, trace-level, quantitative method. In addition to the separation and detection of GTIs, consideration for the sample preparation as well as the type of assay, limit test, or quantitative assay, must also be evaluated. As the analytical challenges extend beyond development and into the QC laboratory, GTI methods must also be robust. In addition, the use of nonconventional approaches in the QC environment, for example, LC/MS or LC/MS/MS, may require new equipment not traditionally utilized and trained personnel, to successfully execute these types of trace-level analyses on a routine basis.

ACKNOWLEDGMENTS

The authors thank Mark Bolgar, Michelle Madden, and Baiming Xiao from Bristol-Myers Squibb Company for data utilized in this chapter.

REFERENCES

1. International Conference on Harmonisation (ICH). Guideline Q3A(R): Impurities in New Drug Substances: Feburary 2002., ICH Website, http://www.ich.org/cache/compo/276-254-1.html.
2. Kasper, P., Kirkland, D., Lebanc, B. and Sjoberg, P. Spindlerp. Acceptability of Low Levels of Genotoxic Impurities in New Drug Substances: Conclusions of the Drug Information Association (DIA)/European Medicines Agency (EMEA) Workshop, October 27–28, 2003. in London, UK.
3. Committee for Proprietary Medicinal Products (CPMP)* CPCP/SWP/5199/02/draft2. London: CPMP December 2002. * Note: CPMP now called Committee for Medical Products for Human Use (CHMP).
4. European Medicines Agency (EMEA), Committee for Medical Products for Human Use (CHMP). Guideline on the limits of genotoxic impurities, 2006. CPMP/SWP/5199/02; London, UK. June 2006. EMEA Website, <http://www.emea.europa.eu/pdfs/human/swp/519902en.pdf>.
5. (a). Rulis, A. M. Food Safety Assessment. ACS Symposium Series 484:132–139, 1992. (b). Cheeseman, M. A., Machuga, E. J. and Bailey, A. B. A tiered approach to threshold of regulation. *Food Chem. Toxicology* 37:387–412, 1999. (c) CRF 170.39, 1998. Threshold of Regulation for substances used in food contact articles. FDA Website <http://www.access.gpo.gov/nara/crf/waisidx_98/21cfr170_98.html>.
6. Kroes, R., Renwick, A. G., Cheesman, M., Kleiner, J., Mangelsdorf, I., Piersma, A., Schilter, B., Schlatter, J., van Schothorst, F. and Vos, J. G. Structure-based thresholds of toxicological concern

(TTC): guidance for application to substances present at low levels in the diet. *Food Chem. Toxicology* 42:65–83, 2004.

7. Muller, L., Mauthe, R. J., Riley, C. M., Andino, M. M., De Antonis, D., Beel, C., DeGorge, J., De Knaep, A. G. M., Ellison, D., Fagerland, J. A., Frank, R., Fritschel, B., Galloway, S., Harpu, E., Humfrey, C. D. N., Jacks, A. S., Jagota, N., Mackinnon, J., Mohan, G., Ness, D. K., O'Donovan, M. R., Smith, M. D., Vudathala, G. and Yotti, L. A rationale for determining, testing, and controlling specific impurities in pharmaceuticals that possess potential for genotoxicity. *Regul. Toxicol. Pharmacol.* 44:198–211, 2006.
8. Bos, P. M. J., Baars, B. J. and van Raaij, M. T. M. Risk assessment of peak exposure to genotoxic carcinogens: a pragmatic approach. *Toxicol. Lett.* 151:43–50, 2004.
9. Question and Answers on the CHMP Guideline on the limits of genotoxic impurities, EMEA/CHMP/SWP/4311994/2007, revision 1; European Medicines Agency (EMEA), London, UK, June 2008. EMEA Website, <http://www.emea.europa.eu/pdfs/human/swp/43199407en.pdf>.
10. FDA Draft Guidance, Guidance for Industry Genotoxic and Carcinogenic Impurities in Drug Substances and Products: Recommended Approaches. FDA Web site, http://www.fda.gov/downloads/Drugs/GuidanceComplianceRegulatoryInformation/Guidances/ucm079235.pdf.
11. Jacobson-Kram, D. and Jacobs, A. Use of Genotoxicity Data to Support Clinical Trials or Positive Genetox Findings on a Candidate Pharmaceutical or Impurity . . .Now What? *Int. J. Toxicology* 24:129–134, 2005. (b). McGovern, T. and Jacobson-Kram, D. Regulation of Genotoxic and Carcinogenic Impurities in Drug Substances and Products. *Trends Anal. Chem.* 25:790–795, 2006.
12. (a). Kirklan, D. and Snodin, D. Setting Limits for Genotoxic Impurities in Drug Substances: Threshold-Based and Pragmatic Approaches. *Int. J. Pharm. Med.* 18:197–207, 2004. (b). Kasper, P. Assessment and Acceptance of Thresholds of Genotoxic Impurities in New Drug Substances: A Regulatory Perspective. *Int. J. Pharm. Med* 18:209–214, 2004. (c). Leblanc, B., Charuel, C., Ku, W. and Ogilvie, R. Acceptability of Low Levels of Genotoxic Impurities in New Drug Substances: Case Reports. *Int. J. Pharm. Med.* 18:215–220, 2004. (d). Humfrey, C. H. Recent Developments in the Risk Assessment of Potentially Genotoxic Impurities in Pharmaceutical Drug Substances. *Toxicol. Sci.* 100:24–28, 2007. (e). Bower, N. A review of the new draft FDA guidance on genotoxic and carcinogenic impurities. *Regulatory Rapporteur* 6:13–15, 2009.
13. (a). Delaney, E. J. An impact analysis of the application of the threshold of toxicological concern concept to pharmaceuticals. *Regul. Toxicol. Pharmacol.* 49:107–124, 2007. (b). Snodin, D. EU Guideline on Genotoxic Impurities Needs Updating. *The Regulatory Affairs J.: Pharma.* 19:1–8, 2008. (c). Snodin, D. EU Guideline on Genotoxic Impurities Needs Updating - Part II. *The Regulatory Affairs J. Pharma.* 19(10):1–8, 2008. (d). Snodin, D. and Vudathala, A. case for a regulatory rethink GAPS News Magazine, pp. 20–27, February 2009. (e). Munro, I. C. and Renwick, A. G. Danielewska-Nikiel, B., The Threshold of Toxicological Concern (TTC) in risk assessment. *Toxicol. Lett* 180:151–156, 2008.
14. EMEA, Press Release: European Medicines Agency Announces Recall of Viracept, London, June 6, 2007. <http://www.emea.europa.eu/emea/251283/2007>.
15. EMEA, Letter to Marketing Authorizatyion Holders, London, 27TH February 2008. <http://www.emea.europa.eu/emea/CMDh/98694/2008>.
16. (a). Elder, D. P., Teasdale, A. and Lipczynski, A. M. Control and analysis of alkyl esters of alkyl and aryl sulfonic acids in novel active pharmaceutical ingredients (APIs). *J. Pharm. Biomed. Anal.* 46:1–8, 2008. (b). Snodin, D. J. Residues of genotoxic alkyl mesylates in mesylate salt drug substances: Real or imaginary problems? *Regul. Toxicol. Pharmacol.* 45:79–90, 2006. (c). Teasdale, A., Eyley, S. C., Delaney, E., Jacq, K., Taylor-Worth, K., Lipczynski, A., Reif, V., Elder, D. P., Facchine, K. L., Golec, S., Oestrich, R. S., Sandra, P. and David, F. Mechanism and Processing Parameters Affecting the Formation of Methyl Methanesulfonate from Methanol and Methanesulfonic Acid: An Illustrative Example for Sulfonate Ester Impurity Formation. *Org. Proc. Res. Dev.* 13:429–433, 2009. (d). Chan, L. C., Cox, B. and Sinclair, R. S. Selective Hydrolysis of Methanesulfonate Esters. *Org. Proc. Res. Dev.* 12:213–217, 2008.
17. Paulson, B. Viracept recall Implicates API GMPs. *International Pharmaceutical Quality* 1:1–5, 2007.
18. Muller, L. and Singer, T. EMS in Viracept—The course of events in 2007 and 2008 from the non-clinical safety point of view. *Toxicol. Lett.* 2009. doi:10.1016/j.toxlet.2009.02.005.

19. (a). Pierson, D. A., Olsen, B. A., Robbins, D. K., DeVries, K. M. and Varie, D. L. Approaches to Assessment, Testing Decisions, and Analytical Determination of Genotoxic Impurities in Drug Substances. *Org. Process Res. Dev.* 13:285–291, 2009. (b). Dobo, K. L., Green, N., Cyr, M. O., Carson, S. and Ku, W. W. The application of structurebased assessment to support safety and chemistry diligence to manage genotoxic impurities in active pharmaceutical ingredients during drug development. *Regul. Toxicol. Pharmacol.* 44:282–293, 2006. (c). Snodin, D. Genotoxic Impurities in Drug Substances. *Org. Process Res. Dev.* 13:409, 2009.
20. DEREK software, <https://www.lhasalimited.org/index.php/derek/>.
21. MulitCASE software, <http://www.multicase.com>.
22. Ames, B. A. and McCann, J. Detection of carcinogens as mutagens in the Salmonella/microsome test: assay of 300 chemicals: discussion. *Proc. Nat. Ach. Sci.* 73:950–954, 1976.
23. International Conference on Harmonisation (ICH). Guideline Q2(R1): Validation of Analytical Procedures: Text and Methodology: November 1996, ICH Website, <http://www.ich.org/LOB/media/MEDIA417.pdf>.
24. Miller, S. A., Ju, J., Cucolo, M., Dai, J. Y., Jia, J., Narra, K., Smith, A. M. and Aubrey, A.-F. Poster Presentation at The 52nd ASMS Conference on Mass Spectrometry and Allied Topics, 2004. Nashville, TN.
25. Liu, D. Q., Chen, T. K., McGuire, M. A. and Kord, A. S. Analytical control of genotoxic impurities in the pazopanib hydrochloride manufacturing process. *J. Pharm. Biomed. Anal.* 50:144–150, 2009.
26. Taylor, G. E., Gosling, M. and Pearce, A. Low level determination of p-toluenesulfonate and benzenesulfonate esters in drug substance by high performance liquid chromatography/mass spectrometry. *J. Chrom. A* 1119:231–237, 2006.
27. Elder, D. P., Lipczynski, A. M. and Teasdale, A. Control and analysis of alkyl and benzyl halides and other related reactive organohalides as potential genotoxic impurities in active pharmaceutical ingredients (APIs). *J. Pharm. Biomed. Anal.* 48:497–507, 2008.
28. Yuabova, Z. Y., Holschlag, D. R., Rodriguez, S. A., Qin, C., Papov, V. V., Qiu, F., McCaffrey, J. F. and Norwood, D. L. Genotoxic impurities: a quantitative approach. *J. Liq. Chrom. Relat. Tech.* 31:2318–2330, 2006.
29. An, J., Sun, M., Bai, L., Chen, T., Liu, D. and Kord, A. A practical derivatization LC/MS approach for determination of trace level alkyl sulfonates and dialkyl sulfates genotoxic impurities in drug substances. *J. Pharm. Biomed. Anal.* 48:1006–1010, 2008.
30. Borman, P. J., Chatfield, M. J., Crowley, E. L., Eckers, C., Elder, D. P., Francey, S. W., Laures, A. M.-F. and Wolff, J.-C. Development, validation and transfer into a factory environment of a liquid chromatography tandem mass spectrometry assay for the highly neurotoxic impurity FMTP (4-(4-fluorophenyl)-1-methyl-1,2,3,6-tetrahydropyridine) in paroxetine active pharmaceutical ingredient (API). *J. Pharm. Biomed. Anal.* 48:1082–1089, 2008.
31. Vanhoenacker, G., Dumont, E., David, F., Baker, A. and Sandra, P. Determination of arylamines and aminopyridines in pharmaceutical products using in-situ derivatization and liquid chromatography–mass spectrometry. *J. Chrom. A* 1216:3563–3570, 2009.
32. Raman, N. V. V. S. S., Reddy, K. R., Prasad, A. V. S. S. and Ramakrishna, K. Development and validation of RP-HPLC method for the determination of genotoxic alkyl benzenesulfonates in amlodipine besylate. *J. Pharm. Biomed. Anal.* 48:227–230, 2008.
33. Soman, A., Jacob, S. and Swanek, F. Low Level Determination of 4-Amino-2-Ethoxy-Cinnamic Acid and its Ethyl Ester in Drug Substance and its Formulation Prototypes by HPLC–UV-DAD. *J. Chrom. Sci.* 47:315–319, 2009.
34. Jacq, K., Delaney, E., Teasdale, A., Eyley, S., Taylor-Worth, K., Lipczynski, A., Reif, V. D., Elder, D. P., Facchine, K. L., Oestrich, R. S., Sandra, P. and David, F. Development and validation of an automated static headspace gas chromatography–mass spectrometry (SHS-GC–MS) method for monitoring the formation of ethyl methane sulfonate from ethanol and methane sulfonic acid. *J. Pharm. Biomed. Anal.* 48:1339–1344, 2008.
35. Alzaga, R., Ryan, R. W., Taylor-Worth, K., Lipczynski, A. M., Szucs, R. and Sandra, P. A generic approach for the determination of residues of alkylating agents in active pharmaceutical ingredients by in situ derivatization–headspace–gas chromatography–mass spectrometry. *J. Pharm. Biomed. Anal.* 45:472–479, 2007.

36. Sun, M., Bai, L. and Liu, D. Q. A generic approach for the determination of trace hydrazine in drug substances using in situ derivatization-headspace GC–MS. *J. Pharm. Biomed. Anal.* 49:529–533, 2009.
37. Colon, I. and Riholl, S. M. Determination of methyl and ethyl esters of methanesulfonic, benzenesulfonic and p-toluenesulfonic acids in active pharmaceutical ingredients by solid-phase microextraction (SPME) coupled to GC/SIM-MS. *J. Pharm. Biomed. Anal.* 39:477–485, 2005.
38. Huybrechts, T. Genotoxic Impurities 2007, September 24th, 2007. Hilton Prague.
39. Hansen, S. H. and Sheribah, Z. A. Comparison of CZE, MEKC, MEEKC and non-aqueous capillary electrophoresis for the determination of impurities in bromazepam. *J. Pharm. Biomed. Anal.* 39:322–327, 2005.
40. Jagota, N. K., Chetram, A. J. and Nair, J. B. Determination of trace levels of hydrazine in the penultimate intermediate of a novel anti-infective agent. *J. Pharm. Biomed. Anal.* 16:1083–1087, 1998.
41. Megoulas, N. C. and Koupparis, M. A. Twenty years of evaporative light scattering detection. *Crit. Rev. Anal. Chem.* 35:301–316, 2005.
42. Jiranek, I., Peckova, K., Kralova, Z. and Moreira, J. C. Barek, The use of silver solid amalgam electrode for voltammetric and amperometric determination of nitroquinolines. *J. Electrochimica Acta* 54:1939–1947, 2009.
43. NIST Webbook, http://webbook.nist.gov.
44. Ackermann, B. L., Berna, M. J. and Murphy, A. T. Recent advances in use of LC/MS/MS for quantitative high-throughput bioanalytical support of drug discovery. *Curr. Top. Med. Chem.* 1:53–66, 2002.
45. Zeiger, E., Anderson, B., Haworth, S., Lawlor, T. and Mortelmans, K. Salmonella mutagenicity tests: V. Results from the testing of 311 chemicals. *Environ. Mol. Mutagen.* 19(Suppl. 21):2–141, 1992.
46. Haworth, S., Lawlor, T., Mortelmans, K., Speck, W. and Zeiger, E. Salmonella mutagenicity results for 250 chemicals. *Environ. Mutagen.* 5(Suppl. 1):3–142, 1983.
47. Handley, A. J. Extraction Methods in Organic Chemistry (Sheffield Analytical Chemistry Series), Taylor & Francis, Inc., 1999. ISBN 0849397405.
48. Knapp, D. R. Handbook of Analytical Derivatization Reactions, J. Wiley and Sons, 1979. ISBN 047103469X.

5 QUALITY BY DESIGN AND THE IMPORTANCE OF PAT IN QbD

ROBERT W. BONDI, Jr. AND JAMES K. DRENNEN, III

Mylan School of Pharmacy, Duquesne University, Pittsburgh, Pennsylvania, USA

I. INTRODUCTION

With a distinctively rapid pace and uniquely evolutionary scope, regulatory bodies around the globe have, with the new millennium, promoted a significantly more modern philosophy regarding pharmaceutical manufacturing. This paradigm involves numerous new concepts that modernize drug development and manufacturing while simultaneously minimizing risk to consumers.

On August 21, 2002, the United States Food and Drug Administration (FDA) announced a new initiative to enhance and modernize FDA's regulations of pharmaceutical manufacturing and drug product quality. This initiative is known as "Pharmaceutical CGMPs for the 21st Century: A Risk-Based Approach." The primary objective of the FDA is to minimize the risk to consumers through application of scientific risk-based approaches to pharmaceutical development and manufacturing. The goals of the initiative are to incorporate risk management and quality systems concepts, ensure that the latest scientific advances in pharmaceutical manufacturing and technology are adopted by manufacturers, and ensure the efficient use of resources to address the most critical health risks.[1] The guidance, released in final form in September 2004, outlined a paradigm shift to take place within the FDA in the coming years and direction for continuous improvement of the pharmaceutical industry through innovation in pharmaceutical manufacturing. In the document, the FDA recognized that reduction in variability through process understanding produces improvements in quality and productivity, rewarding both the consumer and the manufacturer.

The initiative proposed major change in the regulation of pharmaceutical manufacturing. Historically, the regulatory environment has been perceived by industry to be one that hindered the adoption of modern methods of manufacturing that would enable frequent process optimization, requiring submission of documentation even for what some consider to be insignificant changes to a process. As a result, pharmaceutical companies were reluctant to implement changes to their manufacturing processes. The initiative encouraged the use of the latest advances in manufacturing science, quality management systems, and risk management, which in turn results in the most robust manufacturing processes. Such robust processes enable an environment where process adjustments allow the manufacturer to consistently produce the highest quality pharmaceutical products. Manufacturing processes that are well characterized and which utilize such modern approaches as described in the initiative will, within the "risk-based" approach to regulation, be less scrutinized than "traditional" processes.[2]

Introduced in September 2003 and released in final form in September 2004, in parallel with "Pharmaceutical CGMPs for the 21st Century: A Risk-Based Approach," the FDA's guidance for industry titled "PAT—A Framework for Innovative Pharmaceutical Development, Manufacturing, and Quality Assurance"[3] was intended to describe a regulatory framework that encourages the implementation and application of innovative techniques to pharmaceutical development, manufacturing, and quality assurance. Goals of process analytical technology (PAT) include enhancing understanding of the product and process, and controlling the manufacturing process. The desired state of the pharmaceutical industry was also outlined, with the following desired traits:

- assurance of pharmaceutical quality and performance through the design of effective and efficient manufacturing processes;
- specifications based on scientific understanding of how formulation and process variables affect product performance and quality;
- continuous real-time quality assurance;
- adoption of risk-based regulatory policies that recognize the capability of process control strategies to mitigate the risk of poor quality products.

While the dawn of quality by design (QbD) is sometimes credited to International Conference on Harmonization (ICH) Q8(R1), the theme of QbD was introduced in the PAT guidance with the concept that "quality cannot be tested into products; it should be built-in or should be by design." Building quality into products is achieved through a comprehensive understanding of the product and manufacturing process and development of relationships that link material attributes, the manufacturing process, and environmental variables to final product quality. Another central tenet of QbD that was introduced in the PAT guidance and presented as a desired goal of PAT was the design and development of well-understood processes that consistently ensure a predefined quality of the manufactured product. A drug product should have target quality characteristics that can be predicted based on the relationship between the manufacturing process, material attributes, and quality characteristics. This, in turn, enables integration of process monitoring and control strategies that continually monitor the state of a process with the capability of manipulating the process to the desired state, thereby ensuring that the final product achieves its predefined quality characteristics. The concepts of building quality into products and achievement of predefined quality upon completion of the manufacturing process form the core of QbD in the pharmaceutical industry.

While the foundation of QbD was laid in the FDA guidance documents, the paradigm shift continued to evolve through the release of three ICH documents (described later): "ICH Q8(R1), Pharmaceutical Development," "ICH Q9, Quality Risk Management," and "ICH Q10, Pharmaceutical Quality Systems."

The global pharmaceutical industry has been slow to adopt, on a wide scale, many of the modern manufacturing methods that other industries have employed successfully for decades. This lack of innovation has been due, in part, to the restrictions of operating in a heavily regulated industry. During the past decade, however, the FDA recognized the potential benefits to be gained through an overhaul of the regulations governing pharmaceutical manufacturing. The modern approach to pharmaceutical development and manufacturing provides the industry an opportunity to redefine pharmaceutical quality. The definition of "pharmaceutical quality" was recast by Janet Woodcock in 2004[4] where quality was defined in terms of products that are

- consistently and reliably delivering the clinical performance and other characteristics stated in the label;
- not contaminated;
- available.

This new definition of pharmaceutical quality places greater emphasis on the consumer's needs, specifically regarding the performance of the product, and implores the question—Does the drug product deliver the therapeutic benefit indicated by the label claims? However, as opposed to other industries, it is often difficult for a consumer to rigorously assess the quality of drug products[4]; equal responsibility is placed on the regulatory bodies and industry to ensure that high-quality drug products which consistently deliver the desired therapeutic benefit are reaching the consumers.

With the introduction of the current good manufacturing practices (CGMPs) for the twenty-first century,[4] the Agency has focused on addressing and mitigating the various risks associated with pharmaceutical development and manufacturing through a modernization of the regulations governing the industry. The 2004 guidance titled "Pharmaceutical CGMPs for the 21st Century—A Risk-Based Approach" outlines the primary objectives of the FDA in ensuring pharmaceutical quality. A goal of the initiative was to facilitate industry's application of modern quality management techniques to all aspects of pharmaceutical production and quality assurance. Additionally, the industry is encouraged to adopt a risk-based approach to the development process, directing attention to the critical areas of pharmaceutical production. For example, complex manufacturing processes with greater probability of posing risk to drug product quality should be under greater scrutiny than simpler processes that pose little risk to quality. Also, manufacturing processes that have demonstrated the assessment and mitigation of risks to drug product quality should also be under less scrutiny than processes with little or no risk mitigation strategies.[2]

To achieve risk-based drug product development and manufacturing, it is imperative to apply modern science-based principles throughout the entire lifecycle of a drug product. Such modern approaches include the utilization of quality risk management (QRM) principles, PAT, and a quality systems approach. In its entirety, the implementation of modern science-based principles will direct pharmaceutical development in a new direction that will build quality into products as opposed to merely testing quality into products. This forms the core of QbD in that risks associated with pharmaceutical development and manufacturing will be addressed and mitigated throughout the lifecycle of the product. Subsequent to the release of the "CGMPs for the 21st Century" guidance, various documents defining the necessary methods for the application of QbD, released by FDA and the ICH, have served to unify both regulators' and industry's direction in improving pharmaceutical quality, providing the driving force and framework for modernizing pharmaceutical development and manufacturing to a state that ensures only high-quality drug products reach consumers.

A. Core Documents

Three ICH documents provide the structure for the application of risk-based approaches during pharmaceutical development and manufacturing; they are "ICH Q8(R1), Pharmaceutical Development," "ICH Q9, Quality Risk Management," and "ICH, Q10 Pharmaceutical Quality System." ICH Q8(R1) provides the overall framework for implementation of QbD in pharmaceutical development. The document highlights the various tools and applications that can be used to achieve QbD. For example, the concept of design space has garnered significant attention in the pharmaceutical industry. Development of a thorough design space exhibits that the manufacturing process has been well characterized and critical sources of risk to quality have been mitigated. PAT is another concept discussed in ICH Q8(R1) as a potential tool that can aid in the achievement of QbD.

ICH Q9[5] describes how QRM principles—used extensively in other industries—can be applied to the pharmaceutical industry throughout the lifecycle of a product. The primary objective of risk management as it applies to the pharmaceutical industry is that risk to quality should be defined based on scientific knowledge and ultimately linked to the protection of the patient.

ICH Q10[6] illustrates a model for a pharmaceutical quality system that can be implemented throughout the lifecycle of a product. This primary objective is achieved through the implementation of three specific goals, including (1) establish, implement, and maintain a system that ensures the delivery of high-quality drug product with the appropriate quality attributes to consumers; (2) develop an effective monitoring and control system that continually assesses process performance and product quality; and (3) continually identify and implement improvements to the product and process to consistently fulfill the quality needs of the final drug product.

The pharmaceutical quality system should be applied throughout all stages of the product's lifecycle, including pharmaceutical development, technology transfer, commercial manufacturing, and product discontinuation.[6]

While QbD principles are explicitly described in ICH Q8(R1), in order to successfully achieve true QbD, the collective application of ICH Q8(R1), ICH Q9, and ICH Q10 is necessary as the principles outlined in these three documents provide, in sum, the risk- and science-based approach to pharmaceutical development set forth by the FDA.

B. Structure of QbD in the Pharmaceutical Industry

To facilitate implementation of a QbD approach to pharmaceutical development and manufacturing, it is important to understand how quality attributes are described for a given drug product and how such quality attributes are affected by process parameters and variables. The initial steps in implementing QbD involve the generation of the design goals for the product which are described in the target product profile (TPP). The TPP should direct the formulation and process development as it contains information regarding route of administration, dosage form, maximum and minimum doses, and target patient population.[7] The TPP summarizes the overall intent of the drug development program, providing information about the drug product at the corresponding stage in development. It is typically organized in parallel with the drug label and links development activities to concepts intended for inclusion in the drug label.[8] Thus, the TPP can be viewed as a patient-centric element of QbD as it serves as the "user interface" of the drug product.[7]

The target product quality profile (TPQP) provides quantitative descriptions of clinical safety and efficacy that can be utilized to design and optimize a drug product. Because it describes aspects of clinical safety, the TPQP should comprise quantitative targets for attributes that are related to product performance. Examples include stability requirements and release profiles as these attributes are specifically related to the product's performance. It must be noted that the TPQP is not a specification because it may include analyses that are not required for batch release.[7] Effectively, the TPQP serves as the interface between the product/process and the patient, where performance characteristics of the product relevant to the clinical performance are quantified. Providing the link between the clinical performance and the process is essential in the QbD paradigm because pharmaceutical quality is defined such that risk to the consumer is minimized and the product performance is commensurate with the label.

A critical quality attribute (CQA) is defined as "a physical, chemical, biological or microbiological property or characteristic that should be within an appropriate limit, range, or distribution to ensure the desired product quality."[9] In general, CQAs of solid oral dosage forms are characteristics that affect the purity, strength, drug release, and stability of the drug product.[9] CQA has been used by some to describe performance characteristics such as dissolution while others have used CQA to describe physical properties of a drug product or intermediate such as particle size. Thus, the TPP can be viewed as a patient-centric element of QbD as it serves as the "user interface" of the drug product. It has been proposed by some that physical material attributes such as radial tensile strength (RTS) be categorized as critical material attributes (CMAs) that can be used to monitor or control the manufacturing process. For example, dissolution is considered a performance characteristic and is part of the TPQP, but it is difficult to directly link dissolution to the manufacturing process, in some cases. The relationship between CMAs that affect dissolution (such as RTS) should be determined, and the identified CMAs should be directly linked to the manufacturing process, as they can be directly related to the input materials (RTS can be directly linked to compression force, excipient components, etc.). Thus, in certain applications, CMAs can be used to provide the mechanistic link between elements of the TPQP and the manufacturing process.

A critical process parameter (CPP) is defined as "a process parameter whose variability has an impact on a CQA and therefore should be monitored or controlled to ensure the process produces the desired quality."[9] Process parameters can also be classified as noncritical, meaning variability in that parameter does not change the desired quality of the product, or unclassified in that the criticality of the parameter is unknown. Critical and noncritical process parameters can be distinguished by the sensitivity of the CQAs or CMAs to a change in a given process parameter.[7] If a small change to a process parameter causes failure to meet the desired quality, CQAs and CMAs are sensitive to that process parameter and should be classified as critical. Process parameters classified as critical should be part of the process design space and the control strategy to ensure they remain in a state of control that ensures the product meets its desired quality.[9] Classification of process parameters is primarily achieved through formal experimental designs but can also be classified based on results of a risk assessment or prior knowledge.

To summarize, the TPP is the overall summary of the drug product and product development activities, serving as the user interface of the drug product. Because the TPP is a qualitative descriptor, a category specific for quantifying the product performance is required. The TPQP provides quantitative targets for product performance, such as targets for impurities or release profiles. It should also be noted that the TPQP can contain product-specific performance characteristics, such as adhesion in the case of a transdermal patch. CQAs serve as a general category for quality attributes of the drug product and intermediates and can be defined in terms of physical material attributes or describe parts of the TPQP. Process parameters can be classified as critical, noncritical, or unclassified. Process parameter criticality is best described in terms of the sensitivity of a CQA or CMA to a process parameter.

II. THE ROLE OF PAT IN QbD

PAT is less than a decade-old concept to the pharmaceutical industry, yet it has been known for decades as process analytical chemistry (PAC) in other industries. Callis *et al.* defined the goal of PAC as "supplying quantitative and qualitative information about a chemical process to monitor and control a process, as well as optimize its efficient use of energy, time, and raw materials."[10] The guidance on PAT was released in final form by the FDA in September 2004 as a framework for innovative pharmaceutical development, manufacturing, and quality assurance. The overarching goal of the PAT initiative is to enhance process understanding, leading to improved control of the manufacturing process. While the ultimate goal of QbD is to build quality into products by design,[3] without PAT, QbD in the pharmaceutical industry cannot be realized. PAT tools and principles provide the means to achieve the objectives set forth by QbD. Extensive application of PAT as a means to achieve QbD is encouraged by FDA representatives[11] and it is in the authors' opinion that utilization of PAT is required for the complete achievement of QbD.

The traditional pharmaceutical manufacturing approach leads to more rejected product[12] than is acceptable for many industries and provides limited efficiency. Process variability arising from multiple factors including environmental variability, process parameters, and operational variability often lead to failure in reaching the desired quality characteristics.

Manufacturing under the QbD paradigm follows a process that is understood and controlled, which in turn enhances the likelihood that the final product will meet the predefined quality characteristics. In such a situation, the influence of formulation and process variables on product quality is understood, and when such critical variables are under control, final product quality is guaranteed.[7] The significance of PAT to QbD can be fully realized only when such relationships exist. ICH Q8(R1)[9] made clear that PAT is useful to ensure a process remains within the defined design space. However, an extension of that would be to continuously monitor CMAs and provide active feedback control to CPPs that affect such attributes. Such a system would manage input variability of CPPs to ensure CMAs meet predefined specifications. Lionberger *et al.*[7] identified the primary drawback of defining a design space in terms of multiple univariate process parameters as limiting flexibility because combinations of CPPs outside a univariate design space may give a final product that meets the TPQP. In an ideal situation, the design space is developed based on the definition in ICH Q8(R1)[9] as the "multidimensional combination of input variables and process parameters" and is coupled with PAT in that input variables and process parameters constituting the design space are manipulated during the process based on CMAs monitored by way of PAT tools. Such an approach guarantees final product quality by assuring material attributes monitored during the process are maintained at their target values.

One area where the PAT initiative seeks to improve traditional pharmaceutical manufacturing is through the elimination of time-defined endpoints. This does not imply that time is not significant in most processes; but, using time as a sole indicator of process completion does not allow consideration of effects such as temperature, humidity, and raw material variability.[3] The PAT and QbD paradigms seek to define a process "endpoint" based on mechanistic understanding of the process related to CMAs of the product (or intermediate). Leuenberger *et al.* discussed the pitfalls of attempting to define an explicit endpoint for the manufacture of pharmaceutical granules.[13] While various techniques such as power consumption and torque measurements provide information as to unfolding the various stages of the granulation process, such techniques should not explicitly define the endpoints; rather, they should be used to control the endpoint.

Leuenberger states that an intrinsic endpoint to granulation processes in general does not exist. Rather, the granulation endpoint should depend on critical granule properties such as particle size distribution or moisture content. Power and torque measurements may be used to provide a preliminary signal to the endpoint, followed by process parameter adjustment to reach the desired CMAs. Defining endpoints in terms of CMAs linked to CPPs is one example of the movement away from testing quality into products to building quality into products.

A. Multivariate Techniques for PAT

The complex, multivariate nature of pharmaceutical products and processes warrants the use of multivariate techniques such as partial least squares (PLS) and principal component analysis (PCA).[3] One of the benefits of utilizing such multivariate techniques is that the dimensionality of the variable space can be considerably reduced, enabling a thorough understanding of the effects that process variables impart on the product and/or intermediate quality attributes. The ultimate benefit that multivariate methods offer to the pharmaceutical industry is the ability to predict final product quality from process data, enabling the implementation of advanced control strategies.

Multivariate techniques are essential for the successful implementation of PAT. Most pharmaceutical products are complex multifactorial systems manufactured through multifaceted processes; therefore, to thoroughly understand the product and process, complex multivariate techniques are paramount.[3] Also, most PAT tools are multivariate in nature. For example, NIR sensors are a common PAT tool used to analyze multiple processes, but due to the nonspecific and nonselective nature of NIR spectra, multivariate techniques are required to generate meaningful relationships. Chemometrics refers to "the application of multivariate, empirical modeling methods to chemical data"[14] (typically spectroscopic data) and is an essential tool in unlocking the important relationships between data collected through process analyzers and product quality. The two general functions (applied to analytical chemistry) of chemometrics are instrument specialization in which multivariate models are generated to make a multivariate analyzer selective to the defined product quality attributes or analyte of interest, and information extraction in which tools are used to unlock multivariate information already contained in the instrument to improve process understanding.[15] As with spectroscopic data, complex process data are also multivariate in nature, requiring multivariate mathematical techniques to understand the underlying phenomenon. Because PCA and PLS will be referred to in later examples, the fundamentals of these methods will be presented here.

The most common algorithms used in the analysis of multivariate process and spectroscopic data are principle component analysis (PCA) and PLS. PLS and PCA are similar in that they are both factor analysis methods,[16] and they both significantly reduce the dimensionality of the variable space. This is done by representing the data matrix (**X**) with a few orthogonal variables that explain most of the variance. The main difference between PLS and PCA is that PLS can be referred to as a supervised technique that maximizes the covariance between the response (**Y**) and input variables (**X**) in as few factors as possible while PCA simply aims to maximize the covariance of **X**. Mathematically, PCA reduces the data matrix using eigenvector decomposition of the covariance matrix of the data matrix.[17] Essentially, the data matrix is broken down into principal components (PCs), represented by pairs of scores (**t**) and loadings (**p**). The loading vectors are equivalent to the eigenvectors of the covariance matrix of **X**, and the corresponding eigenvalues (λ) represent the variance of each corresponding PC. Suppose **X** is composed of n samples on q variables. The first PC is defined as $\mathbf{t}_1 = \mathbf{X}\mathbf{p}_1$ and explains the greatest amount of variance, while the second PC is defined as $\mathbf{t}_2 = \mathbf{X}\mathbf{p}_2$ having the next greatest amount of variance and so on. Up to q PCs can be defined, but only the first few (M) are significant in explaining the main variability of the system. Selection of optimal number of PCs can be accomplished in various ways and is well described in the chemometrics literature.

PLS, however, works to achieve two goals: capture variance and correlate **X** and **Y**.[18] The first latent variable (LV) $\mathbf{t}_1 = \mathbf{X}\mathbf{w}_1$ is a linear combination of the **X** variables that maximizes the covariance of **X** and **Y**, where $\mathbf{w}_1$ is the first eigenvector (called weight vector) of the covariance matrix. The columns of **X** are then regressed on $\mathbf{t}_1$ to give a regression vector $\mathbf{p}_1$. The original X matrix is then deflated as follows: $\mathbf{X}_2 = \mathbf{X} - \mathbf{t}_1\mathbf{p}_1^T$. $\mathbf{X}_2$ is the resultant data matrix after removing the element of the original data matrix (X) that was most correlated with **Y**. The second LV is then computed from $\mathbf{X}_2$, $\mathbf{t}_2 = \mathbf{X}\mathbf{w}_2$, where $\mathbf{w}_2$ is the first weight vector of the covariance matrix of $\mathbf{X}_2$ and **Y**.[19] These steps are repeated until q LVs are computed. As with PCA, the optimum number of LVs may be chosen via cross-validation methods. While PLS and PCA are the most common algorithms in use, other algorithms have been developed and are used. For a detailed description of other algorithms and the use of chemometrics in PAT, the reader may refer to Miller.[15]

PLS is essentially a regression tool[20] and may be used to relate process variables to product quality attributes. PCA can also be used as a regression tool in that the significant PCs may used to generate a regression model that relates process variables to product quality attributes (when PCA is used in such a way it is referred to as principal component regression, or PCR). Identifying the link between critical process variables and product quality attributes is critical to successful implementation of a QbD approach to product development.[9] The ability to generate such relationships in turn decreases the variation in final product quality by adjusting CPPs to ensure final product quality. When final product quality can be predicted from process data, corrective action strategies can be moved upstream to ensure that out-of-specification product will not be produced.[21]

B. Process Understanding Through Process Monitoring

A primary goal of the FDA's PAT initiative is to enhance understanding of the product and process, allowing quality to be built into the final product. Predefined quality attributes of the final product defined during development can only be achieved upon completion of the manufacturing process if the process is well understood. Achievement of enhanced process understanding does not occur at a particular phase during drug development or manufacturing; rather, continuous learning over the life cycle of the product is essential. Beginning at small scale, process analyzers capable of collecting data in real time may be used to gain initial insight as to how various process and formulation variables influence final product quality. Information regarding optimization, technology transfer, and control may also be attained at this stage. Once a product reaches the production phase, process understanding is enhanced when new variables, for example, environmental factors that influence the process, are introduced. According to the FDA's guidance on PAT, a process is considered to be understood when critical sources of variability are identified and explained, variability is managed by the process, and quality attributes can be accurately predicted over the established design space.[3] Enhanced process understanding is significant in that risk of manufacturing a product that does not meet the defined quality attributes is at a minimum when process and product understanding is at a maximum.[21]

Process monitoring is an integral part of successful implementation of QbD. One of the first steps toward QbD is the establishment of a process monitoring system that can provide warning of potential failure. Process monitoring enables the control of critical steps in a manufacturing process, thereby enabling the control that will ultimately avoid failure.[11] The pharmaceutical industry relied previously upon statistical quality control (SQC), involving the measurement of important quality attributes over time. The overarching goal of SQC is to maintain a given quality attribute within specified limits. Unfortunately, SQC alone may often provide inadequate evidence that the process is under control[22,23] and does not adequately facilitate the advancement of process knowledge. Progressing to statistical process control (SPC), where critical process variables are monitored on univariate control charts, the industry applies statistical techniques to determine if a process is stable over time. While SPC offers more than SQC, because the root cause for critical process variability can be more easily traced, data collected during a pharmaceutical process (e.g., spectroscopic, process variables, etc.) are multivariate and usually highly correlated,[23] often rendering univariate techniques such as SPC to be less than adequate in most pharmaceutical systems.[24] Multivariate SPC (MSPC) refers to the use of multivariate algorithms such as PLS and PCA to mine information from extensive multivariate data sets.[20] Due to the high correlation of process variables, data sets comprising hundreds of variables can be reduced to a significantly smaller dimension, usually less than 10 LVs.[23] Utilization of such techniques captures the complex multivariate nature of pharmaceutical processes but in a reduced, interpretable variable space.

MSPC in its simplest form may involve the analysis of score plots. Score plots have been described by Martens as "a map of the landscape of samples."[25] Essentially, scores from PCA provide information on how samples relate to one another. Every sample subjected to PCA or PLS has a corresponding score value for each PC (or LV) which provides information on that sample. Score plots are generally constructed as biplots where scores from the first PC (or LV) are plotted against the scores from the second PC (or LV). Confidence limits can be generated for the score plot, and when an observation falls outside the confidence limit, the sample can be further evaluated to determine the source of failure. While observing score plots is more efficient than traditional SPC, plotting all combinations of scores on a biplot is still inefficient when monitoring a process in real time. For example, if a data matrix of process variables was reduced to four PCs (or LVs), six score plots would still be required to monitor all combinations.[22]

A more efficient extension of the score plot is the generation Hotelling's T^2 and squared prediction error (SPE) statistics. The Hotelling's T^2 statistic is a powerful singular value that can be used to monitor the overall performance of the main events of a system.[23] The T^2 statistic is desirable for charting multivariate data because it is flexible, easy to apply, sensitive to small process changes, and offers sufficient signal interpretation. In general, the Hotelling's T^2 statistic is similar to the univariate t statistic in that it is a measure of statistical distance between an observed sample mean and its corresponding population mean.[24] Hotelling's T^2 is determined from the scores (t) and eigenvalues (λ) calculated from PCA or PLS by the following equation: $T^2 = \mathbf{t}(\mathbf{t}^\mathrm{T}\mathbf{t})^{-1}\mathbf{t}^\mathrm{T} * \mathrm{df}$ where $\mathbf{t}$ represents the scores for M PC's and df represents the degrees of freedom.[17] It is an ideal statistic for monitoring a process because it contains information from multiple process variables but is a singular value for a given sample. A critical threshold can be determined for Hotelling's T^2 and in a situation where a new sample gives a value outside the acceptable limit, that sample can be further investigated by observing the variable contribution plot to Hotelling's T^2. Essentially, a variable contribution plot gives a list of the process variables that were most responsible for the sample above the T^2 threshold.[22] This enables real-time diagnosis of an abnormal situation that leads to the source of deviation. Depending on the control strategy in place, corrective action may be taken immediately, or the process can be shut down to correct the abnormal event.[23] The Hotelling's T^2 statistic and the corresponding contribution plots provide advantages over SQC and SPC because of the diagnostic capability presented when all process variables are monitored using one chart. A visual description of Hotelling's T^2 statistic and the corresponding contribution plots are depicted in Figures 1–4.

As previously stated, the first M PCs or LVs capture the main source of variance in a system. However, a small portion of variance that was not modeled still remains in the system, and this variability constitutes the residuals, commonly referred to as the SPE. The SPE is calculated as follows: $\mathrm{SPE} = \sum_{i=1}^{q}(x_{\mathrm{new},i} - \hat{x}_{\mathrm{new},i})^2$, where $\hat{x}_{\mathrm{new},i}$ is determined from the reference model (note: $\hat{x}_{\mathrm{new},i} = \mathbf{t} * \mathbf{p}^\mathrm{T}$). Note that the term $(x_{\mathrm{new},i} - \hat{x}_{\mathrm{new},i})^2$ for an individual sample is referred to as the Q residual, which is the contribution of a single sample to the SPE. It is critical to ensure that the unmodeled variability in the system is still within a critical threshold; variance that was not modeled initially does not mean it is not significant to the system. As with T^2, SPE can be monitored to ensure it does not exceed a critical threshold value. Contribution plots may also be utilized for SPE values to diagnose abnormal situations in the manufacturing process.[23] As opposed to Hotelling's T^2, SPE gives a squared perpendicular distance from the plane described by the model while Hotelling's T^2 gives a distance to the center of the model.[19] The mathematical difference between Hotelling's T^2 and SPE is that SPE gives the squared perpendicular distance from the plane described by the model while Hotelling's T^2 gives the distance to model center. The combination of Hotelling's T^2 and SPE may also be used to assign a process signature because these two values give the combination of all variables affecting the process.[23]

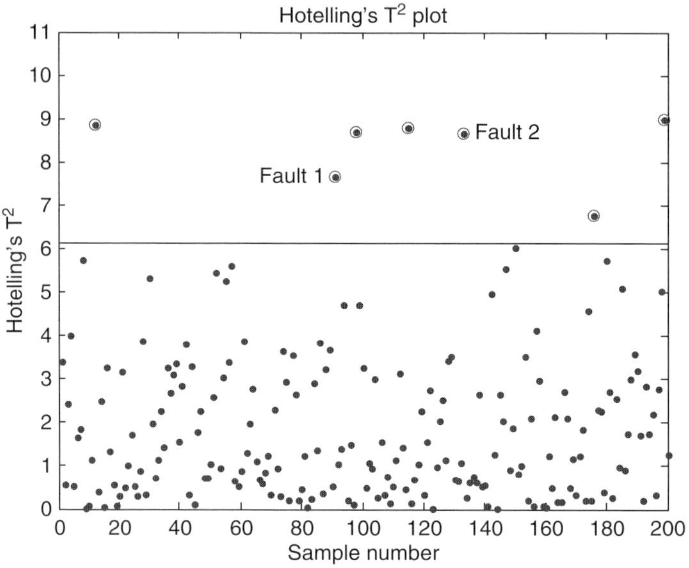

FIGURE 1 Hotelling's T^2 plot for a given process. The horizontal line corresponds to the critical threshold of the Hotelling's T^2 statistic. The points circled in red correspond to samples that are outside the acceptable limits for the process. See Figures 2 and 3 for contribution plots of samples marked Fault 1 and Fault 2. (For interpretation of the references to color in this figure legend, the reader is referred to the Web version of this chapter.)

Process variable (or parameter) classification is essential to express a thorough understanding of the process. An unclassified process parameter (UPP) is defined as a process parameter in which its influence on quality attributes is undefined.[11] In general, the number of UPPs gives an estimate of the level of process understanding at a given moment in time. If the number of UPPs is large, the process is not well understood because the relationship between those variables and final product quality has not been achieved. Thus, process variable classification is a critical step in expressing process understanding. Recall that at the outset of product development, the number of UPPs is relatively large, mainly due to the fact that a thorough evaluation of the product and process has not been completed. As development progresses, relationships between process variables and final product quality are established, which in turn means a reduction in the number of UPPs. The complexity of most pharmaceutical processes warrants the use of multivariate techniques such as MSPC for such classification. If process data exist from normal operating conditions that produced final product with acceptable quality, Hotelling's T^2 statistics can be generated for those batches. When a batch is produced with unacceptable product quality, resulting in a Hotelling's T^2 or SPE value above the threshold, that sample can be further observed using contribution plots. Once the T^2 value is deconstructed to exhibit the contribution of each variable, the process parameter(s) giving the greatest influence can easily be identified. These parameters may now be deemed critical; as a realistic change resulted in a failure to meet the final product quality attributes.[7]

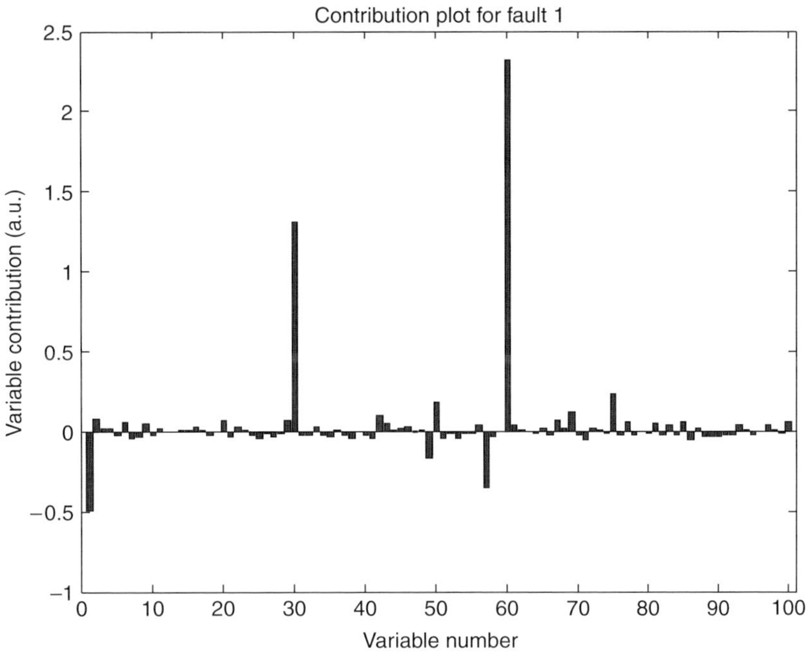

FIGURE 2 **Contribution plot corresponding to Fault 1 in Figure 1. Variable number corresponds to the process variables measured during the process. From this contribution plot, one can observe variables 1, 30, and 60 are the main contributing variables for the fault.**

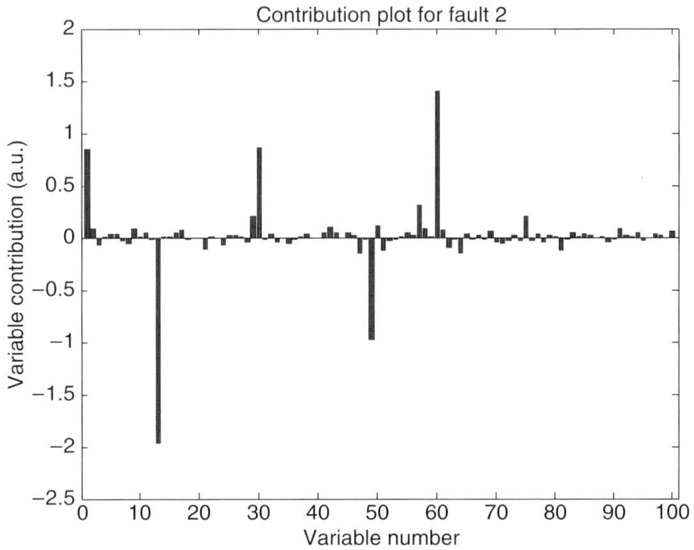

FIGURE 3 **Contribution plot corresponding to Fault 2 from Figure 1. Variable number corresponds to the process variables measured during the process. From this contribution plot, one can observe variables 1, 13, 30, 49, and 60 are the main contributing variables for the fault.**

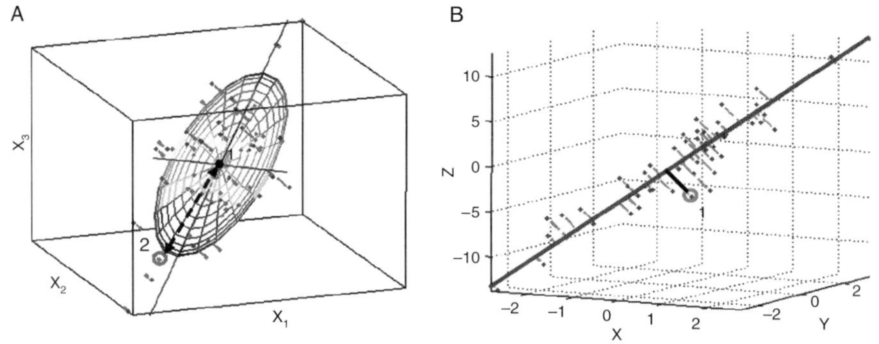

FIGURE 4 **(A) Representation of Hotelling's T^2. Point 1 represents the center of the model plane while point 2 represents an individual sample. Hotelling's T^2 for the sample represented by point 2 is given by its distance to the center of the model, within the model's plane, which is represented by the broken line. (B) Representation of Q residual and SPE. The Q residual for the sample marked as point 1 is defined as the squared perpendicular distance from point 1 to the plane of the model (blue solid line) which is represented by the solid black line. The SPE would be determined by summing the Q residual for all samples shown in the figure. (For interpretation of the references to color in this figure legend, the reader is referred to the Web version of this chapter.) (See Color Insert.)**

Gabrielsson *et al.*[26] combined spectroscopic profiles and process variables in one MSPC model to detect and diagnose process deviations in the conversion of nitrobenzene to aniline. UV spectra and various process variables were used as inputs while reaction time was the output. The approach was somewhat unique in that spectral and process data were concatenated into one data matrix. Score plots were observed and any extraneous deviations from normal operating conditions were further evaluated through contribution plots. In one instance, contribution plots exhibited that a higher reactor temperature caused an increase in reaction rate to a point where the batch was ahead of schedule. In another instance, the reaction rate was significantly slower than under normal conditions, caused by a low reaction temperature and inadequate starting material. Spectroscopic data also showed significant contribution to the reaction rate deviating from its normal conditions. It is also important to note that evaluation of the process using the concatenated data matrix identified process deviations that were not recognized using the spectral and process data independently, meaning MSPC was able to detect an interaction between process and spectral data that was unseen using a more traditional technique. Such an approach enabled thorough process understanding because interaction between chemical and physical properties was established and isolation of process conditions owing to decreased reaction rates was demonstrated.

Process and product understanding can also be linked to the definition of a process signature. The FDA presented a definition of process signature based on measurements from various process sensors.[3] Process signature is defined by ASTM as "a single or multidimensional signal indicative of the attributes of the process."[27] Near-infrared spectroscopy (NIRS) has been demonstrated as an appropriate tool for the detection of a process signature. In terms of NIRS, a process signature can be defined as the spectral characteristic imparted to a material based on the processing steps encountered. While it may be risky to consider process signatures as a qualitative aspect of the material, spectral signatures may alert the manufacturer to potential quality issues. More often, unique process signatures provide challenges to the implementation of nondestructive spectroscopic calibrations across manufacturing scales or sites. Spectral decomposition methods such as PCA can be used to treat the data quantitatively, reducing the complications of calibrating/predicting in such situations. This was exemplified by Cogdill *et al.* in a series of papers presented in 2005. Experimentation was used to determine the level of interbatch spectral variation and the possibility of pooling tablets from multiple lots and production scales for one calibration set. PCA was used to deconstruct the NIR data, with two PCs explaining > 97% of the spectral variation. Upon observation of the PC scores plot, two data clusters were evident, each representing multiple batches (Figure 5). Each cluster in the PCA score plot corresponded to a unique production site and scale; for example, compression using a rotary tablet press versus a single station tablet press. The spectral characteristic imparted to the material was unique to the method and scale of compression and was clearly observable through PCA. It is also important to note that the investigators were able to identify differences between production facilities.[28,29] This not only proves the feasibility of NIRS for the identification of a process signature, but also demonstrates the importance of monitoring and controlling the process rather than relying solely on end-product testing. While certain final product quality attributes may be similar between the batches corresponding to each data cluster, the final products will not be identical. Other product properties that may not be considered during the final product testing can potentially be significantly different. To obtain uniformity in all product properties, the process conditions must also be monitored

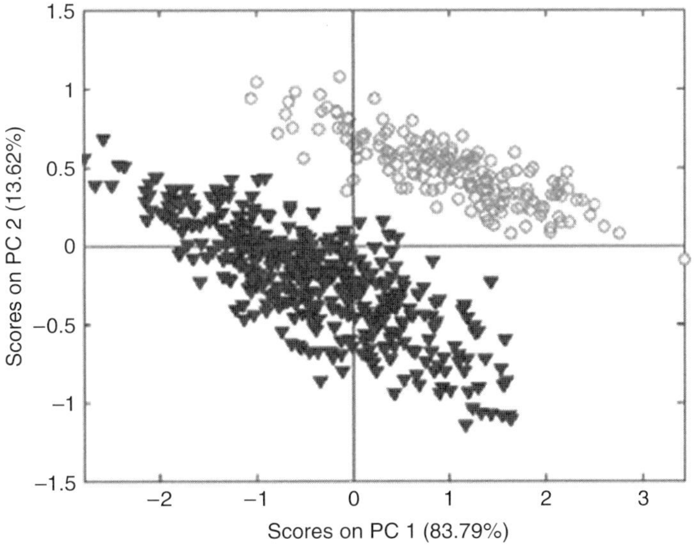

FIGURE 5 **Scores plot of first and second PCs from PCA of production samples using reflection data (reprinted with permission from Cogdill *et al.*[29]). The two clusters of samples correspond to unique production sites and equipment.**

and controlled. The process path may be as significant as the final product quality attributes[23] in that an alteration in the process path may lead to deviations in product performance when administered to patients.

C. Technologies for Process Monitoring

Nondestructive analytical techniques are essential in developing a system to increase product and process understanding. Such measurements contain information related to biological, physical, and chemical attributes of the materials being processed. However, the major advantage is realized through real-time measurements as materials are being processed, enabling enhanced understanding of the product and process. Nondestructive measurements may be collected in three ways: the sample is physically removed from the process (at-line); the sample is diverted from the process stream, analyzed, and returned to the process stream (on-line); or the sample is directly analyzed while in the process stream (in-line).[3] All traditional laboratory analyses are static in that a sample is pulled from the process stream and analyzed. Real-time process measurements collected in-line enable dynamic analysis of samples which may contain critical information about the process that is not observable through a static measurement. Spectroscopic techniques, particularly NIRS and Raman spectroscopy, are well suited for real-time nondestructive analysis of a process. NIRS offers flexibility in that NIR sensors and instrumentation is capable of collecting information at-, on-, and in-line.[30] Another major advantage of NIRS that makes on- and in-line analysis feasible is that solid samples do not require extensive preparation, allowing analysis to be carried out directly.[31] However, NIR spectra typically consist of broad, overlapping bands causing potential challenges in regard with selectivity. To mitigate selectivity issues, various data preprocessing techniques are used to enhance relevant chemical and physical features while reducing physical and chemical interference.[30] While Raman spectra contain sharp, discrete peaks which may offer a greater degree of chemical selectivity,[32] preprocessing may still offer benefits, depending on the application. Detailed descriptions on preprocessing techniques are beyond the scope of this discussion and are well described in the chemometric literature. NIRS and Raman spectroscopy are examples of technologies that provide fast, nondestructive means of analysis that can be used to enhance process knowledge and understanding.

Blending of solids is one of the most common unit operations in development of a solid dosage form. Blending has been extensively studied using at-, on-, and in-line spectroscopic techniques. De Beer *et al.*[33] used an in-line Raman spectrometer setup to monitor the blending process of diltiazem HCl, Avicel PH 102, lactose, and silicum dioxide. A two-level full factorial design with two additional experiments at the mid-point (10 experiments total) was performed to evaluate the effects of API content (%, w/w), blender speed, and loading (kg) on blending time required to reach homogeneity. SIMCA was used to develop a classification method to determine the blending endpoint in real time based on the orthogonal distance (OD) and Mahalanobis distance (MD) to the model center. The last seven spectra from each experiment were used as the reference spectra. The threshold for separation of classes was based on ± 3 times the standard deviation of the MD and OD values for the reference spectra. The remaining spectra (test set) exhibited large values of MD and

OD, well outside the specified threshold giving indication that the model may be used for real-time endpoint determination. Shi *et al.*[34] utilized two in-line NIR sensors (one on top and one on side) to characterize a powder blending process in a bin blender. A ternary mixture of acetaminophen, lactose, and MCC was used as the model system. NIPLS (multi-Y PLS algorithm) was used to generate predictive models for all three constituents. An endpoint determination algorithm containing root mean square from nominal value (RMSNV) and two-tailed Student's *t*-test that was developed from PLS predicted concentrations of all three constituents was used to determine the blend endpoint. The endpoint algorithm proved to be more robust than the moving window standard deviation algorithm that is traditionally used as an endpoint indicator. The significance of using two sensors was also exhibited as each sensor gave significantly different endpoints.

Unknown solid-phase transformations during processing can be devastating to a pharmaceutical process as downstream processing, bioavailability, and physicochemical properties can be substantially affected. Not only is full classification of all solid forms crucial, but also monitoring the solid form during the whole production cycle may be critical (See Chapter 2). Raman spectroscopy has been used to monitor solid form conversions in multiple pharmaceutical unit operations. High-shear wet granulation has great potential to induce phase transformations due to addition of binder solution (typically aqueous solution) and the input of significant mechanical energy. Wikstrom *et al.*[35] demonstrated the feasibility of Raman spectroscopy for monitoring the transformation kinetics of anhydrous theophylline (AT) to theophylline monohydrate (MT) in a drug formulation containing mannitol, microcrystalline cellulose, and water (binder liquid) as the excipients. A calibration model using three Raman peaks unique to theophylline (as opposed to the excipients) was generated to predict the ratio of AT to MT which was further used to determine the conversion kinetics. NIRS was also investigated in this case, but Raman proved to be superior as water dominated the NIR spectra. Hu *et al.*[36] demonstrated the utility of in-line Raman spectroscopy to monitor the crystallization of flufenamic acid (FFA), which commonly exists as one of two polymorphs: form I or form III. Raman spectroscopy was sensitive to solution concentration and selective to both polymorphs.

It must also be noted that multiple PAT analyzers can be used on a single unit operation to understand the process. For example, Aaltonen *et al.*[37] used in-line Raman spectroscopy and in-line NIRS to monitor the dehydration of MT in a lab-scale fluid bed dryer. A PLS model was generated to quantify the solid-state transformation using Raman spectra while water removal was monitored using NIRS. This proved to be an effective combination because physical and chemical information were monitored and extracted by way of NIRS and Raman spectroscopy, respectively. Such combinations of process analyzers offer enhanced capabilities to improve process understanding and control.

Thermodynamic parameters and spectroscopic data can also be monitored simultaneously during a process. For example, during fluid bed granulation, thermodynamic parameters such as temperature, relative humidity, and differential pressure in the vessel may be critical to the process. Parameters derived from in-line sensors may include moisture content (via NIRS) and particle size (via focused beam reflection measurement, FBRM). Figure 6 represents a schematic of the

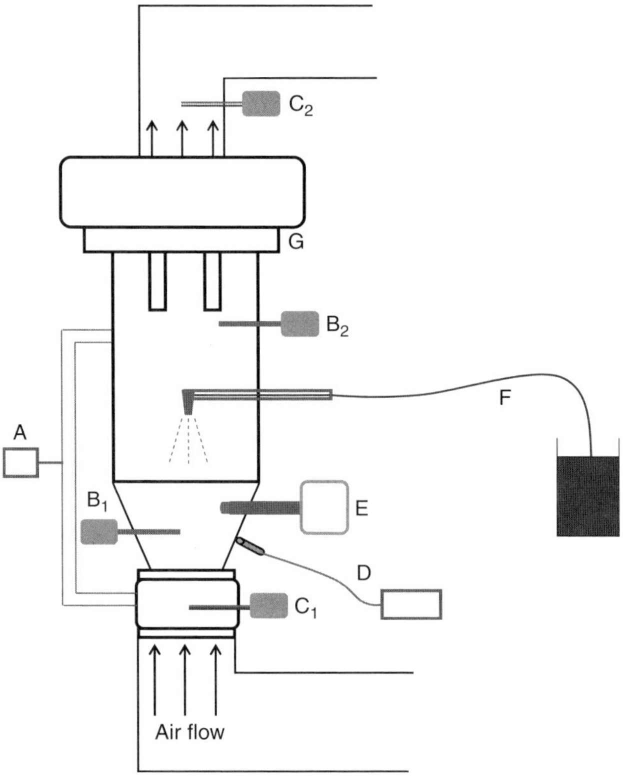

FIGURE 6 **Fluid bed granulator with multiple PAT sensors measuring critical variables online. (A) Digital manometer, (B) thermocouple, (C) thermocouple with humidity sensor, (D) NIR spectrometer with fiberoptic probe, (E) particle size analyzer, (F) liquid addition, and (G) filter bag apparatus.**

various thermodynamic and analytical sensors on a fluid bed system. Inlet and outlet temperature and humidity (measured via C_1 and C_2, respectively) can be monitored for determination of process endpoint. The product vessel temperature (measured via B_1 and vessel temperature B_2) can also be used to facilitate the maintenance of a constant temperature in the vessel, thereby controlling the mechanism of drying. The pressure drop across the bed can be measured using a digital manometer (A in Figure 6), used to control bed height during granulation and drying. The moisture content of vessel contents can be monitored via NIRS for the purpose of controlling liquid addition and to determine drying endpoint. Particle size can also be measured in-line to gather information regarding the growth of granules during liquid addition and attrition during drying. Aside from process monitoring, these process measurements can also be used in conjunction with control software to actively manipulate the process. Active process control will be discussed in subsequent sections.

While NIRS and Raman are two of the most common PAT analytical technologies, other advanced analytical methods are also capable of PAT application. One such method is terahertz spectroscopy. The terahertz region of the electromagnetic spectrum falls between the mid-IR and microwave regions, specifically 3 and 100 cm^{-1}. Unlike IR radiation, terahertz photons are not energetic enough to make molecules vibrate, but they can excite torsional low-energy motions. More important, the weak forces that hold molecules together in a crystal lattice are capable of being excited by terahertz radiation, meaning crystalline materials have a defined terahertz spectrum while amorphous materials are essentially invisible in the terahertz region.[38] Presently, the majority of pharmaceutical applications of terahertz spectroscopy involve time-domain spectroscopy, meaning terahertz radiation is generated using ultrashort laser pulses on the order of femtoseconds (time-domain terahertz spectroscopy is referred to as terahertz pulse spectroscopy (TPS)).[39] TPS is advantageous because it is suitable for imaging applications and has superior signal-to-noise ratio, as well as the ability to measure the appropriate data for the determination of the absorption coefficient and/or refractive index of a given material.[40,41]

Strachan *et al.*[39] utilized TPS to quantify different polymorphs of four drugs: carbamazapine, enalapril maleate, indomethacin, and fenoprofen calcium. Quantitative models generated using PLS were capable of predicting polymorph content of each drug in physical mixtures from corresponding TPS spectra. The authors suggested TPS is capable of quantifying crystalline and amorphous forms of most organic molecules. Terahertz pulsed imaging (TPI) was used by Fitzgerald *et al.*[42] to analyze tablet coatings from a brand ibuprofen product and generic ibuprofen product. The philosophy behind TPI for measuring coating thickness is based on unique interactions of terahertz radiation at each interface of the coated tablet. The terahertz pulse is reflected back to the detector, and the amplitude of the reflected terahertz radiation is recorded as a function of time. Once the refractive index of the coating material is derived, the coating thickness can be determined from the terahertz pulse duration. In the Fitzgerald study, it was demonstrated with TPI that the brand tablets contained several coating layers with a thickness of 436 μm while the generic tablets contained only one layer that was 314 μm thick. The difference in structure of each coating was also evident through TPI. All results were confirmed using microscope photographs. The value of integrating terahertz spectroscopy and chemometrics for quantifying tablet concentrations was exemplified by Wu *et al.* using theophylline tablets as a model system.[43] It was demonstrated that PLS and PCR gave quick and reliable predictions for major components in the formulation from terahertz spectra using derivative methods for preprocessing. It was also demonstrated that when multiplicative scatter correction was applied to the spectra as a preprocessing technique, tablets manufactured through wet granulation were differentiated from tablets produced via direct compression. It was concluded that terahertz spectroscopy may be a valuable area for future research concerning the quantification of components in solid dosage forms.

Chemical imaging (CI) may also be implemented for PAT applications. CI is a technique that combines imaging and vibrational spectroscopy (IR, NIR, and Raman) to derive spatial and spectral information from an object.[44] Two general categories of instrumentation exist for chemical imagers: staring and push-broom. The staring configuration is more popular among NIR imagers and works by collecting an image at one wavelength at a time, keeping the sample and camera stationary. Spectral information is provided by discrete filters, tunable filters, or an imaging Fourier-transform spectrometer.[30] The push-broom configuration is more common with Raman imagers. Push-broom imagers function by acquiring the desired spectrum at one point on the sample, then the sample is moved and another spectrum is recorded. This is repeated until spectrum from every point on the object is acquired. In both cases, a chemical image, known as a spectral hypercube, is generated. Essentially, a spectral hypercube is a three-dimensional data block consisting of two spatial and one wavelength dimension.[44] Figure 7 represents a schematic of a spectral hypercube.[45]

A major advantage of CI over point spectroscopy methods is the capability of deriving spatial distributions (as well as ingredient concentration) of heterogeneous solid forms (e.g., tablets). While application of CI techniques to pharmaceutical products and processes is limited when compared to traditional spectroscopic techniques, CI offers a viable solution to the challenging analytical needs of the pharmaceutical industry.[30] Roggo *et al.*[46] utilized IR-CI to deduce the effects of process conditions of melt granulation on dissolution rates of tablet cores. "Bad" batches were produced in which the time–temperature gradient varied in the melt granulation step. Chemical images revealed a substantial increase in magnesium stearate on the surface of the tablet core. It was determined that the process change modified the physical properties of the material prior to compaction, causing magnesium stearate to migrate to the surface upon compaction, which in turn decreased the dissolution rate. Lee *et al.*[47] utilized NIR-CI as a high-throughput method for the determination of content uniformity (CU) of

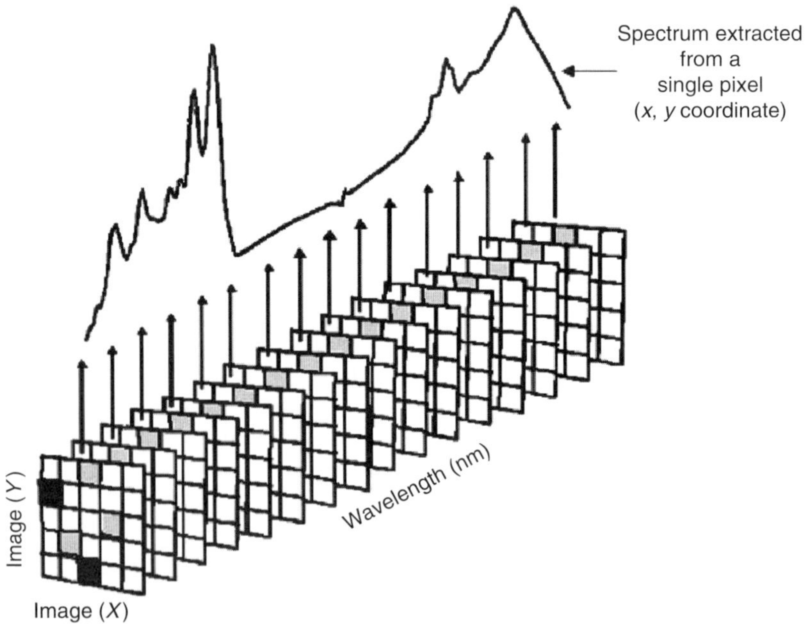

FIGURE 7 **Schematic of spectral hypercube. Each pixel is spatially located in the x–y image plane while a representative NIR spectra for each pixel is represented in the third dimension. Reprinted with permission from Lyon *et al.*[45]**

pharmaceutical tablets. A large field of view was used to simultaneously image 20 tablet samples at 61 wavelengths that captured a spectral characteristic specific to the API. Of the 20 tablet samples, five were calibration samples and 15 were unknown samples. This was advantageous because a calibration curve was generated for every image, negating calibration transfer issues. CU results were in agreement with the traditional destructive UV method, proving the method was a viable alternative to the traditional CU testing method. The significant advantage over the traditional method was time of analysis; the traditional method required half a working day while the NIR-CI method required less than 2 min. Sasic[48] utilized a push-broom Raman imaging system to map the API content of a low-dose (<1%, w/w) formulation containing alprazolam. Reliable detection of the drug was only feasible through PCA which was further used to determine the spatial distribution of Alprazolam. This study demonstrated CI is not limited to dosage forms containing a relatively substantial amount of API, but spatial distribution of API for low-dose formulations can also be derived.

One of the primary goals of the PAT initiative is to enhance product and process understanding. Enhancing the level of process understanding is significant in that the relationships between process variables and parameters, and final product quality cannot be deduced until the product and process are thoroughly evaluated and understood. Due to the complexity of most pharmaceutical products and processes, multivariate methods are essential to rigorously assess the system. While traditional methods such as SQC and SPC offer some utility in enhancing process knowledge, they may hinder the advancement of complete process understanding by ignoring critical interactions between various process variables. Thus, MSPC tools must be utilized to completely evaluate a system. Various spectroscopic techniques such as Raman spectroscopy and NIRS offer advantages over other analytical techniques for enhanced process understanding because they are nondestructive in nature and can be used to collect information in-line and in real time. Such multivariate and spectroscopic methods provide the tools necessary to evaluate and classify UPPs. As previously suggested, the number of UPPs for a given process gives some indication of the level of understanding at that moment in time. Multivariate methods and the nondestructive techniques discussed provide excellent tools to reduce the number of UPPs and ultimately classify those variables and parameters critical to the process that warrant inclusion to the control strategy. The advancement of product and process understanding is essential because when such knowledge is at a maximum with appropriate control strategies in place, the risk of manufacturing product of poor quality is at a minimum.[21]

D. Process Control, Design Space, and Real-Time Release

A fundamental goal of PAT is to enable monitoring and control of the manufacturing process. Process control is established through enhanced process knowledge and understanding, meaning a process cannot be under control until it is thoroughly understood.[49] The process monitoring and control strategies should be used to continuously monitor the state of the process through timely measurements and provide a means to actively change the process to a desired state when necessary.[3] As previously mentioned, the first step in implementing

QbD is to develop a process monitoring system, while QbD in its entirety should include the control of critical sources of process variability to ensure that final product quality is met. In the ideal case, when the process monitoring system detects an out of control event, the process control system automatically adjusts the process to a state of control.[11] The two general categories of process control are feed forward and feedback process control. Feedback control refers to reactive actions in that manipulations to the process are made based on the output (product or intermediate quality attributes), whereas feed forward control refers to proactive actions in that manipulations to the process are made based on deviations of the input conditions or materials.[23] The process engineering community defines the purpose of feedback process control as "to operate the plant such that the net return is maximized in the presence of disturbances and uncertainties, exploiting the available measurements."[50] This implies that process control should be focused toward ensuring final product quality in the presence of process disturbances through manipulation of the process, not simply monitoring the process. A common misconception in the pharmaceutical industry, stemming from the traditional manufacturing paradigm, has been that a process is considered under control if the process variables are continually monitored and kept within their validated limits. Historically, this has been achieved by testing quality attributes upon completion of the manufacturing process. However, under the PAT paradigm, the process control strategy uses real-time analysis to adjust the critical process variables to ensure the final product quality attributes are constant.[49]

As with product and process understanding, multivariate methods play a crucial role in process control. To achieve the successful development of a control strategy, a reliable relationship between final product quality attributes, critical intermediate material attributes, and process parameters must be developed. These relationships are typically multivariate due to the complexity and multivariate nature of pharmaceutical products and processes. It is advantageous to know the overall "health" of the manufacturing process at all stages to provide corrective action if the process moves to a state where probability of fault is high. The development of multivariate models linking CMAs, process parameters, and final product quality attributes, early in the development process, will ensure that the final product will meet the desired attributes. The ability to predict final product quality from material attributes early in the process stream allows for corrective action to be moved upstream, reducing final product variability and enhancing the likelihood that the final product will meet its specification.[21] Such a proactive approach is the foundation for feed forward control. Material attributes of intermediates created early in the process stream allow for adjustment in downstream operations to ensure the final product will be consistent and in specification. Feedback control should also be applied to bring the material attribute of the intermediate to a value that ensures final product quality is met. Such dynamic processes adjust for variability of input materials as well as significant deviations in final product quality. Under the traditional manufacturing paradigm, it was assumed that consistent product would come as a result of a constant process. However, it is *impossible* to produce a consistent product from a fixed process with variable input. The process must be adaptable to the variability of input materials to consistently produce product with high quality[51,52]; thus, feed forward and feedback control is essential in ensuring final product quality.

An increase in variability in the process stream directly increases the variability of the final product. The ability to observe variation in the process and predict variation in final product quality from process variation allows immediate corrections if the process deviates from a state where the probability of producing out of specification product is at a minimum.[21] As with predicting final product quality from material attributes and process parameters, the ability to link critical sources of variability with downstream process status and final product quality also enables critical aspects of the control strategy to be shifted upstream. Such relationships ultimately enable the implementation of a real-time release (RTR) strategy in lieu of extensive end product testing. RTR is defined as "the ability to evaluate and ensure the quality of in-process and/or final product based on process data, which typically includes a valid combination of measured material attributes and process controls."[9] Such a system is advantageous over the traditional end product testing approach because once production of a batch is complete it can immediately be shipped to market, dramatically decreasing QC/QA efforts.[53] RTR can only be achieved through the continuous monitoring and control of material attributes, process parameters, and process variation leading to the prediction of final and intermediate quality attributes, implying the significance of multivariate analysis and real-time measurements.

Under the QbD paradigm, the process model that relates process inputs (i.e., CPPs) to process outputs (i.e., CQAs) is referred to as the *design space*. Specifically, design space is defined as "the multidimensional combination and interaction of input variables (e.g., material attributes) and process parameters that have been demonstrated to provide assurance of quality."[9] The design space is part of the process control strategy in that CPPs must be kept under control to remain in the design space which in turn ensures the final product meets all relevant quality attributes.[9] This is advantageous to industry considering post-approval submissions are costly and time consuming. Essentially, the design space is the foundation of the process control strategy because it defines the multidimensional space of input variables that ensures a desired level of product quality.

In most cases, the results of pharmaceutical processes are path dependent; thus, the design space of input parameters is as significant as the final product CQAs.[54] The design space is the platform in which product quality can be defined as a function/model of input parameters (raw material attributes and process parameters) as follows: Quality $= f$ (raw material attributes, process parameters), where *Quality* refers to the set of CQAs for a given product. When such a relationship is defined, raw material variability can be compensated for by adjustment of the process parameters. However, if a process is fixed, variability in raw material attributes will be conveyed in final product CQAs. The appropriate combination of process parameters can be determined from the observed raw material attributes and desired set of final product CQAs, facilitating the achievement of final product quality. Thus, the design space is the model that relates raw material attributes and process parameters to final product quality, and by some descriptions, defines the range of process parameters for which the model has been confirmed. This implies that the design space cannot be exclusively described as a range of process parameters because the process path (the multidimensional combination of raw material attributes and process parameters) compensates for variability expressed in raw material attributes. The traditional approach to pharmaceutical manufacturing utilized fixed processes, resulting in a single range of process parameters corresponding to a specific set of raw material attributes for which final product quality is assured. Such an approach limits manufacturing flexibility by utilizing only one of multiple combinations of raw material attributes and process parameters, overlooking raw material variability and various process disturbances.[55]

While the design space/process model describes the relationship between CPPs and CQAs, the control model functions to ensure that the CPPs and other process variables are maintained in a state of control within the working range that corresponds to the desired set of CQAs. There are multiple avenues to ensure that the process remains within the desired set of processing conditions, such as following identical manufacturing routes for a given set of raw material attributes or implementation of a multivariate control model that updates the process in real time through utilization of PAT tools.[56] With a control mode capable of adjusting the process in real time, process disturbances that cause deviations from the target set of CMAs can be compensated for during process operation. Such a system ensures intermediate CMAs are within the appropriate limits, and the product will meet the target CQAs as processing adjustments made in real time through the control model bring the process back to the desired state. Figure 8 represents a

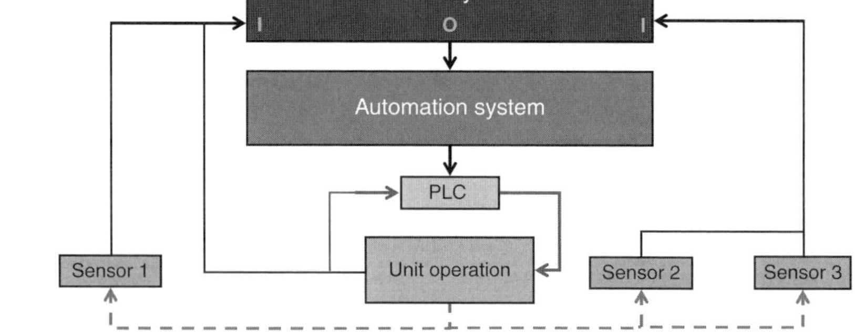

FIGURE 8 General schematic of a control system. I and O represent input and output, respectively; sensor 1, 2, and 3 represent various on- and in-line sensors (i.e., PAT tools); unit operation represents an individual unit operation (e.g., fluid bed granulation); PLC refers to programmable logic controller; response 1, 2, and 3 represent responses from the various on- and in-line sensors; CPP_1, CPP_2, and CPP_3 represent critical process parameters for the given unit operation.

schematic of a general control strategy that depicts how various in-line measurements can be used to accomplish process adjustments for an individual unit operation in real time. At the center of the control system is the data management and control software that functions to collect and organize all pertinent process data. Multiple sensors collect data in real time. The real-time data are sent to the data management and control system where it is organized and sent to the appropriate analytical and process models (e.g., NIR spectra sent to the appropriate analytical model to predict moisture content). The responses from the analytical and process models are then sent to the data management and control system where they are reorganized and sent to the appropriate control models. The control models then execute the appropriate adjustments for the corresponding process parameters. The adjusted process parameters are then organized in the data management and control system and sent to the automation system. The automation system communicates with the programmable logic controller (PLC) which in turn sends the adjusted process parameters to the processing equipment (defined as "unit operation" in Figure 8). Thus, process parameters can be adjusted according to data collected from in-line sensors using a similar control system.

It is most desirable to continuously monitor CMAs associated with a given process via PAT tools and actively apply feedback control based on the real-time analysis of those CMAs using a control system similar to that depicted in Figure 8. Such a system allows manipulation of the process based on variation observed in the CMAs caused by any number of disturbances (e.g., raw material variability, environmental changes, etc.).[7] The concept of design space plays a critical role in such a philosophy because the design space would provide the model that links the process parameters to the CMAs, enabling the proper adjustments that will bring the CMAs back to their target values. However, it must also be noted that the relationship between the CMAs and CQAs must be established and understood. For example, a CQA common to most solid oral dosage forms is the dissolution profile, and particle size is a CMA that may have a significant effect on the final products dissolution profile. In such a scenario, the design space can provide the functional relationship between process variables and particle size in that if the particle size falls outside its specified range, the process can be adjusted so the particle size meets the desired range which in turn gives the appropriate dissolution profile for the final product. Thus, as previously suggested, final product CQAs should be based on the product's performance while CMAs should be those material attributes that critically affect the performance characteristics described by the CQAs. The design space would ultimately describe the relationship between the CMAs and process variables to provide the platform by which process parameters can be adjusted in real time to achieve the desired CMAs and corresponding CQAs.

Nondestructive analytical techniques are required for the implementation of an advanced process control system and achievement of RTR because they provide a means of measuring CMAs and/or CMAs in real time. Skibsted *et al.*[57] used nondestructive analytical techniques (NIRS), multivariate analysis, and MSPC to implement a RTR strategy that implemented feed forward and feedback control. The RTR strategy was applied to a typical tablet manufacturing process containing mixing, granulation, drying, mixing, and tabletting. Four models were generated, one MSPC model and three regression models to predict quality indicators.

The MSPC model utilized NIR spectra collected subsequent to granulation to detect particle size deviations that may potentially exacerbate future processing. The first regression model made quality predictions on intermediate material to provide a means of feedback control. Water content was predicted using a PLS model generated from in-line NIRS data. The model was capable of providing real-time control of the drying time as it was determined that drying was complete when the granule temperature reached 34 °C. The second was a pair (models I and II) of PCA models that made predictions on mean disintegration time for the tablets from process variables and NIR spectral scores while manufacturing was ongoing. Model I used the first three scores from the PCA model derived from average NIR spectrum from the first mixing step concatenated with mixing time and granulation liquid flow. For model II, the first three scores from the PCA models derived from the average NIR spectrum of the granulation, the last spectrum from drying, and the average NIR spectrum from the glidant mixing step were concatenated with process variables (drying temperature, drying time, and upper punch force) to predict disintegration time. It was demonstrated that both models enabled feed forward process control; model I was used to set the granulation liquid flow while model II was used to determine the upper punch force to give the desired disintegration time. The final model utilized NIR transmission spectroscopy to determine the content of API in the final dosage form to give the final quality prediction. This provides an excellent example of how multivariate analysis couple with NIRS enables a RTR strategy through enhanced process understanding and control.

Traditionally, end product testing for solid dosage forms involved off-line destructive analytical techniques that measured various properties such as CU and RTS. Cogdill *et al.*[28,29,58] presented a series of papers that demonstrated the potential of NIRS in replacing destructive analytical techniques used for end product testing of tablets. The first part of the series demonstrated the feasibility of on-line NIRS for quantitative methods. The on-line method utilized a conveyor system that positioned tablets from a tablet press over NIR reflectance and transmission instruments. It was demonstrated that positioning error can be mitigated by preprocessing methods (also reflectance measurements were less sensitive than transmission); therefore, the conveyor system was feasible for on-line analysis. It was also demonstrated that samples produced from a range of manufacturing scales can be pooled for quantitative calibration development. In the second paper, a model was generated using PLS that predicted API content (mg) from spectral data. The calibration set consisted of tablets from production batches as well as laboratory batches to enhance variability in API content and RTS (the product was manufactured to tight specifications). A second model was generated using PLS that predicted tablet RTS from spectral data. Consistent with API content, the calibration set contained tablets from production and laboratory batches. In both cases, the standard error of the model using NIRS was consistent with that of the standard reference method (HPLC for CU and diametral crushing strength for RTS). The final paper of the series was based on the development of a calibration monitoring and transfer methods. Calibration monitoring was achieved through observing instrument performance parameters and Hotelling's T^2 and Q residuals. Calibration transfer was achieved using baseline subtraction with as few as 15 transfer samples. This series of papers demonstrated the feasibility of NIRS for the development of a quantitative method to measure API content and RTS on-line and in real time in lieu of inefficient destructive techniques.

The quantitative NIRS method developed by Cogdill *et al.* was implemented as part of a RTR strategy for a marketed product, specifically an immediate release tablet.[53] The process included wet granulation, fluidized bed drying, compression, and application of a nonfunctional film coat. The RTR strategy was based on the prediction of dissolution, and through formal risk assessment exercises, it was demonstrated that the input variable that had the greatest effect on dissolution was particle size. A process model was developed that predicted dissolution from raw material attributes and process parameters and was used to provide feed forward control at the process outset. Thus, if the raw material attributes were outside the given range that generated the appropriate dissolution profile, CPPs could be adjusted accordingly to give the appropriate dissolution profile. API content (CU) and RTS were determined via the method developed by Cogdill *et al.* prior to film coat application. CU and RTS were determined in real time using the NIRS method, eliminating the need for traditional, time-consuming CU analyses. The RTR and control strategy was suitable for use in full-scale production and proved traditional end product testing was redundant.

III. SUMMARY AND CONSIDERATIONS REGARDING RETURN ON INVESTMENT FOR QbD/PAT

With the current economic and political climate posing new risks to the pharmaceutical industry, and considering both the industry's and regulators' desire to provide quality pharmaceutical products to the public at a reasonable price, there is no logical choice but for pharmaceutical manufacturers to seek the efficiency and quality that is achievable through the philosophy of QbD, enabled by PAT.

The practical example cited in Ref. [53] provides evidence that achievement of advanced process control and RTR is possible in the pharmaceutical industry. That innovative system proved to be of value when compared to traditional testing and control procedures; because the strategy was applied to an existing product, the manufacturer was able to determine, quantitatively, the advantages of the RTR approach. It was reported that cycle time from initial dispensing of material to availability to market was reduced from about 12 days to 4, and QC/QA work was reduced from 8 days to 8 h. "This reduction in time leads to a need for less working capital, which is a major cost-saving and given the work to confirm process robustness as part of the scheme, there is additional assurance that product will pass specification, giving a more predictable supply chain."[53] This exemplifies the ultimate advantage of utilizing PAT and QbD in the pharmaceutical industry.

While QbD/PAT offers significant opportunity for the pharmaceutical industry to enhance product quality, the industry will not realize a return on such investments without an appropriate correlated effort to capitalize on the manufacturing efficiencies that are possible. Achieving the desired return on investment (ROI) requires a synergy between the QbD/PAT approach and the operational efficiencies provided through a philosophy of Lean manufacturing. Methods for estimating the ROI that can be achieved have been discussed by Cogdill *et al.*[59]

New approaches to integrating performance-based product specifications into the QbD/design space philosophy will provide a link between clinical performance and manufacturing variability.[60] The removal of operational, technological, and regulatory roadblocks that reduce quality and prohibit maximum process efficiency will strengthen our industry, enabling the scientists and engineers who develop and manufacture pharmaceuticals to better serve the public.

REFERENCES

1. Nasr, M., Migliaccio, G., Allen, B., Baum, R. and Branning, R. FDA's pharmaceutical quality initiatives. *Pharm. Tech.* 32:54–72, 2008.
2. Food and Drug Administration. Final Report on Pharmaceutical cGMPs for the 21st century—A Risk Based Approach. http://www.fda.gov/cder/gmp/gmp2004/CGMP%20report%20final04.pdf. Updated 20042008
3. Guidance for Industry: PAT—A Framework for Innovative Pharmaceutical Development, Manufacturing, and Quality Assurance, 2004.
4. Woodcock, J. The concept of pharmaceutical quality. *Am. Pharm. Rev.* 7(6):10–15, 2004.
5. Q9: Quality Risk Management. *ICH Harmonized Tripartite Guidelines.* International Conference on Harmonization of Technical Requirements for Registration of Pharmaceuticals for Human Use, 2006.
6. Q10: Pharmaceutical Quality System. *ICH Tripartite Guidelines.* International Conference on Harmonization of Technical Requirements for Registration of Pharmaceuticals for Human Use, 2007.
7. Lionberger, R. A., Lee, L. S., Lee, L., Raw, A. and Yu, L. X. Quality by design: Concepts for ANDAs. *AAPS* 10:268–276, 2008.
8. FDA Guidance for Industry and Review Staff: Target Product Profile – A Strategic Development Process Tool (Draft Guidance).
9. Q8 (R1): Pharmaceutical Development, Revision 1. *ICH Harmonized Tripartite Guidelines.* International Conference on Harmonization of Technical Requirements for Registration of Pharmaceuticals for Human Use, 2007.
10. Callis, J. B., Illman, D. L. and Kowalski, B. R. Process analytical chemistry. *Anal. Chem.* 59:624A–637A, 1987.
11. Yu, L. X. Pharmaceutical quality by design: Product and process development, understanding, and control. *Pharm. Res.* 25:781–791, 2008.
12. Munson, J., Gujral, B. and Stanfield, C. F. A review of process analytical technology (PAT) in the U.S. pharmaceutical industry. *Curr. Pharm. Anal.* 2:405–414, 2006.
13. Leuenberger, H., Puchkov, M., Krausbauer, E. and Betz, G. Manufacturing pharmaceutical granules: Is the granulation end-point a myth? *Powder Technol.* 189:141–148, 2009.
14. Miller, C. E. Chemometrics and NIR: A match made in heaven? *Am. Pharm. Rev.* 2:41–48, 1999.
15. Miller, C. E. Chemometrics in process analytical chemistry. In *Process Analytical Technology,* (K.E. Bakeev, ed.), Blackwell Publishing, pp. 226–325, 2005.
16. Haaland, D. M. and Thomas, E. V. Partial least squares methods for spectral analyses. *Anal. Chem.* 60:1193–1202, 1988.
17. Eigenvector Research Inc. (Manson, WA). PLS_Toolbox Version 3.0, 2003.
18. Wise, B. M. and Gallagher, N. B. The process chemometrics approach to process monitoring and fault detection. *J. Process Control* 6(6):329–348. doi:10.1016/0959-1524(96)00009-1, 1996.
19. Kourti, T. Optimisation and control. In *Design and Analysis in Chemical Research.* 3rd ed., (R. Tranter, ed.), CRC Press, Boca Raton, FL, pp. 314–364, 2000.
20. AlGhazzawi, A. and Lennox, B. Model predictive control monitoring using multivariate statistics. *J. Process Control* 19(2):314–327. doi:10.1016/j.jprocont.2008.03.007, 2009.
21. Wold, S., Cheney, J., Kettaneh, N. and McCready, C. The chemometric analysis of point and dynamic data in pharmaceutical and biotech production (PAT)—Some objectives and approaches. *Chem. Intell. Lab. Syst.* 84:159–163, 2006.

22. Kourti, T. The process analytical technology initiative and multivariate process analysis, monitoring and control. *Anal. Bioanal. Chem.* 384:1043–1048, 2006.
23. Kourti, T. Process analytical technology beyond real-time analyzers: The role of multivariate analysis. *Crit. Rev. Anal. Chem.* 36:257–278, 2006.
24. Mason, R. L. and Young, J. C. Multivariate Statistical Process Control. SIAM, Philadelphia, PA, 2002.
25. Martens, H. M. and Martens, M. Multivariate Analysis of Quality: An Introduction. Wiley, West Sussex, England, 2001.
26. Gabrielsson, J., Jonsson, H. and Trygg, J. Combining process and spectroscopic data to improve batch modeling. *AIChE* 52(9):3164–3172, 2006.
27. ASTM. International Standard E2363-04 Standard Terminology Relating to Process Analytical Technology in the Pharmaceutical Industry, 2004.
28. Cogdill, R. P., Anderson, C. A., Delgado, M., Chisholm, R., Bolton, R., Herkert, T., Afnan, A. M. and Drennen III, J. K. Process analytical technology case study: Part II. Development and validation of quantitative near-infrared calibrations in support of a process analytical technology application for real-time release. *AAPS PharmSciTech.* 6(2):E273–E283, 2005.
29. Cogdill, R. P., Anderson, C. A., Delgado-Lopez, M., Molseed, D., Chisholm, R., Bolton, R., Herkert, T., Afnán, A. M. and Drennen III, J. K. Process analytical technology case study Part I: Feasibility studies for quantitative near-infrared method development. *AAPS PharmSciTech.* 6(2): E262–E272, 2005.
30. Reich, G. Near-infrared spectroscopy and imaging: Basic principles and pharmaceutical applications. *Adv. Drug Deliv. Rev.* 57(8):1109–1143. doi:10.1016/j.addr.2005.01.020, 2005.
31. Rasanen, E. and Sandler, N. Near infrared spectroscopy in the development of solid dosage forms. *JPP* 59:147–159, 2007.
32. Jayawickrama, D., El-Hagrasy, A. and Chang, S. Raman applications in drug manufacturing processes. *Am. Pharm. Rev.* 9:10–17, 2006.
33. De Beer, T. R. M., Bodson, C., Dejaegher, B., Walczak, B., Vercruysse, P., Burggraeve, A., Lemos, A., Delattre, L., Vander Heyden, Y., Remon, J. P., Vervaet, C. and Baeyens, W. R. G. Raman spectroscopy as a process analytical technology (PAT) tool for the *in-line* monitoring and understanding of a powder blending process. *J. Pharm. Biomed. Anal.* 48(3):772–779, 2008.
34. Shi, Z., Cogdill, R. P., Short, S. M. and Anderson, C. A. Process characterization of powder blending by near-infrared spectroscopy: Blend end-points and beyond. *J. Pharm. Biomed. Anal.* 47 (4–5):738–745. doi:10.1016/j.jpba.2008.03.013, 2008.
35. Wikstrom, H., Marsac, P. J. and Taylor, L. S. In-line monitoring of hydrate formation during wet granulation using Raman spectroscopy. *J. Pharm. Sci.* 94(1):209–219, 2005.
36. Hu, Y., Liang, J. K., Myerson, A. S. and Taylor, L. S. Crystallization monitoring by Raman spectroscopy: Simultaneous measurement of desupersaturation profile and polymorphic form in flufenamic acid systems. *Ind. Eng. Chem. Res.* 44:1233–1240, 2005.
37. Aaltonen, J., Kogermann, K., Strachan, C. J. and Rantanen, J. In-line monitoring of solid-state transitions during fluidisation. *Chem. Eng. Sci.* 62:408–415, 2007.
38. McGoverin, C. M., Rades, T. and Gordon, K. C. Recent pharmaceutical applications of Raman and terahertz spectroscopies. *J. Pharm. Sci.* 97:4598–4621, 2008.
39. Strachan, C. J., Taday, P. F., Newnham, D. A., Gordon, K. C., Axel Zeitler, J., Pepper, M. and Rades, T. Using terahertz pulsed spectroscopy to quantify pharmaceutical polymorphism and crystallinity. *J. Pharm. Sci.* 94(4):837–846, 2005.
40. Han, P. Y., Tani, M., Usami, M., Kono, S., Kersting, R. and Zhang, X. A direct comparison between terahertz time-domain spectroscopy and far-infrared Fourier transform spectroscopy. *J. Appl. Phys.* 89:2357–2359, 2001.
41. Taday, P. F., Bradley, I. V., Arnone, D. D. and Pepper, M. Using terahertz pulse spectroscopy to study the crystalline structure of a drug: A case study of the polymorphs of rantidine hydrochloride. *J. Pharm. Sci.* 92:831–838, 2003.
42. Fitzgerald, A. J., Cole, B. E. and Taday, P. F. Nondestructive analysis of tablet coating thicknesses using terahertz pulsed imaging. *J. Pharm. Sci.* 94:177–183, 2005.
43. Wu, H., Heilweil, E. J., Hussain, A. S. and Khan, M. A. Process analytical technology (PAT): Quantification approaches in terahertz spectroscopy for pharmaceutical application. *J. Pharm. Sci.* 97:970–984, 2008.

44. Gowen, A. A., O'Donnell, C. P., Cullen, P. J. and Bell, S. E. J. Recent applications of chemical imaging to pharmaceutical process monitoring and quality control. *Eur. J. Pharm. Biopharm.* 69:10–22, 2008.
45. Lyon, R. C., Lester, D. S., Lewis, E. N., Lee, E. N., Yu, L. X., Jefferson, E. H. and Hussain, A. S. Near-infrared spectral imaging for quality assurance of pharmaceutical products: Analysis of tablets to assess powder blend homogeneity. *AAPS PharmSciTech.* 3(3):1–15, 2002.
46. Roggo, Y., Jent, N., Edmond, A., Chalus, P. and Ulmschneider, M. Characterizing process effects on pharmaceutical solid forms using near-infrared spectroscopy and infrared imaging. *Eur. J. Pharm. Biopharm.* 61(1–2):100–110. doi:10.1016/j.ejpb.2005.04.005, 2005.
47. Lee, E., Huang, W. X., Chen, P., Lewis, E. N. and Vvilecchia, R. V. High-throughput analysis of pharmaceutical tablet content uniformity by near-infrared chemical imaging. *Spectroscopy* 21:24–32, 2006.
48. Sasic, S. Raman mapping of low-content API pharmaceutical formulations. I. Mapping of alprazolam in alprazolam/xanax tablets. *Pharm. Res.* 24:58–65, 2007.
49. Guidance for Industry: Process Validation: General Principles and Practices *DRAFT GUIDANCE*, 2008.
50. Engell, S. Feedback control for optimal process operation. *J. Process Control* 17(3):203–219. doi:10.1016/j.jprocont.2006.10.011, 2007.
51. Morris, K. R. Risk-based development and CMC question-based review: Asking the right questions for process understanding, control and filing. http://www.fda.gov/ohrms/dockets/ac/04/slides/2004-4052S1_12_Morris.ppt, 2009, Updated 2004. Accessed May 26, 2009.
52. Morris, K. R. Risk-based development and CMC question-based review: Asking the right questions for process understanding, control, and filing. FDA Advisory Committee for Pharmaceutical Science. http://www.fda.gov/ohrms/dockets/ac/04/slides/2004-4052S1_12_Morris.ppt, 2009, Updated 2004. Accessed 5/19.
53. Potter, C. PQLI application of science- and risk-based approaches (ICH Q8, Q9, and Q10) to existing products. *J. Pharm. Innov.* 4:4–23, 2009.
54. Moore, C. M. V. Establishing, Presenting, and Maintaining a Design Space—An FDA Perspective. *Drug Information Association 44th Annual Meeting*, 2008.
55. Kourti, T. Quality by design in the pharmaceutical industry: The role of multivariate analysis. *Am. Pharm. Rev.*, 2009.
56. Bolton, R. and Tyler, S. PQLI engineering controls and automation strategy. *J. Pharm. Innov.* 3:88–94, 2008.
57. Skibsted, E. T. S., Westerhuis, J. A., Smilde, A. K. and Witte, D. T. Examples of NIR based real time release in tablet manufacturing. *J. Pharm. Biomed. Anal.* 43(4):1297–1305. doi:10.1016/j.jpba.2006.10.037, 2007.
58. Cogdill, R. P., Anderson, C. A. and Drennen III, J. K. Process analytical technology case study, Part III: Calibration monitoring and transfer. *AAPS PharmSciTech.* 6:E284–E295, 2005.
59. Cogdill, R. P., Knight, T. P., Anderson, C. A. and Drennen III, J. K. The financial returns on investments in process analytical technology and lean manufacturing: Benchmarks and case study. *J. Pharm. Innov.* 2:38–50, 2007.
60. Cogdill, R. P. and Drennen, J. K. Risk-based quality by design (QbD): A Taguchi perspective on the assessment of product quality, and the quantitative linkage of drug product parameters and clinical performance. *J. Pharm. Innov.* 3:23–29, 2008.

6 NOVEL DOSAGE FORM ANALYSIS

JOSEPH ETSE

Pharmaceutical and Analytical Development, Novartis Pharmaceuticals Corporation, East Hanover, New Jersey, USA

I. INTRODUCTION

The key aspect of dosage form formulation is the addition of excipients to the active pharmaceutical ingredient (API) or drug substance (DS) to make it amenable for processing operations into a final dosage type.[1] Novel pharmaceutical dosage forms cover a broad range of formulation delivery platforms such as tablets, capsules, cachets, sustained release dosage forms, parenteral dosage forms, transdermal dosage forms, metered-dose inhalants, solutions, emulsions, and suspensions that are designed to improve patience acceptance and/or to assure maximum absorption following administration.

Excipients in the pharmaceutical processing operations may originate from a wide variety of sources, including plants (e.g., starches, sugars, celluloses), animals (gelatin, lactose, shellac), minerals (e.g., dicalcium phosphate dehydrate, magnesium stearate), and from synthetic origin (e.g., polyvinylpyrolidone, polysorbates, polyethylene glycols, EUDRAGIT®). Similar to the API or DS, excipients exhibit a wide range of solubilities which influence the type of sample preparation technique that may be employed.[2] Since the goal of any sample preparation technique is to get the DS into solution to achieve quantitative recovery (>98%) of the API, the number of sample treatment steps and the choice of solvents used become a critical consideration. Sample treatment steps whether manual or automated involve laboratory sampling, weighing, sample extraction/dilution, and subsequent injection of an aliquot of the sample for analysis by HPLC.[3] Unlike excipients, the API or DS is typically well characterized and may be organic acids or bases which exhibit different solubility behavior in water.[4,5]

Solubility of the dosage form is a key factor when it comes to analysis. Solubility is expressed as the concentration of a substance in a saturated solution at a defined temperature. The United States Pharmacopoeia (USP) defines solubility as shown in Table 1. The nature of the API may be such that it may be nonionizable and so poorly soluble in water but its corresponding salts may be ionizable and as a consequence become water soluble.[4,5] Depending on the nature of the dosage form, the first step in the sample preparation technique to employ is to determine what type of solvent system will be best to achieve maximum solubility of the API in the presence of excipients. Maximum solubility of the API ensures that its recovery from the sample preparation steps will be an accurate measure of drug content in the pharmaceutical dosage form.

In view of the importance of API solubility in the HPLC analysis of pharmaceutical dosage forms, this chapter will examine some of the strategies that may be employed to assure optimum and quantitative recovery of the API during the sample preparation steps prior to HPLC analysis.[6,7] In addition, emphasis on sample preparation strategies used for tablets and capsules will frequently be referred to and also serve as examples throughout the chapter because tablets and capsules are the most commonly used novel pharmaceutical dosage form delivery platforms.

TABLE 1 Solubility Definitions by USP

Descriptive term	Parts of solvent required for one part of solute
Very soluble	Less than 1
Freely soluble	From 1 to 10
Soluble	From 10 to 30
Sparingly soluble	From 30 to 100
Slightly soluble	From 100 to 1000
Very lightly soluble	From 1000 to 10,000
Insoluble	From 10,000 and over

Source: *USP 32-NF-27*, U.S. Pharmacopeial Convention: Rockville, MD.

II. TABLETS

These are solid dosage forms that contain drug with or without suitable diluents and are prepared either by compression or compaction methods.[8] Tablets made by compression or compaction methods may be manufactured using three general methods, namely, wet granulation; dry granulation; and direct compression as film-coated, enteric-coated (e.g., diclofenac sodium delayed release tablets, USP), or as multiple-layered tablets. Other tablets made by compression or compaction include buccal and sublingual tablets which are usually small, flat, oval shaped and are administered by inserting the tablets into the buccal cavity to dissolve or erode slowly.[8]

In the analysis of any tablet dosage form, sample preparation may involve a number of unit operations that reduce the dosage form into a suspension or solution form that is amenable to analysis.[3,7] Though a number of analytical techniques such as titrimetric and gravimetric techniques can be used for analysis of solid dosage forms, the focus of this chapter will be limited to the use of HPLC and in particular reversed-phase HPLC (RP-HPLC).

The biggest challenge for RP-HPLC analysis is the accurate measurement of drug concentration in tablets that are formulated with excipients that exhibit poor aqueous solubility. Since the primary goal of any sample preparation technique is the effective solubilization of the drug to achieve a solution sample that is amenable to RP-HPLC analysis, careful selection of solvents or solvent mixtures is a prerequisite that will avoid sample precipitation or emulsification issues.[3,7] At the same time, chromatographic conditions selected should also be such that it allows for the positive retention and identification of the API and avoids its elution in the void volume.[9,10]

A. Sample Preparation and Choice of Solvents

The primary aim of any sample treatment technique is to provide a reproducible and homogenous solution suitable for HPLC analysis. In that sense, an aliquot of the prepared sample needs to (1) be free of any interference, (2) be stable in the solvent/solvent mixtures, (3) dissolve in the mobile phase without affecting retention or resolution of the API, and (4) be compatible with the intended HPLC method.

Usually, the type of dosage form dictates the kind of sample preparation technique(s) to employ. Regardless of the sample preparation technique employed, appropriateness of the technique(s) is measured by the quantitative recovery of the API from the sample matrix. If quantitative recovery of the API is achieved, the sample preparation technique employed is judged to be successful.

The initial step in any sample preparation technique is to reduce the sample (if it is a solid) into finely divided powder sample that will dissolve faster and will be easier to extract due to increase surface area.[3,7] The next step is to determine what kind of solvent system will be best to extract the API from the sample matrix.

B. Choice of Solvents

Since majority of APIs contain one or more acidic or basic functional groups, solvents used for HPLC sample preparation may require pH control to influence the ionizable state of the compounds in a similar fashion to what is done for most

RP-HPLC mobile phases.[3–5,9–12] When the pH of an acidic compound in solution is more than 2 pH unit above its pK_a, the compound will be >99% ionized and will be >99% nonionized below its pK_a. Conversely, for basic compounds, they are ionized below their pK_a and nonionized above their pK_a. The nonionized form of either an acidic or basic compound will be less polar (i.e., more hydrophobic or nonpolar) and strongly retained in the RP-HPLC chromatographic system to require the addition higher amounts of organic modifiers (i.e., less polar solvents such as methanol or acetonitrile) to effect their elution (see Table 2). Similarly, in the sample preparation solvent system, the nonionized form of the analytes being more hydrophobic may require the addition of organic modifiers to assure their higher solubility in the sample matrix. Thus, the effect of pH on either the RP-HPLC mobile phase or the sample preparation solvent can drive the analyte equilibrium concentration to either extreme (i.e., neutral, nonionized, or ionized) to ultimately influence the extent of the analyte solubility in either the sample preparation solvent and/or affinity to the HPLC stationary phase.[3,4] For this reason, buffers are widely used in HPLC analysis to preferably force the analytes to remain predominantly (>90%) in a single ionization state (see Table 3).

The commonly used sample solvent mixtures in RP-HPLC are hydro-organic mixtures of which methanol, acetonitrile, and/or combination of these two are the predominant organic solvents used. Other organic solvents such as tetrahydrofuran (THF), isopropyl alcohol (IPA), ethyl acetate, and acetone have also been used (see Table 4). However, the primary consideration for choice of solvents include miscibility of the solvents, compatibility with the RP-HPLC mobile phase, solubility of the API in the sample solvent, polarity, light transmission, viscosity, stability, and pH.[9,10] Compatibility of the sample solvent used with the HPLC

TABLE 2 Properties of Some Common Solvents Used in RP-HPLC Analysis

Solvent	Refractive index (20 °C)	Viscosity (cP)	Miscibility number (*M*)	Polarity (*P*′)
Acetone	1.3587	0.36	15, 17	5.1
Acetonitrile	1.3441	0.38	11, 17	5.8
Ethyl acetate	1.3724	0.45	19	4.4
Ethyl ether	1.3524	0.24	23	2.8
Isopropyl alcohol	1.3772	2.40	15	3.9
Methanol	1.3284	0.55	12	5.1
Tetrahydrofuran	1.4072	0.55	17	4.0
Trifluoroacetic acid	1.2850	0.926	–	–
Water	1.3330	1.00	–	10.2

Missing values in table indicate data not available. For miscible solvents, *M* numbers differ by 15 units or less. For immiscible solvents, *M* numbers differ by 17 or more units. Polarity values (P') range from very low to high values. For nonpolar solvents, P' is very low (e.g., hexane, $P'=0$). For very polar solvents, P' is high (e.g., water, $P'=10.2$).
Source: See Ref. 3.

TABLE 3 Common Buffers Used in RP-HPLC Analysis

Buffer	pK_a (25 °C)	pH range	UV cutoff (nm)	LC-MS compatible
Trifluroacetic acid	0.3	1.5–2.5	210 (0.1%, v/v)	Yes
Phosphoric acid (pK_1)	2.15	1.1–3.1	<200 (10 mM)	No
Phosphoric acid (pK_2)	7.20	6.2–8.2	<200 (10 mM)	No
Phosphoric acid (pK_3)	12.33	11.3–13.3	<200 (10 mM)	No
Citric acid (pK_1)	3.13	2.1–4.1	230 (10 mM)	Yes
Citric acid (pK_2)	4.76	3.7–5.7	230 (10 mM)	Yes
Citric acid (pK_3)	6.40	4.4–6.4	230 (10 mM)	Yes
Formic acid	3.75	2.8–4.8	210 (10 nm)s	Yes
Acetic acid	4.76	3.8–5.8	210 (10 mM)	Yes
Trisamino methane (TRIS)	8.1	7.1–9.1	205 (10 mM)	No
Borate	9.2	8.3–10.3	<200 (10 mM)	Yes
Dimethylamine	10.7	9.7–11.8	<200 (10 mM)	Yes
Triethylamine	10.8	9.8–11.8	<200 (10 mM)	Yes
Ammonium hydroxide	9.3	8.3–10.3	<200 (10 mM)	Yes
Pyrrolidine	11.3	10.3–12.3	<205 (10 mM)	No
N-Methyl pyrrolidine	10.3	9.3–11.3	<205 (10 mM)	No
Piperidine	11.1	10.1–11.1	<205 (10 mM)	No
Ammonium acetate		3.8–5.5	210 (10 mM)	Yes
Ammonium bicarbonate (pK_1)	6.4	5.4–7.4	<200 (10 mM)	Yes
Ammonium bicarbonate (pK_2)	10.3	9.3–11.3	<200 (10 mM)	Yes
Ammonium	9.3	8.3–10.3	<200 (10 mM)	Yes

Sources: See Refs. 9,11.

TABLE 4 UV Cutoff for Some Organic Solvents Used in RP-HPLC Analysis

Solvent	UV cutoff (nm)
Acetonitrile	190
Isopropyl alcohol	205
Methanol	205
Ethanol	205
Tetrahydrofuran	215
Ethyl acetate	256

Source: See Ref. 9.

mobile phase composition is necessary to avoid precipitation of the API on the column and in the sample solvent. Miscibility is important to avoid the unfavorable partitioning of the analytes between immiscible solvent mixtures. Light transmission is important when using UV detection because solvents with high UV cutoffs such as acetone and ethyl acetate cannot be used for analyses at low UV wavelengths such 210 nm.[9] Acetonitrile has very low UV cutoff (< 190 nm) and is one of the contributing factor toward its common use as sample solvent. Methanol, ethanol, and IPA have UV cutoff >205 nm and is generally not recommended for work at UV wavelength <210 nm with these solvents. Viscosity is also another important consideration because highly viscous solvents such as methanol and IPA (see Table 2) can lead to reduced diffusion rate resulting in peak broadening as well as excessively high backpressures in the column.[9,12] Solvents such as THF and other ethers are prone to oxidation to form peroxides which can react with the analyte or with mobile phase components causing the appearance of spurious peaks.[9,11,12] Therefore, the choice of sample solvent is critical for ensuring reproducible chromatographic conditions and maintaining high solubility of the API in solution.[3]

C. Sample Preparation for Tablet Dosage Forms

It is recommended to start the sample preparation for tablets by first adding an acidified aqueous solvent to the tablet in a suitable container to disintegrate the tablet. This approach is highly recommended if the API contains one or more basic functional groups or basic salts. However, for tablets that contain acidic or neutral functional groups, use of acidified aqueous solvent is not necessary and the addition of plain water may be all that is required. Next is the addition of the organic solvent to achieve a final aqueous to organic solvent ratio that can range from about 10:90 to 50:50 (v/v).

Sample agitation to effect extraction of the analyte from the tablet matrix is the step that follows after the addition of the hydro-organic solvent. Sample agitation is accomplished through a variety of techniques which include grinding, shaking, and sonication.[3,7,13,14] Grinding requires the use of a mortar and pestle to reduce the tablet into finely ground powder and from which an aliquot equivalent to the tablet weight is weighed and extracted. Grinding as a sample preparation step is not recommended because of potential desegregation that can occur from the grinding process and cause nonhomogenous powder mixtures that may no longer be truly representative of the original tablet content. In addition, the frictional rigor of the grinding process can generate heat that may cause heat labile compounds to degrade.[13] To minimize potential occurrence of sample nonhomogeneity due to grinding, the recommended sample preparation technique is shaking or sonication of the sample after the addition of a well-defined volume of a hydro-organic solvent. When shakers or sonicators are used, it is important to ensure that initial volume of the sample solvent that is added to the container is not more than 70% of the total volume of the container. The required duration of shake or sonication time necessary to achieve quantitative recovery of the API from the tablet matrix is established through series of recovery experiments (see Table 5). Instead of using only shaking or sonication, combinations of shake and sonication steps are usually used to promote the rapid extraction of the API from the tablet matrix. Whether by shake, sonication, or combination of shake/sonication steps, efficient extraction of the API from the tablet matrix is evaluated by estimating the amount of API recovered in an accuracy or recovery study.[7,10] After the defined shaking or sonication period, the sample is diluted to the

TABLE 5 Recovery of a Spiked Known Amount of an API from a Placebo Tablet

Sample prep no.	Target amount (%)	Spike amount (μg/mL)	Recovered amount (μg/mL)	Recovered (%)
1	50	50.1	50.1	100.0
2		50.0	50.5	101.0
3		49.5	49.8	100.6
1	100	101.1	100.9	99.8
2		99.7	99.4	99.7
3		99.9	99.6	99.7
1	150	150.8	149.9	99.4
2		149.6	149.1	99.7
3		149.7	147.3	98.5
			Overall mean	*99.8*
			% RSD	*0.7*

Experiment consisted of spiking percentage of target known amounts of the API into a placebo matrix of the tablet dosage form. Samples were extracted and analyzed as directed in the method. Results show quantitative recovery of API which indicated reliability of the sample preparation procedure.

final volume of the container with the hydro-organic solvent and mixed well to ensure a homogenous sample solution that is devoid of concentration gradients is achieved.

Sonicators are widely used for sample preparation of solid dosage forms such as tablets and capsules because it causes solids in a solid–liquid mixture to quickly disintegrate into finer particles and thereby increase the total solid surface in contact with the solvent. Sonicators achieve this rapid disintegration because of the generation and propagation of ultrasound waves in the liquid.[14–16] However, propagation of the ultrasound waves is not homogeneous and unavoidably produces heat that may lead to the creation of local hot spots within the ultrasonic bath.[15,16] In addition, continuous ultrasonication often leads to warming up of bulk liquids due to endothermic reactions.[14] The combined effects of warming up of bulk liquids and the creation of local hot spots could potentially cause degradation of heat labile compounds to occur. Usually, the local hot spots are also the position of highest intensity of the ultrasound waves. So simple methods that rely on the use of aluminum foil in what is termed "aluminum foil test" have been developed to locate the position of local hot spots.[14] The method requires placing aluminum foil sheets inside the ultrasonic bath and operating the unit. The zones within the bath where the most intense ultrasound waves occur are quickly identified by perforations that appear in the aluminum sheet.[15,16] To minimize the warming up of bulk liquid and its potential impact on the chemical stability of dosage forms during the sonication step, water cooling recirculation systems can be attached to the ultrasonic bath to maintain the temperature at levels that will not be detrimental to the chemical stability of samples. For these reasons, it is imperative that explicit instructions be provided if the sample preparation step requires the use of sonicators.

A prerequisite for any successful sample preparation technique is the demonstration of the reliability of the analytical procedure. This is accomplished through the measurement of accuracy of the sample. Accuracy is defined as the agreement between the expected content of the API in the dosage form and the actual amount found after analysis.[10,17] Table 5 shows a recovery experiment performed to demonstrate the quantitative recovery of an API from a tablet dosage form. In this example, accuracy was demonstrated by the quantitative recovery of API after it had been spiked into a placebo sample of the drug product. Accuracy is typically assessed in a recovery experiment that uses a minimum of nine determinations at a minimum of three different concentration levels as shown in Table 5. Because the type of sample preparation technique adopted can greatly influence the outcome of the recovery experiment, careful consideration must be given to the selection of the particular sample agitation technique such as grinding, shaking, sonication, filtration, and centrifugation. Besides, accuracy, completeness of extraction of the sample using any of the sample agitation techniques discussed above should be evaluated. This can be done either by kinetic or thermodynamic means.[10] If a kinetic approach is used, the experiments can consist of taking the sample left over from the initial recovery experiment and reshaking or resonication at longer shake or sonication times and determining the amount of material still recoverable after the additional shake or sonication time. Alternatively, the initial sample can be used by shaking and/or sonication at different longer times. For a thermodynamic approach, different extraction volumes at a fixed shake or sonication time may be used. A generalized procedure for either the kinetic or the thermodynamic approach for evaluating completeness of extraction is shown in Table 6. If results from either set of experiments suggest incomplete extraction because additional

TABLE 6 Extraction Study to Evaluate Kinetic and Thermodynamic Effects

	Kinetic approach: extraction volume is kept constant V (mL) = V_1		Thermodynamic approach: extraction time is kept constant T (min) = T_1	
Experiment no.	**Extraction time (min)**	**Amount recovered (%)**	**Extraction volume (mL)**	**Amount recovered (%)**
1	T_0	99.3	V_0	99.4
2	T_1	99.6	V_1	99.5
3	T_3	99.1	V_3	99.2
	Mean % recovered	99.3	*Mean % recovered*	99.4
	% RSD	*0.3*	*% RSD*	*0.2*
	Mean % diff (K vs. T)	*0.1*		

Note: Recovery experiment was evaluated by (1) keeping the sample extraction volume constant and shaking at longer times and (2) keeping the shake time constant and increasing the extraction volumes. Less than 0.1% difference between the kinetic and the thermodynamic techniques (1) indicates robustness of the adopted sample preparation procedure and (2) justifies use of either technique.

API was recovered or less than quantitative recovery was obtained, the sample preparation steps would require modification by either increasing the duration of the sonication or shake time, or increasing the sample volume in order to achieve a more efficient and complete recovery of the API from the sample matrix.

Prior to filling an aliquot of the prepared sample into vials for HPLC analysis, the sample should be free of particulates because the presence of particulates can cause damage not only to the HPLC column but also to the HPLC system.[12] The most common methods for removing particulates are filtration, centrifugation, and sedimentation.[17–20]

1. Sedimentation

This is the simplest technique for removing particulates. It relies on gravitational pull on samples to cause heavier particles to settle first. In this approach, the sample suspension is allowed to stand over time at room temperature till all particulates in the sample solution settle to the bottom of the container. The clear solution that forms at the top is carefully removed and filled into suitable vials for HPLC analysis.

2. Filtration

Filtration is the process of separating liquids from solids with the purpose of obtaining optically transparent liquid for HPLC analysis.[17,18] The ability of a filter medium to eliminate solid matter from a liquid is termed "retention." Different filter types that may be used for sample filtration are given in Table 7. Regardless of the filter type, the filtering medium is composed of countless channels which impart porosity to the medium. Both the retentive ability and filtration rate of a liquid through the medium are dependent on the porosity of the medium. Selecting the correct filter type for an application depends on the conditions and objectives of the analysis as indicated in Table 7. Paper filtration is a relatively straightforward technique.[17] However, for many HPLC analyses, disposable syringe filters equipped with Luer fittings are commonly used.[18] Here, the sample is placed in a syringe and filtered through the membrane using gentle pressure. Variety of membrane materials with different porosities and dimensions are available for HPLC analysis (see Tables 8 and 9). In general, the lower the porosity of the filter medium, the cleaner the filtrate and the longer the filtering time. For most HPLC samples, the recommended filter membrane porosity lies between 0.25- and 0.45-μm nominal porosity range. Nominal or absolute porosity rating refers to a filter membrane's ability to retain 60–98% of the specified particle size.[18] Membrane filter with 0.25-μm porosity or pore size means it will remove particulates that are ≥ 0.25 μm and larger macromolecule. If the sample contains colloidal material or large amount of fines, considerable pressure may be required to force the liquid sample through the filter. Difficult-to-filter solutions may require use of a larger diameter syringe filter or a syringe filter with a 1.0-μm glass prefilter to prevent membrane plugging (see Table 9). If an inappropriate solvent is used, the filter may dissolve (or soften) and the sample becomes contaminated. Therefore, in determining which filter membrane to use, the sample matrix, chemical resistance, and wet strength should be carefully considered.

TABLE 7 Different Filtration Approaches in HPLC

Filtration media	Typical filter types	Comment
Filter paper	Cellulose	Filter paper fibers can get into samples and will cause sample contamination. Ensure sample solvent compatibility.
Membrane	Nylon	This is a general purpose membrane for HPLC sample filtration. It is hydrophilic and suitable for aqueous samples. It has limited chemical resistance to acids and bases and tends to bind to proteins.
	Polytetrafluroethylene (PTFE)	This is hydrophobic and is chemically resistant to most solvents, acids, and bases. It is ideal for HPLC solvent and sample filtration.
	Polyvinylidene difluoride (PVDF)	This is a hydrophilic, solvent-resistant membrane that exhibits low levels of UV absorbing extractables. It is ideal for HPLC sample filtration and general biological sample filtration.
	Polypropylene (PP)	This is a hydrophilic membrane that exhibits a wide range of chemical compatibility for organic solvents. It has low protein bind and is suitable for HPLC sample and biological sample filtration.
	Polyethersulfone	This is a durable membrane with high-temperature resistance and has good chemical compatibility. It is suitable for biological sample filtration where protein loss is critical.

Source: *Technical Note CS35*; Guide to Syringe Filter Selection, Chromatographic Specialties Inc., Brockville, ON.

TABLE 8 Different Filter Pore Sizes

Filter pore size	Suggested application
0.2 μm	Removes fine particles (<10 μm)
0.45 μm	Removes small particles and is generally used for clarification of HPLC samples
1.0 μm and larger	Removes coarse particulates and is suitable for processing viscous samples

Source: See Ref. 18.
Difficult-to-filter solutions may require the use of a large diameter filter coupled with a glass prefilter to prevent plugging of the membrane.

3. Centrifugation

Centrifugation is the separation process that relies on the action of centrifugal force to separate particles in a solid–liquid mixture into two distinct phases consisting of the sediment and centrifugate (also called supernatant liquid).[19,20] During centrifugation, the separation is caused by the effects of centrifugal force operating on the solid–liquid mixture at the selected speed or revolutions per minute (RPM) that moves solid particles radially away from the axis of rotation. The force on the particles (compared to gravity) is called the relative centrifugal force

TABLE 9 Filter Diameter and Its Application

Diameter of filter (mm)	Application
4	Used for sample volumes < 1 mL
13	Used for sample volumes from 1 to 10 mL
17	For sample volumes up to 30 mL
25	For samples volumes from 10 to 100 mL
30	For sample volumes up to 150 mL

Source: See Ref. 18.

(RCF) and the magnitude of this force is influenced by the type of rotors in the centrifuge (see Figure 1 and Table 10). Rotors in centrifuges can broadly be classified into three common categories: namely, swinging-bucket rotors, fixed-angle rotors, and vertical rotors (see Table 10). In swinging-bucket rotors, the sample tubes are loaded into individual buckets that hang vertically while the rotor is at rest. When the rotor begins to rotate, the buckets swing out to a horizontal position. This rotor is particularly useful when samples are to be resolved in density gradients. In fixed-angle rotors, the sample tubes are held fixed at the angle of the rotor cavity, and this causes the solution in the tubes to reorient when the rotor begins to rotate and is most commonly used for pelleting applications. In vertical rotors, the sample tubes are held in vertical position during rotation and it is not suitable for pelleting

FIGURE 1 An example of a laboratory tabletop centrifuge with fixed-angle rotors.

TABLE 10 Types of Centrifuge Rotors and Their Applications

Type of rotor	Pelleting
Fixed angle	Excellent
Swinging-bucket	Inefficient
Vertical	Not suitable

Source: See Ref. 19.

TABLE 11 Chemical Compatibility of Popular Centrifuge Tube Materials

Tube type	Clarity	Chemical resistance
Polypropylene (PP)	Opaque	Good
Polyallomer (PA)	Opaque	Good
Polycarbonate (PC)	Clear	Poor
Polyethylene terephtalate	Clear	Poor

Source: See Ref. 19.

(i.e., hard-packed concentration of particles in a tube after centrifugation) applications but is most efficient for density separations due to the short pathlength. Depending on the objective and the nature of the centrifuge tubes, consideration should be given to the type of rotors and the type of centrifuge tube to use (see Table 11). For most HPLC analysis, the goal of centrifugation is to obtain samples that (1) readily separate into clear supernatant in tubes that prevent sample leakage or loss, (2) are not chemically reactive with the sample, and (3) allow for easy sample recovery. To ensure reproducible sample preparation, it is important to specify the rotor type and speed at which the centrifuge should be operated at. In general, it is better to achieve the pelleting efficiency when the *K*-factor is small because longer pathlength of rotors permits better separation of individual particle types from a mixture.[19,20] *K*-factor is calculated using the formula:

$$K = \frac{2.53 \times 10^{11} L_n(R_{\max}/R_{\min})}{(\mathrm{RPM})^2},$$

where RPM is the revolutions per minute (speed); $R_{\max}$, the maximum radius from the axis of rotation in centimeters; $R_{\min}$, the minimum radius from the axis of rotation in centimeters; RCF, the relative centrifugal force; RCF $= 11.17 \times R_{\max}$ $(\mathrm{RPM}/1000)^2$.

If the time to pellet a sample in an "old" rotor is known, the time it would take for the same sample to pellet in a "new" rotor can be determined using the formula shown below:

$$\frac{T_1}{K_1} = \frac{T_2}{K_2} \rightarrow T_1 = T_2\left(\frac{K_1}{K_2}\right),$$

where T_1 is the time to pellet in the "new" rotor; T_2, the time to pellet in the "old" rotor; K_1, the K-factor of the "new" rotor; K_2, the K-factor of the "old" rotor.

For example, for an old rotor (Beckman® JA-10) with $K_2 = 3610$ and pelleting time $(T_2) = 20$ min and a new rotor (Sorvall® SLC-1500) with $K_1 = 1676$, the new pelleting time is calculated to be $(T_1) = 9.2$ min.[19]

Finally, the choice of using either filtration or centrifugation for sample preparation will depend on the outcome of recovery experiments. If comparative data from the recovery experiments show higher relative recoveries using either filtration or centrifugation, the technique that yields the highest recovery becomes the method of choice for sample preparation (see Table 12).

III. CAPSULE DOSAGE FORMS

Capsules are solid dosage forms in which the API is enclosed in either a hard or soft soluble container or shell of a suitable form of gelatin that is made from collagenous material by hydrolysis.[8,23] Capsule shells consist of two sections known as the "cap" that slips over the other section called "body." Capsules are supplied in a variety of sizes that are numbered from "000" (the largest size) to "5" (the smallest size) with approximate total fill capacity that ranges from about 615 to 60 mg for "000" to "5," respectively (see Table 13). There are two main types of capsules, the hard gelatin capsule (HGC, also referred to as dry-filled capsules) and soft elastic capsule (SEC, also called soft gelatin capsule, Liqui-Gels®, Softgels).[8,23]

Manufacture of HGC dosage forms compared with tablets requires minimum formulation efforts and usually involves granulation of the API powders with diluents such as lactose, mannitol, calcium carbonate, and lubricants such as magnesium stearate. Prior to filling the capsules with the granulated API powder, the body and cap are separated. The cap is replaced after completion of the filling

TABLE 12 Filtration Versus Centrifugation Study

Experiment no.	Target amount (%)	Amount recovered after filtration (%)	Amount recovered after centrifugation (%)	Difference between filtered and centrifuged samples (%)
	100	99.6	99.4	0.2
2	100	99.8	99.8	0
3	100	100.2	100.2	0
4	100	101.9	101.6	0.3

Note: Since the difference between filtration and centrifugation technique is less than 2.0%, the results indicate either technique can be used for sample clarification.

TABLE 13 Capsule Size Chart

Size	Outer diameter (mm)	Height or locked length (mm)	Actual volume (mL)	Typical fill weight (mg)[a]
000	9.97	26.14	1.37	615–1370
00	8.53	23.30	0.95	430–950
0	7.65	21.7	0.68	305–680
1	6.91	19.4	0.50	225–500
2	6.35	18.0	0.37	165–370
3	5.82	15.9	0.30	135–300
4	5.31	14.3	0.21	95–210
5	4.91	11.1	0.13	60–130

Reprinted from *Torpac, Inc*, Fairfield, NJ, USA.
[a]Fill weight depends on density of the powder.

operation. To avoid separation of the cap and body after the capsules have been filled, capsules shells can be spot-welded by means of heating process that fuses the cap to the body or may be banded with molten gelatin laid around the joint in a strip and dried.[8]

Softgels, unlike HGC, offer the possibility of delivering a liquid in a solid oral dosage form. The shells of soft gelatin capsules are somewhat thicker, softer, elastic, and pliable than that of HGC because of the addition of plasticizers. Typically, the ratio of plasticizer to shell polymer determines the hardness of the shell.[8] Large groups of liquids that can be encapsulated into softgels fall into two main categories, namely, (1) water-miscible and (2) water-immiscible liquids.[21] Water-miscible liquids include polyethylene glycol (PEG) and nonionic surfactants such as polysorbates. Water-immiscible liquids include hydrogenated vegetable oils, aromatic oils, aliphatic, aromatic and chlorinated hydrocarbons, ethers, esters, high-molecular-weight organic acids, and some alcohols. The fill liquid is typically either a solution or a suspension of the API that is designed to fit in the smallest possible softgels with acceptable physical stability and production efficiency. The cap and body of softgels are sealed through a plasticization process by the addition of glycerin, sorbitol, or similar polyol.[21] The essential difference between a filled softgel and HGC is that softgels have a seam at the point of closure of the cap and body that is heat sealed. Commercially, filled softgels come in a wide variety of sizes and shapes.[8,21]

A. Sample Preparation Procedures for HGC Dosage Forms

Sample preparation considerations for HGC follow the same steps as previously discussed for tablets (see Sections II.A–II.C). HGC can be ground using a mortar and pestle followed by weighing of an aliquot of the finely ground powder equivalent to the unit dose of the capsule into a suitable container.[7,17] Extraction of the API from the powdered mixture requiring the use of solvents, sample agitation using either shakers or sonication, and sample filtration or centrifugation follow the same steps as previously discussed for tablets. Because of concerns for potential sample desegregation as previously discussed in Section II.C, the practice of grinding capsules is not recommended. Instead, two alternative sample preparation approaches can be followed.

In the first approach, the whole capsule is weighed and placed into a suitable container containing a fixed volume of a suitable hydro-organic solvent mixtures that can cause the capsule shell to swell and to open up to release its contents into the solution. While this approach is better than the grinding method, the presence of capsule shell constituents and fragments in the sample solution can cause interference in the sample chromatogram because of the presence of unwanted peaks that may originate from the endogenous capsule shell materials. The presence of capsule shell materials can also make filtration of the sample solution or suspension difficult.

The second approach, which is preferred, involves opening the capsule shells by carefully removing the cap and very gently emptying the capsule powder contents directly into a suitable container. This approach (1) avoids the presence of unwanted interference peaks due to capsule shell fragments and endogenous materials in the sample solution and (2) makes it easy to filter the solution. A potential drawback of this method is that emptying of the capsule contents requires very careful opening of the capsule to avoid lose of material due to accidental spillage. Accidental spillage of capsule contents will lead to low assay results that will not be representative of the actual dosage content of the capsule. However, if the emptying of the capsule content is done carefully, results that are more reliable and reflective of the true content of the capsule can be obtained. An example of this sample preparation approach is described for a 25-mg HGC product that was manufactured by a dry blend granulation process. In this example, 10 capsules are first weighed to obtain the average weight of the capsules. Next, contents from the 10 capsules are emptied into a 250-mL volumetric flask. The empty capsule shells are collected into a separate 30-mL beaker to which 5 mL of acetonitrile is added to completely submerge all the empty capsule shells. The beaker is gently swirled to promote rinsing of the inside of the empty capsule shells so as to remove any residual powder that may be trapped inside. The rinse solution from the beaker is carefully transferred, using a transfer pipette, into the volumetric flask containing the capsule contents. This rinsing step is repeated three times. Finally, about 150 mL of a hydro-organic solvent mixture (i.e., water/acetonitrile (50/50, v/v)) is added to the flask. The flask is first sonicated for about 5 min and then shaken on a mechanical shaker for 45 min. After the flask is removed from the shaker, the flask is allowed to warm to room temperature. The flask is diluted to the 250-mL mark with the water/acetonitrile (50/50, v/v) sample solvent and mixed well. A portion of this solution is filtered through an Acrodisc PFTE syringe filter into a suitable container. Ten milliliters of the filtered solution from this container is pipetted into a 100-mL flask and then dilute to volume with the water/acetonitrile (50/50, v/v) sample solvent and mixed well. An aliquot of this sample solution is filled into a vial for HPLC analysis.

B. Sample Preparation Procedures for Soft Gelatin Capsule Dosage Forms

Analyses of softgels tend to be more complicated than HGC because they contain liquids and grinding as a sample preparation step cannot be employed. Instead, analysis of softgels always require as the initial step, the dissolution of the capsule shell. The uniqueness of this sample preparation method that characterizes softgel capsule analysis have been the subject of a number of publications and the one reported by Shah *et al.* [22] serves as a good example. In this publication, a rapid and simple HPLC method is described for the assay of docusate sodium in soft gelatin capsules. This method, which is an improvement over the existing pharmacopoeial method,[23] involves the placement of the softgel capsules containing 100 mg docusate sodium into a 100-mL volumetric flask and the addition of about 10 mL of water to the flask. This is followed by heating the flask till the capsule shell dissolves. Methanolic solution containing about 2 mg/mL of an internal standard is added to the flask and the solution is finally diluted to 100 mL with methanol. The addition of methanol to the flask causes the gelatin material to precipitate. The precipitate is allowed to settle and then after an aliquot of the clear solution is filtered through a 0.45-μm nylon syringe filter into a vial for HPLC analysis.[22]

Another example of how complicated the analysis of softgel capsules can be is described in the method developed by the University of Chicago and University of Ottawa for the analysis of softgel capsules containing ginseng extract.[24] In this method, the soft gel capsules are first dissolved in a hydro-organic solvent mixture that contains hexane/methanol/water (20/15/10, v/v/v). The sample solution is evaporated to dryness under vacuum at 45–50 °C and the resultant residue dissolved in a mixture containing methanol/0.05% trifluroacetic acid (4:1, v/v). Finally, an aliquot of this solution is filtered using a 0.45-μm nylon Acrodisc 13 filter into a vial for HPLC analysis.

IV. CACHET DOSAGE FORMS

Cachets are related to capsules because they provide a means of oral administration of unpleasant tasting powders. Powder mixtures used for cachets are the same as those used for capsules.[8,25] Cachets are made from two concave edible wafers prepared from flour and water that vary in size from ¾ to ⅛ in. in diameter. There are two types of cachets, the wet- and dry-seal types. The filling process is similar for both types. Wet-seal cachets after filling with powder are closed by wetting the flanges of each half of the wafers and applying pressure to join the two halves. With the dry-seal cachet, the two halves are pushed together by applying pressure. Cachets were taken by holding one between the thumb and first finger, dipping in water to soften, then placing on the tongue and swallowing with a draught of water.[25] Because cachets become soft, elastic, and slippery when moistened with water, they are analyzed using the same approaches described for HGC (see Section III.A).

V. SUSTAINED RELEASE DOSAGE FORMS

These are dosage forms that are designed to slowly release a drug over an extended period of time. There are different types of sustained release dosage types of which the most common types include enteric-coated tablets (where timed release is achieved by barrier coating), delayed release capsules (which contain drug particles coated with polymers to achieve slow release of drug), and transdermal systems (popularly known as patches).[26–28]

Transdermal patches, unlike tablets and capsules, consist of an impermeable backing material, an inner layer (containing the drug in a matrix of release-controlling excipients and materials), an adhesive layer to hold the patch in place on the skin and a protective cover that is peeled away before applying to the patch to the skin.[28,29] The impermeable backing is the visible surface of the patch after it has been applied to the skin and it is made of a polyester film, ethylene vinyl alcohol copolymer, or polyurethane film. The impermeable backing remains when the patch is applied. The removable strip, which is removed before applying the patch, is the protective liner and is often composed of a polyester fabric. The adhesive which keeps the patch attached to the skin is generally a complex mixture of acrylic polymers (of which polyisobutane is a frequent component) and permeation enhancers (such as solvents (e.g., alcohols) and surfactants). Most patches belong to one of two general types: the reservoir system and the matrix (or drug in adhesive) system.[28,29] For both the reservoir and the matrix types, the adhesive and drug reservoir matrix usually contain the same acrylic polymer material (see Figure 2). When several dosage strengths of a transdermal patch are marketed, the formulation used is identical but the patches will have different surface areas to differentiate among the different strengths.

The key consideration in the analysis of any sustained release dosage form as previously discussed (see Sections II.A, II.B, II.C.1, and II.C.2rg) is to determine what solvent or solvent system will be most appropriate to assure the dissolution of the drug and its excipients to make it amenable to HPLC analysis. Aqueous solubility of weak acids and bases is governed by the pK_a of the compound and the pH of the medium. In an acidic or low pH medium, weak acids will be unionized and will be more soluble in organic solvents. The reverse is the case for basic compounds as previously discussed in Section II.B. Because the formulation of sustained release dosage forms tend to rely on the use of insoluble plastics (i.e., methyl acrylate–methyl methacrylate, polyvinyl chloride, and polyethylene), hydrophilic polymers (i.e., methylcellulose, hydroxypropyl-methylcellulose, sodium carboxymethyl cellulose, and carbopol 934), and fatty compounds (i.e., waxes such carnauba wax and glyceryl tristearate), similar hydro-organic solvents and sample preparation steps that have been discussed for tablets and capsules can also used for their analysis by HPLC (see Sections II.A, II.B, II.C.1, and II.C.2).

The first step in the analysis of any sustained release dosage form is to get the dosage form into solution.[3–5,7,10,17] This is achieved by shaking and sonication steps as previously discussed in Sections II.A, II.B, II.C.1, and II.C.2. Next, the resulting solution is subjected to the same clarification steps of filtration,

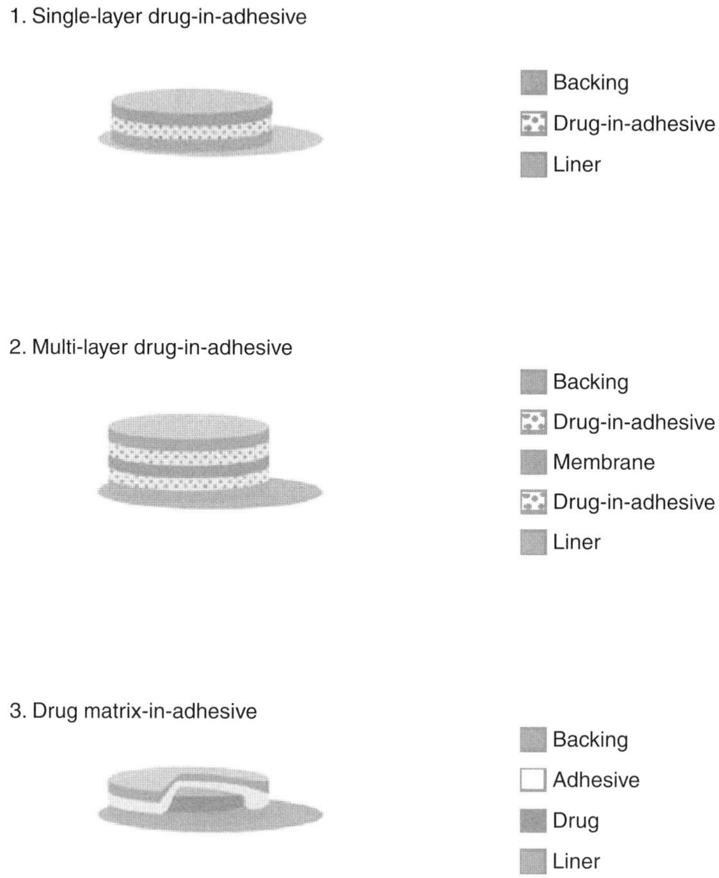

FIGURE 2 **Schematic of different types of transdermal systems. Reprinted from Ref. 29. (See Color Insert.)**

centrifugation, and sonication steps as previously discussed for tablets and capsules (see Sections II and III).

A. Sample Preparation Procedure for a Sustained Release Capsule Dosage Form

A modified release capsule containing immediate and delayed release beads is analyzed using a sample solvent containing 20 mM of sodium acetate solution (adjusted to pH 4.60 ± 0.05 with glacial acetic acid) and methanol, and a chromatographic mobile phase containing sodium acetate solution (adjusted to pH 4.60 ± 0.05 with glacial acetic acid), methanol, and acetonitrile (900:60:40, v/v/v). The first step involved weighing 20 of the capsules to determine the average capsule fill weight. Next, contents of the 20 capsules are carefully emptied, with the aid of a funnel, into a suitable volumetric flask. Methanol is added to the flask to about 80% of the total volume of the flask and is sonicated for about 15 min. After the flask is allowed to cool to room temperature, the solution is diluted to volume with methanol as the sample stock solution. Next, 10 mL of the sample stock solution is pipetted into a 100-mL volumetric flask, and 50 mL of the sodium acetate solution is added to the flask. The resulting solution is sonicated for another 15 min. After the solution is allowed to cool to room temperature, it is diluted to volume (i.e., the 100-mL mark) with the sodium acetate solution and mixed well as the sample solution. An aliquot of this solution is filtered using a Millex-HV hydrophilic PVDF 0.45 μm filter into a suitable vial for HPLC analysis by discarding the first 2 mL to waste.

B. Sample Preparation Procedure for a Sustained Release Tablet Dosage Form

An enteric-coated tablet made by coating the tablet cores with hydrophilic polymers is analyzed using a 50 mM disodium hydrogen phosphate 2-hydrate ($Na_2HPO_4 \cdot 2H_2O$) solution that is adjusted to pH 8.45 with ortho-phosphoric acid (85%) and a mobile phase composition containing 50 mM disodium hydrogen phosphate solution (adjusted to pH 8.45 with conc. phosphoric acid) and methanol in a ratio of 42:58 (v/v). The sample preparation steps involve the addition of 150.0 mL water (using a volumetric pipette) to 20 tablets in a 200-mL volumetric flask and stirring the flask for approximately 15 min using a magnetic stirrer or until all tablets are completely disintegrated. This is made to volume (i.e., 200 mL) with methanol. A portion of the resulting suspension is centrifuged for approximately 10 min at 2000 rpm ($r = 14$ cm) or until the supernatant solution is clear. Ten milliliters of the clear supernatant solution is pipetted into a 20-mL flask, and this is diluted to volume (i.e., 20 mL) with water and mixed well. An aliquot of this solution is filtered using a Millex-HV hydrophilic PVDF 0.45-μm filter into a suitable vial for HPLC analysis by discarding the first 2 mL to waste.

In another example, a delayed release tablet made by coating the tablet cores with a hydrophobic polymer is analyzed by adding methanol to 20 tablets in a suitable flask to about 80% of the total volume of the flask. The flask is sonicated for about 15 min or until all the tablets are completely disintegrated. The flask is allowed to cool to room temperature and the resulting suspension is diluted to volume (i.e., total volume of the flask) with methanol. This is the sample stock solution and is sonicated for approximately another 15 min. A portion of this stock solution is centrifuged at 4000 rpm for about 10 min in order to obtain a clear supernatant. Ten milliliters of the clear supernatant solution is pipetted into a 20-mL flask, and this is diluted to volume (i.e., 20 mL) with methanol. An aliquot of this solution is filtered using a nylon 0.45 μm filter into a suitable vial for HPLC analysis by discarding the first 2 mL to waste.

C. Sample Preparation Procedures for Transdermal Systems

An estradiol transdermal system made to contain 90.0–110.0% of the labeled amount of estradiol ($C_{18}H_{24}O_2$) is analyzed using a sample solvent containing acetonitrile and water in a ratio of 50/50 (v/v) and mobile phase composition containing acetonitrile/water (55/45, v/v).[30] Analysis is performed by first cutting 10 transdermal systems into small pieces. The protective release liner, if any, is removed from the cut pieces and discarded. The cut pieces are placed in a stoppered flask of suitable size and an accurately measured volume of the sample solvent is added to the flask. The flask is shaken by mechanical means for about 3 h, and then sonicated for 15 min. The flask is diluted to volume with the sample solvent and mixed well. An aliquot of the solution is filtered using a suitable syringe filter into a suitable vial for HPLC analysis.

In another example, a transdermal system is analyzed by using methanol as the extraction solvent and an isocratic RP-HPLC mobile phase composition containing acetonitrile (ACN), triethylamine (TEA), and 20 mM potassium dihydrogen phosphate (KH_2PO_4) solution in a ratio of 75/0.2/25 (v/v/v). The mobile phase is prepared by mixing acetonitrile, TEA, and the pH adjusted 20 mM KH_2PO_4 solution (adjusted to pH 5 ± 0.1 with glacial acetic acid). The protective release liner is carefully removed and the patch with the adhesive side facing up is placed into a suitable wide-mouth, screw-capped container using a pair of tweezers. This is followed by the addition of an accurately measured volume of methanol up to about 80% of the total volume of the container. The sample is stirred for about 16 h with a magnetic bar. Next, the sample is sonicated for about 30 min and then made to volume (i.e., final volume of the container). This sample solution is then mixed well. About 10 mL of this sample solution is filtered into a suitable vial for HPLC analysis using a suitable membrane filter (e.g., glass fiber Pall Acrodisc prefilter combined with a Millex 0.2-μm syringe filter).

VI. SOLUTION, SUSPENSION, EMULSION, AND PARENTERAL DOSAGE FORMS

These are dosage forms made by dissolving the API in an aqueous or nonaqueous solvent or by suspending the API in an appropriate medium or incorporating the API into one or two phase solvent system comprising oil and water.[31] Because the API is already incorporated into some kind of solvent system, sample preparation steps for these dosage forms tend to be less tedious than that for tablets and capsules. The main consideration is to ensure that an aliquot of the dosage form taken can easily be dissolved in a suitable solvent system that is amenable for HPLC analysis. The same considerations that determine the choice of solvents that can be used is similar to what has already been discussed in Section II.B. Thus, the same sample preparation techniques of dissolving, shaking, sonication, filtration, and centrifugation are applied. Compared with tablets and capsules, and transdermal systems, analysis of solutions, suspension, emulsions, and extracts are the simplest.

VII. HOMOGENIZATION

Because of the need to increase sample throughput during the sample preparation steps for solid dosage forms, especially during release testing in pharmaceutical research and development and in commercial operations, homogenization methods have emerged as alternatives to the traditional shaking-and-sonication

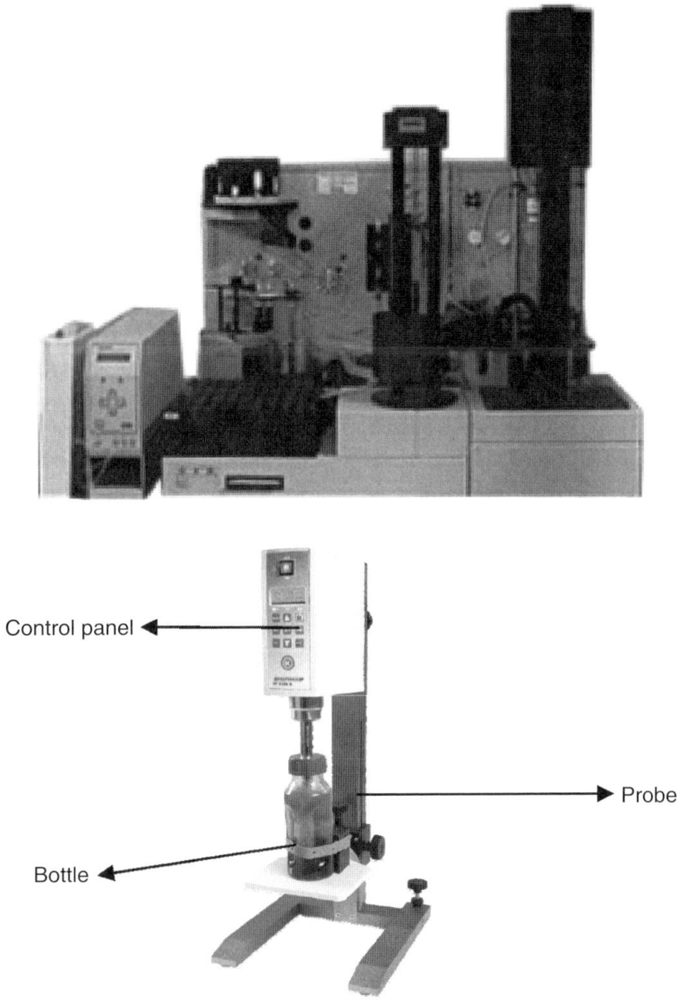

FIGURE 3 (A) An example of Zymate® TPWII sample preparation robot. Reprinted from Cat#58911/4, *American Laboratory Trading*, Inc, East Lyme, CT 06333. (B) A benchtop POLYTRON® PT 3100 lab homogenizer. Source: Kinematica Website, *Kinematica AG*, Dispensing and Mixing Technology.

sample preparation techniques.[32,33] Two homogenization platforms that are commonly used are the fully automated robotic system such as Zymark TPWIITM® (see Figure 3A) and the benchtop lab homogenizers of which the newly designed POLYTRON® PT 3100 is an example (see Figure 3B).

The main advantages that the homogenization platforms offer over the traditional shake and sonication sample preparation methods are speed and reproducibility. Homogenizers equipped with motorized blades, combine wet grinding/shredding/shearing of the sample matrix to achieve sample preparation very rapidly and in much shorter times. Sample preparation steps involve gravimetric addition of solvents instead of volumetric addition, and this contributes to greater efficiency, accuracy, and ultimately better reproducibility of the sample preparation procedure. Additionally, use of laboratory robotic platforms has the advantage of consistently preparing each sample in the same analytical historical manner because robots lack the inherent day-to-day variations and incidental errors that are associated with human behavior.[32,33]

The increasing use of robotic platforms for sample preparation is reflected by the large number of articles that continue to be reported in the literature. In a recent literature publication, an automated Zymate® TPWII sample preparation method was developed for the analysis of drugs in plasma in which theophylline was used as a probe drug. This method yielded better results than the conventional manual sample preparation method.[34] In another literature publication, a manual sample preparation for a controlled release capsule formulation was converted to an automated Zymate® TPWII sample preparation method. This conversion achieved comparable results to the manual method in terms of recovery, precision, carryover, and filter selection.[33]

The benchtop lab POLYTRON® PT 3100 homogenizer has also been used in the sample preparation of HGC. In this particular example, 10 capsules were first weighed to the nearest 0.1 mg and placed into a special wide-mouth, screw-capped bottle. Next, about 212 g (equivalent to 250 mL) of a hydro-organic sample solvent containing methanol/acetonitrile/water/THF (40:40/20/0.05, v/v/v) was weighed and added to the bottle. The homogenizer probe was lowered into the bottle (see Figure 3B) so the tip of the probe head was approximately 3 cm from the bottom of the bottle. The distance from the tip of the probe head to the bottom of the bottle is such that it allows the capsules to fit nicely between the tip of the probe head and the bottom of the bottle and all the capsules were completely submerged in the sample solvent. The capsules were homogenized at a probe speed of approximately 24,000 rpm for 1 min. The speed was then reduced to 14,000 rpm and the homogenization continued for another 5 min. After the resulting suspension was inspected to ensure the absence of intact capsules that failed to homogenize, the probe was removed and carefully cleaned to avoid cross-contamination with the next sample. A portion of the homogenized suspension was centrifuged at 4000 rpm for 5 min. A 5.0 mL aliquot of the clear supernatant was pipetted into a 50-mL flask and diluted to volume with the sample solvent. An aliquot of the solution was filtered into a suitable vial for HPLC analysis.

Compared to the traditional shake and sonication methods which would have required took more than 45 min to prepare the samples, the homogenization method only about 6 min to complete, a near 10-fold reduction in time and speed. This example in conjunction with the other examples reported in the literature, consistently show that robotic sample preparation methods achieve about a 10-fold reduction in sample preparation time, and a doubling of sample throughput.

VIII. METERED-DOSE INHALERS

These are special dosage forms that may be applied as an aerosol or dry powder of the API filled into capsules to deliver the API directly to the patient in very specialized devices.[35] They have been used primarily for the treatment of asthma and in some cases for migraine headache. Three primary categories that can be used to describe these dosage forms are (1) topical aerosols, (2) nasal sprays, and (3) metered-dose inhalers.

Topical aerosols and nasal sprays rely on pump systems to deliver the API in the form of finely dispersed mist or as coarse spray or stream of semisolid. Because of their design, aerosol and nasal products are usually under very high pressure, and this makes it difficult to obtain suitable samples for analysis. In spite of that, a number of very specialized tests which include pressure test, minimum fill, number of discharges per container, delivered dose uniformity, delivery rate, leak test, spray pattern and plume geometry test, assay, and microbial enumeration test (MET) have been devised to test the integrity of the aerosol package and the device used to deliver the dose.[36]

Meter-dose inhalers which can either be dry power inhaler or metered-dose inhalers require a greater amount of testing because the metered valve, oral adapter, and the formulation are collectively responsible for the delivery the API to the appropriate site in the respiratory passages.[35,36] Similar to aerosols and nasal sprays, very specialized tests which include leakage rate, number of dose per container (i.e., emitted dose test), particle distribution, and value delivery tests have been devised to test the suitability of the metered-dose inhalers.

In addition to the specialized tests, there are many other tests that are similar to those used for tablets and capsules. However, complexity of the sampling procedures required to perform the tests and details of the tests are beyond the scope of this chapter. So for further information, the reader is directed to consult the general chapter[36] in the *US Pharmacopeia* on "Aerosols, nasal sprays, metered-dose inhalers, and dry powder inhalers."

IX. CONCLUSIONS

The goal of any dosage form analysis is, first and foremost, the effective solubilization of the API to achieve a solution sample that is amenable to HPLC analysis. Second, is to show that results obtained from the analysis is a true reflection of the actual drug content in the dosage form. Consequently, the choice of solvents used and the sample treatment techniques employed, whether manual or automated, are important considerations that must be carefully evaluated if the desired outcome of verifying the true drug content in the dosage form can consistently and reliably be achieved. Additionally, careful selection of solvents or solvent mixtures is a prerequisite that will guarantee the outcome of the analysis as truly reflective of the actual drug content in the dosage form. Closely aligned with this requirement is the proper selection of chromatographic conditions to ensure a positive retention and identification of the API and avoid its elution in the void volume. Incorrect identification and elution in the void volume will lead to outcomes that may not be representative of the actual drug content in the dosage form.

The increasing trends in the use of homogenization methods that demonstrate good correlation with manual methods are encouraging and will eventually become important considerations in novel dosage form analysis. If one takes into account the reported 10-fold reduction in sample preparation time, comparability to manual methods, a doubling of sample throughput, and the increased reliability and reproducibility with the deployment of homogenization methods, it is not unreasonable to speculate an eventual replacement of the prevailing traditional methods of shake and sonication with homogenization methods.

As long as the lofty goals of high sample throughput, improved efficiencies and increase productivity continues to dominate the strategic imperatives of pharmaceutical companies, the future wider use of homogenizers for sample preparation in novel dosage form analysis looks increasingly promising.

REFERENCES

1. Radebaugh, G. W. and Ravin, L. J. Preformulation. In *Remington: The Science and Practice of Pharmacy.* (A. R. Gennaro, ed.), 19th ed., Mack Publishing Company, Easton, PA, Vol. II, pp. 1495–1522, 1995.
2. Rowe, R. C., Sheskey, P. J. and Quinn, M. E. *Pharmaceutical Excipients*, 6th ed., The Pharmaceutical Press, 1 Lambeth High St., London SE1 7JN, UK, 2009.
3. Majors, R. and Slack, G. Sample preparation. In *Practical HPLC Method Development.* (L. R. Synder, J. J. Kirkland and J. L. Glajch, eds.), 2nd ed., John Wiley and Sons, Inc., 605 3rd Avenue, NY, NY 10158-0012, USA, pp. 100–170, 1997.
4. Pudipeddi, M., Serajuddin, A. T. M., Grant, D. J. W. and Stahl, P. H. Solubility and dissolution of weak acids, bases, and salts; In International Union of Pure and Applied Chemistry (IUPAC). In *Handbook of Pharmaceutical Salts, Properties, Selection and Use* (P. H. Stahl and C.-G. Wermuth, eds.), Wiley-VCH, Zurich, Switzerland, pp. 19–39, 2002.
5. Stahl, P. H. and Wermuth, C. G. Improving Solubility (2002); In International Union of Pure and Applied Chemistry (IUPAC). In *Handbook of Pharmaceutical Salts, Properties, Selection and Use* (P. H. Stahl and C. G. Wermuth, eds.), Wiley-VCH, Zurich, Switzerland, pp. 1–2, 2002.
6. Kazakevich, I. Role of HPLC in preformulation. In *HPLC for Pharmaceutical Scientists* (Y. Kazakevich and R. LoBrutto, eds.), John Wiley and Sons, Inc., 111 River St., Hoboken, NJ 07030, USA, pp. 139–195, 2007.
7. Main, K. B. Analysis of medicinals. In *Remington: The Science and Practice of Pharmacy.* (A. R. Gennaro, ed.), 19th ed., Mack Publishing Company, Easton, PA, Vol. II, pp. 1660–1675, 1995.
8. Rudnic, E. Oral solid dosage forms. In *Remington: The Science and Practice of Pharmacy.* (A. R. Gennaro, ed.), 19th ed., Mack Publishing Company, Easton, PA, Vol. II, pp. 1615–1648, 1995.
9. LoBrutto, R. and Kazakevich, Y. Reversed-phase HPLC. In *HPLC for Pharmaceutical Scientists* (Y. Kazakevich and R. LoBrutto, eds.), John Wiley and Sons, Inc., 111 River St., Hoboken, NJ 07030, USA, pp. 139–195, 2007.
10. LoBrutto, R. and Patel, T. Method validation. In *HPLC for Pharmaceutical Scientists* (Y. Kazakevich and R. LoBrutto, eds.), John Wiley and Sons, Inc., 111 River St., Hoboken, NJ 07030, USA, pp. 461–499, 2007.
11. Dolan, J. *A Guide to HPLC and LC-MS Buffer Selection*, MAC-MOD Analytical Inc., Chadds Ford, PA 19317-9927.
12. Dolan, J. *HPLC Troubleshooting Guide*, MAC-MOD Analytical Inc., Chadds Ford, PA 19317-9927.
13. Etse, J. The role of HPLC in technical transfer and manufacturing. In *HPLC for Pharmaceutical Scientists* (Y. Kazakevich and R. LoBrutto, eds.), John Wiley and Sons, Inc., 111 River St., Hoboken, NJ 07030, USA, pp. 753–754, 2007.
14. Santos, H. M., Lodeiro, C. and Capelo-Martinez, J.-L. The power of ultrasound. In *Ultrasound in Chemistry: Analytical Applications* (J.-L. Capelo-Martinez, ed.), Wiley-VCH, Verlag GmbH and Co KGaA, Weinheim, 2009.

15. Wibetoe, G., Takuwa, D. T., Lund, W. D. and Sawula, G. Coulter particle analysis used for studying the effect of sample treatment in slurry sampling electrothermal atomic absorption spectrometry. *Fresenius' J. Anal. Chem.* 363:46, 1999.
16. Capelo-Martinez, J.-L., Ximenez-Embun, P., Madrid, Y. and Camara, C. Advanced oxidation processes for sample treatment in atomic spectrometry. *Trends Anal. Chem.* 23:331, 2004.
17. Peck, G. E. Separation. In *Remington: The Science and Practice of Pharmacy.* (A.R. Gennaro, ed.), 19th ed., Mack Publishing Company, Easton, PA, Vol. II, pp. 628–638, 1995.
18. Technical Note CS35; *Guide to Syringe Filter Selection*, Chromatographic Specialties Inc., Brockville, ON.
19. Cole-Palmer Technical Library; *Basics of Centrifugation*, Cole-Parmer Instrument Company, 625 East Bunker Court, Vernon Hills, Illinois 60061-1844.
20. Ford, T. C. and Graham, J. M. *An Introduction to Centrifugation.* BIOS Scientific Publishers, Ltd 1 872748 40 6, 1991, 118pp.
21. Bergstrom, D. H., Tindal, S. and Dang, W. Capsules, soft. In *Encyclopedia of Pharmaceutical Technology.* (J. Swarbrick, ed.), 3rd ed., pp. 419–425, 2006.
22. Hogue, D. R., Zimmardi, J. A. and Shah, K. A. High-performance liquid chromatographic analysis of docusate sodium in soft gelatin capsules. *J. Pharm. Sci.* 81(4):359–361, 1992.
23. USP Monograph: Docusate Sodium Capsules (2009); *USP 32-NF-27*, U.S. Pharmacopeial Convention: Rockville, MD.
24. Yat, P. N., Arnason, J. T., Lu, Z. Z., Fitzloff, J. F., Fong, H. H. S. and Awang, D. V. C. *Ginseng Evaluation Program; ABC's Method for Analysis for Ginseng (*Panax *spp).* American Botanical Council, HC# 021091 159, Austin, TX, 1999.
25. Homan, P. G. Powders and Cachets; Information sheet: 8, pp. 1–2, Royal Pharmaceutical Society of Great Britain, 1 Lambeth High Street, London SE1 7JN, 2002.
26. Chiao, C. S. L. and Robinson, J. R. Sustained-release drug delivery systems. In *Remington: The Science and Practice of Pharmacy.* (A. R. Gennaro, ed.), 19th ed., Mack Publishing Company, Easton, PA, Vol. II, pp. 1660–1675, 1995.
27. Lee, V. H.-L. and Robinson, J. R. Drug properties influencing the design of sustained or controlled release drug delivery system. In *Sustained and Controlled Release Drug Delivery Systems; Drug and Pharmaceutical Sciences, Vol.* 6 (J. R. Robinson, ed.), Marcel Dekker, Inc., New York and Basel, pp. 76–78, 1978.
28. Scheindlin, S. Transdermal drug delivery: Past, present, future. *Mol. Interv.* 4(6):308–312, 2004.
29. Chopda, G. Transdermal drug delivery systems: A review. *Latest Reviews* 4(1):1–19, 2006.
30. USP Monograph: Estradiol Transdermal System (2009); *USP 32-NF-27*, U.S. Pharmacopeial Convention: Rockville, MD.
31. Nairn, J. G. Solutions, emulsion, suspension and extracts. In *Remington: The Science and Practice of Pharmacy.* (A. R. Gennaro, ed.), 19th ed., Mack Publishing Company, Easton, PA, Vol. II, pp. 1676–1692, 1995.
32. North, N. and Smith, S. Laboratory automation: Some perspectives on the challenges in the implementation of the technology in pharmaceutical development. *J. Automat. Chem.* 20(4):111–115, 1998.
33. Han, S. M. and Munro, A. Transfer from manual to automated sample preparation: A case study. *J. Pharm. Biomed. Anal.* 20(5):785–790, 1999.
34. Hempenius, J., Wieling, J., Jonkman, J. H. G., de Noord, O. E., Coenegracht, P. M. J. and Doornbos, D. A. Development of a laboratory robotic system for automated bioanalytical methods. The determination of theophylline in human plasma: A comparison between the robotized and manual method. *J. Pharm. Biomed. Anal.* 8(4):313–320, 1990.
35. Sciarra, J. J. Aerosol. In *Remington: The Science and Practice of Pharmacy.* (A.R. Gennaro, ed.), 19th ed., Mack Publishing Company, Easton, PA, Vol. II, pp. 1676–1692, 1995.
36. USP General Chapters. *USP <601> Aerosols, nasal sprays, metered-dose inhalers and dry powder inhalers,* USP 32-NF-27. U.S. Pharmacopeial Convention, Rockville, MD, 2009.

7

METHOD DEVELOPMENT FOR PHARMACEUTICAL CHIRAL CHROMATOGRAPHY

GREGORY K. WEBSTER* AND LAILA KOTT†

*Global Analytical Research and Development, Abbott Laboratories, Abbott Park, Illinois, USA
†Analytical Development Small Molecules Millennium Pharmaceuticals, Cambridge, Massachusetts, USA

I. INTRODUCTION

In Chapter 10, "Method Development," in the first edition of the *Handbook of Modern Pharmaceutical Analysis*, the basics for pharmaceutical chromatographic separations were discussed. These separations are called "achiral chromatography" for the separations are not stereospecific. In today's pharmaceutical industry, our separation needs do not end simply in the achiral realm. The active pharmaceutical ingredients (APIs) are not only resolved from impurities and degradants but are also required by compendial guidelines to be characterized in terms of its enantiomeric purity. For example, warfarin, a common anticoagulant, can be synthesized in two enantiomeric forms: R-(+) and S-(−) (Figure 1).

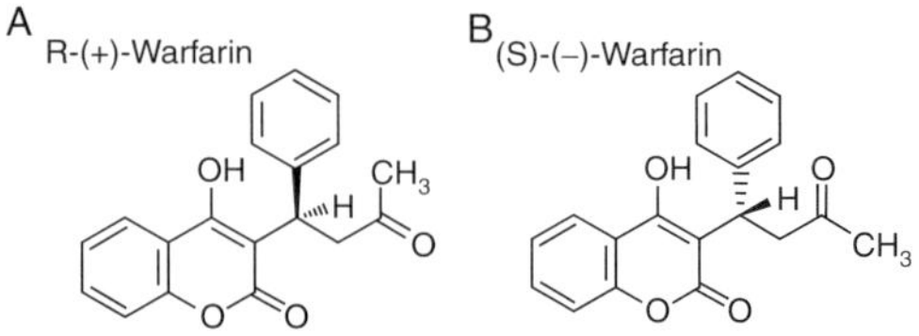

FIGURE 1 **Stereoisomers of warfarin. (A) *R*-(+)-warfarin. (B) *S*-(−)-warfarin.**

The form for pharmaceutical application is the R-(+) form. Thus, the pharmaceutical lab must evaluate how much, if any, of the S-(−) form is present. Analytical chemistry has progressed to a point where it is used to distinguish and separate two isomers that are identical except for their ability to rotate a plane of light. Over the years, techniques such as polarimetry, circular dichroism (CD), and spectroscopic techniques have been used to determine enantiomeric purity and establish stereoselective identity. In today's pharmaceutical industry, the vast majority of this work is now done by chiral chromatography.

Today's pharmaceutical industry pays much attention to stereoselectivity. Historically, the majority of the drugs on the market have been racemic. Yet, we know the human body only uses L-amino acids and the only bioactive form of Vitamin D_3 is (+)-cholecalciferol. It was simply more economical over the years to synthesize racemic forms and adjust the intake to the enantiomeric purity. Tragically, in the 1960s a drug called thalidomide caused significant deformities in fetuses from mothers who took this drug during their pregnancy.[1,2] The drug was prescribed as a sedative. The cause of this tragedy traced to a specific isomer of thalidomide, (S)-stereoisomer, was found to be a teratogen. Since this episode, regulatory agencies have rightfully required the pharmaceutical industry to characterize the stereochemistry of their drugs in development. Today, thalidomide is approved for use in the United States solely as the (R)-stereoisomer. Technology has developed to a point where the single stereoisomer of pharmaceutical interest is to be developed and marketed under current guidelines, not a racemic or mixture of isomers. More than 40%, the new chemical entities currently under development in the pharmaceutical industry are chiral and subject to these guidelines. The pharmaceutical industry addresses its need for enantiomeric selectivity through stereospecific synthesis, preparative chiral isolations, or a combination of both. The need for stereospecific chiral separations propelled the chromatography industry and academia to deliver chiral stationary phases (CSPs) to meet these requirements. This ultimately yielded the cyclodextrin, macrocyclic glycopeptide, polysaccharide, and Pirkle-type stationary phases available today. Chiral chromatography differs from achiral chromatography in that the preparative scale drives the technique instead of analytical applications.

II. OVERVIEW OF CHIRAL SEPARATIONS

Unlike achiral column chromatography, chiral chromatography does not follow a direct and predictive "Snyder and Kirkland" method development strategy.[3] This is due to the need for chiral method development to first identify stationary

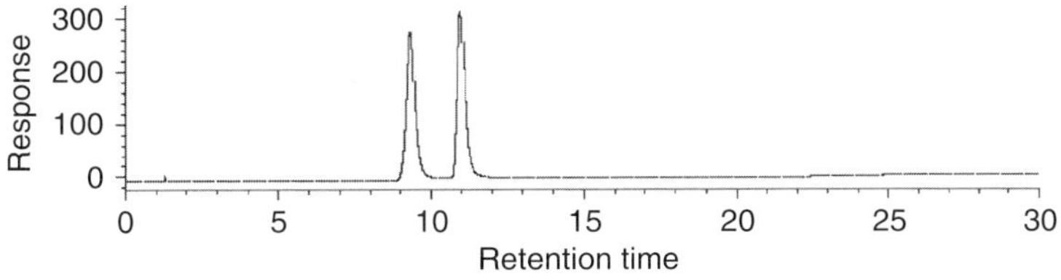

FIGURE 2 **Chiral resolution of warfarin using a Cogent EE column.**

phases that exhibit stereoselective recognition for the molecule of interest. In other words, the stationary phase must have a capability to induce partition coefficient differences between enantiomers. Traditionally, chromatographers consider the stationary phase needing to meet the "three-point rule" promoted by Pirkle. In this rule, a stationary phase will show chiral selectivity if the target API has three simultaneous interactions between the chiral selector and one of the enantiomers. One of these interactions must be stereochemically dependent. Thus, for a CSP to be useful for the chiral separation of the intended target, a pseudodiastereomeric complex must be formed as the target API elutes down the column. This reversible complex formation can be described in terms of the partition coefficient:

$$K_R : API_R + CSP \Leftrightarrow API_R - CSP_{Complex}$$

$$K_S : API_S + CSP \Leftrightarrow API_S - CSP_{Complex}$$

As the retention faction for the separation, k', is the distribution of the enantiomer in the mobile and stationary phase, the k'_R becomes

$$k'_R = \frac{[API_R]}{[API_R - CSP_{Complex}]}$$

and for the successful resolution of stereoisomers, $k'_{R\ \text{and}}$ k'_S must differ significantly from each other. Further thermodynamic explanation of this partitioning leading to chromatographic efficiency is found in Ref. 4. As illustrated in Figure 2, k'_R and k'_S for warfarin differ significantly enough in magnitude to easily allow the chiral resolution of warfarin enantiomers with polysaccharide CSPs.

III. COMMERCIALLY AVAILABLE CHIRAL COLUMNS FOR LIQUID CHROMATOGRAPHY/ SUPERCRITICAL FLUID CHROMATOGRAPHY

Column technology for liquid and supercritical fluid chromatography (LC–SFC) has been improving at a rapid pace. Currently, there are five broad categories of chiral columns which include polysaccharide-, cavity- (i.e., using cyclodextrins, antibiotics, and macrocyclic glycopeptides), π–π interaction- ("Pirkle" columns), protein-, and ligand exchange-based columns. In addition to the variety of chromatographic stationary phases, column technology is also moving away from the 250- and 10-μm particle size columns to the more efficient 5-μm particle size and shorter columns. In some cases, the 3-μm particle size is available and in many dimensions including the "rapid resolution" small sizes, ranging from 2.0 mm × 50 mm to the more traditional 4.6 mm × 250 mm. The smaller particle sizes and shorter column lengths allow for shorter screening runs, less consumption of solvents, and less generation of waste. Tables 1–5 summarize the currently available chiral column phases and their uses. The most commonly used columns are the Diacel columns (CHIRALPAK, CHIRALCEL), which have been used to separate many pharmaceutical compounds and, as discussed below, are well represented in the current chiral screening systems due to their versatility.

IV. COMMERCIALLY AVAILABLE CHIRAL COLUMNS FOR GC

Chiral gas chromatography (GC) is a useful tool when analyzing chiral raw materials and early intermediates because they are generally small molecules, have low molecular weights, are volatile, and typically do not have a chromophore for UV detection. As the majority of the pharmaceutical compounds available today are nonvolatile, the GC chiral screening is less developed than that of its liquid and supercritical fluid column counterparts.

Several of the GC chiral columns used in the pharmaceutical industry are listed in Table 6. Chiral GC typically uses bonded cyclodextrins to trap the volatile chiral components and resolve the stereoisomers based on their stereospecific affinity for the bonded cyclodextrin. The mechanism would be the same as seen with chiral capillary electrophoresis (CE), except cyclodextrin phases is bonded to the GC column support; the diastereomeric complex does not elute. As the diastereomeric complex partitions between the cyclodextrin stationary phase and carrier gas mobile phase, stereoselectivity is achieved.

V. PREPARATIVE CHIRAL CHROMATOGRAPHY

The focus of preparative chiral chromatographer differs from that of the analytical chemist, where the end product is often more important than the separation. In early development, it may be uneconomical to synthesize the off-enantiomer, which is likely required for analytical development, and it is therefore easier to collect it via preparative chromatography. Alternatively, an early synthetic process may yield only a racemic mix, but since this is not acceptable in a lead compound, as discussed earlier, the chemist is faced with separating the enantiomers by column chromatography.

There are both HPLC and SFC preparative chromatography systems and provided the HPLC is run under normal phase, the columns and column packings are, for the most part, interchangeable. There is a trend more toward SFC, since the desired end result of preparative chromatography is the compound, there is less solvent to remove when using SFC. Typical bulk preparative column packings are 16–20 μm, which are used when packing one's own columns; however, prepacked columns ranging in size from 21.2 mm × 150 mm to 50 mm × 250 mm

TABLE I Polysaccharide-Based Chiral Columns Commonly Used in Pharmaceutical HPLC Method Development Screening Systems

Column	Chiral phase	Chiral group bonding	Chromatography
CHIRALCEL AD-H[a] (or AD-RH)	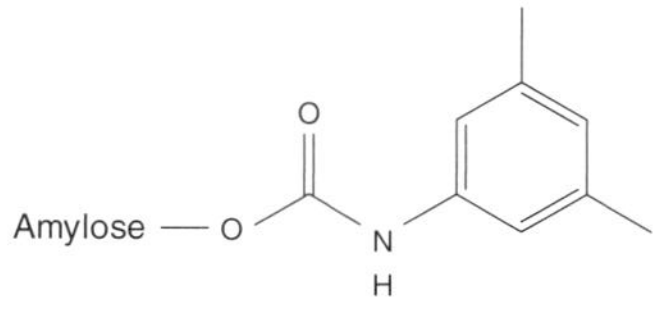	Coated	AD-H: normal phase AD-RH: reverse phase
CHIRALCEL AS-H[a] (or AS-RH)	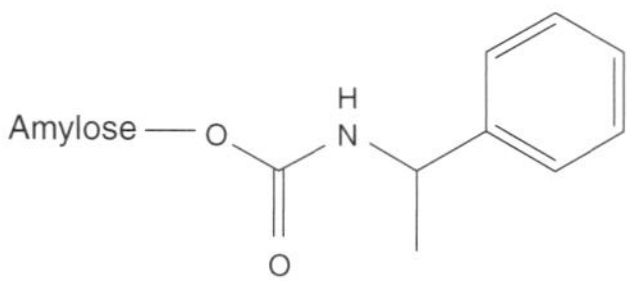	Coated	AS-H: normal phase AS-RH: reverse phase
CHIRALCEL OB-H[a]	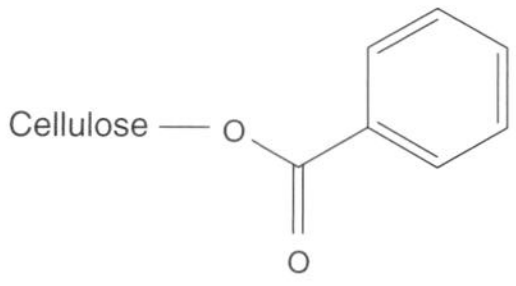	Coated	Normal phase
CHIRALCEL OC-H[a]	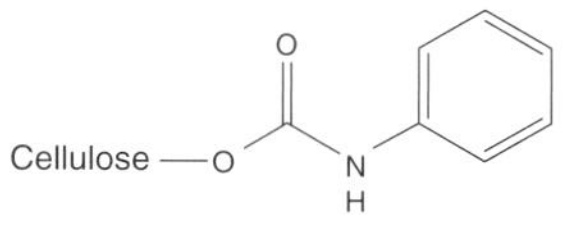	Coated	Normal phase
CHIRALCEL OD-H (or OD-RH)[a]	Cellulose 3,5-dimethylphenylcarbamate 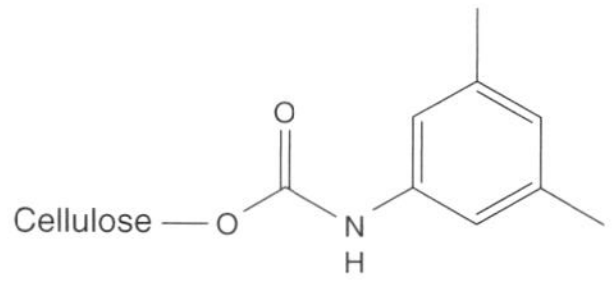	Coated	OD-H: normal phase OD-RH: reverse phase

(Continued)

TABLE I *(Continued)*

Column	Chiral phase	Chiral group bonding	Chromatography
CHIRALCEL OF	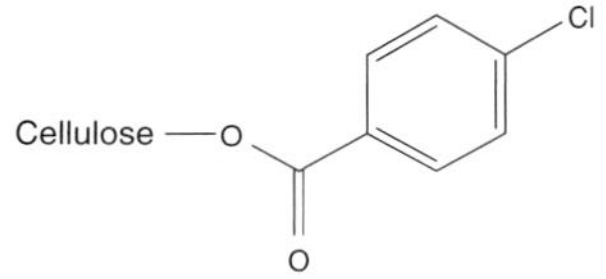	Coated	Normal phase
CHIRALCEL OG	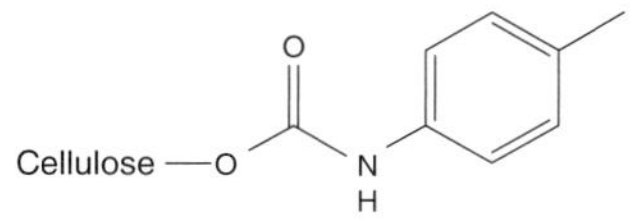	Coated	Normal phase
CHIRALCEL OJ-H[a] (or OJ-RH)	Cellulose tris(methyl benzoate) 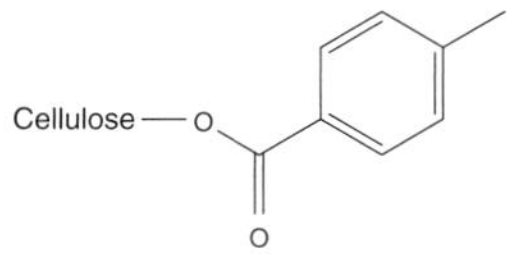	Coated	OJ-H: normal phase, polar organic OJ-RH: reverse phase
CHIRALCEL OK	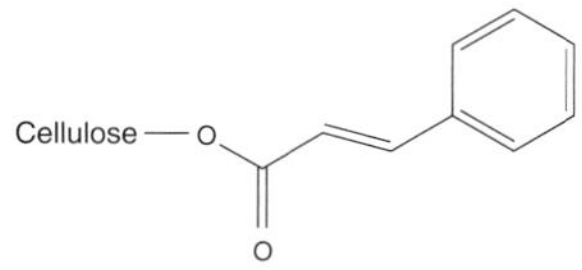	Coated	Normal phase
CHIRALPAK IA	Amylose 3,5-dimethylphenylcarbamate 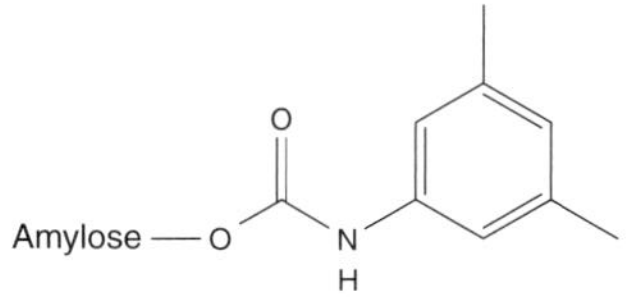	Immobilized	Normal phase

CHIRALPAK IB	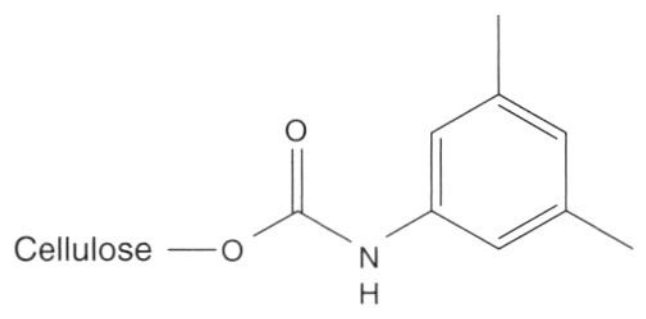	Immobilized	Normal phase
CHIRALPAK IC	Cellulose tris 3,5-dichlorophenylcarbamate 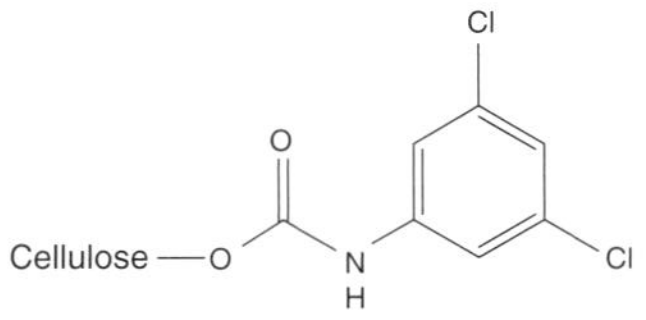	Immobilized	Normal phase
Kromasil Amycoat	Adsorbed polyamylose	Coated	• Normal phase, SFC • Useful for analytical, not for scale up
Kromasil Cellucoat	Adsorbed polycellulose	Coated	• Normal phase, SFC • Useful for analytical, not for scale up
Lux Amylose-2(Phenomenex)	Amylose tris(5-chloro-2-methylphenylcarbamate) 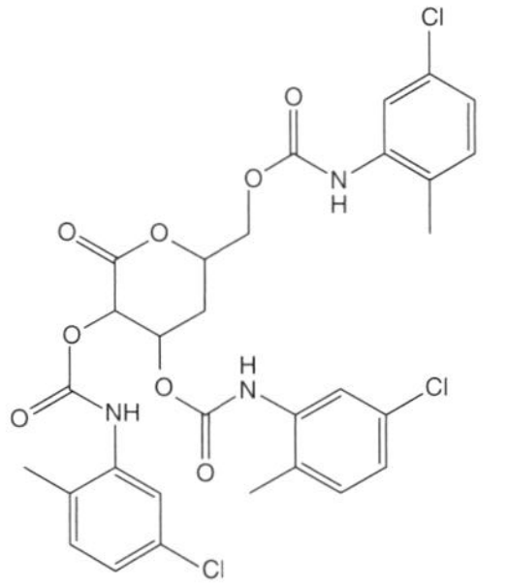	Coated	Normal/polar organic

(*Continued*)

TABLE I (Continued)

Column	Chiral phase	Chiral group bonding	Chromatography
Lux Cellulose-1 (Phenomenex)	Cellulose tris (3,5-dimethylphenylcarbamate) 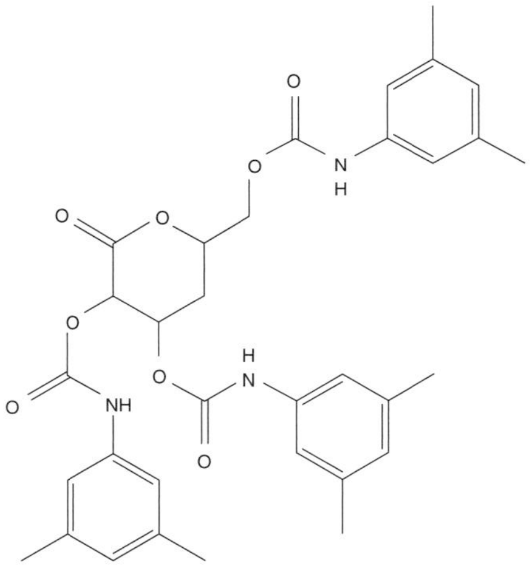	Coated	Normal/polar organic
Lux Cellulose-2 (Phenomenex)	Cellulose tris(3-chloro-4-methylphenylcarbamate) 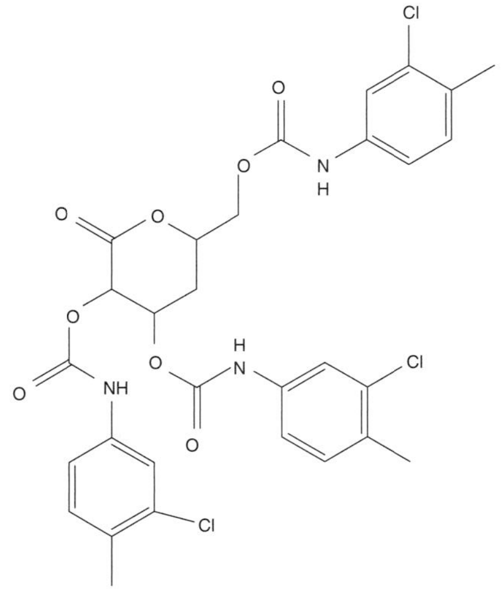	Coated	Normal/polar organic

[a]Higher efficiency 5 μm versions of OJ, OD, AS, AD, OB, and OC phases.

TABLE 2 Cavity-Based Chiral Columns Commonly Used in Pharmaceutical HPLC Method Development Screening Systems

Column	Chiral phase	Chiral group bonding	Chromatography	Uses
Chirobiotic R (Astec)	Macrocyclic glycoprotein Ristocetin A	Covalently bonded		Anionic analytes; α-hydroxy acids, substituted aliphatic acids, chiral alcohols, secondary and tertiary amines
Chirobiotic T (Astec)	Macrocyclic glycoprotein amphoteric glycopeptide, Teicoplanin	Covalently bonded	Reverse, normal and polar organic	Underivatized amino acids, N-derivatized amino acids, that is, FMOC, CBZ, and hydroxy-carboxylic acids, acidic compounds including carboxylic acids and phenols, small peptides, neutral aromatic analytes and cyclic aromatic and aliphatic amines
Chirobiotic T2 (Astec)	Macrocyclic glycoprotein amphoteric glycopeptide, Teicoplanin	Covalently bonded, improvements for preparative separations	Polar organic, polar ionic	
Chirobiotic TAG (Astec)	Macrocyclic glycoprotein amphoteric glycopeptide, Teicoplanin, where the sugars are removed from the macrocyclic glycopeptide to produce an aglycone structure	Covalently bonded	SFC, Polar organic, polar ionic, reverse, and normal	Amphoteric molecules like the amino acids; neutral molecules include hydantoins, oxazolidinones, and diazopenes; sulfur containing compounds using normal phase
Chirobiotic V (Astec)	Macrocyclic glycoprotein Vancomycin	Covalently bonded	• Reverse, normal, and polar organic • 0–100% organic modifier	Neutral molecules, amides, acids, esters, and cyclic amines
Chirobiotic V2 (Astec)	Macrocyclic glycoprotein Vancomycin	Covalently bonded, improvements for preparative separations	Polar organic, polar ionic	

(Continued)

TABLE 2 (Continued)

Column	Chiral phase	Chiral group bonding	Chromatography	Uses
ChiroSil [RCA(+) and SCA(−)] (Regis)	Trifunctional bonding of (+) or ()-(18-Crown-6)-tetracarboxcylic acid 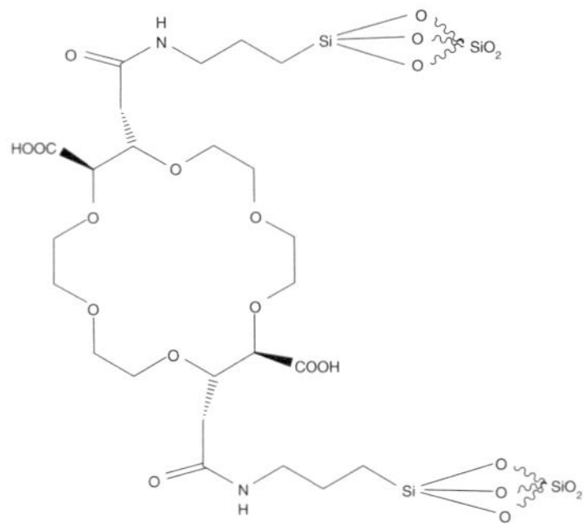	Covalently bonded	Reverse and normal phase	Amino acids and compounds containing primary amines
Cyclobond I 2000 (Astec)	β-Cyclodextrin	Covalently bonded		
Cyclobond I 2000 AC (Astec)	β-Cyclodextrin—peracetylated product of the native beta-cyclodextrin	Covalently bonded		Aromatic alcohols or amines that are chiral on the alpha or beta carbon
Cyclobond I 2000 DM (Astec)	β-Cyclodextrin—dimethylated product of the native β-cyclodextrin	Covalently bonded	Reverse phase only	
Cyclobond I 2000 DMP (Astec)	β-Cyclodextrin—3,5-dimethylphenyl carbamate	Covalently bonded		Useful for derivatized amines, like amphetamine ACQ
Cyclobond I 2000 DNP (Astec)	β-Cyclodextrin—dinitrophenyl functionality bonded through an ether linkage to the secondary	Covalently bonded	Reverse, normal and polar organic (reversed phase yields the greatest	Aromatic rings, carbonyl

	hydroxyl positions of the β-cyclodextrin		number of separations and highest selectivity)	
Cyclobond I 2000 HP RSP (Astec)	High performance β-cyclodextrin, *R*, *S*-hydroxypropyl ether	Covalently bonded		Most beneficial for basic and neutral compounds
Cyclobond I 2000 RSP (Astec)	β-Cyclodextrin—*R*,*S*- hydroxypropyl ether	Covalently bonded		Nonaromatic structures such as t-boc amino acids
Cyclobond I 2000 SP (Astec)	β-Cyclodextrin—*S*-hydroxypropyl ether	Covalently bonded		Extends hydrogen-bonding capabilities to accommodate greater distances of the chiral center from an aromatic ring structure
Cyclobond II (Astec)	γ-Cyclodextrin—8 glucopyranose units arranged in truncated cone shape	Covalently bonded	Polar organic	Multi-ring structures; steroids, porphorins, FMOC amino acids
Cyclobond II AC (Astec)	γ-Cyclodextrin–hydroxyl groups in the 2 and 3 position acetylated	Covalently bonded		Steroids and sterols
Kromasil TBB	*O*,*O*′-bis(4tert-butylbenzoyl)-*N*, *N*′-diallyl-L-tartar diamide	Bonded copolymer	SFC, polar organic, polar ionic, reverse, and normal	
Kromasil DMB	*O*,*O*′-bis(3,5dimethylbenzoyl)-*N*, *N*′-diallyl-L-tartar diamide	Bonded copolymer	SFC, polar organic, polar ionic, reverse, and normal	

TABLE 3 π–π **Interaction-Based ("Pirkle" Columns) Chiral Columns Commonly Used in Pharmaceutical HPLC Method Development Screening Systems**

Column	Chiral phase	Chiral group bonding	Chromatography	Uses
DACH-DNB (Regis)	Contains the 3,5-dinitrobenzoyl derivative of 1,2-diaminocyclohexane 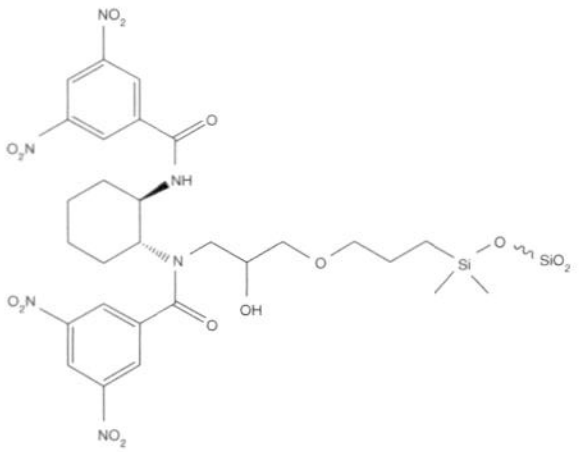	Covalently bonded	Reverse and normal	Amides, alcohols, esters, ketones, acids, sulfoxides, phosphine oxides, selenoxides, phosphonates, thiophosphineoxide, phosphineselenide, phosphine-borane, beta-lactams, organometallics, atropisomers, and heterocycles
Leucine (Regis)	3,5-dinitrobenzoyl leucine, covalently bonded to 5 μm aminopropyl silica 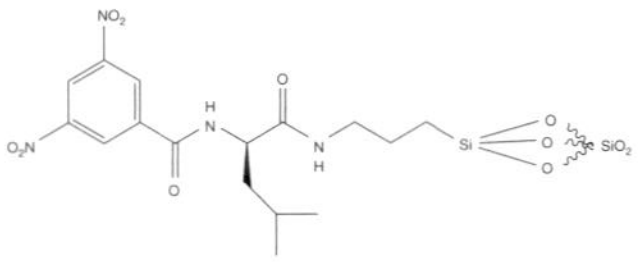	Covalently bonded	Reverse and normal	Benzodiazapines
Phenylglycine (Regis)	3,5-dinitrobenzoyl phenylglycine, covalently bonded to 5 μm aminopropyl silica 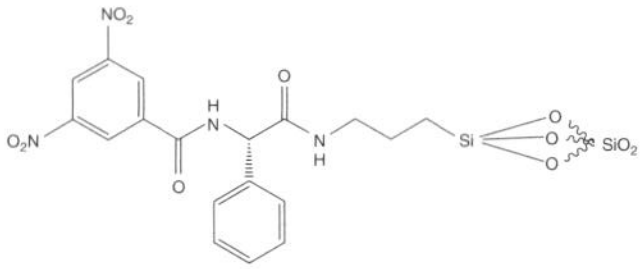	Covalently bonded	Reverse and normal	π-Basic groups, including: aryl-substituted cyclic sulfoxides, bi-β-naphthol, and its analogs, α-indanol, and α-tetralol analogs, and aryl-substituted hydantoins.

Pirkle-1 J (Regis)	CSP contains an unusual - lactam structure 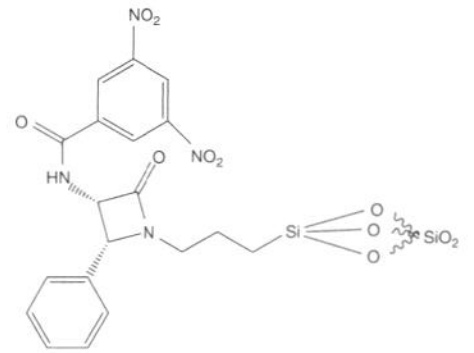	Covalently bonded	Reverse and normal	β-Blocker enantiomers; arylpropionic acid NSAIDs
ULMO (Regis)	3,5-dintrobenzoyl derivative of diphenylethylenediamine 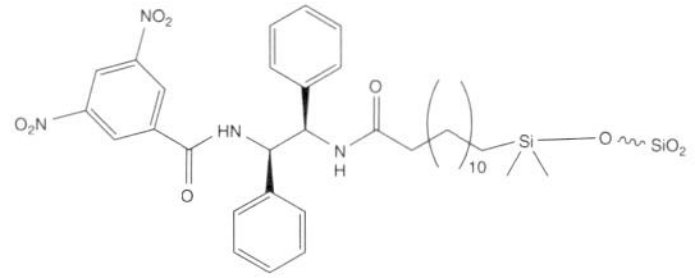	Covalently bonded	Reverse and normal	Particularly good at separating the enantiomers of aryl carbinols
Whelk-O1 (Regis)	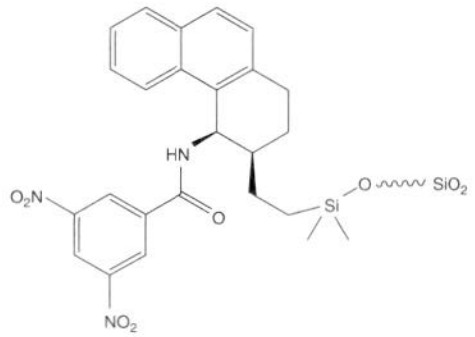	Covalently bonded	Reverse and normal	Amides, epoxides, esters, ureas, carbamates, ethers, aziridines, phosphonates, aldehydes, ketones, carboxylic acids, alcohols, and nonsteroidal anti-inflammatory drugs (NSAIDs).
Whelk-O2 (Regis)	Trifunctional version of the Whelk-O 1 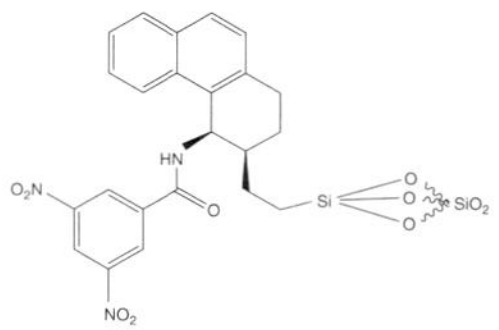	Covalently bonded	• Reverse and normal phase (better resistance against hydrolysis) • Resistant to strong organic modifiers like TFA	

(Continued)

TABLE 3 *(Continued)*

Column	Chiral phase	Chiral group bonding	Chromatography	Uses
α-Burke 2 (Regis)	Dimethyl N-3,5-dinitro-benzoyl-amino-2,2-dimethyl- 4-pentenyl phosphonate covalently bound to 5 μm silica 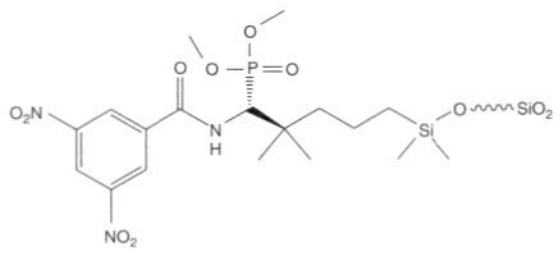	Covalently bonded	Reverse and normal	β-Blocker enantiomers
β-Gem 1 (Regis)	Covalently bonded *N*-3, 5-dinitrobenzoyl-3-amino-3-phenyl-2-(1,1-dimethylethyl)-propanoate, to 5 μm silica through an ester linkage 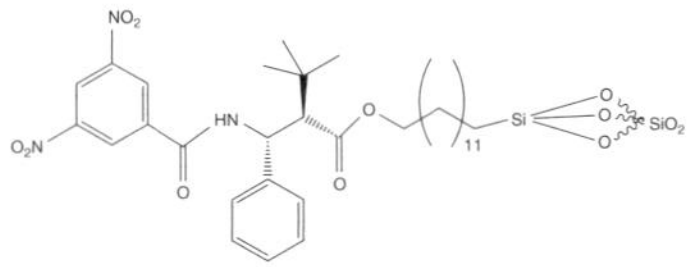	Covalently bonded	Reverse and normal	Outperforms the Phenylglycine analog; anilide derivatives of chiral carboxylic acids, including nonsteroidal anti-inflammatory agents

TABLE 4 Protein-Based Chiral Columns Commonly Used in Pharmaceutical HPLC Method Development Screening Systems

Column	Chiral phase	Chiral group bonding	Chromatography	Uses
Chiral AGP (Regis)	α1- acid glycoprotein	Immobilized on 5 μm silica	• Reverse phase • Enantioselectivity, retention regulated by pH, buffer concentration, and organic modifier	Amines, (primary, secondary, tertiary, and quaternary ammonium), acids, esters, sulphoxides, amides, and alcohols
Chiral CBH (Regis)	Cellobiohydrolase enzyme	Immobilized on 5 μm silica	• Reverse phase • Enantioselectivity, retention regulated by pH, buffer concentration, and organic modifier	Enantiomers of basic drugs from many compound classes
Chiral HSA (Regis)	Human serum albumin	Immobilized on 5 μm silica	• Reverse phase • Enantioselectivity, retention regulated by pH, buffer concentration, and organic modifier	Many carboxylic acids and amino acids can be resolved directly, without derivatization

TABLE 5 Copper Ligand Exchange-Based Chiral Columns Commonly Used in Pharmaceutical HPLC Method Development Screening Systems

Column	Chiral phase	Chiral group bonding	Chromatography	Uses
Davankov ligand exhange chiral (Regis)	A form of octadecyl silane (ODS) coated with copper	Coated	• Reverse phase • Mobile phase required to contain aqueous methanol and copper (II) acetate	Underivatized amino acid enantiomers
Astec CLC-D Chiral	An enantiomeric form of an amine coupled to a proprietary derivative creating an appropriate distance for copper coupling		Reverse phase	Any compound that can coordinated with copper, for example, lactic, malic, tartaric, and mandelic acids

TABLE 6 Commonly Used Chiral Columns Commonly Used in Pharmaceutical GC Method Development Screening Systems

Manufacturer	Astec	Restek	Agilent (J&W)	Supelco	Varian
Columns (stationary phase)	CHIRALDEX G-TA (Gamma trifluoroacetyl) CHIRALDEX B-DP (Beta dipropionyl)	Rt-BDEXsm (Beta dimethyl, mono-t-butyl)	Cyclosil-B (Beta dimethyl, mono-t-butyl)	A-DEX 120 (Alpha methyl) B-DEX 120 (20% Beta methyl) G-DEX 120 (Gamma methyl)	Chrompak Cyclodextrin-β-2,3,6-M-19 (Beta trimethyl) Chrompak Chirasil-Dex CB (Beta unknown bonded phase)

are commercially available with a 5-μm particle size. Not all column packings used for analytical separations are available for preparative work and up until recently two column packings, CHIRALPAK AY-H and CHIRALCEL OZ-H, were *only* available for preparative separations.

Preparative chromatography is not generally considered a scalable process; therefore its use is usually limited to early development compounds and only rarely is used past Phase I in the lifetime of the drug. This means that method development work is often limited and is usually related to scaling up of the analytical method, if there is one in place. Perhaps, the most important aspect in preparative chromatography method development is in the column loading, for the higher the loading, the fewer runs are required to isolate the compound of interest. Therefore, when developing a preparative chiral method, it is most convenient to start with the conditions and column afforded by the analytical method, scale the parameter to the size of the system being used, followed by determining the maximum column loading.

VI. CHIRAL METHOD DEVELOPMENT SCREENING

Chiral method development historically began with a trial and error approach based on matching the structure of the target of interest with the mechanism and functional groups found in Tables 1–5. This was a slow and costly process which oftentimes did not yield a suitable chromatographic separation for regulatory agency approval. The industry as a whole has moved to replace this approach through the development of efficient chiral method development screening strategies. Using dedicated chiral method development screening systems, pharmaceutical laboratories can quickly select suitable CSPs for further optimization. As the chiral screening strategies have been effectively implemented and optimized, oftentimes one of the actual methods from the screening protocol is suitable for early phase drug development. Many chiral screening approaches have been applied for the major separation techniques available to the laboratory chemist. Primarily, the focus has been using normal phase LC[5–10] and SFC.[11–17] CE has been studied for chiral applications but is used to a lesser extent in the pharmaceutical industry, primarily due to cost and sensitivity concerns.[18] Chiral GC is used principally for raw materials and early intermediates as most drug entities are not volatile.[18]

The basis of chiral method development screening strategies is largely dictated by whether the method is to be applied to analytical or preparative applications. Preparative chromatography tends to use volatile modifiers and mobile phases for easy removal after sample collection. Sample isolation capabilities have led to the advancement of chiral SFC since CO_2 is easily driven off. Chiral separations in preparative chromatography only need to be optimized to a point where pure samples can be collected. Conversely, while analytical chromatography can use a greater variety of modifiers and mobile phases, the separation needs to be optimized to a point where suitable resolution ($R \geq 1.5$) is achieved between the enantiomers of interest.

Whether for analytical or preparative application, finding a stationary phase that exhibits enantiomeric recognition for the molecule of interest is the point of emphasis for chiral screening. There are three modes employed in all LC and SFC chiral screens: the presence of an unmodified, acidic-modified, and basic-modified mobile phase. Common modifiers for these studies are listed in Table 7. The actual mobile phase recipes differ from researcher to researcher. The stationary phases and chromatographic system tend to be those which historically have been successful at each company. Commercial systems are available to efficiently evaluate various chiral LC columns and mobile phases in an unattended mode. In general, these screens will likely have polysaccharide phases equivalent to the OD, AD, AS, and OJ CSPs from Diacel and be supplemented with Pirkle and macrolide antibiotic phases. Typical commercial chiral LC columns found on pharmaceutical LC and SFC chiral method development screens are listed in Table 8 and typical chromatographic screening conditions are found in Table 9. For normal phase LC and SFC, these column combinations have illustrated a stereospecific resolution success rate of 80% or better in the major screening studies found in the literature. Column chiral screening in the pharmaceutical industry primarily centers on normal phase separations for LC and SFC method development to enable preparative chromatographic collection of the enantiomers.

Generally, better chromatographic performance is found with chiral separations in the normal phase for most column manufacturers. It is also likely that the easier solvent removal after collecting the isolated enantiomer, is what drove the industry to normal phase chromatography for chiral applications. It is advantageous to the chiral chromatographer that the majority of the commercially available normal phase LC CSPs and modifiers can be used on both LC and SFC instrumentation. This flexibility allows methods developed using one mode to be transferred to the other

TABLE 7 Commonly Used Modifiers in Chromatographic Chiral Method Development Screening

Acidic modifiers	Neutral/basic modifiers
Trifluoroacetic acid	Diethlyamine
Ethanesulfonic acid	Isopropylamine
Acetic acid	Ammonium acetate
Potassium hexaflourophosphate/phosphoric acid	

TABLE 8 Chiral Columns Commonly Used in Pharmaceutical Method Development Screening Systems

Normal phase			Reverse phase		
Polysaccharide	Pirkle	Antibiotic	Polysaccharide	Pirkle	Antibiotic
CHIRALCEL OD	Whelk-O	Chirobiotic V	CHIRALCEL OD-RH	Whelk-Ulmo	Chirobiotic V
CHIRALPAK AD	Ulmo	Chirobiotic T	CHIRALPAK AD-RH	(*R*)-NGLY/DNB	Chirobiotic T
CHIRALCEL OJ	(*R*)-NGLY/DNB		CHIRALCEL OJ-R		
CHIRALPAK AS					

TABLE 9 Representative Normal Phase LC Column Screening Conditions

Parameter	Condition		
Detection	UV at 254 nm		
Flow Rate (mL/min)	1.0		
Column (°C)	Ambient		
Injection (μl)	10		
Mobile Phase A	Heptane		
Mobile Phase B	Ethanol: neat or with 0.1% modifier		
Gradient	Time (min)	%A	%B
	0	95	5
	5	95	5
	25	40	60
	30	40	60
	(Reequilibrate 5 min)		

TABLE 10 Representative SFC Column Screening Conditions

Parameter	Condition
Detection	UV at 254 nm
Flow rate (mL/min)	2–3
Column (°C)	40
Injection (μl)	10
Pressure (bar)	150
Mobile Phase A	CO_2
Mobile Phase B	MeOH: neat or with 0.1% modifier
Gradient	2–30% linear gradient

with minor modifications. LC and SFC can both operate in a normal phase mode and use the same columns. Tables 9 and 10 illustrate suggested chromatographic conditions employed in normal phase and SFC chiral screening. A significant difference in chromatography to a point where only one mode shows method stereoselectivity using the same column is not generally expected.[19]

Reversed phase chiral separations are desired simply for efficiency in generating results from laboratories whose instrumentation is routinely configured to run in reversed and not normal phase modes.[20] Normal phase conditions are less attractive to the analytical chemist for this reason and deter laboratory efficiency. Typical commercial chiral LC columns found on pharmaceutical reversed phase LC chiral method development screens are listed in Table 8. Table 11 shows suggested chromatographic conditions employed in reversed phase chiral screening.

The use of polysaccharide based stationary phases (Diacel AD-RH, OD-RH, AS-RH, and OJ-RH) that are slightly modified to be stable in aqueous mobile phases have been very effective in developing suitable chiral reverse phase

TABLE 11 Representative Reversed Phase LC Column Screening Conditions

Parameter	Condition		
Detection	UV at 254 nm		
Flow rate (mL/min)	1.0		
Column (°C)	Ambient		
Injection (μl)	10		
Mobile Phase A	Water: neat or with 0.1% modifier		
Mobile Phase B	CH_3CN: neat or with 0.1% modifier		
Gradient	Time (min)	%A	%B
	0	95	5
	5	95	5
	25	40	60
	30	40	60
	(Reequilibrate 5 min)		

TABLE 12 Representative GC Column Screening Conditions

Parameter	Condition
Column dimensions	30 m × 0.25 mm
Injector temp	250 °C
Split ratio	50:1
Carrier gas	He
Oven ramp	40–180 °C @ 2 °C/min
Flow—constant pressure	23 psi ~1.5 mL/min
Detection	FID (temp 250 °C)

methods for pharmaceuticals. While these stationary phases have been shown to be useful, it is important to continually evaluate these phases as molecular entities continue to evolve. Thus, as with normal phase screens, pharmaceutical laboratories often supplement their reversed phase column selections with macrolide antibiotic and Pirkle stationary phase platforms as well.

The column and phase selection for chiral GC is considerably less than that of liquid chromatography (see Table 6), which lends itself to simpler screening campaigns. Table 12 shows suggested chromatographic conditions employed in GC chiral screening.

A. Orthogonal Approach to Chiral Method Development

Figure 3 illustrates an orthogonal approach to chiral method development. This tactic resulted from the combined efforts and experience supplied by many pharmaceutical vendor and development laboratories.[18] Each of the sections can

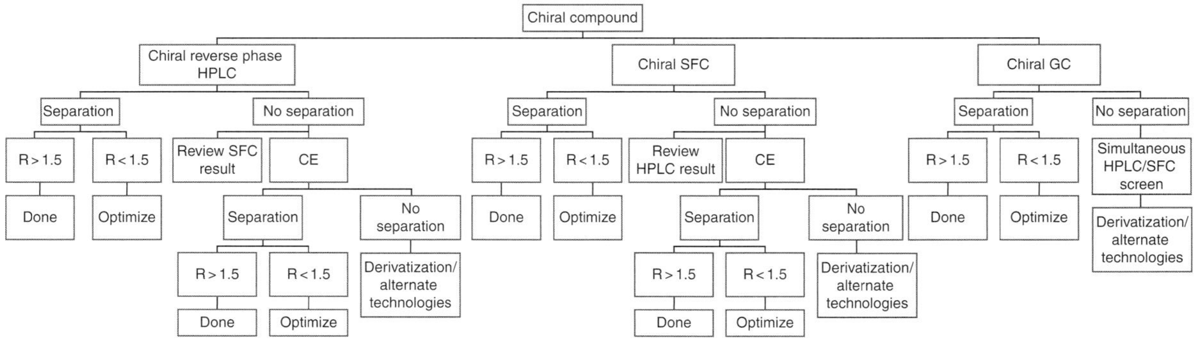

FIGURE 3 Unified chiral screening approach to chiral methods development.[18]

be run independently or combined, as individual laboratory setup and instrumentation availability allow. In addition, for laboratories without the capability for SFC, the use of normal phase LC with the same modifiers as listed for SFC will still enable a systematic and orthogonal approach when combined with reversed phase screening. The principle role of CE is as a complement to SFC and reversed phase LC for water soluble compounds where neither of these techniques provides a satisfactory separation.

VII. CHIRAL DETECTORS FOR LC/SFC SYSTEMS

Choosing the correct column and chromatographic conditions for separation has been shown to be challenging. One tool available to aid in chiral analyses is the availability of certain detectors to show the difference between enantiomeric pairs. In the absence of a detector that can differentiate between enantiomers, the analysts are left with only a diode array detector to identify chiral pairs. The danger, however, in relying solely on the UV spectral data is that this type of data is not unique to one compound. Many compounds can give a similar UV spectrum, therefore trying to discern enantiomeric pairs can be challenging in the presence of other impurities of similar structure.

There are several detectors that can differentiate between enantiomers and this is highlighted in the chromatographic traces by the presence of positive and negative peaks (Figure 4). Currently on the market there are two types of detectors that can be incorporated into chromatographic systems (i.e., HPLC or SFC): Polarimeters and CD detectors. Both detector types use Drude's equation, but in different ways.

A. Polarimeters

Normal light waves vibrate in many planes; however, plane polarized light is generated when normal light is passed through an optical polarizing filter. This effect results in a light beam emerging that vibrates in a single plane (linearly polarized). A compound is optically active if linearly polarized light is rotated when passing through it. The degree of rotation is dependent on both the concentration of a chiral compound and its molecular structure. Figure 5 illustrates the path of light in a polarimeter. Every optically active substance has its own specific rotation (degree of rotation in polarized light) as defined by Biot's law:

$$[\alpha]_\lambda^T = \frac{\alpha_\lambda^T}{cl},$$

where, $[\alpha]$ = specific rotation; l = optical path length in dm; λ = wavelength; T = temperature; α = optical rotation, c = concentration in g/mL.The specific rotation of the molecule, not the absorption characteristics, is what determines the signal strength using the polarimeter. Drude's equation highlights the effect of wavelength on the specific rotation, such that

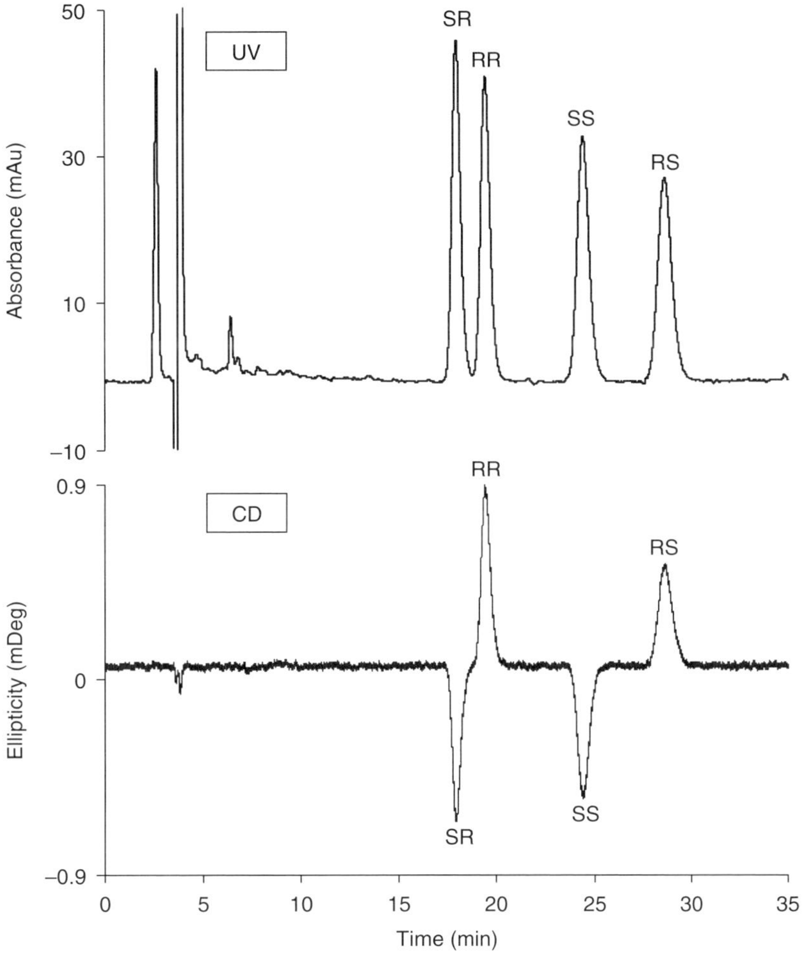

FIGURE 4 **Example of UV and CD chromatograms showing both positive and negative peaks.**[26]

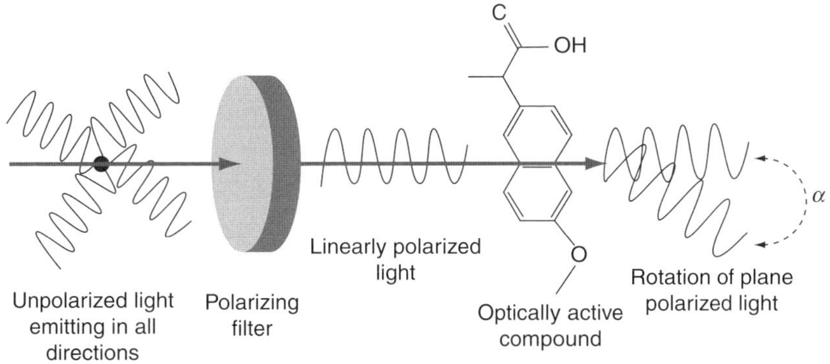

FIGURE 5 **Picture of light in a polarimeter.**[23]

$$[\alpha]_\lambda = \frac{\sum A_i}{(\lambda_2 - \lambda_i)},$$

where A_i is a molecular constant and λ_i is a constant wavelength.[21] The equation shows the normal behavior of *optical rotation dispersion* (the dependence of rotational strength of optically active molecules on the wavelength of light used for the measurements) in the absence of chromophores or in spectral regions that are distant from absorption bands. The equation also points out that the angle of rotation, as a function of wavelength, is greatest at shorter wavelengths. Therefore, to optimize the chiral response in a molecule, lower source wavelengths should yield stronger responses.

However, at lower wavelengths right- and left-handed circularly polarized light* propagate at different velocities and are absorbed by molecules to a different extent. (This phenomenon is known as CD; see below.) When this happens, it causes a deviation from Drude's equation, known as the "Cotton effect." Figure 6[22] shows how the plain optical rotation dispersion curve (or *Plain curve*) is affected by the Cotton effect. Figure 6 also highlights the area where there is maximized specific rotation [α] and minimized Cotton effects, which is in the range of 400–460 nm.

There are currently three commercially available polarimeters: the PDR chiral advanced laser polarimeter, the IBZ Chiralyzer, and the Jasco optical rotary dispersion (ORD) detectors. The PDR chiral polarimeter uses a diode laser at 670 nm as the light source. PDR chiral has chosen this wavelength as it is in the flat region of the *Plain curve* (section of the curve where there is little change in [α], see Figure 6), a region with little optical interferences. Experimentally, it has been found that this polarimeter is the chiral detector most susceptible to scatter, possibly due to the use of this long wavelength.

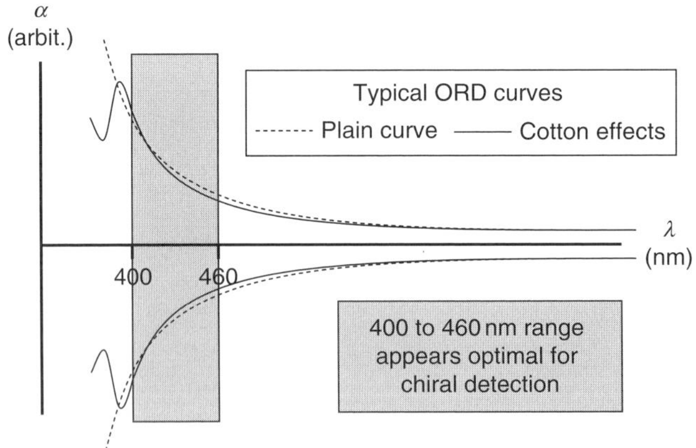

FIGURE 6 **Picture of *Plain curve*.[23]**

*Circularly polarized light: The two beams of linearly polarized light that are of equal amplitude and are a quarter wave out of phase.

The IBZ Chiralyzer detector is similar in design and function to the PDR chiral polarimeter, with the exception of a light emitting diode (LED) at 426 nm being the light source and having a second polarizing filter present in-line after the sample. Based on the *Plain curve* discussion earlier, the Chiralyzer uses a blue light from an LED as the light source to optimize the specific rotation observed.

The ORD detector is similar in design and function to the PDR chiral polarimeter; however, the light source for this detector is an Xe–Hg lamp, which is readily available and utilizes the strong line emissions of Hg at 365 nm. This detector utilizes the lowest wavelength of the polarimeters (365 nm vs. 426 and 670 nm) and therefore one would expect that this detector would give the strongest signal based on Drude's equation; however, it also runs the risk of interference from the Cotton effects.

B. Circular Dichroism Detector

When an optically active compound preferentially absorbs right or left circularly polarized light, the difference between the right and left absorbances $[A_{(r)} - A_{(l)}]$ (often a very small value) is recorded as the CD signal. Figure 7 illustrates the path of light in a CD detector. As with UV absorbance, the CD signal is wavelength dependent. From the earlier discussion of the Cotton effect, it is expected that a molecule should have a chromophore with absorption in the range of 200–420 nm to have strong CD signal. This broad generalization was somewhat disproven by Kott *et al.*[23] but considering that CD is an absorbance-based technique, chromophores will only add to the signal. There are molecules that have CD spectra greater than 420 nm such as some porphyrin rings or retinoic acid; however, for method development purposes, most small APIs fall within this range.

Similar to a UV spectrum, molecules have a distinctive CD spectrum and for the best results the CD detector should be set to the enantiomer pair maximum/minimum wavelength, again similar to choosing the most appropriate wavelength for UV detection in LC. Figure 8 is the representation of CD spectra showing the spectra of the four benzoxathiepin stereoisomers.

If both the UV and CD data are collected, the anisotropy or *g*-factor can be used in achiral chromatography to determine the enantiomeric purity of a compound. Enantiomers do not have different responses in the UV, therefore a 0.2-mg/mL

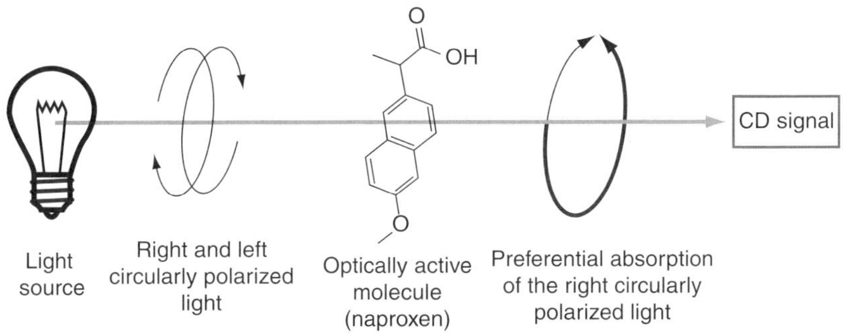

FIGURE 7 Picture of light in a CD detector.[23]

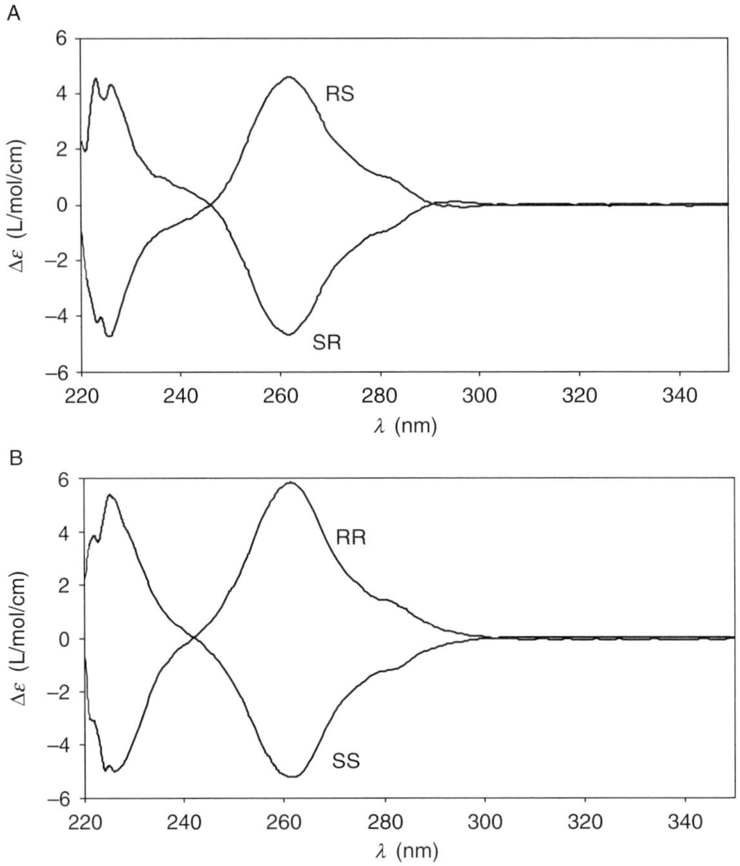

FIGURE 8 **CD spectra of the four benzoxathiepin stereoisomers.**[26]

sample of an S-enantiomer (solution A), 0.2 mg/mL of an R-enantiomer of the same compound (solution B), and a 0.2-mg/mL solution of a racemic mix (solution C) yield the same UV signal intensity at a given wavelength. However, the CD signal for solution A gives the maximum positive signal at the optimum CD wavelength, solution B gives the maximum negative signal, and solution C gives no signal at all. It is the difference in these signals versus the UV signal that can allow for percent enantiomer determinations.

Beer's law states that the absorbance of a compound depends on its molar absorptivity (ε), solution concentration (c), and path-length of the cell (l):

$$A = \varepsilon \times c \times l,$$

and it has been discussed that the CD signal is the difference between the absorbed right, $A_{(r)}$, and left, $A_{(l)}$, circularly polarized light

$$\text{CD} = A_r - A_l.$$

The anisotropy or *g*-factor is the ratio of the two signals:

$$
\begin{aligned}
g\text{–factor} &= \mathrm{CD}/A \\
&= \frac{A_r - A_l}{\varepsilon \times c \times l} \\
&= \frac{\varepsilon_r \times c \times l - \varepsilon_r \times c \times l}{\varepsilon \times c \times l}, \\
&= \frac{(\varepsilon_r - \varepsilon_l) \times c \times l}{\varepsilon \times c \times l} = \frac{\Delta\varepsilon}{\varepsilon}
\end{aligned}
$$

which is independent of concentration and path-length, provided they are constant for both measurements, which would be the case for the HPLC detector. A *g*-factor of zero would then be indicative of a racemic mix.

Currently on the market there is one CD detector that can be easily added to an HPLC system and that is the Jasco CD-2096 CD detector. This detector is capable of collecting both the UV and the CD signals and can be used to determine the enantiomeric purity of a compound.

C. Detector Capabilities

These four detectors have been compared by Kott *et al.*[23]. Parameters of interest for the commercial detectors are discussed below.

1. Gradient versus Isocratic Elution

Early chiral method development started out by using normal phase conditions and usually employing an isocratic mixture of hexane(s) and an alcohol. However, as discussed earlier, with more and more chiral columns currently available, screening a compound with multiple columns is becoming commonplace. As time is something the pharmaceutical industry is always lacking, using gradient chromatographic screens allows the scientist to "test" many isocratic conditions within one chromatographic run.

In chromatography, when using low UV wavelength detection, a change in the gradient results in a change in the baseline, usually sloping up or down with addition of more organic or upon a change of acid concentration. Therefore, for the chiral detectors to be useful, they should also be able to handle gradients.

Gradients work well on the CD, PDR, and ORD polarimeters; however, the Chiralyzer has more trouble with gradients. To be able to use a gradient, the detector needs to be zeroed at the midpoint of the gradient, followed by adjusting the sensitivity such that the early and late portions of the gradient remain on scale. This makes the Chiralyzer a little more cumbersome for gradient screens, unless the same gradient is used with many different columns. Blank runs can be programed in a sequence for changing the gradient and rezeroing the system. This detector, however, is best used with isocratic elution runs.

2. Precision

System precision was determined using six consecutive injections of the target concentration of the analyte and comparing the percent relative standard deviations (%RSDs). The target concentration was ~0.5 mg/mL, which was adjusted based upon higher or lower detector sensitivity. The CD detector yielded the best precision across all compounds studied and for 63% of the experiments, showed a %RSD of less than 5. The Chiralyzer showed a slightly better response over the PDR polarimeter, with the ORD showing the worst precision data.

The USP outlines system suitability for precision of replicate injections for HPLC as five injections for a %RSD of $\leq$2.0% and six replicates for %RSD of $\geq$2.0%.[24] In general, for most procedures, a %RSD of $\leq$2.0% is the benchmark for chromatographic precision. This being the case, only the CD detector shows the standard HPLC detector specifications. In general, the polarimeter responses are low at concentrations of 0.5 mg/mL and lower (typical analytical concentrations), which affects the precision %RSD and makes them less useful as standard HPLC detectors.

3. Linearity

To better investigate the capabilities of the different types of detectors, linearity was run on what would be considered a standard concentration range expected in a typical drug substance batch (0.5–1.5 mg/mL), as well as on an approximately 10-fold lower concentration set to assess the lower working range of each detector. A 10-fold dilution proved to be an appropriate cutoff for determining the lower linear range for the detectors, as some compounds could not be detected at this level, while others still showed good linearity. The correlation coefficient (r) was used to assess linearity.

The data presented here are based on the response of the polarimeters and a CD analysis wavelength of 230 nm, which was not the optimum wavelength for all the compounds studied in the detector comparison. Since the wavelength was not optimized for any of the compounds, the linear response may be further improved upon proper choice of wavelength. For general screening, 230 nm was chosen as a constant wavelength, as this was the lowest wavelength that would show little interference from the solvents used and because more compounds are UV active at lower wavelengths. Note that when a CD wavelength is optimized, enantiomers yield one positive and one negative signal as is the case with polarimeters (Figure 4). The CD detector showed the most linear response overall. The PDR Chiral detector showed the next best linear response, followed by the Chiralyzer and ORD which showed approximately the same linear responses across the compounds. The correlation coefficients determined for the optical rotation detectors were comparably lower than the CD detector. The response of the polarimeters was lesser (i.e., much lower peak heights) and noisier, than the CD detector responses, making integration more difficult and less consistent.

The lower concentration range data highlight the greater sensitivity of the CD detector, which yields linear responses for all the compounds investigated.

4. Limit of Detection

The third method of assessment of the four chiral detectors was the determination of the limit of detection (LOD). This was done both experimentally and mathematically. The LOD was determined mathematically by the formula

$$\mathrm{LOD} = 6s * \frac{C}{\bar{X}},$$

where s is standard deviation of the sample repeats, C, the concentration in mg/mL of the sample, and $\bar{X}$, the average area count of the sample repeats. When it was possible, the calculated LOD was tested by injecting samples at the calculated concentration, to test the reliability of the calculated value.

Across the entire compound set, the CD detector showed the lowest detection limits. For most compounds, the calculated LOD was the same order of magnitude as the observed. The Chiralyzer showed the next best LOD; however, the calculated values were often higher than the observed values. Since the LOD is calculated using the standard deviation of a sample set, the higher LOD values are indicative of scatter. The Chiralyzer, therefore, is considered a sensitive detector but due to the scatter not precise. The lack of precision, however, does not deter its use of this detector for screening purposes. The PDR Chiral and the ORD are ranked third and fourth in sensitivity, respectively.

5. Limit of Quantitation

Having determined the LOD experimentally or mathematically, the LOQ could also be calculated. The ICH guidelines define LOD based on signal to noise (S/N) as $3 \times$ S/N and LOQ as $10 \times$ S/N.[25] Therefore,

$$\mathrm{LOD} = 6s * \frac{C}{\bar{X}} = 3 \times \frac{S}{N}$$

and LOQ is

$$\mathrm{LOQ} = 10 \times \frac{S}{N} = 3.3 \times \mathrm{LOD}.$$

Using this relationship, the LOQs were calculated from the LOD values measured.

The standard concentration range used (0.5–1.5 mg/mL) was appropriate for all CD analyses, with the exception of the amino acid, as the standard concentration range is close to the calculated LOQs for the two enantiomers. In the case of the polarimeters, only 40% of the compounds studied showed an LOQ lower than the standard concentration range.

6. Achiral Use

Use of chiral detection for achiral liquid chromatographic separations has been reported in literature using CD detectors and the *g*-factor.[26–28] These experiments have shown methods for the simultaneous determination of assay and chiral purity, without the chiral column, just the chiral detector. Using the *g*-factor, validations including linearity, precision, and accuracy per the ICH[25] guidelines have been successful.[29] Work on these types of methods, however, is continuing as the sensitivity of such methods has been able to consistently show a range of 1–50% of the off-enantiomer using this technique. Few, if any, have shown down to the 0.05–0.5% levels that are required in the pharmaceutical industry; however, ranges of 0.3–0.7% have been reported.[30–32] The LOD and of LOQ using the *g*-factor are dependent on the LODs and LOQs of the UV and CD signals.[29] As it has been shown that CD detection is less sensitive than detection by UV,[30] these limits are dependent on the CD signal. Therefore, it can be concluded that wavelength optimization and determination of the chromatographic conditions (temperature, mobile phase optimization including pH, etc.[33]) need to be considered in order to yield strong CD signals which will lead to the desired LODs and LOQs.

VIII. SUMMARY AND CONCLUSIONS

Regardless of whether a traditional normal phase LC, reversed phased LC, or SFC format is used, routine analytical chiral method development screening is very effective and efficient for developing chiral selective methods in the pharmaceutical industry.

Of course, for preparative chromatography, normal and SFC systems pose significant advantages in recovering enantiomers from the mobile phase. However, for analytical operations, analysts overwhelmingly prefer not to have to switch to normal phase modes for chiral analysis. Thus, chiral method development screening has become quite developed in all these chromatographic formats. As an ever diverse stream of chemical entities is being developed for market, the analytical chemist must continually look for ways to optimize the screening system. This is required to ensure the best opportunity for success in yielding a robust analytical method for the new chemical entities in development.

REFERENCES

1. Franks, M. E., Macpherson, G. R. and Figg, W. D. Thalidomide. *The Lancet* 363:1802, 2004.
2. Perri, A. J. and Hsu, S. A review of thalidomide's history and current dermatological applications. *Dermatology Online Journal* 3(5), 2003.
3. Snyder, L. R., Kirkland, J. J. and Glajch, J. L. Practical HPLC Method Development. 2nd ed., John Wiley & Sons, New York, NY, 1997.
4. Kromidas, S. HPLC Made to Measure. John Wiley & Sons, New York, NY, 2006.
5. Perrin, C., Vu, V. A., Matthijs, N., Maftouh, M., Massart, D. L. and Vander Heyden, Y. Screening approach for chiral separation of pharmaceuticals. Part I. Normal-phase liquid chromatography. *J. Chromatogr. A* 947:69, 2002.
6. Perrin, C., Vu, V. A., Matthijs, N., Maftouh, M., Massart, D. L. and Vander Heyden, Y. Screening approach for chiral separation of pharmaceuticals part II. Reversed-phase liquid chromatography. *J. Chromatogr. A* 966:119, 2002.

7. de la Puente, M. L., White, C. T., Rivera-Sagredt, A., Reilly, J., Burton, K. and Harvey, G. Impact of normal-phase gradient elution in chiral chromatography: a novel, robust, efficient and rapid chiral screening procedure. *J. Chromatogr. A* 983:101, 2003.
8. Andersson, M. E., Asclan, D., Clark, A., Roeraade, J. and Hagman, G. Evaluation of generic chiral liquid chromatography screens for pharmaceutical analysis. *J. Chromatogr. A* 1005:83, 2003.
9. Lipka, E., Glacon, V., Mackenzie, G., Ewing, D., Len, C., Postel, D., Vaccher, M. P., Bonte, J. P. and Vaccher, C. HPLC Separation and Determination of Enantiomeric Purity of Novel Nucleoside Analogs, on Cyclodextrin Chiral Stationary Phases, Using Reversed and Polar Organic Modes. *Anal. Lett* 37:385, 2004.
10. Esser, C. K., Black, R. M. and Von Langen, D. Implementation of HPLC automation for the analysis and purification of chiral molecules. *Pharm. Disc* 26:, 2004.
11. Zhang, Y., Watts, W., Nogle, L. and McConnell, O. Rapid method development for chiral separation in drug discovery using multicolumn parallel screening and circular dichroism signal pooling. *J. Chromatogr. A* 1049:75, 2004.
12. Berger, T. A. Packed Column SFC. Royal Society of Chemistry, Cambridge, UK, 1995.
13. Terfloth, G. Enantioseparations in super- and subcritical fluid chromatography. *J. Chromatogr. A* 906:301, 2001.
14. K. K. Anton and C. Berger, eds.), *In* Supercritical Fluid Chromatography with Packed Columns-Marcel Dekke, New York, 1998.
15. Zhao, Y., Woo, G., Thomas, S., Semin, D. and Sandra, P. Rapid method development for chiral separation in drug discovery using sample pooling and supercritical fluid chromatography-mass spectrometry. *J. Chromatogr. A* 1003:157, 2003.
16. Borman, P., Boughtflower, B., Cattanach, K., Crane, K., Freebairn, K., Jonas, G., Mutton, I., Patel, A., Sanders, M. and Thompson, D. Comparative performances of selected chiral HPLC, SFC, and CE systems with a chemically diverse sample set. *Chirality* 15:S1, 2003.
17. Di, L., McConnell, O. J., Kerns, E. H. and Sutherland, A. G. Rapid, automated screening method for enzymatic transformations using a robotic system and supercritical fluid chromatography. *J. Chromatogr. B* 809:235, 2004.
18. Akin, A., Antosz, F. J., Ausec, J. L., Greve, K. F., Johnson, R. L., Magnusson, L. E., Ramstad,, T., Secreast, S. L., Seibert, D. S. and Webster, G. K. An Orthogonal Approach to Chiral Method Development Screening. *Curr. Pharm. Anal* 3:53, 2007.
19. Wong, M. M., Holzheuer, W. B. and Webster, G. K. A Comparison of HPLC and SFC Chiral Method Development Screening Approaches for Compounds of Pharmaceutical Interest. *Curr. Pharm. Anal* 4:101, 2008.
20. Holzheuer, W. B., Wong, M. M. and Webster, G. K. Reversed Phase Chiral Method Development Screening for Compounds of Pharmaceutical Interest. *Curr. Pharm. Anal* 5:346–357, 2009.
21. Snatzke G., Angewandte Circular Dichroism and Optical Rotatory Dispersion-Principles and Application to the Investigation of the Stereochemistry of Natural Products. *Chemie International Edition in English* 7(1):14–25, 1968.
22. IBZ Messtechnik website: http://www.ibzmesstechnik.de/news
23. Kott, L., Holzheuer, W. B., Wong, M. and Webster, G. K. An evaluation of four commercial HPLC chiral detectors: A comparison of three polarimeters and a circular dichroism detector. *J. Pharm. Biomed. Anal* 43:57, 2007.
24. USP 32 published by the United States Pharmacopeial Convention Inc., 2009
25. Validation of analytical procedures: Methodology Q2B. In from the International Conference on Harmonization of Technical Requirements for the Registration of Pharmaceuticals for Human Use, 1996, 1996.
26. Beaufour, M., Morin, P., Ribet, J.-P. and Maurizot, J.-C. HPLC quantitation of the four stereoisomers of benzoxathiepin derivatives with cellulose phenyl type chiral stationary phase and circular dichroism detection. *J. Pharm. Biomed. Anal* 41:544, 2006.
27. Song, S., Lun, S., Yuan, L., Sun, T., Zhoa, Y., Zuo, W., Cong, Y., Li, X. and Wang, Method to determine enantiomeric excess of glucose by nonchiral high-performance liquid chromatography using circular dichroism detection. *J. Chromatogr. A* 1179:125, 2008.
28. Zougagh, M., Aranda, P., Castañeda, G. and Ríos, A. Supercritical fluid extraction–achiral liquid chromatography with circular dichroism detection for the determination of menthone enantiomers in natural peppermint oil samples. *Talanta* 79(2):284, 2009.

29. Lorin, M., Delépée, R., Ribet, J. P. and Morin, P. Validation of a method using an achiral liquid chromatography sorbent and a circular dichroism detector. Analysis of the efaroxan enantiomers. *J. Chromatogr. A* 1141(2):244, 2007.
30. Williams, R. C., Edwards, J. F., Joshi, A. S. and Aubry, A.-F. Chiral analysis of drug substance in clinical plasma extracts using achiral HPLC with circular dichroism detection. *J. Pharm. Biomed. Anal* 25:501, 2001.
31. Bertucci, C., Andrisano, V., Cavrini, V. and Castiglioni, E. Reliable assay of extreme enantiomeric purity values by a new circular dichroism based HPLC detection system. *Chirality* 12:84, 2000.
32. Miller, M. T., Ge, Z. and Mao, B. Development of a nonchiral HPLC method with circular dichroism detection for chiral analysis of molybdenum-catalyzed enantioselective synthesis products. *Chirality* 14:659, 2002.
33. Lecour-Lorin, M., Delépée, R., Ribet, J.-P. and Morin, P. Chiral analysis of milnacipran by a nonchiral HPLC - circular dichroism: Improvement of the linearity of dichroic response by temperature control. *J. Sep. Sci* 31:3009, 2008.

8 CHARACTERIZATION AND ANALYSIS OF BIOPHARMACEUTICAL PROTEINS

YIN LUO,* TATJANA MATEJIC,* CHEE-KENG NG,* BRIAN NUNNALLY,† THOMAS PORTER,* STEPHEN RASO,* JASON ROUSE,* TANYA SHANG,* AND JOHN STECKERT*

**Analytical Research & Development, BioTherapeutics Pharmaceutical Sciences, Pfizer, Andover, Massachusetts, USA*

†Site Technical Services, Pfizer Global Manufacturing, Pfizer, Sanford, North Carolina, USA

I. INTRODUCTION

From 1982 (when the first recombinant product was approved) through 2008, there have been 403 approvals of biopharmaceutical products by FDA, including vaccines, blood products, recombinant proteins (including monoclonal antibodies (mAb)), and other biopharmaceuticals. Among these, 103 are recombinant proteins.[1] The market for therapeutic proteins should show double-digit growth over the next few years.

Compared with small-molecule pharmaceuticals, proteins are structurally complex, simply given the difference in typical molecular mass (Table 1).

TABLE 1 Comparison of Small Molecule and Protein Pharmaceuticals

Parameter	Small molecule	Recombinant protein
Typical molecular mass (kDa)	200–1000	5000–280,000
Example	Acyclovir $C_8H_{11}N_5O_3$ 225.09	rhBMP-2[a] $C_{1142}H_{1752}N_{318}O_{328}S_{18}$ 25,761.6[b]
Typical source	Chemical synthesis, semisynthesis, extraction	Recombinant DNA/cellular expression system
Heterogeneity	Exact structure	Multiple isoforms
Purity	Highly homogeneous (well defined)	Low-level process and product-related impurities
Mechanism of action	Usually receptor binding	Varied
Typical half-life	Hours to days	Days to weeks
Analytical methods	Usually standardized	Specific to molecule, or standardized (mAbs)
Regulatory pathway	New drug application (CDER)	Biologics license application (CDER)
Toxicity	Usually because of metabolites	Usually due to unexpected pharmacology
Cost of DS	Low	Currently high but improving
Dose	Low	Varied

[a]Recombinant bone morphogenetic protein-2.
[b]One of multiple isoforms.

However, this complexity is increased many fold by the presence of microheterogeneity resulting from the production process. Changes in the coding DNA sequence, such as point mutations and deletions, may occur, resulting in an undesirable protein sequence. Posttranslational modifications are expected to be present. It has been estimated that posttranslational modifications, combined with differential mRNA splicing, can amplify the information encoded in a protein's DNA coding sequence by 60-fold resulting in many protein isoforms.[2] The purification and drug product fill finish process can introduce stresses to the molecule that cause degradation, such as deamidation and aggregation.

Despite this complexity, the regulatory authorities recognize that much of the heterogeneity in protein therapeutics is expected, and can be found in native proteins. The desired product can be a mixture of anticipated posttranslationally modified forms (e.g., glycoforms), but the manufacturer should define the pattern of heterogeneity of the desired product and demonstrate consistency with that of the lots used in preclinical and clinical studies (comparability), as well as overall lot-to-lot consistency. If a consistent pattern of product heterogeneity is demonstrated, an evaluation of the activity, efficacy, and safety (including immunogenicity) of individual forms may not be necessary, as long as representative lots containing reasonable levels of these isoforms have been evaluated in animals.[3] Recently, however, the concept of quality by design (QbD) was introduced by FDA to achieve greater understanding of the relationship between the critical quality attributes (CQAs) and clinical properties of the product. In the QbD approach, various isoforms are evaluated by risk assessment as well as structure–function studies, and the resulting knowledge is coupled with systematic characterization of the process design space to ensure the process understanding and control. The QbD implementation in the biotherapeutics industry is still at an early stage, but it is clear that thorough product characterization and appropriate analytical method development will play an essential role in the successful implementation of a QbD program. Details of the QbD approach is outside the scope of this chapter, but a review can be found.[4]

An analytical development program for a protein therapeutic comprises in-depth product characterization, assay development, and routine testing activities. The goals of a characterization effort are to define the structure of the major product substances and product-related impurities, to relate structure to function, and to identify and quantify protein/DNA impurities derived from the host-cell. In addition to fundamental product knowledge, a comprehensive characterization effort allows the future implementation of meaningful comparability protocols that support manufacturing changes. The goals of assay development are to develop and implement relevant, routine assays that address biological activity, safety, efficacy, and quality. These assays support process development efforts, manufacturing investigations, drug substance and drug product release, and stability programs. Many assays arise from the characterization activities. For example, a highly resolving ion-exchange chromatography procedure may be required to isolate product isoforms for structural characterization. A similar procedure, perhaps with a shorter run time that sacrifices some resolution, might be implemented as a routine test to monitor isoform composition.

The objective of this chapter is to provide a brief overview of analytical methods commonly used to assess the structure, purity, safety, stability, and potency of protein drugs produced with recombinant technology. In just one chapter, it is not possible to give comprehensive treatment to each method, nor to include all variations on a theme. However, this chapter should serve as a starting point for further reading. Product specifications, method validation, and transfer to a release testing environment, as well as dosage form, delivery, and formulation, for proteins could be significantly different from that for conventional pharmaceutics. They are beyond the scope of this chapter.

II. STRUCTURE

A. Covalent Structure

1. Primary Structure

All proteins are composed of amino acids that are covalently linked by peptide (amide) bonds into long polypeptide chains. Twenty different amino acids are known to serve as the building blocks in all proteins. The amino acid sequence of a naturally occurring polypeptide (i.e., the type, number, and order of each amino acid in the chain) is genetically determined. Two or more polypeptide chains can be covalently linked together by, most commonly, disulfide bonds.

All the covalently linked amino acids in a given sequence constitute the primary structure of a protein. The primary structure in turn determines the identity, the higher-order structures, basic properties, and functions of a protein. Modifications to the amino acids or variations in the connectivity of the disulfide bonds in a protein can potentially lead to irreversible changes in its primary structure. These changes in primary structure may then trigger changes in the protein's higher-order structures, stability, and functions. Thus, modifications in the primary structure of a pharmaceutical protein may have negative impact on product quality. Therefore, an important goal in the analysis of therapeutic proteins is to detect and characterize both posttranslational and process-induced modifications in their primary structure. This section describes a number of techniques commonly used in the biopharmaceutical industry for this purpose.

a. Peptide Mapping/LC/MS

Peptide mapping is a powerful tool for the analysis of the primary structure of a protein.[5] This method typically takes advantage of one or more specific proteases that cleave the protein into smaller peptides, which are then separated and analyzed by reversed-phase liquid chromatography (RP-HPLC), sometimes in conjunction with mass spectrometry (MS).

When peptide mapping is coupled with MS, the amino acid sequence of each peptide can be individually confirmed. In addition, covalent modifications such as glycosylation, oxidation, or deamidation can be identified in a site-specific manner (see below).

Once a peptide map is characterized using online MS, the chromatographic profile alone can serve as a routine analytical tool to monitor the protein's primary structure and covalent modifications, and is often used for batch release or stability testing of biopharmaceuticals. However, whenever in-depth characterization of a protein is needed, such as that required for comparability studies or reference material characterization, the peptide map should be coupled with MS to ensure a thorough examination of all peptides in the map.

In developing a peptide mapping procedure for characterization and/or as routine assay, many considerations need to be taken at each step. They are discussed as follows.

i. Reduction and Alkylation

Cysteine residues can complicate proteolytic digestion, because of either disulfide scrambling or structural hindrance to proteolytic sites. Therefore, modification of cysteine residues by reduction and alkylation typically increases the efficiency and robustness of the proteolytic digestions. Reduction and alkylation is typically performed in a denaturing buffer (e.g., 6 M guanidine chloride) at slightly alkaline pH (7.5–8.4). Common reductants are dithiothreitol (DTT) or tris(2-carboxyethyl)phosphine (TCEP), and common alkylants include iodoacetate, iodoacetamide, or *N*-ethylmaleimide. The optimal conditions for reduction and alkylation (achieving complete reduction and alkylation, without overalkylating) can be evaluated by LC/MS and SDS-PAGE of the reduced/alkylated protein. After reduction and alkylation, excess alkylant may be neutralized by another addition of reducing agent to quench the alkylation reaction and avoid overalkylation.

ii. Desalt and Dilution

Denaturants such as guanidine chloride used in reduction and alkylation inhibit many proteases, and must be removed or diluted out prior to the addition of the protease. Removal by buffer exchange can be achieved by dialysis, desalting filtration, or a desalting column. Dilution is simpler and less time-consuming; however, residual denaturant typically still inhibits protease activity, leading to higher levels of missed cleavages. On the other hand, the lower protease activity also may lead to less nonspecific cleavages, and a cleaner baseline in the peptide map. In our experience, the level of miscleavages tends to be reproducible, and does not affect the usefulness of the map for monitoring structural integrity and covalent modifications. Therefore, we find the dilution option to be reasonable for a routine peptide mapping method.

iii. Protease Selection

Typical choices for peptide mapping are site-specific proteases, including trypsin, Lys-C, Asp-N, and Glu-C. Due to the different specificities, different proteases are expected to cleave the protein into different peptides. Very small peptides tend to be lost in the flow-through of reversed-phase HPLC (RP-HPLC), reducing the sequence coverage, so proteases that do not generate too many di- or tripeptides are preferred. Very large peptides may have either chromatographic issues (recovery, peak size), or potentially reduce the ability to detect small changes chromatographically or by MS. For example, a large, modified peptide may not fully resolve from its unmodified form. The ability to generate moderate sized proteolytic peptides (<5000 Da) may be of particular interest in certain sensitive regions of the sequence, such as the complementarity determining region (CDR) of an antibody, where small changes may be expected to impact efficacy. On the other hand, larger peptides result in simpler maps that tend to be more robust and less time-consuming to analyze. Therefore, the choice of protease is a balancing act with the ultimate goal of good sequence coverage (>90%), ability to detect small structural changes, and digestion robustness.

iv. Digestion

Digestion is typically performed at the optimal pH of the chosen protease for 2 h to overnight. A time course study is useful to determine optimal digest time that ensures robustness of digestion, minimal artifacts (such as deamidation), and practicality. Another factor that influences completeness of digestion is enzyme:substrate ratio (E:S). Too much protease may lead to nonspecific cleavages from low levels of other protease impurities in the commercial protease preparation, while too little can lead to underdigestion. Both situations can lead to issues with the reproducibility and robustness of the peptide map. Digests are quenched by acidification and/or addition of reductant. Acidification not only stops the protease action (this is needed to ensure reproducibility of digestion time), but also reduces likelihood of deamidation and cyclization of N-terminal Gln as a result of prolonged exposure to high pH of the digestion reaction mixture.

v. Chromatographic Separation

Most peptide mapping methods utilize RP-HPLC to separate the proteolytic peptides, although cation exchange chromatography (CEX) has also been used for this purpose. However, RP-HPLC is the preferred method because of its ability to directly interface with MS, making peak identification much easier. A typical peptide mapping method uses a C18 column and a gradient of water and acetonitrile with ion-pairing agents such as TFA. Initial method development involves screening various columns and gradients to identify parameters that lead to well resolved peaks with good peak area recovery. MS is often needed to aid method development, to ensure important peptides (e.g., those containing stability-indicating modifications, N-terminal and C-terminal peptides, and glycosylated peptides) are well separated from other species, and can be readily monitored.

vi. Evaluation of Reproducibility and Robustness

If peptide mapping will be used for batch release or stability testing,[6] the method needs to be evaluated for reproducibility and robustness (e.g., column and protease lot-to-lot consistency, sample stability over time). If any quantitative acceptance criteria are necessary, such a method also needs to undergo qualification and validation.

vii. Detection and Characterization of Separated Peptides

The chromatographically separated peptides are most commonly detected by UV absorbance, especially for routine testing and method development. With a dual-wavelength detector, UV absorbance at 214 nm is universally applicable for all peptides, while absorbance at 280 nm detects only tryptophan- or tyrosine-containing peptides. The photodiode array (PDA) detector has the ability to measure the UV absorbance spectra of each peak in the chromatogram, and is sometimes useful for troubleshooting. For example, a contaminant peak representing a method-, process-, container closure-, or product-related substance may have an unusual chromophore that leads to a UV absorbance spectrum distinguishable from that of a peptide.

For identification of the peaks in a peptide map, online RP-HPLC/MS (or LC/MS) analysis is needed. Accurate mass determinations with modern high-resolution electrospray (ESI) mass spectrometers often provides the ability to distinguish and confidently assign all separated peptides from a recombinant protein in an efficient manner.[7] Commercial ESI hybrid quadrupole time-of-flight (QTOF) mass spectrometers[8,9] with 10,000 resolving power (i.e., mass/Δmass using bovine insulin) typically afford accurate mass measurements for peptides to within 0.003% of the respective theoretical masses in a stable, reproducible manner through the use of lock mass[10] and/or flight tube temperature monitoring[11] correction strategies, following external mass calibration. For mass spectral data analysis, the protein therapeutic amino acid sequence is proteolyzed *in silico* to generate a sorted list of respective peptides and theoretical molecular masses (monoisotopic values). For each detected peptide, the experimental mass is matched to a particular theoretical value for positive peptide identification, provided that the mass error is less than the 0.003% specified tolerance.

In some cases, where the peptide mass alone is insufficient for definitive assignment, or if site-specific information about a particular amino acid position or modification is required, the peak may be collected (i.e., the eluate containing the peptide of interest) and subjected to further analysis, such as off-line high-resolution tandem MS (MS/MS) sequencing with collision-induced dissociation (CID) or electron-transfer dissociation (ETD), Edman N-terminal sequencing, or amino acid analysis (AAA).[12] The more traditional quadrupole- and ion trap-based mass spectrometers with unit mass resolution generally provide mass accuracies of 0.03% or better, which still is sufficient to distinguish and identify all separated peptides in most protein therapeutics.[13] However, with this group of instruments, more confident peptide assignments are obtained by either online LC/MS in conjunction with in-source CID, provided that the peptides are well separated, or online LC/MS/MS with data-dependent scan functions for more complicated peptide separations. In this latter approach, after each recurring mass analysis, one to five abundant peptide precursor ions are selected by the acquisition software for CID in quadrupole instruments or CID and/or ETD in ion-trap instruments, based on user-defined selection criteria and the results of previous scan functions. In recent years, the use of data-independent LC/MS/MS has gained momentum[14,15]; with this approach, quadrupole ion selection is not used and all peptide precursor ions are subjected to CID. However, the mass spectral fragmentation data of closely spaced precursor ions in the LC separation are multiplexed and specialized, proprietary software is required to extract the respective amino acid sequences for peptide identification.

In recent years, a new generation of ultrahigh-resolution ESI-based mass spectrometers have become available that afford mass accuracies of less than 0.0005%, for even more confident peptide identification via LC/MS and LC/MS/MS.[16–19] Additionally, the newer ultrahigh-resolution ESI-QTOF instruments feature greater sensitivities and faster scan rates (i.e., up to 20 spectra per second), which are well suited for fast chromatography applications with ultrahigh performance liquid chromatography (UHPLC) and capillary electrophoresis (CE).

Last, thorough identification of the peaks across the peptide map separation with MS is a very laborious, multihour, hands-on process, whether LC/MS with accurate mass measurements or LC/MS/MS with gas-phase peptide ion fragmentation is employed. Reliable peptide map informatics software, specific to protein pharmaceutical characterization, such as BiopharmaLynx and ProteinLynx Global Server (Waters Corporation), is becoming available as the biotherapeutics research and development sector matures.[20] BiopharmaLynx interfaces with high-resolution Waters QTOF instruments and utilizes accurate mass measurements from LC/MS to identify the peptides represented by each peak in a peptide map—literally in minutes. Additional specificity for peptide identification is gained from amino acid sequence tags if gas-phase fragmentation is conducted in a second, alternating scan function. The BiopharmaLynx algorithm requires the target recombinant protein sequence, instrument type, protease and cleavage specificity, the number of allowed miscleavages, a specified error tolerance between 0.0005% and 0.003% (or better), and the list of possible posttranslational, storage-induced, and method-induced modifications, in order to create a comprehensive theoretical list of unmodified peptides, modified peptides, miscleaved peptides, and protease-derived peptides, including their masses, for automatically elucidating all of the peptides in an experimental peptide map separation.

b. N-Terminal Sequencing

N-terminal sequencing by Edman degradation[21,22] is frequently used to determine the N-terminal amino acid sequence of a protein. The peptide bonds are sequentially hydrolyzed from the N-terminus, and the released amino acid is derivatized and analyzed by HPLC. The amino acid in each hydrolysis cycle is identified and quantitated based on comparison to amino acid standards.

For analysis of biopharmaceuticals, N-terminal sequencing is typically used to confirm the identity of the protein, as well as to assess N-terminal heterogeneity. It provides complementary information to accurate mass analysis of intact proteins and subunits. It is especially useful in the identification of unknown bands in SDS-PAGE (after blotting), allowing the important distinction of product-related (such as fragments) versus nonproduct impurities (such as host–cell proteins (HCP)). In the case of HCPs, identity can sometimes be assigned based on BLAST searches using the N-terminal sequence.

N-terminal sequencing has some limitations. Chemically modified N-termini, such as cyclization of Glu (pyroglutamic acid) or carbamylation, often cannot be successfully sequenced. It is possible to remove N-terminal pyroglutamic acid enzymatically[23] to obtain the remaining N-terminal sequence; however, the yield of the enzymatic removal is variable.

2. Cysteine and Disulfide Bond Characterization

Cysteine is one of the two sulfur-containing amino acids; its side chain contains a thiol, which is one of the most chemically reactive groups found in proteins. The thiol form (–SH) is in equilibrium with the thiolate ($-S^-$), the relative amounts of each depend on the solution pH. Cysteinyl residues in proteins are most commonly found in the reduced form (thiol/thiolate) or as disulfide bonds.

The disulfide bond is a specific oxidized form of thiol functionality, wherein two sulfur atoms form a covalent bond, dimerizing (two) cysteine side chains. The disulfide form of cysteine is often called cystine; however, it may be useful simply refer to such bonds as "disulfides" to avoid confusion.

Disulfide formation of cysteine side chains in proteins results in the covalent cross-linking of polypetide chains. This cross-linking can occur inter- or intramolecularly. Intermolecular disulfide bonds are often found in multimeric proteins to covalently link two or more subunits, stabilizing the quaternary structure (Section II.B.1.d). It should be noted that not all multimeric proteins contain intermolecular disulfide bonds. Intramolecular disulfide formation can have the effect of stabilizing certain elements of protein tertiary structure, or functional domains (Section II.B.1.c). Since disulfide bonded proteins cannot exist in a reducing environment, and inside the cell is reducing, disulfide bonds are found in extracellular proteins, such as circulating or structural proteins. Under the right experimental or physiological conditions, the disulfide oxidation reaction is fully reversible. In addition, disulfide bonds can exchange, or shuffle: a thiol can react with a preexisting disulfide bond, resulting in a new disulfide bond (Figure 1).

$$R_1\text{-SH} + R_2\text{-S-S-}R_3 \longrightarrow R_1\text{-S-S-}R_2 + R_3\text{-SH}$$

Thiol	Disulfide	New disulfide	Released thiol

FIGURE 1 The disulfide exchange reaction.

In contrast to disulfide bonds, which are relatively inactive, cysteine thiols (or more specifically the thiolates) may play a role in a multitude of chemical processes, using a several different mechanisms. These include nucleophilic attack, hydrogen bonding, and ligand–metal binding. The cysteine thiol may also react with oxygen to form products with increasing oxidation states: sulfenic acid (–SOH), sulfinic acid (–SO_2H), and sulfonic acid (–SO_3H); formation of the latter two is irreversible. Protein cysteinyl side chains can also form mixed disulfides with small thiol-containing molecules, such as glutathione and free cysteine, effectively blocking their chemical reactivity.

a. Detection and Quantification of SH

By taking advantage of their inherent chemical reactivity, the presence and amount of unmodified cysteinyl residues (thiol/thiolate) in a protein can be determined. Briefly, a colorimeteric substrate is reacted with the protein; the amount of signal is proportional to the amount of thiol present. The most commonly used chemical method for the determination of thiols is Ellman's assay. When higher sensitivity is required, fluorescence-based assays can be used.

i. Ellman's Assay

Ellman's assay was first described[24] in 1959 and has undergone many minor modifications over the years.[25–27] The procedure is based on the reaction of thiols with Ellman's reagent or 5,5′-dithiobis-(2-nitrobenzoic acid) (DTNB) to yield a modified protein thiol (mixed disulfide), and a molecule of 2-nitro-5-thiobenzoic acid (TNB) as shown in below (Figure 2).

While the unreacted Ellman's reagent is only faintly colored, the TNB anion has a deep yellow color; this can be detected and quantified easily using a spectrophotometer.

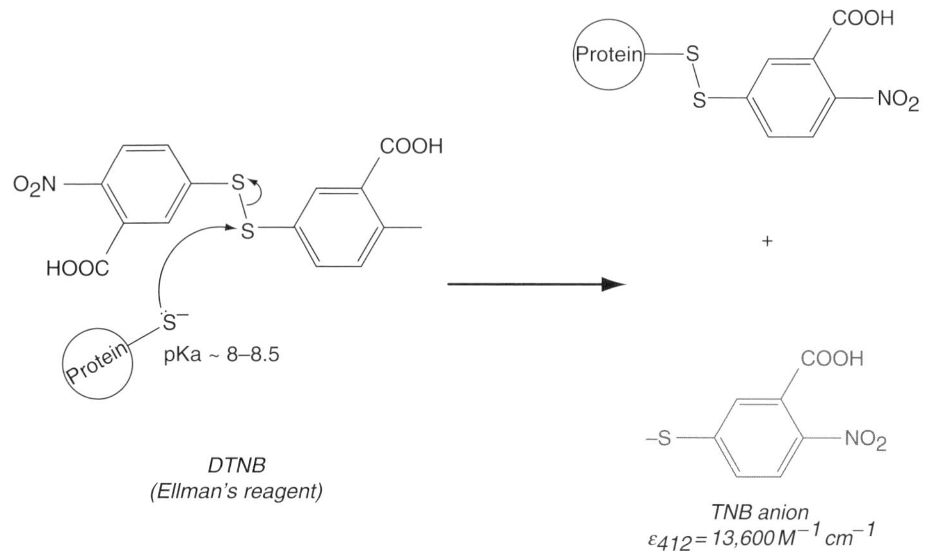

FIGURE 2 **Protein thiol reaction with Ellman's reagent.**

The measurement of absorbance at 412 nm (specified by Ellman) provides a quantitative determination of thiol groups. For every one molecule of free thiol that reacts, one molecule of TNB in produced.

Two major factors that effect this reaction are pH and steric considerations. Since it is actually the thiolate form of cysteine that reacts with the Ellman's reagent, the reaction should be run at a pH where this form is well populated. The pK_a for cysteine thiol deprotonation—the pH where the thiol and thiolate forms are present in equal amounts—is approximately 8.5. If the pH is too far below 8.5, only trace amounts of the thiolate form exist; therefore, the reaction will not proceed at an appreciable rate. At higher pH more thiolate is present; however, DTNB will break down rapidly at pH 9.0, or greater, creating very high background signal. It is necessary to run the reaction under conditions where a significant proportion of the cysteinyl residues are in the thiolate form and the Ellman's reagent is sufficiently stable throughout the course of the experiment. Ellman's original recommendation was pH 8.0, but this value may be adjusted slightly to suit modifications or improvements to the method.

In order to assure that all cysteine thiols in the protein react, it is necessary to perform Ellman's reaction under conditions where the protein is unfolded. This will enable the Ellman's reagent to access thiols that may have been sterically inaccessible in the folded protein. Chemical denaturants, such as urea, guanidine, or detergent are often suitable and convenient. Further, the TNB anion extinction coefficient does not change significantly in the most common denaturants.[28,29] Conducting Ellman's assay under both native and denaturing conditions allows for the additional distinction of solvent-accessible versus buried thiol groups. Any increased DTNB reactivity in the presence of denaturant is assumed to arise from accessibility to thiols which were buried in the native protein.

Lastly, if the number of cysteinyl residues in a protein is known from the amino acid sequence, one can infer how many disulfide bonds, or otherwise modified cysteinyl residues are present. If Ellman's assay cannot account for all of the cysteine residues predicted from the sequence, additional data, like high-accuracy MS, may be necessary to determine if the unreacted cysteinyl residues are involved in disulfide bonds, oxidized, or have undergone other chemical modifications.

It should be noted that Ellman's assay does not require the use of external standards, or the generation of a standard curve, which is very convenient. However, it is limited in its sensitivity. These assay limitations are directly because of the inherent limitations in spectrophotometric detection. For instance, if an Ellman's assay sample shows an absorbance of 0.05 at 412 nm (approaching the lower limit of detection), the thiol concentration is calculated to be 3.7 μM. This may be suitable when assaying proteins in the mg/mL concentration range (1 mg/mL bovine serum albumin = 15 μM), but not for lower concentrations. In some cases, the presence of unmodified cysteine thiols is aberrant or undesirable; a more sensitive assay to detect low levels of such species would be useful.

ii. Fluorescence-Based Assays

In the last 10 years, many, more sensitive, thiol assays have been developed.[30–32] They all exploit the relatively high reactivity of cysteinyl residues, either by forming mixed disulfides (like DTNB), or providing a site for nucleophilic attack by the thiolate sulfur. In each case, the chemically reactive cysteine side chain is coupled with a highly fluorescent molecule. Most of these reagents require the removal of unreacted fluorescent probe from the labeled protein prior to thiol quantification.

Some reagents, however, are not highly fluorescent until they react with the protein; therefore, the unreacted material does not interfere with the signal from the labeled protein.[33,34] Conveniently, this feature eliminates the need for removal of unreacted fluorescent reagent. Much like the breakdown of Ellman's reagent at high pH, one must be careful when using this class of reagent. Reaction of the reagent with a buffer component, or breakdown in solution, will lead to very high background interference and rapid depletion of reagent.

Although these fluorescence-based methods can increase sensitivity by orders of magnitude over Ellman's assay, there are some significant drawbacks. It is necessary to use a suitable model compound, such as DTT or free cysteine, to generate a standard curve with each assay. This can be complicated by the propensity of these model thiol compounds to undergo spontaneous oxidation reactions in solution. In order for the standard curve to provide an accurate value, the model compounds must be 100% in the reduced state. It is not trivial to determine if the model compounds being used have undergone some amount of oxidation, which may lead to systematic overestimation of the thiol content of a protein.

b. Disulfide Structure by Nonreducing Peptide Map-LC/MS

Peptide map analysis of a protein containing two or more cysteine residues typically employs reduction and alkylation chemistry for efficient, reliable proteolysis, reproducible chromatographic profiles, and straightforward characterization by MS. However, when protease digestion is carried out on a nonreduced protein, the disulfide bonds in the protein will maintain the covalent linkage between the peptides that are involved in the disulfide bond. In many cases, it is then possible to choose a protease that will ensure one cysteine residue in each peptide upon cleavage, such that each disulfide bond will associate with a pair of proteolytic peptides.

LC/MS analysis of the resulting peptide map provides identification of the disulfide-linked peptide pairs on the basis of accurate mass, thus revealing the cysteine residues involved in the disulfide bond.[35] The individual peptide components of each disulfide-linked peptide pair are then confirmed by chemical reduction (DTT or TCEP) of the nonreduced peptide digest and subsequent LC/MS. In a more targeted approach, a suspected disulfide-linked peptide pair can be isolated via peak fractionation, reduced chemically, and mass analyzed by LC/MS to confirm individual peptides. Off-line MS techniques such as nanoESI or matrix-assisted laser desorption/ionization (MALDI) may provide the mass of the peptide constituents in rapid manner following reduction, but ion suppression or hydrophobicity effects may prevent clear-cut detection for one of the two peptides. Alternatively, MS/MS approaches involving CID or ETD can yield contiguous amino acid sequence for identification of the disulfide-linked peptides,[36] while ETD has potential to reduce disulfide bonds in the gas-phase, following mild collisional activation, similar to chemical methods.[37]

On the other hand, if a peptide contains more than one cysteine residue, the identification of this peptide component as part of a disulfide bonded peptide cluster will not be sufficient to elucidate the connectivity of the cysteine residues. In this case, the disulfide-linked peptide cluster can be purified and further characterized. If a different proteolytic site exists between the multiple cysteine residues within the purified peptide cluster, this second protease can be used to digest the cluster into smaller peptides with a single cysteine in each peptide. Otherwise, Edman N-terminal sequencing may be used to further elucidate the disulfide connectivity, as well as various MS-based approaches.[38–43]

Trisulfide bonds, originally elucidated in human growth hormone,[44] were detected recently in mAbs at low levels.[45,46] The linkage between light and heavy chain is the predominant location of trisulfide bonds in mAbs. Ironically, nonreducible thioether bonds were detected previously at this same disulfide bond linkage in mAbs via peptide map analysis.[47,48]

3. Posttranslational Modifications

Hundreds of modifications to the protein backbone have been reported and reviewed.[49,50] For many biopharmaceutical proteins, however, only a handful of modifications are usually detected. These include both posttranslational modifications that are results of intracellular enzymatic processes, and covalent modifications that occur during or after the manufacturing process, either induced by process conditions or resulting from degradation (discussed in Section II.A.4).

During product and process development of a biopharmaceutical protein, it is necessary to characterize materials at different stages of development, such as materials generated using varying cell lines, cell culture conditions, purification process parameters, or under stress conditions. Characterization of this wide range of samples leads to the understanding of the relative propensity of the protein biopharmaceutical to undergo various modifications, and the need to monitor them.

Most of the posttranslational modifications can be readily detected by peptide mapping. The nature of the modifications can then be identified by coupling peptide mapping with MS. The exceptions modifications that may be lost or altered during the peptide mapping procedure (disulfide reduction and prolonged incubation at neutral to high pH). Therefore, orthogonal methods, such as intact or subunit LC/MS analyses, are often necessary.

Once modifications are identified, assays can be chosen to monitor relevant modifications for routine testing. Peptide mapping is usually used for this purpose. Even modifications that are unforeseen (not identified during initial characterization of release testing assays) can typically be detected by peptide mapping as a new peak or atypical profile, triggering an investigation. For modifications that are deemed to be particularly important to a protein (e.g., a modification known to impact efficacy or safety, or the predominant degradation product), more specific methods may be developed in addition to peptide map, along with appropriate acceptance criteria to control it. Comprehensive reviews of posttranslational modifications of biopharmaceutical proteins have been published.[50] Below, brief discussions are given for select modifications that are most frequently encountered.

a. Glycosylation

Glycosylation is a common and complex posttranslational modification of recombinant therapeutic glycoproteins expressed in eukaryotic expressions systems.[51–53] The glycosylation biosynthetic pathway consumes considerable energy, is highly conserved evolutionarily and requires a set of well-defined cellular machinery that span from the endoplasmic reticulum to the *trans*-Golgi network. Although the general glycosylation machinery is fundamentally similar across all eukaryotic expression systems, differences that are distinctive of or unique to a given host expression system may affect potential immunogenic responses, pharmacokinetic, or drug stability outcomes.[54–61] As such the application of appropriate analytical methods in monitoring and structural characterization of glycosylated forms are critical in the understanding of drug structure and function, and manufacturing process consistency.

In the mammalian expression systems, there are two major forms of protein glycosylation; N- and O-linked. The addition of N-linked oligosaccharide to the amide side group of Asn occurs at a well-defined amino acid consensus sequence, Asn-X-Thr/Ser (where X is not a Pro), typically located near the beta-bend on the protein backbone, with threonine favored over serine.[62–64] In contrast, there is no known consensus site for the initiation of O-linked glycosylation other than the initial addition of a sugar moiety predominantly occurs on Thr/Ser residues (linkage to the hydroxyl side group) in a mostly Pro rich region, with Thr favored over Ser.[65] Glycosylation may also occur on hydroxylysines and hydroxyprolines. While both N- and O-linked glycosylation form the vast majority protein-glycosidic linkages, *C*-mannosylation of tryptophan, which requires highly specific consensus peptide sequences are known to occur too.[66]

There are three major classes of N-linked oligosaccharides: high mannose (or oligomannose), hybrid and complex, and they share a common core structure—$Man_3GlcNAc_2$. In hybrid and complex glycoforms, terminal galactose units are often capped with sialic acids, and depending on the host type and culture conditions, the core amide-linked GlcNAc may be fucosylated via an α1-6 linkage. Terminal sialylation is an important factor influencing the pharmacokinetics of glycoprotein biotherapeutics.[67,68] Poorly sialylated N-linked oligosaccharides are typically removed from the circulation via cell surface lectins such as asialoglycoprotein receptors in the liver, and/or mannan binding receptors found in various immune surveillance cells (e.g., macrophages, NK cells).[69–71]

The basis for N-linked oligosaccharide heterogeneity is largely determined by a series of innate and/or environmental factors such as attachment site accessibility to initiating glycosyltransferases, relative enzymatic activities of competing glycosidases and glycosyltransferases that are present in a given cell line, and culture conditions.[57] Site occupancy of both N- and O-linked oligosaccharides can be determined through distinct mass shifts of glycopeptides in peptide mapping on RP-HPLC following MS. NMR analysis is of limited use as the degree of heterogeneity and molecular masses of many biotherapeutic glycoproteins typically exceed the capability of the method.

Two major forms of O-glycosylation are known; GalNAc-based (mucin-type) and the O-xylose linked glycosaminoglycans found in proteoglycans.[65] There are up to six different core forms of O-GalNAc-based modifications. Other forms of O-glycans include, O-mannose, O-fucose, O-GlcNAc, O-glucose, and O-galactose.[72] All O-glycosylated initiating core sugar units are often further modified, elongated, or branched by cell-type specific glycosyltransferases to form heterogeneous structures. For instance, terminal sialylation or fucosylation may occur at the GalNAc and/or galactose units. These modifications not only depend on the availability of the appropriate glycosyltransferases within the host-cell expression system, but are greatly affected by culture conditions (e.g., pH, temperature, media composition) and steric accessibility.

The following sections will focus on analytical methods that are commonly used in the biotechnology industry to isolate and characterize the various heterogeneous glycoforms that are found in mammalian (including yeast) or plant-based expression systems. As a rule, *Escherichia coli* and other bacteria-derived proteins are not glycosylated. For a given glycan profile analysis, four key criteria should be considered: glycan release, labeling of glycan (optional), separation of the different glycans forms, and structural characterization.

i. Glycan Release

N-linked glycans are typically released from the protein backbone by hydrolytic endoglycosidases such as peptide-N^4-(*N*-acetyl-glucosaminyl) asparagine amidase F (PNGase F) or endo-*β*-*N*-acetyl glucosamindases (e.g., Endo H). PNGase F hydrolyzes the linkage between the chitobiose units and asparagine, while Endo H will cleave between the two *N*-acetylglucosamine residues of the chitobiose.[73] The use of recombinant endoglycosidases is highly recommended in order to avoid contaminating extraneous endoglycosidase activities that will quantitatively affect the final glycoprofile.[74] It should be noted that PNGase F will not release plant or insect N-linked glycans that contain α1,3-linked fucose to the core amide-linked GlcNAc residue. In this instance, PNGase A is recommended.[75] Unfolding of the protein backbone with *β*-mercaptoethanol and SDS or digestion of the protein with trypsin promotes endoglycosidase accessibility of the glycans.[76] In all instances, the presence of ammonium (e.g., NH_4HCO_3) or substantial amounts of primary amines in the digestion buffer will lead to the generation of glycosylamines.

N-linked glycans may be released with chemical cleavage (e.g., alkaline-based *β*-elimination, hydrolysis with hydrazine, or trifluoromethansulfonic acid (TFMS).[77] Some method optimization will be needed, and care must be taken to avoid chemical destruction of the reducing end. Hydrazinolysis, introduced in the 1960s, is the most commonly used chemical method of choice though the harshness of the method often leads to chemical deacetylation of sialic acids and *N*-acetylhexosamines.[78,79] Nearly all methods will result in a reactive aldehyde reducing end that can be derivatizated (see below) with the appropriate amine containing molecules.

O-linked oligosaccharides are typically released by reductive *β*-elimination or hydrazinolysis. Typically, 0.1 or 0.05 M NaOH containing 1 M $NaBH_4$ at up to 37 °C is used, where the cyanoborohydride converts the reducing end of the O-linked glycan to an alditol thereby preventing additional side products that result from alkali-induced peeling reactions. O-linked oligosaccharides from hydroxyproline may be released by alkaline hydrolysis in $Ba(OH)_2$.

Currently, there is no known universal PNGase F equivalent for the general release of O-glycans though a recombinant O-glycanase from *Streptococcus pneumoniae* (endo-α-*N*-acetylgalactosaminidase), with narrow structural specificity is available for Gal-*β*1-3GalNAc-Thr/Ser (Core I).[80] Substitutions with sialic acid, fucose, or *N*-acetylhexosamine residues will prevent enzymatic cleavage. Likewise, the absence of terminal galactose will inhibit enzymatic cleavage of the glycosylamine (GalNAc-Thr/Ser) linkage.

Regardless of the release method of choice, the release conditions must be optimized to ensure complete (or near complete) release. This can be experimentally confirmed, preferably through orthogonal series of highly sensitive methodologies such as LC/MS, SDS-CE, or MALDI-TOF MS. The method of choice should take into account of the chemical stability and completeness of the released glycoform. For instance, some chemical release methods may result in partial degradation of the reducing terminus sugar (chemical peeling) with concomitant loss of *N*-acetyl groups of amino sugars (e.g., if hyadrazine is used). And the use of Endo H will specifically release all high mannose and certain hybrid structures. To reduce the possibility of chemical peeling by hydrazine, the reaction must be kept anhydrous, and the temperature at up to 37 °C. The *N*-acetyl groups are restored by reacetylation with acetic anhydride in cold sodium bicarbonate prior to further analysis.

ii. Postrelease Glycan Analysis (Nonderivatization)

Postreleased N- or O-linked glycans may be cleaned up, and directly analyzed on a pellicular anion resin column in strong alkaline conditions (e.g., 0.1 M NaOH), a method commonly known as high pH anion-exchange chromatography (HPAEC).[81–83] The basic or cationic nature of the resin promotes efficient binding of oxyanions of sugar hydroxyl groups at a very high pH. Picomole level detection of the glycans is achieved by pulsed electrochemical detection (PED) on a gold electrode, under alkaline conditions. Since its introduction in the early 1990s, HPAEC-PED has been widely adopted by the industry and accepted by drug regulatory agencies as a method of choice by many for the separation (and profiling) of a complex glycan mixture. The identity and oligosaccharide linkage elucidation may be done in conjunction with MS, coeluting standards, or sequential enzymatic digestion. However, HPAEC-PED is difficult to quantify and requires dedicated expertise in the maintenance and proper usage of instruments and the gold electrode. For instance, the high pH mobile phase requires a mostly PEEK retro-fitting of stainless steel lines found in a typical HPLC unit, and the baseline noise originating from the highly sensitive gold electrode can be affected by contaminants or impurities from the water source, drug formulation components, or mobile phase. In some instances, the complexity of the HPAEC profile may be reduced by converting the reducing ends of the oligosaccharides to alditols with a reductant such as cyanoborohydride. In addition to HPAEC-PED, underivatized *N*- and *O*-oligosaccharides may be analyzed on porous graphitized carbon columns with either MS or charged aerosol detection.

Fine structural elucidation of glycan species can be determined or confirmed by sequential glycosidase digestion followed by distinct elution position shifts in the HPLC profile and/or determined with tandem MS techniques such as CID. Elucidation of unknown glycan structures often begins with specific glycosidase treatments that reduce the complexity of the unknown or resolve isomeric structures. Careful accurate mass analysis combined with structural information from CID helps dissect and assign branching patterns, monosaccharide positions and linkages of a given glycan structure.

iii. Postrelease Glycan Analysis (Derivatization)

In recent years, the biopharmaceutical industry and regulatory agencies have embraced oligosaccharide profiling through fluorescence labeling of released *N*-and *O*-glycans followed by separation on various liquid chromatography platforms or CE with fluorescence detection.[84] Conjugation of the reducing end of released oligosaccharides with fluorophores greatly increased the detection capabilities of these methods and enabled relative ratio quantification of a major glycoforms within a given heterogeneous oligosaccharide mix. The typical fluorophores used in chromatrography methods are simple arylamines such as 2-aminobenzamide (2-AB), 2-aminobenzoic acid (2-AA), or pyridine-based 2-aminopyridine (2-AP), and in the case of CE, APTS (8-aminopyrene-1,3,6-trisulfonic acid) is commonly used. The conjugation of the reducing end with the primary amine occurs through reductive amidation in mild acid at elevated temperature (typically up to 65 °C) to form a Schiff's base, and the addition of sodium cyanoborohydride or an amine–borane complex resulting in a stable secondary amine complex. A drawback in labeling of oligosaccharides is potential desialylation of complex glycoforms in the presence of heat and mild acid. During method development, the effect of temperature and labeling time on desialylation must be undertaken and a careful design of experiment approach may be applied to balance the effect of desialylation against optimizing the labeling yield.

After derivatization, the pool of fluorescence-labeled oligosaccharides is cleaned up, and profiled on a chromatographic or electrophoretic method of choice. Many clean up protocols are available, however, we prefer organic precipitation and the more traditional, though laborious paper–solid phase approaches. Both of these methods provide average to excellent recovery of all labeled glycoforms and in our experience thus far, both methods do not speciate one form of oligosaccharide over another.

A common chromatographic profiling method is hydrophilic interaction or normal phase chromatography analysis on an amide-bonded stationary phase.[85] Other chromatographic forms include reversed-phase chromatography on C18 columns, or anion-exchange chromatography. APTS-labeled oligosaccharides can be separated on capillary or gel electrophoresis with laser-induced fluorescence detection,[86] and on a reversed-phased mode using amide-bonded matrices. The identity of the separated labeled oligosaccharide can be determined by online MS or coelution/migration with known and well-characterized commercially available labeled-oligosaccharide standards coupled with confirmatory sequential digestion with known glycosidases (recombinant forms are preferred). It is common to use a combination of these key characterization methods to determine the form and linkage elucidation of N- and O-linked oligosaccharides released from biotherapeutic glycoproteins.

iv. Monosaccharide Composition Analysis

Monosaccharide composition analysis is often done to detect the presence of uncommon or atypical monosaccharides that may be present on the protein backbone, often in O- or C-linked forms. Typically, monosaccharides are released in strong acid at high temperature.[87] However, it must be noted that hexose linkages are usually more stable to heat and strong acid than those of fucose or sialic acids. For instance, sialic acids are known to decompose after prolonged exposure to heat and acid. Optimizing the release of monosaccharides require careful sample preparation, and if possible, removal of all trace of exogeneous sugars, either by dialysis or buffer exchange with spin columns.

After acid hydrolysis, the released monosaccharides are analyzed on liquid column chromatography or CE. Underivatized monosaccharides may be directly analyzed using HPAEC-PAD or ion-moderated partition chromatography (e.g., Supelcogel Pb or Bio-Rad Aminex). Analysis of 2-AA or 2-AB derivatized monosaccharides may be done on reversed-phase or anion-exchange chromatography with online fluorescence detection. APTS-labeled monosaccharide analysis is done on the CE with LIF detection. Given the many orthogonal analytical methods that are used to characterize the biotherapeutic glycoprotein, along with the inherent variability of assay, the use of monosaccharide composition analysis as a drug substance or product release assay has been drastically curtailed.

Sialic acid assay is typically applied to sialylated glycoproteins which oligosaccharides are shown to affect the pharmacokinetics or biological activity of the biotherapeutic.[88] It is also a surrogate product attribute marker in bioprocess manufacturing consistency assessment. Upward to 80 different forms of sialic acids are present in nature.[89] Depending on the expression host and cell culture conditions, the most common mammalian sialic acid forms are *N*-acetylneuraminic acid and *N*-glycolylneuraminic acid. Both forms may be further modified with O-acetylation, sulfation, or methylation by the expression host-cell. Unlike other monosaccharide linkages, sialic acid glycosidic linkages are highly labile to mild acid and heat treatment. As such the sialic acid assay is often done separately from monosaccharide composition analysis. Sialidases with specific specificity for certain linkages may be used to release sialic acids and thereby determine the glycosidic linkage. However, care must be taken as substrate accessibility may be hindered by the protein backbone, and the substrate specificity affected by O-acetylation.

Once released, sialic acids may be analyzed in its underivatized form by HPAEC-PED or derivatized with DMB (1,2-diamino-4,5-methylenedioxybenzene)[90] or OPD(O-phenylene diamine)[91] followed by reversed-phase chromatography with fluorescence detection. If detection and quantification of O-acetylated forms are required HPAEC is not recommended as deacetylation will occur in alkaline conditions. In our experience, the DMB method is highly sensitive, robust, and the chromatographic conditions are amenable to online MS analysis.

b. γ-Carboxylation and Cysteine Modifications

γ-Carboxylation refers to the post-translational modification that converts certain Glu side chains into γ-carboxyglutamate (Gla). γ-Carboxylation occurs in the ER compartment in the cell secretory systems by vitamin-K dependent enzymatic mechanisms. The conversion of Glu to Gla creates the ability to chelate Ca^{2+}, which is essential to the function of the γ-carboxylated proteins. Several biopharmaceutical proteins (such as recombinant factor IX and recombinant factor VIIa) undergo γ-carboxylation within certain highly conserved sequences in the N-terminal domain (Gla domain). Multiple Gla sites within the Gla domain cooperatively bind metal ions, upon which conformational changes occur that allows membrane binding and downstream signal transduction.[92] Therefore the extent, consistency and specificity of γ-carboxylation are important quality attributes for Gla-containing biopharmaceutical proteins.

The identification of Gla sites in the protein can be achieved by mass spectrometric analysis of peptides from the peptide map. Peptide map can also detect relative occupancy of each site, therefore revealing sites of under-carboxylation. The Gla-related isoforms of the intact protein can be separated on the basis of charge using anion-exchange HPLC (AEX-HPLC); confirmation of their identities can be obtained by amino acid analysis of base hydrolysate for Gla content, and peptide mapping with mass analysis. Individual Gla isoform can also be isolated by AEX-HPLC and tested by bioassay to determine the effect of under-carboxylation on activity.

The sulfhydryl group of the Cys residue can form a covalent disulfide bond with another sulfhydryl. The other sulfhydryl may be from another Cys residue in the same protein, or from a small molecule such as glutathione or free cysteine (glutathionylation or cysteinylation). The formation of disulfide bonds in specific configurations is often important to the structure, function, or stability of the biopharmaceutical protein. Therefore, the confirmation of the disulfide connectivity is necessary. The analysis of disulfide bond structure is discussed in more detail in Section II.A.2.

4. Stability- and Process-Related Modifications

A protein biopharmaceutical may encounter various environments during and after the manufacturing process that lead to its degradation. Proteins degrade by chemical and physical pathways. Chemical pathways can include deamidation and related reactions, oxidation, and peptide bond hydrolysis (chemical and enzymatic). Physical pathways can include aggregation, precipitation, denaturation, and adsorption to surfaces.

Many potential degradation products are not observed in protein pharmaceuticals, primarily because much care is taken in the choice of formulations, lyophilization, and storage conditions in order to maintain protein stability. Thus, degradation is minimized and usable shelf lives are on the order of years. In order to study the degradation pathways of a biopharmaceutical protein, and to evaluate the stability-indicating ability of the analytical methods, it is sometimes necessary to perform forced degradation studies, where the biopharmaceutical protein is subjected to a variety of stress conditions, such as varying pH, elevated temperature, or the addition of oxidants.

A comprehensive survey of the degradation products of 73 protein pharmaceuticals indicates that the primary chemical pathway of degradation is succinimide formation at Asn and Asp residues to yield Asp and isoAsp at both residues.[93] Deamidation at Gln in Gln-Gly sequences, hydrolysis at Asp-Pro bonds and Met oxidation were also observed at a lesser extent. These reactions are briefly discussed below.

A discussion of physical degradation products (related to higher-order structural changes) can be found in Section II.B.2.

a. Deamidation and Related Reactions

Asn and Asp residues are involved in the predominant protein degradation pathway, which is spontaneous nonenzymatic hydrolysis of the side chain via a succinimide intermediate[94] (Figure 3).

The initial reaction is the nucleophilic attack of the peptide bond nitrogen of the adjacent amino acid residue (C-terminal) on the carbonyl carbon of the Asn or

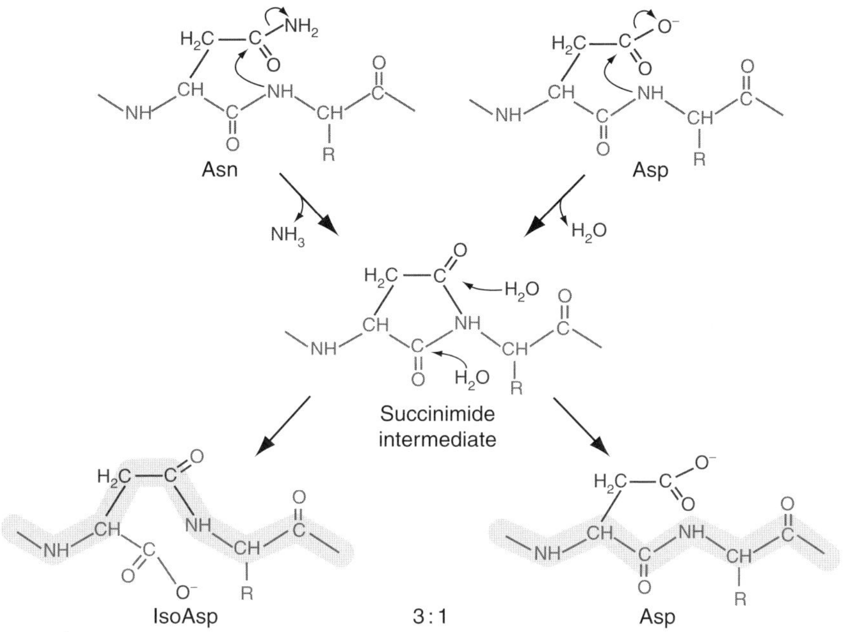

FIGURE 3 **Deamidation of asparagine by succinimide formation. L-aspartic acid and L-isoaspartic acid residues are the products of L-asparagine degradation via a succinimide intermediate. Aspartic acid residues can also form the succinimide intermediate by a similar mechanism with a loss of water. Figure based on that of Clarke et al.[94]**

Asp side chain. This results in the formation of a five-membered succinimide ring and the loss of ammonia or water from Asn or Asp, respectively. The succinimide typically hydrolyzes to Asp and isoAsp in a 1:3 ratio. Cleavage of the peptide bond can also occur at Asn residues.

Solution conditions and protein structure determine the rates of these reactions. Nucleophiles, including phosphate and carbonate anions as well as Tris, and basic pH can accelerate the reaction. The structure of the amino acid side chain in the amino acid residue C-terminal to the Asx residue (Asx = Asn or Asp) has a large effect on the rate of degradation. Asx followed by Gly, Ser, or His are the most reactive, while Asx with large bulky side chains on the C-terminal side are the least reactive.[95] Reactions at Asx are also more prevalent in flexible regions of protein sequence[96] and higher-order structure can protect certain Asx from degradation. For example, deamidation of Asn[67] in native ribonuclease A is 30 times slower than the rate of deamidation of this residue in reduced and denatured protein.[97]

Glutamine can also undergo deamidation in a reaction analogous to that of Asn, with a six-membered glutarimide intermediate. The rate of deamidation at Gln residues is significantly slower than that at Asn residues, probably because of the relative instability of the six-membered ring intermediate when compared with the succinimide intermediate.[98] In contrast to the low rate of deamidation of internal Gln residues, deamidation of N-terminal Gln is much faster. The product of this irreversible reaction, pyroglutamic acid, has a stable five-membered ring[99] (Figure 4).

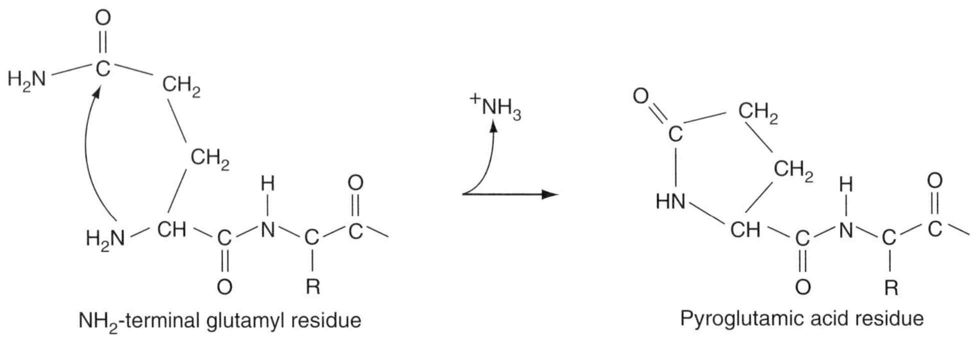

FIGURE 4 **Pyroglutamate formation occurs when a glutamine at the NH_3-terminus reacts with amine groups.**

Pyroglutamate formation should not strictly be categorized as a degradation event, since this amino acid residue imparts aminopeptidase resistance to the protein[93] and the presence of glutamine or pyroglutamic acid in the N-terminal position of many proteins is a naturally occurring source of heterogeneity.

Peptide mapping is usually capable of detecting deamidation products as new peaks (see Section II.A.1.a.). Most of the deamidation products (Asn → Asp, Gln → Glu, N-terminal pyroglutamate) also lead to a change in the net charge of the protein, and can be detected by ion-exchange chromatography.[39] Asp isomerization can be detected by a commercially available kit IsoQuant (Promega), which uses the enzyme isoaspartyl methyltransferase to detect isoAsp and generate a stochoimetric coproduct quantifiable by HPLC. Hydrolysis products can usually be detected by RP-HPLC or SE-HPLC.

b. Oxidation

Oxidation of methionine (Met) to methionine sulfoxide (MetO; Figure 5) is the major oxidative degradation pathway in proteins, however, MetO was found in only 11 of 73 proteins in a survey of protein degradation[93] and is much less prevalent than degradation at Asx.

Oxidation of Met residues has significantly affected biological activity in some instances (subtilisin, *E. coli* ribosomal protein L12), but has shown no effect in others (ribonuclease, Kunitz trypsin inhibitor). As is the case with degradation at Asx, the rate of oxidation at select residues within a protein is dependent upon higher-order structure, presumably because of solvent inaccessibility or steric hindrance.[100] However, unlike the case of Asx degradation, there are no primary sequence or other structural motifs found to strongly correlate with oxidation of Met.[93] Other amino acid residues susceptible to oxidation are Cys, His, Trp, and Tyr.

Given the sensitivity of Met to oxidation compared with that of other amino acid residues, it is important to be able to routinely identify Met oxidation during manufacture and stability studies of active substance and vialed protein, and to understand its effect on bioactivity. In forced degradation studies, the protein biopharmaceutical may be exposed to low levels of oxidants, such as hydrogen peroxide, peracetic acid, or bleach, to evaluate the sensitivity of the protein in question.

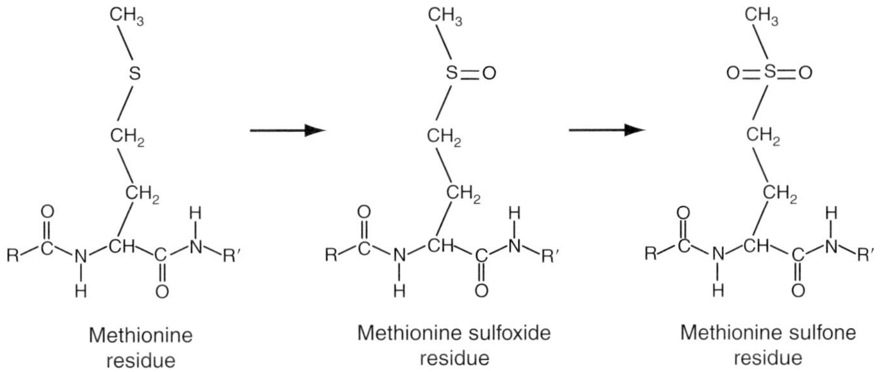

FIGURE 5 Oxidation of methionine leads to methionine sulfoxide. Methionine sulfone is rarely found in protein pharmaceuticals.

Oxidation is usually detectable by peptide map and MS (see Section II.A.1.a). In some cases, the oxidized protein can be separated and quantitated using RP-HPLC without enzymatic digestion.[101]

c. Glycation and Other Modifications

The manufacturing process, modificationsas well as the formulation, handling and storage of a biopharmaceutical protein can lead to various other modifications. Some examples include:

- Misincorporation of amino acids during translation of the recombinant DNA.[102,103] Misincorporation of amino acids often leads to the mass change that is detectable by MS analysis of the intact protein or peptide mapping.
- Urea used in purification steps leading to carbamylation of amino groups of the protein, especially in the N-termini.[104] Carbamylation can also be detected by the peptide map, as a change in retention time of the modified protein as well as a mass change.
- Hydrolysis of peptide bonds, especially if low pH is encountered in process or during storage. The resulting fragments can be detected by the mass analysis of the intact protein, and depending on the size may also be detectable as low molecular weight (LMW) species in size-exclusion chromatography (SEC).
- Glycation (nonenzymatic glycosylation) as a result of glucose in cell culture medium,[105,106] or glucose from hydrolysis of sucrose in the formulation buffer.[107,108] Glycation can usually be detected by a combination of intact protein or subunit MS analysis and peptide mapping (see II.A1.a). Boronate chromatography has been used to separate glycated protein from unglycated protein.[105,106]

B. Higher-Order Structure and Folding

In order to perform their functions, the linear polypeptide chains of the majority of proteins must assume specific and stable three-dimensional structures. This process by which this conformational transformation takes place is referred to as protein folding. Typically, a folded protein has certain general characteristics: the hydrophobic amino acid side chains cluster in the interior of the protein (sequestered away from aqueous solvent) and the charged/polar residues populate the surface of the folded protein. This increases protein solubility while providing a hydrophobic core, which is likely to remain tightly packed under physiological conditions.

There are several structural "tiers" that exist in a folded protein, sometimes referred to as the hierarchy of protein structure. Although different types of proteins can fold in very different ways, this structural hierarchy can be applied generally to all proteins: primary, secondary, tertiary, and quaternary structure.

1. Structural Hierarchy of Proteins

a. Primary Structure

As discussed in Section II.A.1, primary structure is simply the order in which various amino acids are covalently linked (via peptide bonds) to create a large linear molecule, sometimes called the polypeptide chain. The primary structure also has an element of directionality, wherein the first amino acid is at the amino (NH_2) terminus and the last is at the carboxy (COOH) terminus.

b. Secondary Structure

This refers to the three-dimensional arrangements of the polypeptide chain, which are primarily stabilized by local, peptide backbone interactions. There are two very commonly occurring secondary structural elements: α-helix and β-sheet.

In the α-helix the polypeptide chain forms a right-handed helix, stabilized by a network of peptide backbone hydrogen bonds which run roughly parallel to the helical axis (Figure 6).

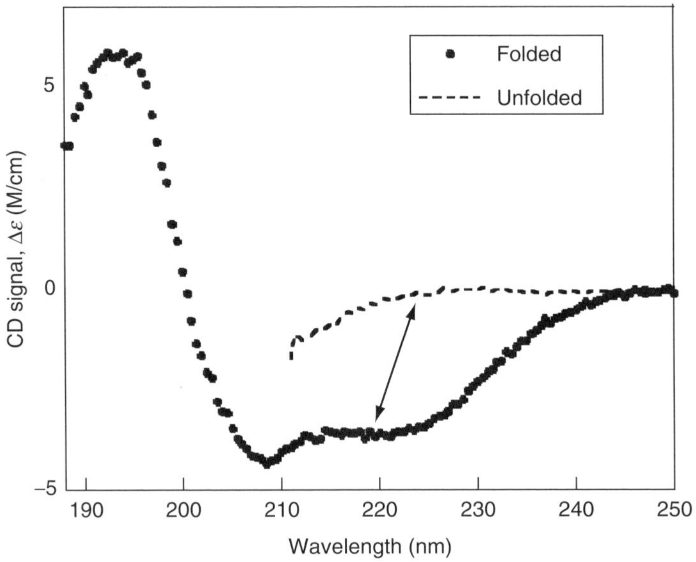

FIGURE 6 Far-UV CD spectra of an α-helical protein. Far-UV circular dichroism (CD) is a convenient method to assess a protein's secondary structure in solution. The spectra are of the folded and unfolded states or the protein.

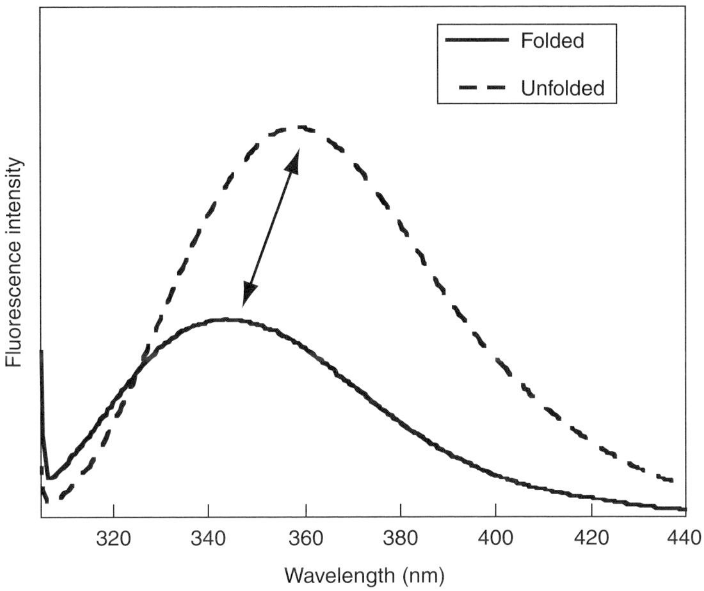

FIGURE 7 **Fluorescence emission of a folded and unfolded protein. Fluorescence emission of tryptophan side chains can be to used monitor a protein's tertiary structure in solution. Tryptophan residues often have very different emission maxima depending on whether they are buried in a folded protein or exposed to solvent (unfolded protein).**

In the β-sheet the polypeptide chain assumes a significantly extended conformation; two or more segments of the peptide chain (called β-strands) are aligned next to each other and stabilized by a network of hydrogen bonds which run perpendicular to the direction of the polypeptide segments. β-Sheets can exist in two major subclasses, parallel and antiparallel, referring to the relative direction (N- to C-term) of adjacent strands.

It's worth noting that several other forms of protein secondary structure exist, but none are as well defined or commonly observed as the α-helix and β-sheet.

c. Tertiary Structure

Tertiary structure describes the ultimate three-dimensional conformation of a folded polypeptide chain. Major elements of tertiary structure typically involve longer range backbone interactions: various elements of secondary structure are associated, resulting in the formation of structural or functional domains, and the amino acid side chains are packed in a well-defined manner. The side chains may stabilize the protein structure through a variety of interactions, including the formation of cystinyl disulfide bonds (Figure 7).

d. Quaternary Structure

By definition, only multimeric proteins have quaternary structure; it refers to the arrangement of two or more folded polypeptide chains. The interactions stabilizing protein quaternary structure may be covalent or noncovalent. Although other forms of covalent side chain interactions have been observed, the most common is the cystinyl disulfide bond.

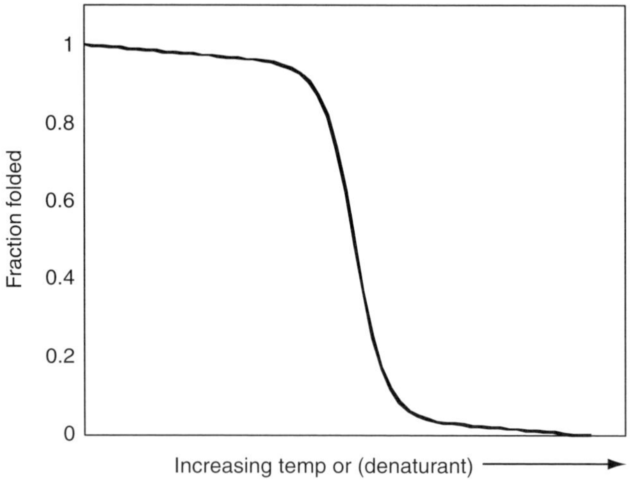

FIGURE 8 **Idealized protein denaturation curve. A discrete unfolding transition is often observed with increasing stress of a folded protein. The position of this transition (temperature or denaturant concentration) provides a measure of stability of the folded state, relative to the unfolded form.**

There are many ways to assess the structure of a folded protein. Many of them are spectroscopic, such as circular dichroism (CD), fluorescence, infrared, and Raman. These methods can be combined with the use of stress to measure the stability of the folded structure by perturbing it with heat or chemicals (Figure 8). High resolution structural information (determined in atomic detail) is often obtained by X-ray crystallography or nuclear magnetic resonance (NMR). These data are routinely deposited in the RSCB (Research Collaboratory for Structural Bioinformatics) protein database for public access.[109]

In addition to biologically relevant self-association processes (quaternary structure), proteins are prone to undergo the aberrant process of aggregation which is often linked to folding and misfolding. Aggregation can have profound effects on protein production and the final product. Many methods have been developed to assess and understand these processes and they are discussed in more detail below.

2. Stability-Related Structural Changes

a. Unfolding

A pharmaceutical protein is constantly exposed to stresses in the manufacturing process that may lead to unfolding of the protein molecule to a nonnative state. Examples of these stresses are extremes of pH and ionic strength, elevated temperature, shear strain, and adsorption to surfaces. Stress-induced unfolding is highly undesirable since it can lead to degradation and destabilization of the protein molecule. Unfolding can initiate degradation by increasing in the accessibility of buried peptidyl, methionyl, tryptophanyl, and disulfide bonds to the aqueous environment. In their new environments, these residues are more susceptible to a number of chemical modifications (such as hydrolysis of peptide bonds, oxidation of methioninyl and tryptophanyl side chains, and cysteinyl-assisted scrambling of disulfide bonds) than in the native protein. Protein unfolding can also uncover hydrophobic and cysteinyl residues and increase their accessibility to the aqueous environment. The exposed hydrophobic and cysteinyl residues can then participate in the formation of noncovalent contacts and disulfide cross-links, respectively, between molecules of the unfolded protein. This is considered to be a key event in the formation of some but not all types of protein aggregate during the manufacturing process.[110]

b. Protein Aggregation

Protein aggregation is a major concern of the biotechnology industry since it may lead to adverse changes in the potency, activity, and safety of a therapeutic protein. The detection and quantification of aggregates in protein drug substances and drug products are therefore of paramount importance to the development of protein pharmaceuticals. SEC is routinely employed by the industry as a quantitative release assay for protein aggregates. However, SEC has a number of drawbacks that may lead to an inaccurate assessment of the quantity of aggregates in a therapeutic protein.[111] For instance, one of the weaknesses of SEC is that protein aggregates may be filtered out of the flow stream by the column bed packing or column frits. In a case such as this, SEC would fail to detect all the aggregates that were present in the sample. This limitation of SEC has not escaped the attention of the regulatory agencies who now request that matrix-free orthogonal methods, such as analytical ultracentrifugation-sedimentation velocity (AUC-SV), batch-mode dynamic light scattering (DLS), or asymmetric flow field flow fractionation (afFFF), be used to detect aggregates that SEC might have missed. AUC-SV, DLS, and afFFF cover a wider range of aggregate sizes than SEC and are capable of detecting irreversible aggregates. AUC-SV and DLS are, however, better suited to detecting reversible aggregates than either SEC or afFFF. Some of the attributes of these three orthogonal methods are summarized in Table 2. The papers referenced in the table provide a fuller description of these methods along with examples illustrating their application to pharmaceutical proteins.

III. CONCENTRATION

A. Introduction

Protein concentration is the most fundamental and one of the most important analytical end-points for a protein pharmaceutical. It is used for dosage determination, as a basis for specific biological activity calculations, to normalize sample concentrations for many release assays, to establish process yields, and it can be stability-indicating.

Of primary concern is consistency of dosing at all stages of development, from animal toxicology studies through clinical trials and on to the commercial market. The dosage for most protein pharmaceuticals is based on mass of protein delivered. Therefore, having a precise, routine protein assay is paramount for

TABLE 2 Methods Orthogonal to SEC for the Measurement of Protein Aggregates

Method description	Method strengths	Method limitations
AUC in the sedimentation velocity mode (AUC-SV). AUC-SV measures the rate at which proteins sediment in a rapidly spinning rotor. The rate of sedimentation is expressed as a sedimentation coefficient, which depends on both molecular mass and shape. Proteins form concentration gradients, known as boundaries, as they sediment toward the base of a sector-shaped centrifuge cell. An optical detection system based on absorbance, Rayleigh interference, or fluorescence is used to record the movement of the boundaries at regular intervals of time. The data from an AUC-SV experiment are then used to calculate a continuous distribution of sedimentation coefficients known as $c(s)$. The resulting $c(s)$ distribution provides a measure of the number, quantity, and size of the sedimenting proteins.[111–114]	• Matrix-free, solution-based method, no protein-column interactions as in SEC. • AUC-SV is an absolute method that is based on physical first principles. Protein size markers are not needed. • Can assess aggregates over a wider range of buffer conditions than SEC. AUC-SV can be performed in low ionic strength formulation buffers as well as in higher ionic strength SEC mobile phase. AUC-SV is therefore useful for detecting differences in protein size distribution between different buffer conditions. • Can assess a wider range of protein sizes in a single analysis than SEC. For example, AUC-SV can potentially detect a 12 MDa aggregate (80-mer mAb) at a single rotor speed of 40,000 rpm. Even larger aggregates can be detected by employing multi-speed AUC-SV.[115]	• The c(s) distribution is calculated with a curve-fitting method. The precision of the information contained in the resulting distribution is therefore highly dependent on the precision of the fitted parameters. • Low accuracy and precision in measuring low levels of small-sized aggregates (dimer–octamer). • AUC-SV is time-consuming and labor intensive and sample throughput is low. • High level of expertise is required. The precision and reliability of the results is highly dependent on understanding subtleties in the theory and method. • Dynamic density and viscosity gradients resulting from the sedimentation of formulation excipients, such as sucrose, may limit the ability of AUC-SV to detect and quantify minor protein species. • AUC-SV equipment and data analysis software are difficult to validate for use in a quality control laboratory.
Dynamic light scattering (DLS) in batch mode A solution of particles is illuminated with light from a laser and the scattered light is detected. The intensity of the scattered light fluctuates on a time scale that depends on the rate of diffusion of the scattering particles. The intensity fluctuations are measured by fast photon counting and processed into an autocorrelation function. This function relates the intensity fluctuations to the time scale of the particle motions. The autocorrelation function can then be analyzed to produce a distribution of the hydrodynamic or Stokes' radii of the scattering particles.[111,112,116]	• Matrix-free, solution-based method, no protein-column interactions as in SEC. • Minimal or no sample preparation required. High concentrations and turbid samples can be analyzed. • Low sample volume requirements ($<20\ \mu L$). • DLS does not involve dilution or separation of the protein species in a sample. Thus, dissociable as well as nondissociable aggregates can be assessed. • DLS can assess aggregates over a wider range of buffer conditions than SEC. However, the distribution of aggregates in one buffer may differ from that in another. • DLS is considerably more sensitive to trace levels of large aggregates than SEC or AUC-SV.	• Size distributions are calculated by curve-fitting mathematical models to the autocorrelation function data. • High level of expertise is required. The precision and reliability of the results is highly dependent on understanding subtleties in the theory and method. • Low size and mass resolution. Species must differ by a factor of 2 in radius or by a factor of 8 in mass in order to be resolved by DLS. • Dust or other particulates are interferences that may be mistaken for large protein aggregates. • Weight fractions of protein species determined from the intensity distribution are neither accurate nor precise.

	• DLS is capable of detecting protein species that vary in size from a few nanometers to a few microns in a single analysis. • High throughput can be achieved with a DLS instrument equipped with a microtiter plate reader.	• Strong scattering from large aggregates may mask the presence of smaller species that scatter light weakly.
Asymmetric flow field flow fractionation (afFFF) Proteins are separated on the basis of size in a narrow channel formed between two flat, parallel plates. The upper plate is impermeable to liquid whereas the bottom plate consists of a porous frit covered by a semipermeable membrane. Mobile phase buffer enters the channel through an inlet and exits through an outlet. Prior to sample injection, the flow of buffer from the outlet is reversed to oppose the flow of buffer from the inlet. The sample is then injected into the channel and focused into a narrow band by the two opposing flows. Protein molecules within the band are concentrated near the surface of the membrane by the perpendicular cross flow of buffer. The movement of the protein molecules toward the membrane is, however, opposed by diffusion. The net effect is that the smaller molecules will partition farther from the membrane than the larger ones. A continuous flow of buffer is then restored to elute the protein molecules along the channel and into a detector. Because the flow of buffer in the channel is laminar (i.e., flow is faster near the center of the channel than at the edges), the smaller molecules will elute ahead of the larger ones. The elution profile can then be monitored with an absorbance, fluorescence, or refractive index detector either alone or in tandem with a light scattering detector. The latter is used to determine the molecular masses of the eluted proteins.[114,117,118]	• Matrix-free, solution-based method, no protein-column interactions as in SEC. • afFFF does not use frits which may trap protein aggregates. • afFFF has a broader size separation range than SEC. Molecules ranging in size from a few nanometers up to 0.5–1 μm can be assessed. Minimal disruption of protein aggregates due to shear.	• Development of afFFF methods involves the optimization of many operating parameters. • Adsorption of proteins and their aggregates to the membrane will result in poor recovery. • Low ionic strength formulation buffers may promote the adsorption of protein onto the membrane or other surfaces. This would then necessitate the use of a higher ionic strength mobile phase buffer. The drawback to this is that the distribution of aggregates determined in the high ionic strength buffer may not be the same as that in the lower ionic strength formulation buffer. • afFFF is not well suited for assessing aggregates whose size distribution depends strongly on protein concentration. Two reasons for this are (1) aggregates may be generated as the protein is concentrated in the focusing step; and, (2) aggregates become increasingly dilute as they elute along the channel. Thus, they may dissociate into smaller species. In either case, afFFF will not yield an aggregate size distribution that is representative of the original sample.

safety and efficacy of the product. Accuracy of the protein concentration determination, both for establishment of product reference material and day-to-day standard curves, if needed, is of lesser importance. In other words, it is more important to maintain consistency of dosage throughout the development program to link early nonclinical results to clinical results and maintain assay relevance.

It is important to establish a clear, logical protein concentration determination strategy at the beginning of the development program covering both reference material and routine assays. For example, the protocol and analytical procedure used to establish the protein concentration of reference materials for a particular product should be clearly defined by subject matter experts, and efficiency can be gained by making this a platform for all products. Typically, the mass extinction coefficient is determined for a product reference material by obtaining the slope of the plot of absorbance at 280 nm versus protein concentration (in mg/mL) determined by AAA.[119] This absorptivity is then used in routine protein concentration determinations using absorbance at 280 nm.

A seemingly simple decision can actually be somewhat complex if the product comprises multiple posttranslational modifications and protein sequence isoforms where there is no predominant species and lot-to-lot variations in isoform content are significant. Molecular mass is used to calculate the mass extinction coefficient, and can be based on just the amino acid backbone, or, the amino acid backbone plus posttranslational modifications. It might be best to ignore the posttranslational modifications and calculate molecular mass solely on the amino acid backbone if the modifications vary greatly from batch to batch. For a product with significant lot-to-lot variation in posttranslational modifications or protein sequence isoform content, where there are multiple major isoforms, a weighted average molecular mass might be used if the isoform distribution can quantitatively be determined. It is important to determine acceptable concentration errors up front, as the variation in isoform content might be larger than the protein assay variability. A useful exercise would be to calculate the protein concentrations of hypothetical mixtures of isoforms that represent reasonable extremes and the protein concentration calculated using the molecular mass of one "average" isoform, and determine if the difference from the concentration calculated using the exact weighted average molecular masses of the mixtures could be tolerated. This simplifies routine protein concentration determinations.

B. Methods

All protein concentration methods have sources of error. Colorimetric methods rely on protein standards that may not be representative of the target product protein's response in the assay. AAA and absorbance methods are more direct than colorimetric methods, however, AAA does rely on amino acid standards, and absorbance methods rely on the correspondence of extinction coefficients of model compounds with those of certain amino acid residues within the target protein. Additionally, AAA may suffer from environmental contamination from the laboratory, whereas absorbance methods may suffer from contaminating compounds that absorb light at the wavelength(s) used. Colorimetric results can be influenced by contaminating protein or formulation excipients.

1. Amino Acid Analysis

AAA affords an absolute quantitative measure of protein content of a sample independent of an external protein reference standard, based on the content of the test article's constituent amino acids. The method is not dependent on the protein's charge or higher-order protein structure, or dye-binding capacity. Automated Edman sequencing and MS techniques have largely supplanted the use of AAA for protein identification. The major application of AAA in the modern biotechnology laboratory is in the determination of protein absorptivity constants, to serve as a basis for simple, routine measurements of protein concentration at 280 nm. However, there are instances in which AAA can serve as a routine concentration assay, such as in cases where a peptide or protein pharmaceutical contains little or no 280 nm-absorbing residues (Trp, Tyr, cystine), or where a particular modified amino acid needs to be quantified in chemically modified peptides, proteins, or conjugates.[120] In addition, AAA can be useful for protein quantitation in cases where UV-absorbing prosthetic groups (covalent or noncovalent) or formulation excipients interfere with UV measurements. Perhaps a less labor intensive strategy is to initially establish a concentration for the peptide or protein reference material batch by AAA, and then use this reference material in the standard curve for a routine protein concentration assay, such as absorbance, or a colorimetric assay.

There have been many AAA procedures developed over the years.[121] The gold standard remains the postcolumn labeling method of Stein and Moore.[122] This method involves protein hydrolysis, cation-exchange chromatography of the released amino acids, followed by in-line postcolumn labeling with ninhydrin. Postcolumn methods are less affected by sample buffer constituents and sample handling variability than procedures based on precolumn labeling. The major drawback of the ninhydrin postcolumn labeling AAA for a lab that does not routinely perform the procedure is the additional cost and upkeep, and lab space requirements of the method-dedicated equipment. In contrast, AAA procedures based on precolumn labeling techniques can be performed on standard HPLC instruments, which allows flexibility in labs that do not perform AAA on a regular basis.

We have found that with high concentration protein products, typically containing 10–50 mM amino acid buffer excipients, the large dilution of the drug substance with water prior to acid hydrolysis results in sufficiently low buffer concentration as to not impact the stoichiometry of the labeling reaction. Of course, the particular amino acid used as a buffer is ignored when calculating the protein concentration, but this is a small price to pay for the convenience of the method. The only sample manipulation is dilution (likely required for any AAA procedure), and sample desalting is avoided.

We find that the precolumn labeling AccQ-tag (Waters, Corp.) method is a good compromise for labs with a need to determine a reference material protein concentration on an infrequent basis. The labeling procedure is relatively simple, with very little sample manipulation and a short reaction time. This, combined with the high resolution and rapid analysis time of a UHPLC system, allows for a relatively rapid, accurate, precise, and convenient method (See Figure 9).

2. Intrinsic Protein Absorbance

As stated above, typically methods for determining a protein's absorbtivity involve the use of model compound extinction coefficients, and assuming those

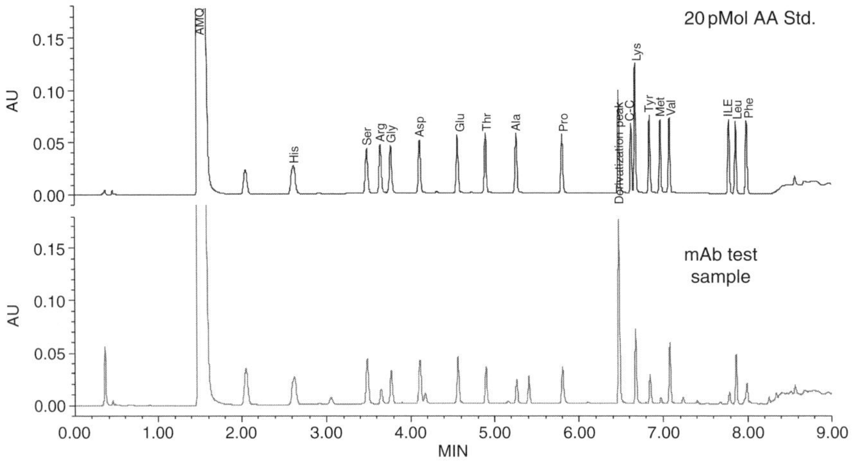

FIGURE 9 Amino acid analysis of a monoclonal antibody (mAb) using liquid-phase acid hydrolysis (6 M HCl), followed by AccQ-tag labeling and UHPLC separation. The top panel shows a chromatogram of amino acid standards. (See Color Insert.)

absorbance properties are similar to the corresponding amino acid residues in an unfolded protein.[123,124] The specific amino acids monitored for concentration determination are tryptophan, tyrosine, and to a lesser extent, cystine (disulfide bonds). Proteins that contain tryptophan or tyrosine residues absorb light significantly in the near-UV region, around 280 nm.

If the amino acid sequence is known, the theoretical molar extinction coefficient at of an unfolded protein ($\varepsilon_{\mathrm{unfold}}$) at 280 nm can be calculated as follows:

$$\varepsilon_{280,\ \mathrm{unfold}} = x(5690) + y(1280) + z(120)$$

where 5690, 1280, and 120 are the molar extinction coefficients (in units of $M^{-1}\,cm^{-1}$) of tryptophan, tyrosine, and cystine disulfide residues, respectively; x, y, and z represent the number of each residue in the protein. The amino acid extinction coefficients were determined in 6 M guanidine-HCl, unfolding conditions for most proteins.[123] If identical dilutions of a protein are made, in 6 M guanidine-HCl and in native buffer, the absorbance at 280 nm (A_{280}) can be measured and used to calculate the extinction coefficient of the native protein using the relationship below[124]:

$$\varepsilon_{280,\mathrm{fold}} = (A_{280,\mathrm{fold}})\ (\varepsilon_{280,\mathrm{unfold}})/(A_{280,\ \mathrm{unfold}})$$

where $\varepsilon_{280,\mathrm{fold}}$ and $A_{280,\mathrm{fold}}$ are the extinction coefficient and absorbance, respectively, of the folded protein. $A_{280,\mathrm{unfold}}$ is the absorbance of the unfolded protein (typically in 6 M guanidine-HCl).

In practice, however, $\varepsilon_{280,\mathrm{fold}}$ is usually within 10% of the calculated $\varepsilon_{280,\ \mathrm{unfold}}$; therefore, $\varepsilon_{280,\mathrm{unfold}}$ often provides a very good approximation for the extinction coefficient of the native protein.[124,125] Based on empirical data, Pace et al.[125] put forth a set of corrected extinction coefficients to represent tryptophan, tyrosine cystine disulfide bonds in a folded protein. These values may also be used to estimate a protein's extinction coefficient from its sequence, but ultimately this value is still an approximation and the values determined by Edelhoch remain the more widely used convention.

When dealing with proteins it is often useful to deal with concentrations in terms of mg/mL as opposed to mol/L. The molar extinction coefficient, ε_{280}, which has units of M/cm can be converted a specific absorptivity (a_{280}), with units of mL/mg cm^{-1}, to be used for concentration determination in mg/mL. To convert a protein's ε_{280} to a_{280} simply divide by its molecular weight (Mr), as shown below:

$$a_{280} = \varepsilon_{280}/\mathrm{Mr}$$

Consider the example of lysozyme: the protein has a mass of 14,308 g/mol, and contains six tryptophans, four tyrosines, and four disulfide bonds.

$$\text{The } \varepsilon_{280} \text{of lysozyme} = 6(5690) + 4(1280) + 4(120) = 38,460\,\mathrm{M}^{-1}\mathrm{cm}^{-1}.$$

If a lysozyme solution has an $A_{280} = 1$, the molar concentration is 1/38,460 = 26 μM. The a_{280} for lysozyme is 38,460/14,308 = 2.7 mL/mg/cm, so for the same lysozyme solution ($A_{280} = 1$), the concentration is 1/2.7, or 0.37 mg/mL.

Though many have a much broader range, most spectrophotometers work optimally within the range of 0.1 to just over 1 absorbance unit. And it is conventionally assumed that the sample path is 1 cm when collecting absorbance data. This optimal spectroscopic range, and the use of conventional path-length cells, has often limited the working range of the absorbance method. However, recent instrumental developments allow the use of droplets (a few μl, instead of 100 s μL to mL volumes) to accurately measure protein concentration when sample volumes are limited. Additionally, instruments have been developed to easily change path length "on the fly," effectively increasing the dynamic concentration range of accurate absorbance measurements from very low (10s of μg/mL) to very high (100s of mg/mL).

Although it is often more convenient, and certainly faster, to only collect absorbance data at 280 nm, it is informative to collect a wavelength spectrum (from 250 to 350 nm) to help assess whether the protein contains UV-absorbing contaminants, protein aggregates, or other particulates.

Some proteins, such as collagen and HPr, do not contain tryptophan or tyrosine residues;[126,127] therefore they do not intrinsically absorb light at 280 nm. In these cases, it will be necessary use either AAA, refractive index (n) detection, backbone absorbance (<240 nm), or a colorimetric method (see below). When using refractive index detection, the incremental change in n with concentration (dn/dc, a constant analogous to a protein's extinction coefficient) must be known. The determination of dn/dc requires the generation of a standard curve (n vs. concentration); the concentration must be previously determined by different method. It quickly becomes evident how circular the determinations can be, with one method relying on the outcome of another.

The absorbance at 280 nm is convenient for monitoring protein concentration because it is broad, relatively insensitive to protein conformation and not overly prone to background interference. In contrast, the protein backbone absorbs strongly in the far-UV region (below 240 nm), but these absorbance bands are highly dependent on protein conformation and very susceptible to background interference (many molecules, such as salts and common buffer excipients, absorb significantly in the far-UV). Use of the absorbance method assumes that the protein is not only free of UV-absorbing contaminants, but is analyzed in a solvent with little or no absorbance. However, it is possible to remove the interference from excipients by chromatography which separates protein and small-molecule excipients (SEC and RP-HPLC are often used), and to monitor the backbone absorbance in the far-UV region. The concentration can then be determined from the total peak area based on a standard curve using a reference material whose concentration is determined by a primary method such as AAA (see above).

3. Colorimetric Methods

There are many cases when it may be more convenient, or even necessary to use a colorimetric assay. These methods can be used when a protein's extinction coefficient is not known (or has no Trp/Tyr residues), or is not free of absorbing contaminants. These methods are also useful when trying to nonspecifically determine total protein concentration. In general, they all rely on an exogenous molecule forming a complex with the protein in solution, which leads to an absorbance/fluorescence signal proportional to the protein concentration.

The most common of these assays are the Bradford, Lowry, and bicinchoninic acid (BCA) methods. In addition to these classical absorbance-based methods, convenient fluorescence-based methods are becoming increasingly more popular.[128] Some salient aspects of these assays are summarized in Table 3.

The Bradford assay monitors a color change associated with the binding of Coomassie brilliant blue G-250 dye (CBBG) to protein in solution; the unbound dye absorbs at 470 nm, while the complex absorbs at 595 nm. Unfortunately, not all proteins bind to CBBG equally, because of differing amino acid compositions. Therefore, not all proteins show the same response in a Bradford assay.

The Lowry and BCA methods are related, in that the first step of the assay is the reduction of copper ion, Cu^{2+} to Cu^{1+}, by protein amides under alkaline conditions. In the Lowry method the reduced copper—and to a lesser extent some protein side chains—react with the Folin-Ciocalteu reagent (phosphomolybdate/phosphotungstate) to produce a blue color. In the BCA assay the BCA complexes with the Cu^{1+}, which absorbs strongly at 562 nm. Although both the Lowry and BCA methods must be carefully timed and are subject to interference from buffer components, the BCA method is less susceptible to interference, especially by detergents.

TABLE 3 **Common Colorimetric Protein Quantitation Assays**

Assay method	Suggested range	Wavelength analyzed	Linear response	Approximate test volume
Bradford assay	1–10 μg	595 nm	No	200 μL
Lowry assay	1–20 μg	660 nm	Yes	200 μL
BCA method	0.5 to >200 μg	562 nm	Yes	200 μL
Fluorescent dye-binding	1 ng to 10 μg	Varies with dye	Depends (often no)	0.2–2.5 mL

Fluorescence-based dye-binding assays work by complexation of the dye to unfolded proteins; once bound, the dye's fluorescence increases greatly. Although usually not linear in response to protein concentration, the fluorescent dyes provide the advantages of procedural ease and a long-lived signal. Proteins may respond differently based on relative hydrophobicity (the more hydrophobic, the more dye binding/signal), but there is not a specific bias with respect to amino acid composition.

All of the colorimetric methods described above, share the disadvantage that they require a standard curve to be generated for data interpretation. Since all of these reagents will react differently with different proteins, it is difficult to know if some generic protein (bovine serum albumin is commonly used) will accurately reflect the response of the protein one is attempting to quantify. With this in mind, one should endeavor to use a protein similar to the protein being analyzed when possible.

A common advantage that all of these methods have is the widespread commercial availability of kits, that include all reagents, reference standards, and optimized protocols. In addition, these methods are easily adapted for use in high throughput, automated fashion. Therefore, many samples and corresponding standard curves can be prepared on a multiwell plate using robotics, and then read quickly using plate-reading spectrophotometers.

IV. PURITY

When developing release tesing and release specification of a protein pharmaceutical, appropriate methods must be identified to determine the purity of the drug substance. In addition, specific impurity species that are likely to be present in the drug substance must be controlled with acceptance criteria. The ICH guidance for impurities in biopharmaceuticals is found in Q6B (Section 4.1.3); impurities in drug substances are classified as either process-related or product-related.[3] The former group of impurities include any nondrug substance material which originated from cell culture (e.g., HCPs, DNA), microorganisms, viruses, purification chromatographic media, solvents, and buffer components. The latter is product-related, namely, "the molecular variants with properties different from those of the desired product formed during manufacture and/or storage."

This section will outline commonly encountered impurities, and the various analytical methods used to monitor their levels in the drug substance.

A. Product-Related Impurities

Product-related variants may arise from the manufacturing, handling, and storage of a biological drug substance.[129,130] These molecular variants of the desired product include precursors, degradation products that may or may not have biochemical or biophysical properties comparable to those of the desired product with respect to biological activity, clinical efficacy, and safety. Conversely, product-related substances are molecular variants or isoforms of the desired produced form, and are not considered impurities.

In defining product variants, separation of the product based on physiochemical properties (e.g., charge heterogeneity profiling on ion-exchange chromatography, electrophoretic mobility in gel/CE, or hydrodynamic size distribution in SEC) must be done. In each of these instances, quantitative and sensitive analytical methods using various separation modes of HPLC, gel, and CE are applied to identify and characterize product-related variants.[131–134]

To classify a particular variant as product isoform versus product-related impurity, further functional characterization is often required during the development phases. Evaluation of a variant includes assessment of impact on biological activity and/or efficacy and safety. Determination of comparable activity depends mainly on the selectivity of the potency assay to distinguish active from inactive variants and its relevance to the physiological activity of the product-related variant. Care must be taken to use the appropriate biological assay or a combination of biological assays, for instance, a binding assay (e.g., SPR), which may be biased toward binding kinetics of different variants be used in combination with a physiological relevant functional assay. Other considerations in assigning product-related impurities include potential immunogenicity, and impact on stability. Further examples of product-related impurities and the corresponding analytical methods are given below.

Product-related aggregates and oligomers are often deemed as impurities as they are removed by the purification process and drug formulation and storage conditions are optimized to reduce its formation.[135,136] The method of choice is SEC with analytical ultracentrifugation (AUC) as an orthogonal verification method (for more details, see Section II.B.2.b). The integrity of the protein structure can be affected by chemical or enzymatic degradation or fragmentation during the manufacture, handling and/or storage. Methods such as SDS-CE or PAGE, N-terminal sequencing or peptide mapping in combination with MS can determine the changes in primary structure, such as cleavage site.[137–139] Charge isoforms, such as that which result from sialylation, carbamylation, deamidation, N-terminal glutamic acid cyclization, or incomplete C-terminal lysine processing by the expression host system can be determined via weak ion-exchange chromatography and/or peptide mapping with MS.[132]

B. Process-Related Impurities

1. Leachables from Affinity-Based Chromatography

Protein A chromatography is the most often used affinity-based chromatography in the purification of therapeutic mAbs, and Fc-containing fusion proteins.[140] It is one of the most efficient and costly chromatographic purification method in the biologics manufacturing process.[141,142] Despite the high cost, the speed and effectiveness of protein A chromatography in impurity removal (typically >95%) is truly remarkable. In combination with other appropriate chromatographic modes, protein A chromatography purified drug substances often achieve pharmaceutically acceptable purity levels.[143] Indeed, the versatility and efficiency of protein A chromatography capturing product from a broth of cell debris and culture additives, makes it the purification method of choice by the industry.[144]

Protein A, a cell wall component of *Staphylococcus aureus* (recombinant forms or native), binds specifically and tightly to the Fc region of mAbs and is often covalently immobilized as a ligand to a variety of support matrices (Sepharose or other equivalent nonprotein sorbent matrices).[145,146] The protein A ligand, despite its covalent immobilization and stability in low pH conditions, is susceptible to proteolysis from nonspecific protease degradation (from cell culture) and matrix shedding as the result of low level acid hydrolysis (from exposure to low pH release of the captured Fc-containing biologic or high pH during cleaning-in-place procedure) or chaotropes.[147] The released protein A fragments are seen as a process-related impurity and must be removed.

The health effects of protein A as an impurity, and potential immunocomplexes resulting from protein A induced IgG aggregation in humans are not fully known or well documented, though past reports have indicated that protein A may have immunogenic[148] and mitogenic effects.[149] Current FDA guidelines (Points to Consider in the Manufacture and Testing of Monoclonal Antibody Products for Human Use, 1997) require that mAb drug substance be monitored using a highly sensitive analytical method to ensure a safe level of contaminating recombinant protein A. In addition, effective methods are essential to demonstrate the removal of protein A during process development. The test method should be direct, highly sensitive, relatively quick, and robust.

Protein A, by virtue of its smaller size and charge, may be detected by common electrophoretic or hydrodynamic separation. However, the detection modes are often insensitive to very low levels of protein A impurity. Indeed, the protein A fragment(s) or whole molecule, may be hidden from separation and subsequent detection by its specific interaction with the Fc (or the variable region).[150,151] The most commonly used method in measuring protein A is the conventional immunoassay such as sandwich enzyme-linked immunosorbent assay (ELISA) where less than 1 ppm (1 ng/mg of protein product) can be detected with reliable accuracy.[152] The detection method may be either colorimetric or fluorospectrometric. To ensure accuracy, the method must contain a sample preparation step that dissociates protein A from the product. To achieve this separation, the test article is subjected to incubation at low pH (typically, pH <4), with or without detergents or heat treatment to remove IgG. For the most part, the presence of low level SDS appears to be sufficient in separating protein A from the Fc-containing product.[153] Another consideration is the source of protein A ligand and its multitude fragments as a reliable control, and the antibody is able to recognize with good specificity the spectrum of leached protein A fragments. In general, chicken antiprotein A polyclonal antibody (IgY) is often used in the ELISA,[154] where the chicken antibody is immobilized in wells of microtiter plates.[155,156] Unchallenged chicken IgY from serum does not bind nonspecifically to protein A. Control recombinant protein A is used to generate the assay standard curve, and test samples containing leached protein A are preincubated in low pH buffer to separate residual/leached protein A that may be associated with the Fc-containing product. Following low pH incubation, the samples are then placed on the protein A antibody coated ELISA microtiter plate. Protein A is then detected by subsequent additions of biotinylated chicken antiprotein A antibody, and may be detected by avidin-conjugated with horse radish peroxidase or fluorescence tags, followed by the appropriate detection mode. Highly sensitive methods have been reported to detect trace levels (sub ng/mL level) of protein A in drug substance, and methods are often optimized to suit the needs of the manufactured Fc-containing glycoprotein.[152]

2. Host-Cell Proteins

Purified recombinant proteins expressed in cell cultures of mammalian, plant, or bacterial origins will have trace amounts of HCPs in the final drug substance.[157] The levels of residual HCPs are often used by regulatory agencies as markers of potential process-related impurities and efficiency of the purification process.[129] These proteins are either secreted by the cells in culture, or derived from lysed or ruptured cells during culture media harvest, and HCPs are by nature, a complex mixture of proteins with highly diverse physiochemical properties. HCPs are often cited as a source of process-related impurities and a potential cause of immunogenicity of drug substance, and the reduction of HCP levels is an important consideration in process development.[158] Specification for HCP is often set based on knowledge of the specific product gained from clinical experience, production process capability to remove HCPs, and the effect of impurities on the efficacy and safety of the product.

The International Conference on Harmonization (ICH) quality guidelines for biological products is set out in section, Q6B, where process-related impurities are defined as those derived from the manufacturing process, to include host–cell DNA, cell culture chemical inducers, antibiotics, media components, and nonproduct protein removal by downstream processing. Given the heterogeneous nature of HCPs, there is no one analytical assay to detect, identify, and quantify with good certainty of their levels in a given sample containing a biologic.[159,160] This analytical limitation is well-understood by the regulatory agencies, and "[c]onsequently, the purity of the drug substance is usually estimated by a combination of methods. The choice and optimization of analytical procedures should focus on the separation of the desired product from product-related substances and from impurities."

A combination of orthogonal analytical methods are often used to detect the presence of HCPs in a given drug substance or in-process drug substance.[160] Such methods and their employment in detection depend on protein sample quantity, nature of impurity tested, accuracy of estimation, and sensitivity.[158] A commonly used electrophoresis method is SDS-PAGE or SDS-CE, and depending on the mode of detection, sub- ng to -μg level of HCPs may be detected. Other methods may include ELISAs or Western blots of drug substance using polyclonal antibodies raised against host-cell-type specific HCPs. Some less used methods such as size exclusion or RP-HPLC may be used to detect non-coeluting μg-level HCPs. Each of these methods on its own may give rise to technical and data interpretation problems or pitfalls. For instance, composition-based estimates such as ELISAs may only be responsive to a limited number of HCPs being assessed, and provide very little structural information of the impurities. Likewise, the sensitivity of some methods, such as SE-HPLC or other chromatographic methods, may be dependent on the type of detection mode and on the interaction HCPs with the separation matrices, or that some HCPs may coelute with the desired drug substance. Therefore, a combination of methods is the preferred approach in determining the levels of HCPs.

3. Host–Cell DNA

Host–cell DNA is an upstream-derived process-related impurity in drug substances derived from a cell culture process, often the result of cell lysis, or rupture resulting from physical exertion (e.g., shear forces, excessive air bubbling, etc.) in a culture vat or cell pelleting or removal during harvest. The guidelines outlined by the WHO suggest that up to 10 ng of residual host-cell DNA per purified dose is considered acceptable.[161] It has been suggested that whenever possible, the level is no more than 100 pg of cellular DNA per dose.

One of the most commonly used methods to detect host-cell DNA is quantitative polymerase chain reaction (qPCR), using random oligonucleotides (typically hexanucleotides) that will amplify the presence of low level host–cell genomic DNA fragments.[162] This method may be used to detect and quantify host–cell DNA in in-process samples and the final drug substance.

4. Cell Culture Medium-Derived Impurities

The cell culture media in many current bioprocesses are well defined in composition, and in many cases, may be supplemented with soy hydrosylates or calf serum to enhance cell culture performance. When supplemented with these poorly defined additives, the ICH guidelines require the manufacturer to show their removal.[3] For instance, if bovine serum is used, a proxy assay such as a bovine IgG assay is needed to show serum protein removal.

A bioprocess that utilizes mammalian cell lines (e.g., CHO or NSO) is known to contain noninfectious retrovirus-like particles (RVLPs) and adventitious viruses may be introduced during cell culture.[163] These viruses are removed by an optimized purification process by $\log_{10}$ reduction value (LRV) of >4.[164] Remaining virus(es) can be detected with highly sensitive and specific real-time quantitative PCR of viral DNA/RNA isolated from postpurified samples. In the cases where bovine or human serum is used in the bioprocess or source of therapeutic protein (e.g., blood and plasma products), infectious agents (e.g., prions) that may cause transmissible spongiform encephalopathy must be removed. Tests for the abnormal prion protein, PrP, often serve as a surrogate for infectivity assays.[165]

Cell lines used in bioprocesses are also known to release cytokines (e.g., IL-1, IL-6, etc.) and growth factors (e.g., G-CSF) during the production phase in the manufacture of biologics. In addition, when biologics are isolated and purified from blood or plasma products, presence cytokines and/or growth factors contamination that may contaminate the product after purification, and can be assessed by highly sensitive and specific ELISAs for these proteins.

Lipopolysaccharides (LPS) or endotoxin may be introduced as the result of low level bacterial contamination of the cell culture or during the fill of the final drug product, and can be assessed by the *Limulus* amebocyte lysate (LAL) assay. Similarly, antifoaming agents that are introduced during cell culture are removed during purification, but verification of removal through quantitative HPLC assays is required. Lastly, purification process impurities such as debris from resin or membrane filter, and leachates such as silicone-based lubricants are tested by macroscopic size determination and the appropriate HPLC-based quantitative analysis for leachables.

V. FUNCTION

A. Introduction

The efficacy, or functionality, of conventional small-molecule drugs in humans and animals is believed to be preserved as long as the uniformity of the chemical structure can be guaranteed by the production process. Since most of the small-molecule drugs are produced by chemical synthesis, the production processes can be highly automated and well controlled. The production of biomolecules, however, is based on processes involving the use of living organisms or cultured cells. Because these processes are more difficult to control than chemical synthesis, they yield products of lower homogeneity. Furthermore, biomolecules, especially proteins, are marginally stable outside the living bodies from which they were isolated and purified. In this unprotected state, pharmaceutical proteins may undergo irreversible changes in structure and stability as a result of exposure to stresses in the manufacturing process or storage (refer to Section II.B.2). Because the function of a protein is intimately linked to its structure, an irreversible structural change in a pharmaceutical protein will ultimately have a negative impact on function and product efficacy. Therefore, functionality is an essential quality attribute of a pharmaceutical protein. To satisfy regulatory requirements, *in vitro* functional assays must be performed with appropriate controls and specifications prior to releasing a protein product as a drug.

Without exception, proteins carry out their functions in living organisms via direct interactions with themselves or through interactions with other proteins or biomolecules. The processes are likely to involve multiple binding partners in multiple binding steps, and may or may not take place across membrane barriers between cells or cell organelles. Because these processes involve complex biological pathways, they and the phenotypes they produce are rarely understood in detail.

The ultimate test of the quality of a pharmaceutical protein is to determine its safety and efficacy by dosing patients in a clinical trial. Thus, an *in vitro* functional test to assess product quality is meaningful only if it can reflect at least part of the *in vivo* mode of action of the drug. Cell-based assays meet this requirement since they produce a quantitative signal in response to interactions of the biotherapeutic molecule with living cells known to exhibit the phenotype of the target disease. This section describes several examples of these assays, which have been successfully developed and applied to pharmaceutical proteins already on the market or in clinical trials.

The design and development of cell-based assays for pharmaceutical proteins are complex and dependent on many factors including the degree of understanding of the target diseases and the availability of necessary reagents. If an interaction between the protein and target ligand is identified as a critical element in the mode of action of the protein drug, then an *in vitro* binding assay that quantitatively measures such an interaction is often used as a surrogate functional assay in place of, or in parallel with, a cell-based assay. Methods that are commonly used in the biopharmaceutical industry to measure protein–ligand binding are described in Section C below. This section also discusses the assay of pharmaceutical proteins that function as enzymes.

B. Functional Bioassays

1. Definition and Selection

The aim of this section is to cover approaches to assay selection, to provide an overview of the possible assay formats and to give a description of the most frequently used assay formats for potency determinations. Bioassays are developed and used for different purposes during drug discovery and development. Even though they can have the same assay format they have a different purpose in a product development cycle starting from target selection to clinical biomarkers.

The potency assays that are described in this chapter are bioassays used as product quality monitoring tools to verify the functional integrity of a product during the product development phase. These assays are an integral part of quality control testing necessary for product characterization, comparability, lot release and stability. After changes in the manufacturing process of a product, comparability testing of the drug substance and drug product is required to demonstrate that the changes do not impact product identity, purity, safety and efficacy. Structural differnces in the product pre- and postprocess change may necessitate functional testing to establish the biological and sometimes even clinical relevance of the change. Bioassays for this purpose have to be able to detect changes in the protein structure because of process/product changes (comparability) and degradation (stability) that impact biological function. In order to serve its intended purpose it is necessary to relate a bioassay for product quality to biological function (mechanism of action) and protein structure (physicochemical characterization) throughout product development to evaluate any impact of structural changes (oxidation, deamidation, aggregation, glycosylation) on biological function of the molecule.

Functional bioassays for product quality (characterization, release, and stability) of therapeutic proteins can be performed in many different formats. The most frequently used functional bioassay formats are *in vitro* cell-based, ligand-binding, or enzyme activity configurations, however, animal-based formats are sometimes employed if necessary. Cell-based assays can measure responses to product based on changes in cell activation markers or end-point function. If the product is an enzyme or proenzyme, enzymatic activity can be measured. If the product possesses binding activity, this can be measured in a binding assay such as an ELISA or surface plasmon resonance (SPR) type assay (see V.C.2. and V.C.3.). The selection of an assay format should be based on the intended use and appropriateness of the assay regardless of the assay format. The choice of an animal-based assay would be a last resort, after all *in vitro* formats have been proven to be inadequate, because of the very high variability of animals (even inbred strains), the high cost of animal maintenance, and technical difficulty.

Understanding the product's biological function is an important first step in choosing the assay format that best mimics the mechanism of action (MoA). In addition, for product release and stability monitoring the assay also has to have properties of a routine quantitative analytical method such as: quantitative readout, reproducible dose response curve with a wide dynamic range, relatively high sample throughput, high precision and accuracy, technically simple to perform, and stable, well-characterized reagents. These assay characteristics indicate the method is appropriate to be routinely used in the regulatory compliant environment of a quality control laboratory. The selection of bioassays for routine lot release is limited by the degree of variability of biological systems, which is generally very high for animal models, less-so for cell-based bioassays, and least variable with ligand-binding assays. Among the several characterization bioassays the one that is relevant to the MoA with properties of a quantitative analytical method is generally selected for a lot release potency assay.

2. Different Stages of Cell Activation as Readouts for Bioassays

Table 4 presents different stages during a cell activation cascade leading to a specific cell function and lists the assay formats commonly used to detect these events. The methods listed in Table 4 are separated into two categories to distinguish more reproducible quantitative assays that are used in industry as quality control assays for a lot release/stability potency determination and the assays that are less suitable for routine use but can be used for characterization of the therapeutic protein and comparability assessment. In some cases the same bioactivity method can satisfy the requirement of both categories, to be representative of MoA and suitable for routine use (quantitative and reproducible).

The basic principle of cell activation is recognition of cell surface receptors by the extracellular ligand, which induces a cascade of biochemical events within the cell signaling pathway leading to induction of gene(s) in the nucleus. As a result of the gene activation the cell reacts by changing its metabolic activity or differentiation state, which ultimately results in exhibition of one or more cellular function(s): activation, proliferation, cytotoxicity, secretion of biological mediators such as chemokines or cytokines.

The majority of the cell-based assays use established cell lines. The cell line of choice would naturally express receptors of interest that respond to the therapeutic protein product, and can be obtained from commercial sources, American Type Culture Collection (ATCC), for example. Alternatively, if an appropriate commercially sourced cell line cannot be found, one can be engineered to respond to a therapeutic protein (see b.Reporter Gene Assay). Most frequently the cells are stably transfected with a vector containing the desired receptor gene or responsive element of a gene of interest. In rare instances when the cell lines are not available, primary cells separated from blood or tissues (human or animal) can be used to develop bioassays. Primary cells are the last resort because of donor-to-donor variability, accessibility of material, and limited practicality of such assays.

A cell-based bioassay procedure generally includes cell growth in suspension or adherent cultures for several days/weeks. In a typical assay, cells are activated/inhibited with the therapeutic protein of interest in a multiwell assay

TABLE 4 Assay Formats Used to Monitor Major Events During Cell Activation

Stages of cell activation	Characterization methods	Methods used as routine potency assays
Receptors		
Receptor binding	Flow cytometry cell surface staining	
Receptor phosphorylation	• Flow cytometry intracellular staining • cAMP • β-arrestin	KIRA ELISA
Activation of signaling pathways		
MAPKinases ERK AKT STAT Smad	• Flow cytometry intracellular staining • Western blot from cell lysate	Reporter gene assay (RGA)
Gene activation		
mRNA	Polymerase chain reaction (PCR)[a]	
Reporter gene		Reporter gene assay (RGA)
Protein secretion		
Cytokine/ chemokine	Flow cytometry intracellular staining	ELISA from cell culture supernatant
End-point function		
Proliferation	BrdU incorporation	• [^{3}H] Thymidine incorporation • Metabolic activity assays that detect live cells-viability dyes and kits (see Table 5 for details).
Differentiation	Flow cytometry, cell surface, and intracellular staining	ELISA for secreted proteins
Cytotoxicity ADCC and CDC Apoptosis Necrosis		• ^{51}Cr release • Metabolic activity assays that detect dead cells (see Table 5 for details)

[a]PCR is a complex molecular biology method that includes several techniques. It is not typically used for potency assays and is not covered in this section.

plate (typically 96-well plate) with multiple concentrations of a test sample along with an assay reference standard. After several hours or days, depending on the end-point, the end-point reaction is detected (see Table 4). Different cell activation events may require different incubation times. The early stages of the cell activation take place within minutes of stimulation (e.g., some stages of the signaling pathways), while the other cellular responses, proliferation, for example, take two to three days for many cell types.

There is a wide variety of readouts and detection systems that are used to detect responses to therapeutic proteins in cell-based assays. Each stage of a cell activation cascade listed in Table 4 can be used as a readout for an assay. The chosen readout representing the extent of the biological response to each concentration of the therapeutic protein can be fited into a curve-fitting equation. Most of the biological responses can be modeled by a 4-parametric equation. However, it is important to establish the best fit for the data. The potency of the therapeutic protein is calculated relative to a representative and well-characterized lot of therapeutic protein (reference material) that is typically used to generate the standard curve for the assay. Testing the entire standard curve and test article curve for parallelism is an important prerequisite for the potency calculation. In addition, having standard and appropriate control(s) on each assay plate is highly recommended.

a. Receptor Phosphorylation and Signaling Pathways

In general, monitoring early events in cell activation as readouts for an assay provides quicker assays with the possibility for higher sample throughput. The specificity of these reactions often needs confirmation by another method because of the fact that many biological molecules can activate several signaling pathways.

Phosphorylation of intracellular proteins followed by transmission of signals through designated signaling pathways are an important regulator of cellular responses to many extracellular stimuli. Many signaling pathways use protein tyrosine kinases, therefore detection of these intracellular events can be very useful in monitoring cellular activation. Bioassays based on the receptor phosphorylation as a readout are typically in a format of an intracellular staining using flow cytometry[166] or "kinase receptor activation" (KIRA).[167] Both methods use specific antibodies that recognize proteins that have phosphorylated tyrosine residues. These assays are utilized to demonstrate receptor agonist or antagonist activity of a therapeutic protein. They are frequently used to monitor agonist-mediated changes associated with G-Protein coupled receptor's (GPCRs) transmembrane signaling. Except for the KIRA ELISA, the receptor phosphorylation and signaling pathway assays are rarely used as routine potency assays, however they are invaluable complementary characterization assay tools in conjunction with a receptor binding assay to demonstrate the ability of the therapeutic protein to engage in cell activation.

b. Gene Activation: Reporter Gene Assays (RGA)

Gene expression profiling is often done in the early discovery phase to identify genes (*in vivo* and *in vitro* models) that are upregulated under the influence of the therapeutic protein. The gene(s) of interest can be used as a base for engineering a cell line. Reporter gene assays (RGA) are based on engineered cells that are transiently or stably transfected with the reporter gene and the gene of interest placed in the same DNA construct (expression vector/plasmid).[168]

For bioassays covered in this section, it is important to use stably transfected cell lines of human or animal origin. In situations where the MoA of the therapeutic protein has an impact on the signaling pathway it is very important to select the appropriate cell line for transfection that has the ability to be activated through the relevant signaling pathway. The expression of the reporter gene is used as a marker to monitor upregulation of the gene of interest. Commonly used reporter genes that encode the proteins that generate measurable fluorescent readouts are green fluorescent protein (GFP),[169] red fluorescent protein (DsRed2 reporter gene) and the enzymes that with appropriate substrate can generate luminescent products: the enzyme luciferase and the *lacZ* reporter gene (in bacteria), which encodes the protein β-galactosidase.[170] Luciferase is detected using the chemiluminescent substrate luciferin[171] and β- galactosisdase is detected using either the colorimetric substrate o-nitrophenyl β-D-galactopyranoside (ONPG) or the more sensitive 1,2-dioxetane-based chemiluminescence substrate. The reporter gene assays are relatively easy to perform and standardize, which makes them suitable for potency assays. The basic principle of these assays is that the cultured transfected cells are incubated with a therapeutic protein for a few hours, usually less than a day, and the signal is generated using the appropriate substrate. There are many different types of RGAs, two examples are described in the referenced papers.[172,173]

c. Protein Secretion

Two methods are typically used to detect secreted protein from stimulated cells: intra cellular staining by flow cytometry V.B.e.iii.(1) or binding assays (described in Section V.C.2) from the cell culture supernatant.

d. Differentiation

Cell activation can lead to differentiation of certain cells into more mature cells that exhibit different structural and functional form. For example, immature hematopoietic cells can be stimulated by some proteins to mature, by changing cell morphology and by inducing expression of different intracellular and surface markers. Expression of differentiation markers can be detected and monitored by flow cytometry using specific antibodies against cell surface or intracellular markers.

Another example is the osteoinductive proteins that, when administered *in vivo*, induce formation of new bone tissue at the site of administration as the result of differentiation of surrounding mesenchymal cells from the soft tissues into bone. Cell lines with the capability to differentiate into bone forming cells *in vitro* can serve as models for *in vivo* differentiation. A bone forming cell line when stimulated *in vitro* with osteoinductive protein demonstrate the sequence of events characteristic for osteoblast differentiation: alkaline phosphatase (ALP) secretion, collagen and osteocalcin production. ALP can be detected by addition of ALP substrate and collagen and osteocalcin can be detected by standard ELISA assays (see Section V.C.2.).

e. End-point Function

i. Proliferation

Proliferation is one of the fundamental methods to measure biological responses to therapeutic protein.

(1) Proliferation Assays Based on Nucleoside Incorporation into DNA: Historically, proliferation was monitored using incorporation of [^{3}H] thymidine into DNA of dividing cells.[174] This assay format was widely used and it is still used in some labs in industry for lot-release potency assays. Even though the incorporation of thymidine into newly synthesized DNA of dividing cells represents a

direct and accurate measure of the cell proliferation, the assay measures proliferation in the entire cell population without any information on how many cells were activated to proliferate. Preparation of samples for radioactivity counting and issues associated with the use and disposal of radioactive material made this method undesirable and obsolete in many laboratories.

The bromodeoxyuridine (BrDu) incorporation (an analog of the DNA precursor thymidine) is another direct method that monitors incorporation of the nucleoside into DNA of dividing cells.[175] Availability of commercial mAbs that recognize the nucleoside analog BrdU enable the implementation of this type of nonradioactive proliferation assays. A flow cytometry based assay where BrdU is detected by fluorescein labeled anti-BrdU mAb, using intracellular staining procedure provides the information about proliferation of individual cells, which is an additional advantage over [^{3}H] thymidine incorporation assay. The ELISA assay capturing BrdU from the cultured dividing cells is another assay format that is based on the BrdU principle. The BrdU methods are not suitable for routine use mainly because of assay complexity and high variability.

(2) Proliferation/Viability Assays Based on Metabolic Activity of Cells: Besides the incorporation of nucleosides into DNA, cell-viability dyes and kits have become valid options to monitor cell viability, activation and the increasing number of viable cells during cell proliferation. Table 5 lists the assays that are based on the metabolic activity of cells. They are also used as an indirect measure of cytotoxicity. The viability/metabolic reagents based on redox indication dyes like tetrazolium (MTT, XTT, MTS WST-1) and resazurin (Alamar Blue) are popular substitutes for radioactive compounds. The tetrazolium assays are colorimetric assays based on bioreduction of tetrazolium salt into a either insoluble (MTT) or soluble (XTT, MTS,WST-1) colored substance formazan by mitochondrial succinate dehydrogenase.[176,177,178] The reaction occurs only in live cells, and the amount of formazan increases in activated and dividing cells. The colorimetric assays mentioned above suffer from limited sensitivity and a relatively small dynamic assay range because of the use of the optical density measurements. The Alamar Blue assay is a fluorescent assay based on intracellular reduction of dark blue Alamar Blue (resazurin) into red fluorescent product (resorufin) by cellular enzymes.[179] Under optimal conditions Alamar Blue can be reliably used to detect live cells. However, Alamar Blue is not suitable for kinetic monitoring of the cell growth.

The nucleotide adenosine triphosphate (ATP) is an essential intracellular metabolite for maintaining viability, while dead cells lose ATP. There are many commercial cell-viability assay kits that are based on an ATP marker combined with the detection (luciferin/luciferase) reagents with a luminescent readout.[180]

The assays based on metabolic activity using the vital dyes are simple, sensitive and, in most cases, provide an accurate measure of the number of live cells. In recent years the metabolic activity methods have been improved with introduction of the commercial reagents and kits, which were used to monitor cell viability and proliferation in simple "add-mix-measure" assay protocols. These simple assay formats allow for a wider cell-based assay application and automation. However, one should exercise caution when developing any assay based on metabolic activity of cells because there are several drawbacks with these types of

TABLE 5 Methods to Monitor Cell Viability and Cell Death

Marker	Detection reagent	Reaction	Readout
Cell viability			
Mitochondrial enzymes activity	MTT (3-(4,5-dimethylthiazol-2-yl) 2,5-diphenyl tetrazolium bromide) (colorimetric)	Tetrazolium reduction into formazan in metabolically active cells	Optometric density
	XTT[a]		
	MTS[b]		
WST-1[c]			
Intracellular enzymes activity	Alamar Blue	Reduction of resazurin in metabolically active cells	Fluorescence
Adenosine triphosphate (ATP)	Luciferin/luciferase	Intracellular ATP upregulation in metabolically active cells	Chemiluminescence
Intracellular and cell surface labeling of live cells	Carboxyfluorescein diacetate (CFSE)	Binds to both intracellular and cell surface proteins by reacting with lysine side chains and other available amine groups.	Fluorescence
Intracellular labeling of live cells	Calcein AM	The enzymatic (intracellular esterase) conversion of the nonfluorescent cell-permeant calcein AM to the intensely fluorescent calcein in live cells.	Fluorescence
Cell death			
Apoptosis			
Activation of the intracellular signaling pathway leading to cell death	Caspase-3	Aminoluciferin cleavage of the luminogenic Z-DEVD-aminoluciferin substrate	Chemiluminescence
Phosphatidylserine (PS)	Annexin V	PS translocates to the cell surface and binds annexin V	Fluorescence
DNA fragmentation	Biotin conjugated dUTP using terminal deoxynucleotidyl-transferase (TdT)	TUNEL reaction terminal deoxynucleotidyl-transferase dUTP nick end labeling of broken ends of the double stranded DNA	Optometric density
Intracellular proteins associated with apoptosis	Flow–cytometry based bead assay	Detects proteins from cells associated with different stages of apoptosis	Fluorescence
Necrosis			
Lactate dehidrogenase (LDH) release	LDH substrate	LDH release from the cells with the ruptured cell membrane	Optometric density
Nucleic acid labeling	7AAD (7 amino-actinomycin D)	Staining of the nucleic acid of the cells with the ruptured membrane	Fluorescence
	Propidium iodide	Staining of the nucleic acid of the cells with the ruptured membrane	Fluorescence
	Ethidium bromide	EthD-1 enters cells with damaged membranes and produces red fluorescence upon binding to nucleic acids of dead cells	Fluorescence
AK (adenylate kinase)	ADP	AK (adenylate kinase) released from dead cells	Chemiluminescence

[a]XTT (sodium 3′-[1-(phenylaminocarbonyl)- 3,4-tetrazolium]-bis (4-methoxy-6-nitro) benzene sulfonicacid hydrate).
[b]MTS [3-(4,5-dimethylthiazol-2-yl)-5-(3-carboxymethoxyphenyl)-2-(4-sulfophenyl)-2H-tetrazolium, inner salt
[c]WST-1 (4-[3-(4-Iodophenyl)-2-(4-nitrophenyl)-2H-5-tetrazolio]-1,3-benzene disulfonate).

assays. Knowing a metabolic reductive capacity of the particular cell line is an important factor in determining the appropriate kind of the metabolic readout.

ii. Cytotoxicity

Cytotoxicity is an end-point function, which is the result of a cell activation leading to death of a specific cell population targeted by a therapeutic protein. The ability of some therapeutic proteins to induce cell death (cytotoxicity) is an important attribute of product quality and safety profiles. Targeted killing of tumor cells by a toxin-conjugated therapeutic mAb is one example of cell killing by a toxin where the mAb is used only to recognize tumor cell targets.[181]

The therapeutic protein can cause cytotoxicity of different cell types without a conjugated toxin by two basic mechanisms: direct and Fc receptor (FcR)-mediated killing. Direct cell killing is when a therapeutic protein directly induces cell death by apoptosis or necrosis (methods for detecting apoptosis and necrosis are listed in Table 5 and described in iii). Therapeutic proteins can cause FcR- mediated depletion of a specific subset of the disease-causing cells by three different mechanisms.

The first is the antibody-dependent cellular cytotoxicity (ADCC), the mechanism by which the antigen binding site of a therapeutic protein, for example, a mAb or a molecule with a similar structure, binds to its receptors on the target cells and the Fc portion of the antibody binds to FcR on the effector cells, mainly natural killer (NK) cells. As a result the effector cells are activated to release cytotoxic factors, which induce rupture of the membrane of the target cells.[182]

The second mechanism of cell death is the complement-dependent cytotoxicity (CDC), where a mAb or a molecule with a similar structure binds to its specific receptors on the target cells and the Fc portion of the antibody binds the complement proteins. Complement binding results in the activation of a complement cascade leading to formation of pores in the cell membrane compromising membrane integrity.

The third mechanism is the FcR-mediated phagocytosis upon antibody binding to its cellular receptor targets on macrophages. Assays based on phagocytosis are very difficult to standardize and use for potency determination.

Cytotoxicity assays based on the Fc-mediated killing can be used as bioactivity/potency assays to evaluate product quality when the desired MoA of the therapeutic protein is to induce death of certain types of cells (e.g., tumor cells). It is well known in the industry and in the literature that some therapeutic proteins (mAbs or molecules of similar structure) with an intact Fc region can exhibit the nontargeted Fc-mediated functions (ADCC, CDC, phagocytosis), which can cause undesired effects, thereby compromising patient safety. The same types of assays can be used to evaluate product's safety profile by testing the ability of the therapeutic protein to exhibit the undesired, nontargeted Fc- mediated functions. In those cases when the Fc function is not a part of the MoA of the therapeutic protein, cytotoxicity assays should be used only as a characterization tool.

iii. Methods to Monitor Cell Death/Cytotoxicity

The classical assay that has been used for years to monitor cytotoxicity is the ^{51}Cr release assay. Briefly, the target cells in either ADCC or CDC assay are labeled with radioactive chromium. In an ADCC assay the target cells with the bound therapeutic protein are incubated with the effector cells at various effector: target ratios and the ^{51}Cr is released upon the target cell death and monitored using a γ scintillation counter.[183]

In the CDC assay the ^{51}Cr release was monitored after incubation of the labeled targets with therapeutic protein and exogenous source of complement (human or baby rabbit serum).

FcR-mediated cytotoxicity assays are variable and difficult to perform, especially the ADCC assay, which uses primary cells from peripheral blood. However, recently these assays have been simplified and used as a bioactivity assays for characterization of therapeutic proteins and as a routine potency assays. Some laboratories in industry are able to use cloned NK cells as effector cells and the established cell lines or transfected cells as target cells in simplified ADCC assays. The CDC assay is used as a potency release assay for several therapeutic proteins. The use of cell lines instead of the primary cells in ADCC assays and nonradioactive detection methods, adenylate kinase (AK) or ATP with luminescent readouts, for example, have a potential to improve the performance of these assays and make them more suitable for the routine use.

As mentioned above there are different mechanisms by which therapeutic molecules can exhibit cytotoxic effects. Apoptosis and necrosis are two mechanisms of cell death. Apoptosis involves well-characterized signaling pathways that lead to cell death through series of events characterized by biochemical markers typical of a stage of cell death. Intracellular signaling molecules, including the family of cysteine proteases, caspases,[184] and phosphatidylserine (PS) exposed at the surface of the cell membrane are early markers of apoptosis followed by DNA fragmentation at the later stage. The series of apoptotic events ultimately lead to rupture of cell membrane and necrosis of cells. The description of different methods to detect apoptosis can be found in reference 185. The decision to select one assay for apoptosis over the other should be based on the stage of apoptosis, and simplicity and reproducibility of the assay.

Cell death by necrosis can be induced directly by the therapeutic agent or it can be a result of apoptotic process. The release of lactate dehydrogenase (LDH) seems a good simple choice for detection of cells with ruptured membranes.[186] However, LDH is not a good choice for every cell type and in some cases is not suitable for a precise quantitative assay. Measuring AK released from the dead cells is a simple and more reliable approach for direct detection of necrotic cells. The 7-amino-actinomycin D (7AAD) and propidium iodide (PI) are good choices for flow cytometry-based detection of necrotic cells especially in combination with other markers in double or multicolor analysis. Both dyes penetrate ruptured membranes and intercalate with DNA of the necrotic cells generating red fluorescence.[187]

The choice of an assay in quantifying cell death is critical. The selection and timing of an appropriate end-point to measure is based on knowledge of the underlying mechanism and kinetics of cell death. The first decision to be made is whether to monitor the marker of dead cells or the marker for surviving (live) cells. Advancement in understanding and techniques to detect the biochemical markers typical of cell death led to the development of relatively simple and direct cell death detection assays, which in many cases represent a better choice of assay, especially for the methods used to measure the loss of membrane integrity. Table 5 lists the choices of dyes and detection reagents used to detect apoptotic and necrotic cells. The choice of reagents depends on the type of cells, cell culture supplements, and the therapeutic protein.

The use of additional methods during bioassay development is highly recommended to verify suitability of the method based on metabolic activity of cells.

Simultaneous use of an additional assay(s) that measures different end-points (e.g., flow cytometry methods) is of great value. The flow cytometry methods, based on single cell measurements, are very useful for selection and optimization of the cell-viability and cell-death assays and for selection of the readout reagents. The double staining method can compensate for the inability of simultaneous detection of live and dead cells by metabolic activity/cell-death kits and verify their accuracy and suitability. It is highly recommended to use double staining of live and dead cells from the same assay wells. For example, using Annexin V and 7AAD[188] allows for simultaneous detection of populations of apoptotic cells and necrotic cells, respectively, and the remaining unlabeled live cells.

For simultaneous detection of live and dead cells, the appropriate combination of two (or more with the more advanced flow cytometers) markers can be used in a flow cytometry assay.

(1) Flow Cytometry: The flow cytometry is a powerful, widely used technology for defining and characterizing different cell types by analyzing cell size, cell surface, and intracellular markers on individual cells. The method measures intensity of fluorescence generated by fluorochrome-labeled antibodies that specifically bind to the target molecules on or inside the cells. Preparation of samples for analysis includes staining of the cell suspension with specific antibodies that are either directly conjugated with a fluorochrome (direct staining) or require secondary antibody conjugated to fluorochrome to generate a signal (indirect staining). Detection of the intracellular molecules (cytokines, phosphorylated receptors, signaling pathways) requires fixation of cells and permeabilization of the cell membrane.[189] Flow cytometry has the capability of simultaneous detection of several cell markers using multiple antibodies conjugated to fluorochromes that after laser excitation emit light at different wavelengths producing readouts of different colors. Even though the flow cytometer instruments have improved and are simplified to be user friendly, analysis, and interpretation of data still require intensive training, scientific knowledge, and experience.[190]

The assays based on flow cytometry are not routinely used as potency assays, however, flow cytometry has a broad range of applications that makes it a valuable tool for laboratories working with cells. Methods based on flow cytometery can be used for characterization of therapeutic protein to understand its function. It is also a valuable tool to be used during development of a routine potency assay in order to understand the assay; verify its appropriateness; confirm the results and characterize cells.

C. Non Cell-Based Binding Assays

A range of methods that are not based on cellular responses are also used to measure protein–ligand interactions. The ligand can be another protein, small-molecule drug, carbohydrate, nucleotide, etc. Some of these methods are more often used in research and drug discovery to identify/screen targets. Examples include immunoprecipitation followed by 2D electrophoresis and MS, yeast or bacterial two-hybrid, phage display, Western blot, glutathione *S*-transferase (GST) pulldown, NMR, atomic force microscopy, etc.[191] The focus of this section is analytical methods that are quantitative and quality-indicating, have been or have the potential to be, used routinely as release or characterization assays in drug development. Robustness, low-cost, and ease of use are among the essential features. For these reasons, some of the more established methods, such as sedimentation equilibrium of analytical centrifugation, fluorescence anisotropy, fluorescence correlation spectroscopy are not frequently used in biopharmaceutical industry. The following section describes some popular methods that are used in protein drug development.

1. Homogeneous Solution Methods

a. Föster Resonance Energy Transfer

The Föster resonance energy transfer (FRET)[192] technique has a long history of being used to characterize the conformation of macromolecules in solution. It is also referred to as fluorescence energy transfer in the literature. However, this name may be misinterpreted to mean that the energy is transferred between molecules via fluorescence radiation. In a FRET measurement, a donor probe absorbs light and is then excited to a higher electronic energy state. The excitation energy may then be nonradiatively transferred from the donor to an acceptor probe via a long-range dipole–dipole interaction. This transfer of energy can only occur if the emission spectrum of the donor overlaps the absorption spectrum of the acceptor. As a result of radiationless energy transfer, the donor relaxes from the excited state to the ground state and its fluorescence decreases. The acceptor on the other hand is excited to a higher energy state and sensitized to emit fluorescence if it is also a fluorophore.

The energy transfer efficiency E is highly dependent on the distance between the donor and the acceptor, r, following an inverse 6th power law

$$E = 1/(1 + (r/R_0)^6)$$

where R_0 is the Förster distance, characteristic of a given donor–acceptor pair, defined as the distance at which $E = 0.5$. This dependence, with R_0 typically in the range of 20–60 Å, enabled the technique to be used as a tool to measure the distances between donor and acceptor probes attached to specific sites on a macromolecule.[192,193] This unique capability of FRET has been taken advantage of in many studies of macromolecular conformation. It is also used to detect intermolecular interactions when the donor and acceptor probes are separately attached to different macromolecules and when their attachment points can be brought sufficiently close by the interaction of those macromolecules.[194] Its application to the study of intermolecular interactions became more practical following the development of chelated lanthanide ions as energy transfer donors. These probes enable FRET to assess molecular distances of ~100 Å.[193,194]

FRET measurements can be performed with many methods, using different types of instruments, all with the same objective of determining, or monitoring changes in, the interprobe distance *r*. The distance is usually calculated from the measured energy transfer efficiency *E* using the above equation. *E* is defined as the ratio of a measurable fluorescence property of the donor, or the acceptor, with and without the influence of energy transfer. Since energy transfer can modulate all properties of a fluorophor, theoretically any property can be used, but in reality, some suffer more difficulties or uncertainties than others. The methods are mainly classified as steady-state methods[192] and time-dependent methods.[193] Measurement of fluorescence intensity (for either the donor or the acceptor) and donor fluorescence depolarization are among the steady-state methods. These methods are relatively straightforward in principle and operation, and require simpler instrumentation than the time-dependent methods. Consider the case of donor fluorescence intensity as an example, $E = 1 - I_{DA}/I_D$, where I_{DA} and I_D are the fluorescence intensities of the donor in the presence and absence of an acceptor, respectively. However, the intensity measurement is affected by many factors that are not simple to control or correct, such as the degree of labeling and concentration of donor- and acceptor-bound molecules, photobleaching, light scattering, inner-filter effect, and background fluorescence including the emission of the acceptor if it is also a fluorophore.

The time-dependent methods, which are unaffected by or less sensitive to the complications that intensity measurements suffer, include the time domain (time-resolved) technique, in which the fluorescence decay lifetime τ is measured, and the frequency domain technique, in which the frequency response of donor emission is used to obtain interprobe distances (distribution). Each requires highly specialized, high-cost, instruments, often custom-built for nanosecond lifetime measurement before the lanthanide probes were developed. In the time-resolved method, $E = 1 - \tau_{DA}/\tau_D$, where τ_{DA} and τ_D are donor fluorescence lifetime in the presence and absence of an acceptor, respectively. For the purpose of studying structure–function relations of macromolecules, the accurate measure of absolute distance is of less importance and interest than the changes in distance in response to condition changes.

Despite the long history, traditional time-resolved FRET has been used in a limited number of laboratories, mostly in academia where there is the expertise to build the necessary instruments. Time-resolved instruments comprise high-cost components such as picosecond-pulsed light sources and nanosecond gating electronics for detecting the fluorescence decay of fluorophores with lifetimes in the nanosecond range. The application of the technology to the study of interactions between macromolecules was further limited by the R_0 values of the traditionally used reagents. This situation has changed since the development of chelated lanthanide ions as energy transfer donors. These probes enable FRET to assess molecular distances of 70 to 100 Å.[195,196] The longer lifetime of lanthanide emission permitted the development of low-cost, commercially available, time-resolved instruments for FRET measurements of intermolecular interactions.

The new generation of FRET utilizes a scheme of two-tiered energy transfer in two different time scales. First, a regular fluorophore which absorbs ultraviolet light for excitation, and emits with nsec lifetime, transfers its energy in the nsec timeframe to a lanthanide ion, for example, $Eu(3+)$, via the nonradiative Föster mechanism. Second, the excited lanthanide, which emits at longer wavelengths (typically >600 nm) with μs–ms lifetimes, transfers its energy in the μs–ms timeframe to the acceptor which is a regular fluorophore that absorbs and emits at wavelengths of >600 nm (e.g., cyanine dyes Cy3, Cy5). All three components have to be in close proximity for the energy transfer to occur. The measurement readout can be either the decay rate (or lifetime) of the lanthanide ion emission or the intensity of the acceptor fluorescence. Both allow calculation of the energy transfer efficiency and, thereafter, the distance between the lanthanide ion and the acceptor, using the same Föster equations. The critical element in this technology is the very long-lifetime emission of lanthanide ions. This is because the emission of lanthanide ions is not fluorescence but phosphorescence, meaning that the transition between their ground (singlet) and the excited (triplet) state is "forbidden" in quantum mechanics, therefore, it is more difficult and takes a longer time to occur. Likewise, their excitation by directly absorbing photons is also more difficult. Nonradiative transfer of excitation energy from a suitable fluorophore from a very close distance proved to be efficient in enhancing the excitation of lanthanide ions. The development of synthesized reagents that link a chelator with a suitable fluorophore, therefore holding the lanthanide ion in an optimal distance from the fluorophore, also played a critical role in the development of the technology. The third essential component of those reagents is a moiety that enables conjugation to macromolecules utilizing a variety of chemical pathways.[196]

The major advantage of the very long lifetime of the lanthanide ion emission is to allow the data collection to be delayed by ~μs following each excitation pulse. This not only leads to less expensive instrumentation[197] (μs rather than ns gating device), but also higher data quality because of the elimination of most of the background signals that decay on the ns time scale (e.g., emission of the fluorophore linked to the chelator, light scattering, fluorescence from solution components, etc.). The degree of labeling of the donor (chelator + lanthanide ion) and the acceptor becomes less important when the emission of the acceptor is measured.

In the application of FRET to detect and quantify protein–target interactions, the protein and the target are each labeled with either a donor or an acceptor prior to being brought into contact. The probes can be conjugated to the macromolecule via primary amine or sulfhydryl or other appropriate chemistries. In general, the unattached probes should be carefully removed by dialysis or other means of buffer exchange. This requirement can be eased when a chelated lanthanide ion is used as the donor.

The largest R_0 values for the known donor and acceptor pairs are in the range of 70–100 Å, similar to the sizes of many types of macromolecules, for example, globular proteins. Depending on the sites of conjugation on the surface of macromolecules and the relative orientation of these molecules in their interactions, the interprobe distance may still not be close enough for an energy transfer to be detectable even when the binding of the two macromolecules takes place. This may lead to an erroneous claim of no binding. Such a conclusion would have to be verified with orthogonal methods during assay development. Attaching multiple probes to the macromolecules may help to increase the chance for energy transfer to occur when they interact. However, caution must be taken that one or more of the probes may interfere with the interaction if it is bound at or near the site of contact. This is a general concern regarding all methods that utilize external probes to report intermolecular interactions, unless the attachment is restricted by specific chemistry at a carefully chosen location.

When using FRET to study interactions between two molecules, one labeled with the donor and the other labeled with the acceptor, a fluorescence property of either the donor or the acceptor (if it is a fluorophore) has to be chosen for detecting the energy transfer. Then the detection is performed with a fixed concentration of the component that is being detected and varying concentrations of the other component, from zero for background correction to a level that the detection signal shows saturation. The saturation is easily confirmed with a log–log or semilog plot of signal versus concentration. The concentration corresponding to the half-maximum signal reflects the affinity of the interaction in terms of a dissociation constant K_D. However, this is only true when the interaction follows a simple 1:1 binding mechanism. Nonlinear fitting software can be used to fit the curves to determine binding affinities data based on prior knowledge of the binding mechanisms.

More precise FRET results can often be obtained by performing a competitive binding experiment. In such an experiment, the maximum signal of the acceptor emission at the maximum energy transfer is first established under optimal binding conditions for the donor- and the acceptor-labeled components. Donor-labeled molecules are then allowed to be displaced by the unlabeled molecules added to the solution by means of competition. The energy transfer will decrease as the concentration of the unlabeled molecule increases, as illustrated in Figure 10.

Quantitative measurement using FRET requires solutions free of scattering particles. The concentration range to be used in the measurement is dictated by the dissociation constant of the binding reaction, which is often close to the competitor concentration corresponding to half-maximum signal. For very strong binding (K_D in pM range, for example), the concern is signal sensitivity, while for weaker bindings (K_D in μM range, for example), sample consumption, aggregation, or the inner-filter effect because the concentration is relatively high, are among the usual experimental concerns.

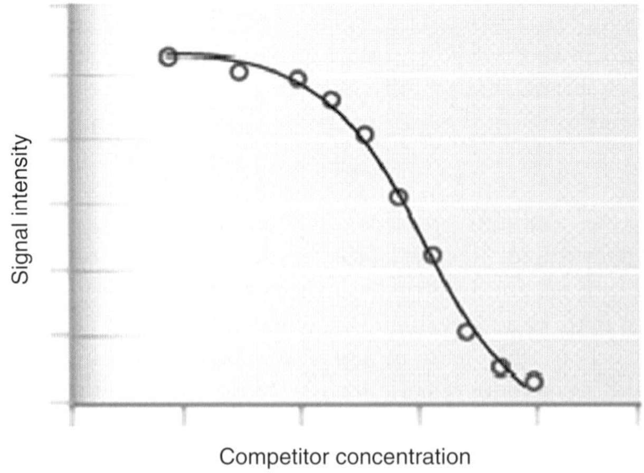

FIGURE 10 **Exemplary binding isotherm of a FRET competitive binding assay using HTRF®.[197]**

b. Isothermal Titration Calorimetry

Calorimetry measures the total heat release or uptake during a chemical and/or physical process. If a process involves two molecules interacting with each other, the measured heat release or uptake (under constant temperature and pressure) can be used to calculate thermodynamic parameters, heat capacity ΔCp, and enthalpy ΔH of the interaction. If the Gibbs free energy ΔG, or the equilibrium constant K, is known, the entropy ΔS can also be calculated.

$$\Delta H = T\Delta Cp$$

$$\Delta G = -RT \ln K = \Delta H - T\Delta S$$

where R is the gas constant and T is the absolute temperature. Carrying out the heat measurements during the course of a titration, that is, successively injecting aliquots of one binding component into the calorimeter cell containing a known amount of the other component, allows for the determination of the equilibrium constant K. Therefore, isothermal titration calorimeter (ITC) is the only practical technique that allows for the determination of the entire thermodynamic profile for a reaction within a single experiment.

The heat that is measured at each injection step after reaching equilibrium is proportional to the increment of complex formation at each step (Figure 11). As the reaction in the cell approaches saturation, the increment diminishes until eventually only the heat of dilution is measured (used for baseline correction in data analysis). At the end of the titration, an isotherm is constructed by plotting the net heat after equilibrium (peak area) versus the calculated molar ratio of the two reactants in the cell at the end of each titration step. The equilibrium constant K, the reaction stoichiometry, and the enthalpy ΔH can then be determined by fitting an appropriate model to the isotherm[199] (see Figure 11).

Under ideal conditions, when the binary reaction in the cell involves only the binding of the two reactants, the measured equilibrium constant is equivalent to the equilibrium binding constant K_A. The range of measurable binding constants by ITC is 10^3–10^8 M^{-1} although higher affinities (10^9–10^{12} M^{-1}) can be assessed with competitive binding and other techniques.[200,201] The binding enthalpy ΔH_b is directly calculated from the measured heat capacity and the temperature. The free energy ΔG_b and entropy ΔS_b of the binding can be calculated using the above equations.

Major advantages of ITC include that the measurements are performed in solution without the need to label or immobilize the reagents. Furthermore, ITC is not affected by optical interferences, such as color or turbidity, and is insensitive to molecular mass. The last feature makes ITC especially useful for measuring the binding of small molecules to macromolecules. However, like FRET, conventional ITC is an equilibrium method and thus is incapable of providing kinetic information about the binding reaction. ITC also suffers from the following drawbacks that need to be considered carefully when carrying out the measurements or analyzing and interpreting the data:

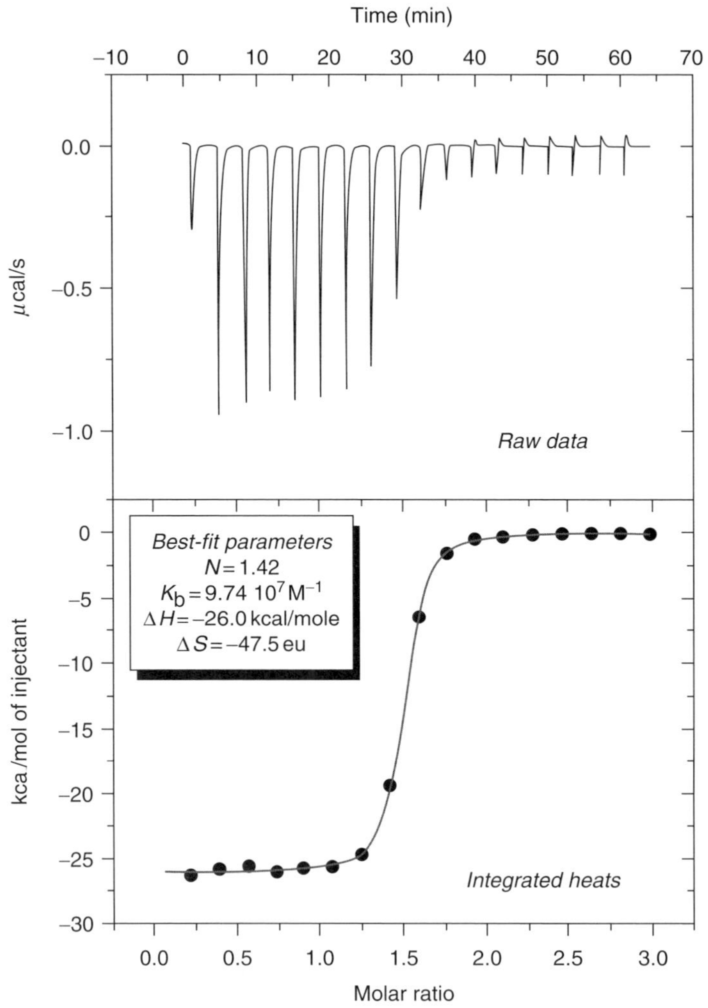

FIGURE 11 **Typical ITC data. Top: raw ITC data. Bottom: binding isotherm.**[198]

1. The development of ultrasensitive calorimeters enabled ITC to perform microscale measurements of biological samples, such as protein and DNA. However, even the most sensitive models currently available still require milliliter or hundreds of microlitter volumes of reagents in micromolar to millimolar concentration for each titration experiment. Thus, the application of ITC may be limited by the availability of reagents.
2. Since the injection volume should be much smaller than the solution volume in the cell, the concentration of the ligand in the syringe has to be much higher. Thus, the ligand must remain soluble at these higher concentrations. An additional complication arises if the ligand in the syringe is a protein that reversibly self-associates. Upon dilution after injection into the calorimeter cell, the protein may dissociate. The degree of

dissociation depends on the concentration of the ligand or target molecule in the cell, as well as the difference between the equilibrium constants of the self-association and the hetero-molecular interactions. Consequently, the heat of dilution measurable by injections at the end of the titration may or may not equal that at the beginning. The latter may lead to a shift in the binding isotherm since the heat of dilution must be subtracted from the total heat measured at each titration step.

3. In order to carry out their function, some proteins undergo a conformational change in response to the binding of certain ligands. The conformational change may take place at a location on the protein molecule that is distant from the ligand-binding site. Motor molecules and transmembrane molecules are examples of these proteins. In an ITC measurement, the heat associated with a conformational change will be indistinguishable from the heats of other processes involved in a binding event, such as changes in hydration or charge in and outside of the contact areas. Thus, the measured heat, and the thermodynamic parameters derived from the heat, should be regarded as apparent values that include contributions from all the microscopic events involved in and following the binding. This is obviously a disadvantage of ITC if the goal is to solely determine the equilibrium parameters for the binding. However, it has been used in an ingenious manner to characterize conformational changes distant to the contact areas following the binding of two molecules. In this unique approach, the thermodynamic information for the events outside of the contact areas are obtained by subtracting orthogonally (and independently) determined thermodynamic information for the binding from the total value obtained by ITC measurement.[202]
4. The interactions involving macromolecules that contain multiple binding sites in different domains or subunits, such as antibodies binding to multiepitope antigens (e.g., VEGF, BMP, TNF, etc.), may be kinetically and thermodynamically complex. For instance, the antibody–antigen immune complexes, or concatemers, may form as antibody and antigen interact. The size of such complexes is concentration dependent, and are often large enough to precipitate out of solution. The stoichiometry of such a binding event is variable and the rate of dissociation is lowered by the avidity effect. Therefore, the ITC results for this kind of binding reactions reflect the averaged avidity (over the size distribution of the complexes), rather than the affinity of a single binding event at an individual site. The effect of avidity is not limited to ITC but is encountered in all affinity measurements made with multivalent macromolecules in homogeneous solutions. Furthermore, as large complexes precipitate out of the solution, the total heat measured by ITC will also include the heat of solubilization. Contributions from this and other interfering processes may be difficult to eliminate from the measurements or account for in data analysis. Without thorough corrections, binding equilibrium models may fail to fit the isotherm. In some cases, different experimental configurations using different technologies may help to reduce certain artifacts. For example, immobilizing one of the binding components on a solid surface or a biosensor, helps to prevent the formation of large concatamers. Therefore, the avidity effect is reduced in comparison to measurements performed in free solution. However, the results of those methods

may be confounded by other complexities, as discussed below. It may be naïve to regard any of the methods as the gold standard for macromolecule interactions, or to assume a value to be absolutely correct even if obtained with more than one method.

2. Solid-Surface Methods

The ELISA is a solid surface binding method widely used for the detection, quantification, and determination of the binding characteristics of different proteins and other biological molecules. ELISA methods can be used as a potency assay in cases when the function of a therapeutic protein is only to bind a target molecule *in vivo* or as a second release assay in addition to the functional potency bioassay to evaluate ability of the therapeutic protein to bind to a designated target. Binding assays are often used as a characterization assay in structure–function studies.

Advances in generation of antibodies especially mAbs enabled the development of wide variety of the ELISA formats. For many applications, the ELISA replaced the older radioimmunoassay (RIA) method that is based on a detection of antibody–antigen reaction in solution using radioisotope. In the ELISA methods the radioactive label of an RIA is replaced by an enzyme conjugate.

The fundamental component of ELISA type of assays are the antibodies. The basic principal of the method is antigen–antibody binding reaction. A capture reagent, bound to the solid surface of the high protein binding 96-well polystyrene plate, initiates the binding to form antigen–antibody complexes that are detected using enzyme-conjugated detection reagent. The plate surface is washed to remove the unbound reagents. The amount of bound conjugate is visualized by adding an appropriate substrate that generates optometric density (OD; colorimetric) read out when hydrolyzed by the enzyme conjugate.

Many enzymes have been used for ELISAs, alkaline phosphatase (ALP) and horseradish peroxidase (HRP) are the most commonly used enzymes. They can be stably conjugated to antibodies generating colored products upon enzymatic reaction with the appropriate substrate. The colored reaction is detected by spectrophotometer (OD plate reader). The detailed protocols for different types and formats of ELISA assays can be found in the reference.[203]

There are several basic formats of ELISA depending on intended purpose and availability, specificity, and affinity of the antibody reagents. The most common types of ELISA are the direct binding, the sandwich assay and the competitive assay.

The direct binding assay design typically has a therapeutic protein of interest (antigen) that is being measured coated on the surface of the assay plate. The specific antibody that recognizes and binds the coated molecule is then added to the plate. The antibody is either conjugated to biotin in which case the avidin linked to the colorimetric enzyme is used as a detection reagent or the antibody is already conjugated with the colorimetric enzyme (HRP or ALP).

The sandwich assay is commonly used for detection of proteins. In the sandwich assay the antibody specific to the molecule of interest (antigen) is coated on the surface of the assay plate. The molecule of interest is then added to bind to the coated antibody forming the antibody–antigen complex. The unbound reagents are washed away; the antibody–antigen complex is detected by another antigen-specific antibody conjugated to the colorimetric enzyme. Because of the potential steric hindrance of the epitopes, this type of ELISA assay is best suited for an antigen that has multiple epitopes that are spatially separated. The sandwich ELISAs are typically performed in stepwise sequential addition of reagents. However, some assays may be performed with simultaneous addition and coincubation of antigen and antibody conjugate.

The signal in both direct and sandwich ELISA is directly proportional to the amount of analyte (protein of interest) present in the test sample.

The competitive assay design is mainly used to detect or quantitate soluble antigens. In the direct competitive assay the purified (molecule of interest) reagent is coated on the plate. The antigen-specific antibody conjugated to an enzyme and the test sample containing the antigen in solution are added concurrently. The antigen in solution competes with the antigen coated on the plate for the antibody binding sites. Since the binding of the antigen-specific antibody-enzyme conjugates to antigen coated on the plate is inhibited by the soluble antigen the signal is inversely proportional to the concentration of the antigen in the sample solution.

All these ELISA can be designed in different and more complex formats depends on the binding properties and nature of the molecule to be detected and availability of the antibody reagents. For example, some molecules may require more complex sandwich assay design consisting of three or more layers of antibodies. Due to the complexity of binding these assays are less sensitive then two antibody sandwich assays.

The final ODs readouts at each concentration of the therapeutic protein can be fit into a curve-fitting equation. Most of the binding immunoassays can be fitted with the four-parametric fit. It is important to determine the best fit for the data. The standard curve typically consists of known concentrations of the purified protein of interest. When ELISA is used as a potency assay the potency of the therapeutic protein is calculated relative to the representative well-characterized lot of therapeutic protein (reference material) that is used to generate the standard curve for the assay. For potency assays, it is important to include the entire standard and test sample curves in the calculation. Testing the entire standard curve and sample curve for parallelism is an important prerequisite for the potency calculation. Having standard and appropriate control(s) on each assay plate is highly recommended.

The ELISA assays have many advantages when assaying soluble antigens and antibodies. The new technologies for mAb generation increased availability of specific antibodies for many therapeutic proteins and other antigens of interest. High affinity and avidity of specific mAbs to the antigen enable strong, long-lasting immuno complexes, which attribute to high sensitivity of the assay and stability of detection and measurement. ELISAs are sensitive and reproducible. The sensitivity of ELISAs to detect and reliably quantitate the antigen in the test sample is at picomolar to nanomolar ranges. Another advantage of ELISAs is that they are relatively short duration assays and easy to perform, which makes them suitable for high throughput testing and automation. Development time for these assays is also relatively short. In addition, a simple OD plate reader is the only necessary equipment, and it does not require extensive training or special skills to use.

However, ELISA assays have technical challenges, pitfalls, and disadvantages to keep in mind while developing a assay and selecting the assay format and reagents. One of the biggest disadvantages of the solid phase assays in general is that the immobilized antibody or antigen on the plate surface can change a conformation often altering the antigenic structure of proteins through unfolding, therefore a capture ELISA in which a protein antigen is coated onto a polymeric surface of a plate may be unsuitable for testing the specificity of antibodies directed against native proteins. Similarly, an orientation of the coated molecule may expose the lower number of binding site influencing the antigen–antibody binding characteristics and affecting detection/quantification of the protein of interest. The presence of interfering factors in the assay either from a test sample matrix (buffers, presence of competing biological molecules, or process or product-related impurities) or from cross-reactivity of reagents represent common technical challenges when selecting a format and reagents for ELISAs.

3. Biosensor-Based Methods

a. Surface Plasmon Resonance

i. Basics of the Technology and Instruments

The most popular biosensor application for macromolecular interactions so far is based on SPR technology. Biacore series have the longest history, and still represent the mainstream, of the commercialized SPR instruments for biomolecules. The instrument is basically composed of an optical system including a light source and response detector; a flow cell chamber that houses a sensor containing a two-sided thin gold film that interacts with the optical system on one side and the sample on the other; and a fluidic system to deliver the samples across the sensor at a constant flow rate. Automated operations can be programmed for multiple samples, allowing higher throughput.

On the gold surface of a sensor chip that is in contact with the liquid in flow cells, carboxylated dextran strands of $\sim$100 nm in length are attached. Molecules of interest can be covalently linked to the dextran via chemical reactions of specific moieties, such as primary amines, sulfhydryls, etc., with the carboxylic acid on the dextran. Specialized sensor chips (e.g., ones precoated with straptavidin for immobilizing biotinylated molecules) are also available. Unlike ELISA and other solid phase attachment methods, all molecules involved in a Biacore study are attached to dextran strands via single chemical bonds or via specific binding to immobilized molecules, and are totally immersed in solution at all times. Therefore, the higher-order structures of the macromolecules are likely to remain intact. However, whether or not the covalent attachment to dextran will result in homogeneous orientation depends on the number, and the relative reactivity, of the active moieties on the molecule that are available for the chemical coupling. In kinetic studies of binding with the immobilized molecules, data result from fitting the experimental curves to kinetic models. Heterogeneity in orientation may lead to poor fitting because no model can account for varied degrees of steric hindrance by chemical linkages which may or may not be near the specific binding site.

The physical principles of SPR phenomenon are discussed and graphically illustrated in biotechnology handbooks[204] and literature.[205,206] The incident light and the sensor chip are at such an angle that the entire beam is reflected (total reflection) at the contact surface of the optic component and the gold film on the sensor. The optical detector covers the whole range of the reflected beam. The incident light induces, by interacting with the electrons in the gold film, an evanescence wave that enters the solution phase from the opposite side of the gold film. This wave is reflected at the contact surface of the gold film and the solution phase, passes through the gold film, and interferes with the reflected incident light. This interference causes the reflected light to be dimmed at a certain angle, detectable by the optical device. This angle varies with the refractive index of the solution within ~200 nm of the gold surface, which in turn, varies with the density of the solution.

In the application to intermolecular interactions, one of the components is first immobilized (see below) to the dextran matrix, and the other is injected when a measurement begins. As the injected molecules bind to the immobilized molecules and accumulate in the dextran matrix, the density of the layer increases, inducing a shift in the angle of interference. Likewise, when the injection is switched to blank buffer, the bound molecules dissociate, causing a change in the opposite direction. The detected change in the interference angle, that is, SPR response, is converted to an arbitrary unit, RU (1000 RU corresponds to 10 mg/mL in the dextran matrix of a CM5 or similar chip); increasing values correspond to an increase in density near surface, and vice versa. A plot of response versus time is the basic form of raw data, called sensorgram. A single curve of the increase and decrease in SPR signal per flow cell is the only output, which reflects all concurrent events in the flow cell, including the binding process of interest and potential artifacts. Some of the artifacts will be discussed below. In-depth understanding of the principle of SPR, and the physical configuration of the instrument, is very helpful in the ability to recognize and correct for artifacts, and optimize the experimental conditions (this often involves consideration of many interplaying factors).

ii. Ligand Immobilization and Experimental Design

The most basic procedure of a Biacore binding study, referred to as direct coupling method, starts from covalently immobilizing one of the binding components, followed by the injection of its binding partner. The immobilized and injected components are referred to as the *ligand* and *analyte*, respectively, in Biacore terminology. The most commonly used reaction for ligand immobilization couples primary amines of proteins with the carboxyl group on carboxylated dextran strands to form peptide-like linkage. It is often obvious which binding component should be chosen to be the ligand. For example, when multiple samples of one binding component are to be tested against a common target, only one flow cell is needed if the target is immobilized, while multiple flow cells, each for one sample (there are four flow cells in commonly used chips), will be used if the opposite is chosen. After the immobilization, nonreacted ligand molecules should be thoroughly washed and a stable baseline should be established before proceeding to binding measurements. It is also a standard practice to have a blank flow cell (treated with conditions identical to ligand immobilization but without exposure to the ligand) as a reference to correct signals not because of specific binding to the ligand. The sources of such nonspecific signals include refractive index spike when an analyte is injected.

Biacore is believed to be a real-time and label-free method for studying intermolecular interactions. The advantage of real-time measurement, coupled with the control over the interaction phases via the control of analyte and buffer injections, is obvious: the binding kinetics can be studied, providing valuable information about binding mechanisms. It should be noted that both association and dissociation take place during the analyte injection; but dissociation dominates after switching to buffer injection. If the experiment is properly designed, the association and dissociation rate constants can be obtained by fitting appropriate kinetic models to the time course of the analyte injection (signal increases) and the subsequent buffer injection phase (signal decreases), together termed sensorgram in Biacore technology. The equilibrium binding affinity can be calculated using these kinetic parameters based on the mechanism. Fitting the kinetic parameters simultaneously to a group of sensorgrams of the same binding reaction, obtained with different analyte concentrations, is recommended for reducing the experimental errors. The practice is termed global fitting. However, the ability to obtain multiple high quality sensorgrams reusing the same ligand surface for high quality global fitting depends on many factors, including complete removal of analyte molecules bound to the ligands in the previous injection. This is typically done by the injection of a cleaning reagent (also called regeneration solution) following the dissociation phase, but care must be taken not to damage the immobilized ligands. These conditions are required for reproducible sensorgrams, but not always easy to achieve.

A very useful alternative design is the capturing method, in which the ligand is noncovalently captured, via specific binding, by another molecule (capturing molecule) which is covalently coupled to the dextran surface. The binding of the capturing molecule with the ligand should have very high affinity so that only minor dissociation occurs during the measurement of the ligand–analyte interaction, while not affecting the binding of the ligand with the analyte. After a measurement of ligan–analyte binding is completed, the surface with only the capturing molecules should be regenerated by removing the all the captured ligand, with or without an analyte bound. The cleaning conditions have to be very effective in dissociating the ligand, however, benign to the capturing molecule. One such example is the capture of mAbs by either protein A or anti-Fc antibodies. The Fc domain of the mAb (ligand) interacts with the capturing molecule on the surface, leaving the Fab domains free to bind its antigen (analyte). Protein A is known to also bind to the CH1 domain of some IgG heavy chains, but does not interfere with the antigen binding in the variable region. Short injections (30 s) of low pH buffer (2.5–3.5) can completely dissociate antibodies from both protein A and anti-Fc IgG without damaging them. The inclusion of salt in the regeneration solution may help with cleaning the surface and the fluidic lines. In general, higher density of the capturing molecules on the surface helps to increase the stability of the ligand capturing. Since protein A is smaller and has higher affinity to Fc domains than anti-Fc IgGs, protein A is more advantageous to be used for capturing Fc-containing ligands. Other useful capturing strategies include using engineered tags at specific locations on the ligands. For example, it is popular to engineer a tag, containing six to ten sequential histidine residues, attached to either the N- or C terminus of a recombinant protein to facilitate the purification. Anti-His antibodies are commercially available from many sources. Some of them have sufficiently high affinity to the His-tag on ligands, therefore, can be used for capturing in a binding assay. Again, a low pH solution is effective and benign for regeneration.

The advantages of the capturing method compared to the direct coupling include the following: (1) Higher efficiency. The need to search for suitable regeneration conditions for new binding studies is largely eliminated; and a chip can be used for multiple projects if all the ligands bind to the same capturing reagent. A protein A-coated chip can be used for >1000 cycles in a month. (2) Better ligand quality. The quality of the ligand molecule is better ensured for all repetitive measurements of ligand–analyte binding because the captured ligand molecules are removed by the regeneration solution after each measurement, and fresh ligand molecules are captured for the next binding cycle. (3) Enhanced homogeneity in ligand orientation. In contrast to randomly coupling ligand with one of the reactive amino acids, which may or may not impede the ligand–analyte binding, the binding of a capturing reagent to ligand for the immobilization is specific in location on the ligand, resulting in a more uniform relative orientation of the ligands from the standpoint of subsequent analyte binding. This is important to the quality of data analysis, since the kinetic models used to fit the sensorgrams assume well-defined binding affinities for one ligand–analyte binding event, and do not account for heterogeneity.

Is Biacore truly a label-free method? The major concern with labeling probes for the purpose of monitoring intermolecular interactions is the possibility of modifying or interfering with the interactions by the attachment of the probes. For protein, this can result from structural alteration upon reaction and/or steric blocking if a probe is in or close to the binding site. The effect can be variable if multiple attachment locations are accessible. In the Biacore technique, similar chemical reactions are used to react with primary amines or free thiols in direct coupling of a protein–ligand, and the attachment to long dextran strands only amplifies the effect of steric blocking if the analyte-binding site is nearby. Therefore, only when specific locations far away from the analyte-binding site are used for ligand immobilization (via engineered cysteines or by using capturing method as described above) the technology may offer the benefit of label-free measurement for intermolecular interactions.

The dextran strands cannot only block or slow down the analyte-binding of the attached ligands by direct steric interference, they can also reduce the accessibility of the ligands attached to other dextran strands nearby. When analyte molecules are fluxed through the dextran matrix, they search for ligand to bind like searching for targets in a dense forest. It is conceivable that smaller analyte molecules can penetrate faster and deeper in the dextran layer (i.e., closer to the solid surface), and larger analyte molecules are less likely to reach all the ligands if the ligand is much smaller. This is demostrated by the inconsistent maximum responses (R_{max}) observed when the choice of ligand and analyte are switched, especially if the two binding components are very different in size/mass. Biacore response is proportional to mass change; the theoretical R_{max} for analyte binding can be calculated given the immobilized ligand baseline, the molecular masses of the two binding molecules and their binding stoichiometry. In reality, however, it is not uncommon that the fitted R_{max} differ when the ligand and analyte are switched.

In our studies of antibody–antigen binding, when the ~150 kDa antibodies are captured by protein A as the ligands at relatively low density (~200 RU), and the analyte molecules are $\leq$15 kDa, the theoretical R_{max} values are obtained. Lower R_{max} values are observed if the antigens, as the analyte, are larger. The R_{max} can be lower than 50% of the theoretical value when the binding components are switched and small antigens are immobilized as ligands. These results indicate that, while R_{max} may theoretically be used to calculate the binding stoichiometry, or to indicate the binding activity when compared with the theoretical R_{max}, caution needs to be taken. The dependence of experimental R_{max} on the method and conditions may not be directly related to the binding properties of the biomolecules analyzed. It is more reliable to compare the results against that of known control samples under identical conditions. It is worth mentioning that using the smaller molecule as the analyte may not always be advantageous because the response for the binding event is lower (proportional to mass). The total analyte-binding signal can be increased by raising the ligand density. However, high ligand density may cause mass transport issues, that is, the rate of analyte transport limits the rate of association.

Most of the condition-related caveats mentioned above tend toward underestimation of binding affinity via negative effects on the analyte-to-ligand association rate. There are, however, situations where the measured binding affinities represent overestimation because of a slower dissociation rate. For example, if an analyte contains two or more binding sites for the ligand, for example, IgG antibody, a bivalent, or multivalent binding event may take place if the ligand density is high. The off-rate of an analyte molecule with double-attachment to surface is lower than that with single attachment. In other words, the binding is stabilized by the avidity. Even though a kinetic model corresponding to this case is available in the Biacore data analysis software, the fitted parameters, reflecting the avidity rather than the affinity of binding to an individual site, may not be the goal of the study. To achieve single-site binding when studying multivalent molecules, one needs to keep the ligand density low, or to switch the analyte–ligand arrangement. Despite this avidity concern, the biosensor method is still more advantageous than the solution phase methods when both binding components are bi- or multivalent. That is because multimolecular complexes (concatamers in general, or immune complexes specifically in the case of antibody–antigen binding) can form upon mixing the two components in solution. This is avoided in a biosensor method as one of the molecules is immobilized.

It is apparent that, while Biacore is a very useful tool for studying intermolecular interactions, designing a good experiment to obtain valid, high-quality results is not a trivial task. It requires extensive experience and thorough understanding of all physicochemical processes that could potentially be involved. The choices of design and conditions are always limited by the properties of the molecules and the configuration of the instrument. Every choice is a compromise between different benefits and artifacts. It has been shown that for exactly the same reaction using reagents from the same sources, experimenters in different laboratories using different experimental designs and conditions do not always produce the same results.[205]

iii. Data Acquisition and Interpretation

Based on the range of the binding affinity, two major strategies of data acquisition and fitting, that is, kinetic and steady-state, can be used. The kinetic measurement, suitable for stronger binding, with a dissociation constant (K_D) in the 10^{-13} to 10^{-7} M range, requires generating nonsaturating sensorgrams with low ligand density and low analyte concentrations (similar in order of magnitude with the K_D). Analysis of the data requires fitting a chosen kinetic model for the kinetic parameters, such as association rate constant (on-rate, k_a), dissociation rate constant (off-rate, k_d), and R_{max}. The binding affinity, either in terms of association constant K_A or dissociation constant K_D, can be calculated using the fitted on-rate(s) and off-rate(s) based on the relations determined by the chosen kinetic mechanism. Simultaneous (global) fitting of multiple sensorgrams generated with analyte concentrations ranging from 20- to 100-fold is a common practice to reduce experimental error. An example is shown in Figure 12.

BiaEvaluation software reports Chi^2 to reflect fitting quality. The value may be helpful in choosing a kinetic model and/or adjusting fitting parameters for the analysis of one experiment. However, to compare data quality between different binding reactions, the Chi^2 values need to be normalized by the absolute response level of each experiment. Practically, when the sensorgrams exhibit reasonable association rates (not too fast; adjustable with analyte concentration) and dissociation rates (not too slow; data collected long enough for an appreciable signal change), it may be more reliable to judge the fitting quality by visually comparing the experimental and the calculated curves. Especially, when reporting or publishing results, sensorgrams overlaying the experimental and fitted curves are far more meaningful than nonnormalized Chi^2 values.

Finding the best-fit kinetic model can help to elucidate the reaction mechanism. However, good fitting does not guarantee the correctness of either the experimental conditions or the model. Further testing with other Biacore methods (e.g., Injection

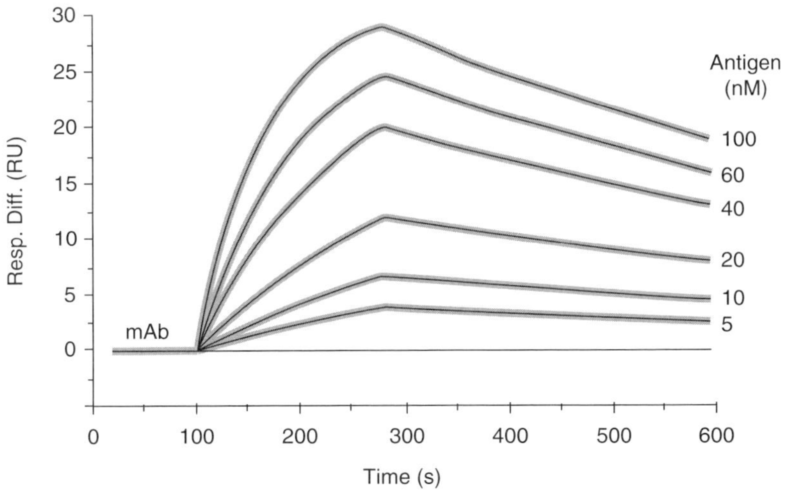

FIGURE 12 Typical sensorgrams (gray) for an antibody–antigen binding are overlaid for global fitting (black).

Time test, see Biacore manual) and/or other orthogonal technologies, as well as structural information of the molecules, should always be considered when making mechanistic interpretations.

When the binding is weak, that is, a K_D in the $10^{-7}$to 10^{-4} M range, the dissociation is fast. To gain an adequate signal, higher analyte concentrations need to be employed, which lead to fast association. The fast-on, fast-off kinetics gives rise to rectangle-shaped sensorgrams even with analyte concentrations far from required to saturate the ligand surface. In this situation, the signal at the plateau of each sensorgram represents a steady-state binding condition (not equilibrium because the system is open with constant flow). It is no longer practical or accurate to fit kinetic models for the on-rate and off-rate of such systems. In these cases a steady-state method needs to be employed, which fits a binding isotherm, that is, the plot of steady-state signal versus the analyte concentration, to the steady-state binding equation (see BiaEvaluation manual) with the dissociation constant K_D as a parameter. One example is shown in Figure 13.

Theoretically, the K_D value equals to the analyte concentration giving rise to the signal half of the maximum. Therefore, experimentally reaching, at least approaching, the maximum (saturation) is critical for obtaining a reliable K_D by curve fitting. Using a log scale for analyte concentration in the isotherm helps to ensure meeting the condition. Switching to steady-state fitting, after finding relatively flat sensorgrams obtained with low analyte concentrations aimed for kinetic measurement, is erroneous because of the large uncertainty in half-maximum when the maximum (saturation) is ill defined.

b. Bio-Layer Interferometry

The application of bio-layer interferometry (BLI) technology to studying *in vitro* interactions of biomolecules was commercialized in more recent years, first by

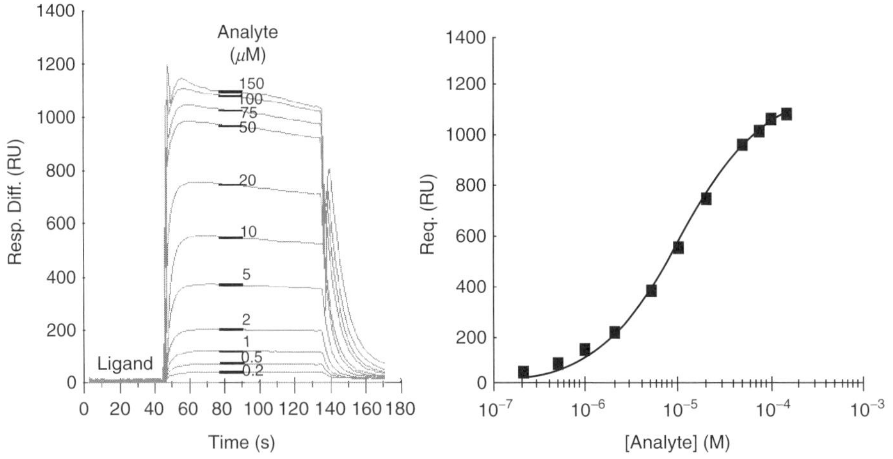

FIGURE 13 **Typical sensorgrams for a steady-state binding (left panel) and the binding isotherm (right panel) derived from the sensorgrams. The dissociation constant K_D is obtained by fitting the measured steady-state response (R_{eq}, horizontal lines in left panel and squares in right panel) with the steady-state binding equation $K_D = (R_{max} - R_{eq})C/R_{eq}$, where C is the analyte concentration, R_{max} is the response at true saturation, also a fitting parameter.**

ForteBio with the Octet series. As an alternative biosensor-based technology, these instruments are gaining popularity fast, mainly because of significant advantages in the ease of operation/maintenance and the higher throughput compared to current models of Biacore.[208,209] The following features contribute to these advantages: (1) "Single-use" disposable sensors. A fiber optic in a needle-shaped plastic sensor mounted on the robotic arm transmits the incident and reflected light between the biolayer at the tip and the main optical device in the instrument. The robotic arm picks up new sensors at the initiation of a binding cycle and ejects them after the dissociation phase is complete. Unlike Biacore, finding a regeneration condition to reuse the sensor is not necessary for the Octet. (2) Simultaneous measurements. Up to eight tips can be mounted on the robotic arm, therefore, up to eight binding events can be measured in one experiment. (3) Batch-mode solution contact. The tips of the sensors on the robotic arm are dipped in solutions contained in individual wells of a 96-well plate. To proceed along the steps of a binding cycle, the sensor tips are moved from one set of wells to another by the robotic arm instead of flowing different solutions through a stationary sensor chip as in Biacore. The elimination of the fluidic system significantly simplifies the mechanics and eliminates one major source of high maintenance.

BLI differs from SPR in the optical detection of binding in biolayer near the sensor surface. BLI measures interference of light reflected from the solution back to the sensor with the back-reflected light directly from the tip surface. The readout of this measurement varies as the density of the biolayer changes, but does not have linear correlation with the solution density. The intensity of the solution-reflected light, dampens with the distance from the tip surface. The farther away molecules bound in the biolayer, the weaker the reflection, resulting in less interference. For example, if a capturing strategy as described in the Biacore section is used to avoid heterogeneous ligand orientation, then the analyte-binding, which is the interest of the study, will be in the third molecular layer from the sensor tip surface. The drastic decrease in detection sensitivity is evident in the cases when the ligand and the analyte are of similar molecular mass. The maximum signal resulting from the analyte binding to the immobilized ligand is much lower than that from ligand capturing by the capturing reagent. Such variation in sensitivity limits the technology to be applied to binding studies with complex configurations, for example, capturing ligand with preimmobilized molecules or "sandwich" binding formats, etc. It also renders the method not suitable for quantitaion without using reference standards. Other drawbacks of Octet instruments include (1) the versatility and flexibility of softwares for both experimental operation and data handling are limited; (2) kinetic fitting is limited to the simplest 1:1 model; (3) lower precision/reproducibility because of $\geq 10\%$ tip-to-tip variability; (4) a more limited range of measurable association rate because of lower maximum data acquisition rate.

Despite the many attractive features of BLI, some of them complement the weaknesses of SPR technology, there is not yet a trend to replace Biacore as the major biosensor-based tool for *in vitro* binding measurements. It's not uncommon that research and development laboratories are equipped with both, with Octet more often used for quick feasibility explorations and Biacore for more accurate kinetic measurements. One particular area that Octet enjoys increased popularity is protein titer estimation, thanks to the high throughput and low requirement for clean solutions. In this application, the concentration of a sample is determined by interpolating the BLI readout, either the steady-state binding signal or the initial rate of binding, against a standard curve, generated using a purified sample of a known concentration.

4. Enzymatic and Other *in Vitro* Functional Assays

As indicated above, for many therapeutic proteins, the eventual biological outcome is initiated via intermolecular interactions with a receptor or target. To quantitatively measure and compare the therapeutic molecule's activity, cell-based bioassays are available if the protein–target interaction induces some kind of cellular response. *In vitro* binding assays based on a variety of physicochemical techniques can also be used if the binding of a protein with its target is required for its mechanism of action. However, some therapeutic proteins are enzymes or proenzymes, and their biological effect is catalysis. There are a variety of catalytic reactions, such as synthesis, hydrolysis, isomerization, and many others. A common feature of catalytic reactions is that only a transient contact between the enzyme and the target molecule (substrate) is required for the enzyme to catalyze the desired change in the substrate molecules, and the products formed dissociate from the enzyme quickly, allowing it to participate in another reaction cycle. The entire process of association–catalysis–dissociation can complete in a very short time (μs to s). Apparently, the cell-based assays and most of the *in vitro* binding methods described in the above sections are not suitable to be functional assays for enzymes.

The behavior and reaction activity of enzymes, measured by the rate of either the substrate consumption or the product generation, comprise the study of enzyme kinetics.[210,211] Fast and accurate quantitation of either a substrate or a product is the key in developing enzymatic assays. If possible, instantaneous physical readout, such as absorbance in visible or UV range, is advantageous (see Section III), especially for determining rates of fast reactions. However, the substrates and products of enzymatic reactions occurring *in situ* (or occurring in biological systems) are very diverse in their physicochemical and biochemical properties, reflecting their diverse functions. Assay methods have to be customized to individual proteins based on knowledge of their function and properties.[212]

Genetic enzyme deficiencies can have severe, even fatal, consequences. Replacement therapy employing a recombinant enzyme has been demonstrated as an effective treatment. A few examples of enzyme or proenzyme products for replacement therapies are discussed below, with the purpose of demonstrating the diversity and specificity of their functional assays.

Gaucher disease is caused by a deficiency of ß-glucocerebrosidase activity, resulting in accumulation of glucocerebroside. Long term enzyme replacement therapy with recombinant ß-glucocerebrosidase is available for pediatric and adult patients. As a typical enzyme, glucocerebrosidase catalyzes hydrolysis of glucocerebroside to glucose and ceramide, but with some degree of cooperativity in the absence of an activator.[213] The enzyme activity assay involves the incubation of the enzyme with a natural substrate glucocerebroside under optimized conditions. The freed α-glucose (product) can be captured and monitored with gas–liquid chromatography. However, pure and well-characterized natural glycocerebroside is difficult to obtain in large enough quantities to meet the needs of the biotechnology industry. Therefore, synthetic surrogates such as aryl β-D-glucoside is more often used as the substrate in routine β-glucocerebrosidase activity assay. A technically more convenient approach is the use of chromogenic synthetic substrate, *p*-nitrophenyl β-D-glucopyranoside (*p*NPGlc), of which the hydrolyzed product *p*NP absorbs at 400 nm. In this instance, the unit activity of the enzyme corresponds to the release of 1 nmol of *p*NP from *p*NPGlc per hour at 37 °C.

Hemophilia A and B are caused by deficiencies in blood coagulation factors VIII (FVIII) and IX (FIX), respectively. Plasma-derived and recombinant FVIII and FIX are commercially available for replacement therapy. FIX is a zymogen (proenzyme), which is converted to an active enzyme after targeted proteolysis by activated factor XI. FVIII, after proteolytic activation to FVIIIa, is a cofactor for activated FIX (FIXa) in enzymatic activation of factor X (FX) by FIXa. FVIIIa does not possess enzymatic activity itself, but greatly enhances the catalytic activity of FIXa. Despite the fact that FVIII and FIX have different functions at the molecular level, they are both essential components of the intricate cascade of blood coagulation, giving rise to the possibility of applying the same assay to both if the readout can be taken at steps downstream of that involving FVIII and FIX. The simplest method to measure the function of FVIII and FIX is to record the clotting time obtained after mixing FVIII in FVIII-deficient plasma, or FIX in FIX-deficient plasma.[214] The occurrence of clotting is judged by visual inspection or viscosity increase. This simple method has been widely used for longer than half a century, despite suffering from high variability because of a lack of standardization of the biologically sourced assay components, including pooled blood plasma.[214]

Alternatively, a biochemically defined chromogenic substrate method can be used to assess the function of FVIII and FIX. The chromogenic substrate method is performed using purified reagents representing the macromolecules associated with the specific reactions involving FVIII and FIX. Because the function of the enzymatic complex of FVIIIa and FIXa is the generation of FXa, a synthetic chromogenic substrate of FXa is then used to detect the FXa produced. The assay is configured to readout FXa units which are directly proportional to the limiting FVIII or FIX added. Since the effectiveness of a replacement treatment correlates with the dosage of the therapeutic, the functional assays are in practice also used as quantitation methods to determine the dosage strengths of the therapeutics. For FVIII, both the clotting time and chromogenic substrate methods have been exploited as quantitation methods, whereas for FIX, the chromogenic substrate method still suffers from reagent instability issues.[215,216]

VI. SUMMARY AND CONCLUSIONS

This chapter provides a brief overview of analytical methods commonly used to assess the structure, purity, safety, stability, and potency of recombinant protein pharmaceuticals. These methods are technically challenging to develop because the protein products are large in size, complex in structure and function, and marginally stable in aqueous solution. The basic biochemical, biophysical, and cell biology principles behind the methods are described in the chapter.

The properties and functions of all proteins are determined to a large extent by their amino acid sequences, as well as the posttranslational modifications by cellular machinery. Section II describes some basic methods commonly used in analytical laboratories in the biopharmaceutical industry to characterize primary structure. It also covers characterization of higher-order structure, which is also a determinant of protein properties and functions. Section III then describes methods used to measure protein concentration, while Section IV describes methods used to assess product- and process-related impurities. Lastly, the development of functional bioassays, non cell-based functional assays, and *in vitro* binding assays is described in Section V. Our experience indicates that the development of high quality, high efficiency analytical methods benefits from collaboration between highly skilled analysts with strong backgrounds in biochemistry, biophysics, and cell biology.

ACKNOWLEDGMENTS

The authors would like to extend their gratitude to Denise O'Hara, Karen Blank, Andrew Hanneman, Mark Hardy, Mike Jankowski, and Bruce Tangarone for their technical input and review of the manuscript.

REFERENCES

1. http://www.pharmamanufacturing.com/articles/2009/058.html?page=full.
2. Jensen, O. N. Modification-specific proteomics: Characterization of post-translational modifications by mass spectrometry. *Curr. Opin. Chem. Biol.* 8:33–41, 2004.
3. ICH Harmonised Tripartite Guideline Specifications: Test Procedures and Acceptance Criteria For Biotechnological/Biological Products Q6b.
4. Rathore, A. S. Roadmap for implementation of quality by design (QbD) for biotechnology products. *Cell* 27:546–553, 2009.
5. Judd, R. C. Peptide Mapping by High-Performance Liquid Chromatography. 2nd ed., Protein Protocols Handbook, 559–561, 2002.
6. US Pharmacopeia 1055 Biotechnology-Derived Articles—Peptide Mapping. 2009.
7. Mann, M., Højrup, P. and Roepstorff, P. Use of mass spectrometric molecular weight information to identify proteins in sequence databases. *Biol. Mass Spectrom.* 22:338–345, 1994.
8. Blackburn, R. K. and Moseley, M. A. Quadrupole time-of-flight mass spectrometry: A powerful new tool for protein identification and characterization. *Am. Pharm. Rev.* 2:49–59, 1999.
9. Morris, H. R., Paxton, T., Dell, A., Langhorne, J., Berg, M., Bordoli, R. S., Hoyes, J. and Bateman, R. H. High sensitivity collisionally-activated decomposition tandem mass spectrometry on a novel quadrupole/orthogonal-acceleration time-of-flight mass spectrometer. *Rapid Commun. Mass Spectrom.* 10:889–896, 1996.
10. Charles, L. Flow injection of the lock mass standard for accurate mass measurement in electrospray ionization time-of-flight mass spectrometry coupled with liquid chromatography. *Rapid Commun. Mass Spectrom.* 17:1383–1388, 2003.
11. Rouse, J. C., McClellan, J. E., Patel, H. K., Jankowski, M. A. and Porter, T. J. Top-down characterization of protein pharmaceuticals by liquid chromatography/mass spectrometry application to recombinant Factor IX comparability—A case study. *Methods Mol. Biol.* 308:435–460, 2005.
12. Johnson, K. A., Paisley-Flango, K., Tangarone, B. S., Porter, T. J. and Rouse, J. C. Cation exchange-HPLC and mass spectrometry reveal C-terminal amidation of an IgG1 heavy chain. *Anal. Biochem.* 360:75–83, 2007.
13. Porter, T. J., Rathore, S., Rouse, J. C. and Denton, M. Biomolecules in tissue engineered medical products (TEMPs): A case study of recombinant human bone morphogenetic protein-2 (rhBMP-2). *Tissue Eng. Med. Prod.* 1:150–171, 2004.

14. Geromanos, S. J., Vissers, J. P. C., Silva, J. C., Dorschel, C. A., Li, G.-Z., Gorenstein, M. V., Bateman, R. H. and Langridge, J. I. The detection, correlation, and comparison of peptide precursor and product ions from data independent LC–MS with data dependant LC–MS/MS. *Proteomics* 9:1683–1695, 2009.
15. Li, G.-Z., Vissers, J. P. C., Silva, J. C., Golick, D., Gorenstein, M. V. and Geromanos, S. J. Database searching and accounting of multiplexed precursor and product ion spectra from the data independent analysis of simple and complex peptide mixtures. *Proteomics* 9:1696–1719, 2009.
16. Olsen, J. V., de Godoy, L. M. F., Li, G., Macek, B., Mortensen, P., Pesch, R., Makarov, A., Lange, O., Horning, S. and Mann, M. Parts per million mass accuracy on an orbitrap mass spectrometer via lock mass injection into a C-trap. *Mol. Cell. Proteomics* 4:2010–2021, 2005.
17. Haas, W., Faherty, B. K., Gerber, S. A., Elias, J. E., Beausoleil, S. A., Bakalarski, C. E., Li, X., Villen, J. and Gygi, S. P. Optimization and use of peptide mass measurement accuracy in shotgun proteomics. *Mol. Cell. Proteomics* 5:1326–1337, 2006.
18. Yates, J. R., Cociorva, D., Liao, L. and Zabrouskov, V. Performance of a linear ion trap-orbitrap hybrid for peptide analysis. *Anal. Chem.* 78:493–500, 2006.
19. Asa, D. High performance mass spectrometry for small molecule and protein applications. *Curr. Trends Mass Spectrom.* 30–32, 2009. (supplement to spectroscopy).
20. Xie, H., Gilar, M. and Gebler, J. C. Characterization of protein impurities and site-specific modifications using peptide mapping with liquid chromatography and data independent acquisition mass spectrometry. *Anal. Chem.* 81:5699–5708, 2009.
21. Edman, P. Sequence determination. *Mol. Biol. Biochem. Biophys.* 8:211–255, 1970.
22. Heinrikson, R. L. Application of automated sequence analysis to the understanding of protein structure and function. *Ann. Clin. Lab. Sci.* 8:295–301, 1978.
23. Mozdzanowski, J. Deblocking of proteins containing N-terminal pyroglutamic acid. *Methods Mol. Biol.* 211:365–369, 2003.
24. Ellman, G. L. Tissue sulfhydryl groups. *Arch. Biochem. Biophys.* 82:70–77, 1959.
25. Sedlak, J. and Lindsay, R. H. Estimation of total, protein-bound, and nonprotein sulfhydryl groups in tissue with Ellman's reagent. *Anal. Biochem.* 25:192–205, 1968.
26. Habeeb, A. F. S. A. Reaction of protein sulfhydryl groups with Ellman's reagent. *Methods Enzymol.* 25:457–464, 1972.
27. Bulaj, G., Kortemme, T. and Goldenberg, D. P. Ionization-reactivity relationships for cysteine thiols in polypeptides. *Biochemistry* 37:8965–8972, 1998.
28. Riddles, P. W., Blakeley, R. L. and Zerner, B. Ellman's reagent: 5, 5′-dithiobis(2-nitrobenzoic acid)—A reexamination. *Anal. Biochem.* 94:75–81, 1978.
29. Riddles, P. W., Blakeley, R. L. and Zerner, B. Reassessment of Ellman's reagent. *Methods Enzymol.* 91:49–60, 1983.
30. Hatsuo, M., Matsuno, H., Ushida, M., Katayama, K., Saeki, K. and Itoh, N. 2, 4-Dinitrobenzenesulfonyl fluoresceins as fluorescent alternatives to Ellman's reagent in thiol-quantification enzyme assays. *Angew. Chem. Int. Ed.* 44:2922–2925, 2005.
31. Pullela, P. K., Chiku, T., Carvan 3rd, M. J. and Sem, D. S. Fluorescence-based detection of thiols in vitro and in vivo using dithiol probes. *Anal. Biochem.* 352:265–273, 2006.
32. Hansen, R. E., Ostergaard, H., Norgaard, P. and Winther, J. R. Quantification of protein thiols and dithiols in the picomolar range using sodium borohydride and 4, 4′-dithiodipyridine. *Anal. Biochem.* 363:77–82, 2007.
33. Wright, S. K. and Viola, R. E. Evaluation of methods for the quantitation of cysteines in proteins. *Anal. Biochem.* 265:8–14, 1998.
34. Yi, L., Li, H., Sun, L., Liu, L., Zhang, C. and Xi, Z. A highly sensitive fluorescence probe for fast thiol-quantification assay of glutathione reductase. *Angew. Chem. Int. Ed.* 48:4034–4037, 2009.
35. Mhatre, R., Woodard, J. and Zeng, C. Strategies for locating disulfide bonds in a monoclonal antibody via mass spectrometry. *Rapid Commun. Mass Spectrom.* 13:2503–2510, 1999.
36. Mikesh, L. M., Ueberheide, B., Chi, A., Coon, J. J., Syka, J. E. P., Shabanowitz, J. and Hunt, D. F. The utility of ETD mass spectrometry in proteomic analysis. *Biochim. Biophys. Acta* 1764:1811–1822, 2006.

37. Wu, S.-L., Jiang, H., Lu, Q., Dai, S., Hancock, W. S. and Karger, B. L. Mass spectrometric determination of disulfide linkages in recombinant therapeutic proteins using online LC–MS with electron-transfer dissociation. *Anal. Chem.* 81:112–122, 2009.
38. Qi, J., Wu, J., Somkuti, G. A. and Watson, J. T. Determination of the disulfide structure of sillucin, a highly knotted, cysteine-rich peptide, by cyanylation/cleavage mass mapping. *Biochemistry* 40:4531–4538, 2001.
39. Zhang, W., Marzilli, L. A., Rouse, J. C. and Czupryn, M. J. Complete disulfide bond assignment of a recombinant immunoglobulin G4 monoclonal antibody. *Anal. Biochem.* 311:1–9, 2002.
40. Chelius, D., Wimer, M. E. H. and Bondarenko, P. V. Reversed-phase liquid chromatography in-line with negative ionization electrospray mass spectrometry for the characterization of the disulfide-linkages of an immunoglobulin gamma antibody. *J. Am. Soc. Mass Spectrom.* 17:1590–1598, 2006.
41. Wypych, J., Li, M., Guo, A., Zhang, Z., Martinez, T., Allen, M. J., Fodor, S., Kelner, D. N., Flynn, G. C. and Liu, Y. D., *et al.* Human IgG2 antibodies display disulfide-mediated structural isoforms. *J. Biol. Chem.* 283:16194–16205, 2008.
42. Zhao, L., Almaraz, R. T., Xiang, F., Hedrick, J. L. and Franza, A. H. Gas-phase scrambling of disulfide bonds during matrix-assisted laser desorption/ionization mass spectrometry analysis. *J. Am. Soc. Mass Spectrom.* 20:1603–1616, 2009.
43. Zhang, B. and Cockrill, S. L. Methodology for determining disulfide linkage patterns of closely spaced cysteine residues. *Anal. Chem.* 81:7314–7320, 2009.
44. Canova-Davis, E., Baldonado, I. P., Chloupek, R. C., Ling, V. T., Gehant, R., Olson, K. and Gillece-Castro, B. L. Confirmation by mass spectrometry of a trisulfide variant in methionyl human growth hormone biosynthesized in *Escherichia coli*. *Anal. Chem.* 68:4044–4051, 1996.
45. Pristatsky, P., Cohen, S. L., Krantz, D., Acevedo, J., Ionescu, R. and Vlasak, J. Evidence for trisulfide bonds in a recombinant variant of a human IgG2 monoclonal antibody. *Anal. Chem.* 81:6148–6155, 2009.
46. Gu, S., Wen, D., Weinreb, P. H., Sun, Y., Zhang, L., Foley, S. F., Kshirsagar, R., Evans, D., Mi, S. and Meier, W., *et al.* Characterization of trisulfide modification in antibodies. *Anal. Biochem.* 400:89–98, 2010.
47. Tous, G. I., Wei, Z., Feng, J., Bilbulian, S., Bowen, S., Smith, J., Strouse, R., McGeehan, P., Casas-Finet, J. and Schenerman, M. A. Characterization of a novel modification to monoclonal antibodies: Thioether cross-link of heavy and light chains. *Anal. Chem.* 77:2675–2682, 2005.
48. Cohen, S. L., Price, C. and Vlasak, J. β-elimination and peptide bond hydrolysis: Two distinct mechanisms of human IgG1 hinge fragmentation upon storage. *J. Am. Chem. Soc.* 129:6976–6977, 2007.
49. Delta Mass http://www.abrf.org/index.cfm/dm.home.
50. Walsh, G. Post-translational modifications in the context of therapeutic proteins: An introductory overview. In *Post-translational Modification of Protein Biopharmaceuticals*, (G. Walsh, ed.), Wiley-VCH, Weinhaim, pp. 1–76, 2009.
51. Hirschberg, C. B. and Snider, M. D. Topography of glycosylation in the rough endoplasmic reticulum and Golgi apparatus. *Annu. Rev. Biochem.* 56:63–87, 1987.
52. Reuter, G. and Gabius, H. J. Eukaryotic glycosylation: Whim of nature or multipurpose tool? *Cell. Mol. Life Sci.* 55:368–432, 1999.
53. Weerapana, E. and Imperiali, B. Asparagine-linked protein glycosylation from eukaryotic to prokaryotic systems. *Glycobiology* 16:91–101, 2006.
54. Schellekens, H. Bioequivalence and the immunogenecity of biopharmaceuticals. *Nat. Rev. Drug Dis.* 1:457–462, 2002.
55. Brooks, S. A. Appropriate glycosylation of recombinant proteins for human use. *Mol. Biotechnol.* 28:241–255, 2004.
56. Mahmood, I. and Green, M. D. Pharmacokinetic and pharmacodynamic considerations in the development of therapeutic proteins. *Clin. Pharmacokinet.* 44:31–347, 2005.
57. Butler, M. Optimisation of the cellular metabolism of glycosylation for recombinant proteins produced by mammalian cell systems. *Cytotechnology* 50:57–76, 2006.
58. Morrow, K. J. Advances in antibody manufacturing using mammalian cells. *Biotechnol. Annu. Rev.* 13:95–113, 2007.

59. Dingermann, T. Recombinant therapeutic proteins: Production platforms and challenges. *Biotechnol. J.* 3:90–97, 2008.
60. Mukovozov, I., Sabljic, T., Hortelano, G. and Ofosu, F. A. Factors that contribute to the immunogenicity of therapeytic recombinant human proteins. *Throm. Haemost.* 99:874–882, 2008.
61. Kawasaki, N., Itoh, S., Hashii, N., Takakura, D., Qin, Y., Huang, X. and Yamaguchi, T. The significance of glycosylation analysis in development of biopharmaceuticals. *Biol. Pharm. Bull.* 32:796–800, 2009.
62. Kornfeld, R. and Kornfeld, S. Assembly of asparagines-linked oligosaccharides. *Annu. Rev. Biochem.* 54:631–664, 1985.
63. Shakin-Eshleman, S. H., Spitalnik, S. L. and Kasturi, L. The amino acid at the X position of an Asn-X-ser sequon is an important determinant of N-linked core-glycosylation efficiency. *J. Biol. Chem.* 271:6356–6363, 1996.
64. Jones, J., Krag, S. S. and Betenbaugh, M. J. Controlling N-linked glycan site occupancy. *Biochim. Biophys. Acta* 1726:121–137, 2005.
65. van der Steen, P., Rudd, P. M., Dwek, R. A. and Opdenakker, G. Concepts and principles of O-linked glycosylation. *Crit. Rev. Biochem. Mol. Biol.* 33:151–208, 1998.
66. Furmanek, A. and Hofsteenge, J. Protein C-mannosylation: Facts and questions. *Acta Biochim. Pol.* 47:781–789, 2000.
67. Sinclair, A. M. and Elliott, S. Glycoengineering: The effect of glycosylation on the properties of therapeutic proteins. *J. Pharmaceut. Sci.* 94:1626–1635, 2005.
68. Bork, K., Horstkorte, R. and Weidemann, W. Increasing the sialylation of therapeutic glycoproteins: The potential of the sialic acid biosynthetic pathway. *J. Pharm. Sci.* 98:3499–3508, 2009.
69. Henkin, J., Dudlak, D., Beebe, D. P. and Sennello, L. igh sialic acid content slows prourokinase turnover in rabbits. *Throm. Res.* 63:215–225, 1991.
70. Rice, K. G. and Lee, Y. C. Oligosaccharide valency and conformation in determining binding to the asialoglycoprotein receptor hepatocytes. In *Advances in Enzymology and Related Areas of Molecular Biololgy* 66, (A. Meister, ed.), John Wiley & Sons Inc., New Jersey, pp. 41–83, 1993.
71. Jones, A. J. S., Papac, D. L., Chin, E. H., Keck, R., Baughman, S. A., Lin, Y. S., Kneer, J. and Battersby, J. E. Selective clearance of glycoforms of a complex glycoprotein pharmaceutical caused by terminal *N*-acetylglucosamine is similar in humans and cynomolgus monkeys. *Glycobiology* 17:529–540, 2007.
72. Spiro, R. G. Protein glycosylation: Nature, distribution, enzymatic formation, and disease implications of glycopeptide bonds. *Glycobiology* 12:43–56, 2002.
73. Plummer, T. H. J., Elder, J. H., Alexander, S., Phelan, A. W. and Tarentino, A. L. Demonstration of peptide:*N*-glycosidase F activity in endo-beta-*N*-acetylglucosaminidase F preparations. *J. Biol. Chem.* 259:10700–10704, 1984.
74. Tarentino, A. L., Quinones, G., Changhien, L.-M. and Plummer, T. H. J. Multiple endoglycosidase F activities expressed by *Flavobacterium meningosepticum* endoglycosidases F2 and F3. *J. Biol. Chem.* 268:9702–9708, 1993.
75. Fan, J.-Q. and Lee, Y. C. Detailed studies on substrate structure requirements of glycoamindases A and F. *J. Biol. Chem.* 272:27058–27064, 1997.
76. Nuck, R., Zimmermann, M., Sauvageot, D., Josic, D. and Reutter, W. Optimized deglycosylation of glycoproteins by peptide-*N*-(*N*-acetyl-β-glycsoaminyl)-asparagine amidase from flavobacterium meningosepticum. *Glycoconjug. J.* 7:279–286, 1990.
77. Edge, A. S. Deglycosylation of glycoproteins with trifluoromethanesulphonic acid: Elucidation of molecular structure and function. *Biochem. J.* 376:339–350, 2003.
78. Takasaki, S., Mizuochi, T. and Kobata, A. Hydrazinolysis of asparagines-linked sugar chains to produce free oligosaccharides. *Methods Enzymol.* 83:263–268, 1982.
79. Patel, T. P. and Parekh, R. B. Release of oligosaccharides from glycoproteins by hydrazinolysis. *Methods Enzymol.* 230:57–66, 1994.
80. Kobata, A. Use of endo- and exoglycosidases for structural studies of glycoconjugates. *Anal. Biochem.* 100:1–14, 1979.
81. Townsend, R. R. and Hardy, M. R. Analysis of glycoprotein oligosaccharides using high-pH anion exchange chromatography. *Glycobiology* 1:139–147, 1991.
82. Hase, S. High-performance liquid chromatography of pyridylaminated saccharides. *Methods Enzymol.* 230:225–237, 1994.

83. Hardy, M. R. and Townsend, R. R. High-pH anion-exchange chromatography of glycoprotein-derived carbohydrates. *Methods Enzymol.* 230:208–225, 1994.
84. Anumula, K. R. High-sensitivity and high-resolution methods for glycoprotein analysis. *Anal. Biochem.* 283:17–26, 2000.
85. Guile, G. R., Rudd, P. M., Wing, D. R., Prime, S. B. and Dwek, R. A. A rapid high-resolution high-performance liquid chromatographic method for separating glycan mixtures and analyzing oligosaccharide profiles. *Anal. Biochem.* 240:210–226, 1996.
86. Evangelista, R. A., Guttman, A. and Chen, F.-T. A. Acid-catalyzed reductive amination of aldoses with 8-aminopyrene-1, 3, 6-trisufonate. *Electrophoresis* 17:347–351, 1996.
87. Hardy, M. R., Townsend, R. R. and Lee, Y. C. Monosaccharide analysis of glycoconjugates by anion exchange chromatography with pulsed amperometric detection. *Anal. Biochem.* 170:54–62, 1988.
88. Byrne, B., Donohoe, G. G. and O'Kennedy, R. Sialic acids: Carbohydrate moieties that influence the biological and physical properties of biopharmaceutical proteins and living cells. *Drug Dis. Today* 12:319–326, 2007.
89. Kelm, S. and Schauer, R. Sialic acids in molecular and cellular interactions. *Int. Rev. Cytol.* 175:124–137, 1997.
90. Hara, S., Yamaguchi, M., Takemore, Y. and Sakamura, M. Highly sensitive determination of *N*-acetyl- and *N*-glycolylneuraminic acids in human serum and urine and rat serum by reversed-phase liquid chromatography with fluorescence detection. *J. Chromatogr.* 377:111–119, 1986.
91. Anumula, K. R. Rapid quantitative determination of sialic acids in glycoproteins by high-performance liquid chromatography with a sensitive fluorescence detection. *Anal. Biochem.* 230:24–30, 1995.
92. Brown, M. A. and Stenberg, L. M. Biopharmaceuticals: Post-translational modification carboxylation and hydroxylation. In *Posttranslational Modifcation of Protein Biopharmaceuticals*, (G. Walsh, ed.), Wiley-VCH, Weinheim, pp. 210–252, 2009.
93. Powell, M. F. A compendium and hydropathy/flexibility analysis of common reactive sites in proteins: Reactivity at Asn, Asp, Gln, and Met motifs in neutral pH solution. In *Formulation, Characterization, and Stability of Protein Drugs*, (R. Pearlman and Y. J. Wang, eds.), Plenum Press, New York, NY, 1996.
94. Clarke, S., Stephenson, R. C. and Lowenson, J. D. Lability of asparagine and aspartic acid residues in proteins and peptides: Spontaneous deamidation and isomerization reactions in stability of protein pharmaceuticals, part a. In *Chemical and Physical Pathways of Protein Degradation*, (T. J. Ahern and M. C. Manning, eds.), Plenum Press, New York, NY, 1992.
95. Brennan, T. V. and Clarke, S. Deamidation and isoaspartate formation in model synthetic peptides: The effects of sequence and solution environment. In *Deamidation and Isoaspartate Formation in Peptides and Proteins*, (D. W. Aswad, ed.), CRC Press, Washington DC, pp. 65–90, 1995.
96. Chazin, W. J. and Kossiakoff, A. A. The role of secondary and tertiary structures in intramolecular deamidation of proteins. In *Deamidation and Isoaspartate Formation in Peptides and Proteins*, (D. W. Aswad, ed.), CRC Press, Washington, DC, pp. 193–206, 1995.
97. Wearne, S. J. and Creighton, T. E. Effect of protein conformation on rate of deamidation: Ribonuclease A. *Proteins* 5:8–12, 1989.
98. Robinson, A. J. and Rudd, C. J. Deamidation of glutaminyl and asparaginyl residues in peptides and proteins. *Curr. Top. Cell. Regul.* 8:2247–2295, 1974.
99. Fukawa, H. Changes of glutamine-peptides on heating in aqueous media. *J. Chem. Soc. Jpn.* 88:459–463, 1967.
100. Pearlman, R. and Bewley, T. A. Stability and characterization of human growth hormone. *Pharm. Biotechnol.* 5:1–58, 1993.
101. Yokota, H., Saito, H., Masuoka, K., Kaniwa, H. and Shibanuma, T. Reversed phase HPLC of Met58 oxidized rhIL-11: Oxidation enhanced by plastic tubes. *J. Pharm. Biomed. Anal.* 24:317–324, 2000.
102. Bogosian, G., Violand, B. N., Dorward-King, E. J., Workman, W. E., Jung, P. E. and Kane, J. F. Biosynthesis and incorporation into protein of norleucine by *Escherichia coli*. *J. Biol. Chem.* 264:531–539, 1989.
103. Muramatsu, R., Misawa, S. and Hayashi, H. Finding of an isoleucine derivative of a recombinant protein for pharmaceutical use. *J. Pharm. Biomed. Anal.* 31:979–987, 2003.

104. Stark, G. R. Reactions of cyanate with functional groups of proteins. IV. Inertness of aliphatic hydroxyl groups. Formation of carbamyl- and acylhydantoins. *Biochemistry* 4:2363–2367, 1965.
105. Quan, C., Alcala, E., Petkoska, I., Matthews, D., Canova-Davis, E., Taticek, R. and Ma, S. A study in glycation of a therapeutic recombinant humanized monoclonal antibody: Where it is, how it got there, and how it affects charge-based behavior. *Anal. Biochem.* 373:179–191, 2008.
106. Zhang, B., Yan, Y., Yuk, I., Pai, R., McKay, P., Eigenbrot, C., Dennis, M., Katta, V. and Champion Francissen, K. Unveiling a glycation hot spot in a recombinant humanized monoclonal antibody. *Anal. Chem.* 80:2379–2390, 2008.
107. Fischer, S., Hoernschemeyer, J. and Mahler, H. C. Glycation during storage and administration of monoclonal antibody formulations. *Euro. J. Pharm. Biopharm.* 70:42–50, 2008.
108. Banks, D. D., Hambly, D. M., Scavezze, J. L., Siska, C. C., Stackhouse, N. L. and Gadgil, H. S. The effect of sucrose hydrolysis on the stability of protein therapeutics during accelerated formulation studies. *J. Pharm. Sci.* 98:4501–4510, 2009.
109. http://www.rcsb.org/pdb/home/home.do.
110. Philo, J. S. and Arakawa, T. Mechanisms of protein aggregation. *Curr. Pharm. Biotechnol.* 10:348–351, 2009.
111. Philo, J. S. A critical review of methods for size characterization of non-particulate protein aggregates. *Curr. Pharm. Biotechnol.* 10:359–372, 2009.
112. Arakawa, T., Philo, J. S., Ejima, D., Tsumoto, K. and Arisaka, F. Aggregation analysis of therapeutic proteins, Part 2. Analytical ultracentrifugation and dynamic light scattering. *Bioprocess. Int.* 5:36–47, 2007.
113. Berkowitz, S. A. Role of analytical ultracentrifugation in assessing the aggregation of protein biopharmaceuticals. *AAPS J.* 8:E590–E605, 2006.
114. Liu, J., Andya, J. D. and Shire, S. J. A critical review of analytical ultracentrifugation and field flow fractionation methods for measuring protein aggregation. *AAPS J.* 8:E580–E589, 2006.
115. Stafford, W. F. and Braswell, E. H. Sedimentation velocity, multi-speed method for analyzing polydisperse solutions. *Biophys. Chem.* 108:273–279, 2004.
116. Philo, J. S. Is any measurement method optimal for all aggregate sizes and types? *AAPS J.* 8: E564–E571, 2006.
117. Arakawa, T., Philo, J. S., Ejima, D., Sato, H. and Tsumoto, K. Aggregation analysis of therapeutic proteins, Part 3. Principles and optimization of field-flow fractionation (FFF). *Bioprocess Intern.* 5:52–70, 2007.
118. Cao, S., Pollastrini, J. and Jiang, Y. Separation and characterization of protein aggregates and particles by field flow fractionation. *Curr. Pharm. Biotechnol.* 10:382–390, 2009.
119. Anders, J. C., Parten, B. F., Petrie, G. E., Marlowe, R. L. and McEntire, J. E. Using amino acid analysis to determine absorptivity constants: A validation case study using bovine serum albumin. *Biopharm. Int.* 16:30–37, 2003.
120. Noble, J. E., Knight, A. E., Reason, A. J., Di Matola, A. and Bailey, M. J. A. A Comparison of protein quantitation assays for biopharmaceutical applications. *Mol. Biotechnol.* 37:99–111, 2007.
121. Cooper, C., Packer, N. and Williams, K., eds., *In* Amino Acid Analysis Protocols. Humana Press, Totowa, NJ, 2001.
122. Macchi, F. D., Shen, F. J., Keck, R. G. and Harris, R. J. Amino acid analysis, using postcolumn ninhydrin detection, in a biotechnology laboratory. *Methods Mol. Biol.* 159:9–30, 2001. (Totowa, N. J).
123. Edelhoch, H. Spectroscopic determination of tryptophan and tyrosine in proteins. *Biochemistry* 6:1948–1954, 1967.
124. Gill, S. C. and Von Hippel, P. H. Calculation of protein extinction coefficients from amino acid sequence data. *Anal. Biochem.* 182:319–326, 1989.
125. Pace, C. N., Vajdos, F., Fee, L., Grimsley, G. and Gray, T. How to measure and predict the molar absorption coefficient of a protein. *Protein Sci.* 4:2411–2423, 1995.
126. Traub, W. and Piez, K. A. The chemistry and structure of collagen. *Adv. Protein Chem.* 25:243–352, 1971.
127. Herzberg, O. and Klevit, R. Unraveling a bacterial hexose transport pathway. *Curr. Opin. Struct. Biol.* 4:814–822, 1994.

128. Olson, B. J. S. C. and Markwell, J. Assays for the determination of protein concentration. *Curr. Protoc. Protein Sci.* 3.4.1–3.4.29, 2007.
129. Guideline, I.H.T. Specifications: Test Procedures and Acceptance Criteria For Biotechnological/Biological Products. Q6b.
130. Herman, A. C. Purity of biological products: Related and unrelated impurities. *Dev. Biol. Stand.* 16:57–62, 1998.
131. Perkins, M., Theiler, R., Lunte, S. and Jeschke, M. Determination of the origin of charge heterogeneity in a murine monoclonal antibody. *Pharmaceut. Res.* 17:110–1117, 2000.
132. Harris, R. J., Kabakoff, B., Macchi, F. D., Shen, F. J., Kwong, M., Andya, J. D. and Shire, S. J., *et al.* Identification of multiple sources of charge heterogeneity in a recombinant antibody. *J. Chromatogr. B Biomed. Sci Appl.* 752:233–245, 2001.
133. Beck, A., Bussat, M.-C., Zorn, N., Robillard, V. and Klinguer-Hamour, C., *et al.* Characterization by liquid chromatography combined with mass spectrometry of monoclonal anti-IGF-1 receptor antibodies produced in CHO and NS0 cells. *J. Chromatogr. B* 819:203–218, 2005.
134. Ahrer, K. and Jungbauer, A. Chromatographic and electrophoretic characterization of protein variants. *J. Chromatogr.* 841:110–122, 2006.
135. Berthold, W. and Walter, J. Protein purification: Aspects of processes for pharmaceutical products. *Biologicals* 22:135–150, 1994.
136. Cromwell, M. E. M., Hilario, E. and Jacobson, F. Protein aggregation and bioprocessing. *AAPS J.* 8:E572–E579, 2006.
137. Bayol, A., Bristow, A., Charton, E., Girard, M. and Jongen, P. Somatropin and its variants: Structural characterization and methods of analysis. *Pharmeuropa. Bio.* 2004:35–45, 2004.
138. Yan, B., Vallier-Douglass, J., Brady, L., Steen, S., Han, M. and Pace, D., *et al.* Analysis of post-translational modifications in recombinant monoclonal antibody IgG1 by reversed-phase liquid chromatography/mass spectrometry. *J. Chromatogr.* 1164:153–161, 2007.
139. Xie, H., Gilar, M. and Gebler, J. C. Characterization of protein impurities and site-specific modifications using peptide mapping with liquid chromatography and data independent acquisition mass spectrometry. *Anal. Chem.* 81:5699–5708, 2009.
140. Roque, A. C. A., Lowe, C. R. and Taipa, M. A. Antibodies and genetically engineered related molecules: Production and purification. *Biotechnol. Prog.* 20:639–654, 2004.
141. Hober, S., Nord, K. and Linhult, M. Protein A chromatography for antibody purification. *J. Chromatogr.* 848:40–47, 2007.
142. Low, D., O'Leary, R. and Pujar, N. S. Future of antibody purification. *J. Chromatogr.* 848:48–63, 2007.
143. Huse, K., Bohme, H.-J. and Scholz, G. H. Purification of antibodies by affinity chromatography. *J. Biochem. Biophys. Methods* 51:217–231, 2002.
144. Follman, D. K. and Fahrner, R. L. Factorial screening of antibody purification processes using three chromatography steps without protein A. *J. Chromatogr.* 1024:79–85, 2004.
145. Hahn, R., Schelgel, R. and Jungbauer, A. Comparison of protein A affinity sorbents. *J. Chromatogr.* 790:35–51, 2003.
146. Ghose, S., Hubbard, B. and Cramer, S. M. Binding capacity differences for antibodies and Fc-fusion proteins on protein A chromatographic materials. *Biotechnol. Bioeng.* 96:768–779, 2007.
147. Boschetti, E. Advanced sorbents for preparative protein separation purposes. *J. Chromatogr.* 658:207–235, 1994.
148. Gomez, M. I., Lee, A., Reddy, B., Muir, A., Soong, C., Pitt, A., Cheung, A. and Prince, A. *Staphylococcus aureus* protein A induces airway epithelial inflammatory responses by activating TNFR1. *Nat. Med.* 10:842–848, 2004.
149. Bensinger, W. I., Buckner, C. D., Clift, R. A. and Thomas, E. D. Clinical trials with staphylococcal protein A. *J. Biol. Resp. Modif.* 3:347–351, 1984.
150. Hoffman, W. L., Ruggles, A. O. and Tabarya, D. Chicken anti-protein A prevents *Staphylococcus aureus* protein A from binding to human and rabbit IgG in immunoassays and eliminates most false positive results. *J. Immunol. Methods* 198:67–77, 1996.
151. Carter-Franklin, J. N., Victa, C., McDonald, P. and Fahrner, R. Fragments of protein A eluted during protein A affinity chromatography. *J. Chromatogr.* 1163:105–111, 2007.

152. Zhu-Shinomi, J., Gunawan, F., Thomas, A., Vanderlaan, M. and Stults, J. Trace level analysis of leached protein A in bioprocess samples without interference from the large excess of rhMAb IgG. *J. Immunol. Methods* 341:59–67, 2009.
153. Steindl, F., Armbruster, C., Hahn, R., Armbruster, C. and Katinger, H. W. D. A simple method to quantify staphylococcal protein A in the presence of human or animal IgG in various samples. *J. Immunol. Methods* 235:61–69, 2000.
154. Larsson, A., Wejaker, P.-E. and Sjoquist, J. Chicken anti-protein A for the detection and capturing of protein A from *Staphylococcus aureus* in the presence of absence of mammalian IgG. *Hybridoma* 11:239–243, 1992.
155. Lucas, C., Nelson, C., Peterson, M. L., Frie, S., Vetterlein, D., Gregory, T. and Chen, A. B. Enzyme-linked immunosorbent assays (ELISAs) for the determination of contaminants resulting from the immunoaffinity purification of recombinant proteins. *J. Immunol. Methods* 113:113–122, 1988.
156. Godfrey, M. A. J., Kwasowski, P., Clift, R. and Marks, V. A sensitive enzyme-linked immunosorbent assay (ELISA) for the detection of staphylococcal protein A (SpA) present as a trace contaminant of murine immunoglobulins purified on immobilized protein A. *J. Immunol. Methods* 149:21–27, 1992.
157. Shukla, A. A., Jiang, C., Ma, J., Rubacha, M., Flansburg, L. and Lee, S. S. Demonstration of robust host-cell protein clearance in biopharmaceutical downstream process. *Biotechnol. Prog.* 24:615–622, 2008.
158. Wang, X., Hunter, A. K. and Mozier, N. M. Host cell proteins in biologics development: Identification. Quantitation and risk assessment. *Biotechnol. Bioeng.* 103:446–458, 2009.
159. Eaton, L. C. Host cell contaminant protein assay development for recombinant biopharmaceuticals. *J. Chromatogr.* 705:105–114, 1995.
160. Hoffman, K. Strategies for host-cell protein analysis. *Biopharm* 13:38–45, 2000.
161. WHO Guideline requirements for the use of animal cells as in vitro substances for the production of biologics. *Requirements for biological substances* 50:, 1997.
162. Flatman, S., Alam, I., Gerard, J. and Mussa, N. Process analytics for purification of monoclonal antibodies. *J. Chromatogr.* 848:79–87, 2007.
163. Andersen, K. P., Low, M. A., Lie, Y. S., Keller, G. A. and Dinowitz, M. Endogenous origin of defective retroviruslike particles from a recombinant Chinese hamster ovary cell line. *Virology* 181:305–311, 1991.
164. Strauss, D. M., Lute, S., Tebaykina, Z., Frey, D. D., Ho, C., Blank, G. S., Brorson, K., Chen, Q. and Yang, B. Understanding the mechanism of virus removal by Q sepharose fast flow chromatography during the purification of CHO-cell derived biotherapeutics. *Biotechnol. Bioeng.* 104:371–380, 2009.
165. MacGregor, I. R. Screening assays for transmissible spongiform encephalopathies. *Vox Sang* (Suppl. 2):3–6, 2004.
166. Krutzik, P. O. and Nolan, G. P. Intracellular phospho-protein staining techniques for flowcytometry: Monitoring single cell signaling events. *Cytom. A* 55:61–70, 2003.
167. Sadick, M. D., Intintoli, A., Quarmby, V., McCoy, A., Canova-Davis, E. and Ling, V. Kinase receptor activation (KIRA): A rapid and accurate alternative to end-point bioassays. *J. Pharm. Biomed. Anal.* 19:883–891, 1999.
168. Brasier, A. R. and Fortin, J. J. Nonisotopic assays for reporter gene. *Curr. Protoc. Mol. Biol.* 9:12–21, 1995.
169. Zhang, W.-W., Labrecque, S., Azoulay, E., Dudley, R. and Matlashewski, G. Development of a p53 responsive GFP reporter: Identification of live cells with p53 activity. *J. Biotechnol.* 79–86, 2000.
170. Zhang, X. and Bremer, H. Control of the *Escherichia coli* rrnB P1 promoter strength by ppGpp. *J. Biol. Chem.* 270:11181–11189, 1995.
171. Fan, F. and Wood, K. V. Bioluminescent assays for high-throughput screening. *Assay Drug Dev. Technol.* 5:127–136, 2007.
172. Caserman, S., Menart, V., Gaines Das, R., Williams, S. and Meager, A. Thermal stability of the WHO international standard of interferon alpha 2b (IFN-α2b): Application of new reporter gene assay for IFN-α2b potency determinations. *J. Immunol. Methods* 319:6–12, 2007.

173. Zhang, J., Chen, D., Gong, X., Ling, H., Zhang, G., Wood, A., Heinrich, J. and Cho, S. Cyclic-AMP response element-based signaling assays for characterization of Trk family tyrosine kinases modulators. *Neurosignals* 15:26–39, 2006.
174. Gillis, S., Ferm, M. M., Ou, W. and Smith, K. A. T-cell growth factor: Parameters of production and a quantitative microassay or activity. *J. Immunol. Methods* 120:2027, 1978.
175. Porstmann, T., Ternynck, T. and Avrameas, S. Quantitation of 5-bromo-2-deoxyuridine incorporation into DNA: An enzyme immunoassay for the assessment of the lymphoid cell proliferation response. *J. Immunol. Methods* 82:169–172, 1985.
176. Mosmann, T. Rapid colorimetric assay for cellular growth and survival: Application to proliferation and cytotoxicity assays. *J. Immunol. Methods* 65:55, 1983.
177. Buttke, T. M., McCubrey, J. A. and Owen, T. C. Use of an aqueous soluble tetrazolium/formazan assay to measure viability and proliferation of lymphokine-dependent cell lines. *J. Immunol. Methods* 157:233, 1993.
178. Scudiero, D. A., Shoemaker, R. H., Paull, K. D., Monks, A., Tierney, S., Nofziger, T. H., Currens, M. J., Seniff, D. and Boyd, M. R. Evaluation of a soluble tetrazolium/formazan assay for cell growth and drug sensitivity in culture using human and other tumor cell lines. *Cancer Res.* 48:4827, 1988.
179. Nakayama, G. R., Caton, M. C., Nova, M. P. and Parandoosh, Z. Assessment of the Alamar Blue assay for cellular growth and viability in vitro. *J. Immunol. Methods* 204:205–208, 1997.
180. Crouch, S. P. M., Kozlowski, R., Slater, K. J. and Fletcher, J. The use of ATP bioluminescence as a measure of cell proliferation that the ADP:ATP ratio may reflect mitochondrial and cytotoxicity. *J. Immunol. Methods* 160:81–88, 1993.
181. Pagliaro, L. C., Liu, B., Munker, R., Andreefff, M., Freireih, E. J., Schneiberg, D. A. and Rosenblum, M. G. Humanized M195 monoclonal antibody conjugated to recombinant gelonin: An immunotoxin with anti-leukemic activity. *Clin. Cancer Res.* 4:1971, 1998.
182. Whiteside, T. L. Measurement of cytotoxic activity of NK/LAK cells. *Current Protocols in Immunology* , 2001.
183. Brunner, K. T., Mauel, J., Cerottini, J. C. and Chapuis, B. Quantitative assay of the lytic action of immune lymphoid cells on 51Cr-labelled allogeneic target cells in vitro: Inhibition by isoantibody and by drugs. *Immunology* 14:181, 1968.
184. Riedl, S. J. and Shi, Y. Molecular mechanisms of caspase regulation during apoptosis. *Nat. Rev. Mol. Cell Biol.* 5:897–907, 2004.
185. Muppidi, J., Porter, M. and Siegel, R. M. Measurement of apoptosis and other forms of cell death. *Current Protocols in Immunology* 1–36, 2004. Immunology-Cell Activation.
186. Korzeniewski, C. and Callewaert, D. M. An enzyme-release assay for natural cytotoxicity. *J. Immunol. Methods* 64:313, 1983.
187. Malcolm, A. and King, M. A. Detection of dead cells and measurement of cell killing by flow cytometry. *J. Immunol. Methods* 243:155–166, 2000.
188. Lecoeur, H., de Oliveira-Pinto, L. M. and Gougeon, M. L. Multiparametric flow cytometric analysis of biochemical and functional events associated with apoptosis and oncosis using the 7-aminoactinomycin D assay. *J. Immunol. Methods* 265:81–96, 2002.
189. Holmes, K. L., Lantz, M., Fowlkes, B. J., Schmid, I. and Giorgi, J. V. Preparation of cells and reagents for flow cytometry; Immunofluoresence and cell sorting. *Current Protocols in Immunology Immunology* 1–24, 2001. Immunology—Immunofluoresence and Cell Sorting.
190. Holmes, K. L., Otten, G., Wayne, M. and Yokoyama, W. M. Flow cytometry analysis using the becton dickinson FACS calibur. *Current Protocols in Immunology* 1–22, 2001. Immunology—Immunofluorescence and Cell Sorting.
191. Golemis, E. A. and Adams, P. D.e. Protein–Protein interactions: A Molecular Cloning Manual. Cold Spring Harbor Laboratory Press, New York, 2005.
192. Lakowicz, J. R. Energy transfer. *Principles of Fluorescene Spectroscopy.* 3rd ed., Springer, US, pp. 443–475, 2006. (online).
193. Lakowicz, J. R. Time-resolved energy transfer and conformational distributions of biopolymers. *Principles of Fluorescence Spectroscopy.* 3rd ed., Springer, New York, pp. 477–506, 2006. (online).
194. Li, Z., Gergely, J. and Tao, T. Proximity relationships between residue 117 of rabbit skeletal troponin-I and residues in troponin-C and actin. *Biophys. J.* 81:321–333, 2001.

195. Selvin, P. R., Rana, T. M. and Hearst, J. E. Luminescence energy transfer. *J. Am. Chem. Soc.* 116:6029–6030, 1994.
196. Heyduk, E. and Heyduk, T. Thiol-reactive, luminescent europium chelates: Luminescence probes for resonance energy transfer distance measurements in biomolecules. *Anal. Biochem.* 248:216–227, 1997.
197. http://www.htrf.com/technology/htrfassay/Immunoassays.
198. http://www.microcal.com/technology/itc.asp.
199. Velazquez-Campoy, A., Leavitt, S. A. and Freire, E. Characterization of protein–protein interactions by isothermal titration calorimetry. In *Methods in Molecular Biology: Protein–Protein Interactions. Methods and Applications*, (H. Fu, ed.), Humana Press Inc., Totowa, New Jersey, pp. 35–54, 2004.
200. Sigurskjold, B. W., Berland, C. R. and Svensson, B. Thermodynamics of inhibitor binding to the catalytic site of glucoamylase from *Aspergillus niger* determined by displacement titration calorimetry. *Biochemistry* 33:10191–10199, 1994.
201. Bjelic, S. and Jelesarov, I. A survey of the year 2007 literature on applications of isothermal titration calorimetry. *J. Mol. Recognit.* 21:289–311, 2008.
202. Ladbury, J. E. and Williams, M. A. The extended interface: Measuring non-local effects in biomolecular interactions. *Curr. Opin. Struct. Biol.* 14:562–569, 2004.
203. Hornbeck, P. Assays for antibody production. Enzyme-Linked Immunosorbent Assays. *Current Protocols in Immunology Immunology* 2.1.1–2.1.22, 1991. Antibodies—Enzyme-Linked Immunosorbent Assays.
204. http://www.biacore.com.
205. Sambrook, J. and Russell, D. W. Molecular Cloning: A Laboratory Manual. 3rd ed., Cold Spring Harbor Laboratory Press, Cold Spring Harbor, New York, 2001.
206. Myszka, D. G. Kinetic analysis of macromolecular interactions using surface plasmon resonance biosensors. *Curr. Opin. Biotechnol.* 8:50–57, 1999.
207. Rich, R. L., Papalia, G. A., Flynn, P. J., Furneisen, J., Quinn, J., Klein, J. S., Katsamba, P. S., Waddell, M. B., Scott, M. and Thompson, J., *et al.* A global benchmark study using affinity-based biosensors. *Anal. Biochem.* 386:194–216, 2009.
208. Rich, R. L. and Myszka, D. G. Higher-throughput, label-free, real-time molecular interaction analysis. *Anal. Biochem.* 361:1–6, 2007.
209. Do, T., Ho, F., Heidecker, B., Witte, K., Chang, L. and Lerner, L. A rapid method for determining dynamic binding capacity of resins for the purification of proteins. *Protein Expr. Purif.* 60:147–150, 2008.
210. Danson, M. and Eisenthal, R. Enzyme Assays: A Practical Approach. Oxford University Press, Oxford, 2002.
211. Copeland, R. A. Enzymes: A Practical Introduction to Structure, Mechanism, and Data Analysis. Wiley-VCH, New York, NY, 2000.
212. Lundblad, R. Using in vitro assays for therapeutic enzyme characterization. *Bioprocess International* 40–44, December 2009.
213. Vanderjagt, D. J., Fry, D. E. and Glew, R. H. Human glucocerebrosidase catalyses transglucosylation between glucocerebroside and retinol. *Biochem. J.* 300:309–315, 1994.
214. Langdell, R. D., Wagner, R. H. and Brinkhous, K. M. Effect of antihemophilic factor on one-stage clotting tests. *J. Lab. Clin. Med.* 41:637–647, 1953.
215. Barrowcliffe, T. W., Mertens, K., Preston, F. E. and Ingerslev, J. Laboratory aspects of haemophilia therapy. *Haemophilia* 8:244–249, 2002.
216. Peetz, D. Factor VIII methods: Which assay principle for which indication? In *36th Hemophilia Symposium Hamburg 2005*, (I. Scharrer and W. Schramm, eds.), Springer Medizin Verlag, Heidelberg, pp. 71–74, 2007.

9 PREFORMULATION AND EARLY PHASE METHOD DEVELOPMENT

KEVIN COLEMAN BYNUM

Boehringer Ingelheim Pharmaceuticals, Inc., Ridgefield, Connecticut, USA

I. INTRODUCTION

The transition from discovering a new therapeutic agent to developing a new pharmaceutical product begins with an investigation of the physical and chemical properties of the active pharmaceutical ingredient (API), a process commonly referred to as "preformulation." This is a key milestone in the drug development process, as the focus shifts from looking at the therapeutic agent as a molecular entity to how it can be effectively made into a product. Preformulation studies, when effectively conducted, can be accurate predictors of the challenges that will be encountered in combining the API with a suitable system that will deliver this new therapeutic agent in a safe and effective manner to a patient. Following preformulation, formulation development activities will begin, with the primary goal of developing a robust bioavailable, efficacious, and safe dosage form that can be manufactured at a suitable scale for commercialization.

This chapter will discuss the role of the analytical chemist in the preformulation process and highlight the methods that will need to be developed, validated and utilized to support these studies. An assortment of analytical techniques is needed to measure a number of critical quality attributes of the new molecular entity such as solubility, purity, and crystalline habit. The methods will be used to make important decisions such as the choice of salt form, or which solid oral dosage form has the best probability of providing adequate exposure in an early phase clinical study. These analytical tools are critical to the decision that a pharmaceutical development organization uses to evaluate the first round of solid dosage form development prior to the availability of clinical data.

The pharmaceutical analytical chemist is a key collaborator with the formulation development scientist who is charged with generating and providing the information about the dosage form(s) under study which will guide the formulation scientist in his/her quest to transform a chemical entity into a medical product.

One key distinction that needs to be made is the difference between information (or knowledge) and data. Data are a collection of numbers and/or values, for example, an assay value generated by high-performance liquid chromatography (HPLC). The individual datum may be accurate and correct, but in and of itself it provides little value. It may be evaluated against a predetermined acceptance criterion, to serve as a release mechanism for a test article, but it does not, by itself provide analytical information about the drug product. Only when that individual datum is placed into the correct context does it allow for the generation of knowledge about the behavior or attributes of the product. One example of this approach is to compare this individual datum to other data points collected under different conditions, such as that done during the performance of a stability study. Now the individual datum is placed in a context of time under a fixed storage condition (such as the stress condition of 40 °C/75%RH). When compared with other assay values at different times under a constant storage condition, a trend analysis can be performed. For example, one can apply such a strategy to potency and see if this critical attribute changes with time. This trend can then be extrapolated to later time points to predict where the potency value will be at a future time which is a key attribute in determining the shelf life of a product. The data, placed in the correct context, will generate useful information which will allow for informed decisions and potential prediction of future outcomes. Further context can be added such as the formulation of the drug product and additional sets of data generated from different formulations so that, for example, the stability of one formulation relative to another can be derived from these assay value(s). The reader is referred to Chapter 12 devoted to stability for a further discussion on this topic as it relates to early method development.

The key point in such a situation is context. Analytical chemistry is no longer considered a "service" function but rather is a collaborative venture with those scientists developing formulations or an API process. Communication and partnering with other departments is key to success. If there is a clear and open transfer of information about the problem that needs to be solved, then the best and most efficient analytical methodology can be applied to the task at hand.

This will be the common trend that will be addressed in this chapter, the utilization of the expertise of the analytical chemist to generate useful knowledge in addressing challenges associated with the testing of pharmaceutical samples.

II. PREFORMULATION API PHYSICOCHEMICAL PROPERTIES

Prior to developing a drug product, a number of experiments need to be conducted that will fully characterize the API. This information will then be used to help guide formulation development efforts. Each of the critical pieces of information that are needed for formulation development will be discussed below. These include a screening process where an ideal solid-state salt form is selected and the generation of a suite of analytical methods necessary to probe the critical quality attributes of the new chemical entity, to support synthetic optimization scale-up, stability studies, and the release of material for GLP and experimental use.

A. Salt Screening

The selection of a suitable salt form for development is a critical decision that needs to be made before an API is formulated for clinical development. The salt form chosen will have a significant impact on many critical factors of the API, such as solubility, stability, and physical processing parameters. Unstable or metastable forms cannot be readily manufactured into suitable dosage forms.[1]

As part of the salt screening process, the potential formation of polymorphic forms and solvates should be evaluated to determine which salt form is most appropriate for development based on target attributes (see also Chapter 2). Each potential salt and/or polymorphic form should be evaluated to ascertain its stability at accelerated conditions (temperature and humidity). The water sorption of each should be evaluated to determine if the material will likely be incompatible with excipients that contain moisture. Following the synthesis of different salt forms each must be evaluated by a battery of chemical and physical tests.[2] Given the importance of this work, an entire chapter in this text is devoted to solid-state characterization and the reader is encouraged to peruse them for further details.

1. Solubility

The solubility of the compound will typically be studied by creating a saturated solution, adding an excess of material under study to a known amount of solvent until dissolution ceases. The sample will then be shaken under controlled room temperature (20–25 °C) typically for 24–48 h. After this time, the solution will be assumed to have achieved a thermodynamic equilibrium, and the liquid will be saturated. It is important to be certain that some solute remains, to ensure that the liquid is completely saturated. The liquid is then assayed to determine the exact amount of material that went into solution. This value is the thermodynamic solubility of the molecule. This is typically called the "shake flask method."[3]

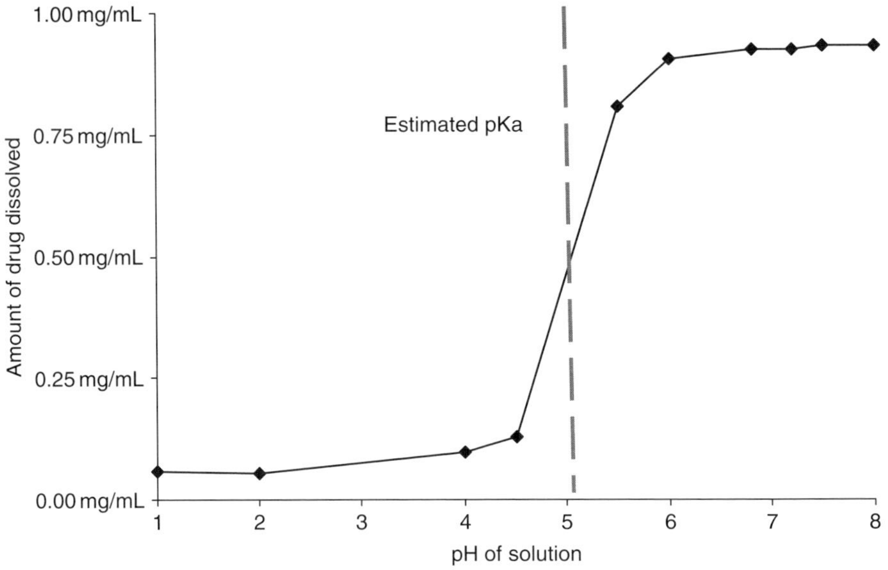

FIGURE 1 The determination of a compound's pK_a.

The aforementioned process will need to be repeated for a number of different solvents. Typically, the solubility will be observed in water, as well as a number of different aqueous solutions which are buffered to a standard set of pH values. This profile of pH versus thermodynamic solubility is one of the most important pieces of information that is needed throughout drug product development. The inflection point in this pH profile will be a good method to measure the pK_a of the material. An example of this technique is shown in Figure 1.

The pH should also be measured for each solution after a thermodynamic equilibrium has been established. If the pH of the solution is different from the pH of the initial buffer, this will indicate that the ionized drug substance has overcome the buffering capacity of the buffer solution and has changed the solution pH, which can lead to incorrect assumptions about solubility at a given pH.[4]

For salt forms of our therapeutic compound, we will have other considerations. The strong ionic interaction between the organic moiety and counter ion within the crystal lattice of a salt will increase the potential energy of salt form, relative to the base. This difference in potential energy will lead to a different apparent solubility for the salt form. Typically, the solubility of a salt will be much higher when it is first introduced into an aqueous medium, compared to the equilibrium (thermodynamic) solubility measured for the base. This higher solubility is called apparent solubility. Since this supersaturation condition represents a high energy state, the drug will slowly precipitate until the thermodynamic solubility is reached. Since the dissolved fraction of material changes overtime, it is often called kinetic solubility. This apparent solubility is very important for the salt selection process.[5,6]

Each salt form under study may have a different apparent solubility. In fact, for most poorly soluble drug substances, the driving force of the salt screening process will be to increase the apparent solubility of the compound in an attempt to improve the exposure of the compound *in vivo*. Since this supersaturation effect observed with salts occurs and persists on a timescale relevant to biological processes (tens of minutes to hours), it is a desirable process that needs to be controlled and measured for each salt form under consideration.

The most common manner to study apparent solubility will be to conduct the shake flask method as described above, but instead of waiting for 24 h the sample will be assayed at earlier time points (1, 2, 4 h...) to generate a plot of apparent solubility versus time. This profile should be measured at multiple pH values to see how the apparent solubility will be affected by different pH values. Finally, the use of surfactants will typically be studied. This has some biorelevance, since there are a number of bile salts in the intestine that will have a surfactant effect. Also this will provide valuable technical information for solubilizing the API for sample preparation, dissolution testing, as well as utilizing surfactants in the drug product.

Conducting these solubility experiments will require a great deal of resources. To help reduce the amount of time and labor required for these experiments, there are a number of automation options for these studies. There are robotic platforms that can be used to conduct the entire salt screening study, and solubility specific units that will utilize *in situ* UV to measure apparent solubility overtime. Regardless of how this measurement of apparent solubility is made, it is a critical parameter to measure for salt selection.

Another "flavor" of solubility that is typically studied for salt screening is intrinsic solubility. This experiment typically utilizes a compress pellet of pure API in a rotating die. This die is then placed into a USP dissolution apparatus, and the amount of drug released into a relevant media, such as water, simulated gastric fluid (SGF), or simulated intestinal fluid (SIF) is studied with time. This experiment provides information about the dissolution rate that can be expected for the pure drug substance.

Recent enhancements in fiberoptic dissolution technology has allowed for a modification of this technique to observe the rapid intrinsic dissolution of a free flowing powder, instead of a compress pellet.[7,8] This experiment is conducted in a very small volume, typically 2–10 mL, so a much smaller amount of API is needed for this measurement. The use of a fiberoptic optrode, placed in the small volume apparatus, measures the amount of drug dissolved in real time, at a rate of up to one measurement per second. This fast sampling is critical, since, the free powder typically dissolves in a minute or two. This technique has been shown to be effective at observing particle size, and surface area differences in pure drug substance.

2. Polymorphic Purity

Each initial salt needs to be studied for polymorphic purity, and the presence of solvates. Ideally, only one polymorphic form will be present. This polymorphic purity will be measured by one of the solid-state technique(s) listed in Section II.D. Initial screening will typically be conducted with less labor intensive and less expensive methods such as differential scanning calorimetry (DSC) and TGA.

Once these techniques indicate that there may be multiple polymorphs or solvates present, more complex and involved techniques such as X-ray powder diffraction or solid-state NMR (SSNMR) can be employed to confirm the initial test and provide more information about the exact polymorphic species in the API.[9]

3. Stability of Salt Form

The chemical and physical stability of the API will be another critical factor in salt selection.[10,11] Each salt from can exhibits a different stability profile over time. Both physical and chemical changes can lead to a stability failure overtime. The chemical stability must be tracked with a stability indicating assay/degradation product method to monitor structural changes (typically covalent changes) in the organic component of the drug substance. This is most commonly done with a stability-indicating HPLC or UPLC method. Typically, UV detection is utilized, but mass spectrometry, combined with UV detection will provide much more information if a new degradant is observed. In addition to chemical degradation, the polymorphic form must also be studied. It is important that this polymorphic form should not change on stability under long term storage conditions (25 °C/60%RH) or ideally even under accelerated storage conditions. Techniques such as X-ray powder diffraction and thermo gravimetric analysis (TGA) are useful to track the polymorphic form and are discussed in Section II.D in greater detail.

The end result of this process will be the identification of one salt form that will then be moved into drug product development. It should be noted that this decision is a critical one, and that the methodology selected to provide this critical GO-NOGO decision on each salt candidate must be appropriate. If a nonoptimal salt form is selected, then this could compromise the entire project and result in the rejection of a viable therapeutic compound, or the continuation of a candidate that has undesirable physical or chemical properties.

B. Assay and Impurities by HPLC/UPLC

After the salt form is selected the chemical development group will begin a campaign to manufacture a sufficient quantity of drug substance to fuel the drug development process. This may include synthetic optimization activities, and scale-up of the drug substance synthesis. All of these activities will require an analytical methodology that will be able to measure the purity of the drug substance that is produced.

The most common methodology for this purpose is HPLC, or more recently UPLC.[12–14] Modern HPLC and UPLC methods are able to accurately and robustly separate complex mixtures and quantitate all of the components with UV detection in a fast and efficient manner. In addition, this system can be coupled to a mass spectrometer, which can increase the utility of this technique significantly, so that unknown components can be tentatively identified, and peak purity established.

There are many method development strategies for the development of HPLC methods. One such method that has been successfully utilized with "great success" in the authors' laboratory is described below.

The starting point will be a method which will utilize an aqueous mobile phase that is buffered to a pH that is at least two pH units away from the pK_a of the active ingredient. Ideally, this pH will be on the nonionized "side" of the pK_a so that the analyte under study is as hydrophobic as possible (not ionized) and has a great affinity for the stationary phase (for reverse phase HPLC/UPLC). Establishing that the pH of the mobile phase is at least two pH units away from a pK_a ensures that the API is in only one state (preferably a nonionized state). Current hybrid stationary phases, which are stable across a wide pH range (typically from pH values 12-1), facilitate this approach for acidic, basic, and neutral compounds. For basic compounds, typically a basic mobile phase is a good starting point, where the inverse is true for acidic mobile phases. The salt selected to buffer the mobile phase should be volatile so that it can be utilized with mass spectrometry. This aqueous mobile phase will then be utilized with a broad gradient (10% aqueous to 90% organic) to elute all of the components off of the column. Typically, either methanol or acetonitrile are used as an organic component for gradient elution. When a high quality stationary phase is utilized, with the smallest particle size possible (2.5–3.0 μm for HPLC) a very high plate count can be realized with relatively fast separations. With these smaller particle sizes, separations can further be accelerated by increasing column temperature which will allow for greater flow rates. Again current hybrid technology columns are able to tolerate temperatures in excess of 50 °C even at pH extremes.

Combining this optimized chromatography with mass detection will significantly speed the method development, and allow for the tentative identification of degradation products and processing impurities. Such a system can be utilized initially for small scale-up batches, and as experience is gained about the process, and impurity process, the method can be tweaked to improve the separation of the observed impurities and degradation products.

Once the synthetic route is well established, and known impurities can be controlled, a series of forced degradation experiments will be very useful to ensure that the analytical methodology will be stability indicating. The drug substance will be exposed to acidic, basic, oxidative conditions, as well as thermal and photonic stress (light) to determine what degradation products may form, and if this chromatographic methodology can separate the degradation products. Again the mass detection will provide to be invaluable in this process. If the chromatographic process is not stability indicating, and a co-elution does occur, then the mass detection will likely resolve this new component. If only UV detection were utilized, it is unlike that coelutions will be obvious.

C. Chiral Purity

If the molecule being studied possesses one or more chiral centers, then methods must be developed to control any potential chiral impurities (also see Chapter 7 in this volume). It has been shown that different optical isomers of the same molecule can have dramatically different pharmacological effects that vary from being nonefficacious to having toxicological concerns.[15] A chiral impurity could have less pharmacological activity, compared to the parent compound, or could even cause a toxic effect that is not evident in the other optical isomer.

Therefore, the chiral nature of the drug substance must be known and controlled. This must be done throughout the synthesis of the drug substance, so that the exact chiral species is consistently generated. Also, once an enantiopure compound has been generated, it must be shown that the chirality does not change on stability. Clearly, a chiral metric is required.

In order to characterize the chiral nature of a drug substance, and to then track its optical purity on stability, a method which can resolve the subtle chiral differences between the main chiral form, and its optical isomer is required. The most robust way to measure chiral purity is to separate the two enantiomers by means of a chiral chromatographic technique. Once the species are separated, they can easily be quantitated by UV detection. In order to achieve a separation of two enantiomers, they must interact in some way with a second enantiomeric species. There are two general method of how to accomplish this chromatographic separation.[16]

The first is to utilize a stationary phase, which has a chiral nature to it. This is most commonly accomplished by bonding a chiral sugar to a silica base support. Since the sugar moiety is chiral in nature, the two different enantiomers under study will have a different affinity for the chiral stationary phase, and one will interact with it to a greater degree, and therefore be retained greater. In this manner a chiral separation can be affected. It should be noted that such an arrangement requires a normal phase separation mechanism, since the polar silica with bonded sugar will not remain intact in an aqueous environment. In fact, water must be actively excluded from the system, including the sample preparation, to ensure that the column is not adversely affected by the presence of water. Also, the chiral columns utilized are not only fragile, but also very expensive; it is not uncommon for a chiral column to cost thousands of dollars.

An alternate chromatographic approach is to utilize a standard reverse phase HPLC method, which incorporates a chiral modifier in the aqueous mobile phase. This methodology typically uses a chiral cage type molecule, such as cyclodextrin as a chiral modifier, and a standard achiral reverse phase stationary phase. The method requires that the enantiomeric analyte will be able to interact with and become enclosed within the chiral cage molecule. This chiral interaction will be different for the two different enantiomeric species, which will result in a difference in retention on the reverse phase stationary phase. This methodology has the advantage that it does not require a dedicated normal phase system, or a different water free sample preparation. Also, it does not require an expensive and fragile chiral HPLC column. The key disadvantage is that the use of a chiral modified method requires a good deal of method development, and unless a good understanding of the interaction of the chiral analyte with the chiral modified is realized, this approach can take a great deal of empirical trial and error to facilitate a separation.

D. Solid-State Analysis

Prior to formulation development, it is critical to evaluate the solid-state crystalline nature of the API (also see Chapter 3 in this volume).[17] Different polymorphic forms have shown to have different solubility and stability characteristics, and as such can impact on the bioavailability of the final therapeutic compound. The polymorphic form of the API that has been synthesized and crystallized needs to be fully characterized and controlled. This form can be impacted by not only the intrinsic properties of the organic compound, but also the method of crystallization which is typically the last step in the synthetic process. In addition, any post crystallization steps, such as milling, could also impact the crystal form.

As part of the solid-state characterization of the new API, a series of experiments should be conducted which try to determine if it is possible to form other polymorphic states from the new compound. The compound should be recrystallized from a number of different solvents to determine if other polymorphic forms exist. These different polymorphic forms can then be analyzed by a number of solid-state tests to demonstrate that the main polymorphic form of the API can be differentiated from these different forms. These samples can be used as controls to demonstrate the appropriateness of the solid-state technique chosen to monitor the solid-state nature of the API. If these other polymorphic forms are not observed on during stability testing of the API, it can be assumed that the main polymorphic form is stable.

A number of solid-state techniques are currently utilized to support drug product development. X-ray powder diffraction and SSNMR are two powerful techniques which will be discussed. Thermal techniques such as DSC and TGA can be used as well to observe the thermal behavior of a crystalline substance. Scanning electron microscopy and polarized microscopy are two very useful microscopic techniques that can be a critical tool for solving solid-state problems.

The key information that will be obtained from a solid-state technique will describe the crystal form of the API. Once a salt form is selected, the crystalline nature of the API will be fully characterized. Any potential polymorphic forms, hydrates, or cocrystals will be identified. The drug substance itself should not convert from the initial crystalline habit, into any different form during storage at normal storage condition, or ideally, during accelerated stability. This is critical preformulation information that must be established before drug product development.

There are many different techniques that could be used to evaluate the solid-state properties of the API under study. Typically, cheaper and less technically challenging techniques such as thermal techniques (TGA and DSC) are utilized to screen materials. If these initial techniques show a multitude of species, or are inclusive, then typically a more advanced technique such as X-ray powder diffraction or SSNMR can be employed, if available to the project team. A brief description of the most commonly used solid-state techniques will follow.

1. Thermal Methods

There are two types of thermal methods that are routinely utilized to evaluate the physical properties of a pharmaceutical solid, DSC and TGA.[18] DSC is a technique which measures heat flow with changing temperature, and TGA measures weight change with changes in temperature. Both techniques have the advantage that they are very mature and robust techniques, and require the expenditure of very little sample. These techniques can be used to identify polymorphic transitions, in order to determine if multiple polymorphic forms are present in a sample of material. The USP has recognized the utility of these techniques and in USP <891> describes a compendial methodology for the determination of polymorphic purity, for samples which exhibit a true melt. In addition to this utility, hydrates can also be detected by DSC and TGA. Due to the robust nature of this technique, and ease of operation, it is usually employed as a first tool to evaluate a solid-state material.

2. X-ray Powder Diffraction

Since pharmaceutical API is typically composed of fine "powdered" crystals, X-ray powder diffraction is a useful tool to evaluate the crystal nature of both pure API and API in a drug product.[19] This technique utilizes radiation in the X-ray region of the spectrum to observe the scattering angle of a powdered sample. These observed scattering angles are representative of the crystal structure of the sample. Chemically identical molecules which have different crystalline structures generate a different pattern of scattering angles. This allows for the direct and conclusive differentiation of crystal forms. It is also useful to determine the degree of crystallinity, in the presence of amorphous impurities. In this role, it is often used as a confirmatory test following an initial indication that a second polymorphic form is present.

3. Solid-State NMR

SSNMR requires the use of special techniques to overcome fundamental challenges in the analysis of solid samples.[20] Cross polarization (CP) and magic-angle spinning (MAS) is used to reduce/eliminate the line broadening that is observed in SSNMR. This allows for the high resolution acquisition of NMR spectra from samples that are in the solid state. This provides for a valuable tool that can be used to observe the NMR spectrum of an API in its native form within the actual dosage form. Polymorphic forms can be differentiated and identified based on the resolved differences in chemical shift of the individual resonances. For solvates, the exact solvent that is incorporated into a solvate can be identified based on the chemical shift of the solvent. With this tool and the information that it can generate comes a cost. The SSNMR hardware requires a significant cost in both capital instrumentation, as well as infrastructure. Also SSNMR is a very specialized area of spectroscopy, and SSNMR experts are in high demand.

4. Microscopy

Microscopic techniques play a critical role in trouble shooting physical problems that are often observed with other techniques.[21] Microscopic techniques provide morphological observations on the microscale, which can provide a great deal of information about the material under study. There are two very powerful microscopic techniques that can provide information about a drug substance.

Scanning electron microscopy can provide very detailed images with a great deal of magnification to demonstrate the morphology and habit of crystalline material. This technique can provide a great deal of information about the surfaces of the crystals, as well as their overall shape. When combined with energy dispersive X-ray spectroscopy (EDAX) elemental information can be generated from the SEM field of view.

A complementary technique is polarized light microscopy (PLM). Since this technique utilizes light as opposed to electrons to visualize the specimen under study, the magnification where effective resolution can be realized is not as great, when compared to SEM. Despite this resolution limitation, it does provide a great deal of information about the crystalline nature of the drug substance under study. The PLM utilizes a pair of crossed polarizers oriented at 90 degrees to one another one below the sample and one above it. In such an arrangement, the polarized light from below the sample, passing through a noncrystalline (amorphous) material, will be blocked by the second polarizer and the sample will appear black. Crystalline material which has a refractive index, when oriented with the optical axis parallel to one of the polarizers will appear dark, but will be bright when the sample is rotated (facilitate by the round rotating stage which is the most noticeable feature of a PLM) and the optical axis lies at 45° between the polarizers. This technique can be used to characterize, and differentiate crystalline material, based on their unique optical axes that are the result of the crystalline structure. This technique can be very valuable in identifying trace noncrystalline impurities in samples of crystalline API.

E. Stability Testing

Once the new API is synthesized, a batch of this material will need to be placed on stability to demonstrate that the chemical and physical properties do not change with time. This is commonly achieved by packaging the material in a container type that will be utilized for storage of the material (such as a polyethylene bag in a fiber drum). Additionally, other container types can also be used such as bottles. It is often common to place the material in an open container, such as a petri dish, to see if it is stable when exposed to ambient conditions, as a worse case stability condition. The samples will then be stored under different environmental conditions for a given period of time. Typically, samples are stored at 25 °C/60%RH, as at normal room temperature (recommended or long term conditions). In addition, samples can be stored under accelerated conditions, typically 40 °C/75%RH. The accelerated samples can be used to try to predict the stability behavior of the material that is stored at the normal room temperature sample, if a sample is stable under accelerated conditions for *X* months, it will likely be stable for at least 2*X* months at recommended conditions. This assumption is based on the Arrhenius equation[22] (first derived by van't Hoff in 1884) which states that, in general, for many chemical reactions, the rate of reaction doubles for every 10 °C increase in temperature. If no chemical or physical changes are noted in a material after storage, in an open petri dish for 3 months at 40 °C/75%RH, it will be very safe to assume that the material will be stable for 6 months in a closed container at normal room temperature and humidity.

The key to any stability analysis will be to select the proper stability indicating methodology to ensure that the critical properties of the material under study do not change overtime. It is crucial that we monitor both the chemical stability as well as the physical stability for a duration that will be long enough so that the material can be utilized for its intended purpose. This length of time could be short if this is a small batch that will be completely consumed in a toxicological study that will last only 1 month. One month of storage at room temperature and accelerated conditions would be sufficient of this purpose. If instead the material will be used to produce a batch (or perhaps multiple batches) of drug product that will be used in a lengthy late phase clinical study, it may be desirable to generate stability data that will justify a much longer retest period.

The techniques chosen for the analysis of the stability samples must be able to indicate if changes have occurred on stability, the methods must be stability indicating. For a chromatographic method, the technique should be able to separate and measure all of the degradation products observed from stress resulting from heat, light, acid, base, and oxidative conditions. This is typically done through a process of forced degradation. The chiral method should be able to separate the main chiral form from the other optical isomers, which will require the synthesis of the other pure form, or at least a racemic mixture of both. Additionally, a suitable solid-state technique(s) must be chosen to evaluate if the crystal structure of the API changes on stability overtime.

It is very critical that the proper techniques be investigated to provide the proper information required to demonstrate conclusively that the API is stable under the given test conditions. For example, if a chiral method is not utilized to track the conformation of a chiral center, and on stability the R form converts to the S form at room temperature, this change will be detected by the normal reverse phase method. At best this missed degradation pathway will be discovered at a later stage (perhaps in the drug product) and could result in a large expenditure in development resources and could slow or even result in the need to repeat a clinical study with pure enantiomeric material. At worst this error could result in a significant safety issue in a clinical or commercial setting.

The suite of technique must also be developed to have the necessary specificity that will be able to resolve a stable sample from one that has changed on stability. For example, if a "stability indicating" assay/degradation products method is developed without the aid of mass detection, and the method does not resolve a major degradation product from the main component, again the information derived from the data will be in error, and the project can be led down a path which will cause significant development delays.

F. Particle Size

One final critical factor in drug substance preformulation analysis, particle size, should be discussed briefly. The particle size of a drug substance can have a significant impact on a number of factors in our drug product. Larger particles sized material tends to have slower intrinsic dissolution rate. This slower intrinsic dissolution rate can significantly slow the rate of dissolution of the API from a final dosage form. This slower dissolution rate could also have an impact on the bioavailability of the drug product. Particle size and morphology can also have an impact on critical drug product processing parameters.[23]

It is clear that the particle size should be at least controlled for consistency, if not optimized for both processing efficiency and rapid dissolution of the drug substance. These controls can take two general forms. The particle size can be controlled by the synthetic process, typically with the parameters of the final recrystallization step. The other method is by means of milling the material to a certain particle size distribution.

This particle size control methodology should be monitored with an appropriate analytical technique. This is typically done with either a laser scattering based approach, by means of a sieve analysis of the milled material, or by means of a microscopic examination. It should be noted that each technique can result in a very different measured particle size distribution, because of the different fundamental properties that are observed by each technique. The differences of each technique are outlined elsewhere.

It is critical that particle size be evaluated, and controlled prior to the development of a drug product, in order to ensure consistent drug product performance.

III. DRUG PRODUCT

Once a drug substance is promoted to clinical development, and a salt form is selected, the process to develop a solid dosage form for clinical studies begins. There are many different approaches to delivering a therapeutic compound, resulting in a wide variety of solid dosage forms. Depending on the chemical and physical properties of the API, some forms of drug delivery will be preferred over others. The delivery system can take many forms, tablet, capsule, beads, transdermal patches, parental, powder in bottle. The selection of a delivery vehicle is outside of the scope of this discussion, but there are a number of common factor that apply to all dosage forms and will be discussed here.

A. Excipient Compatibility

The first step in developing a dosage unit is to find a set of pharmaceutical excipients that will be compatible with the API. The purpose of the excipients will be to act as a vehicle to deliver the drug substance. Some will be simply a bulking agent that will have physical properties that are desirable for the dosage form (compressibility). Other excipients or can take an active role in delivering the drug substance (disintegrants, permeating agents, and surfactants). Regardless of the purpose and role of each excipient, each must be compatible with the API. Each excipient must be chemically compatible with the active ingredient, it cannot react with the excipients. Any such reaction will reduce the shelf life of a given drug product which will at best create difficulties during clinical development, and reduce its commercial potential, or at worst derail the clinical development to the point project termination.

In addition, the solid state of the API must not be impacted by the presence of the excipients. The polymorphic crystal form must not change while in contact with the excipients in the dosage unit. Such a change could result in a lower solubility of the new polymorph, reducing the overall bioavailability of the compound.

The question then arises how do we select excipients that will be compatible with our new drug candidate? There are two basic studies that can be conducted.

Both rely on combining the API with and individual or a mixture of excipients and then studying the chemical and/or physical properties of the API to determine if a change has occurred.

In the first method, a numbers of samples are made consisting of the API and a single or mixture of excipients. The samples are created at a relevant drug loading that will represent a worst case formulation (lowest drug/excipients ratio). The solid material is simply blended (usually by hand) and stored in small bottles. The samples are then stored under relevant accelerated storage conditions. Typically, samples are stored under 40 °C/75%RH in open and closed bottles. In addition to study a worst case scenario, a small amount of water can be added directly to the samples, either using a micropipette, or by placing the water in a small micro capillary tube, and adding the tube to the vial.[24] The vials are then sealed and stored under the accelerated condition.

The vials are then removed and tested to determine if the accelerated storage condition has resulted in a potency loss, degradation, or if any solid-state changes have occurred. Typically, the samples are tested using an HPLC method. The HPLC method developed to characterize the drug substance can be utilized. This method should be stability indicating, and many degradation products can usually be resolved with the HPLC method utilized for the drug substance characterization and stability analysis. Many new degradation products may be present, which may result from a chemical interaction between the active component and the excipients. These new degradation products may not be resolved from the API, and could go unnoticed, unless a labor intensive method development strategy is undertaken. In addition, the excipients themselves could cause additional difficulties, by coeluting with the main component or with new degradation products.

To overcome these challenges, mass spectrometry can be utilized in addition to UV detection for our HPLC method. A relatively inexpensive single quadrupole mass spectrometer can be utilized in series after a standard UV detector. Our drug substance HPLC method can then be utilized, and many of the degradation components can be separated and measured with the UV detector. For each sample, the mass spectrum can then be observed, and the purity for each of the peaks can be evaluated. Additionally, the mass spectrum of the initial sample can then be compared to the sample stored under accelerated conditions. If the mass spectrum under the main component changes between the initial and the accelerated sample, then it is likely that a degradation product has been formed that is not resolved. The single ion chromatogram of that mass can then be observed and used to estimate the amount of degradant formed. It should be noted that this will be an estimate of the amount of degradant formed, since the ionization of the degradation product and main component could be different. Since only a rough estimate of the level of degradation is needed to eliminate a possible excipient (do we really care if 14% or 15% of API is degraded by an excipient in a week either value indicates that a stability concern exists?) this is not a significant problem. If a large excipient peak is observed, then the mass spectrum can again be used to look for degradant components masked by the interfering component.

One additional complication is that the recovery of the API will not be known with this methodology. Each powder sample will be extracted in a solvent which is able to solubilize the API. It is quite possible that the recovery of the API from these excipients will not be 100% even though an optimal extraction solvent is used. Again the analysis of the initial samples will be useful to overcome this challenge. The time = 0 samples which are simple mixtures of the API and excipients will be analyzed with no storage under accelerated conditions. This will provide a good estimate of the amount of drug substance recovered under optimal conditions. If greater than say 80% of the API is recovered, then the extraction conditions are likely sufficient for this excipient. This extraction efficiency can then be applied to later samples, stored under accelerated conditions, and if after the extraction efficiency correction is applied, a large drop in potency is seen, it is likely that the potency drop is a result of degradation, especially if degradation components are observed that increase as the potency decreases.

This process should be sufficient to exclude excipients that cause chemical degradation, A drop in potency (compared to the initial extraction of the $T=0$ sample) combined with an increase in degradation products will be a clear indication that a chemical degradation is occurring with that given excipient. The question of what happens to the solid-state properties of the drug component still remains. The samples which were stored under accelerated conditions for the chemical compatibility study can be tested by a solid-state test such as X-ray powder diffraction. This will demonstrate that the crystal form of the API will remain unchanged under these stressed conditions, and in contact with the excipient under consideration.

This process will require a significant amount of labor to construct all of the samples, and then conduct extensive LC–MS and solid-state analysis to determine compatibility with a given excipient. Automation can be used to speed and reduce the amount of labor required for this process. One such robotic technique utilizes a robotic platform to construct the excipient and drug mixtures. There are a number of automated platforms that can dispense and mix powders in an accurate and reproducible manner. The platform can then utilize a fluid handling system to extract and inject the samples into an HPLC system for chemical analysis, or the samples can be removed and tested by a solid-state analysis system (XRPD or SSNMR). The samples can then be stored under accelerated conditions, and returned to the automation platform and tested to determine the stability when in the presence of the excipient. Most robotic platforms are able to track the samples using barcodes, to ensure that the sample identity and storage condition are linked properly to the data generated by the analytical systems.

A different approach to utilizing HPLC to measure the loss of chemical potency and monitor degradation products is to directly monitor the thermal activity, using a technique called thermal activity monitoring (TAM). In this process, excipients and API are mixed in a manner similar to the HPLC–MS approach above. The samples then are monitored with an isothermal microcalorimeter for a length of time, usually a few days. The microcalorimeter monitors the reaction enthalpy that will be evident if a chemical reaction occurs. A change in enthalpy will also be evident if a solid-state change affects the solid-state properties (crystal form) of the API. If the excipient is compatible, then the microcalorimeter will indicate that no heat flow was evident from the sample. If instead, there was a change in heat flow with time, then it is very likely that a chemical or physical change has occurred with the excipient–API mixture. The sample can then be tested with HPLC–MS and/or anointer solid-state analytical technique to determine what physical or chemical change has occurred.

Regardless of what technique is utilized, it is very important to ensure that the excipients chosen for a solid dosage form are compatible with the API. A mistake at this stage of development, the inclusion of an excipient that will react with the API, could result in the development of a solid dosage form which will not have optimal, or even acceptable stability. This could result in a waste of valuable resources, expensive API, could slow development, or delay early phase clinical studies. The excipient compatibility must exclude inappropriate excipients. In order to do this, the analytical team must provide this information, which can only be derived if the experiments are designed properly, which must generate accurate data to provide relevant information about the compatibility of each excipient under study. This is not simply a process of collecting data. Proper controls ($t = 0$ samples) must be generated to determine extraction efficiencies, the proper excipients must be selected in relevant drug loads, and finally the proper analytical tools must be utilized to derive the necessary information. The data once collected (in an accurate manner) must be evaluated over time, peak purity must be evaluated, degradation products identified, solid-state changes observed. These data are then trended with time and stability condition, resulting in the necessary information, relative stability of the API with each excipient mixture.

B. Formulation Support

Once a set of compatible excipients has been selected, formulation development activities will begin. Depending on the complexity of the drug product, the process of formulation development may be simple and straightforward, or difficult, and complex. Typically, there will be a formulation strategy behind the solid dosage form. This could be as simple as getting the API into a hard gelatin capsule. For many dosage forms, this is a good rapid formulation for first in man studies. For other formulation strategies, tablets may be chosen for over all development speed (first in man formulation could become the commercial product) or to serve as a special delivery system (delayed release, control release, enhancement of solubility). Regardless of the formulation strategy, it is critical that the pharmaceutical analysis group be made aware of the formulation strategy, and the details of the actual formulation under development. The formulation strategy will help the analytical group develop an appropriate suite of methods to test the formulation. For example, a delayed release enteric coated tablet would require a more complex two step dissolution test to demonstrate the integrity of the enteric coat. Such a test would not be required for a simpler dosage unit such as hard gelatin capsule. Again, a close partnering of analytical sciences with the pharmaceutics group is critical. There should be a complete sharing of information related to the project, and formulation under development. The analytical group should not be treated as an internal contract lab, where samples are dropped off in a sample submission box with a request form, in fact, the sample drop box and request for analytical support form are sure signs communication are not taking place.

Before samples are ready for testing, the pharmaceutics group should discuss the formulation goals with the analytical group, and a set of tests needed to support the formulation development activities developed prior to the initiation of formulation development. The analytical group should take part in this discussion in an active manner. In this way, the most appropriate tests, and methodologies can be utilized to generate the required information in the shortest time possible.

Once the formulation work begins, it is a good idea to continue with this close patterning between the analytical and formulation support group. The analytical group should not be kept in the dark as to the progress of the formulation work, simply waiting for the generation of samples. The best practice is to have an analytical person present in the formulation development lab, observing and if possible helping with the sampling. In this role, the analytical scientist will act as both a liaison and will have a detailed knowledge about the formulation process, and any issues that may have occurred during the formulation development.

There are two requirements of the analytical group during formulation development. The first requirement is that the analytical tools chosen must provide accurate analytical data about the formulation. An incorrect analytical result could result in an incorrect decision being made about the formulation under study. Such an error, if not caught quickly, could derail the formulation development process, sending the drug development process down a dead end. One example would be a "stability indicating" assay method which is not capable of tracking chemical degradation during stability testing. Such a method, which did not resolve a major impurity from the parent peak, may indicate that a formulation is stable. When the error is discovered during a later clinical phase, an unplanned redevelopment of the formulation may need to be undertaken, delaying the clinical development process.

The second requirement is that information be generated quickly, and immediately fed back to the formulation group, preferably in real time. A great deal of trust and collaboration is needed between the two groups in order to make this happen. Since the analytical group is the "eyes and ears" of the formulation group, the analytical information is quickly needed to help the formulation group determine if their process is working properly. Typically, the formulator will make a number of batches at a small scale, and then request that samples from these batches be tested for critical parameters (uniformity of dose, dissolution, hardness...). Based on the information that is developed from these data, a decision will be made as to which formulation strategy is working properly. Any delay in the analytical information, will directly impact on the timeline of the formulation development. For this reason, it is very important that the analytical tool selected for this support provide real-time or near real-time information.

The risk of real-time information is that lengthy data checking is not possible, when data are rapidly evaluated, and information generated. Both groups need to be aware of this risk, and mitigate it whereever possible. Some analytical systems lend themselves to real-time data generation (fiberoptic dissolution) and are designed to perform all of the calculations in real time, using hard coded software, as long as the initial data inputs are correct (standard mass, dose, volume, dilutions) the result will be correct. If experienced chemists support this analytical work stream, the risk of incorrect data being generated is low. Additionally, the data checking process can still be conducted, but it should be decoupled from the information stream. The data can be generated by our experienced chemist, using real-time techniques, and the information can then be reported just after the completion of the experiment as "unverified results" so that decisions can be made as rapidly as possible. The data can then be checked for errors, and if a gross error is discovered, the error can be corrected, and a determination made to see if the data error impacts the information and therefore the decision. It is the authors experience that many analytical errors made generally do not impact the information generated from the data, for example, a calculation error may shift an assay value from 100.1% to 100.3%. This is clearly an analytical error, and could be a significant issue for a GMP sample, but the information for a formulator is the same in either case... all of the drug weighed out for a batch made it's way into the solid dosage form, the potency is correct.

During formulation development, this information will not be used for GMP purposes (clinical release, stability...) so the regulatory burden on these data is not great. As such, the documentation and data verification can be greatly reduced, compared to a GMP sample analysis. The data should be documented in a manner needed to reproduce the experiment, and data checking should be done to ensure accuracy, but lengthy data review processes and GMP documentation are not needed, and will slow the generation of information needed by the formulation group.

C. Analytical Tools for Formulation Support

As previously stated, the complexity, and specifics of the formulation will drive the suite of analytical tools needed to support formulation development. Despite this, there are standardized experiments that are generally required to support the formulation development process. These experiments control the uniformity of dosage units, the dissolution rate of the formulation, and the solid-state properties of the API within the formulation.

1. Uniformity of Dosage Units

One of the most critical properties of an experimental formulation is the uniformity of the dosage units. Unless the formulation has the correct potency, and the tablet-to-tablet variability is adequate, all of the other dosage unit attributes are not relevant. The uniformity of the dosage units will be largely dependent on the formulation process chosen. This process needs to be designed and implemented in a manner that will provide for the generation of solid dosage units with the correct potency and little tablet-to-tablet variability. There are a number of analytical controls that can be used to ensure that the final drug product is produced with a uniform and correct level of drug substance.

Most formulations are a multistep process. At a minimum, components are combined in a blender, mixed, sieved, and tableted/filled into capsules. Additional steps such as granulation, drying are also included as demanded by the formulation strategy. Since there are many steps in this process, there are many points at which the material can be tested in process to determine assay, and drug uniformity. The obvious place to look for uniformity will be the final product, typically the individual tablet or capsule. An accurate and rapid uniformity of dosage unit's method for the final dosage unit will be required. This individual tablet/capsule uniformity method will be essential to ensuring the quality of the final product.

Even if the final product is determined to be uniform, the point at which the uniformity is accomplished during the process may not be known. Ideally, the entire process will be understood, and each step optimized so that a uniform product is developed in a robust manner, with no unnecessary effort. This understanding of the process is critical to troubleshooting any problems that will be encountered in the utilization of this manufacturing process later in development. In order to understand how the uniformity of the formulation is impacted by the process, we will want to look at the uniformity of critical steps in the process. In doing so, we will ensure that we impart quality into the design of our process (quality by design-QbD).

One obvious place to look at homogeneity is during the blending step. The purpose of this step is to combine the API with the excipients, and make a homogenous mixture. There are a number of techniques used for the analysis of blend uniformity. The most straightforward approach is to utilize a thief to sample the blend during the blending process, to remove samples for external chemical analysis for the active ingredient.[25,26] This process requires that the blending process be stopped at set intervals during the blending process, the blender opened, and samples taken from multiple points and levels of the powder bed (typically at least 5–10 samples per time point, depending on the size of the batch). This sampling procedure is then repeated for a number of time points throughout the blending process. The end result is many samples (typically 50–100) taken from various portions of the blend bed, at different blend times. These samples must then be individually weighed, extracted, and assayed for drug substance content, typically using an HPLC method. This process can take a great deal of time and resources, even when fast HPLC and automation are utilized. The end result is a set of assay values throughout the blend bed, at different time points. The variability (%RSD) of the assay values (weight normalized) at each time point can then be calculated. The trend of this %RSD measurement throughout the blend process will give us the information we are looking for: at what point in the blend process are all of the components mixed?

This is the most direct and straightforward methodology in determining the uniformity of a blend. There are a number of disadvantages to this technique. The first is that it is labor intensive methodology. The sampling process is tedious, and requires opening the blender many times during the blend process. This tends to be a messy operation and has the potential to alter the blending process, by disturbing the blend bed with sampling. Each of the samples collected must be weighed to determine the weight normalized assay value. This weighing step alone could take a chemist a number of hours to accomplish. The next step, the extraction process will likely take a number of hours as well, with the final HPLC step requiring multiple hours to complete. The total time needed for this process can easily run into days, during which the formulator may be sitting idle, waiting for the information necessary to optimize the process.

Another disadvantage to this technique is that the number of time points that can be sampled is finite. As more time points are sampled, the more the blend process is disturbed. Sampling the blend even once every 5 min for a 30-min blend may be considered excessive by some, and will not be desirable. Even with 5-min sampling, the exact time when blending is optimized is only approximate. At best the optimal blend time can be estimated within about 10 min with 5-min sampling, since we are looking at the trend and want to find the time when the RSD of the HPLC assay levels off, and typically three time points will be needed to make this determination. In this case, if the blend is optimized at say 6 min, we would have sampled at 5, 10, and 15 min. We would have observed a flattening of the RSD at 10 and 15 min. This means that the optimal blend time will lie between 5 and 10 min. This data set would likely result in a 10-min blend time for a blend that only requires 6 min, better than just over blending for 45 min, but certainly not optimized.

Finally, the off line blend uniformity measurement is typically made once each time the process is optimized, this may only happen a few times during the development of a drug product. Once the blend time is set, it will be blended blindly. This is due to the effort and expense of an offline blend uniformity study.

What is clearly needed is a metric that could monitor the blend uniformity, in real time as the blend occurs. Ideally, such a process would be noninvasive to the blend process, as accurate as the offline measurement, and nonlabor intensive. There are a number of techniques that have been developed to serve as a real-time blend uniformity monitoring. Near infrared (NIR) spectroscopy is probably the most mature technology that has been applied to blend uniformity analysis and will be discussed in detail below.

2. Blend Uniformity by NIR Spectroscopy

NIR has been successfully utilized to monitor blend uniformity for a number of years.[27] A number of approaches have been used, from manually probing the blend bed with a hand probe[28] to collect NIR spectra, to real-time monitoring of the blend using a remote wireless sensor attached to the blender. NIR imaging has even been utilized to observe the blend uniformity during the blend process.[29]

The big advantage to NIR spectroscopy for the application of blend uniformity is the ability to observe the blend process as the mixing takes place, in real time. This is most commonly accomplished utilizing a blender equipped with a transparent window, usually made of sapphire. An NIR reflectance probe is attached to the window on the outside of the blender, so that broadband radiation can be imparted into the blender, illuminating the contents. Some of the radiation will be diffusely reflected back to the window, and will be received by the sensor, and measured by a detector, to record the spectra of the contents of the blender that is in front of the spectral sensor window at a given time. Typically, the spectral sensor has a solid-state monochromator, a number of commercially available systems utilize either an AOTF or a PDA spectrometer to collect full spectra in a rapid manner. The careful reader should see a problem with this arrangement. The spectrometer is mounted on a *rotating* blender. Spectrometers need a means to draw power for operation, and unless data are stored locally, a data connection is also required. The fact that the spectrometer is mounted on a rotating blender makes a hard-wired power and data connection impractical. To overcome this challenge, one of the two mechanisms is utilized. Either the spectrometer is wireless, using a battery to power the unit, and a wireless transmitter to send data to a remote computer, or there is some sort of slip ring arrangement, at the axis of rotation of the blender, which couples the power and data into the spectral sensor. One final and novel approach was to place the NIR reflectance probe into the axis of rotation of the blender,[30] this approach has been patented, and does tend to limit the placement of the spectral sensor.

Regardless of how the technical issues of spectra acquisition are overcome, there are some common traits, requirements, and challenges to monitoring blend uniformity by NIR spectroscopy.

In its simplest form, NIR can be used to monitor blend uniformity with no chemometric model built. The system is simply utilized to collect NIR spectra, and then monitor the variability (%RSD) at a single, or at multiple wavelengths in the spectra. This provides a very good indication of the amount of spectral change that is occurring in the area of the blender immediately in front of the spectral sensor window. Based on the assumption that once the spectral variability stops the contents of the blender have reached a point of homogeneity, this will indicate when the contents are blended. Unfortunately, this assumption does not always hold true. This methodology will not pick up clumps of pure component (API or excipient) that will cause uniformity issues, since the clumps will likely be coated with bulk blend, and cannot be differentiated from the bulk material. Also material can become stuck to the blender window, which will have a constant spectrum, but will not reflect the bulk contents of the blender. In these cases, a chemometric model which generates an accurate analytical value for the component(s) in the blender in real time is needed to ensure that the end point that is measured is in fact a well mixed powder bed.

Model Construction

In order to obtain meaningful information from near infrared spectroscopy one must construct a model, or training set, consisting of spectra collected from known samples, and mixtures of samples. This training set is necessary because the NIR spectra of most compounds contains many overlapping broad peaks that are overtones and combination bands of the fundamental peaks observable in a typical mid-IR spectra. If there is only one component to be measured, this is not an issue, but when monitoring a mixture of components, such as is encountered in a pharmaceutical blend, complex mathematical models need to be utilized to generate accurate analytical information from an NIR spectrum.

This process consists of presenting the system with a number of pure components, and known mixtures of the components that will be encountered during the blend process. These data are often called a training set. Typically a spectral pretreatment is applied to the raw spectra, and then the data are fed into a multivariate model. The details of NIR model building are beyond the scope of this chapter. Most commercially available NIR spectrometers contain software that makes the generation of these models easier, but knowledge of this model building process and chemometrics is necessary, so specialized training and the assistance of an expert in this area is required. Once the training set is built, and "validated" (one cautionary note, validation in the NIR spectroscopy world has a different meaning from the world of pharmaceutical development) unknown mixtures can then be presented to the system (spectrometer and data model) and if the model was built correctly, accurate concentration data derived from the overlapping spectra.[31] Before a blend is conducted, the model should be challenged with what is called an external validation set. This set consists of a number of mixtures of components that were not included in the training set. The ability of the model to then extrapolate these "unknown" samples, based on the spectral properties of the training set is observed. The actual value of the concentration of API in the validation sample will then be compared with the measured value generated by the NIR system. The difference between the two is a good measure of the accuracy of the system. This exercise should be conducted across the expected concentration range that is expected (100–0% for a typical blend) of the component to be studied (typically the API). Once the model shows the ability to measure unknown samples that are not in the training set, then the model is valid to use to monitor blend uniformity samples.

This model building process can be followed for just the key component, such as the API, or can be applied to ALL of the components simultaneously. This approach, monitoring all of the components in a blend simultaneously was termed "the mass balance approach"[32,33] for blend uniformity. This has the advantage of better characterizing the blend, in that not only the API are tracked, but also functional excipients such as disintegrants, and lubricants. It will also resolve pure clumps of material in the blender, which may be coated with homogenous blend, which have an identical spectral response as the bulk blend. This is the result of the limited depth of penetration of the NIR beam (a few microns). Since the pure clump of say filler, such as lactose, will cause a drop in the lactose level in the bulk blend and a slight increase in the relative concentration of the other excipients and API. If all components are not monitored, this may not be evident. When all of the components are monitored this is much more obvious, and can be corrected (by sieving) to avoid a solid clump of excipient being tableted, which could result in a failure in the uniformity of the dosage form, and a product quality issue.

Once the model is constructed, the contents of a rotating blender can then be assayed in real time. A number of things can be done with this new information stream. As stated above, the blend uniformity can be directly monitored by looking at the variability of the assay value with time. When the variability of the assay value approaches zero, it is likely that the blend is homogenous. Checking the observed assay value from the NIR unit and comparing it with the theoretical amount of material loaded into the blender will provide a confirmation of the blend homogeneity, and that it is likely that there are no large clumps of material which are not being observed by the spectral sensor. In addition, this measurement confirms the identity and actual amount of material added to the blender. This check could catch an error that may occur if the wrong type, or wrong amount of material is filled into the blender.

The real-time data also allow for a direct feed back into the manufacturing process. The real-time monitor can be configured to automatically stop the blender when the system indicates that the blend is homogenous. In this manner, the system will only blend for the needed time, and the data to justify the blend time will be present to demonstrate the product quality.

This benefit does not come without a cost. The spectroscopic system represents a significant capital cost. More significant is the cost of the skilled and specialized scientists needed to maintain the equipment, build the models, and conduct the real-time analysis. Also a significant amount of work will need to be conducted to prove the concept, qualify the system, and then validate the individual assay. As with any other analytical system the cost versus the benefit must be weighed, and the decision made to implement the technology, only if the benefits will outweigh the cost.

3. Other Areas to Monitor Homogeneity

The previous discussion has pointed out various process analytical tools that can be used to monitor uniformity during the blending process. The blending step is one of the more critical points where dosage uniformity can be monitored, but it is not the only location. The same technologies can be used to measure homogeneity during other pharmaceutical steps such as, granulation, melt extrusion, tableting (monitoring demixing in a hopper). Most of the same limitations, constraints, and benefits of real-time blend uniformity exist for other process analytical tools that could be used to speed formulation development. Since these applications are not as mature as blend uniformity monitoring, there are likely more challenges and difficulties in a successful implementation of this technology to these new applications.

4. Solid Dosage Form-Assay/CU/Degradation Products

This brings us back to the uniformity of our final drug product. Even with the application of process controls, and real-time monitoring during manufacturing, the final dosage form will have to be measured for uniformity of drug content. Again this piece of information is time critical since the formulation group will be dependent on this information to monitor and guide their formulation development strategy.

The most labor intensive portion of the uniformity test is the sample extraction. The extraction process must fully disrupt the dosage unit, and fully dissolve the API. For immediate release tablets, the dosage unit is generally designed to disintegrate in an aqueous environment. A buffered aqueous media can quite often cause an immediate release tablet to disintegrate completely and rapidly. The difficulty arises when the API is poorly soluble in water. In such a case, it may be possible to utilize an extraction solvent which has a pH that favors the dissolution of the API. Quite often this is still not enough to effectively dissolve the API in a reasonable volume of solvent, and an organic solvent must be utilized. A common technique is to use a portion of aqueous buffer to cause the tablet to disintegrate, and then add a sufficient volume of organic solvent to complete solvate the API. Techniques such as sonication, shaking, and mechanical homogenization can be utilized to speed this process considerably. A full discussion of the process of developing robust and effective extraction methods is discussed elsewhere.[34]

HPLC/UPLC will be the method of choice for the initial uniformity of dosage units assay method. At this early stage, the method can serve multiple purposes. Ten individual dosage units (or perhaps more) can be tested, each in a small volume[a], to evaluate the uniformity of the batch under study, and the average of the multiple units can be used as a composite assay. In addition the stability indicating drug substance method can be utilized (possibly modified based on the experiences from the excipient compatibility study) to provide information of the purity of the API in the dosage unit. This stability indicating assay and purity information will demonstrate that the formulation process did not cause any degradation in the dosage unit. For this purpose, the HPLC system can also be coupled to a mass spectrometer to allow for the detection of impurities that may not be fully resolved from the main component. This is a very good methodology to improve the stability indicating nature of the assay method.

A great deal of efficiency can be realized by utilizing this one test for assay/ purity/ and dose uniformity. The one down side is that a longer HPLC run time will be needed for the stability indicating assay/degradation product analysis, compared to a simple uniformity measurement, which does not necessary resolve all of the degradation products. This down side will be mitigated with fast HPLC methods (15-min stability indicating methods are common with high efficiency HPLC columns) or better yet UPLC analysis, which can execute a stability indicating chromatographic run in well under 10 min.

5. Dissolution Testing

Once the uniformity of the dosage units in controlled, the next critical parameter to investigate for a new formulation is the dissolution rate. The dissolution test procedure is a physical test that will seek to demonstrate how the solid dosage form will disintegrate and solubilize the API. There are many factors that can impact the rate at which a tablet disintegrates and the drug substance dissolves. These factors can be grouped into two broad categories, drug substance and drug product factors. For drug substance factors, salt form, polymorphic form, particle size, and surface area all play a role in the dissolution of the final drug product. For drug product factors, the dissolution test will be effected by the formulation process (granulation, melt extrusion, etc.) and excipients selected. All of these drug substance and drug product factors have the potential to impact the dissolution rate of the tablet. When we observe a resultant dissolution curve we are in essence looking at the combined sum of all of these factors.

6. Types of Dissolution Methods

There are three basic classes of dissolution methods, all three of which have a role to play in early phase drug product development. These three classes are

[a]One word of caution, if a small volume preparation is utilized, it is usually better to pipet the known amount of solvent into an open container, such as a test tube, using a class A volumetric pipette. This approach is superior to placing a dosage unit into a small 10 or 20-mL volumetric flask and bringing the flask to volume, since this undissolved excipients will displace a good amount of volume relative to the small volume of the flask resulting in a significant analytical error.

intrinsic dissolution methods for drug substance, biorelevant dissolution methods, and finally drug product release methods. Intrinsic dissolution has already been discussed in the previous section, and is typically utilized to characterize a given drug substance prior to drug product development. For drug product development, the use of both biorelevant and drug product release methods are important.

7. Biorelevant Dissolution Methods

In the effort to develop an oral solid dosage form the formulation group generally has a strategy to provide the maximum exposure of the therapeutic compound to the patient. This strategy is usually developed to overcome some sort of technical challenge posed by the drug substance. The most common challenge is to provide adequate exposure for a poorly soluble drug substance. In order for a drug substance to be absorbed into the human body, it must first be solubilized. The challenge for the formulation scientist is to design a formulation (s) that will be able to fully solubilize a drug substance so that adequate exposure can be observed. Since human and animal clinical trials are expensive and time consuming, an *in vitro* test is needed as a metric that will represent the environment of the human body. This *in vitro* test method can then be used to screen experimental formulations, to determine if the formulation will provide adequate exposure. Currently, there is no perfect *in vitro* methodology that can determine exposure prior to the clinical study. There are a number of dissolution tests that can determine if it is likely that a formulation will have enhanced solubilizing power, compared to other formulations.

In general, biorelevant dissolution tests try to mimic the conditions found in the human body as best as possible. In order to do this, the tablet or capsule is typically exposed to a small volume (~200 mL) solution of SGF (dilute HCl with sodium chloride) for a period of time representing the expected gastric emptying time (20 min–2 h). This test is typically executed in a standard USP apparatus I or II system. Other apparatus such as apparatus III, IV, and custom apparatus have been used to try to mimic the human digestive system. Regardless of the apparatus chosen, the solid dosage form will first be observed in this gastric fluid, and both visual observations of the disintegration, as well as chemical analysis can be made to determine the amount of drug that can be solubilized in this acid media. Following the acid exposure, a solution of phosphate buffer, with enough sodium hydroxide to shift the pH to ~6.8 will be added to simulate the dosage unit moving from the stomach to the intestine. Typically, this buffer will contain a low level of surfactants to simulate the natural bile salts found in the human digestive system. The rest of the test will be conducted at this higher pH and volume to simulate exposure to the human intestinal tract, the location with absorption generally takes place.[35]

In order to quantitate the amount of drug released by the dosage form in our biorelevant method, we need an accurate analytical technique. The ideal technique for this analysis is *in situ* UV analysis, facilitated with a fiberoptic optrode. Fiberoptic dissolution has been applied to the analysis of solid dosage forms for over 15 years.[36,37] The technique has been applied successfully to a whole host of dosage forms, and has been shown as a fast and effective technique, when applied to single component dosage forms, which contain an active ingredient with a strong chromophore.[38,39] One of the big advantages of this technique is that it can "sample" the contents of a dissolution many times a minute, generating a continuous dissolution profile. In addition, the results are generated in real time, so that the content of the drug released in the vessel can be observed as the dissolution test is conducted. This technique when applied to our biorelevant dissolution test, can resolve the fate of our drug substance in the biorelevant dissolution media.

This biorelevant methodology can be used to observe the enhancement in supersaturation, which is a desirable state to generate *in vivo* for a poorly soluble drug substance. A common formulation strategy is to utilize functional excipients to enhance and prolong the intrinsic nature of some drug substances to solubilize at a level greater than the thermodynamic solubility. This test utilizes the biorelevant experiment above, where the solid dosage form is exposed in a low volume of acid for a period of about an hour. Some of the drug will dissolve in this low volume of acid, depending on the pH solubility profile. Following the addition of the buffer, it is likely that much more of the drug will dissolve, because of the greater aqueous volume, and also because of the surfactants, which simulate the natural bile salts found *in vivo*. Weakly basic compounds will be very useful to study with this experiment, since the drug will be less soluble in the basic media, which simulates conditions in the intestine. Generating and maintaining a certain level of supersaturation for a weakly basic API can lead to increased levels of absorption. An example of such an experiment is shown in Figure 2. During the acid step, a small amount of drug substance is dissolved as the dosage unit disintegrates. The continuous UV fiberoptic measurement is able to measure this process, for each of the six samples studied in real time. With the addition of the buffer, the pH is shifted to 6.8, the volume is increased to 500 mL, and the dosage unit is exposed to a low level of surfactant which is considered to be biorelevant.[40] Under these conditions, we observe a large jump in the level of dissolved drug, which is much greater than the thermodynamic solubility in this media. This observed supersaturation, is a highly desirable state to observe for a formulation, and different formulations can be compared and screened based on this property. This information can then be utilized by the formulation scientist, and the whole drug product development team, to make decisions about which formulation is most likely to perform best *in vivo*.

A word of caution should be noted here: This methodology is simply a "best guess" which can be used to rank order different formulations. This methodology is in no way a complete simulation of the human digestive tract. One (of many) factor that is not considered in this experimentation is drug permeability. Ideally such a technique would observe both drug dissolution under biorelevant conditions, and observe the permeability of this dissolved drug in real time. Some attempts have been made to observe both critical factors simultaneously,[41,42] but to date no robust and routine techniques have been developed.

8. Drug Product Release Method

Once our experimental formulations have been ranked based on our biorelevant method, it is important to observe how long it takes for the drug

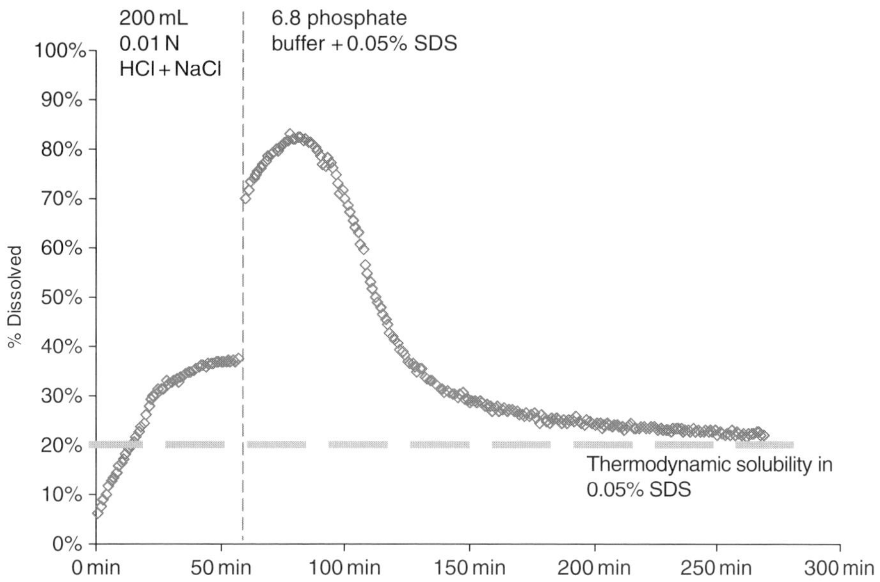

FIGURE 2 Biorelevent dissolution method. The dosage unit is exposed to 200 mL of 0.01N HCl with NaCl for 1 h, followed by the addition of a concentrated buffer containing SDS.

substance to be completely released from the dosage unit. This is an important technical and regulatory requirement[43] we need to make sure that our dosage unit will completely disintegrate and release the full potency of the API. It should be noted that the biorelevant method that studied what may occur (we hope) *in vivo*, does not generally dissolve all of the drug substance. This is because we are studying the drug product in a media where we do not have "sink condition."[44] Sink condition is defined as three times the thermodynamic solubility of the API. Our biorelevant method does not require that we dissolve all of the drug substance to generate the information required, and in fact we need to conduct this study in a media where supersaturation can occur. Our drug product release method will be conducted under sink conditions, and will usually achieve this level of drug substance solubility (for a poorly soluble API) with the use of synthetic surfactants (SDS, Tween80, etc.). Once we are at sink conditions, as the drug substance becomes exposed to the media it should quickly and quantitatively dissolve. The observed dissolution profile will then reflect the ability of the formulation to either quickly release, in the case of an immediate release tablet or capsule, or retard the release, in the case of a control release formulation. A drug product dissolution curve is shown in Figure 3. The time that the formulation takes to reach 100% dissolved will be observed, and since we are at sink condition for our API, this time will be most directly impacted by the drug product process, and excipients. The purpose of this test will be to control the release characteristics of the drug product, to ensure that different batches are similar, and that overtime no physical changes occur in the dosage form which will change the release rate of the drug product (stability testing). It is important that the method be developed in

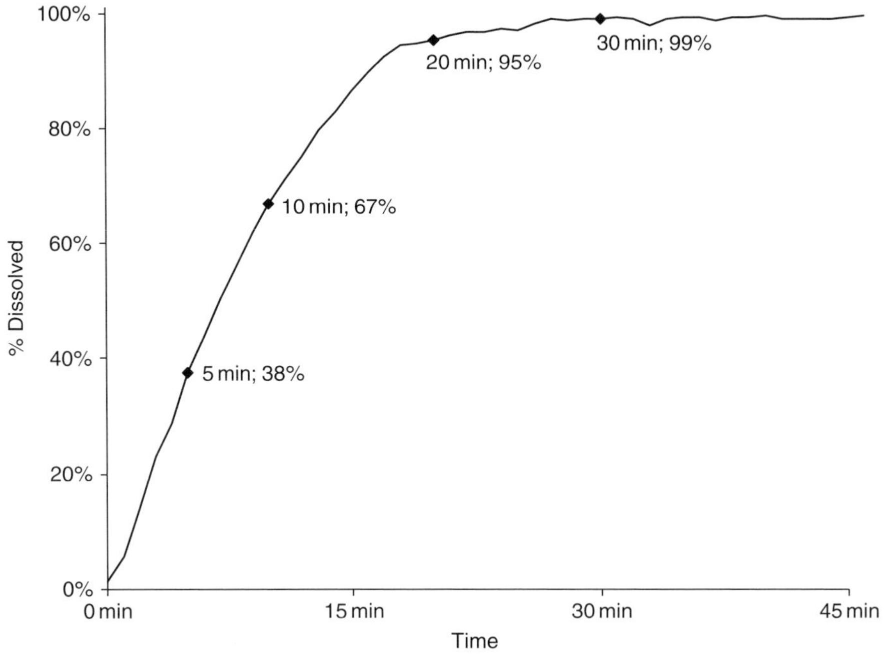

FIGURE 3 **Drug product release method. USP apparatus II (paddle) at 75 RPM, pH 6.8 phosphate buffer + 0.75% SDS.**

a manner that is discriminating enough to identify batches which will have different release profiles, which could impact bioavailability, but not be too discriminating, in which case it will pick up on small manufacturing changes, which have no impact *in vivo*. This aspect of dissolution method development is a very critical one, and is outside the scope of this brief communication, the reader is directed elsewhere for more detail on this topic.[45–47]

One final note on dissolution testing for drug product. The two classes of methods above should be considered two different tools, both of which will be needed throughout drug product development. These two methods will provide different information about our drug product. Neither method can completely stand alone. The biorelevant method can shed some light as to the *in vivo* performance of our drug product. It can be used during formulation development to evaluate different formulations, or following a significant change such as a change in particle size of the API. It looks at both the properties of the drug substance and drug product, so it can provide information about both. Our drug product release method gives us a different information set. It focuses much more on the drug product factors (process and excipients), and how they impact on the time it takes to release 100% of the drug. As such it is very valuable for monitoring process changes, such as process scale-up, process transfer and stability, where physical changes can be expected. The drug product release method will also be used to demonstrate the stability of the final drug product that is filed with the relevant regulatory agency. It is important that this difference is understood, and that each method is utilized for the proper purpose that is required to generate the information that is needed at that specific time in the development process.

9. Notes on Laboratory Automation

Many of the techniques described above, excipient compatibility, blend uniformity by HPLC dissolution, and content uniformity/assay by HPLC can be effectively automated by robotic sample preparation. Each of these techniques requires that the sample under study be dissolved in an appropriate solvent and fully extracted from any excipients. There are a number of commercially available products that have proven to be effective and robust in this sample preparation role. This robotic process can reduce both the analyst hours required to prepare a number of samples, and turnaround time on the sample analysis, since the robotic systems will operate unattended over night and on weekends. There is of course a cost to pay for laboratory automation. There is a significant capital cost, and then an ongoing maintenance cost for the continued operation of the system. Also it is critical that a specialist be available in-house to care for the system, develop the methods, and troubleshoot any issues with the system. The cost of the system and specialist must be weighed against the advantages of speed and lab capacity enhancement realized with a successful automation implementation.

10. Solid-State Analysis of Drug Product

Now that we have control over the content, degradation products and dissolution rate of our new drug product under development, we need to address the solid-state properties of the formulation. It is likely that we may have already collected some information about the solid-state nature of the drug product. Our dissolution method may be sensitive to some of the solid-state properties (hardness, crystal form, particle size) and if any of the critical parameters have changed, it is likely that a properly designed dissolution method will pick up on this change. The root cause of this dissolution change will not be evident. So our discriminating dissolution method will indicate that a change has occurred, but this technique cannot point to the root cause of the problem. A more sophisticated solid-state technique will be needed to fully characterize the solid-state change that has changed our dissolution rate.

Once the drug product development efforts are initiated, the question now arises, have we changed the crystal form with our formulation processing? Any of the processing steps could have an impact on the crystal form of the API. Additionally, the question now arises; will this crystal form be stable during the shelf life of the drug product?

These are important questions to answer since solid-state changes to the API in the drug product can lead to changes in the observed dissolution rate of the material, and potentially the *in vivo* performance of the drug product.

All of the techniques used to monitor the solid-state properties of our API prior to formulation, can be utilized to look at the crystal form of the API in the drug product.

X-ray powder diffraction can be utilized to study the API crystallinity even in the presence of excipients. The scattering pattern of the pure API must be compared to the pattern of the interfering excipients to determine which angles represent the API. These peaks can then be observed to compare different batches of material, different manufacturing processes, and different stability conditions/times to evaluate if the polymorphic form is changing with the varying condition.

Due to the high resolution nature of SSNMR spectroscopy, the spectra of the API can be resolved from that of the excipients in the formulation. It has been demonstrated that the spectra of different polymorphic forms and solvates are different, and the level of these forms can be quantitated. This technique will provide both a detailed chemical analysis of all of the components in the formulation, as well as information about their crystalline state. This combination of attributes makes this technique one of the most powerful tools available for the analysis of solid dosage forms.

Both SEM and PLM can still be utilized to solve drug product challenges. It is not uncommon to observe spots or blemishes on the surface of solid dosage form. Most chemists have encountered the infamous "black spots" that tend to appear when O-rings begin to wear in the manufacturing area. Changes can also occur in solid dosage forms during stability testing, resulting in a change in the physical surface of our dosage unit under study. Microscopy can provide a very valuable problem solving tool for these issues.

Other more specialized microscopic tools can be very valuable to identify the root cause in drug product problems. FTIR and Raman microscopy can provide detailed spectral information about very small areas in a solid dosage form, to help us identify the origin of the blemish or foreign matter.

Regardless of the technique utilized, the root cause of many problems observed with our more routine tools is often found with the resolving power of a microscopic technique.

IV. CLINICAL DOSAGE FORMS

Once a drug product formulation(s) has been successfully developed, the next step will be to put in places all of the GMP and analytical controls necessary to manufacture a clinical batch that can be used in a phase I clinical study.

We should already have a functional suite of analytical methods available to us from the formulation development support activities, and preformulation drug substance activities. We should have a stability indicating HPLC/UPLC chromatographic method that is capable of measuring all of the known degradation products in our dosage form. We also have a dissolution method that will be able to demonstrate the full release of our drug product. If a chiral center exists in our compound, we should have a drug substance chiral method which can resolve and measure the enantiomeric forms of the drug substance. We should also have a full suite of solid-state methods that have already been used to control the polymorphic form in the drug substance, and during formulation development.

With most of the method development work accomplished, now the focus will be on optimization of the methods and sample preparation to ensure that they are efficient, validatable, and appropriate for use in GMP testing.

A. Assay/Degradation Products Method

Since we have a stability indicating HPLC method that has been developed for formulation support, we should be able to apply this tool directly to our first official GMP method. We may want to modify the extraction process to make it more robust, and it would be a good idea to investigate the completeness of extraction, by preparing a composite assay as per the test method with the target extraction period, sampling the extraction flask, and then continuing the extraction process for a longer period of time. After the extended extraction a second test sample will be taken and assayed with the sample taken at the target extraction time. The assay values of these two samples, one at the target extraction time and one at the extended extraction time, should be identical, if the extraction is complete by the end of the target extraction period. Once we know we have a complete extraction that matches the theoretical dose of the drug product, it is very likely that our method is able to fully extract the drug substance from our drug product matrix.

The method should also be evaluated to see if degradation products can be quantitated at a level that is consistent with ICH guidance.[48] Ideally, the method is developed to initially control the expected level of degradation products in the dosage unit, but if the dose of the drug has changed or if impurities need to be controlled at a level lower that the ICH guidelines, because of a suspected toxic impurity, the method may have to be modified. If the analytical group is part of the project team and is aware of the shifting requirements of the analytical needs on a project, this change can be made. If the analytical group is just receiving samples and blindly testing them, they may not be aware of these new and more stringent requirements.

The sensitivity of an HPLC method can generally be increased by either increasing the injection volume or increasing the concentration of the sample extraction. Care must be taken in either approach. As injection volume is increased, it is possible to observe volume and/or mass overload for the main component. Also care must be taken that the new injection volume does not generate absorbance values that are in a nonlinear region of our UV detector. Many of the same considerations exist with the approach of sample concentration, with the additional concern, that the increase in sample concentration may decrease the extraction efficiency and result in a lower sample recovery.

B. Uniformity of Dosage Units

For our formulation screening activities, we were utilizing our full stability indicating HPLC/UPLC method for the measurement of uniformity of dosage units, since this allowed us to look at the content uniformity, assay, and gain insight into any degradation that had occurred during the processing of the drug product. This allowed us to extract as much information as possible from one experiment. This will likely not be a desirable approach for a GMP compliant method, where a separate stability indicating composite assay/degradation products method will be needed. Since we will now establish purity and level of decomposition products in a separate method, this information will no longer be needed from our uniformity of dosage unit's procedure. In order to increase efficiency, we can now reduce the run time of our uniformity of dosage units HPLC method. We can adjust the gradient so that the run time is much shorter, which will allow for faster turnaround of the uniformity data. The method should still separate all known degradation products from the main component, but it is not critical that these degradation components be separated from each other, and can elute in the void volume or in a postcolumn wash, since they do not need to be measured. With a bit of method development work, a fast method which chromatographs the main peak, with no interference can be developed for most compounds.

C. Chiral Purity Method

If the drug substance has been shown to have a chiral center, and on stability conversion to the other enantiomer is observed in the drug substance than it is likely that this chiral HPLC method will need to be modified to test the drug product for the presence of this chiral impurity. This may still be needed even if the conversion is not observed in the drug substance, since the drug product may facilitate this chiral conversion. This new method will require an aqueous free extraction of the drug product which can be difficult for some solid dosage forms. Most immediate release drug products are designed to disintegrate in the stomach or intestine, which is an aqueous environment. It is likely that a normal immediate release tablet will not disintegrate as readily in a pure organic solvent, and it may have to be disrupted by mechanical means, by either grinding followed by extraction, or by placing in an organic solvent and then break the dosage unit apart by either sonication or mechanical homogenization.

Once the extraction is facilitated the method must be checked for specificity against the excipients in the dosage unit, sensitivity, and linearity for the dosage unit under study.

D. Dissolution Method

Our formulation development activities we developed both a biorelevant dissolution method, and a drug product release method. At this point in development, all of our formulation development activities are complete, making the biorelevant method less important. The formulation(s) is fixed, and we are now concerned with demonstrating that we can consistently manufacture batches of drug product. We will also want to show that the release characteristics of the dosage unit do not change on stability. Our drug product release method will fulfill this role and will be the primary dissolution method, until the method conditions need to be revisited when *in vivo* data become available from our first clinical studies. At that time, the method should be revisited to evaluate how biorelevant our *in vitro* data are.

For formulation development support, we relied heavily on the use of *in situ* UV monitoring with a fiberoptic sensor as an analytical technique to measure the amount of drug released from our formulation overtime.[49] This provides accurate real-time dissolution data that will help speed formulation development activities. For our GMP method, we may want to move away from this real-time monitoring toward either offline HPLC or UV measurements. The offline measurements will use an autosampler to collect a vial or test tube of sample from the dissolution vessel at set points during the dissolution test. The samples will then be analyzed by either a semiautomated procedure (moved to a UV or HPLC by hand), or directly tested in an automated fashion (dissolution sample is collected into a vial and injected by an automated dissolution HPLC system). The reason for this is simple, for a GMP sample, it is always desirable to have a retained sample at every time point taken. If we then observe an analytical problem with the analytical portion of the analysis, we have a sample that can be retested as part of an investigation. As long as the auto sampler worked properly this will allow us to determine the root cause (bad injection, wrong pathlength used) and reanalyze the sample and obtain an accurate analytical measurement.

The real-time *in situ* measurement does not allow for sample reanalysis. If an error occurs, no retained sample exists for retesting. This is an acceptable risk during non-GMP formulation support activities. If an error occurs, we only lose efficiency, since the test may have to be repeated. For a GMP analysis, such as a clinical batch release, the analytical failure could bring into question the quality of the clinical material, and could delay a pivotal clinical study.

For our new automated or semiautomated analytical method for dissolution, one key experiment needs to be run. We must show that our automated sipper apparatus is equivalent to manual sampling. This is a requirement in the USP dissolution chapter <711>. The experiment, in its simplest form, consists of sampling dissolution vessels with both our automated and manual methodology. The vessels will contain a known amount of drug substance and a relevant level of placebo excipient. Both the automated and manual samples are then assayed by an appropriate method. If the two values compare, within the accuracy of the analytical method, then the automated method is validated. The other considerations that should be made are autosampler carryover, system cleaning, and maintenance. The system should not carryover drug substance from sampling point to sampling point, or more importantly from experiment to experiment. To evaluate this, one should run a high level standard followed by the standard autosampler purge and/or wash process normally used. A blank sample will then be tested, and evaluated for carryover. This test may lead to a modification to either the inter sample purge time, or the postrun cleaning process, if excessive carryover is noted.

Once our methods have been fully developed, and procedures drafted, a protocol driven, ICH compliant method validation should be conducted. This validation should be appropriate for this early stage of development. Typically, since only one lab is running this method, and likely one chemist, intermediate precision and robustness do not need to be evaluated. Other critical attributes, such as specificity, linearity, and accuracy will need to be evaluated.

V. CONCLUSIONS

After following the selection of a new therapeutic drug candidate, a great deal of effort must be expended to develop the analytical methodology that will be utilized to support the development of a safe and effective medicine. A suite of analytical techniques must be selected that will be utilized to guide the development team in making the key decisions necessary to make a successful delivery system for the new compound. These analytical tools are a critical analytical foundation that will be relied upon in later stages of clinical development. This suite of analytical methods will first characterize the new compound, to provide the development team with critical information needed to make key decisions regarding salt selection and formulation strategy. These methods will then be built upon to take on a new role in supporting the development of the new drug product. Finally, these methods will be modified to serve as a suite of analytical methods to demonstrate the quality, purity, and potency of the new drug product in clinical development.

REFERENCES

1. Gibson, M. Pharmaceutical preformulation and formulation: A practical guide from Candidate Drug Selection to Commercial Dosage Form. Interpharm/CRC press, Boca Raton, 2004.
2. Engel, G. L., Farid, N. A., Faul, M. M., Richardson, L. A. and Winneroski, L. L. Salt form selection and characterization of LY333531 mesylate monohydrate. *Int. J. Pharm.* 198:239–247, 2000.
3. Adeyeye, M. C. and Brittan, H. G. Preformulation in Solid Dosage Form Development. Informa Healthcare USA Inc., Location, 2008.
4. Sarfaraz, N. Handbook of Bioequivalence Testing. Informa Healthcare USA, Inc, New York, 2007.
5. Bhattachar, S., Deschenes, L. and Wesley, J. Solubility: It's not just for physical chemists. *Drug Discov. Today* 11(21/22), 2006.
6. Swarbrick, J. and Boylan, J. Encyclopedia of Pharmaceutical Technology, Vol I. 2nd ed., Marcel Dekker Inc., New York, 2002.
7. Avdeef, A., Voloboy, D. and Foreman, A. Dissolution and Solubility, Comprehensive Medicinal Chemistry II Vol. 5. Elsevier, Amsterdam, 2007.
8. Berger, C., Tsinman, O., Voloboy, D., Lipp, D., Stones, S. and Avdeef, A. Technical Note: Miniaturized Intrinsic Dissolution Rate (Mini-IDR™) Measurement of Griseofulvin and Carbamazepine. Dissolution Technologies, Leawood, KS, 2007.
9. Giron, D. Monitoring polymorphism of drugs, an on-going challenge—Part 1. *Am. Pharm. Rev.* 11 (1):66, 68–71, 2008.
10. Variankaval, N. and Sheth, A. Physical stability of crystal forms—Implications in drug development. *Am. Pharmaceut. Rev.* 10:96–101, 2007.
11. Qiu, Y., Chen, Y. and Zhang, G. Developing Solid Oral Dosage Forms Pharmaceutical Theory & Practice. Academic Press, New York, 2009.
12. Chesnut, S. and Salisbury, J. The role of UHPLC in pharmaceutical development. *J. Sep. Sci.* 30(8), 2007.
13. Kadav, A. and Vora, D. Stability indicating UPLC method for simultaneous determination of atorvastatin, fenofibrate and their degradation products in tablets. *J. Pharm. Biomed. Anal.* 48 (1):120–126, 2008.
14. Nováková, L., Matysová, L. and Solich, P. Advantages of application of UPLC in pharmaceutical analysis. *Talanta* 68(3):908–918, 2006.
15. Ali, I., Aboul-Enein, H. and Ghanem, A. Enantioselective toxicity and carcinogenesis. *Curr. Pharm. Anal.* 1:109–125, 2005.

16. He, B. and Shi, Y. Comprehensive chiral HPLC method development for pharmaceutical compounds during late stage development. *Am. Pharmaceut. Rev.* 11(1):47–52, 2008.
17. Byrn, S., Ralph, R., Pfeiffer, R. and Stowell, J. Polymorphs, regulations, crystallization, and solid-state chemistry. *Am. Pharmaceut. Rev.* 5(3):92–99, 2005.
18. Redman-Furey, N. and Collins, W. Thermoanalytical characterization of pharmaceutical materials: Small molecule applications. *Am. Pharmaceut. Rev.* 5(3):100–105, 2002.
19. Brittain, H. X-Ray powder diffraction of pharmaceutical materials. *Am. Pharmaceut. Rev.* 5 (1):74–80, 2002.
20. Offerdahl, T. and Munson, E. Solid-state NMR spectroscopy of pharmaceutical materials. *Am. Pharm. Rev.* 7(1):109–112, 2004.
21. Lee, D. and Webb, M. Pharmaceutical Analysis. Taylor & Francis Inc., Philadelphia, 2003.
22. Huynh-Ba, K. Handbook of Stability Testing in Pharmaceutical Development. Springer-Verlag, New York, 2008.
23. Shekunov, B., Chattopadhyay, P., Tong, H. and Chow, A. Particle size analysis in pharmaceutics: Principles, methods and applications. *Pharm. Res.* 24(2), 2007.
24. Serajuddin, A., Thakur, A., Ghoshal, R., Fakes, M., Ranadive, S., Morris, K. and Varia, S. Selection of solid dosage form composition through drug-excipient compatibility testing. *J. Pharm. Sci.* 88(7):696–704, 1999.
25. Guidance to Industry Blend Uniformity Analysis FDA CDER US Oct 1999.
26. Berman, J. and A, J. Planchard "Blend Uniformity and Unit Dose Sampling" *Drug Dev. Ind. Pharm.* 21(11):1257–1283, 1995.
27. Storme-Paris, I., Clarot, I., Esposito, S., Chaumeil, J. C., Nicolas, A., Brion, F., Rieutord, A. and Chamina, P. Near InfraRed Spectroscopy homogeneity evaluation of complex powder blends in a small-scale pharmaceutical preformulation process, a real-life application. *Eur. J. Pharm. Biopharm.* 72:189–198, 2009.
28. Ciurczak, E. W. Following the progress of a pharmaceutical mixing study via near-infrared spectroscopy. *Pharm. Technol.* 15(9):140–145, 1991.
29. Lai, C., Holt, D., Leung, J. C., Cooney, C. L., Raju, G. K. and Hansen, P. Real time and noninvasive monitoring of dry powder blend homogeneity. *AlChE J.* 47(11):2618–2622, 2001.
30. Hailey, P. A., Doherty, P., Tapsell, P., Oliver, T. and Aldridge, P. K. *J. Pharm. Biomed. Anal.* 14:551–559, 1996.
31. Li, W. and Worosila, G. D. Quantitation of active pharmaceutical ingredients and excipients in powder blends using designed multivariate calibration models by near-infrared spectroscopy. *Int. J. Pharm.* 295:213–219, 2005.
32. Bynum, K., Zielinski, J., Patel, S., Wabuuyele, B., LoBrutto, R. and Vivilecchia, R. Blend uniformity: A mass balance approach. *Am. Pharm. Rev.* 10(6):16–27, 2007.
33. Li, W., Johnson, M., Bruce, R., Ulrich, S., Rasmussen, H. and Worosila, G. Mass-balanced blend uniformity analysis of pharmaceutical powders by at-line near-infrared spectroscopy with a fiber-optic probe. *Int. J. Pharm.* 326(1–2):182–185, 2006.
34. Ahuja, S. and Dong, M. Handbook of Pharmaceutical Analysis by HPLC. Elsevier, San Diego, 2005.
35. Dressmann, J. and Kramer, J. Pharmaceutical Dissolution Testing. Taylor & Francis Group, New York, 2005.
36. Cho, J., Gemperline, P., Salt, A. and Walker, D. UV/visible spectral dissolution monitoring by in situ fiberoptic probes. *Anal. Chem.* 67(17), 1995.
37. Aldridge, P. and Kostek, L. In situ fiberoptic dissolution analysis. *Dissolution Technol.* 2(4):10–11, 1995.
38. Toher, C., Nielsen, P., Foreman, A. and Avdeef, A. In situ fiberoptic dissolution monitoring of a vitamin B12 solid dosage formulation. *Dissolution Technol.* 10(4), 2003.
39. Nir, I., Johnson, B., Johansson, J. and Schatz, C. Application of fiberoptic dissolution testing for actual products. *Pharm. Technol.* 25(5), 2001.
40. Zoeller, T. and Klein, S. Simplified biorelevant media for screening dissolution performance of poorly soluble drugs. *Dissolution Technol.* 14(4), 2007.
41. Grundy, J., Anderson, K., Rogers, J. and Fosters, R. Studies on dissolution testing of the niefedipine gastrointestinal therapeutic system. I. Description of a two-phase in vitro dissolution test. *J. Control. Release* 48:1–8, 1997.

42. Grundy, J., Anderson, K., Rogers, J. and Fosters, R. Studies on dissolution testing of the niefedipine gastrointestinal therapeutic system. II. Improved in vitro–in vivo correlation using a two-phase dissolution test. *J. Control. Release* 48:9–17, 1997.
43. Guidance for Industry Dissolution Testing of Immediate Release Solid Oral Dosage Forms, U.S. Department of Health and Human Services Food and Drug Administration Center for Drug Evaluation and Research (CDER) 1997.
44. USP 32-NF 27 <1092> the Dissolution Procedure: Development and Validation.
45. Lagace, M., Gravelle, L., Di Maso, M. and McClintock, S. Developing a discriminating dissolution procedure for a dual active pharmaceutical product with unique solubility characteristics. *Dissolution Technol.* 11(1), 2004.
46. Krishna, R. and Yu, L. Biopharmaceutics Applications in Drug Development. Springer, New York, 2007.
47. Hanson, R. and Gray, V. Handbook of Dissolution Testing. 3rd ed., Dissolution Technologies, Inc., 2004.
48. 48. ICH Q3B(R2): Impurities in New Drug Products 2006.
49. Bynum, K., Roinestad, K., Kassis, A., Pocreva, J., Gehrlein, L., Cheng, F. and Palermo, P. Analytical performance of a fiberoptic probe dissolution system. *Dissolution Technol.* 8(4), 2001.

10
IN-PROCESS CONTROL TESTING

CHRISTOPHER WOOD

Analytical Chemistry Research & Development, Bristol-Myers Squibb Pharmaceutical Development, New Brunswick, New Jersey, USA

I. INTRODUCTION

In-process testing, as defined by many, is the testing of the isolated intermediates that are prepared leading up to the final active pharmaceutical ingredients (APIs). However, in this discussion of in-process testing, we will be covering the testing that is performed during the actual processing step to prepare the intermediate or API material and not the isolated intermediate analysis. Since the isolated intermediate analysis is essentially the same as the API analysis, the sections dealing with the API analysis apply to intermediates also but not necessarily to the actual in-process testing that will be discussed here.

Most pharmaceutical analysts are very familiar with developing methods and analyzing the intermediates, APIs, and drug products of either chemical, biological, or pharmaceutical processes. They are also very familiar with the challenges that working with these samples can provide. For the analyst dealing with in-process testing, there are some key additional challenges that include the many different sample matrices, sample or matrix instabilities, sample reactivities, and, in early development, the uncertainty of the sample matrix from batch to batch. The challenges can lead to modify the way we qualify and validate our methods and how the technical transfer is performed.

Since this chapter deals with analyzing samples taken during a unit operation within a processing step, we will first spend time discussing a typical processing step and how we go about determining what, when, and how we test the samples, followed by the technical transfer of these test methods to a production facility.

II. API PROCESS CONTROL OVERVIEW

In-process control (IPC) methods are part of the quality control (QC) strategy used to monitor the chemical and biochemical process synthesis of the API. These tests are used to determine if the chemical entity is ready to proceed to next step of the manufacturing process. IPC tests are generally used to verify whether the following stages of the processing have been completed.

- Completion of reaction (COR): Test determines that the reaction has reached a predetermined end point by the consumption of starting material (SM), generation of product, or maximal levels of impurities.[1]
- Completion of extraction(s) or impurity removal.[2]
- Completion of solvent exchanges: Displacement of a solvent using distillation to another solvent. Solvent exchanges[3,4] are used to prepare a nonisolated intermediate for the next reaction or the crystallization.[5]
- Completion of crystallization: Product solution concentrations are determined before and after crystallization.[6,7]
- Completion of filter cake washing: Removal of impurities,[8] solvents, or salts.[9]
- Completion of product drying[10,11]: Isolated products are assayed for residual water[12] and organic solvents.[9,13]

If the analysis indicates that the processing is not complete, actions must be taken to drive the process to completion (i.e., adding more reagent to drive the reaction to completion, performing additional extraction(s), or adjusting the pH for impurity removal).

The selected IPC sampling points are typically used to monitor three classes of process impurities listed below.[14,15]

- Organic impurities: SMs, intermediates, reaction by-products and degradents, and enantiomers.
- Inorganic impurities: Reagents, ligands, catalyst heavy metals, and other material (e.g., filter aids, charcoal).
- Solvents: Organic volatile chemicals from the process or generated during processing.[16]

IPC tests are significant in establishing a synthesis and manufacturing control strategy that will reduce process variability, shorten process cycle times, and ensure that quality intermediates and APIs are produced. The analytical control strategy for an API process evolves during development.[17] Critical impurity control, detection, and identification are the primary goals of process development.[16,18]

Once the critical impurities are known, there are two options available to ensure quality API: minimize the formation of impurities or minimize their presence throughout the process synthesis.[19] The QC of process reagents, solvents, and intermediates in conjunction with in process controls helps to ensure a quality API is produced (see Figure 1).

A. Selecting an IPC Test

In-process testing requires robust methods that are fairly quick and easy to perform. The decision to include an IPC sampling point is based on scientific judgment to ensure product quality, maximumS yield, process efficiency, batch

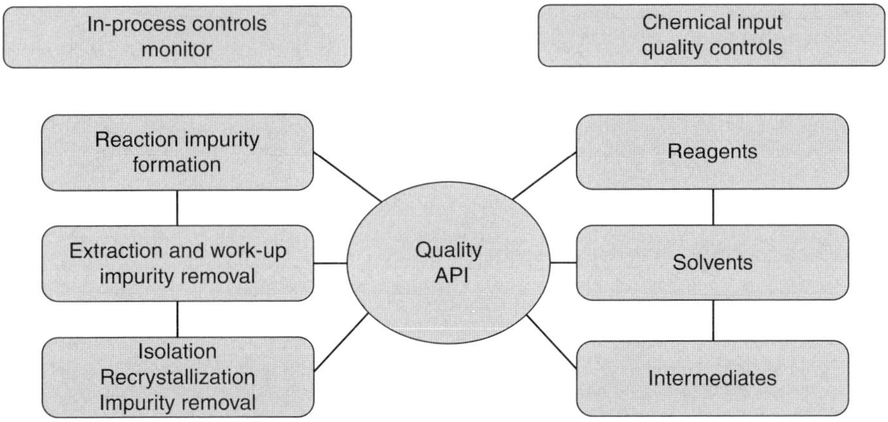

FIGURE 1 Relation between IPCs API quality and chemical input.

consistency, safety, and environmental compliance.[20] Some IPC tests can be simple such as, visual determination of a phase splits or disappearance of color. TLC (thin layer chromatography),[21] HPLC (high-performance liquid chromatography),[22,23] GC (gas chromatography),[24,25] and spectroscopy[5,26] are techniques used to monitor COR, impurity, and volatile component removal. Conductivity and pH[19] are tests that can be used to monitor the completion of a product cake washing process. Loss on drying (LOD), TGA,[27] Karl Fischer (KF) titration, and GC[28] are tests that are commonly used to determine if product drying is complete. The use of these techniques for IPC will be discussed later in this chapter.

B. In-Line Assays

The pharmaceutical industry has increased the development of in-line or on-line spectroscopic analyses for monitoring of chemical and mechanical processes. Most process sample analyses are performed off-line where samples are brought to the lab for testing. This results in a delay created by taking the sample and waiting for the lab results. In-line assays move the analysis closer to the process, which eliminates the sampling, sample transport, and preparation, which can increase the rate of "sampling," thereby providing a greater potential for increased process information. In-line process analytical technology (PAT)[29] is commonly used to monitor pH and temperature. Vibrational spectroscopy techniques such as near-infrared (NIR), Fourier transform infrared (FTIR),[2,30] and Raman[31–33] are a few of the techniques used to monitor process samples that would otherwise be difficult to analyze off-line. An example of PAT used to analyze a difficult sample was demonstrated by Wiss *et al.* using in-line NIR to monitor the formation of a Grignard reagent.[34] The concentration of the added reagent and the reaction initiation of the formed Grignard reagent were determined in real time. The development of the in-line assay improved the safety handling of an exothermic reaction. The assay was able to monitor the cross-coupling reaction and determine the real-time reaction yield. The in-line PAT reduced the cycle times and eliminated the waiting because of an off-line HPLC method. Another example of in-line PAT was demonstrated by Lin *et al.* using in-line FTIR to monitor the mono- and bis-HCL salt formation.[30] The in-line method overcame the limitations of the HPLC method which was unable to differentiate between the bis- and mono-salt. The team was successful in isolating the desired mono-HCL salt.

The common theme for PAT use is greater process understanding,[31] safety,[2,34] and time saving.[3,10] For an in-depth discussion on this topic, the reader should refer to Chapter 5 on PAT in this text.

C. Process Sample Plan Development

An analyst working in a group where in-process testing is performed needs to work closely with the process chemist or engineer to understand the chemistry of the process. This understanding allows the analyst to better help the team in determining where the testing is needed, the best technology to use, and the sensitivity and robustness requirements of the methods. Therefore, in order to have a thorough discussion of in-process testing, an example of a typical API process will first be examined to determine what tests are needed for the process, followed by discussion of the tests.

During the evaluation of the synthetic steps, most analytical methods will be covered, along with a short discussion on the limits. This discussion covers real-world situations, but not necessarily in one processing step as shown here. Scheme 1 is the example that will be evaluated.

The steps in Scheme 1 show a common coupling reaction of compounds A and B to give C. The solvents and reagents used in the coupling reaction are also shown along with the process flow diagram (PFD) in Figure 2.

During the process development of the reaction, the process chemist, the process engineer, and the process analyst work together to best determine where in the process an IPC sampling point is needed. Since excess sampling can cause a burden on the analytical staffing and significantly increase the overall processing time, a careful evaluation of the process needs to be done to evaluate the risk of proceeding at each unit operation of the processing step with and without the analytical data. The risk will depend on the process robustness, process knowledge needed or known, and process operating ranges versus the ability of the plant to control the process within those ranges during the operation. As process development continues and process knowledge is gained, these IPC sampling points will be changed, removed, or added to. This point was demonstrated by Ragan *et al.*, where the water level in the SM made it difficult to drive the acylation reaction to completion without using an excess of acid chloride reagent.[35] The acid chloride was relatively inexpensive and could be purged during the product crystallization. However, the SM's water level varied from 0.5% to 4%. It was found

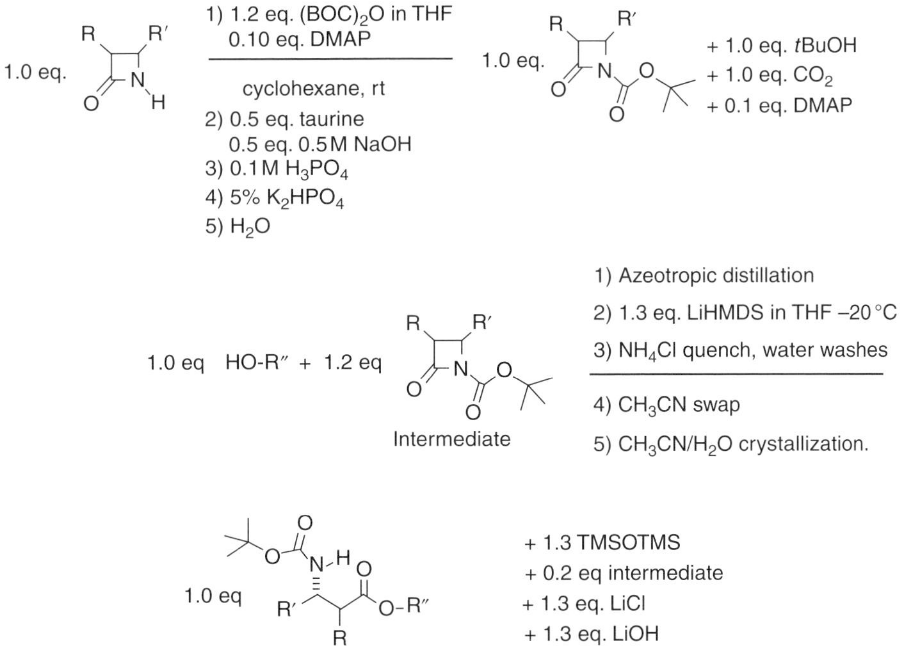

SCHEME 1 Process evaluation.

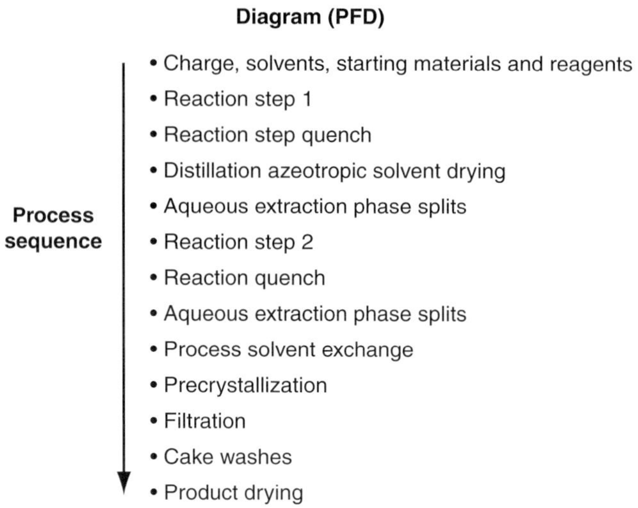

FIGURE 2 **Process flow diagram (PFD).**

that the water level had a significant effect on the reaction yield (65–99%). Laboratory experimentation was used to set a process specification of $<1\%$ water. The in-process specification allowed the product yields to be greater than 90%.

Over time, typically, the number of IPC sampling points will be reduced because of the increase in processing knowledge, the robustness of the process, and the widening of the processing parameters. When the team decides on the IPC sampling points, they will develop a list of required samples, which may be designated as a sample plan. Once the sample plan is determined, the analyst now goes to work determining what methods are needed to ensure that these samples can be analyzed correctly to give a true "picture" of the process. Furthermore, acceptance criterion will be established for each IPC sample in the sample plan based on the current processing knowledge. These limits will also change over time during the development of the process, depending on how well the process can purge the impurities which increase or decrease based on the in-process limit specification

In addition to having IPC samples in the plan, there are–usually "for information only" (FIO) samples included. These samples are used to increase the process knowledge and are not used to control the quality. These samples are pulled and analyzed after the IPC is completed, thereby not slowing the processing of the batch. Typically, these samples cover the overall mass balance of the process and indicate the fate of the impurities during the unit operations. These samples can be very valuable for setting IPC limits during early development of the process. FIO sampling points may be monitored during process validation but are typically not obtained during routine production of the API. Therefore, the technical transfer of these methods to the production facility is more of an informal process performed without transfer protocols.

From the evaluation of the process in Scheme 1, the sample plan was developed by the team, as shown in Figures 3 and 4.

Sample identification	Analytical test
IPC 1 Moisture in solvents	Coulometric titration
IPC 2 BOC protection reaction	HPLC (relative area% of starting material and product)
IPC 3 BOC anhydride in organic phase	GC (wt%)
FIO 1 Aqueous split	HPLC (wt%) pH
FIO 2 Organic phase after washes	HPLC
IPC 4 Azeotropic drying	Coulometric titration

FIGURE 3 Sample plan (BOC protection).

Sample identification	Analytical test
IPC 5 Reaction completion coupling	HPLC
FIO 3 Spent aqueous	HPLC
IPC 6 Solvent swap to acetonitrile **FIO 4** Slurry	GC HPLC
FIO 5 Mother liquor and washes	HPLC (wt%)
FIO 6 Washes **IPC 7** Dry cake	Conductivity Loss on drying

FIGURE 4 Sample plan (coupling).

As can be seen in the sample plan, the plan contains a wide variety of tests that include many common types of analytical techniques such as chromatographic assays, moisture determination, solvent assays, loss on drying, pH, and conductivity measurements.[36] As with tests for the API or drug product, the IPC tests can be categorized as indicated in the list below.

- Quantitative tests: COR, concentration analysis.
- Limit tests: COR, impurity determinations, concentration analysis.
- General or compendial tests: Loss on drying (LOD), KF, residue on ignition (ROI), color, pH, specific rotation, and titrations.

III. ANALYSIS CONSIDERATIONS

During the method development, the analyst needs to consider a number of aspects of the process before proceeding too far down the method development path. These considerations may be determined by asking some of the following questions:

- What is the simplest analytical technique suitable for the test?
- What is the stability of the IPC sample and the standard solutions?
- What is the concentration range of product or analyte in the process streams?
- What technique will give the shortest analysis time?

The simplest analytical technique to use is often the most overlooked. Yates *et al.* used an off-line colorimetric test to determine if a sodium sulfite treatment of a process reaction stream quenched the excess *N*-bromosuccinimide remaining in the reaction.[37] This was done by adding aliquots of the treated reaction solution into a starch/iodine test. The disappearance of the blue color indicated there was no more bromine present. Semiqualitative tests are generally used in the early stages of process development. A series of semiqualitative tests are commercially available impregnated papers from EM Science (e.g., peroxide test and pH papers).

In-process testing requires robust methods that are fairly quick and easy to perform. Samples that arrive at the lab during off-hours may not have an experienced analyst who is usually available during the day shift to perform a complex analysis. The use of new analytical technologies during early process development may be suitable for IPC analysis; however, as the process development continues to the technical transfer, alternative methods should be developed that are compatible with QC lab capabilities. In addition, analysts should not try to optimize the method to generate other information that is seldom used in a normal manufacturing operation.

Furthermore, in-process sample analyses are shared by various team members (engineer, chemist, and analyst) with varying amounts of analytical skills during the early stages of process and IPC development. Complicated analytical techniques can make the test difficult to transfer or share with other team members.[19] Additionally, external instrumentation is another concern with complicated analytical techniques.

Sample and standard solution stability knowledge are essential during the early stages of method development and optimization.[35,37] The solutions must be stable during the trip to the lab, or an alternative procedure should be used to increase sample stability. For example, the sample may need to be immediately cooled or quenched at the time of the sample pull. If quenching the sample is a means of increasing the stability, then additional work may needed to defend the rate of quenching, the quantity of quenching solution needed, and addition order of quench reagents. The importance of reaction quenching was demonstrated by Kopach *et al.* who had obtained some anecdotal information suggesting the reaction quench order was important to minimize stereochemical scrambling.[38] Although the stereochemical scrambling was not noted as a caution for sample preparation, it could be implied from the quench reaction that this could happen in the reaction sample preparation. In some instances, reactions with the mobile phase or sample diluent may cause impurities. This phenomena was reported by Laviana *et al.* who found sultamicillin imines were readily hydrolyzed at acidic pH in the HPLC mobile phase or sample diluent.[39] These examples help to reinforce the idea that an IPC sample stability and mobile interaction must be evaluated before finalizing the in-process method development.

A. Preparation for Analysis

Before an intermediate or API process begins, a tentative IPC sample arrival schedule is produced by a chemist, engineer, or plant manager. This is only a tentative schedule because IPC samples can arrive hours early or late. There are many process factors that affect sample arrival times such as reactor heating rates, distillation speeds, and equipment issues. The analyst and team communicate frequently on the day of the process to obtain updated sample arrival times.

Because of the uncertainty of sample delivery times, rapid/fast IPC chromatographic methods are needed to maintain an efficient analytical laboratory and plant operation.[40–42] Since many samples can be generated during the process development, IPC methods should be developed and optimized for the shortest analysis time possible. The need for a fast IPC method was demonstrated by Wu *et al.* who redeveloped an in-process method to shorten the run time for a reaction conversion HPLC analysis from 30 to 10 min using a monolithic HPLC column.[22] The newly developed method could also be used to determine mother liquor concentrations and perform impurity profiles for the crude product. This optimized method reduced the process cycle time for lab instrumentation. The dual purpose for an IPC method (i.e., COR, impurity profiling, and/or concentration analysis) is another opportunity to improve the process and lab efficiency. Another example of a dual-purpose method use was demonstrated by Nageswara Rao *et al.* who developed a 15-min RP-HPLC method that could determine the COR and impurity profile (isolated product) for two different processes for production of 4-methoxyphenyl acetic acid.[43] The dual-purpose method is a common theme for in-process analyses. It is also important to remember that a reduction in the analysis time and overall manufacturing time can reduce the cost of the API.

IV. HPLC METHOD DEVELOPMENT

The development of a suitable HPLC IPC method requires collaborations with process chemists to help enhance the analyst's knowledge of the potential impurities generated from the reaction or processing. Many people think: Why not use a modified version of the intermediate release method? This sounds simple; however, development of COR and process mass balance samples have been analyzed long before any formal intermediate release methods are developed. In many instances, the in-process methods are the foundation for the intermediate release methods.[37]

The COR IPC method should meet the following requirements before proceeding to validation:

- There should be no interference between blank sample matrix, SM, and product.
- SM must be resolved from reaction impurities and product.
- Linearity of product and SM at target concentration should be confirmed.
- IPC sample preparation and results must be repeatable.
- Accuracy and precision must meet the requirements of the process.
- Reaction solution and standard solutions stability must meet the process requirements.[44]

The required accuracy and linearity may vary significantly during API process development. The individual developing the COR IPC method must continuously check with the process team members (chemists and engineers) to determine if the method is still suitable for its intended purpose.[45]

Developing a suitable HPLC IPC method for a two- to three-component mixture (SM and product) is perceived to be straightforward. However, the addition of process reagents, reaction intermediates, and by-products (which are purged and not tested in an intermediate release method) increases the complexity of the separation. Therefore, the rapid evolution of API processes forces the analyst to move away from the trial- and-error approach.[46] A systematic approach is needed to help make method development more efficient. It is recommended that a computer assisted-method development tool (CAMD)[47] and automated column switching[46,48] be used to perform mobile phase and stationary phase screening followed by LC/MS for peak identification and purity analysis. Chromsorb®,[49] ACD (Advanced Chemistry Development),[49]and Drylab®[50] are a few of the commercially available CAMD software packages. The use of column switching and software optimization was demonstrated by Xiao using an LC Spiderling (column switching value) and Chromsorb® software to develop a 10-min separation of structural epimers of β-methyl epoxide and α-methyl epoxide.[46] Another use of CAMD software for methods development was reported by Argentine *et al.* using Drylab® modeling software. Drylab software can rapidly optimize an HPLC separation for a 30-component mixture of a reaction SMs, intermediates, and process-related impurities within 30 min.[51] LC/MS was used for peak identification during the method development. The Drylab software was also applied to shorten the 30-min method analysis time to 12 min for all 30 components (4.6 mm $\times$ 150 mm, 3 μm HPLC column). The final method analysis time was further decreased to 5 min for the mixture by using sub-2-μm column and high-pressure LC equipment.

The use of orthogonal methods is recommended to confirm the final separations specificity. Examples of this concept have been reported by Pellet *et al.*[52] for drug product impurities and degradants and by Argentine *et al.*[18,51,53] for control of process-related impurities. A flow diagram of method development strategy is provided in Figure 5. For additional information, the reader should refer to method development discussed in various chapters in this text.

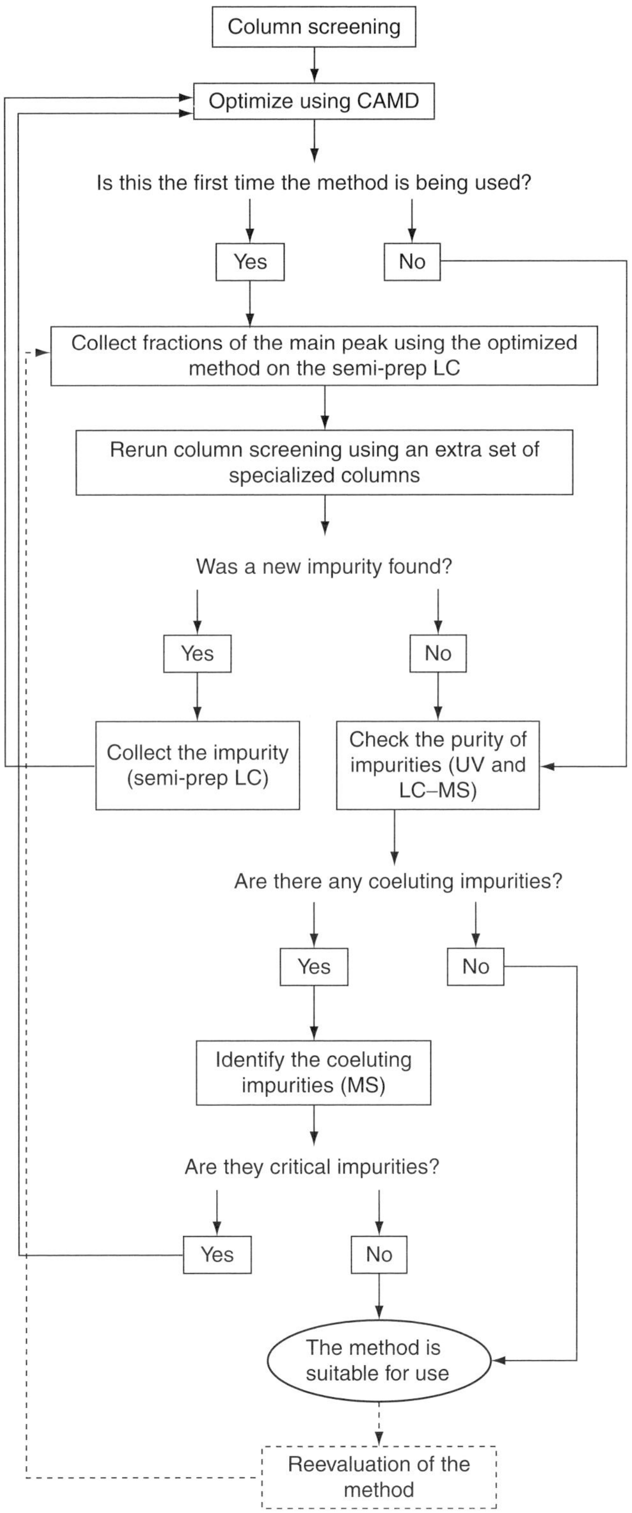

FIGURE 5 **Method development workflow.**

V. IN-PROCESS CONTROL TESTS

In general, the first IPC of a process is the COR assay or test. Process samples are typically analyzed by TLC,[21] HPLC,[22,23,43,46] GC,[24,25] infrared, or Raman spectroscopy.[2,5,26] This assay determines when the reaction has reached it predetermined end point.

TLC is typically for samples used in the early stages of process development for screening and reaction monitoring.[21] A broad range of compounds can be separated by using different plates (i.e., silica, cellulose, or reverse phase) and mobiles phases.[54] TLC has an advantage in that all impurities in principle can be detected because they are applied to sorbents. Everything is shown on the sample plate from origin to the sample front after development. Many of these plates are impregnated with phosphor to detect compounds using short-wavelength (254 nm) UV or by long-wavelength (350 nm) for compounds that are naturally fluorescent.[55] There are also many spray reagents to enhance specificity and/or sensitivity for many compounds when UV detection is inadequate.[55,56] The major hurdles for TLC are the limited resolution, detection, and ease of quantification.[54]

Currently, off-line HPLC has surpassed TLC in most reaction monitoring applications; however, TLC plays a useful role when cost and simplicity are needed.[54]

GC is not generally used for COR analyses because most IPC samples are non-volatile and potentially thermally labile in nature.[24] There are some exceptions however; Müller *et al.* used GC to determine in-process COR reaction of a poorly water-soluble volatile aromatic hydrocarbon with a polymer.[9] GC analysis was also used to monitor reaction progress and follow the formation of a process impurity (vinyl nitrile) by Vaidyanathan *et al.*[57]

NMR is a common tool used by organic process research chemists to monitor reaction progress.[58] An example of in-process COR analysis done by NMR was performed by Kopach *et al.* using ^{1}H NMR to follow a Grignard reaction initiation by a derivatization. The active Grignard itself was not stable enough to analyze by direct measurement.[38] An aliquot of the Grignard reaction mixture was quenched into CD_3OD, which converted any active Grignard to THP (tetrahydropyan). The extent of the reaction could then be determined by comparing the integration of 4.2 ppm 4-chlorotetrahydropyran to THP ^{1}H NMR resonance. For more discussions on NMR analysis, the reader should refer to various chapters in this text.

A. Completion of Reaction (COR) Relative Area Percent (RAP)

Off-line HPLC/UV detection is the most common analysis technique used to monitor the disappearance of SM(s) and formation of product(s).[23,41,42] COR IPC criteria are usually expressed as a relative area percent (RAP) of SM to product.[22,51] The IPC method should separate all the known impurities from the SM, but not necessarily from the product since minor impurities co-eluting with the product will not significantly affect the IPC result. This approach allows the development of shorter methods. The UV λ_{max} of the SM and/or critical impurities should be used as the HPLC detector setting (if possible) for the analysis, not the product.[19] Using the λ_{max} of the product may underestimate the reaction's completion and critical impurity formation. The method should have the ability to handle variable reaction product loads that occur during process optimization and should be able to reach system equilibration with just a few injections (e.g., blank, system suitability, and IPC sample).[44]

A system suitability preparation is made at the reaction criteria before a COR sample is analyzed (e.g., if a reaction IPC criteria is ≤ 1 RAP SM to product). A product to SM mixture standard preparation is made at 99 parts product to 1 part SM in the linear range of the detector. Similar system suitability preparations were performed by Wrezel *et al.* for reaction conversion samples.[44] The analyst needs to know the anticipated concentration of the product or analyte in the process stream (e.g., if the anticipated concentration of the product in reaction solution is 50 mg/mL, then the SM concentration in solution should be <0.5 mg/mL). This tells the analyst the reaction sample dilution factor needed to remain in the linear range of the detector. It is important to remember that during synthesis development and process optimization, the anticipated product, SM, and impurity concentrations may vary significantly. Figure 6 shows an example of a COR IPC chromatogram. It is important to note the number of process-related impurities in the chromatogram.

B. Impurity Determination RAP

Impurity determination IPCs are used to monitor the removal of critical impurities from the product-rich solution layer or by solvent extractions or cake washes. The IPC criteria are usually expressed as RAP of the impurities relative to product in the product-rich solution layer and concentration of the impurity in extracted solution or mother liquor. The use of an impurity profile IPC was

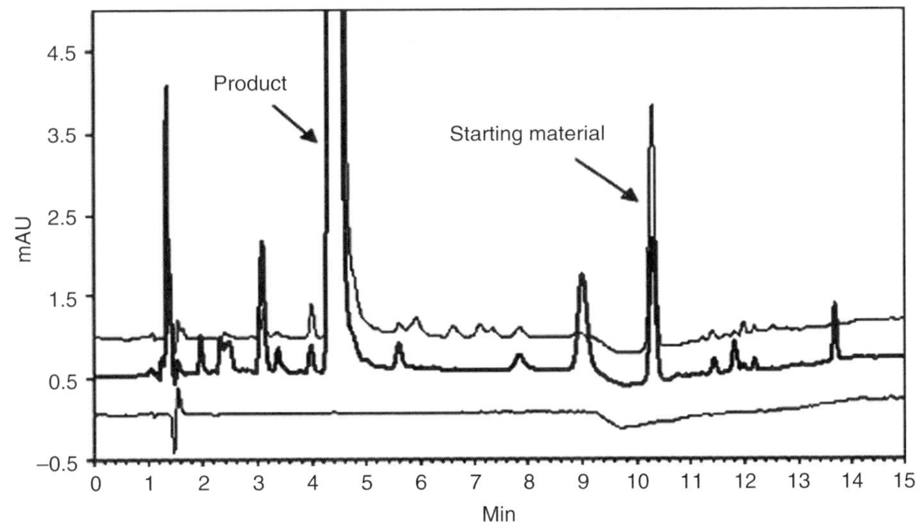

FIGURE 6 Completion of reaction (HPLC/UV). From bottom to top: diluent blank injection, reaction IPC, and reaction system suitability chromatogram.

demonstrated by Hansen *et al.* who used HPLC analysis for product solution to determine the time needed to remove residual HI and complete the free-basing operation. Solid K_2CO_3 was used to remove HI from the process solution. This problem was solved by noting that iodide gave a significant peak at 227 nm in the HPLC chromatogram. The extent of neutralization was monitored by sampling the mother liquors and stirring until the iodide peak was no longer visible by HPLC (1 h time). Zheng also used an impurity IPC to follow the removal of unreacted SM and benzyl alcohol by using isopropyl acetate extractions of an aqueous product layer by HPLC.[59]

The analyst needs to know the concentration of analyte in the product-rich solutions to ensure that sample preparation is within the method's linear range. The primary method development goal should be one method for COR, impurity determination, and product concentrations (if needed).[51] An example of this concept was demonstrated by Wu[22] and Nageswara Rao *et al.*[43] who developed dual-purpose HPLC methods suitable for the monitoring reaction progression and determining the impurity profiling of the isolated product produced from the reaction. A secondary method goal should be the use of the same column and mobile phase, but different gradient conditions for all three IPCs (if possible). One HPLC method or multiple methods using the same column and mobile phase reduces instrument usage and helps the laboratory work more efficiently. This concept was demonstrated by Argentine *et al.* in the development of a single 5-min high-pressure LC/MS that tracked the formation of product fate of starting materials, intermediates, and process-related impurities throughout a four-step process.[51] The fast method improved the laboratory productivity and process by shortening the analysis time between IPC and process development samples.

C. Product Solution Concentration Analysis

Product concentration IPCs are used to measure initial reactor charge concentrations, reagent charges, product losses during phase splits or waste streams, and to determine the start and end of product crystallization. HPLC/UV detection is the technique most often used for this analysis. Off-line NIR[7] and Raman spectroscopy[31] have also been used for these samples. IPC assays are performed similar to product-release assay using qualified reference materials. The HPLC method must resolve all peaks from the component of interest, but not necessarily from each other. As with most IPC methods, a key component is the sample preparation. Many of the concentration analysis samples are heated (reaction samples) and/or cooled (crystallization samples), meaning the product may precipitate or redissolve, affecting the IPC concentration values. The analyst must ensure that the process temperature (heating or cooling samples) is maintained during the sample preparation or adapt a new sampling protocol when performing the analysis. This point was demonstrated by Zhou *et al.* who had difficulty obtaining representative samples for an HPLC analysis to determine the product concentration prior to seeding in a crystallization process.[7] The sampling required handling of a supersaturated pilot-plant reactor solution at elevated temperatures (65 °C). The concern was that the product could prematurely crystallize and change the concentration results. The samples were obtained by taking 10–20 mL of the reactor process solution through sample port after immediately flushing into a tared 100-mL glass screw-cap bottle. The bottles were closed immediately and reweighed. The contents of the bottle were quantitatively rinsed and diluted into a volumetric flask and diluted with solvent. The HPLC analysis was later replaced with in-line NIR spectroscopy because of the sampling uncertainty with off-line LC.

Before starting the analysis, it is important the analyst have a sample description and know the anticipated concentration of product or analyte in the process stream for any IPC sample. The theoretical process sample concentration can tell the analyst if the process is proceeding as expected or if there was an error in the sample preparation. A significant change in the sample concentrations could indicate a process or sampling error (e.g., a wrong layer of a biphasic sample delivered to the laboratory). The method development process in Figure 5 can also be used for concentration HPLC assays.

D. Process Solvent Exchanges

A process solvent exchange is the displacement of a solvent using distillation to another solvent. Solvent exchanges are used to prepare a nonisolated intermediate for the next reaction or for the crystallization step.[4] An incomplete solvent exchange can result in the loss of product or the generation of unwanted impurities in the next step. Zeng *et al.* reported high mother liquor product losses because of a poor solvent exchange from tetrahydrofuran/toleune to acetonitrile. The “lost” material had to be recovered from a rework of the mother liquors.[59] This example shows the importance of an IPC for solvent-exchange analysis.

Solvent-exchange IPC samples are usually analyzed by direct injection capillary GC with flame ionization detection (FID);[2] NIR,[3] and Raman spectroscopy[31,32] have also been used for this assay. The final acceptance criteria are typically expressed in % volume/volume (%, v/v) or % weight/volume (%, w/v) of the solvent being removed relative to the volume of solvent in the process mixture.

Rapid sample analysis is important for each IPC in the manufacturing process. Using generic GC analysis for solvent swaps is an approach to keep the lab more efficient and reduce need for multiple instruments. The literature has examples of generic GC methods used to detect ICH classes 1–3 solvents in the isolated intermediates and the API. This concept was used by Klick *et al.* who validated a generic GC method for use during drug development and for QC in commercial manufacturing.[25] The validated method included 5 class-1 solvents, 17 out of 27 class-2 solvents, 17 out of 27 class-3 solvents and three unclassified solvents according to the ICH guideline.

The limited number of solvents used in the API synthesis and processing suggests that a generic method approach may be applied to the process solvent-exchange analyses.[24] The initial method suitability can be evaluated before samples are submitted for analysis. The need to develop a unique solvent-exchange method may be unnecessary when a generic method is available. The generic method can be optimized later to reduce the analysis time for a faster sample turnaround if needed. This approach was demonstrated by P. Sandra *et al.* for the determination of residual solvents in pharmaceutical formulations.[24] A generic

GC method was used that separates the 20 most common API solvents for the identification of unknown solvent residues in samples. Then a second shorter analysis time method was used for fast throughput and quantitation of samples. An example of a generic GC process solvents exchange method is illustrated in Figure 7, with a mixture of 26 common process solvents listed in the ICH guideline[17] diluted in *N*,*N*-dimethylacetamide. Figure 8 shows an example of actual process solvent-exchange analysis.

Another technique used to determine completion of solvent exchange is Raman spectroscopy.[31,32] Spectroscopic methods offer a simple sample preparation and rapid analysis. An example of an off-line Raman solvent method developed to determine the completion of a solvent exchange from 30% to less than 2% dichloromethane (DCM) in acetone is shown in Figures 9–11.[60] A multiple linear regression was used for quantitative determination of the total DCM concentrations in the reactor samples. The reliable prediction of total DCM concentration was successfully achieved by Raman spectroscopy, as judged by a comparison with GC data.

E. Genotoxic Impurity (GTI)

Genotoxic impurities (GTIs) are compounds that induce genetic mutation, chromosomal breaks and/or chromosomal rearrangements, and have the potential to cause cancer.[61,62] Detailed discussions of GTI impurities and method development are included in Chapter 4 of this text. Therefore, we will look at GTI analyses from an

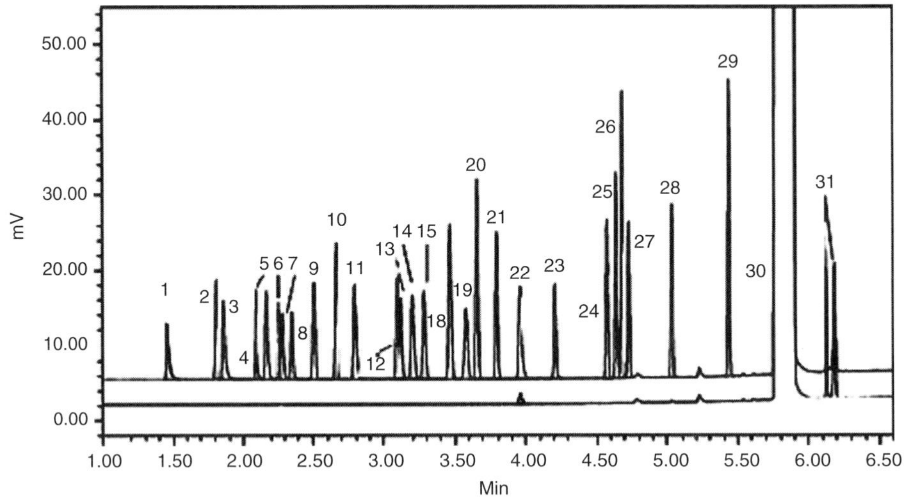

FIGURE 7 **Generic solvent-exchange method, direct injection GC/FID. From bottom to top: blank injection and GC volatiles test solution. Peaks: 1, methanol; 2, *n*-pentane; 3, ethanol; 4, acetone 5 isopropyl alcohol; 6, acetonitrile; 7, methyl acetate; 8, methylene chloride; 9 methyl tertiary butyl ether; 10, *n*-hexane; 11, propanol; 12, methyl ethyl ketone; 13, ethyl acetate; 14, *sec*-butanol; 15, tetrahydrofuran; 16, cyclohexane; 17, hexamethyldisiloxane; 18, benzene; 19, *n*-heptane; 20, butyl alcohol; 21, 1,4-dioxane; 22, methyl isobutyl ketone; 23, pyridine; 24, toluene; 25, isobutyl acetate; 26, *n*-butyl acetate; 27, *p*-xylene; 28, dimethylacetamide; 29, solvent impurities.**

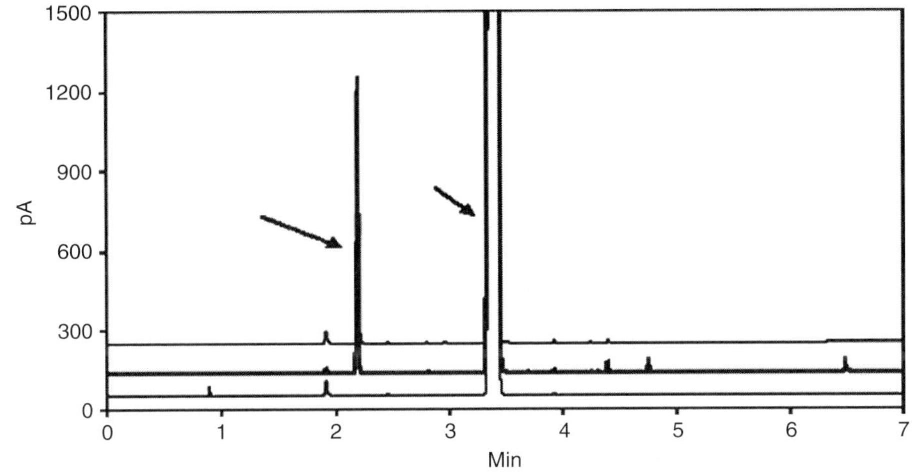

FIGURE 8 Example of process solvent exchange, direct injection limit test: From bottom to top: blank solvent injection, IPC vol% sample solution and standard solution corresponding to 2 vol% of exchanged solvent, Retention time for the exchanged solvent is approximately 2.2 min the new solvent is at 3.5 min.

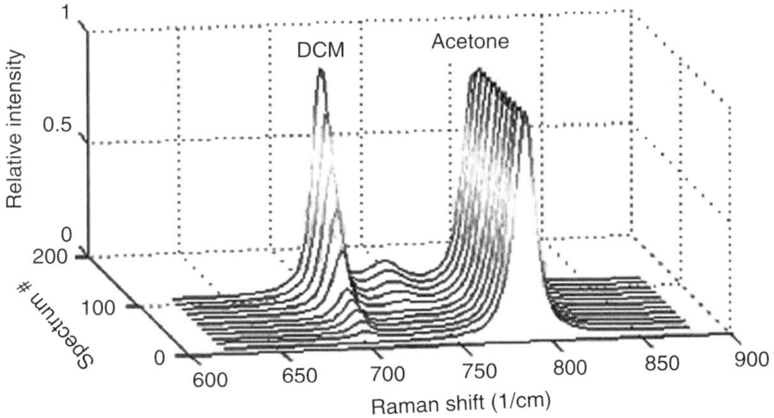

FIGURE 9 Waterfall off-line plot Raman spectra of DCM from 0% to 30% v/v standards in acetone. (See Color Insert.)

IPC perspective. GTIs in the API synthesis may come from reagents, SMs, process intermediates, and/or reaction by-products. A cross-functional team (regulatory, toxicology, process chemistry, and analytical) determines the specification level to which a GTI(s) must be controlled in the isolated API.[63,64] It is preferred to control the GTIs in the SM(s) or process intermediates for process flexibility and cost.[64,65] Process chemist perform spike and recovery laboratory experiments to determine if crystallization or extraction rejects, reduces, or removes the GTI.[18,64] This value forms the basis for the IPC acceptance criteria. The analysis of the isolated intermediate provides a secondary check to confirm that the GTI(s) are reduced

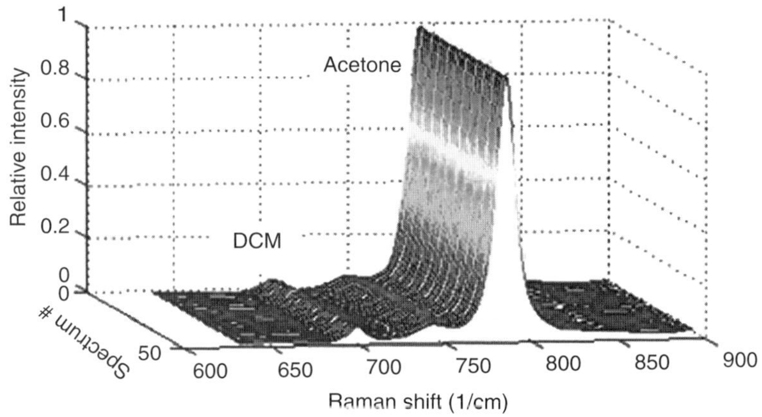

FIGURE 10 **Waterfall plot off-line Raman spectra of reaction plot samples after solvent exchange. (See Color Insert.)**

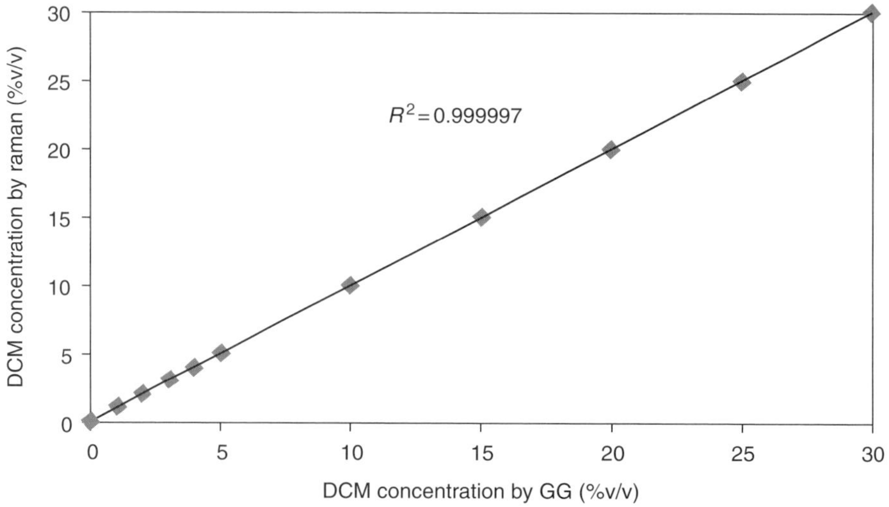

FIGURE 11 **Plot calibration model. DCM% (%, v/v) by Raman calibration results and actual GC results from the sample.**

to an acceptable level. The isolated material may need to be recrystallized if it does not meet the release specification.

GTI IPCs are usually trace analyses quantitated to ppm levels versus the product concentration in solution.[65] The analysis can be either a limit (pass/fail) or quantitative test performed at point(s) where the impurity can be reduced or removed (e.g., product solution extractions, wet cake washes, or crystallizations). Quantitative GTI results are usually needed for process development samples to track the level of GTI removal for the procedure.[64] Limit tests are routinely used for manufacturing IPCs. The method validation follows those suggested by the ICH (specificity, linearity, limit of detection, solution stability, spike, and recovery).

Most of the GTI detection and quantitation must be done by HPLC/MS or GC/MS using a selective ion mode.[64–67] MS gives the greatest opportunity to rule out interference that may occur using HPLC/UV and GC/FID detections; thus, HPLC/UV or GC/FID detection must always be assessed as an alternative. Lui *et al.* demonstrated the development of a hyphenated MS-based analysis to control the level of GTIs for a pazopanib process.[65] The LC/MS techniques were not robust enough to transfer to QC laboratories. By setting proper controls of GTI impurities upstream in the process, HPLC/UV could be used for the analysis. This made the methods more suitable for transfer to a QC lab.

The simplest analysis techniques are the best ones for IPC samples. HPLC/UV and GC/FID can be used for the IPC analyses if the sample has high solubility and the process or sample solvent has no GTI peak interferences. Figures 12 and 13 are examples of HPLC/UV and GC/FID GTI IPC method chromatograms for compounds with high solvent solubilities (~100 mg/mL).

Each GTI matrix brings along its own unique set of challenges for method development and optimization. There are examples in the literature of trace levels analysis of sulfonic esters (alkylating agents) quantitated to low levels.[61,62,66] Because a method works for a particular GTI does not means it can be applied generically for every synthesis containing that compound. Each GTI analyses must be evaluated on a case-by-case basis to determine if it can meet the process IPC criteria for specificity and sensitivity.

F. Water Concentration

Managing and understanding the effect of water on a synthetic process is important. An accurate determination of water maybe important to the success of reaction completion,[28] process yields,[35] azeotropic distillations,[38] crystallizations,[51] form control, and product drying. KF is a rapid analysis technique optimal for water

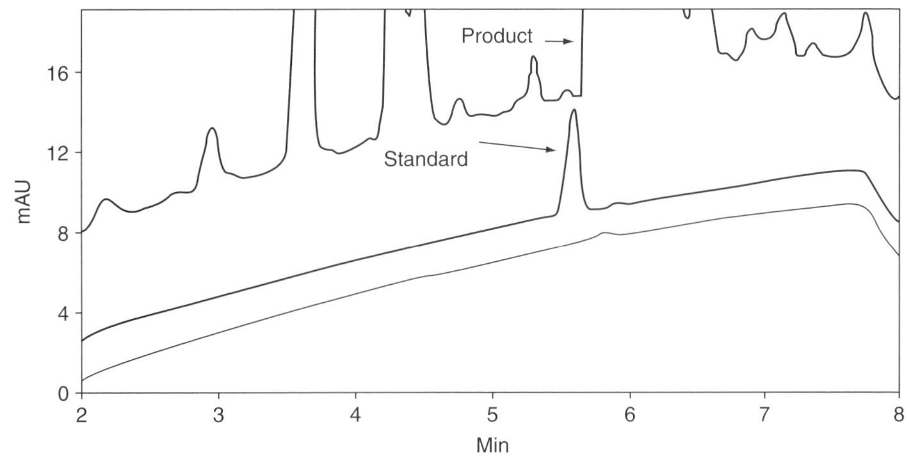

FIGURE 12 **Example of (HPLC/UV) limit test: From bottom to top: blank injection standard solution corresponding (50 ppm) GTI, and IPC sample solution (100 mg/mL). Retention time for the GTI is approximately 5.5 min.**

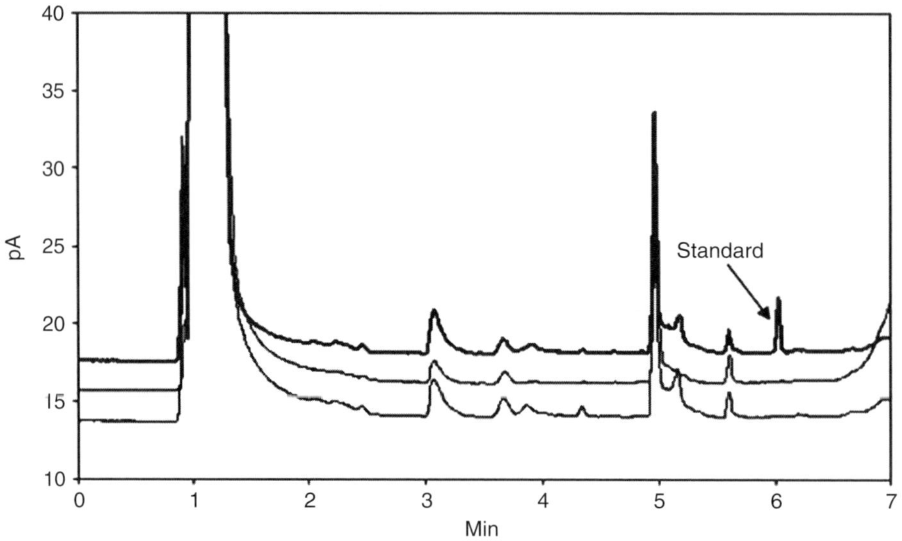

FIGURE 13 **Example of a volatile GTI GC/FID direct injection limit test: From bottom to top: blank solvent injection, IPC sample Solution at 100 mg/mL, and standard solution corresponding 50 ppm GTI. Retention time for GTI is approximately 6.2 min.**

determination in process samples.[7,12,68] The effect of water on a process IPC was demonstrated by Barkalow *et al.* who had to adjust a thionyl chloride charge to >1 equivalent of product in solution to compensate for the water in the process streams.[28] This adjustment for process water ensured that the reaction went to completion (<1% RAP SM). Pesti *et al.*[2] and Kopach *et al.*[38] used KF analysis to determine the water levels of the reaction solution after an azeotropic drying distillation prior to the use and formation of a Grignard reagent. Water levels can affect the Grignard formation, thus hindering the subsequent reactions. Another example of the effect of water was shown by Argentine *et al.*, who had to use an IPC test for water to control and ensure that spontaneous nucleation did not occur during API crystallization.[51] The control of water in the process also increased the product yield.

There are two types of KF titrations available:

- Volumetric KF titrations are used for samples with >1% water.
- Coulometric KF titrations are used for samples <1% water. Or less sample can be used for the analysis of samples with a water content greater that 1%.

The following are the reactions for the potentiometric KF titration[69,70]:

$$CH_2OH + SO_2 + RN \rightarrow [RNH]SO_3$$
$$H_2O + I_2 = [RNH]SOCH3 + [RNH]SO_4CH_3 + [RNH]I$$
$$[RN = Base]$$

Great care must be taken to ensure that the process components do not interfere with the KF titration causing erroneous results. Aldehydes, ketones, strong amines, siloxanes, thiols, phenols, peroxides, and mercaptans can interfere with the KF titration.[69,70] In-process samples are complex mixtures of many compounds and reagents. The analyst needs to look beyond the reaction solvent and into the process chemistry to determine if there will be interference (e.g., a KF titration is requested for the water content of a process stream containing a 3% benzaldehyde impurity). Aldehydes react with methanol to form acetals, generating a high bias water.[69,70] The analyst should review the process chemistry and confer with the team to determine if any compounds or reagents are present that will interfere or cause erroneous results with the KF analyses. If the sample's water quantity cannot be determined by KF, then loss on drying, thermogravimetry analysis (TGA),[12] or GC with thermal conductivity detection (TCD)[71] are orthogonal methods that can be used to verify the results if interference is suspected. O'Keef *et al.* used GC/TCD for the determination of the water content of acetone solutions containing methyl isobutyl ketone (MIBK) and other by-products of the MIBK synthesis. KF was used to confirm the analysis.[72] Hinz used loss on drying, TGA (coupled with DSC), direct KF titration, and the KF oven to determine if the methods gave the same results with six different substances with unknown chemical and thermal behaviors.[12] The substances were piperazine hexadrate, triphenylphosphine, magenesium sulfate, hydranal standard, kromasil, and an API. A combination of direct KF and KF oven method was found to be the optimal technique for determination of water in substances of unidentified chemical and thermal behaviors.

G. Drying End-Point Determination

Once the intermediate or API has been filtered, washed, and isolated from the crystallization slurry, it is placed into a dryer (tray or agitated) under vacuum with nitrogen. Sampling usually involves lowering the oven temperature, breaking the vacuum, removing a sample, and delivering the sample to the laboratory. An analysis is performed in the laboratory to determine if the solvent/water has been removed to a specified level. KF,[12] loss on drying (USP method),[9] loss on drying using infrared light (LOD/IR)[9] thermogravimetric analysis (TGA),[12] and GC for residual solvents are generally the methods used to determine the drying end point.[17,24] If the drying end point fails, the drying process is started again. The sampling and analysis are time-consuming operations. PAT has the potential to eliminate multiple in-process samples thus speeding up the drying cycle time. The use of PAT to monitor the API drying process has been demonstrated by Parris *et al.* (using NIR vapor phase monitoring)[10] and am Ende *et al.* (using MS vapor phase monitoring).[11] The drying profile and endpoint monitoring were achieved using these techniques.

H. Aqueous Solution Conductivity

Aqueous solution conductivity is an IPC test used to determine the removal of dissolved ionic impurities (salt) from filter cakes and extractions of product-rich organic phases.[19,73] The initial water conductivity (μS/cm) is used as the basis for the criteria for IPC. The end point for washes is usually established at a level when the conductivity of the final aqueous extraction(s) and wet cake wash solutions are close to the initial conductivity of the prewash water.

There are many impurities (organic compounds) in the solution that may not contribute to the conductivity reading. Solution conductivity IPC is seen as a qualitative technique.

A study was performed by Briône *et al.* comparing two different methods, LC-ELSD (evaporative light scattering detection) and capillary electrophoresis (CE) with indirect UV detection with solution conductivity, to determine the amount of residual sodium in mother liquor washes of API.[74] The cake washing step was critical because the mother liquor salts affected the API quality attributes (hydroscopicity). The results from the LC-ELSD and CE analysis correlated with solution conductivity measurements, indicating that solution conductivity can be used as a process-monitoring tool. As another example, Barkalow *et al.* used ion chromatography (IC) to determine the completion of the product cakes for removal of ammonium ions. The product was washed until the solution contained less than 5 ppm of ammonium.[28] IC, CE, and HPLC are all quantitative tests that can be used to determine levels of inorganic impurities in cake washes. However, conductivity is a semiquantitative analysis that can be used to rapidly determine completion of the cake-wash unit operation.[19]

I. Aqueous Solution pH

Aqueous solution pH is another test used as an IPC for processes that run in water or have an aqueous component. The pH IPCs can be used to determine the completion of wet cake washing, impurity removal, neutralization of reagents, and process safety. An example of pH IPC was demonstrated by Anderson *et al.* who determined the removal of excess HCl from a wet cake.[75] The pH of the wash solution determines if the washes were sufficient to remove excess HCl. Another example of a pH IPC was demonstrated by Vaidyananthan *et al.* who used high pH to remove hemisuccinate impurities from the filter cake of an isolated product after a single wash.[57] Incomplete phase separations and impurity removal were seen with a lower pH base. A pH IPC can also be used to improve the safe handling of a process. This test was demonstrated by Pesti *et al.* during the pilot plant synthesis of an antifungal agent. The process generated lithium cyanide (LiCN) *in situ* from lithium hexamethyldisilazane (LiHMDS) and acetone cyanohydrohydrin. The evolution of HCN from reaction streams was controlled by maintaining a $pH > 8$. The pH process control minimized plant operators' opportunities for exposure to HCN.

Many in-process solutions are mixtures of organic/aqueous solvents making them difficult to obtain an accurate reading by a pH probe. Therefore, pH paper can be used to determine the alkalinity or basicity range of the organic/aqueous process solution mixtures as a qualitative process test.

VI. CHROMATOGRAPHIC IPC VALIDATION

There is an ICH guideline for method validation of drug substances and drug products.[45] However, there are no official guidelines to follow for the validation of in-process tests. [44]Validation of IPC test methods is proof that the method is suitable for its intended purpose.[45] IPC tests and the process are rapidly changing at the early stage of the API synthesis; a formal validation protocol may not be needed for early method validation. The amount of validation performed for an IPC should be fit for its purpose and stage of development.[37,44,45] However, a formal validation protocol must be approved and executed prior to transferring the methodology to manufacturing. A detailed discussion of methods validation and transfer can be found in other chapters of this book. This section will focus on the chromatographic validation procedures specific to in-process test methods.

IPC method validation is similar to the drug substance and product method validation; however, special consideration should be given to the sample reactivity and stability. IPC method validation requires coordination with the process chemist/engineers to provide fresh reaction and process samples for the analysis. Table 1 has a list of in-process validation parameters that should be evaluated for chromatographic IPC limit and quantitative tests. These parameters are based on the ICH guidelines.[45] The IPC analyses are categorized into RAP (COR and impurity determination) and solution concentration assays (%, w/v; % v/v; and mg/mL).

A. Linearity for Concentration Assays, COR, and Impurity Profiles

Determining the correct sample preparation concentration is a key component to all IPC sample analyses. The target IPC sample concentration is preferred to be around 100% of the anticipated concentration of product or analyte in the process stream; however, this is a moving target during API synthesis development. It is recommended that the analyst broaden the linear range from 25% to 200% (product, SM impurities, and solvent exchanges) to allow a buffer for any future IPC specification changes.

Product linearity: The suggested range is 50–150% of the product anticipated IPC solution concentration (e.g., if the anticipated concentration of the reaction product in the process solution is 50 mg/mL, the linearity range of the method should be established from 25 to 75 mg/mL).

SM and critical impurity linearity: The suggested range is 50–150% of the SM or impurity concentration in the IPC solution (e.g., if the anticipated solution concentration of the SM or impurity is 0.5 mg/mL the linearity range of the method should be established from 0.25 to 0.75 mg/mL). A concentration of 0.5 mg/mL SM to 50 mg/mL product would be equivalent to COR or impurity criteria of 1 RAP. This is assuming that both components have the same response factor. If the IPC involves a transient species that cannot be isolated or qualified, a diluted solution of the product or structurally similar qualified process impurity can be used as a substitute to determine the compound linearity and response factor.[44] The assumption is that both components have comparable response factors.

B. Precision and Repeatability

Sample repeatability: Perfecting the sample preparation and obtaining a representative process sample are also key components to IPC sample analysis. Multiple

TABLE I Validation Parameters for Chromatographic IPCs

Validation parameter	Quantitative tests: Solution concentration assays — Assays	Quantitative tests: Completion of reaction and impurity determination — (RAP)	Limit tests: Solution concentration assay — Assays	Limit tests: Completion of reaction and impurity determination — (RAP)
Specificity	X	X	X	X
Known impurities[a]	X	X	X	X
System suitability	X	X	X	X
Linearity/range[b]	X	X	X	X
Precision				
Repeatability	X	X	X	X
Intermediate	X	X		
Reproducibility	X	X		
Accuracy	X	X		
Sensitivity				
Detection limit	X	X	X	X
Quantitation limit		X		
Robustness				
Solution stability	X	X	X	X
Variation of operating parameters	X	X	X	X

[a]For concentration assay methods, separation of all known impurities from the main component of interest should be demonstrated. It may not be necessary to demonstrate the separation of all known impurities from each other depending on the requirements of the IPC. For completion of reaction methods and impurity determination, separation of all known impurities from the starting material should be demonstrated. It may not be necessary to separate or resolve minor impurities from the product. This requirement will be process- and IPC-specific. Separation of known impurities from each other is not necessary.

[b]For concentration impurity determination methods, linearity of the main component of interest should be demonstrated, not of all known impurities. For volatile component determinations (solvent exchanges, residual solvents, GTIs), only the linearity of the volatile component(s) being measured should be demonstrated. For completion of reaction methods (RAP), linearity of the reaction product, starting material, and any stable impurity used in the COR calculation should be demonstrated.

samples of the process should be requested to verify the process sampling and sample preparation variability during the development of the API synthesis. Typically, six IPC samples from the process are prepared separately and analyzed.

C. Accuracy

Sample recovery: Some reactions will not allow the spiking of the reaction and/or process mixture with SM without rapidly converting it to product. In these cases, a quenched reaction or process simulation mixture may be used as a substitute. Process simulation mixtures are used for IPC samples that are only stable for a few hours, or spike recovery cannot be performed in the reaction or process matrix. A process simulation sample is made using the same chemical reagents in the reaction. The components of the process simulation matrix should be determined in - collaboration with the process chemist. The stability of the process simulation mixture should be determined before the initiation of the formal method validation.

D. Robustness (Sample Solution Stability)

The stability of working sample solutions is assessed under ambient conditions. The analyst should verify the stability of the sample received from the process chemist or plant to determine if any special handling procedures are required. The impact of the sample preparation delay on the IPC preparation should also be evaluated. The length of the sample stability study will be defined by process requirements.[44]

VII. IN-PROCESS CONTROL TEST TECHNOLOGY TRANSFER

It is common during the synthesis of the long-term stability study (LTSS) batches for the quality and manufacturing groups to become involved in the review of the selected IPCs test. It is important during this stage to consider what instrumentation and technical expertise the quality group and/or contract vendor has before the process is transferred. A formal technology transfer of the IPC analytical methods must occur to qualify a laboratory that will perform in-process testing in support of process justification, scale up, validation, and/or commercial manufacture.

A. Early Process Development Transfers

A formal method transfer generally is not required for IPC methods used to support the early process development projects. An informal method transfer is done with chemists, engineers, and/or analysts at the development or contract-manufacturing site. This procedure is done to allow the manufacturing site to give feedback on the IPC analyses. The transfer happens months before the process will be run in the plant. The process-receiving site typically runs small-scale laboratory experiments for familiarization. In general, there is no QC/laboratory involvement in the IPC testing during this small-scale process evaluation. However, the method development analyst should review all the IPC data and confer with the process chemist or analyst on the receiving site to determine how the IPC methods have performed. The information gained from this informal transfer can be used to improve or modify the formal transfer to the QC laboratory.

B. QC Formal Transfers

Transfers using a preapproved protocol are generally used for the transfer of IPC methods to the manufacturing site. The method transfer protocol contains the criteria by which the receiving laboratory is deemed "qualified" to perform the IPC testing. Many process IPC samples are only stable for a few hours making them difficult to ship to the receiving laboratory or to be used in interlaboratory comparisons. In these cases, process simulation samples are utilized as substitutes. A process simulation sample is prepared at the IPC limit using the same chemical reagents at the receiving and expert laboratory. If an interlab comparison cannot be performed with the process simulation samples, spike and recovery can be used as a substitute.

C. Technology Transfer Protocol

A technology transfer protocol should be used to facilitate the transfer of IPCs (e.g., COR, KF, solvents exchange, LOD) QC. The IPC technology transfer protocol is derived from Drug Master File (DMF) or a similar document. The DMF is a document that contains a detailed description of the API and/or intermediates synthesis along with the sample plan and acceptance criteria for all IPCs. The DMF is used to transfer the synthetic process to manufacturing. The IPC technology transfer protocol should contain the following information:

- Introduction
- Description/background (as appropriate)
- Responsibilities
- Tests that must be formally transferred (from the DMF sample plan). All IPC tests in the sample plan must be addressed in the transfer protocol.
- Method performance verification tests and acceptance criteria, as appropriate.
- Detailed IPC sample preparations, testing procedures, and acceptance criteria, as appropriate.
- Sample lots, batches or process simulation samples. If an interlab comparison cannot be done, spike and recovery can be used as a substitute.
- Approval signature page: The transfer protocol must be approved by the expert laboratory, receiving laboratory, and the receiving laboratory QA representative. All approvals must be obtained before laboratory qualification studies can begin.

Suggested Method Transfer Protocol Parameters

- *Quantitative test*: System suitability specificity, repeatability, linearity, interlab comparison, LOQ, and LOD should be evaluated at the receiving lab.
- *Limit test*: System suitability, specificity, LOQ, and LOD
- *General or compendial test*: The receiving laboratory may be considered qualified if they routinely perform similar tests and meet compendial performance criteria. These tests may need to be included in the formal transfer of analytical methodology if verification of comparability in results to the

expert laboratory is needed. A transfer waiver and/or memo from the expert laboratory is required for the exclusion of these IPC tests from the methods transfer.

VIII. FUTURE TRENDS FOR IN-PROCESS ANALYSIS

The FDA's PAT initiative has been embraced throughout the pharmaceutical industry. Most of the applications of PAT in drug substance synthesis have been in the qualitative or trending mode (e.g., monitoring the API drying process).[10] PAT is used to determine when to take a sample because of reactivity, stability, safety (e.g., hydrogenation reactions[33]), or to determine when a process is complete for sampling (product drying). Spectroscopic methods offer a simple sample preparation and rapid analysis. PAT (spectroscopy) is currently being used off-line in development to perform IPC analyses for COR, product dryness, solvent exchanges, and solution concentration before and after crystallizations.[18,31]

Currently, UPLC® or high-pressure HPLC/UV detection using sub-2-μm column packing material has the potential of 2-min gradient samples analysis times for IPCs (e.g., COR, impurity determination, or solution concentration).[51,76] Waters Patrol® UPLC offers in-line and off-line versions. Real-time in-line HPLC has not been used on a regular basis in development enviroments.[77] Most reactors are sampled manually with analysis performed off-line.

Both in-line and off-line spectroscopic PAT or high-pressure HPLC analysis will play important roles in the development of robust control strategies in drug substance synthesis. One would expect to see expanding use of both techniques because of their ability to improve the analytical laboratory and manufacturing productivity.

IX. SUMMARY

IPC tests can effectively reduce process variability and ensure the quality of API produced. An effective IPC involves setting the appropriate process and analytical controls criteria. In-process testing is playing an increasing significant role for the manufacture of the API. This chapter has highlighted the selection, development, validation, and transfer of the IPC test.

ACKNOWLEDGMENTS

The author thanks Dr. Charles Ray for his informative discussions during the preparation of this chapter.

REFERENCES

1. Laporte, T. L., Hamedi, M., Depue, S., Shen, L., Watson, D. and Hsieh, D. Development and scale-up of three consecutive continuous reactions for production of 6-hydroxydbuspirone. *Org. Process Res. Dev.* 12:956–996, 2008.

2. Pesti, J., Chanien-Kang, C., Spangler, L., Delmonte, J. A., Benoit, S., Berglund, D., Bien, J., Brodefuehrer, P., Chan, Y., Corbett, E., Costello, C., Demna, P., Discordia, P. R., Doubleday, W., Zhinong, G., Gingra, S., Grosso, J., Haas, O., Kascur, D., Lai, C., Leng, S., Miller, M., Muslehiddinoglu, J., Nguyen, N., Qiu, J., Olzog, M., Reiff, E., Thoraval, D., Totlben, M., Vanyo, D., Vemishetti, P., Wasylak, J. and Wei, C. The process development of raviconazole: An efficient multikilogram scale preparation of antifungal agent. *Org. Process Res. Dev.* 13:716–728, 2009.
3. Ge, Z., Buchanan, B., Timmermans, J., DeTora, D., Ellison, D. and Wyvratt, J. On-line monitoring of the distillates of a solvent switch process by near-infrared spectroscopy. *Process Control Qual.* 11:277–287, 1999.
4. Yao-En, L., Yubo, Y., Kalthod, V.l and Tyler, M. S. Optimization of solvent chasing in API manufacturing process: Constant volume distillation. *Org. Process Res. Dev.* 13:, 2009. Introduction.
5. Sistare, F., St Pierre Berry, L. and Mojica, A. C. Process analytical technology: An investment in process knowledge. *Org. Process Res. Dev.* 9:332–336, 2005.
6. Feng, S., Shan, N. and Carpenter, J. K. Crystallization of amoxicillin trihydrate in the presence of degradation products. *Org. Process Res. Dev.* 10:1212–1218, 2006.
7. Zhou, G. X., Crocker, L., Jing, X., Tabora, J. and Ge Zhihong, G. In-measurement of drug substance via near infrared spectroscopy to ensure a robust crystallization process. *J. Pharm. Sci.* 95:2337–2347, 2006.
8. Mosely, D. J., Parminder, B., Bowden, S., Couch, M. E., Hubacek, I. and Weinggärtner, G. Trouble with potassium carbonate and centrifuges: Mass transfer and scale-up effects in manufacturing of ZD9331 POM quinacetate. *Org. Process Res. Dev.* 10:155, 2006.
9. Müller, M., Schneeberger, R., Wieckhusen, D. and Cooper, M. Example of finishing technologies as key elements for successful active pharmaceutical ingredient process development. *Org. Process Res. Dev.* 8:376–380, 2004.
10. Parris, J., Christian, A., Escott, R., Rydzak, J. and Crocombe, R. Monitoring the API drying operations with NIR. *Spectroscopy* 20(2):34, 2005.
11. am Ende, J. D., Preigh, J. M., Hettenbach, K., Ajhjanian, P. and Ward, W. H. On-line monitoring of vacuum dryer using mass spectrometry. *Org. Process Res. Dev.* 587–593, 2004.
12. Hinz, D. Evaluation of methods for determination of water in substance with unknown chemical and thermal behavior. *J. Pharm. Biomed. Anal.* 43:779–780, 2006.
13. Cabri, W., Ghetti, P., Giovanni, P. and Alpegiani, M. Polymorphisims and patent market and legal battles: Cefdinir case study. *Org. Process Res. Dev.* 11:64–72, 2007.
14. ICHQ3A Impurities in New Drug Substances (Revised Guideline). www.ich.org, Oct 25 2006.
15. Jiben, R. Pharmaceutical impurities–A mini-review. *AAPS PharmSciTech* 3(2), 2002. http:www.aapspharscitech.org, 2002. (article (6)).
16. ICH Q3C(R4) Impurities: Guideline for Residual Solvents Step 4 version. www.ich.org, Feb. 2009.
17. am Ende, D., Bronk, S. K., Mustakis, J., O'Connor, G., Santa Mari, L. C., Nosal, R. and Watson, J. N. T. API quality by design example from torcetrapib manufacturing process. *J. Pharm. Innov.* 2:71–86, 2007.
18. Argentine, M. D., Owens, P. K. and Olsen, B. A. Strategies for investigation and control of process-related impurities in drug substance. *Adv. Drug Deliv. Rev.* 59:12–18, 2007.
19. Anderson, N. G. *Practical Process Research and Development*, Academic Press, San Diego, California, 2000.
20. Ganzer, P. W., Materna, A. J., Mitchell, B. M. and Wall, K. Current thoughts on critical process parameter and API synthesis. *Pharm. Technol.* 29:46–66, 2003.
21. Vogel, I. and Tatchell, A. R. In *Vogel's Textbook of Practical Organic Chemistry.* (S. B. Furnis, J. A. Hannaford and G. W. G. Smith, eds.), John Wiley & Sons, Inc., New York, NY, 5th ed, pp. 197–198, 1996. (Hardcover by A.I.).
22. Wu, N., Jillian, D., Dempsey, J., Yehl, P., Dovletogolous, A., Ellison, D. and Wyvratt, J. Practical aspect of fast HPLC separation for pharmaceutical process development using monolithic columns. *Anal. Chim. Acta.* 523:149–156, 2004.
23. Anotinucci, V. and Wright, L. Development of practical chromatographic methods for the active esters. *J. Liq. Chromatogr. Relat. Technol.* 24(14):2145–2159, 2001.

24. Sandra, P., David, F. and Szucs, R. Some application of state of the art capillary gas chromatography in pharmaceutical industry. *Trends Anal. Chem.* 21(9–10):665–668, 2002.
25. Klick, S. and Sköld, A. Validation of a generic analytical procedure for determination of residual solvents in drug substances. *J. Pharm. Biomed. Anal.* 36:401–409, 2004.
26. Watson, J. D., Dowdy, D. E., Depue, S., Kotnis, A., Leung, S. and O'Reilly, B. C. Development of safe and scalable oxidation process for preparation of 6-hydroxybuspirone: Application of in-line monitoring for process ruggedness and product quality. *Org. Process Res. Dev.* 8:619–620, 2004.
27. Minli, X., Ziemba, M. T. and Maurin, M. Sublimation characterization and vapor pressure of an HIV nonnucleoisde reverse transcriptase inhibitor using thermogravimetric analysis. *AAPS PharmSciTech.* 4(23):1–9, 2003.www.pharscitech.org, 2003.
28. Barkalow, H. J., Breting, J., Gaede, B., Haight, R. A., Henry, R., Kotecki, B., Mei, J., Pearl, B. K., Tedrow, S. J. and Viswanath, K. S. Process development for ABT-472, a benzimidazole PARP inhibitor. *Org. Process Res. Dev.* 11:693–698, 2007.
29. Hinz, D. Process analytical technologies in the pharmaceutical industry: The FDA's PAT initiative. *Anal. Bioanal. Chem.* 384:1036–1042, 2006.
30. Lin, Z., Zhou, L., Mahajan, A., Song, S., Wang, T., Ge, Z. and Ellison, D. Real-time endpoint monitoring and determination for a pharmaceutical salt formation process with in-line FT-IR spectroscopy. *J. Pharm. Biomed. Anal.* 41:99–104, 2006.
31. Wethman, R., Ray, C. and Wasylyk, J. Development and implementation of an in-line quantitative raman method for in-process pharmaceutical monitoring. *Am. Pharm. Rev.* 8(6):57–63, 2005.
32. Lewis, R. I. and Edwards, G. M. H. *Handbook of Raman Spectroscopy from the Research Laboratory to the Process Line*, Marcel Dekker, Inc, New York, 2001. pp. 275–306.
33. Tumuluri, V., Kemper, S. M., Sheri, A. S., Seoung-Ryoung, C., Lewis, R. I., Avery, A. M. A. and Avery, A. B. Use of Raman spectroscopy to characterized hydrogenation reactions. *Org. Process Res. Dev.* 10:297, 2006. Introduction.
34. Wiss, J., Länzlinger, M. and Wermuth, M. Safety improvement of Grignard reaction using on-line NIR monitoring. *Org. Process Res. Dev.* 9:365–371, 2005.
35. Ragan, J. A., Bourassa, D. E., Blunt, J., Breen, D., Busch, F. R., Cordi,, E. M., Damon, D. B., Do, N., Engtrakul, A., Lynch, D., McDermott, R. E., Mongillo, J. A., O'Sullivan, M. M. and Vanderplas, P. R. R. C. Development of a practical and efficient synthesis of CP-945, 598-01, a CBI antagonist for the treatment of obesity. *Org. Process Res. Dev.* 13:186–197, 2009.
36. Yates, M. H., Koening, M. T., Kallman, J. N., Ley, P. C. and Mitchell, D. An efficient synthesis of multipotent eicosanoid pathway modulator. *Org. Process Res. Dev.* 13:268–275, 2009.
37. Zayas, J., Sanchez, V. and Talley, M. In-process control methods for manufacturing of active pharmaceutical ingredient. *Pharm. Tech.* 29(4):154–162, 2005.
38. Kopach, M. E., Singh, U. K., Kobierski, M. E., Trankle, W. G., Pietz, M. A., Forst, M. B., Stephenson, G. A., Mancuso, V., Giard, T., Vanmarsenille, M. and De France, T. Practical synthesis of chiral 2-morpholine: (4-Benzylmorpholin-2-(s)-yl)-(tetrahydropyran-4-yl) methanone mesylate, a useful pharmaceutical intermediate. *Org. Process Res. Dev.* 13:209–224, 2009.
39. Laviana, L., Fenández-Mari, F., Bayod, M. and Blanco, D. HPLC for in-process control in the production sultamicillin. *J. Pharm. Sci.* 31:321–328, 2003.
40. Sajonz, P., Schafer, W., Gong, X., Shultz, S., Rosner, T. and Welch, C. J. Multiparallel microfluidic high-performance liquid chromatography for high-throughput normal-phase chiral analysis. *J. Chromatogr. A* 1145(1–2):149–154, 2007.
41. Welch, C. J. Microscale chiral HPLC in support of pharmaceutical process research. *Chirality* 21:114–118, 2009.
42. Welch, C. J., Sajonz, P., Biba, M., Gouker, J. and Fairchild, J. Comparison of multiparallel mircofluidic HPLC instruments for high throughput analysis in support of pharmaceutical process research. *J. Liq. Chromatogr.* 29:2185–2220, 2006.
43. Rao Nageswara, R., Nagaraju, D., Parimala, P. and Adapa, R. S. Development and validation of liquid chromatographic method for monitoring reaction involved in the synthesis of 4-methoxyphenylacetic acid. *J. Chromatogr. A* 972:277–282, 2002.
44. Wrezenl, P. and Chion, L. Validation and implementation of in-process control HPLC assays for active pharmaceutical ingredient. *LCGC North Am.* 22:1006–1009, 2004.
45. ICH Q2R1 Guideline Text on Validation of Analytical Procedures. www.ich.org, 1994. (incorporated Nov, 2005).

46. Xiao, P. K., Yuan, X., Fang, Z. L. and Rustum, M. A. Efficient method development strategy for separation of pharmaceutical molecules using advance chromatographic technologies. *J. Chromatogr. A* 1163:145–156, 2007.
47. Garcia-Lavandria, J., Losada, J., Martinez-Pontevedra, J., Lores, M. and Cela, R. Computer-assisted method development in liquid chromatography mass spectrometry: New proposals. *J. Chromatogr. A* 1208:116–125, 2008.
48. Karcher, D., Davies, M., Venit, J. and Delaney, E. Multi-dimensional screening and analysis (MDSA): An automated tool for HPLC. *Am. Pharm. Rev.* , Nov–Dec 2004. (Introduction).
49. Hewitt, E., Lukulay, P. and Galushko, S. Implementation of a rapid and automated high performance liquid chromatography method development strategy for pharmaceutical drug candidates. *J. Chromatogr. A* 1107:80–81, 2006.
50. Schmidt, A. and Molnar, I. Computer-assisted optimization in the development of a HPLC for the analysis of Kava pyrones in *Piper methysticum* preparation. *J. Chromatogr. A* 948:53, 2002.
51. Argentine, M. D., Braden, T. M., Czarnik, J., Conder, E. W., Dunlap, S. E., Fennell, J. W., LaPack, M. A., Scherer, R. B., Schmid, C. R., Vicenzi, J. T., Wei, J. G., Werner, J. A. and Roginski, R. T. The role of new technologies in defining a manufacturing process for PPARα agonist LY518674. *Org. Process Res. Dev.* 13:131–143, 2009.
52. Pellett, J., Lukulay, P., Yun, M., Bowen, W., Reed, R., Ma, M., Munger, R., Dolan, J., Wriley, L., Medwid, K. and Medwid, J. "Orthogonal" separation for reversed-phase liquid chromatography. *J. Chromatogr. A* 1101:122–123, 2006.
53. Argentine, M. D., Owens, P. K. and Olsen, B. A. Strategies for investigation and control of process-related impurities in drug substance. *Adv. Drug Deliv. Rev.* 59(12-18), 2007.
54. Ahuja, S. Assuring the quality of drugs by monitoring impurities. *Adv. Drug Deliv. Rev.* 59:3–11, 2007.
55. Advamovics, A. J. *Chromatographic Analysis of Pharmaceuticals*, Marcel Dekker, Inc, New York, 1997. pp. 57–72.
56. Hellmut, J., Funk, W., Fischer, W. and Wimmer, H. *Thin Layer Chromatography: Reagents and Detection Method Volume 1B*, VCH, P.O. Box 101161, D-69451 Weinheim, Germany, 1994.
57. Vaidyanathan, R., Hesmondhagh, L. and Hu, S. A chemoenzymatic synthesis of an androgen receptor antagonist. *Org. Process Res. Dev.* 11:903–906, 2007.
58. Schafer, W. A., Hobbs, S., Rehm, J., Rakestraw, D. A., Orella, C., McLaughlin, M., Ge, Z. and Welch, C. J. Mobile tool for HPLC reaction monitoring. *Org. Process Res. Dev.* 11:870–876, 2007.
59. Zheng, N., Armstrong, D. J., Eng, K. K., Keller, J., Liu, T., Purick, R., Lynch, J., Hartner, W. F. and Volante, R. P. A convergent asymmetric synthesis of growth hormone secretagoque. *Tetrahedron Asymmetry* 14:3435–3446, 2003.
60. Huang, M., Wethman, R., Wasylyk, J., Ray, C. and Grosso, J. *Oral Presentation # 463, The Role of Raman Spectroscopy in Productivity Enhancement, FACSS*, Louisville, KY, Oct 18–22, 2009. *www.facss.org*, Oct 18–22, 2009.
61. *Guideline on Limit of Genotoxic Impurities CPMP/SWP/5199/02*, European Medicines Agency (EMEA) Committee for Medicinal Products for Human Use (CHMP), London, UK, 2006.
62. *US Department of Health and Human Services, Food and Drug Administration for drug Evaluation and Research.* Guidance for Industry, Genotoxic and Carcinogenic Impurities in Drug Substances and Product: Recommended Approaches. www.FDA.gov, December 2008.
63. Argentine, D. M., Owens, K. P. and Olsen, A. B. Chemical Process and Analytical Approaches for Addressing Genotoxic Impurities throughout Development. www.usp.org/pdf/EN/meetings/ASM, 2008.
64. Pierson, A. D., Olsen, A. B., Robbins, K. D., DeVreis, M. K. and Varie, L. D. Approaches to assessment testing decisions and analytical determinations of genotoxic impurities in the drug substances. *Org. Process Res. Dev.* 13:285–291, 2009.
65. Lui, Q. D., Chen, K. T., Mcquire, A. M., Aleriza, K. and Kord, S. Analytical control of genotoxic impurities in pazopanib hydrochloride manufacturing. *Process J. Pharm. Biomed. Anal.* 50:144–150, 2009.
66. Elder, D. P., Lipczynski, A. M. and Teasdale, A. Control and analysis of alkyl and benzyl halide and other related reactive organohalides as potential genotoxic impurities in active pharmaceutical ingredients. *J. Pharm. Biomedical Anal.* 48:497–507, 2008.

67. Raman, N. V. V. S. S., Ratnakar, R., Prasad, A. V. S. S. and Ramakrisha, K. Development and validation of RP-HPLC method for the determination of genotoxic aklyl benzesulfonates in impurities amlodine besylate. *J. Pharm. Biomed. Anal.* 48:227–230, 2008.
68. Hettenbach, K., David, J., Dias, E., Brenek, S., Laforte, C. and Barnett, S. Microwave assisted vacuum drying and endpoint determination using mass spectrometry, part 1. *Org. Process Res. Dev.* 8: 2004. (Introduction).
69. *Karl Fischer Reagents Technical Manual*, Mitsubishi Chemical Corporation, 2000. pp. 39–41.
70. Scholz, E. Karl Fischer titrations of aldehydes and ketones. *Anal. Chem.* 57:2965–2980, 1985.
71. Nu*ß*baum, R., Lischke, D., Paxmann, H. and Wolf, B. Quantitative GC determination of water in small samples. *Chromatographia* 51:119–121, 2000.
72. O'Keef, W., Ng, T. T. F. and Rempel, G. Validation of gas chromatography/thermal conductivity detection method for the determination of the water content in oxygenated solvents. *J. Chromatogr. A* 1182:113–116, 2008.
73. Seibert, D. K., Sethuraman, S., Mitchell, D. J., Griffiths, L. K. and McGarvey, B. The use of routine process capability for the determination of process parameter criticality in small-molecule API synthesis. *J Pharm. Innov.* 3:105–112, 2008.
74. Briône, W., Herbots, C., Köttgen, C., Loix, S., Gibella, M., Bertrand, M. A. and Ceccato, A. Comparison of liquid chromatography and capillary electrophoresis methods for the quantification of sodium residual. *J. Pharm. Sci. Biomed. Anal.* 44:615–622, 2007.
75. Anderson, G. N., Ary, T. D., Berg, J. A., Bernot, P. J., Chany, Y., Chen, C. K., Davies, M. L., Dimarco, J. D., Dennis, R. D., Desphande, R. P., Do, H. D., Drouhini, R., Early, W. A., Gougousta, J. Z., Grosso, J. A., Harris, J. C., Haas, O. W., Jass, P. A., Kim, D. H., Kodersha, G. A., Kotnis, A. S., LaJeunesse, J., Lust, D. A., Madding, G. D., Modi, S. P., Moniot, J. L., Nguyen, A., Palaniswanny, V., Phillipson, D. W., Simpsom, J. H., Thoraval, D., Thurston, D. A., Tse, K., Polomski, R. E., Wedding, D. L. and Winter, W. J. Process development 5-fluor-3-[3-[4-(methoxy-4-pyrimidymyl)-1-piperazinyl]propyl]-1H-indole dihydrochloride. *Org. Process Res. Dev.* 1:30, 1997.
76. Jenkins, T. Online monitoring of in-process samples waters applies it UPLC technology to obtain QC results on the biomanufacturing floor. *Genet. Eng. Biotechnol. News* 29:2, 29 Jan 2009. www.genengnews.com, 29 Jan 2009.
77. Depalma, A. Stream streamlining HPLC in bioproduction customer needs are evolving beyond simple instrument performance and capabilities. *Genet. Eng. Biotechnol. News* 27(11), 3 Jun 2008. www.genengnews.com, 3 Jun 2008.

11 VALIDATION OF ANALYTICAL TEST METHODS

ALISON E. BRETNALL AND GRAHAM S. CLARKE

Bristol-Myers Squibb, Moreton, UK

I. BACKGROUND AND CHAPTER OVERVIEW

Analytical method validation is the process of demonstrating that analytical procedures are suitable for their intended use. More specifically, analytical method validation is a matter of establishing documented evidence that provides a high degree of assurance that the specified method will consistently provide accurate test results that evaluate a product against its defined specification and quality attributes.[1] The U.S. Federal Register states that "validation data must be available to establish that the analytical procedures used in testing meet proper standards of accuracy and reliability."[2] All analytical procedures require some level of validation, regardless of whether the method is used for stability, in-process analysis, release, or acceptance testing, which is part of the analytical quality system that provides accurate and reliable test data.[3]

This chapter discusses approaches to method validation that will meet these challenges. Chromatography is still the cornerstone of pharmaceutical analysis, and although much discussion in this chapter focuses on the validation of chromatographic methods, fundamentals as described in ICH Q2 and examples in scientific literature can be applied to the validation of most analytical techniques. The chapter also provides a discussion on how the development of a validated method may be staged over the development process for drug substances and products. To fully explore every aspect of method validation is beyond the scope of this chapter. The concept of method validation, applied to both chromatographic and nonchromatographic methods, has been the subject of several classic and recent textbooks devoted to this subject to which the reader is referred for additional information.[4–8]

A. Introduction

In the research-based pharmaceutical industry, it is rare that a fully validated analytical method will be developed *ab initio* for a commercial product. Typically, methods progress through a life cycle of development and validation. Methods will be used in discovery chemistry, which provide necessary information for continued development; however, minimal validation of these procedures is necessary to support such work. As the manufacturing process for the active pharmaceutical ingredient (API) is refined and scaled up, methods will be developed for isolated intermediates and the final API. These methods may be used as the basis for drug product test procedures which are ultimately registered as methods defined in an approved regulatory submission. At each stage of development, an appropriate level of validation should be conducted to demonstrate that the method is fit for its intended purpose. Examples of some exceptions to this process might be for methods used for testing clinical trial comparator products, generic products or reformulated products of an existing API. Method development and validation are considered part of the same continuum; for example, specificity of a method is a critical parameter in determining its suitability for use, whereas robustness is typically conducted late in development, prior to the method being transferred to a quality or routine testing environment.

Analytical method validation requirements for pharmaceutical products have been clearly documented by several regulatory authorities,[9–11] although the protocol to be followed can be customized to meet the demand of the application. The content of this chapter uses the International Conference on Harmonization (ICH) guidelines as its reference; however, for countries outside of ICH jurisdiction, national guidelines may also be applicable. The ICH guidelines were originally written in two documents, Q2A[12] and Q2B,[13] and brought consistency to method validation in the principal economic markets for the pharmaceutical industry. These guidelines have now been revised in their latest iteration and combined into a single document, Q2 (R1).[9]

Since validation requirements differ during the development process of pharmaceuticals, this chapter emphasizes requirements for later stages of development and for methods used to evaluate marketed products while at the same time recognizing that a risk-based approach may be adopted to reduce the validation burden.

A generalized flowchart of the method validation process is detailed in Figure 1.

B. Interrelated Elements to Ensure Reliability of the Analytical Methodology

Government agencies require accurate and detailed information and data be recorded both in regulatory filings as well as day-to-day operations involving pharmaceutical manufacturing and testing. From a pharmaceutical laboratory's

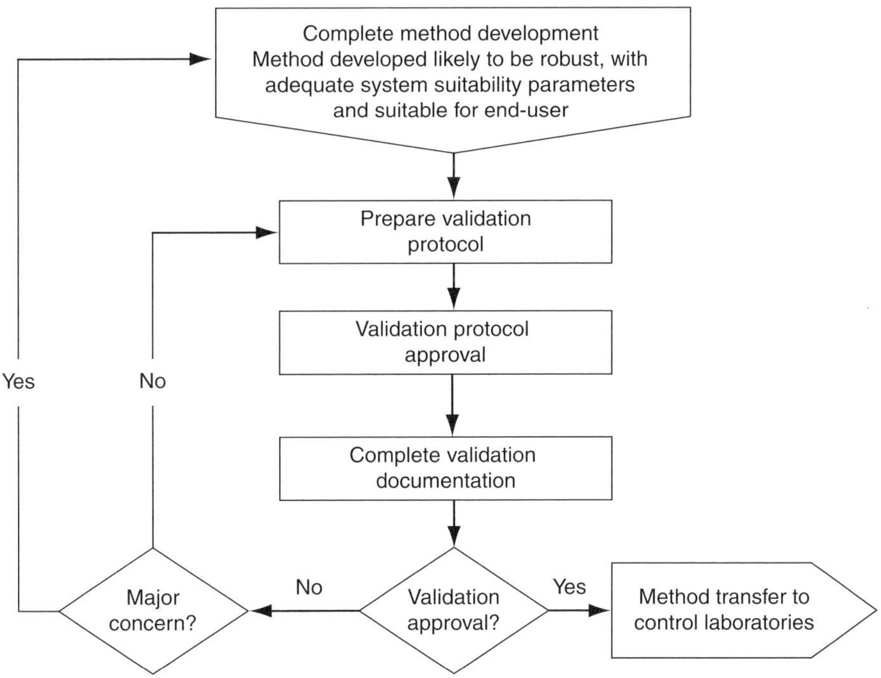

FIGURE 1 Generalized method validation flowchart.

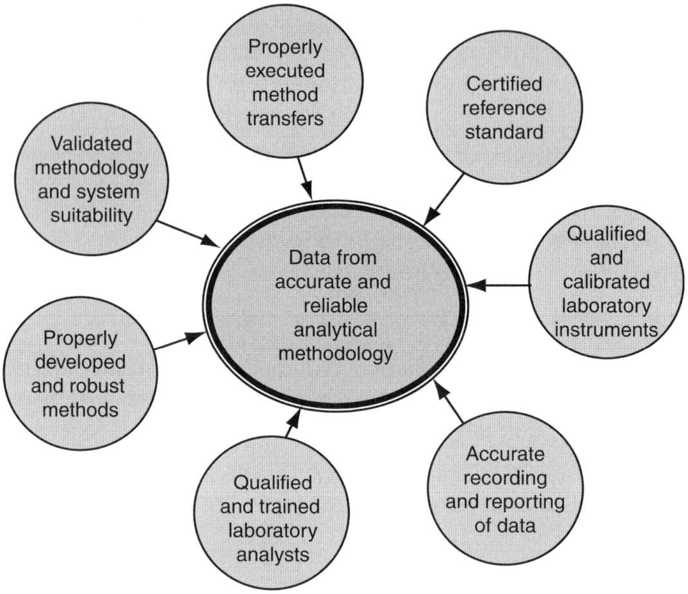

FIGURE 2 **Unified elements that ensure reliability of data from analytical methodology.**

perspective, analysts need to ensure the accuracy and reliability of the data generated by the execution of their test methods. As shown in Figure 2, there are required and fundamental controls that ensure the overall quality of the analytical test data. The interdependent processes represented in Figure 2 correlate to ensure the quality of the reported data.[14]

This chapter, while focusing on the validation process, also elucidates the importance of adequate method development and the availability of suitable reference standards. As stated previously, analytical method development and validation are intertwined and as such coexist as a single iterative process that culminates in successful transfer of the method(s) from the developing laboratory to that which will use them on a routine basis. Method development transfer is discussed in Chapter 13 of this text.

II. VALIDATION TERMINOLOGY AND DEFINITIONS

It is important to define the terms used in regulatory guidelines when discussing method validation. In the following definitions, the italic portions are quoted verbatim from the ICH guideline.[9]

Accuracy: "*The accuracy of an analytical procedure expresses the closeness of agreement between the value which is accepted either as a conventional true value or an accepted reference value and the value found.*" "Method accuracy is the agreement between the difference in the measured analyte concentrations of fortified and unfortified samples . . ., the fortification procedure is called spiking."[15]

Precision: "*The precision of an analytical procedure expresses the closeness of agreement between a series of measurements from multiple sampling of the same homogeneous sample under prescribed conditions. Precision may be considered at three levels: repeatability, intermediate precision, and reproducibility.*"

Repeatability: "*Repeatability expresses the precision under the same operating conditions over a short interval of time.*"

Intermediate precision: "*Intermediate precision expresses within laboratories" variations: Different days, different analysts, different equipment, etc.*

Reproducibility: "*Reproducibility expresses the precision between laboratories.*"

Specificity: "*Specificity is the ability to assess unequivocally the analyte in the presence of components which may be expected to be present.*"

Detection limit: "*The detection limit of an individual analytical procedure is the lowest amount of analyte in a sample which can be detected but not necessarily quantitated as an exact value.*"

Quantitation limit: "*The quantitation limit of an individual analytical procedure is the lowest amount of analyte in a sample which can be quantitatively determined with suitable precision and accuracy.*"

Linearity: "*The linearity of an analytical procedure is its ability (within a given range) to obtain test results which are directly proportional to the concentration (amount) of the analyte in the sample.*"

Range: "*The range of an analytical procedure is the interval between the upper and lower concentration (amounts) of the analyte in the sample (including these concentrations) for which it has been shown that the analytical procedure has a suitable level of precision, accuracy, and linearity.*"

Robustness: "*The robustness of an analytical procedure is the measure of its capacity to remain unaffected by small, but deliberate, variations in method parameters and provides an indication of its reliability during normal usage.*"

III. METHOD DEVELOPMENT AND ITS INFLUENCE ON METHOD VALIDATION

A. Method Development Life Cycle and Phasing Validation

Successful implementation of the method development process requires careful planning and development of the requirements of the method, excellence in laboratory work, qualified instrumentation, and proper documentation from beginning to end. Indeed, the objective of any analytical development project should include the "scope," designing flexibility and robustness into the procedure, understanding the needs and environment of the laboratory where the method will ultimately be used on a routine basis, and taking into account the validation requirements of the method. Those scientists developing the method need to strive to minimize the effort required to use it and simplify processing and interpretation of the data. As many development drug compounds do not reach commercialization for a variety of reasons, it is prudent to only conduct

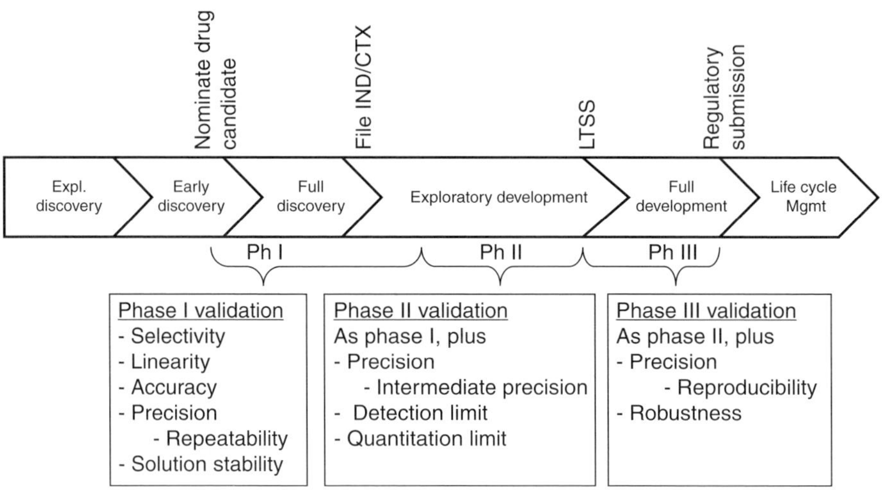

FIGURE 3 **Staged method validation approach during the drug development process.**

method validation that is appropriate to the phase of development. Typically, validation will become more comprehensive as the molecule progresses through the development process. A general development process flow is shown in Figure 3 and demonstrates which validation parameters should be conducted at the stage of product development. Consideration should also be given to any country-specific guidelines.

The method development "life cycle" parallels established approaches to analytical method validation[16] and computer validation.[17] Analytical methods should be reviewed periodically and revised or changed if they have become less suited for purpose, for example, if new technology or column chemistry is available to improve the efficiency of a chromatographic separation. In practice, this review is often neglected until inefficiency in a method becomes intolerable, or a lab investigation following a test failure indicates a method is no longer fit for purpose. One should also consider potential changes to ingredients of the API and drug product such as starting materials and excipients, respectively, as changes in their manufacture and/or purification lead to manifestations in the drug product methods downstream.

I. Method Development Considerations

Before embarking on a method development experimental work, it is advisable to have a clear understanding of the purpose of the method, any prior knowledge of the analyte(s) of interest, and any identified requirements of the end user laboratory. Examples of such considerations include the following:

- The purpose and regulatory requirements of the method, for example, phase of development, stability-indicating requirements, etc.
- Compliance requirements for the method.
- Existing information (technical and safety) of the drug substance, known related substances, and impurities and formulation ingredients.

- Review of the data, deviations, and reports on any earlier method versions, noting that a "crossover" study may be required in the plan to correlate the two methods.
- End user considerations, for example, availability of technology, technique preferences, chromatographic gradient limitations.
- Productivity requirements (e.g., how many samples will need to be analyzed within a given measure of time).
- The requirements and availability of resources (reference standards, instrumentation, consumables, technical staff, etc.).
- Technical requirements of the method (analytical equipment, chemicals used, required sensitivity, selectivity, chirality, etc.).
- Suitability for automation.

2. Quality by Design (QbD) of Analytical Methods

In ICH Q8 (pharmaceutical development),[18] several principles are developed for applying QbD to the pharmaceutical development process which can be applied to the development of validated methodology.[19–21] Methods should be developed with quality "built-in" rather than being validated to demonstrate the quality or suitability of the method. These principles enable a "design space" or acceptable variables within a method to be identified, for example, the composition of HPLC mobile phase constituents could be specified within a predetermined validated range, to enable system suitability criteria to be achieved. This concept has many attributes of ruggedness verification in its application, but by defining a "design space" at registration, would enable method variables to be applied without compromising regulatory compliance. Given the growing importance of QbD in the pharmaceutical industry and analytical chemistry, Chapter 5 is devoted to this subject.

3. *In Silico* Method Development

In recent years, *in silico* tools which are commercially available can be used to provide a starting point for the establishment of analytical conditions, for example, HPLC column chemistry, mobile phase composition, and gradient profile. These tools such as DryLab and Chromsword[22–24] can expedite rapid identification of potentially suitable chromatographic conditions, using structural and physical information (pK_a, log P, log D), and contribute to the QbD paradigm for analytical method development.

4. Robustness Testing

Robustness testing defines the capacity of the method to perform consistently and remain unaffected by small, but deliberate, variation(s) in the method parameters. By means of a limited set of experiments (often using an experimental design or "design space" approach),[25] the critical parameters that may affect the ruggedness of the method can be identified, understood, and improvements made as necessary. It is worth noting that a method will be more robust if its component steps are kept as simple as possible.

5. Method Evaluation and Method Modification

If a method is to be modified or changed, the process should be guided using established change control systems. The revised methodology should be correlated with the existing methodology using a crossover study that evaluates data from a variety of samples analyzed by the two methods.

B. "Deliverables" of the Method Development Process to Support Validation Activities

Prior to beginning the validation activities, the analyst should have a clear understanding of the following deliverables.

1. Method Robustness

The ability of a method to perform effectively in a typical laboratory environment and with acceptable variations is evaluated during robustness testing.

Generally, if robustness is designed into the method development process, the validation results should not provide any unanticipated findings and in many cases should formally confirm what the analyst already knows. In robustness testing, a variety of parameters are evaluated to determine the extent to which they can be varied without affecting the performance of the method. The parameters evaluated will depend on the analytical technique being used, but as an example, an HPLC experiment[26–29] may evaluate the following representative parameters (factors):

- HPLC manufacturer
- Lot-to-lot column variation
- Column supplier
- Flow rate
- Column temperature
- Mobile phase pH
- Ionic strength
- Detector wavelength
- Gradient slope
- Injection size, sample concentration

Methods developed for dissolution testing will have alternative robustness challenges, such as the pH, concentration and deaeration conditions of the dissolution medium, sinker type, and agitation speed.

Examples of robustness validation of spectroscopic methods, for example, NIR systems, are also found in the literature and include evaluation of the partial least squares (PLS) algorithms used in the regression analysis,[30] as well as more conventional parameters such as slit width, wavelength accuracy, and time constants.

The range (levels) over which the various parameters are evaluated should be meaningful; that is, will the method perform successfully if the pH is adjusted ± 0.2 pH units from that specified in the method through HPLC experiments; typical "responses" (i.e., capacity factor, resolution, and peak tailing) are tracked while factors are adjusted. To optimize the evaluation of robustness, these factors can be evaluated simultaneously through an experimental design approach. Success has been achieved using a two-factor Plackett–Burman statistical approach.[31] In this approach, 7 factors can be evaluated in 12 injections.

2. Method Optimization

If the method does not meet operating criteria, for example, run time or robustness requirements, it may require further optimization. Optimization for non-chromatographic methods will require other considerations such as determining sink conditions, use of surfactants or the stability of the drug in the dissolution medium. Method development optimization is typically continued until the objectives outlined in the development plan are met. Experienced method development analysts often have the experience necessary to adequately optimize the methodology; others may benefit from the documented and organized development approach of available software packages. As described earlier, commercial software applications are available that combine classical chromatographic theory with statistical design to predict optimum separation conditions with a minimum number of experiments. There is also the possibility of combining *in silico* development with experimental optimization to arrive at a final method in a more facile manner.

3. Method Specificity and "Stability Indication"

The method specificity and "stability indicating" ability should be demonstrated by the analysis of samples subjected to forced degradation conditions (e.g., pH, light, heat, oxidation). Typically, specificity is determined through peak purity using ultraviolet (UV) diode array or liquid chromatography (LC)/mass spectrometry (MS) analysis. In methods for analysis of drug products, analysis of a placebo formulation (and stressed placebo) must yield "blank" chromatographic baselines. The development and validation of a stability-indicating assay and purity method require that studies be performed to verify the ability of the method to resolve all possible degradation products and synthetic impurities from the active drug substance while accurately measuring their concentration in the presence of product excipients. Such studies are regarded as a minimum requirement for the development and validation of almost any "stability indicating" method.[32,33]

4. System Suitability Test (SST)

Generally speaking, it is anticipated that an analytical test method will perform in an acceptable manner each time it is used. System suitability parameters are most often associated with chromatographic methods; however, for other techniques, consideration should be given to the use of "quality control" check samples which can be used as a reliable reference to demonstrate the method is performing accurately on the day of use. An example of this might be when using Karl Fischer analysis for the determination of water, where the sample analysis can be verified as accurate by the determination of a known addition of water. In considering chromatographic methods, system suitability testing ensures that the total instrument system is functioning at any given time. System suitability testing, coupled with prior instrument qualification, periodic calibration, and method validation, provides assurance that the test method will provide accurate and precise data for its intended use. Properly chosen, system suitability criteria will fail prior to the point where the system will begin to produce less *acceptable* data; however, criteria should not be so strict that acceptable data cannot routinely be used because of the presence of too restrictive criteria. It is the challenge of the method development scientist to develop realistic and meaningful system suitability criteria.[34,35] Table 1 reviews some system suitability tests and the associated validation parameters they monitor from run to run.

As implied from the preceding discussion, only meaningful system suitability criteria and those required by in-house or regulatory policy should be evaluated. During method development and robustness evaluation, marginal performance of the system can be observed. The experienced and attentive scientist will use these circumstances to suggest some of the system suitability criteria; minimum peak tailing and minimum resolution between "critical peak pairs" are typical examples. System suitability should be monitored over time to verify that the criteria remain realistic and achievable while continuing to provide assurance of the suitable performance of the method.

Careful consideration should be given in determining the necessity to include, and subsequently select criteria, for any required system suitability test which demonstrates the analytical technology will operate within the boundaries to provide valid analytical results. This may either include the use of reference materials or standards, as described above, which evaluate performance at the time of analysis. For nonchromatographic methods, this might include determining the accuracy of a wavelength determination, or an assessment of the reproducibility of the response to a test material.

TABLE 1 Typical System Suitability Testing to Continuously Monitor Validation Parameters

Validation parameter	System suitability test
Linearity	Control sample or diluted standard (LOQ sample); response is within acceptable limits compared to 100% standard
Accuracy	Control sample—evaluate standards as "samples"
Precision	RSD of replicate injections (typically 5 or 6) of a standard
Selectivity	Retention factor Capacity factor Number of theoretical plates Tailing factor Resolution (injection of a resolution mixture)
LOQ/LOD	Injection of a dilution of the standard to verify LOD/LOQ
Stability of solutions	RSD (area) of a series of injections or standards throughout a run

RSD, relative standard deviation; LOQ, limit of quantitation; LOD, limit of detection.

The following system suitability parameters are examples which should be considered for use in chromatographic methods to evaluate and monitor performance:

Resolution: The resolution (Rs) is a measure of how well two peaks are separated. For reliable quantitation, well-separated peaks are essential. The separation of all peaks of interest is checked visually using a synthetic sample solution. The resolution factor (R) between the critical peak pair is calculated according to the formulas described in *United States Pharmacopoeia* (USP)[36] and the *European Pharmacopoeia* (EP)[37]

Relative standard deviation: This serves as a daily evaluation of the repeatability of the system. Often, the relative standard deviation for five or six replicate injections of a reference standard is measured at the beginning of each analysis.

Tailing factor: The tailing factor is used during system suitability testing (SST) in the case where there is a tendency to tailing of the peak of the active ingredient or one of the related compounds. This is a critical parameter if peak tailing is exacerbated as HPLC columns age. The tailing factor is calculated according to the formula described in USP[36] and EP.[37]

Limits of quantitation: The system's ability to detect the limits of quantitation (LOQs) should be evaluated with each sample sequence. An injection of the LOQ concentration during SS evaluation may also be used as a rough check of the linearity of the system over the range from LOQ to 100% of the target active concentration.

Additional system suitability parameters: Other parameters for system suitability testing can be considered (e.g., capacity factor, number of theoretical plates, etc.). Capacity factor can be especially important in methods used for stability testing, where adequate resolution of the peak of interest from the solvent front is required so that any degradation products are suitably resolved.

Reference standard check: The operating policies of some laboratories require that a duplicate injection of a separately weighed reference solution should be analyzed as a control to serve as a check on the accuracy of the standard weighing. The expected result for the second standard should be $98.0\% \leq$ reference standard potency $\leq 102.0\%$. Laboratories may also evaluate the change in the system response over time (typically every 10–12 injections) through monitoring variation in the area counts of the reference standard; area counts are anticipated not to change by more than $\pm 2\%$ during a given chromatographic sequence.

5. Method Procedure

The method procedure, description, or STM (standard test method) should contain sufficient information to execute the analysis, and the following should be considered:

- An introductory summary, including the scope and principle of the method.
- A list of reagents and their specifications (HPLC grade, reagent grade, etc.); any precautions should also be included, for example, "test the bottle of THF (tetrahydrofuran) to confirm the absence of peroxides, prior to use."

- A list of required standards.
- A list of supplies, for example, glassware, filters, etc.
- A list of instrumentation and equipment.
- A description of solution preparation (mobile phases, standards, and samples).
- A list of method conditions (flow rate, wavelength, gradient profile, equilibration procedure, etc.)
- Procedure for sample/standard preparation.
- System suitability criteria and how to calculate.
- If there is no company policy, a suggested sample sequence scheme, including order and number of injections of standards, blanks, system suitability, controls, and samples.
- Complete calculations/example calculations, including calibration frequency and how calibration is performed.
- A table of analyte relative retention times and relative response factors (RRFs), including excipients peaks.
- A representative labeled chromatogram; a separate system suitability chromatogram may also be required.

IV. VALIDATION REQUIREMENTS OF THE METHOD

A. Noncompendial Methods

Clearly, all quantitative analytical methods used to support regulatory filings (setting of specifications, etc.), toxicology testing, release of clinical or marketed materials, and methods used in stability studies require some degree of validation; requirements for validation of various analytical methodologies for commercial products are outlined in both USP and ICH guidelines. Drug registration agencies have accepted these guidelines, and Table 2 is taken from ICH Q2A (R).

In summary, many of the requirements of a method development project are dictated through regulation and internationally accepted guidelines, as well as a knowledge of current "good practice" accepted in the pharmaceutical industry.

B. Compendial Analytical Procedures

The methods provided in official monographs have been validated by the laboratory submitting the monograph and/or other laboratories designated by the USP. This validation has been performed with material produced or used by the laboratory and on equipment contained in the laboratory. It is important for all compendia methods that each individual laboratory performs limited validation of the method or verification of the method's suitability in its laboratory.[38] This verification should include any parameters which might be affected by the blinding process, such as specificity at specific wavelength when additional excipients have been used, for example, capsule back fill or colored coatings.

TABLE 2 Validation Parameters Required for Identification, Impurities, and Assay[9]

Type of analytical procedure	Identification	Testing for impurities		Assay: Dissolution (measurement only) Content/potency
Characteristics		**Quantitation**	**Limit**	
Accuracy	–	+	–	+
Precision				
Repeatability	–	+	–	+
Intermediate precision	–	+[a]	–	+[a]
Specificity[b]	+	+	+	+
Detection limit	–	–[c]	+	–
Quantitation limit	–	+	–	–
Linearity	–	+	–	+
Range	–	+	–	+
Solution stability	–	+	+	+

[a]In cases where reproducibility has been performed, intermediate precision is not needed.
[b]Lack of specificity of one analytical procedure could be compensated by other supporting analytical procedure(s).
[c]May be needed in some cases.
–, signifies that this characteristic is not normally validated; +, signifies that this characteristic is normally validated.

C. Automated Methods

There is an increasing availability of analytical technology which is capable of increasing laboratory throughput, using automated equipment such as robotics. Typically, such technologies will automate a process which has previously been developed and validated as a manual method. Some of the sample preparation or chromatographic parameters may not require further validation as they may remain unchanged (e.g., chromatographic specificity); however, contingent on the design on the automated equipment, there may be other validation issues to consider. An example of this would be the Tablet Processing Workstation, previously marketed by Zymark/Caliper and latterly by Sotax. A single vessel is used to disperse a solid dosage unit prior to online analysis, which is then washed and rinsed before the next sample in the series is extracted using the same vessel. In this case, validation should include an evaluation of sample carryover, and robustness testing should include the determination of any degradation of the analyte because of the high energy input from the dispersion probe.

A single comprehensive list of additional validation parameters for automated procedures cannot be developed for all automated equipment; however, the design and functionality of each technology should be carefully considered on a case-by-case basis. General principles for the development and validation of automated methods are available[39]; however, at a minimum, if a manual procedure is available, equivalency should be demonstrated between the two procedures.

V. VALIDATION DOCUMENTATION

The validation documentation typically consists of a protocol, analytical data, and a final report. One approach to simplifying validation documentation is to focus on a thorough protocol with preapproved acceptance criteria, which are necessary to determine whether the validation results demonstrate that the method is fit for purpose. In early development, this protocol may be captured in a lab notebook which could be in either a paper or an electronic environment. There is an ever increasing trend in the use of electronic laboratory notebooks which makes it simple to execute previously developed protocols. Later in development, the protocol should be a unique document, which may have data tables to enter the test results, requiring only a short executive summary to summarize the results and a reference or attachment of raw data. A development lab will often use these well-developed and optimized master method validation protocols[40,41] as templates for subsequent validations. A copy of the method procedure and, where available, a method development report may be appended to the validation protocol. In general, the validation protocol should contain the following:

Validation protocol
1. Method principle/objective
2. List of responsibilities (laboratories involved and their role in the validation)
3. List of reagents/samples (including test lots) and standards required
4. Test procedures to evaluate each validation parameter and proposed acceptance criteria
5. Plan or procedure when acceptance criteria are not met
6. Requirements for the final report

Appendixes
1. Method development report
2. Method procedure

The validation process cannot proceed until the protocol and all parties and their quality unit approves the acceptance criteria, which should be determined based on the required use and capabilities of the method. Although classical acceptance criteria, suitable for regulatory approval, are often available as a template approach, other statistical scientific approaches can also be justified.[42]

Once the extensive validation experiments are complete, minor changes to the method description may be required. Typically, these involve adding validation data (RRFs for the related substances, LOQs, etc.) but may also include slight changes to the system suitability requirements due to data from multiple laboratories. There should not be any fundamental changes that would alter the principles of the methodology or necessitate revalidation unless a portion of the validation failed, suggesting minor method adjustment and repeat of the required validation experiments.

Like many laboratory processes, there are now electronic tools available to simplify and automate the method validation process.[43] These often use standard templates which adhere to the company's standard procedure, and can be used repeatedly to remove the necessity of creating a unique protocol each time a new validation is conducted. Care should be taken, however, to assess that there are no peculiarities to the method being validated which requires specific validation considerations.

VI. VALIDATION EXPERIMENTATION

Depending on the requirements of the validation, there can be a preferred order to efficiently perform the validation experiments, and the ICH Q2 (R1) guidelines provide a rational order for determining validation parameters. Experimental validation characteristics can be considered simultaneously using an experimental design to provide a sound overall knowledge of the capabilities of the analytical procedure.[9] Although these parameters are widely applicable, they are most commonly understood in the context of chromatographic methodology.

A. Determination of Method Specificity

Specificity is one of the most important characteristics of a "stability-indicating" method and should be determined as one of the first validation parameters. A specific method can accurately measure the analyte of interest even in the presence of potential sample components (placebo ingredients, impurities, degradation products, etc.). When criteria for specificity are not met, this often indicates that the method is not sufficiently developed; furthermore, it is likely that criteria for accuracy, precision, and linearity may also not be fulfilled. A major objective of determining specificity is to ensure "peak purity" of the main compound to be determined; in other words, confirm that no related compound or product ingredient co-elutes and interferes with the measurement of the assayed compound. Specificity is required from the sample matrix as well as impurities and potential degradation products.

1. Separation from Impurities and Excipients

Knowledge of synthetic impurities and degradation products can be derived from the historical information that has accumulated for the drug substance/product. Ideally, a library of impurity and degradation compound reference standards is synthesized and characterized and sufficient quantities are made available. These compounds or an "impurity cocktail" can be spiked into the "pure" drug substance and any excipients (placebo sample matrix) to determine if the matrix interferes with the quantitation of the compound(s) of interest by comparing to an unspiked sample and sample diluent.

2. Separation from Potential Degradation Products

Method specificity stress testing of drug substances/product types under several conditions should be performed to support the suitability of the method. Depending on the matrix and packaging, these include extremes of pH, heat, exposure to oxidation, humidity, and light. In the case of drug substances, heat

(80 °C), light (1000 fc), acid (0.1 N HCl), alkali (0.1 M NaOH), and oxidant (e.g., 3% H_2O_2, radical initiators such as AIBN (azobisisobutyronitrile), transition metals, e.g., Cu^{2+} or Fe^{3+}) are often used. For drug products, heat, light, and humidity (85%) are used as stress conditions. Analyte peaks are evaluated for peak purity upon sufficient stress to effect 5–30% degradation of the active compound. Studies can be terminated after a week if insufficient degradation is observed.

B. Evaluation of Peak Purity

The peak purity in these degraded or spiked samples should be determined by using specific detection techniques, such as diode array UV or HPLC-MS. Software to evaluate peak purity is often available on commercial diode array and LC-MS data systems. Although both of these detection techniques provide relatively straightforward peak purity evaluations, there are limitations to the appropriateness of this approach, including the ability to detect low levels of co-eluting species. A less direct, but perhaps more persuasive, approach is to isolate the peak of interest and reinject on a chromatographic system that is based on a different "noncorrelated" or orthogonal separation mechanism, for example, for evaluating a reversed-phase HPLC method, isolate the peak and reinject on an ion-exchange HPLC system. The chromatogram produced from the ion-exchange system is evaluated to observe any secondary peaks that may have eluted under the peak isolate on the reversed-phase system. Capillary zone electrophoresis (CZE) has also been used extensively as a noncorrelated analytical technique to evaluate peaks isolated from reversed-phase methodology.

C. Demonstration of Linearity and Range: Determination of Relative Response Factor

Linearity is the ability to obtain results that are directly or indirectly (by well-defined mathematical transformation) proportional to the concentration of an analyte in a sample within a given range. The range is the interval between the upper and the lower levels of the analytical method that have been demonstrated to obtain acceptable accuracy, linearity, and precision. Hence, the following parameters are typically evaluated during linearity experiments.

The relationship between the sample concentrations and the corresponding instrumental signals for the majority of analytical techniques is one of a straight-line (first order) type. A line that fits best through the coordinates of the measured signals and the corresponding concentrations of the sample represent such a relationship. This line, known as the calibration line, is expressed by an estimated first-order equation:

$$Y = aX + b,$$

where Y is the measured signal; X, the concentration of the sample; and a and b, the linear regression coefficients of the line, a being the slope of the line and b, the intercept.

The calibration lines are usually calculated by ordinary least squares (OLS) regression. A precondition for the application of OLS regression is that the variance of the signal should be independent from the signal itself. This property is also called homoscedasticity. When this is not the case, one is dealing with a heteroscedastic data which can be clearly observed by reviewing a plot that displays residuals (Figure 5).

1. Linearity of the Active Component

Linearity can be demonstrated by analyzing five or more (seven are recommended statistically) concentrations of the active compound as the "pure" drug substance, for example, 70%, 80%, 90%, 100%, 110%, 120%, and 130% of the normal sample concentration for a stability-indicating method, a minimum of one preparation at each level (ideally performing an accuracy check at one level). For a method to determine the content uniformity of the active, linearity can be demonstrated over the range 80–120%. Linearity can also be determined in the presence of sample matrix using a similar approach. There are also added advantages to evaluating the linearity over the whole range from LOQ to 130%, for example, to accommodate future changes to product potency.

Linearity can be established by visual evaluation of a plot of the area as a function of the analyte concentration (Figure 4). Furthermore, the correlation coefficient, *y* intercept, slope should be calculated. The *y* intercept should statistically not differ from 0. The residuals should be calculated and when plotted against concentration, a linear response is indicated by a random scatter of data points (Figure 5). A progressive curvature of the display is indicative of nonlinearity. Low levels of the active compound are also examined to determine its LOQ.

2. Linearity of the Related Compounds

The range for impurities and degradants should minimally encompass five equally spaced concentrations (seven are recommended statistically) in the presence of the matrix: at LOQ, at the specification level, at an upper level above specification, and at least two intermediate concentrations (e.g., for a specification of $\leq$0.1, suggested levels are 0.03%, 0.05%, 0.07%, 0.08% 0.1%,0.12%, and 0.13%). It is recommended to make three replicate injections at each level, and these data can be used to determine precision, RRFs and limit of detection (LOD) and LOQ.

Linearity can be established by visual evaluation of a plot of the area as a function of the analyte concentration (Figure 6). The correlation coefficient, *y* intercept, slope should be calculated. The y intercept should statistically not differ from 0. The residuals should be calculated and plotted exhibiting random arrangement if the response is linear. The slope of the active compound curve (Figure 4) is divided by the slopes of these curves (Figure 6) to determine the RRFs; these are recorded in the method procedure if the method does not prescribe the use of external standards for related compounds.

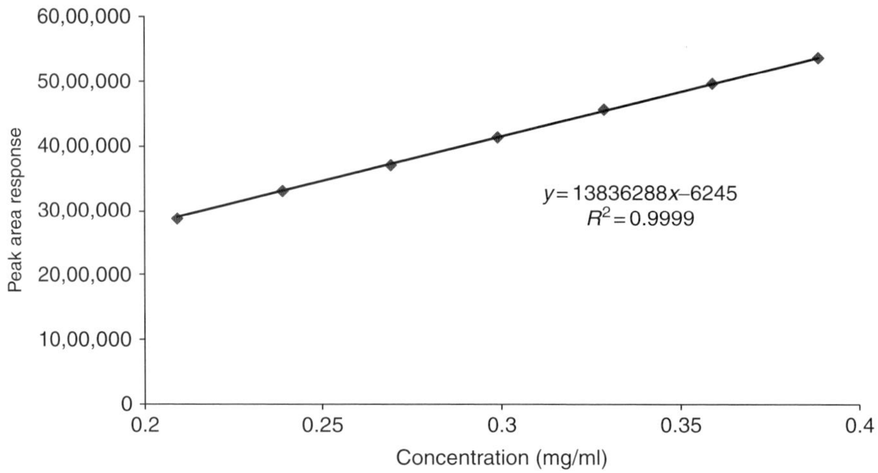

FIGURE 4 **Linearity of the active component (70–130%).**

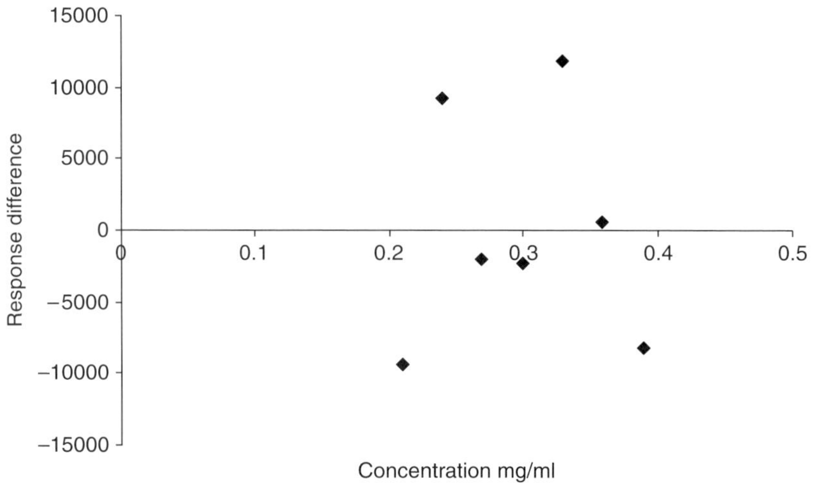

FIGURE 5 **Residuals of the active component (70–130%). Random scatter of data points is indicative of linearity.**

D. Determination of Limits of Detection and Quantitation

As there are different ways to define how these limits are calculated and interpreted, it is important to state clearly how the LOD and LOQ are calculated in the validation protocol. Both the USP and the ICH guidelines similarly define the LOD and LOQ in ways that are widely accepted in the industry. In the most straightforward case, any compound detected with a response at about three times the noise response level is construed to be at its LOD. For the LOQ, the value is commonly taken as 10-fold the noise response level. Therefore, the LOD and LOQ can be simply determined from the known amount (concentration) of an

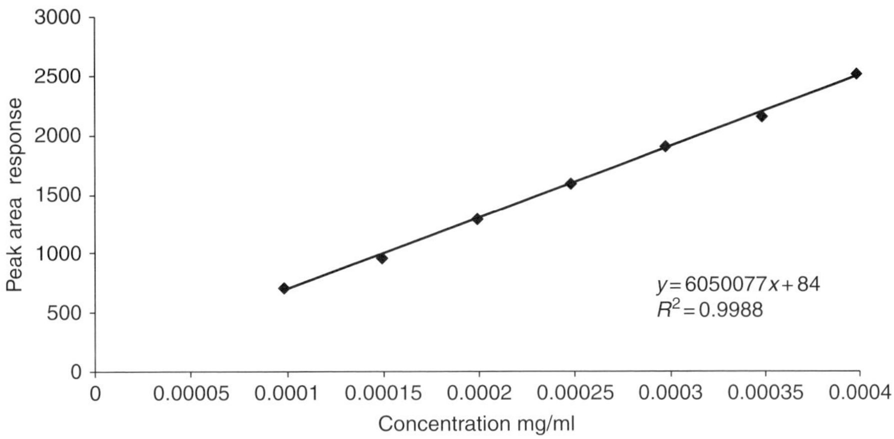

FIGURE 6 **Linearity of specified related substances (0.03–0.13%).**

analyte that produces such responses when the noise level can be easily measured. In fact, some chromatography data systems can be programmed to report baseline noise. An alternative method of determination is described by the ICH guidelines as follows:

$$\text{LOD} = \frac{3.3\sigma}{\text{Slope}} \tag{1}$$

$$\text{LOQ} = \frac{10\sigma}{\text{Slope}}, \tag{2}$$

where σ is the standard deviation.

This method can be conveniently applied to the regression data obtained in linearity studies. However, parameters estimated by this approach are often verified experimentally. The target LOQ and LOD may be stricter for a drug substance than for a drug product. In pharmaceutical analysis of the active drug substance, the target value for the LOQ is typically set at 0.05%.

E. Demonstration of the Accuracy of the Method

This is defined as the closeness of agreement between a test result and the accepted reference value (combination of random and systematic errors). The accuracy is usually examined by determination of the trueness of a test result, which is the closeness of agreement between the average value of a large number of test results and the true result or accepted reference value. The measure of the trueness is expressed by the bias, which is the difference between the expectation of the rest results and an accepted reference value.

The accuracy of a method can be determined by performing recovery experiments, implementing standard addition calibration procedures, testing reference materials, etc. It is also possible to compare the test results of a new method with those of an existing fully validated reference method through "cross-validation" experiments.

Accuracy is most commonly determined by recovery studies in which the analytes are spiked into a solution containing the matrix. The matrix, excipients in formulations including any coating constituents, should be found not to interfere with the assay of the compound(s) of interest. For stability-indicating HPLC methods, it is necessary to determine the accuracy of the active ingredient and that of all known related compounds.

Accuracy of the method to recover the active should be established across a specified range for the analytical procedure, so keeping the level of excipients constant for each formulation being tested (a placebo formulation is ideal to use if available), API from 70% to 130% of the nominal concentration should be spiked into the solutions (for products requiring content uniformity, testing the API should be typically spiked in at the range 80–120%). Make a minimum of one preparation at each level.

It is possible to determine the accuracy of each related compound separately, but it is more efficient to validate these related compounds in a combined spiked mixture of all the related compounds at their appropriate levels. Impurities can be spiked into the active at levels from the reporting limit to typically 120% of the specification. The analyst should be certain that the impurity standards used to spike the solutions are pure and do not contain significant impurities that would affect the results, by running an unspiked sample of the active, as a control.

F. Determination of Method Precision

The precision of an analytical procedure expresses the closeness of agreement between a series of measurements obtained from multiple sampling of the same homogeneous sample under the prescribed conditions. Precision may be considered at three levels.

Injection repeatability: Injection precision is measured by multiple injections, a minimum of $n=6$ is typically recommended, of the reference standard or sample solution at the 100% level and indicates the performance of the HPLC instrument using the chromatographic conditions on one particular day and in one laboratory. The relative standard deviation, RSD (%), as specified here, will determine the lowest variation limit of the analytical results. Repeatability for impurities determination is often assessed by making repeat injections typically $n=6$ of an impurity mixture and a statistical evaluation maybe performed using area response.

Sample repeatability: Sample repeatability expresses the precision under the same operating conditions over a short interval of time. It typically consists of multiple preparations and measurements of a homogeneous sample by the same analyst on the same day, for example, analysis of six samples from the same batch.

Intermediate precision: Intermediate precision expresses the effects of random events on the precision of the analytical procedure within the same laboratory. The investigation typically includes the analysis of three sample preparations from each of three batches on each of three days. Fresh mobile phase, diluents, and standard solutions are prepared on each day. The standard deviation, relative standard deviation, and confidence interval for each precision parameter are reported.

Reproducibility: Reproducibility assesses the closeness of agreement between a series of measurements obtained by the analysis of the same samples by different laboratories and is often performed during analytical method transfer. Precision is analyzed separately for each laboratory.

G. Robustness Solution Stability

The stability of the working standard solution and sample solution should be assessed over a minimum of 24 h, ideally up to at least 7 days, at several conditions, for example, room temperature, room temperature protected from light, and refrigerated. The stability may be assessed versus freshly prepared standard solutions and/or by monitoring total impurities.

H. Target Acceptance Criteria

Table 3 contains an example of validation target acceptance criteria for a stability-indicating HPLC method for the assay of an active drug substance and its related compounds. These criteria or others cited in the literature [33,44,45] can be used as a general guideline when considering acceptance criteria for a validation protocol.

1. Establishment of Acceptance Criteria Plan if Criteria Are Not Met

When one or more items during method validation fail to meet the acceptance criteria, laboratory management should decide whether

1. The results can still be accepted with justification.
2. A retest should be performed on the same sample preparation(s).
3. The test should be repeated (reanalysis).
4. The concern is significant, and the method needs adaptation (through additional method development) after which the test or validation is repeated.
5. As a result of the failure to meet acceptance criteria, a limitation could be put on the method (e.g., range of method limited to 80–120%).

All deviations to the validation procedure should be documented and authorized by laboratory management and the quality assurance (QA) department, and a list of deviations, if any, is included in the final validation report.

VII. METHOD TRANSFER

The transfer of analytical methods is covered in chapters 10 and 13.

TABLE 3 Example Acceptance Criteria for an HPLC Assay/Purity Method

Parameter	Limit for API	Limit for impurities
Linearity		
Correlation coefficient	>0.995	>0.990
y intercept (relative to the active or related compound)	≤2.0%	≤2%
Visual	Linear	Linear
Accuracy		
Active ingredient		
Mean recovery per concentration	98.0–102.0%	
Related compounds (amount spiked%)		
LOQ		±0.02% (absolute)
≤0.20		±0.04% (absolute)
>0.20		±20% (relative)
Precision		
Active ingredient	RSD	
Injection repeatability	≤2.0%	
Sample repeatability	≤2%	≥0.25% RSD ≤20% relative Between 0.10% and 0.25% SD ≤ 0.04
Intermediate precision	Between day ≤3% Within day ≤2%	≥0.25% RSD ≤ 20% relative Between 0.10% and 0.25% SD ≤ 0.04
Robustness	Absolute difference	
Standard solution stability	In total impurities between day 0 and timepoint is <1.0% In purity between day 0 and timepoint is <2.0%	
Sample solution stability	In total impurities between day 0 and timepoint is <1.0% In purity between day 0 and timepoint is <2.0%	No individual impurity ≥0.25% changes by more than 20% (relative) No individual impurity <0.25% changes by more than 0.05 (absolute)
Specificity	Contribution of any peaks from the sample matrix or diluent must be <1% of the sample concentration	All critical impurities must be resolved from peaks from the system, other impurities and placebo matrix
Sensitivity		The LOQ should be less than the reporting limit. %RSD of area response at the LOQ should be ≤15%

VIII. REVALIDATION

According to the validation life cycle,[46] test methods may require additional validation or revalidation when regulatory agencies issue new requirements or when changes are made to the methodology. Revalidation may be necessary after changes in (a) drug substance synthesis, (b) drug product composition, and (c) the analytical procedure, for example, instrumental changes, method modifications. The degree of revalidation required depends on the nature of the changes made.

IX. REFERENCE STANDARDS

Any discussion of the method development/validation process should consider obtaining a sufficient quantity of qualified reference standard and appropriate samples to support the development and validation campaign. In most cases, the limit of the accuracy of the analytical test will be related to the correctness of the standard's assay. There is little consensus[47,48] on the common requirements of reference standards as outlined by the regulatory authority[49]; typical practice favors an approach of thorough analytical characterization of the standard supported by adequate documentation.

A. Types of Standards

Materials selected for use as a standard should be of high purity and stability. Practical and regulatory requirements for reference standards differ, depending on the objective and stage in the drug development cycle. For pragmatic reasons, the reference standard used in the early stages of development may be limited to a well-characterized portion of one of the early lots of material. In later stages, the primary reference standard may be a unique synthesis or a purified version of the material—sometimes in a different salt form to enhance stability or purity. The following sections describe a process of qualification of a reference material:

Primary reference standard of the active compound: The standard of an active compound is typically comprehensively characterized. Table 4 lists the detail required.

Secondary standard (working standard): A material of high purity that has been characterized against a primary or compendial standard. A secondary standard is often used to conserve the amount of primary standard.

Related compound standard: These standards are typically characterized fully, but not to the same extent as the primary reference standard of the active drug substance. Because quantities of these standards are typically limited, response factors for these standards can be assigned to eliminate routine use in HPLC purity determination.

Compendial standard: USP/National Formulary (NF)/American Society for Testing and Materials (ASTM) standards typically do not require further characterization. A certificate of analysis should be available.

TABLE 4 Analytical Tests for Certification of Reference Standards

Characterization requirement	Measurement technique
Physical properties	
Proof of structure	Nuclear magnetic resonance (NMR)
	Mass Spectrometry (MS and MS/MS)
	Fourier transform infrared (FTIR)
	UV spectroscopy
	Crystallinity—isoforms
	Combustion analysis
Physical description	Appearance
Physical properties	Melting point
	pK_a
	Optical rotation
Purity	
Related compounds	HPLC purity
Chiral purity	HPLC, CZE chiral purity
Inorganic impurities	Residue on ignition
Solvents	Inductively cover plasma gas chromatography (ICP GC)
Water	Karl Fischer titration
Counterions	Ion chromatography
Verification—organic impurities	Thin-layer chromatography (TLC)
Assay	
Verification	Titration
	Assay by difference[a]
	CH and N analyses
	Mass balance calculation

[a]Assign assay by difference: 5 assay = 100% (% impurities/degradants, including solvents) − (% water) − (% other, including counterions).

Internal standard: Internal standards require a varying amount of characterization, depending on the application. Often, a qualified commercially available material is chosen.

Other standards: Reagents and chemicals that are commercially available at high purity that have been characterized by the vendor can be suitable for standard materials. Sufficient documentation should be provided by the vendor prior to use.

B. Handling of Standards

1. Storage

When possible, the standard material should be protected from light, heat, and moisture and housed in a controlled environment. Typical storage conditions for a stable solid standard would be at 2–8 °C, protected from light, in a desiccator or cabinet to protect it from humidity. Reference stock solutions and working solutions should also be labeled with the required storage conditions and expiration dates. The method procedure should specify the storage conditions, and the use period, which should be supported by experimental data. There should be controlled access to all standard materials. Although there is a risk of breakage, storage in tightly capped, amber glass bottles is preferred. Containers should be sized to minimize headspace; small changes to assay have been observed in overly large containers as the solvents and water vapors in the solid standard materials equilibrate with the headspace.

2. Handling

Reference materials must be handled to preserve the integrity of the sample. As discussed previously, the solid materials must be protected from heat, light, and high humidity. Standard materials that are stored at subambient temperatures must be equilibrated to ambient before opening to prevent water condensing on the materials, particularly in humid environments. If primary standard materials are used frequently, qualification of a secondary standard may be justified. Alternatively, primary reference standards can be subdivided into several smaller containers to limit contamination and changes in assay. The retest (recertification) date of the standard should be justified through stability monitoring in the actual storage container.

Some laboratories permit selective standard to be oven dried prior to use. If this is the practice, these drying times/conditions should be adjusted to effect drying of the material and minimize any potential decomposition. Warm standards should be equilibrated to ambient conditions prior to weighing. If this practice is followed, attention should be given to ensure that there is no thermal degradation of the standard.

3. Documentation

A reference standard must have documentation to support its use as a standard, establish its assigned assay, and defend the retest date (or expire date for chemical standards).

a. Reference Standard Qualification Report

A detailed qualification report is drafted for a primary reference standard of the active compound; less comprehensive documentation is typical of the other standards. Portions of this report may be submitted to the regulatory authorities. The report references the synthetic route for the standard and any purification schemes that were utilized in preparing the pure standard. Other components of this report may include the following:

- Description or reference to test methods used in characterization of the standard
- Report results of the characterization with representative data (UV, NMR spectra, HPLC chromatograms, etc.)

- Verification of data to establish purity and assay of the standard
- Suggested requirements for requalification interval and requalification tests
- Appendix, detailing additional requalification, and stability data as they are obtained

b. Certificate of Analysis

In general, all standards should have a certificate of analysis generated either from a qualified supplier or through analysis of in-house data. The certificate should contain the following information:

- Standard name
- Lot number and supplier
- Effective data and recertification date purity
- Assay

c. Dispensing

A reference standard log is typically maintained to provide an inventory of the standard, indicting who, when, and how much standard was removed. Care should be exercised when handling to avoid contamination of the standard. The user should consider developing guidance on limiting the number of times an individual vial is sampled, to minimize contamination or deterioration from moisture. Ideally, aliquots contained within a vial should be sufficient for a single test, although it is recognized that this may not be cost effective or practical when reference standards are purchased commercially. Furthermore, observe safety cautions when handling potent drug substance materials.

X. FUTURE PERSPECTIVES

It is clear that validation of analytical methods used in the testing of pharmaceutical products is now well established and formalized, with a wealth of guidelines and regulations issued for the benefit of the method development scientist. It is unlikely that there will be any fundamental change in the requirements or approaches to method validation unless necessitated by new technologies. However, at present, there are changes in the way pharmaceutical products and manufacturing processes are being developed. There is a shift in placing greater emphasis on controlling the process rather than determining that a product meets a required specification.[50] To be effective, this Quality by Design paradigm requires real-time analytical testing, which in itself is best served by online, noninvasive spectroscopic tools, and requires consideration of how to validate such measurements. For example, real-time testing does not necessitate a "sample" which can be evaluated, but requires the development of a model to monitor the process which is being controlled. Clearly, the application of new technologies will require the development of new approaches to validation, which, although different in practical determination, will adhere to the well-established principles discussed herein.

A consequence of this paradigm change is that it signals the phasing out of traditional end-of-process testing on finished drug product and heralds the advent of real-time release (RTR) methods which will require the development of new method validation strategies. The validation of methods used to monitor process control Process Analytical Technology (PAT) and RTR is dealt with in a separate Chapter 14.

XI. SUMMARY AND CONCLUSIONS

Because of the current accuracy and precision in analytical instrumentation, reagents, and capabilities of modern data processing systems, even poor methods may be acceptably validated; however, this does not necessarily certify a method as being "good," robust, or suitable for a control environment. Ideally, method validation should be integral to and performed during the development process as part of the method optimization.

This chapter has covered a systematic and scientific approach to the method development and validation of analytical methods, which should demonstrate to regulatory agencies that methods used in the testing of pharmaceutical products, at each stage of development and ultimately at commercialization, are fit for purpose. To ensure that the data are accurate and reliable, qualified and trained laboratory analysts must always perform methods on qualified equipment, using suitable standards.

It should always be recognized that the validation of such methods is not required, solely to comply with good laboratory practices, or satisfy regulatory guidance. It is to ensure that data used to approve medicinal products is accurate and that patients can receive life enhancing medicines without compromise to their welfare or safety.

ACKNOWLEDGMENTS

The authors wish to acknowledge Dr. Jonathan Crowther as the previous author of this chapter, sections of which remain unchanged in this edition.

REFERENCES

1. *Analytical Procedures and Methods Validation*, U.S. Department of Health and Human Services, Food and Drug Administration, Rockville, MD 20857, 2000. http://www.fda.gov/downloads/Drugs/GuidanceComplianceRegulatoryInformation/Guidances/ucm122858.pdf.
2. Federal Register 21 CFR 211A65 (e) and 211A94 (a)(2).
3. Konieczka, P. The role of and the place of method validation in the quality assurance and quality control (QA/QC) system. *Crit. Rev. Anal. Chem.* 37:173–190, 2007.
4. Chan, Chung Chow, Lee, Y. C., Lam, H. and Zhang, XueMing. *Analytical Method Validation and Instrument Performance Verification*, John Wiley & Sons, Inc., 2004.
5. Riley, C. M. and Rpsanke, T. W. *Development and Validation of Analytical Methods*, Pergamon, Oxford, UK, 1996.

6. McPolin, Oona *Validation of Analytical Methods for Pharmaceutical analysis*, Mourne Training Services, 2009.
7. Ermer, J. and Miller, J. H. *Method Validation Pharmaceutical Analysis: A Guide to Best Practice*, Wiley-VCH Verlag GmbH & Co. KGaA, 2005.
8. Miller, J. H. and Crowther, J. B. *Analytical Chemistry in a GMP Environment: A Practical Guide*, John Wiley & Sons, Inc., 2000.
9. ICH Q2 (R1) International Conference on Harmonisation. Validation of Analytical procedures: Text and Methodology. http://www.ich.org
10. *Reviewers guidance, validation of chromatographic methods*, Center for Drug Evaluation and Research, Food and Drug Administration, Rockville, MD 20857, 1994. http://www.fda.gov/downloads/Drugs/GuidanceComplianceRegulatoryInformation/Guidances/UCM134409.pdf.
11. Chandran, S. and Singh, R. S. P. *Pharmazie* 62:1–14, 2007.
12. ICH Q2A International Conference on Harmonisation. Text on Validation of Analytical Procedures.
13. ICH Q2B International Conference on Harmonisation. Validation of Analytical Procedures: Methodology.
14. Ermer, J. Validation in pharmaceutical analysis. Part I: An integrated approach. *J. Pharm. Biomed. Anal.* 24:755–767, 2001.
15. Jenke, D. R. Chromatographic method validation II. Guidelines for primary validation parameters. *J. Liq. Chromatogr. Relat. Technol.* 19:737–757, 1996.
16. Hokanson, G. C. A life cycle approach to validation of analytical methods during pharmaceutical product development I. The initial validation process. *Pharm. Technol.* 18:118–130, 1994.
17. PMA Computer Systems Validation Committee. Computer systems validation–Staying current: Introduction. *Pharm. Technol.* 13:60–65, 1980.
18. ICH Q8(R1) International Conference on Harmonisation, Pharmaceutical Development, http://www.ich.org.
19. Krause, S. QbD for analytical methods. *Biopharm. Int.* , Feb, 2009.
20. Borman, P., Nethercote, P., Chatfield, M., Thompson, D. and Truman, K. *Pharm. Technol.*, October 2007.
21. Li, Y., Terfloth, G. J. and Kord, A. S. A systematic approach to RP-HPLC method development in a pharmaceutical QbD environment. *Am. Pharm. Rev.* 12(4), May 2009.
22. Baczek, T. Computer-assisted optimization of liquid chromatography separations of drugs and related substances. *Curr. Pharm. Anal.* 4:151–161, 2008.
23. Flieger, J. and Markowski, W. Application of gradient elution optimized by Chromsword software in chromatography of phenothiazines in reversed phase systems controlled by chaotropic effect. *Chem. Anal.* 54:187–202, 2009.
24. Fekete, S., Fekete, J., Molnár, I. and Ganzler, K. Rapid high performance liquid chromatography method development with high prediction accuracy, using 5 cm long narrow bore columns packed with sub-2 μm particles and Design Space computer modeling. *J. Chromatogr. A* 1216:7816–7823, 2009.
25. Vander Heyden, Y., Nijhuis, A., Smeyers-Verbeke, J., Vandeginste, B. G. M. and Massart, D. L. Guidance for robustness/ruggedness tests in method validation. *J. Pharm. Biomed. Anal.* 24:723–754, 2001.
26. Kleinschmidt, G. In *Robustness*, (J. Ermer and J. H. M. Miller, eds.), Wiley-VCH, Weinheim, pp. 120–127, 2005.
27. Suresh Kumar, R., Narasimha Naidu, M., Srinivasulu, K., Raja Sekhar, K., Veerender, M. and Srinivasu, M. K. Development and validation of a stability indicating LC method for the assay and related substances determination of Exemestane, an aromatase inhibitor. *J. Pharm. Biomed. Anal.* 50:746–752, 2009.
28. Somasekhar, V., Gowrisankar, D. and Shivakumar, H. N. Development and validation of a rapid RP-HPLC method for the determination of venlafaxine hydrochloride in pharmaceutical dosage forms using experimental design. *E-J. Chem.* 6(4):1091–1102, October 2009.
29. Nandan, S. R., Reddy, R., Rao, S. and Ravindranath, L. K. Regulatory requirement–Validated, specific, and stability indicating analytical method for zoledronic acid and its related impurities by Ion Pair reversed phase liquid chromatography (IP-RPLC). *J. Liq. Chromatogr. Relat. Technol.* 32:2307–2321, 2009.

30. Xiang, D., Konigsberger, M., Wabuyele, B., Hornung, K. and Cheney, J. Development of robust quantitative methods by near-infrared spectroscopy for rapid pharmaceutical determination of content uniformity in complex tablet matrix. *Analyst* 134:1405–1415, 2009.
31. Jimidai, M., Niemeijer, N., Peeters, R. and Hoogmartens, J. Robustness testing of a liquid chromatography method for the determination of Vorozole and its related compounds in oral tablets. *J. Pharm. Biomed. Anal.* 18:479M85, 1998.
32. Analytical procedures and method validation: Chemistry, manufacturing, and controls documentation. Draft Guidance. http://wwwfda.gov/cber/guidelines.htm, August 2000.
33. Jimidar, I. M., Hartmann, C., Cousement, N. and Massart, D. L. *J. Chromatogr. A* 706:479, 1995.
34. Wiggins, D. E. System suitability in an optimized HPLC system. *J. Liq. Chromatogr.* 14:1045–1060, 1991.
35. Wahlich, J. C. and Carr, G. E Chromatographic system suitability; tests: What should we be using? *J. Pharm. Biomed. Anal.* 8:619–623, 1990.
36. USP 32–NF 27 United States Pharmacopoeial Convention, Inc.: Roskville, MD, 2009.
37. EP. *European Pharmacopoeia*, 6th ed, The European Directorate for the Quality of Medicines within the Council of Europe, Strasbourg, 2010.
38. Miller, J. H. M. In *Validation of Pharmacopoeial Methods. Future Trends in Analytical Method Validation*, (J. Ermer and J. H. M. Miller, eds.), Wiley-VCH, Weinheim, pp. 301–336, 2005.
39. Walsh, A., *et al.* Development and validation of automated methods for finished product testing. *Pharm. Technol. Eur.* , May 2000.
40. DeSain, C. Master method validation protocols. *BioPharm* 6:30–34, 1992.
41. Johnson, J. D. and Van Buskirk, G. E. Analytical method validation. *J. Valid. Technol.* 2:88–105, 1998.
42. Ermer, J. and Ploss, H.-J. Validation in pharmaceutical analysis Part II: Central importance of precision to establish acceptance criteria and for verifying and improving the quality of analytical data. *J. Pharm. Biomed. Anal.* 37:859–870, 2005.
43. Lukulay, P. and Morgado, J. Automation of the chromatographic analytical method validation process. *LCGC North Am.* 24(2):151–156, 2006.
44. Green, J. M. A practical guide to analytical method validation. *Anal. Chem.* 68:305A–309A, 1996.
45. Jenke, D. R. Chromatographic method validation: A review of current practices and procedures II. Guidelines for primary validation parameters. *J. Liq. Chromatogr. Relat. Technol.* 19:737–757, 1996.
46. Hokanson, G. C. A life cycle approach to the validation of analytical methods during pharmaceutical product Development II. Changes and the need for additional validation. *Pharm. Technol.* I8:92–100, 1994.
47. *PMA Workshop on Reference Standards*, PMA Analytical Steering Committee, Washington, DC, 1992.
48. Housepian, P. K. Bulk drug substances and finished products. In *Development and Validation of Analytical Methods*, (C. M. Riley and T. W. Rpsanke, eds.), Pergamon, Oxford, UK, 1996.
49. FDA guidelines for supporting documentation in drug applications for the manufacturing of drug substances, 1987.
50. Rudd, D. In *Future Trends in Analytical method Validation*, (J. Ermer and J. H. M. Miller, eds.), Wiley-VCH, Weinheim, pp. 387–397, 2005.

12 STABILITY STUDIES

JESSICA CHA, TIMOTHY GILMOR, PHILIP LANE, AND JOSEPH S. RANWEILER

Johnson & Johnson Pharmaceutical Research and Development, Raritan, New Jersey, USA

I. INTRODUCTION

Patients taking a pharmaceutical product for a particular malady expect the product to be safe and efficacious. Regulatory agencies worldwide demand that the product retains its identity, quality, purity, and potency for the time that it is commercially available. Consequently, the agencies expect to see stability data supporting the proposed shelf life in the marketing application.

Various stability guidelines[1–3] have been published in the past 10 years describing the types of studies, data and analysis needed to satisfy the expectations of regulatory agencies worldwide. Currently, the guidelines promulgated by the International Conference on Harmonization (ICH) are most commonly accepted in the United States, European Union (EU), and Japan.[1] Based on the ICH, these countries established an agreement that stability data generated in any of the three regions would be mutually acceptable in the other two. In the United States, the Food and Drug Administration (FDA) stability guidance[4] and the section on Pharmaceutical Dosage Forms <1151> in USP 32/NF29[5] reflect the ICH guidelines. All countries, including the United States, EU, and Japan, are covered by the World Health Organization (WHO) guideline on stability testing of active pharmaceutical ingredients (APIs) and finished pharmaceutical products (FPP).[3] This guideline has recently been finalized.

In the broadest sense, stability studies that are conducted must provide evidence of how the quality attributes of the API and drug product change over time when subjected to various environmental conditions, such as temperature, humidity, and light. In addition, the impact of other variables may also need to be studied. Although not an all-inclusive list, examples of such variables are variations in packaging components as well as manufacturing variations, such as changes in process, batch size, equipment train, manufacturing site (for both API and drug product), API synthesis, and excipient vendor. Any proposed modification to the API or drug product should be carefully evaluated to determine the quantity of stability data that will be required to obtain approval. In the US postapproval change (SUPAC[6]) can be classified into different categories and, depending upon the change classification, varying amounts of stability data will need to be provided.

Quality by design (QbD), a component of the ICH Q8(R2) guidance, offers the possibility of improving product quality.[7] QbD is a scientific, risk-based, holistic, and proactive approach to pharmaceutical development. It is a deliberate design effort from product conception through commercialization. The goal of QbD is to fully understand how certain product attributes and process relate to the overall product performance.

A goal of United States Food and Drug Administration's Center for Drug Evaluation and Research (CDER) is to implement the ICH guidances Q8(R2), Q9, and Q10 covering drug product development (through QbD), risk management, and quality management systems.[7–9] The ICH guidances are aligned with CDER's Critical Path initiative, *Moving Manufacturing into the 21st Century*, to develop new manufacturing approaches that improve the sponsors' ability to assess and improve drug product quality. Some of the QbD definition to consider:

1. *Key stability attribute (KSA)*: An attribute which may/will influence API shelf life when stored at fixed conditions. *KSAs are a subset of key quality attributes* (*KQAs*).
2. *Critical stability attribute (CSA)*: An attribute (e.g., particle size) which will primarily determine API shelf life (via SLLA) when stored at fixed conditions and will be part of API release requirements. *CSAs are a subset of critical quality attributes* (*CQAs*).
3. *Shelf life limiting attribute (SLLA)*: An attribute which will breach its limit when stored at fixed conditions at a point predetermined by CSA (s) (e.g., degradation product . . . and would be measured on stability).
4. *Stability design space (SDS)*: Is part of the API "release requirements," a well-understood space built through development knowledge, understanding of KSAs and verified by SLLAs that, allows confident shelf life assignment for a given API, independent of scale, site, process, and route. SDS lies within the QbD space.

QbD approach to stability is an emerging trend in the pharmaceutical industry. The SDS is built using a number of sources and tools listed below:

1. *In silico* modeling and chemistry knowledge of the active molecule
2. Literature and knowledge on other compounds
3. Excipient compatibility studies
4. Forced degradation studies
5. Development and clinical stability studies
6. Microcalorimetry
7. Package selections for moisture sensitive APIs

The potential advantage of the QbD approach to stability studies is to avoid conducting additional stability studies and filings when the scale, site, process, and route are altered from the initial registration stability batches. The barriers of this approach are related to cost, regulations, and guidelines.

In the following sections, the operations that must be considered to conduct viable stability studies for excipients, APIs, and drug products will be presented and discussed.

II. OPERATIONAL ASPECTS OF STABILITY

A. Introduction

This section will discuss the fundamental components for planning and conducting a stability program. Topics that will be covered include the design and implementation of the stability program for the specific studies, the methods and equipment required to administer them, and the documentation needed to effectively plan, execute, and report the results of the studies. The design and outcome of the studies are highly regulated and therefore must follow guidelines that specify the requirements of the agencies to which the results of the studies will be submitted in regulatory filings. These guidelines also provide a degree of consistency and an assurance of the acceptance of the study design. This section will cover aspects of stability planning such as protocol design with the ability to amend the studies based on the resulting data. The protocol is the single most important document for execution of the studies; however, documents such as stability data tables and reports, testing schedules and assay sheets, deviation reports, out-of-specification (OOS) and out-of-trend (OOT) reports, and batch and package documentation are also important and will also be discussed.

I. Phases of Pharmaceutical Product Development

The phases of pharmaceutical development are closely aligned with the progression of the clinical program for the product under development. Although individual organizations may refer to development phases differently, for the purpose of this discussion a product will go through the following five clinical development phases during its life cycle. These five phases are as follows:

1. *Preclinical or Phase 0*: animal toxicological studies in one or more species
2. *Phase I*: safety in healthy volunteers
3. *Phase II*: safety and dose determination in patients
4. *Phase III*: definitive clinical efficacy studies
5. *Phase IV*: postapproval studies for other indications

During the various phases of development, specific needs for stability data can change. Even in the preclinical phase when a chemical compound is used in animal toxicological studies, stability or "end of use" data of the API (denoted as "test article" in preclinical studies) and the dosing formulation are required by Good Laboratory Practice (GLP) regulation in the United States and Organisation for Economic Co-operation and Development (OECD) in the EU. The stability or end of use data is needed to show that the test article and associated dosing formulation are stable during the toxicological study.

To support phase I clinical studies, the evaluation of the physical and chemical properties and stability of the API is the primary focus. Such studies usually include salt and form selection, solid-state polymorph structure, probe (or screening) stability, and stress studies to evaluate relative stability of the API and determine its degradation pathways. The reader is referred to Chapter 3 for a detailed discussion on determination of degradation mechanisms. Stability of the molecule represents a key criterion for selecting a lead candidate for further development. During phase I, a very simple clinical dosage form (such as drug in bottle or drug in capsule) is utilized. During these clinical studies, the stability programs for the dosage forms are short and focused on supporting the limited scope clinical trials.

Stability studies are required to run for a longer duration in support of phase II clinical studies. During this phase, the technical focus shifts to development of a commercially viable formulation. In this regard, excipient compatibility studies, multiple formulations, and packaging materials are generally screened. In most cases, short-term (6-month maximum) probe stability data are sufficient, followed by longer term (2–3 years) studies for the selection of the best formulation candidate. The API should also be monitored for stability with studies lasting up to 5 years in duration.

By phase III, the dosage form and dose(s) have been chosen. During phase III, the formal registration stability batches are manufactured and packaged in the proposed configurations, and the stability program initiated. For these studies, the use of regulatory guidelines is crucial as the resulting data will support the filings in the targeted market countries and to support shelf life and storage recommendation. The importance of conducting longer term studies at stressed and accelerated conditions mimicking the climatic zones where the product will be marketed cannot be overemphasized.

Phase IV typically occurs after product approval and launch, and the stability studies are a part of the marketed product stability program. These studies are conducted to assure that the product continues to meet its approved acceptance criteria up to its approved expiration date. Therefore, these studies are conducted at the label storage conditions only. However, if new strengths, packages, or formulations are being developed, these studies would be conducted at additional storage conditions as well.

2. Regulatory Guidelines

In addition to the guidance given to the industry by the compendia of the various countries or regions and the general requirements of good manufacturing practices (GMPs), several organizations have set standards of practice, including stability, in an attempt to standardize product development. Both the ICH[1] and WHO[3] have issued stability guidelines for API and drug product. The ICH guidelines are agreed to in practice by three marketing regions: Japan, the EU, and the United States. The WHO guidelines are for the development of pharmaceutical products worldwide and include the aforementioned regions. ICH stability guidelines apply to climatic zones I and II, whereas WHO lists requirements for climatic zones I–III, IVa, and IVb (See Table 1).

3. Climatic Zones

The standard climatic zones for use in pharmaceutical product stability studies are shown in Table 1. Reference to these zones and their associated storage conditions is discussed elsewhere.[3,10]

TABLE 1 WHO Definition of the Different Climatic Zones Around the World and Long-Term Testing Conditions

Climatic zone	Definition	Long-term testing conditions
I	Temperate climate	21 °C/45% RH
II	Subtropical and Mediterranean climate	25 °C/60% RH
III	Hot and dry climate	30 °C/35% RH
IVa	Hot and humid climate	30 °C/65% RH
IVb	Hot and very humid climate	30 °C/75% RH

4. Generic and Supplemental Products

A majority of the discussion in this chapter applies to innovator products (new drug products containing APIs that have not yet been approved for marketing). In general, a generic product manufacturer follows these same practices in developing a product for commercialization.

For supplements to an existing product registered in a specific region, the requirements can be different depending on the proposed change. Required stability information can range from no stability study requirements beyond the marketed stability program for a minor change (e.g., batch size) to a complete stability program for a formulation change. Required studies for changes made to an existing product are given in guidance documents such as the FDA SUPAC.[6] In the case of generic products and supplemental changes to existing products, it is important to consult the requirement of each local governing agency and their guidance documents.

B. Protocols

The protocol is a written document that describes the necessary components of a regulated and well-controlled stability study. It details the basic strategy that will be executed; its two major components constitute the matrix of tests to be performed against the schedule of testing. The types of batches that require such a protocol include clinical, formulation development, registration, and commercial products. In addition, compatibility of a product with a vehicle (e.g., an injectable product in an intravenous saline solution) is often studied to support the use of such a product for hospital use. Probe stability studies are generally more experimental in nature and may not follow a formal protocol but may have some documentation for intended purpose documented elsewhere (such as in a laboratory notebook). Regulatory agencies require a study for leachables and extractable for perenteral products packaged with plastic components that contact the product. These studies are part of the package selection process but can be incorporated in the testing of the stability samples.

The protocol stipulates the data that are expected to be obtained from the study that will be used to develop a stability profile for the product. It requires all involved parties (formulators, analysts, management, regulatory affairs) to understand their roles with respect to conducting and supporting the study. The protocol should also reflect the regions where the product will be marketed. For example, if the product is planned to be used in climatic zones I–III, IVa, and IVb, the stability program must include all of these zones. The formulation scientist should be able to determine the requirements of the batch(es) being studied including the batch size, package configurations, and the amount of data needed. The analyst should be able to assess the suitability of the test methods or the need for additional development. It also describes a summary of the overall information to be collected to evaluate the stability of the product. A mechanism (such as change control) must also exist to allow modification of the protocol. The protocol amendment process is discussed in Section II.C.1.

In addition, the stability protocol documents other information and procedures that are needed to organize the study design and ensure that the drug product is in its intended package configuration. Information common to most studies should be documented in standard operating procedures (SOPs) if not already contained in governmental regulations such as the cGMPs. These will be discussed later in this chapter. The protocol should also contain an approval section for the groups involved in the execution of the study, signifying that there is agreement with the design and testing plan as written.

I. Tests, Acceptance Criteria, and Methods

This section of the protocol is a list of the analytical tests to be performed on the stability samples stored at the various environmental conditions. This list may look similar to the list of tests in the product specification that are applied to product at the time of release. An example of a list of tests for stability is given in Table 2. The tests that monitor the quality, purity, potency, and identity which could be expected to change upon storage are chosen as stability tests. Therefore appearance, assay, degradation products, microbiological testing, dissolution, and moisture are standard tests run on stability. The batches used for the stability study must meet all of the release testing requirements; therefore, the tests such as heavy metals, residue on ignition, residual solvents are examples of tests that provide data only at the time of product release but not required to be repeated during stability testing. If leachables or extractable are studied during the stability program, tests to monitor these components in the formulation need to be

TABLE 2 List of Typical Stability Tests for API and Various Drug Products

API	Tablet	Capsule	Oral liquid	Liquid injectable	Topical cream
Assay	Assay	Assay	Assay	Assay	Assay
Degradation products	Degradation products	Degradation products	Degradation products	Degradation products	Degradation products
Identity—IR	Dissolution	Dissolution	pH	pH	pH[a]
Appearance	Appearance	Appearance	Appearance	Appearance	Appearance
Moisture[a]	Moisture[a]	Moisture[a]	Preservatives or antioxidants[b]	Preservatives or antioxidants[b]	Preservatives or antioxidants[b]
			Microbiological tests[c]	Microbiological tests[c]	Microbiological tests[c]
			Precipitate formation[a]	Precipitate formation[a]	Viscosity
Other tests[d]	Other tests[d]	Other tests[d]	Other tests[d]	Other tests[d]	Other tests[d]

[a]Test if moisture is a critical to product stability or performance.
[b]Test in preservatives or antioxidants is included in the formulation.
[c]Microbiological tests may include sterility, preservative effectiveness, and/or microbial count, where appropriate.
[d]Other tests as discussed in ICH guidance Q6A (e.g., enantiomeric purity, particle size, polymorph form, etc.).

included. The stability criteria are included in the limits for each test, thereby defining that a product is stable at the time point and condition tested. Typically, these are the same values as specification limits for a product; however, tighter limits may be employed.

The example found in Table 2, lists the typical tests for the API and dosage form products. For tablets the tests would include appearance, assay, degradation products, and dissolution. Other products might require specialized testing regimes. For example, a hygroscopic product or one that exhibits hydrolytic degradation might warrant a test for moisture content while a parenteral or oral liquid product might need testing for pH, microbiological sterility, and preservative content. For a cream product the list of tests may also include a test for viscosity. The ICH guidance Q6A contains a discussion of tests used for various API and drug products and the rationale for selection of appropriate tests for release and for stability testing.[11]

2. Stability Schedule

A stability schedule is a series of test intervals at which time, all or some of the stability tests will be conducted on a sample in the study. Matrixed against this schedule is the environmental conditions that the sample is exposed to. The combination of results obtained for the various test intervals and environmental conditions creates the stability profile for the product. Data from this profile are used to define the recommended storage conditions, storage temperature, transportation controls, and expiration date. A typical stability schedule for a new product is found in Table 3. This schedule covers climatic zones I–IV.

The primary stability condition for a product to be stored at ambient conditions is controlled at room temperature or 25 °C/60% RH (relative humidity) for climatic zones I and II. Additional stability conditions may be added to cover transportation excursion conditions, for example, 50 °C for 7 days and/or 70 °C for 12 h. ICH condition also calls for the samples to be exposed to light of a designated intensity[12] as part of the stability study. A stability program may also require more intense environments for climatic zones III and IV if the product is to be registered in countries located in these zones.

TABLE 3 Typical Test Schedule for Stability of a New Product

Environment	Intervals	Intended climate zones
25 °C/60% RH	3, 6, 9, 12, 18, 24, 36	Long term for zones I and II
30 °C/35% RH	3, 6, 9, 12, 18, 24, 36	Long term for zone III
30 °C/65% RH[a]	3, 6, 9, 12, 18, 24, 36	Long term for zone IVa, or intermediate condition for zones I and II out to 12 months
30 °C/75% RH	3, 6, 9, 12, 18, 24, 36	Long term for zone IVa or intermediate condition for zones I and II out to 12 months
40 °C/75% RH	3, 6	Accelerated condition for all zones

[a]Test 9 and 12 months only if failure to meet specification at 40 °C/75% RH or assay is different from initial by greater than 5% for zones I and II submission only.

3. Stability Chambers

a. List of Environmental Conditions

The environmental conditions at which stability studies are expected to be performed have evolved during the last several years.[1,3] The guidelines from WHO and ICH have harmonized with respect to the use of mean kinetic temperature[41] and specified climatic zones. The recommended storage conditions proposed in these guidelines are given in Table 4 along with the expected tolerance for temperature (typically ±2 °C) and humidity (typically ±5% RH). The guidelines discuss conditions for room temperature storage, refrigeration, and freezer storage. In addition to these standardized conditions, a typical stability program will have accelerated and stress conditions that are used to probe the degradation mechanisms of the product and define the "edges of failure" for the product and its primary package. Photostability testing of the API and drug product is an element of the stress testing studies included in a stability program.[12] The ICH discusses the need for accelerated testing at 40 °C and 75% RH and stress testing at higher temperature(s). In addition, the effects of temperature cycling (e.g., between freezing and thawing or hot and cool temperature) are discussed by the guidelines. Both the ICH and the WHO recommend that such studies be conducted for liquids and semisolid dosage forms.

For the environmental conditions listed in Table 4, the ICH and WHO provide guidance on the allowable excursions for temperature and humidity. Excursions that exceed the tolerances for more than 24 h should be investigated and documented, including a determination of the impact on the results of the study.

b. Qualification and Calibration of Stability Chambers

Chambers used in the formal stability program must go through qualification and calibration procedures. The chambers should be qualified at the time of purchase to assure a proper installation, a process denoted installation qualification (IQ). This procedure documents that the chamber has been installed properly and that it meets the manufacturer's specifications. The chamber should also be

TABLE 4 Proposed Storage Guidelines

Environment	Tolerance	Guidelines[a]
25 °C/60% RH	±2 °C and ±5% RH	ICH and WHO
30 °C/35% RH	±2 °C and ±5% RH	WHO
30 °C/65% RH	±2 °C and ±5% RH	ICH and WHO
30 °C/75% RH	±2 °C and ±5% RH	WHO
40 °C/75% RH	±2 °C and ±5% RH	ICH and WHO
5 °C	±3 °C	ICH and WHO
−20 °C	±5 °C	ICH and WHO
Light	1.2 million lux	ICH

[a]ICH guidelines from Ref. 1; WHO guidelines from Ref. 3.

qualified to confirm that it is operating within the prescribed parameters and tolerances required for the stability studies. This operational qualification (OQ) should confirm that the chamber maintains the proper conditions of temperature, humidity, etc., within specified ranges. For temperature, this OQ would typically incorporate a "mapping" of the chambers to verify that the specified temperature is maintained consistently and evenly throughout the chamber. For chambers with humidity control, multiple readings for humidity may also be needed. Finally, the continued performance of the chambers should be routinely monitored to address performance qualification (PQ). PQ would ideally be a continuous monitoring of the chamber to verify that the environment is within the specified tolerances. An automated alarm system adds to the reliability of the monitoring system, especially during off-work hours. The use of uninterrupted power supplies (UPS) for the chamber monitoring system and/or the power to the chambers adds to the reliability of the stability program. An important part of the OQ/PQ of the chambers is a routine maintenance program that involves the inspecting, testing, and replacing of worn parts before their failure. In the authors' program, a semiannual schedule is maintained. A discussion of a proposal for IQ/OQ of stability chambers has been published.[13]

c. Inventory of Samples Within Chambers

The stability system must be capable of maintaining an accurate inventory of the samples in each chamber for all studies. This is usually handled by a computerized database system (alternately a logbook system can be used) that tracks the number of samples at all times. Following this procedure, it can be determined whether sufficient samples exist to complete or extend a study. In anticipation of extending studies, it is routine practice to store additional "back-up" or "retain" samples at each storage condition for each study. For liquid samples stored in vials with stoppers or bottles with caps, the stability samples should be stored in both the upright and the inverted position. The inverted samples are tested as the "worst-case" format, and the upright samples typically tested only to determine a difference is observed in the inverted samples. The system should document the total number of samples and should indicate the sample requirement at each testing interval. As samples are withdrawn from the chamber and used for testing, the system must document this change in inventory. The inventory information is also useful if a chamber malfunctions, requiring the documentation of the occurrence for each study involved.

d. Documentation

As the stability chambers are an integral part of the stability program and require continuous adherence to specifications for long-term studies, all aspect of the chambers must be described in detail in an SOP. The SOP should include the procedures and the schedule for calibration of the chambers, the description and operating parameters, a routine maintenance schedule, inventory system, IQ/OQ procedure and a monitoring system (PQ), and emergency procedures for malfunctions or unusual occurrences.

4. Batch Selection

Batch information should be included in the stability protocol. Depending on where the drug being studied is in the development process, a batch may be chosen for a variety reasons. For a batch slated for use in clinical trials, sufficient stability data are required to support the clinical program. Formulation development batches are studied to gather data to select the market image for the product. During these early development phases of a product the guidelines provided from the regulatory agencies are usually less specific and the choices are governed by the developing company. However, it is necessary to choose relevant batches and to document the rationale for their choice.

The selection of the registration batches for stability testing is governed by more stringent requirements, as stipulated by ICH and WHO. These guidelines include the following criteria:

- Typically, three batches of a formulation in each container–closure system are studied.
- Generally, pilot- or production-scale batches are tested. It is recommended that each batch used is no smaller than 10% of the proposed production-scale batch size.
- The process used for all batches should be representative of the final commercial process.
- The container-closure system should be that which will be used for market.
- Matrixing or bracketing of batches or packages may be allowed with justification. Bracketing assumes that an intermediates condition is represented by the batches being tested. Prior agreement with the agency is recommended. Matrixing and bracketing are discussed further in Section V.D.1.

As it pertains to batch selection, site-specific stability has been the subject for discussion between FDA and industry. "Site-specific" means that stability batches are manufactured at the proposed manufacturing site. The "site-specific" stability is to assure that the batches used for expiration date determination will be representative of the batches made for commerce. Although "site-specific" stability testing has typically been implemented for years, it is not an absolute requirement from all authorities. It is recommended that the requirements for "site-specific" stability be determined for each country prior to initiation of the registration stability studies.

5. Responsibilities, Data Evaluation, and Final Reports

The written protocol should not only contain information regarding the batches, packages, tests specifications, and testing schedule, but should also indicate the responsibility of the study coordinator and analytical testing laboratories involved, if not already defined in SOPs. The strategy for data tabulation and evaluation, including the statistical analysis to be used should be defined. The following responsibilities should be defined in the protocol if not documented elsewhere, such as in one or more SOPs.

- Who will provide the samples and what amounts are needed?
- How are the samples to be stored and under what environmental conditions?
- How many samples will be delivered to the testing laboratories and at what intervals?
- Who will approve/validate the test results?
- How will the data be reported?
- How will the data be evaluated?
- What is the projected expiration date?

When this information is complete, the protocol document should be reviewed and signed by the study coordinator and by the testing laboratory management, signifying agreement to initiate the study as designed. Given that many companies are migrating toward totally electronic or "paperless" systems, the signature may be electronic in nature.

C. Documentation

1. Protocols and Protocol Amendments

As has been discussed, the stability protocol is the document that describes the studies to be conducted in detail. The information contained in the protocol will be used to complete the stability report at the conclusion of the study. The information to be included in the protocol is

- purpose of the stability study and any pertinent background information;
- batch information (size, formula number, strength, manufacturer site, date of manufacture, dosage form type, drug substance supplier, lot number, etc.);
- package information (package material, component suppliers, and lot numbers);
- initial sample identifier where the initial, that is, "time zero data" can be found;
- tests and specification (acceptance criteria) to be used for evaluation;
- testing schedule including the intervals to be tested for each environmental condition;
- number of samples at each condition;
- data analysis methods, including the use of any computer programs, if applicable;
- specific reporting requirements, if appropriate;
- approvals of the study coordinator and the testing group.

At the time a study is initiated, the protocol defines the plan to be followed to ensure its successful completion. During the course of the study, it may be necessary to modify the plan to meet new requirements. An amendment to the stability protocol can be used to formally change the plan while the study is on-going. This amendment must be agreed to between the study coordinator and the testing group and approved accordingly through formal change control procedures covered in an SOP. The amendment must contain the protocol number and studies that are affected, the proposed change requested, and the rationale for the change. Each group involved must approve the amendment. An example is given in Figure 1.

Typical changes that would require an amendment are addition or deletion of a test or sample, rescheduling of a sample earlier or later, decision not to market a particular package or strength, remove a manufacturing site, or other change to the protocol that may affect the outcome or interpretation of the results. Minor changes allowed by SOPs may not require a protocol amendment.

2. Deviation Report

In the course of developing pharmaceutical products, there are situations where a conscious decision to not follow written procedures occurs. This may happen in several instances such as the manufacture of the batch, the analytical testing, or other possible reasons. Just as an approved stability protocol has an amendment mechanism to document changes to the plan, the use of a formal deviation report mechanism allows work to be continued and any deviations to be

Stability protocol amendment

Stability study number	
Stability protocol number	
Product	
Proposed change	
Reason for change	

Approved by:

Analytical chemistry Date

Print name

Study coordinator Date

Print name

FIGURE 1 Protocol amendment form.

documented along with a justification and the appropriate corrective action, if needed. The report should list the deviation, why it occurred, who agreed to it, any corrective action plan, and approvals of the initiator, management, and the quality assurance unit.

3. Out-of-Specification and Out-of-Trend Reports

Any time a sample is tested against a predetermined set of acceptance criteria, there is a chance that the sample may fail to meet one or more of its specification limits, a condition termed out of specification (OOS). This is especially true during stability studies conducted on products in development as limited experience with the long-term behavior of the product is available. There is also a chance that the test results will be out of trend (OOT) in comparison with previous data. The ICH Q1A[1] and WHO guidance[3] defines that if assay data is 5% different from the initial data, the results are a significant change and should be considered OOT. It is important to distinguish between results that are OOS or OOT due to instability versus those resulting from errors in sample handling or testing.

If OOS or OOT results are generated, a formal mechanism for investigating these results must be rigorously followed, as defined by the *Barr* decision.[14] This investigation procedure should be defined in SOPs. This procedure must have the ability to evaluate the data collection procedures and, in the case of an error, identify the root cause of the error. The results of the investigation must be documented and included as a formal record in the final report summarizing the stability study. Stability data cannot be dismissed without just cause. The SOPs also formalize the procedure to be followed if a retest of a sample is required. In this case, the following should be documented: the OOS or OOT results, a summary of the investigation together with a conclusion, disposition of the results, and any corrective action required including any effect on other batches/samples in the study matrix. This investigation is approved by the author, management, and quality assurance. This report would be maintained as a record in the stability study. An example of a representative OOS or OOT report format is given in Figure 2.

If the data support the OOS or OOT result as being representative of the sample, it should be clear that the data are consistent with other stability findings. The use of historical data stored in electronic data management system, mass balance considerations, or other changes in the sample should support these findings. This type of data evaluation is made easier by evaluating multiple studies in the stability program for a given product. For the stability report, the results of these studies and the interpretation of the results should be consistent with the stability of the product.

4. Test Results and Raw Data

Test results and the corresponding "raw" data must be maintained in an organized and legally defensible form. The use of notebooks (paper or electronic) or other "document controlled" form is required. Most modern analytical laboratories have some form of laboratory information management system (LIMS) or database system to systematically collect and store the results of the stability

Preliminary laboratory investigation report

Product:	**Tracking number:**
Sample ID #(s):	**Initiation date:**
Lot/batch #:	**Date of discovery of OOS/OOT or suspect result:**

Reason for investigation

Description of analytical results
Actual test result:

Specification:

Preliminary investigation review

- **Equipment/instrumentation review**
- **Reagents, standards, and sample review**
- **Raw data review**
- **Calculations review**
- **Condition of glassware**
- **Method review**
- **Laboratory notebook or testing template**
- **Other areas that may have been reviewed (describe)**

Summary of investigation

Conclusions and rationales

Final results and disposition of results

Corrective action (for confirmed lab errors only)

Documentation reference

Analyst/lab management:

Does material owner need to be informed? (circle one) **Yes / No**

Approvals

______________________/__________
Signature of analyst / Date

______________________ ___/________________
Signature of lab management / Completion date

Material owner:

______________________/__________

Signature of material owner / Date

Formal investigation required? (circle one) **Yes / No**

If "Yes", the material owner must obtain a formal investigation tracking number.

Formal investigation tracking number - ______________

Material owner's comment (if applicable):

FIGURE 2 Example of an OOS investigation report.

studies. The stability database system should be CFR Part 11 compliant, password protected with a revision control feature for data integrity and protection. These databases allow for uniform retention of data and a mechanism to assure that "required information" is captured and offers a tool for a uniform reporting format.

These database systems must have a mechanism to "lock" the data in the system to prevent inadvertent or unauthorized changes from being made. In the event that the data in the system would need to be revised, an authorization procedure should exist, and the database system must create an audit trail of the results, both original and revised. The audit trail is a means for retaining data that have been modified in a historical archive. These historical data are maintained along with a record of why the change was made and who made it. All of this information comprises the resulting audit trail.

5. Stability Report

The culmination of the stability plan design in the protocol and the data generated from the stability studies is the stability report. This may be a formal document in case of clinical and registration stability studies, or a similar presentation of the study plan and results in a regulatory submission. The report presents the results of the data evaluation and the conclusions of proposed expiration date in the various packaging configurations.

A stability report is written to document the retest period, shelf life or expiration dates, and a storage statement for the registration batches for the formal stability studies. It is also used to document the shelf life for formulations that are being used in the clinic. In this case, the expiration date is generally a date when the product usage should be discontinued or that additional data are required to extend the retest period. A more detailed discussion of the information included in the stability report is given in Section II.E.

D. Scheduling

Once a stability study is initiated, the schedule of sample testing is set for a period of about 2–5 years. This presents a significant challenge to the testing laboratory to manage and track the work schedule.

1. Planning and Workload

The tracking of samples and the scheduling of their test dates is easily handled with a modern computer system scheduler. With the computerized scheduling system in place, the workload can be organized on a daily, weekly, or monthly basis. Each sample should be identified with the methods to be used and a key factor for planning is to be able to look ahead for the required period of time to identify the resources needed so that adjustments can be implemented to cover the scheduled workload.

A few alternatives to scheduling the work are possible. All samples can be pulled from the chambers based on their scheduled dates and then retained in the laboratories for testing. The SOP would indicate the maximum time allowed between the time the sample was removed from the chamber and the initiation of testing. Alternatively, based on the schedule of samples, the laboratory can choose when a sample should be tested based on limitations given in the SOP. In the authors' laboratory, all samples are taken from stability chambers on a weekly basis and retained in the testing laboratory. In doing so, laboratory management is aware of the schedule of samples and makes certain that samples are tested within desired time frame. The use of electronic data management systems and contract testing laboratories has aided in the administering of the stability system to handle high- and low-demand periods. The data management system can forecast "peaks and troughs" in the study samples to alert management about the necessary resources needed for the workload. It can also analyze new studies being planned to allow decisions to plan efficient testing of the same products already in the program and to stagger the start of large studies to avoid a future backlog for the testing laboratories.

2. Turnaround Time

The time to complete the stability tests relative to the schedule date depends on the workload of the laboratory. As previously discussed, the more control the laboratory is given for timing the receipt of the samples for testing in the laboratory, the greater the likelihood that the results will be available in a reasonable period of time. Many laboratories have a "sliding scale" where early time points are tested immediately and as the time points are progressively longer, the window for testing is extended. Currently, in the authors' laboratory, the samples are taken on a weekly basis. The sample analysis begins within a week for monthly samples, and within a month for quarterly samples. The maximum time between the scheduled date and the actual assay date is 1 month or less. The actual date of the analysis is recorded with the stability data in the LIMS system.

E. Stability Reports

The primary objective for a stability program is to determine the shelf life or retest period of a drug product or API in its respective package(s). The secondary objective is to determine the proposed storage statement recommended for the product to continue to meet its requirements of quality, purity, potency, and safety in the marketplace. The label storage statement must include storage precautions, if appropriate, for the pharmacists and the patients.

In each case, the report includes a summary of the design given in the protocol, the tests and procedures used in the studies along with the appropriate limits for evaluation, the stability data tables with the necessary batch documentation, and finally the discussion of results and conclusions.

1. Background Information

The stability data tables generated for the stability studies are only a part of the documentation required for a stability program which typically exist as appendices in the final report. The report should begin by describing the objective(s) of the study, a discussion of the batches including their formulation and site of manufacture, the package description and site of packaging, the test methods and the stability criteria, and the testing schedule including any future planned test intervals. This information defines the manufacturing and packaging processes that the data represent. The stability studies would be incomplete without this information. A summary of this information (product strength, batch, formula number, drug substance supplier, package, date of manufacture, manufacturing site, and date on stability) should be included in the stability data tables reported below.

2. Data Tables

The stability data table is the means for reporting the results of the stability study in a concise format for ease of review and interpretation. An example of one representation of a data table format is given in Figure 3. The table consists of a summary of the important batch and manufacturing information as well as the test data. The example given has the data grouped first by storage condition and then by time interval to present the stability as a function of time for each environmental condition studied. With the data in columns across the table, any apparent connection between multiple stability results for a sample can be easily made and interpreted (e.g., a decrease in assay and an increase in degradation products). Each study should be included in a separate table and, in some instances; multiple tables (one table prestorage condition) may be used. In addition, a graphical presentation of stability data versus time for the test data can be used to illustrate trends in the data and may be helpful for data evaluation. A graphical presentation of the data; however, cannot replace the tabular presentation for a regulatory filing.

3. Evaluation and Conclusions

The data from each test should be evaluated and discussed. Any indication of instability should be noted with a discussion regarding the projected cause of the change and the effect on the product's performance. Correlation to previous

Compound
Manufacture lot
Date of manufacture
Manufacturer site
Package information
Initial date placed in chambers

Storage condition	Interval	Appearance	Assay	Impurity RRT1.25	Impurity RRT1.38	Total organic impurities	Moisture
	Initial	White to off-white powder	99.4	0.10	0.41	0.51	0.04
5 °C	3 months	White to off-white powder	99.7	0.09	0.48	0.57	0.04
	6 months	White to off-white powder	99.3	0.09	0.43	0.52	0.05
25 °C/60% RH	3 months	White to off-white powder	99.5	0.09	0.48	0.57	0.04
	6 months	White to off-white powder	99.3	0.09	0.44	0.53	0.06
40 °C/75% RH	3 months	White to off-white powder	99.9	0.09	0.43	0.52	0.04
	6 months	White to off-white powder	99.8	0.09	0.44	0.53	0.06

FIGURE 3 Data table.

chemical stability data (e.g., degradation in the drug product related to the API) should be noted to differentiate between the inherent stability of the drug substance versus "formulation-related" stability behavior.[1] The results of the statistical analysis, where appropriate[1] and the analysis of impurities should also be discussed. ICH Q3A and Q3B provide a discussion on these matters in detail.[15,16] The effect of the stability data on the retest period, shelf life, and expiration date and recommendations for the storage statement should also be presented and be consistent with stability guidance documents relevant to the appropriate regulatory agency.[1–5]

F. Audits and Inspections

Audits of the stability program by a quality unit within the organization should be conducted routinely and should include verification of the data collected for the studies presented in reports.[2] In addition, regulatory agencies will conduct inspections of programs and facilities prior to the approval of a specific product (preapproval inspection or PAI) or as a routine inspection of a facility (general GMP inspection).

1. Internal Audits of Data and Program

The stability program should be audited by the company's quality unit on a periodic basis. In the authors' laboratory, stability data are audited at least once per study by the quality assurance unit and by testing laboratory at the time the report is written or revised. The equipment, processes, and procedures that support the stability program are a part of this audit to verify that appropriate systems exist and that they are being followed. The SOPs are reviewed, the activities and facilities are observed, the data are audited, and an audit report is issued. The use of electronic notebooks for data capture into reports minimizes the need for data transcription review. At the time a stability report is written, the data in the report are audited against the raw data obtained from the laboratories and an audit report is issued. These audit reports are used to provide an objective view of the stability program as well as the data against the current regulations to assist management in the evaluation and possible improvement of the program. This audit further increases the probability of a successful outcome during an inspection from a regulatory agency, such as a PAI.

2. Inspections (Preapproval and Facility Inspections)

PAIs are inspections related to products that under regulatory review. These are scheduled by the regulatory agency during the review process to verify the accuracy of the data, the completeness of the documentation, and the acceptability of the facilities and procedures used to collect the data. Agencies such as FDA and the Medicines Control Agency have programs for initiating these inspections. The stability data, as an important part of a regulatory submission of a product, are often an area of focus during inspections.

Specifically in a PAI, a subset of the stability data is often reviewed by the inspectors to verify the results, to monitor the manufacturing records for the batch, and to review the methods used to collect the analytical test data and the systems used to control manufacturing and testing.

General inspections of facilities are also conducted on a routine basis by regulatory agencies, approximately every 2 years. These inspections are performed to determine general adherence to GMPs and to identify any product problems being experienced by the facility being inspected. Included in the inspection are checks to determine if appropriate systems are in place to effectively produce and control the manufacture of the products. Often a check of the stability program and the product data are a part of such an inspection. Therefore, it is critical for the stability program to have a set of SOPs to follow that adequately control the entire stability process.

III. EXCIPIENTS

A. Formulation Development

Excipients are inert pharmaceutical ingredients that are used in product formulations. Excipients may perform a variety of functional roles in the pharmaceutical product. In the vast majority of cases, excipients have limited (if any) pharmacological activity, unlike the API. Because of this difference in the expected biological activity and the risk/benefit relationship between excipients and active drug substances, the approaches for the safety assessment of excipients and API will differ.[17]

Each excipient serves a specific purpose (e.g., binder, disintegrant, or pH adjustment) for the proper performance of the dosage form. The properties of the final dosage form (i.e., stability) are, for the most part, highly dependent on the excipients chosen, their concentrations and interaction with both the active compound and each other.[18] An experimentally designed study to determine the excipient compatibility with the API should be done prior to phase III to systematically identify which excipients are compatible. These studies should evaluate the excipients planned to be used in the formulation including any colorant, flavors, or preservative required for the formulation. Chapter 9 includes details of excipient compatibility studies that should be done to screen formulations of a new drug candidate.

Most of the excipients used in product formulations are compendial items (United States Pharmacopeia (USP), European Pharmacopeia (EP), Japanese Pharmacopeia (JP), etc.), and therefore, are widely used. Many of the excipients are also used in the food industry. Therefore, there is unusually no requirement to perform stability tests for them as individual excipients.

B. Stability of New Excipients

For noncompendial excipients, there may be a need to generate stability data even though there is no formal stability requirement. New excipients need to be handled on a case-by-case basis. An excipient that is a physical modification of a compendial item but does not meet all of the compendial tests is a case for which stability data would not be needed. An example could be a component that has been physically modified (e.g., milled), but not chemically changed, or an excipient that is a mixture of compendial items to form a "new" excipient. These typically should not require generation of a stability database.

For a totally new chemical that has never been used in a drug product to be used as an excipient, stability data to support its use are highly recommended. In this case, a stability program similar to that for an API could be designed and a stability profile generated. The use of excipients is ultimately the responsibility of the drug product manufacturer; however, sufficient characterization of the excipients, which may include stability data, are typically generated by the manufacturer of the excipient and documented in their Drug Master File (DMF). The drug product manufacturer is required to test and release each excipient used in the drug product and/or set up an on-going certification program to certify that the supplier continues to provide excipients that meet the requirements for the drug product. A totally new chemical excipient may also require a toxicology study to support its use in a drug product. The USP 32/NF 27[5] contains a guideline, USP <1074> that addresses biological safety for new pharmaceutical excipients. A general discussion on the GMPs for manufacturing excipients is detailed in a WHO publication.[2]

C. Probe Stability

It is important to have a good understanding of the excipients used in a formulation to assure proper performance and facilitate the interpretation of stability data generated on the product. The following section will give a few selection examples of stability data of drug products that contain them. For these reasons, a probe stability study may be useful to obtain data to distinguish whether a stability observation for the drug product is due to the excipient or the drug substance. A probe stability study is a short-term, accelerated study to gain insight for evaluating later stability studies. In these cases, a probe stability study would be helpful unless similar information can be obtained from the literature, the excipient supplier, or prior knowledge.

D. Examples of Excipient Stability Information

The following four examples are presented to illustrate the importance of understanding the properties of excipients and their stability. These examples have been encountered in the author's laboratories. They represent likely concerns in many company's future stability studies where these common excipients are used in their formulations. The information can be important to the quality of the pharmaceutical products in which they are used. This information aids in the selection of the right excipient for the product and helps in evaluation of the stability data for the final product.

The parabens are a class of 4-hydroxybenzoic acid esters (primarily methyl, ethyl, propyl, and butyl esters) that are used as biological preservatives in liquid and semisolid dosage forms.[18] Another preservative often used is benzyl alcohol. Parabens and benzyl alcohol are compendial items. These materials are very stable compounds, but in the presence of water the paraben esters will hydrolyze to form 4-hydroxybenzoic acid and the corresponding alcohol. Likewise, benzyl alcohol can easily oxidize in a formulation to form benzaldehyde and benzoic acid. These reactions occur slowly for the unformulated material but can proceed more quickly in a formulation. It is important to know that these reactions do occur in the drug product and that we can distinguish the excipient degradation product from the active ingredient and it degradation products. Knowing that the hydrolysis of the parabens is catalyzed in acid can also help the formulation group select an appropriate pH for the product. Hence, studies to determine the compatibility and reactivity of excipients with the API are critical to the success of the resulting formulation.

Long-chain unsaturated oils such as fatty acid esters are used in a number of semisolid and liquid products. It is known from the use of these oils that the double bonds in their chemical structure can react with oxygen to form peroxides and/or hydroperoxides. Some compendial monographs for these materials will have specifications for peroxide level (i.e., USP/NF, Peroxide Level <401>) for this reason.[5] For APIs that undergo oxidation, these excipients may need to be avoided to minimize the catalytic oxidation of the API and maximize the stability of the drug product.

It has been observed that FD&C Blue #2 (indigo carmine) dye can oxidize which causes its color to fade.[19] In a formulation it can react to form a green to yellow product (mixture of blue and yellow). Therefore, a light blue tablet can change from its blue color to a greenish blue tablet or a green tablet can appear yellowish because of the oxidation of the FD&C Blue #2. However, at higher concentrations of blue dye, this change may not be evident because of greater amounts of the original blue color. Therefore, it can be important to select the proper dye or concentration to minimize a possible change in color.

A special consideration for a formulation and its stability is the effect of water. Carstensen[20] provides a good discussion of the effect of water on solid dosage forms. The effect of water can be a problem in a formulation when the API is moisture sensitive. Shalaev discusses how residual water can affect the solid-state degradation of drugs.[21] When water affects the formulation and causes instability in the product, water levels should be controlled or an excipient that minimizes the association of water should be chosen. A specific example is the difference between tightly bound water (e.g., water is held in the crystal lattice) and water that is adsorbed to an excipient by "weak" hydrogen bonds. For an API that degrades by hydrolysis or is catalyzed by water, an excipient that has loosely bonded moisture will promote degradation while water of hydration in the crystal structure or an anhydrous excipient will not facilitate degradation. Excipients containing water held by hydrogen bonds tend to absorb moisture easily, and this water is usually more readily available for hydrolysis. An example of a crystal hydrate is the hydrated form of lactose, whereas microcrystalline cellulose tends to have loosely adsorbed moisture.

E. Conclusions

Stability studies are generally not required to support the use of excipients for product development. These materials generally follow compendial standards and are suitable for their intended use. Since the excipients are chosen to perform an important function in the drug product, it is important to understand the chemistry of the excipients, the effect that the excipient has on bioavailability and the stability of the active ingredient, so that the proper excipient can be used in the product.

IV. ACTIVE PHARMACEUTICAL INGREDIENT

Information on the stability of the API is fundamental to the entire drug development process. Although there are examples of the instability of a drug substance being improved by proper formulating and packaging,[22] the inherent stability of the API will ultimately influence the design and expiration date of the commercial product. As a result, regulatory agencies expect stability data on the pure drug substance to be part of any product dossier.

A. Developmental Stability Testing

The foundation of any viable stability program for a drug product begins with a thorough understanding of the chemistry of its active ingredient(s). Information on the mechanism(s) of degradation is essential to understanding how the API will behave under various environmental changes and conditions. This information is usually determined by conducting stress degradation studies under varying conditions of acidity, basicity, light, heat, and oxidation. Other conditions may be chosen depending upon the particular drug substance being studies. From these stress studies, knowledge of the inherent stability and mechanisms of degradation can be obtained. The propensity of the API to preferentially degrade under specific conditions will be the area to which particular attention will have to be paid in the storage and formulation of the drug product, as well as the choice of excipients to formulate the product. If the drug substance is prone to hydrolysis, the effect of humidity on storage will necessarily be a focal point in the stability protocol. The choice of excipient can affect the stability of the API in the formulated product. The excipient compatibility studies will be discussed in Section V. The reader is referred to Chapter 3 devoted to an in-depth discussion of degradation studies.

Whereas the stress conditions will usually be more severe than the API will experience under its normal storage and processing conditions, the distribution and kinetics of formation of degradation products that are obtained in these studies can be extrapolated to predict the exact degradation pathway(s) that would be expected at each environmental condition. Identification of the degradation products at this stage is invaluable in establishing the degradation pathway. Although degradation pathways under stress conditions are usually much more complex than those at room temperature, the mass distribution of the degradation products will be an indicator of which degradation products will be seen on long-term storage. The ability to separate all of the degradation products will also be important for developing and validating the stability-indicating properties of any proposed analytical methods. With an understanding of the degradation chemistry, specification can be established that will ensure that the drug substance will retain its identity, quality, purity, and potency over its expected shelf life when properly stored, which is typically up to 5 years at the labeled storage condition.

The ICH guideline on "Stability Testing of New Drug Substance and Products"[1] suggests that for stress testing with regard to temperature, 10 °C increments at accelerated temperature (50, 60 °C, etc.) should be studied. If appropriate, the API should also be exposed to humidity conditions above 75% RH. The effect of pH on the susceptibility of the molecule to hydrolysis should be studied across a wide range, in solution or suspension. Photostability stress testing should be performed in accordance with the conditions specified in ICH Q1B.[12] Oxidative stress should always be included in any assessment of degradation pathways.

The regulatory agencies recognize that the synthetic scheme, specifications, testing regimen, and methods will evolve as the drug development process continues through the various clinical phases. As the database of knowledge regarding the API increases, stability data requirements also evolve. In phase I clinical studies, stability data to support the toxicological assessment and proposed clinical plan must be collected. Adequate stability data should be gathered to demonstrate that the API remains intact during the course of the toxicological and phase I clinical trials. Depending upon the stability of the API, impurities that increase during storage may be qualified by reference to these early human and animal studies. In phase II, the stability studies should support the investigational formulations used in the clinical trials to aid in selection of the final formulation, container, and closure system.

The information obtained from the stress studies and the phases I and II studies should be reflected in the protocol(s) or reported for the primary stability studies to support the final phase III clinical trials.

B. Batch Selection

Since API stability data are needed to support the three clinical phases, the selection of batches at each phase reflects the synthetic scheme and batch size used at that stage of development. In phase I studies, the batches used in the toxicological and clinical trials should be monitored. As the synthetic scheme and batch size evolve, batches made by the new process should be selected for stability testing and monitored for the duration of the phase II clinical trials. Knowledge about the API gathered in phases I and II should form the basis for the stability protocol for the API manufactured by the final synthetic process that will be used to support the phase III clinical trials and the primary registration stability studies.

Stability data from a minimum of three batches of API exposed to both accelerated and long-term storage conditions are required for filing by most regulatory agencies. Long-term data should include information from a minimum of 12 months of storage. These batches should, at the very least, be manufactured by the same synthetic route and simulate the final process to be used on a commercial scale. The quality of the batches tested for stability should be of a purity representative of the batches used in preclinical and clinical studies or better. The batches should also be packaged in containers with closure systems that are representative of the packaging to be used on the commercial product. Data from "laboratory scale" or "glass plant" batches may be submitted to supplement the data from the registrational studies.

The first three production batches of API slated for conversion to marketed drug product, sometimes termed "validation batches," should be monitored on long-term stability using a similar protocol as that submitted in the approved drug application.

C. Specifications and Test Methods

The properties of the drug substance that influence quality and safety/efficacy should be the primary focus of the stability study. Prior to the initiation of primary stability studies it is not unusual for the specifications and test methods to evolve in like fashion as the synthetic route, manufacturing scale, and impurity profile. By the onset of primary registration stability studies, both specifications and test methods should be finalized and "locked" so that as testing occurs over time monitors the key attributes of the API for significant changes from their established limits. The specifications should include, where applicable, physical, chemical biological, and microbiological properties. The established limits should be numerical values, ranges, or other acceptance criteria for individual tests. The test methods used to measure the various properties should be validated according to ICH guideline Q2(R1)[23] to show that they accurately and precisely detect and measure any changes that might occur over the duration of the stability study.

In the ICH guideline on "Chemical Substances" (Q6A), API tests are classified into two categories, universal tests and specific tests.[11] In the universal test category are tests for appearance, identification, assay, and impurities. These tests are generally applicable to all new molecular entities. The appearance test should include sufficient detail and instructions to allow the analyst to make a qualitative statement about the physical state and color of the drug substance. The identification test should be able to distinguish between closely related compounds. Normally, infrared (IR) spectroscopy is used. Use of a single chromatographic system is not generally acceptable; the use of two chromatographic systems using different separation mechanisms, such as thin-layer chromatography (TLC) and high-performance liquid chromatography (HPLC), or a combination of a chromatographic system and another technique, such as HPLC/mass spectrometry is generally acceptable. If the new molecule is a salt, an identity test for the specific counter-ion is required. A specific, stability-indicating method is necessary for determining the assay content of the API. Although HPLC is the most common technique used for assay, a combination of a nonspecific assay, for example, a nonaqueous titration, and a method for impurities, such as another chromatographic procedure specific for the degradation products, may also be justifiable. Impurity tests for inorganic, organic, and residual solvents usually employ chromatographic techniques such as TLC, HPLC, and gas-liquid chromatography. Generally, the tests for content and for organic impurities are combined in one chromatographic test. Although process analytical technology (PAT) is gaining favor as a substitute for release testing of API (as well as drug product), a slate of methods that effectively probe the stability of the API are needed.

In the specific test category, there are tests for physicochemical properties, particle size, polymorphic form, chirality, water content, inorganic impurities, and microbial limits. These tests should be used when the specific attribute can have an impact on the quality of the drug substance. Physicochemical tests include pH, melting point/range, and refractive index. Particle size testing is an important test for solid and suspension-type formulations. The effect of particle size on dissolution, bioavailability, and stability cannot be overemphasized. Methods to determine particle size include the Coulter method, laser-light scattering, microscopy with image analysis, and time-domain analysis.[24,25] Many APIs can exist in different polymorphic forms that have vastly different properties. Polymorphic forms may include hydrates, solvates, crystalline, and amorphous forms. The differences in the properties of the various forms can also have a significant impact on the quality of the finished product, its bioavailability, and ultimately its stability. As a result, the correct solid-state form must be specified. Common techniques to differentiate between forms include solid-state IR, nuclear magnetic resonance (NMR), melting point (including hot-stage microscopy), X-ray powder diffraction, thermal methods (differential scanning calorimetry, thermogravimetric analysis, and differential thermoanalysis), optical microscopy, and Raman spectrophotometry. The reader is referred to Chapter 2, which is devoted to the various techniques for the study of solid-state characteristics of APIs.

If the API is hygroscopic, susceptible to hydrolytic degradation, or exists as a stoichiometric hydrate, testing for water content is required. Common methods for measuring water content are loss on drying (LOD) and Karl Fischer (KF) titration, although GC with thermal conductivity detection (TCD) has also been employed. KF titration is the generally preferred for water analysis; however, it can be subject to interference from certain types of compounds.[26,27] For the initial stability time point, if tests for inorganic impurities are required, (as from a catalyst), compendial tests for heavy metals or sulfated ash may be used or other techniques, such as atomic absorption spectroscopy, may be required for the analysis.

For APIs being developed as a single enantiomer, control of the other enantiomer is required on stability if the desired enantiomer is prone to conversion to the undesired form. A suitable method should be used to test the API. The combination of an achiral method and an appropriate method for controlling enantiometric purity is also acceptable. Chapter 1 details the various techniques and approaches to the determination of chirality. An achiral method may also be acceptable when the drug chemistry degradation studies have shown that recemization is not a significant degradation pathway.

Tests for microbial limits may also have to be considered depending on the nature of the API, its method of manufacture, and its intended use. Sterility testing may be appropriate for APIs that will be used in parenteral or oral solution products that are sterile. Endotoxin testing may be needed for APIs intended for injectable products. Testing for the total count of aerobic microorganisms, yeasts, and molds and the absence of specific objectionable bacteria, such as *Staphylococcus aureus*, *Escherichia coli*, *Salmonella*, and *Pseudomonas aeruginosa*, may also be required. The tests described in the various compendia are generally used.

For detailed information on establishing specifications and test methods, interested readers are referred to the ICH guidelines on specification (ICH Q6A[11]) and on method validation (Q2 (R1)[23]).

D. Storage Conditions

As a general principle, an API should be investigated for stability under storage conditions that reflect the temperature, humidity, and length of storage in the area of the world where the drug product will be manufactured, stored, shipped, and marketed. The long-term storage conditions will generally be reflected in the labeling.

Currently, the world is divided into four climatic zones based on the mean kinetic temperature. The definition of room temperature (i.e., long-term testing conditions) in the four zones is given in Table 5, as is the RH. The countries that fall into these zones can be found in the WHO guidance.[3]

A minimum of 12 months of real-time stability data should be available at the time of submission. The studies should continue through the proposed expiration date. Additional data collected while the submission is under review may be submitted with the prior agreement of the regulatory agency. Data from accelerated storage conditions (Tables 6–8) should also be part of the submission to support the long-term data and to simulate any short-term temperature and humidity excursions that the API may experience during shipment or storage that go beyond the label conditions.

Other conditions may be used; however, they must be justified. Agreement with the regulatory agencies prior to the initiation of studies at other conditions is highly recommended.

For an API, a "significant failure" is defined by the ICH[1] and WHO[3] as a failure of the test article to meet preapproved specification(s). This is the reason that specifications are established prior to the initiation of registration stability studies so that the criteria for comparison are clearly identified. If a significant failure occurs in the first 6 months at accelerated conditions (typically 40 °C and 75% RH),

TABLE 5 International Climate Zones

Climatic zone	Definition	Long-term testing conditions
I	Temperate climate United Kingdom Northern Europe Russia	21 °C/45% RH
II	Subtropical and Mediterranean climate United States Japan Southern Europe	25 °C/60% RH
III	Hot and dry climate Iraq	30 °C/35% RH
IVa	Hot and humid climate Iran Egypt	30 °C/65% RH
IVb	Hot and very humid climate Brazil Singapore	30 °C/75% RH

TABLE 6 General Drug Substance Storage Conditions

Study	Storage condition	Minimum submission interval
Long term (I, II)	25 °C/60% RH	12 months
Long term (III)	30 °C/35% RH	12 months
Long term (IVa), or intermediate condition[a]	30 °C/65% RH	12 months as long-term, or 6 months as intermediate condition
Long term (IVb), or intermediate condition[a]	30 °C/75% RH	12 months as long-term, or 6 months as intermediate condition
Accelerated	40 °C/75% RH	6 months

[a]Whether long-term stability studies are performed at 25 ± 2 °C/60 ± 5% RH or 30 ± 2 °C/65 ± 5% RH or 30 ± 2 °C/75 ± 5% RH is determined by the climate zone in which the product is intended to be marketed. Testing at more severe long-term condition can be an alternative to storage at 25 ± 2 °C/60 ± 5% RH or 30 ± 2 °C/65 ± 5% RH.If 30 ± 2 °C/65 ± 5% RH or 30 ± 2 °C/75 ± 5% RH is the long-term conditions, there is no intermediate condition.If 25 ± 2 °C/60 ± 5% RH is the long-term storage condition, the intermediate condition is tested at 9 and 12 months when the 40 °C/75% RH is failure to meet specification or assay is different from initial by greater than 5%.

TABLE 7 Drug Substance Requiring Refrigeration

Study	Storage condition	Minimum submission interval (months)
Long term	5 °C	12
Accelerated	25 °C/60% RH, or 30 °C/75% RH	6

TABLE 8 Drug Substance Requiring Storage in a Freezer

Study	Storage condition	Minimum submission interval (months)
Long term	−20 °C	12

at least 6 months of data at the intermediate storage condition from a 12-month study must be included in the original submission.

If a significant failure occurs within the first 3 months at 40 °C/75% RH, the proposed retest date should be based on real-time long-term data. In addition, if a failure does occur within the first 3 months, data will have to be presented to allow use of the API if any excursions occur outside of the proposed labeling.

For APIs requiring freezer storage, the retest date is based on the real-time long-term stability data. To cover excursions beyond the labeling, short-term testing on a single batch of API at 5 °C or 25 °C/60% RH is suggested. APIs requiring storage below −20 °C should be considered on a case-by-case basis.

E. Testing Frequency

Testing should be conducted often enough to define the stability characteristics of the API relative to the proposed specifications. The ICH[1] and WHO[3] recommendations for climatic zones I–IV are as follows:

- For a 12-month retest date, the samples stored at the proposed long-term condition should be tested every 3 months for the first year, every 6 months for the second year and then annually.
- For accelerated storage conditions, a minimum of three test points including the initial and end points (e.g., 0, 3, 6 months) should be obtained.
- For intermediate storage conditions, a minimum of four test points including the initial and end points (e.g., 0, 6, 9, and 12 months) should be obtained. There is no intermediate condition for climatic zones III and IV.

F. Stability Commitment

The magnitude of the stability database included in the submission influences the stability commitment that must be made postapproval.

- If the submission includes long-term storage data on three production batches through the proposed retest period, no postapproval commitment is needed. Even though no further commitment is required, it is recommended that one production-sized batch be selected for stability testing annually to monitor stability of future batches over time.
- If the submission does not include long-term storage data on the registration batches though the proposed retest date, the studies should be continued through the postapproval phase to establish an appropriate retest period.

In the latter case, several options are available:

- If the submission includes stability data on at least three commercial batches, a commitment to continue testing through the proposed retest period must be made.
- If the submission includes stability data on less than three commercial batches, a commitment to continue testing through the proposed retest period must be made and additional production batches to total at least three must be selected for stability testing and tested though the proposed retest period.
- If the submission does not include stability data on any production batches, a commitment to select the first three production batches for long-term stability testing and to continue testing through the proposed retest period must be made.

The same stability protocol used for the registration batches must be used for the postapproval stability commitment unless scientifically justified otherwise. In this case prior agreement with the regulatory agencies is recommended.

G. Stability Studies for API Starting Materials and Process Intermediates

API intermediates, which include starting materials and isolated reaction intermediates, may be held for certain period of time from the date of manufacturing without retest before use. In the clinical phase when only small batches of API are being manufactured, stability is not typically conducted for synthetic starting materials and intermediates. The retest period of the starting materials and intermediates is based on the scientific evaluation and information that may be available from vendors. In some instances, warehouse samples are tested in support of the retest period for intermediates and starting materials. During commercial scale up, stability studies on isolated intermediates used to manufacture the primary registration batches are conducted. The frequency of stability testing for an intermediate is related to the length of the holding time for the material and follows the ICH[1] and WHO[3] guidelines on testing of finished API. Where practical, testing should be done at a minimum of three time points after the initial testing. All criteria parameters should be evaluated at release of an intermediate and immediately before its use in the manufacture of the finished drug substance. As many starting materials may be expensive and/or have long lead times, if the stability information is not available from the supplier, many companies conduct long-term studies on such materials for up to 5 years.

H. Data Evaluation

The ultimate goal of the stability study is to establish a suitable retest period that can be applied to all future API batches manufactured in a similar manner to that of the proposed process. The retest period will be established using a minimum of three batches by assessing the stability information collected on the API's physical, chemical, biological, and microbiological properties as a function of time, temperature, humidity, and light. The degree of certainty that the stability of a specific future lot of API will remain within specifications until the next retest period is directly influenced by the variability of the materials studied on stability.

Studies that show virtually no degradation or variability will usually not require any formal statistical analysis. Under these circumstances the requested retest period is normally granted when a complete justification for not needing a statistical analysis is part of the submission. The analysis should include an examination of all key attributes of the API.

The evaluation of the data should begin with an examination of the individual data points that have been collected over the course of the stability study. In a normal data set, some variability is to be expected. The variability can arise from variability in the product, the method, and the analysts performing the testing. The specifications for API assay generally ranges from 98.0% to 102.0% on an anhydrous basis. If the result must be calculated on the anhydrous basis, the variability in the water content will also influence the results. Thus, over the course of the stability study the individual data points will most likely increase and decrease within the specification range. One question to consider is whether the variability in the data can be explained by the expected variability of the determination. The data points for individual impurities and total impurities should be examined as the study continues to ascertain if the specification limits are being approached and if a failure is imminent. A comparison to the impurity levels in the preclinical studies and clinical trials should be made to be certain that the levels of impurities being seen in the stability study have been qualified in animal studies or clinical trials. Any unknown peaks in the chromatograms will need to be investigated, discussed in the submission, and depending on their level, isolated, identified, and qualified. For APIs that coexist in several polymorphic states, the data must be carefully evaluated to see if a change in crystal form is occurring on stability. This change may be very subtle, and a trained spectroscopist armed with suitable tools may have to analyze the data to detect such a change. Upon examining, one must ask if all of the individual data points remain within specification or are some OOS. If some are OOS, they will have to be investigated and documented. When an OOS result is obtained, the data should be investigated immediately for an assignable cause. A delay in investigating may prevent the determining of the correct cause. If the cause can be ascertained quickly, the same sample may then be retested and the correct result reported. It is strongly recommended that the data be evaluated as they are being collected and not at time of study completion or at the time of the submission. Waiting until the end may make it impossible to resolve or explain anomalous data points.

According to ICH Q1E,[28] an appropriate approach to retest period estimation is to analyze a quantitative attribute (e.g., assay, degradation products) by determining the earliest time at which the 95% confidence limit for the mean intersects the proposed acceptance criterion. For an attribute known to decrease with time, the lower one-sided 95% confidence limit should be compared to the acceptance criterion. For an attribute known to increase with time, the upper one-sided 95% confidence limit should be compared to the acceptance criterion. For an attribute that can either increase or decrease, or whose direction of change is not known, two-sided 95% confidence limits should be calculated and compared to the upper and lower acceptance criteria. If the data show that the batch-to-batch variability is small, it may be worthwhile to combine the data into the overall estimate. The appropriate statistical modeling is used to analyze the data.[29]

Extrapolation of the real-time data from the long-term stability study to support a retest period longer than that of the real-time data can be proposed. The proposal should be supported by accelerated data, knowledge of the degradation mechanism, additional supportive stability data from batches other than the primary registration batches that might be available, and the goodness of fit of the mathematical model used in the statistical analysis.

It is strongly recommended that the API sponsor and reviewers confer and agree on the adequacy of the proposed stability protocol before implementation so that both parties know what is expected and what will be submitted, thus avoiding any unnecessary delays in dossier approval.

V. DRUG PRODUCT

One key attribute of a pharmaceutical drug product is its stability over time. In the pharmaceutical industry, manufacturers are expected to conduct appropriate studies of the product in the container and closure system in which it will be marketed, sold, and dispensed. The objective of the stability studies is to provide evidence on how the quality of the formulated product varies with time under the influence of a variety of environmental factors such as temperature, humidity, and light. These studies enable recommended storage condition, retest period, and shelf life to be established. The ICH and WHO developed several guidelines covering regulatory requirements concerning efficacy, quality, and safety aspects of the finished drug product.[1–3] These guidelines are intended to be comprehensive documents that provide information on all aspects of stability data generation and use. It provides recommendations regarding the design, conduct, and use of stability studies and the resulting data that should be performed to support investigational new drug applications, new drug applications, abbreviated new drug applications, new dosage forms supplements and annual reports, and license applications for biologics, and Common Technical Documents (CTD). The following ICH guidelines set out the stability requirements for a registration application within the EU, Japan, and the United States: (1) Stability Testing of New Drugs and Products [Q1A(R2)][1]; (2) Stability Testing: Photostability Testing of New Drug Substances and Products [Q1B][12]; (3) Stability Testing for New Dosage Forms [Q1C][30]; and (4) Bracketing and Matrixing Designs for Stability Testing of New Drug Substance and Products [Q1D].[38] The ICH stability guideline Q1A (R2) was revised in February 2003. The changes to ICH Q1A (R2) were as follows. (1) The intermediate storage condition was changed from $30\pm2\,^{\circ}$C/60% RH ±5% RH to $30\pm2\,^{\circ}$C/65% RH ±5% RH for API, drug product, and drug products packaged in semipermeable container. (2) The intermediate storage condition $30\pm2\,^{\circ}$C/65% RH ±5% RH was added as a suitable alternative long-term storage condition to $25\pm2\,^{\circ}$C/60% RH ±5% RH for both drug product and API. (3) The storage condition $30\pm2\,^{\circ}$C/35% RH ±5% RH was added as a suitable long-term storage condition to $25\pm2\,^{\circ}$C/60% RH ±5% RH for drug products packaged in semipermeable containers.

The WHO[31–33] has established several separate parent guidelines on stability testing that covers the countries around the world. The WHO guidelines are aligned with the ICH stability guidelines for countries that fall within the climatic zones I and II. The WHO guidelines establish guidelines for the zones III, IVa, and IVb countries around the world.

A. Stability Testing During Drug Development—Phases I–III

Stability data are required in all phases of drug development to demonstrate that the formulated products remain within acceptable chemical and physical limits for the planned duration of the proposed clinical investigation. Sufficient information should be available to ensure that the proper identification, quality, purity, and potency of the investigational formulation products is constant for the duration of the clinical study. The amount of information needed to achieve such assurance will vary with the phase of the investigation. If a very short-term clinical trial is proposed, the supporting stability data can be correspondingly very limited. All stability data should be generated under cGMP conditions.

1. Phase I Studies

Studies to support the stability of the formulated product during the toxicological investigations and the very early clinical trials are conducted in the proposed container/closure system at the recommended storage conditions. In phase I, the testing frequency, storage conditions, and storage periods indicated in the ICH guidelines need not be followed. In most cases, the dosage forms used in such early trials are not representative of the final presentation and could be drug in bottle or drug in capsule formulations.

2. Phase II Studies

In similar fashion to that done in phase I, information on the stability of the finished product is collected in the proposed container/closure system to support phase II clinical studies. At this stage of development, protocol-driven stability studies are initiated and the conditions indicated in the ICH guidelines[1] are recommended. The testing frequency is every 3 months for the first year and every 6 months in the second year. The storage conditions recommended in the ICH guideline such as room temperature (25 °C/60% RH) and heat/humidity (40 °C/75% RH) are followed. In phase II, the shelf life of the formulated product is estimated and provided in support of the clinical investigation. Shelf life is based on the on-going phase II stability studies, the phase I stability studies, and the data generated for the API. The expiration dating can vary based on the stability of the API and its formulated product. For example, with limited phase II drug product stability data, the expiration dating may be extended beyond the available stability data if the stability of the drug substance has shown no degradation under stress conditions and the phase I formulation also showed no degradation on stability. Typically, the 3-month stability data can be used to estimate and provide expiration dating of 9–12 months for the formulated product.

Furthermore, additional stability investigations are conducted to facilitate the development of a final formulation. Examples include excipient compatibility studies for potential interactive effects between the API and the excipient of the formulation and selection of the most appropriate container and closure systems. Typically, these investigations (sometimes termed "probe stability" studies) are conducted under extreme conditions over a very limited period of time, such as 60 ± 2 °C/75% RH $\pm 5\%$ RH, 80 °C or 1000 foot-candles of light for less than 3 months. The data generated at this stage are part of the drug development process and are used to select excipients, primary packaging components, and storage conditions for the formulated final product.

If an overencapsulated comparator is used in the clinical trial, a stability investigation of the overencapsulated comparator also needs to be conducted. The overencapsulated comparator is sometimes packaged in a configuration that is different from its commercial presentation. This is usually done to preserve the "blinding" of the study. The test frequency for the overencapsulated active comparator may be less frequent than that recommended in the ICH guidelines and is based on the commercial expiration date of the product.

3. Phase III Studies

During phase III studies, the emphasis in stability testing should be on studying the final formulation in its proposed market packaging. The expiration date is based on the on-going phase III stability studies, the phase II drug product stability studies, along with any on-going studies being conducted on the API. The expiration dating varies with the stability of the phase II drug product stability studies and the API. For example, if the phase II stability studies show the product to be stable, the API is stable and degradation does not take place under stress conditions, a long expiry date can be proposed. Although the phase III formulated product is potentially different from that of phase II, with limited phase III stability data, expiration dating can be extended beyond the available stability data of the phase III product, based on the phase II experience. In many cases, the formulation used to initiate phase III trial may be different from the marketed formulation; however, the use of "bridging" in the clinical trials allows real experience to be gained with this formulation prior to registration and commercialization. Typically, the 3-month accelerated stability data can be used to estimate and provide expiration dating of 9–12 months for the phase III formulated product.

4. Primary Registrational Stability Studies

Because of the general rule that a minimum of 12 months of real-time stability data is generally required at the time of regulatory submission, it is typical for the product to be manufactured and packaged approximately 15 months before submission. Futscher and Schumacher[34] proposed that the world be divided into four climatic zones based on temperature and humidity: zone I (temperate), zone II (Mediterranean, subtropical), zone III (hot dry), and zone IV (hot humid). The ICH and WHO guidelines on stability are applicable. The principles listed in the ICH guidelines have established that stability information generated in any one of the three zone I or zone II climatic regions (Europe, Japan, and the United States) would be mutually acceptable to the other two regions, provided the information is consistent with the ICH Q1A(R2)[1] guideline and the labeling is in accord with national/regional requirements. The WHO establishes guidelines for all four of the different climatic zones.[3] The WHO stability requirements for zones I and II countries are in alignment with ICH Q1A (R2). The WHO guidelines establish long-term storage conditions for the zones III and IV climatic zones.

The objective of the stability studies is to provide evidence on how the quality of the finished product varies with time under the influence of a variety of environmental factors such as temperature, humidity, and light and to enable recommended storage conditions and shelf life to be established.

5. Test Attributes, Procedures, and Acceptance Criteria

Analytical testing should cover those attributes susceptible to change during storage and those most likely to influence quality, safety, and/or efficacy of the product. All analytical methods must be validated according to the ICH guideline Q2(R1), "Validation of Analytical Procedures: Text and Methodology."[23] For stability studies, the analytical methods must be capable of detecting and quantitating degradation products attributable to the API and products from interaction of the active molecule with excipients and/or the packaging components. The analytical methods should also be able to demonstrate where the synthetic impurities elute in the chromatogram. For a solid dosage form, the dissolution method should be validated to detect significant manufacturing variations in the process or product. This method must be "discriminating" to subtle changes in the product that may occur over time such as tablet hardening or capsule shell cross-linking.

Specifications for the finished product that apply through the shelf life of the product should be proposed and followed during the course of the study. The specifications should include physical, chemical, and microbiologic characteristics of the product. The acceptable limit(s) for degradation product after storage under recommended conditions should be established before the initiation of the stability study. These acceptance criteria should include individual and total upper limits for degradation products. ICH guideline Q3B(R2), "Impurities in New Drug Products"[16] addresses degradation products in new drug formulations. The degradation products of the active or interaction products (sometimes termed "adducts") from the active and excipient and/or active and container component should be reported, identified, and/or qualified when the proposed thresholds are exceeded.

In ICH Q3B(R2) the threshold for the drug product for reporting, identification, and qualification of the degradation products are specified. The suggested thresholds are shown in Tables 9–11. The final proposed marketed product specifications should be verified by the conducting of the long-term drug product registration batch stability studies.

Higher thresholds may be proposed with justification and qualification if the target-reporting threshold cannot be achieved. For genotoxic impurities, the proposed specification limit should be much lower than that recommended in the

TABLE 9 Reporting Thresholds for Impurities

Maximum daily dose (g)	Threshold (%)
≤1	0.1
>1	0.05

TABLE 10 Identification Thresholds for Impurities

Maximum daily dose	Identification threshold
<1 mg	1.0% or 5 μg total daily intake, whichever is lower
1–10 mg	0.5% or 20 μg total daily intake, whichever is lower
>10 mg to 2 g	0.2% or 2 mg total daily intake, whichever is lower
>2 g	0.10%

TABLE 11 Qualification Thresholds for Identification of Impurities

Maximum daily dose	Qualification threshold for identification
$<$10 mg	1.0% or 50 μg total daily intake, whichever is lower
10–100 mg	0.5% or 200 μg total daily intake, whichever is lower
$>$100 mg to 2 g	0.2% or 3 mg total daily intake, whichever is lower
$>$2 g	0.15%

ICH guideline to ensure the safety of the product. See the Chapter 4 by Parsons and Miller for further discussion on the subject of genotoxic impurities.

6. Selection of Batches

As recommended in the ICH Q1A(R2) guideline,[1] data from primary stability studies are to be collected for at least three batches of the drug product. The primary batches should be of the same formulation and packaged in the marketed container–closure system. The manufacturing process used for the primary batches should simulate that of the production batches and should provide product of the same quality and meet the specifications proposed for marketing. Two of the three batches should be at least pilot scale batches and the third one can be smaller, if justified. Where possible, batches of the drug product should be manufactured using different batches of API.

B. Packaging/Containers

Primary packaging materials play an important role in the stability of a formulated product since such materials are in direct contact with the drug at all times until use by the consumer. Stability testing should be conducted on the dosage form stored in the package proposed for marketing. Additional testing of unprotected drug product can also be useful with regard to stress testing and package evaluation for studies carried out on other related packaging materials in support of the primary packages.

The primary packaging materials are tested and qualified for medical product use. As stated above, the term "primary" implies a package component that is or may be in direct contact with the dosage form. The FDA has issued several guidance documents on packaging components "Container Closure Systems for Packaging Human Drugs and Biologics."[35] The type and extent of information that should be provided in the regulatory filing will depend on the dosage form and the route of administration. More detailed information usually should be provided for a liquid-based dosage form than for a powder or a solid, as liquid-based dosage form is more likely to interact with the package components. A recent review article from the FDA focused on the evaluation of the safety and quality impact that extractable/leachable could have on biologic protein products.[36] In addition, a recent article was written focusing on the identification of leachables in packaged liquid formulations.[37]

C. Testing Frequency

In ICH Q1A(R2),[1] it is recommended that frequency of testing should be sufficient to establish the stability profile of the new drug product. For products with a proposed shelf life of at least 12 months, the testing frequency at the long-term storage condition will occur at every 3 months over the first year, every 6 months over the second year and annually thereafter through to the proposed shelf life expiration date.

In the case of accelerated storage conditions, a minimum of three test points, including the initial and end points, for example, 0, 3, and 6, months is recommended. When testing at the intermediate storage condition is necessary as a result of significant change at the accelerated storage condition, a minimum of four test points, including the initial and final time points, is recommended, for example, 0, 6, 9, and 12 months.

D. Storage Conditions

Stability studies on the finished product should be sufficient to cover storage, shipping and subsequent use, including reconstitution or dilution of the product if appropriate. The storage condition at which long-term testing is conducted will be reflected in the labeling and expiration date. As discussed in Section 4 (Primary Registration Stability), long-term storage conditions will depend on which climatic zone the product is expected to be approved in Table 12 outlines the different climatic zones defined by the WHO.[31–33]

Table 13 describes the ICH guideline Q1A(R2)[1] recommended storage conditions for the following zones I and II countries: United States, Europe, and Japan. It is recommended in the ICH guideline that long-term testing should be conducted at a minimum of 12 months at the time of submission and should be continued for a sufficient period to cover the proposed shelf life. Additional data accumulated during the assessment period of the registration application should be submitted to the regulatory agencies if requested. Data from the accelerated storage condition or from the intermediate storage condition as appropriate may be used to evaluate the impact of short-term excursions outside the

TABLE 12 WHO Definition of the Different Climatic Zones Around the World

Climatic zone	Definition	Criteria: Mean annual temperature measured in the open air/mean annual partial water vapor pressure
I	Temperate climate	$\leq$15 °C/$\leq$11 hPa
II	Subtropical and Mediterranean climate	>15–22 °C/>11–18 hPa
III	Hot and dry climate	>22 °C/$\leq$15 hPa
IVa	Hot and humid climate	>22 °C/>15–27 hPa
IVb	Hot and very humid climate	>22 °C/>27 hPa

TABLE 13 ICH Drug Product Storage Conditions

Intended storage condition	Test	Conditions	Minimum period (months)
Room temperature	Long term	$25 \pm 2\,^\circ$C/60 $\pm 5\%$ RH	12
	Intermediate[a]	$30 \pm 2\,^\circ$C/65 $\pm 5\%$ RH	6
	Accelerated	$40 \pm 2\,^\circ$C/75 $\pm 5\%$ RH	6
Refrigerated	Long term	5 °C/ambient	12
	Accelerated	$25 \pm 2\,^\circ$C/60 $\pm 5\%$ RH	6
Freezer	Long term	−20 °C/ambient	12

[a]Intermediate condition tested if significant change is seen at $40 \pm 2\,^\circ$C/75 $\pm 5\%$ RH.

TABLE 14 WHO Drug Product Storage Conditions

Test	Conditions	Minimum period (months)
Long term[a]	$25 \pm 2\,^\circ$C/60 $\pm 5\%$ RH or $30 \pm 2\,^\circ$C/65 $\pm 5\%$ RH or $30 \pm 2\,^\circ$C/75 $\pm 5\%$ RH	12
Intermediate[b]	$30 \pm 2\,^\circ$C/65 $\pm 5\%$ RH	6
Accelerated	$40 \pm 2\,^\circ$C/75 $\pm 5\%$ RH	6

[a]Whether long-term stability studies are performed at $25 \pm 2\,^\circ$C/60 $\pm 5\%$ RH or $30 \pm 2\,^\circ$C/65 $\pm 5\%$ RH or $30 \pm 2\,^\circ$C/75 $\pm 5\%$ RH is determined by the climate zone in which the drug product is intended to be marketed. Testing at more severe long-term condition can be an alternative to storage at $25 \pm 2\,^\circ$C/60 $\pm 5\%$ RH or $30 \pm 2\,^\circ$C/65 $\pm 5\%$ RH.
[b]If $30 \pm 2\,^\circ$C/65 $\pm 5\%$ RH or $30 \pm 2\,^\circ$C/75 $\pm 5\%$ RH is the long-term conditions, there is no intermediate condition.

label storage conditions, such as those that might occur during shipping and handling of the product.

The WHO guideline titled *Expert Committee on Specification for Pharmaceutical Preparation*[31] describes the storage conditions for the different climatic zones. The WHO divided the zone IV countries into two different subgroups: zone IVa (hot and humid) and zone IVb (hot and very humid). Table 14 describes the WHO proposed storage conditions for the different climatic zones around the world. The WHO long-term storage conditions for zones I and II countries are harmonized with ICH Q1A(R2).[1]

It is recommended to submit a minimum of 12 months stability data at the time of product registration in all countries. Additional data accumulated during the assessment period of the registration application should be submitted to the regulatory agencies if requested.

It would be normally expected that data would be generated at the long-term storage condition and at 40 °C/75% RH, but alternative (and justified) conditions could be accepted. For example, for products with normal storage conditions that would be lower than 25 °C/60% RH, such as a 5 °C condition, the accelerated condition would be 25 °C/60% RH. The intermediate condition would be used when "significant change" for the product occurs. In general, "significant change" is defined as:

1. a 5% change in assay from its initial value; or failure to meet the acceptance criteria for potency when using biological or immunological procedures;
2. any degradation product exceeding its acceptance criterion;
3. failure to meet acceptance criteria for appearance and physical attributes, and functionality test (e.g., color, phase separation, resuspendability, caking, hardness, dose delivery per actuation); however, some changes in physical attributes (e.g., softening of suppositories, melting of creams) may be expected under accelerated conditions;
4. failure to meet the acceptance criterion for pH; or
5. failure to meet the acceptance criteria for dissolution for 12 dosage units.

I. Design of Matrixing and Bracketing Studies

Bracketing is defined by ICH Q1D: *Bracketing and Matrixing Designs for Stability Testing of New Drug Substances and Products*[38] as a design of a stability study such that only samples at the extremes of certain design space factors (e.g., strength, container size, and/or fill) are tested at all time points as in a full design. Although bracketing has been in use for many years, the current move toward the QbD paradigm has generated increased interest in its utilization. Matrixing or bracketing can be applied to formal stability studies for most types of drug products although additional justification should be provided for certain complex drug delivery systems where there are a large number of potential drug–device interactions. For API, matrixing is of limited utility and bracketing is generally not applicable. Consultation with regulatory agencies may be advisable before a drug product complex matrix design can be utilized.

Design factors are variables to be evaluated in a study design for their effect on product stability. Examples of some of the acceptable variables are:

1. Strength changes with no or a small change in the proportion of ingredients.
2. Container–closure sizes and/or fills.
   (a) Care should be taken to select the extremes by comparing the various characteristics of the container–closure system that may affect product stability
      i. Container wall thickness
      ii. Closure geometry
      iii. Surface area to volume ratio
      iv. Headspace to volume ratio
      v. Water vapor permeation
      vi. Rate of oxygen permeation rate

Table 15 provides an ICHQ1D[38] example of a bracketing design. This example is based on a product available in three strengths and three container sizes. In this example, it should be demonstrated that the 15 and 500 ml high-density polyethylene container sizes truly represent the extremes of such a design.

Another set of examples that the ICH Q1D[38] provides is matrixing designs with regard to time points for a product of two strengths (S1 and S2). Table 16 depicts a one-half reduction in the time points needed to be tested and Table 17 depicts a one-third reduction in the number of time points needed to be tested.

Before proposing a matrixing design, the following list of factors should be considered. This list is not exhaustive.

1. Knowledge of data variability
2. Expected stability of the product
3. Availability of supporting data
4. Stability differences in the product within a factor or among factors and/or
5. Number of factor combinations in the study.

TABLE 15 Bracketing Example

Strength (mg)		50			75			100		
Batch		1	2	3	1	2	3	1	2	3
Container size (ml)	15	T	T	T				T	T	T
	100									
	500	T	T	T				T	T	T

T, sample tested.

TABLE 16 One-Half Reduction in Time Point Testing

Time points (months)			0	3	6	9	12	18	24	36
Strength	S1	Batch 1	T	T		T	T		T	T
		Batch 2	T	T		T	T	T		T
		Batch 3	T		T		T	T		T
	S2	Batch 1	T		T		T		T	T
		Batch 2	T	T		T	T	T		T
		Batch 3	T		T		T		T	T

T, sample tested.

TABLE 17 One-Third Reduction in Time Point Testing

Time points (months)			0	3	6	9	12	18	24	36
Strength	S1	Batch 1	T	T		T	T		T	T
		Batch 2	T	T	T		T	T		T
		Batch 3	T		T	T	T	T	T	T
	S2	Batch 1	T		T	T	T	T	T	T
		Batch 2	T	T		T	T		T	T
		Batch 3	T	T	T		T	T		T

T, sample tested.

In general, a matrixing design is applicable if the supporting data indicate predictable product stability. The supporting data for a matrix study should exhibit slight, if any, variability. If the data exhibits moderate variability, a matrixing design should be justified through the use of statistics. If the supportive data shows large variability, matrixing should be avoided.

There are potential risks to using a matrix model for registrational stability studies. A matrixing design on factors other than time points is generally less precise in shelf life estimation and yields a shorter shelf life than a corresponding full design. In addition, the matrix design might have insufficient statistical power to detect certain main or interaction effects which will lead to incorrect pooling of the resulting data.[38]

2. Bulk Holding Studies for Drug Product Intermediates

Drug product intermediates such as blends, triturates, cores, extended release beads, or pellets may be held for 30 days from the date of manufacturing without retest prior to continued manufacturing. An intermediate that is held for longer than 30 days or manufactured in another facility should be monitored for stability under controlled, long-term storage conditions for the length of the holding period. Stability data should be generated to cover shipping and transportation conditions. In addition, the finished product manufactured with such an intermediate should be monitored for long-term stability for the remainder of the shelf life. The frequency of stability testing for an intermediate is related to the length of the holding time. Where practical, testing should be done at a minimum of three time points after the initial testing. All criteria parameters should be evaluated at release of an intermediate and immediately before its use in the manufacture of the finished drug product.

E. Data Evaluation

ICH Q1E, *Evaluation for Stability Data*[28] recommends the ways and means to use stability data generated in accordance with the principles in the ICH guideline Q1A(R2) *Stability Testing of New Drug Substances and Products.*[1] A summary of ICH Q1E is described below.

The scope of ICH Q1E addresses the evaluation of stability data that should be submitted in registration applications for new molecular entities and their associated drug products. A systematic approach should be adopted in the presentation and evaluation of the stability information. The stability information should include, as appropriate, results from physical, chemical, biological, and microbiological tests. Mass balance should be assessed and the factors that can cause an apparent lack of mass balance; the mechanism of degradation and the stability-indicating capability; and inherent variability of the analytical procedures.

Data from formal registration stability studies, and as appropriate, supporting data should be evaluated to determine CQAs that will likely influence the quality and performance of the drug product. The retest period or shelf life proposed should not exceed that predicted for any single CQA.

A statistical analysis and/or graphical presentation of the long-term data should be included in an appropriate format.

For products intended to be stored at room temperature, the assessment of the data should begin with the accelerated conditions and, if appropriate, at the intermediate condition. If there is no change at the accelerated condition, the retest period or shelf life would depend on the nature of the long-term and accelerated data. If there is little or no change over time and little or no variability of the long-term and accelerated data, a statistical analysis is normally considered unnecessary but justification for the omission should be provided.

If the long-term or the accelerated data shows change over time and/or variability within a factor or among factors, statistical analysis of the long-term data can be performed to establish a retest or shelf life. If the data is not amenable to statistical analysis, the proposed retest period or shelf life can be up to one-and-a-half times, but should not be more than 6 months beyond the period covered by the long-term data. If the data is amendable to statistical analysis, the retest or shelf life can be twice but not more than 12 months beyond the period covered by the long-term data. In case of significant change at the accelerated condition, the intermediate condition will dictate the retest or shelf life. If the long-term data is not amenable to statistical analysis, the proposed retest or shelf life can be up to 3 months beyond the period covered by the long-term data. If the long-term data is amenable to statistical analysis, the proposed retest period or shelf life can be up to one-and-a-half times, but should not be more than 6 months beyond the period covered by the long-term data. If there is a significant change at the intermediate condition, the retest or shelf life should not exceed the period covered by the long-term data.

For drug products intended to be stored at refrigerated conditions, the same principles can be applied as for room temperature storage. If there is little to no significant change at the accelerated condition, the proposed refrigerated retest period or shelf life can be up to one-and-a-half times but should not exceed more than 6 months beyond, the period covered by the long-term data. If change does occur at the accelerated condition, the proposed retest period or shelf life can be up to 3 months beyond the period covered by the long-term data if the following criteria are met:

(1) The long-term data are amenable to statistical analysis but a statistical analysis in not performed.
(2) The long-term data are not amenable to statistical analysis but relevant supporting data are provided.

If the refrigerated long-term or accelerated data does show change over time and/or variability, the proposed retest period or shelf life can be up to one-and-a-half times but should not be more than 6 months beyond the long-term data if:

(1) The long-term data are amenable to statistical analysis and a statistical analysis is performed.
(2) The proposed retest or shelf life is supported by the results of the statistical analysis and any relevant supporting data.

If there is significant change at the accelerated refrigerated conditions between 3 and 6 months, the proposed retest period or shelf life should be based on the long-term data. Extrapolation is not appropriate in this situation.

If the drug product is intended to be stored frozen, the retest period or shelf life should be based on long-term data. If accelerated conditions are not studied, a single batch should be evaluated at an elevated temperature for an appropriate time to address the effects of short-term excursions outside the proposed label storage condition.

To evaluate the stability data for a quantitative attribute and establishing a retest period or shelf life, regression analysis is considered an appropriate approach. To estimate the retest period or shelf life it is acceptable to analyze a quantitative attribute (e.g., assay, degradation product) by using the earliest time at which the 95% confidence limit for the mean intersects the proposed acceptance criterion.

For an attribute that is known to decrease over time, the lower one-sided 95% confidence limit should be compared to the acceptance criterion. For an attribute that is known to increase over time, the upper one-sided 95% confidence limit should be compared to the acceptance criterion. If the attribute increases or decreases over time or whose direction of change is not known, a two-sided 95% confidence limit should be calculated and compared to the upper and lower acceptance criteria. Other examples of statistical approaches to the analysis of stability data from single or multifactor, full or reduced design studies can be found in ICH Q1E *Evaluation for Stability Data*.[28]

F. Statements and Labeling

ICH Q1A(R2)[1] and the WHO *Expert Committee on Specification for Pharmaceutical Preparation*[31] provide similar recommendations for storage statements and labeling.

The ICH Q1A(R2) recommends that a storage statement should be established for the labeling in accordance with relevant national/regional requirements. The storage statement should be based on the stability evaluation of the drug product. Where applicable, specific instruction should be provided, particularly for drug products that cannot tolerate freezing. Storage terms such as "ambient conditions" or "room temperature" should be avoided.

The stability data of the drug product should be a direct link to the label storage statement. An expiration date should be displayed on the container label.

G. Postmarketing Changes

Even after an application has been approved and the product has been marketed, consideration should be given to the effects of changes in the manufacturing formula or manufacturing process. Applications for the changes to the manufacturing formula or process will need to be submitted as necessary. However, with QbD approach to stability studies, postapproval and the additional stability studies may be reduced or eliminated when the scale, site, process, and route are altered from the initial registration stability batches.

H. QbD Approach to Stability Studies

QbD on stability studies are science and risk-based approach to understanding API physical and chemistry properties. The API's KSAs are typically moisture, solvent content, impurities, polymorph, particle size, stoichiometry, and other attributes specific to the individual API. Risk-based criticality analysis of these KSAs can address their potential influence on the API stability, therefore, help to define the API stability control strategy.[39,40]

The same methodology and principle of QbD for API stability studies can also be applied to drug product stability studies but have additional factors from the formulation, packaging, and manufacturing to consider. Examples of QbD type of stability are published. A review article by Lee *et al.* summarizes the current trend of the pharmaceutical industry.[39]

VI. SUMMARY AND CONCLUSIONS

Conducting a dynamic, compliant stability program requires more than merely knowing and adhering to regulatory requirements. It also requires management of the stability samples, the environmental chambers, and all of the associated documentation. SOPs, processes for protocol amendments/deviations, OOT or OOS investigations are all key elements of a compliant stability program. This chapter describes how to manage both the operational elements and the regulatory compliance issues to ensure a successful application and inspection by any regulatory agency. The ICH and the WHO guidelines related to stability have been discussed in detail.

REFERENCES

1. Q1A(R2): Stability Testing of New Drug Substances and Products (Second Revision), International Conference on Harmonization, issued on February 2003. http://www.ich.org/cache/compo/276-254-1.html.
2. *Quality Assurance of Pharmaceutical: A compendium of Guidelines and Related Materials*, Vol. 2, 2nd updated edition: *Good Manufacturing Practices and Inspection,* World Health Organization, Geneva, 2006.
3. Stability testing of active pharmaceutical ingredients and finished pharmaceutical products, Annex 2, World Health Organization, Expert Committee on Specification for Pharmaceutical Preparation, 43rd Report, Technical Report Series No. 953, p. 87, 2009. http://apps.who.int/bookorders/MDIbookPDF/Book/11000953.pdf.
4. *Guidance for Industry Q1A(R2) Stability Testing of New Drug Substances and Products*, Food and Drug Administration, issued November 2003.
5. USP 32/NF 27, United Pharmaceutical Convention, Inc., Rockville, MD, 2009.
6. Immediate Release Solid Oral Dosage Form. Scale-Up and Postapproval Changes (SUPAC); CM&C, In-Vitro Dissolution Testing, and In Vivo Bioequivalence Documentation.
7. Q8(R2): Pharmaceutical Development, International Conference on Harmonization, issued on August 2009.
8. Q9: Quality Risk Management, Current Step 4 version International Conference on Harmonization, issued on November 2005.

9. Q10: Pharmaceutical Quality System, Current Step 4 version International Conference on Harmonization, issued on June 2008.
10. Grimm, W. *Drug Dev. Ind. Pharm.* 19:2795–2830, 2003.
11. Q6A: Specifications: Test Procedures & Acceptance Criteria for New Drug Substances & New Drug Products: Chemical Substances, International Conference on Harmonization, issued October 1999.
12. Q1B: Photostability Testing of New Drug Substances and Products, International Conference on Harmonization, issued November 1996.
13. Cloud, P. A. *BioPharm. Nov.* 30–42, 1997.
14. United State v. Barr Laboratories, Inc. Civil Action No. 92–1744, U.S. District Court for the District of New Jersey; 812 F. Supp. 458; 193 U.S. Dist. Lexis 1932; 4 February 1993, as amended 30 March 1993.
15. Q3A(R2): Impurities in New Drug Substance (Revised Guideline), International Conference on Harmonization, issued October 2006.
16. Q3B(R2): Impurities in New Drug Products (Revised Guideline), International Conference on Harmonization, issued June 2006.
17. *Excipient Biological Safety Evaluation Guidelines*, General Chapter <1074>: USP and NF Excipients, in USP 32-NF27. U.S. Pharmacopeil Convention, Rockville, MD, 2009.
18. The Handbook of Pharmaceutical Excipients. 5th ed., Pharmaceutical Press, London, 2007.
19. Merck Index. 14th ed., Merck & Company, Rahway, NJ, 2006.
20. Cartensen, J. T. Pharmaceuticals of Solids and Solid Dosage Forms. John Wiley & Sons, New York, 1977. Chapter VII-2.
21. Shalaev, E. Y. and Zografi, G. J. *Pharm. Sci.* 85:1137–1141, 1996.
22. Effect of packaging on stability of drugs and drug products. (S. C. Gad, ed.), John Wiley & Sons, Inc., pp. 641–686, 2008.
23. Q2(R1): Validation of Analytical Procedures: Text and Methodology, International Conference on Harmonization, issued November 2005.
24. Iacocca, R. Particle size analysis in the pharmaceutical industry. *Am. Pharm. Rev.* 10(4), 2007.
25. Shekunov, B. Y., Chattopadhyay, P., Tong, H. H. and Chow, A. H. L. Particle size analysis in pharmaceutics: Principles, methods and applications. *Pharm. Res.* 24(2):203–227, 2007.
26. R. E. Wrolstad, ed., In *Karl Fischer Titration*, John Wiley & Sons, Inc., 2005.
27. Stern, M. Reagent selection in the Karl Fischer analysis of pharmaceutical products. *Am. Lab.* 38(4), 2006.
28. Q1E: Evaluation of Stability Data, International Conference on Harmonization, issued February 2003.
29. Chen, J., Zhong, J. and Nie, L. *Stat. Med.* 27:2361, 2008.
30. Q1C: Stability Testing for New Dosage Forms, International Conference on Harmonization, issued November 1996.
31. World Health Organization. Expert Committee on Specification for Pharmaceutical Preparation. 2009. 43rd Report, Technical Report Series No. 953.
32. World Health Organization. WHO guidelines on stability testing. *WHO Drug Information* 16(1):35, 2002.
33. Stability testing for hot and humid climates. *WHO Drug Information* 18(2):113, 2004.
34. Fuitscher, N. and Schumacher, P. *Phar. Ind.* 34:479–483, 1972.
35. Container Closure Systems for Packaging Human Drugs and Biologics, Food and Drug Administration: Questions and Answers, issued May 2002.
36. Markovic, I. *Expert Opin. Drug Saf.* 6(5):487–491, 2007.
37. Pan, C., Harmon, F., Toscano, K., Liu, F. and Vivilecchia, R. *J. Pharm. Biomed. Anal.* 46:520–527, 2007.
38. Q1D: Bracketing and Matrixing Designs for Stability Testing of New Drug Substances and Products, International Conference on Harmonization, issued February 2002.
39. Lee, S. L., Raw, A. S. and Yu, L. *Drugs Pharm. Sci.* 178:571–583, 2008.
40. Yin, S. X. and Grosso, J. A. *Curr. Opin. Drug Discov. Devel.* 11:771–777, 2008.
41. Haynes, J. D. *J. Pharm. Sci.* 60:927, 1971.

GLOSSARY

Bracketing The design of a stability schedule such that only samples at the extremes of certain design factors, for example, strength and package size, are tested at all time points as in a full design. The design assumes that the stability of any intermediate levels is represented by the stability of the extremes tested. Where a range of strengths is to be tested, bracketing is applicable if the strengths are identical or very closely related in composition (e.g., for a tablet range made with different compression weights of a similar basic granulation, or a capsule range made by filling different plug fill weights of the same basic composition into different size capsule shells). Bracketing can be applied to different container sizes or different fills in the same container–closure system.

Developmental stability Stability studies conducted in the early stages of product development where an assessment of various formulations is being made with the intent of selecting the formulation that will have the longest shelf life. Other factors, such as tablet compressibility, and target product profile, may influence these studies. These studies may or may not be protocol driven, depending upon the stage of development.

Current good manufacturing practices (cGMPs) Practices or procedures that the regulatory agencies currently consider to be the standard, accepted manner of operation for products made for human consumption either in a clinical or commercial venue.

International Conference on Harmonization (ICH) A consortium formed to provide an opportunity for tripartite harmonization initiatives to be developed with input from both regulatory and industry representatives. The ICH is sponsored by the European Commission, the European Federation of Pharmaceutical Industry Associations, the Japanese Ministry of Health and Welfare, the Japanese Pharmaceutical Manufacture's Association, the United States Food and Drug Administration (FDA), and the Pharmaceutical Research and Manufactures of America (PhRMA). The objective of the ICH is to harmonize the technical requirements for the registration of pharmaceutical products in the European Union, Japan, and the United States.

Laboratory information management system (LIMS) A computerized database that allows the tracking of samples, their various attributes and their analytical test data. Using appropriate software, the data can be manipulated to facilitate the development process, for example, to predict shelf life.

Matrixing The statistical design of a stability study so that only a fraction of the total number of samples is tested at any specified time interval. At a subsequent time, different sets of samples of the total number would be tested. The design assumes that the stability of the samples tested represents the stability of all samples. Matrixing can cover reduced testing when more than one variable is being evaluated. Thus, the design of the matrix will be dictated by the factors being evaluated. In every case, however, the initial and end-point samples of all batches must be tested. Whenever matrixing is to be used for a complex study, prior agreement with the appropriate regulatory agency is strongly recommended.

Mean kinetic temperature A single derived temperature which, if maintained over a defined period, would afford the same thermal challenge to an API or drug product as would have been experienced over a range of both higher and lower temperatures for an equivalent defined period. The mean kinetic temperature (MKT) is higher than the arithmetic mean temperature and takes the Arrhenius equation into account.

New drug application (NDA) A formal submission to the United States Food and Drug Administration (US FDA) requesting permission to market a new product within the United States and its territories.

Common technical dossier (CTD) CTD is a set of specification for application dossier for the registration of Medicines and designed to be used across Europe, Japan, and the United States. It was developed by the European Medicines Agency (EMEA, Europe), the Food and Drug Administration (FDA, USA), and the Ministry of Health, Labour and Welfare (Japan).

Out-of-specification (OOS) value A measured result obtained from a sample that does not comply with one or more predetermined acceptance criteria. When an OOS value is confirmed, an investigation of the specific batch and related batches is required to determine the cause and impact of the failure.

Out-of-trend (OOT) value A measured result obtained from a sample that does not appear consistent with the stability data trend. When an OOT value is identified, an investigation of the sample is made to determine the validity of the OOT result and the impact on the overall stability study.

Preapproval inspection (PAI) An inspection by the FDA of the facilities that are proposed as the manufacturing, packaging, and testing sites in the NDA. The PAI normally occurs during the review process. Besides inspecting the facilities, the documentation will be reviewed for compliance to GMPs, SOPs and to verify the integrity of the data in the submission.

Probe stability See the definition of developmental stability above.

Retest date The date after which an active API should be reexamined to ensure that the material is still in compliance with the specification and thus is still suitable for use in the manufacture of a drug product.

Standard operating procedures (SOPs) Documented procedures and practices that are the standard way of conducting business within an organization. The regulatory agencies expect that SOPs for critical functions and manner of behavior by employees are in place. The organization must also be able to document the training of its employees on all SOPs that apply to them. Any deviation from SOPs must be documented and be available for inspection during a PAI. The SOP system should be a viable, flexible system that is constantly under review and modified to reflect the actual practices of the organization in practice at that time.

13 ANALYTICAL METHODOLOGY TRANSFER

STEPHEN SCYPINSKI AND JOEL YOUNG

Analytical Research & Development, Bristol-Myers Squibb Company, New Brunswick, New Jersey, USA

I. INTRODUCTION

The process for the transfer of analytical methodology is, on the surface, a relatively simple operation. In its most common form, analytical method transfer is the verification that a method or test procedure works in an equivalent fashion at two or more different sites or laboratories. This is evidenced by the results of the transfer meeting all predefined acceptance criteria. The process is driven by compliance and governed by a statistical treatment of the resulting data. This interlaboratory transfer aspect of the overall transfer process has been covered comprehensively by McGonigle, who stressed that successful transfers are linked to the method validation process. Method transfer was defined in this case as "the introduction of a validated method into a designated laboratory so that it can be used in the same capacity for which it was originally developed."[1] The second portion of the technology transfer process concerns the transfer of technical ownership from one laboratory to another. This latter type of transfer is usually associated with the movement of drug development projects from research and development to the commercial release environment; however, it can also be associated with transfer to (and from) Contract Research Organizations (CROs) or Contract Manufacturing Organizations (CMOs). In many case studies involving technical transfer, the process of ensuring that the receiving laboratory is sufficiently familiar with all scientific aspects of the project is not carried through to fruition. In such cases, problems with the operation of the methods can manifest themselves later during testing of commercial products. Indeed, rapid and complete transfers are crucial to the success of process validation experiments for pharmaceutical dosage forms. The importance of analytical transfers is underscored by the incidence of 483 observations laboratories have received from the Food and Drug Administration (FDA).[2] While there is no technology transfer guidance available from either the FDA or from the International Conference on Harmonization (ICH), guidance documents have been provided by some working groups such as the International Society of Pharmaceutical Engineers (ISPE),[3] the French Society of Pharmaceutical Science and Technology (SFSTP),[4] and the Pharmaceutical Research and Manufacturers of America (PhRMA).[5] Both facets of technology transfer will be dealt with in this chapter. The transfer of analytical methods will be discussed first, followed by the associated documentation requirements.

II. THE DRUG DEVELOPMENT PROCESS

Large pharmaceutical companies develop products by using similar general processes, as depicted in Figure 1 and described in detail elsewhere in this text. Although there are a plethora of internal initiatives targeted at reducing the timeline and cost of drug development and hence the overall time to market, the basic process of drug development has remained unchanged. Pharmaceutical analysts working in research and development (R&D) develop and refine the methods that will eventually be used to test the identity, quality, purity, potency, and composition of marketed products. It is not unusual for analytical methods to undergo multiple iterations during the development life cycle of a pharmaceutical product. Method alterations are a result of changes to any number of product parameters, including, but not limited to, active pharmaceutical ingredient (API) synthesis, formulation composition, and dosage form manufacturing processes. It is important to keep the needs of the quality unit in mind when refining the methods for which

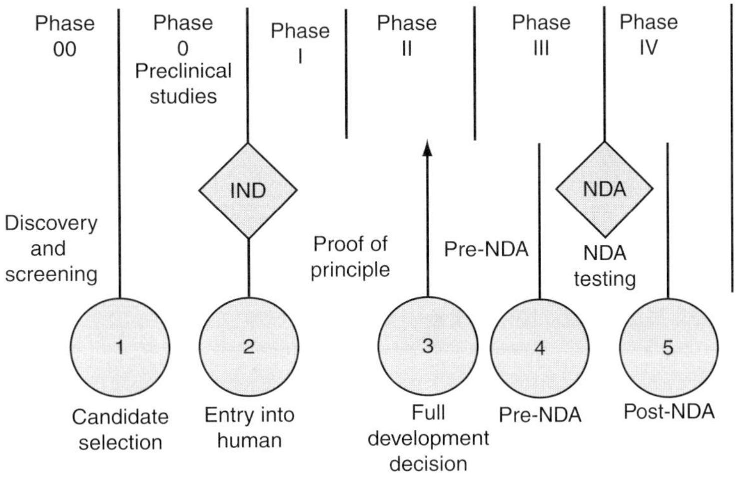

FIGURE 1 **The phases of drug development from a clinical perspective.**

they will eventually have technical ownership, as the method requirements during development may not be the same as those for commercial release. For example, during development, sensitive, accurate, and precise methods are required to monitor genotoxic impurities. The data acquired during development will be used to improve the process and minimize formation of these impurities. Method results will demonstrate product quality and safety and will be used to establish suitable acceptance criteria for the test article. At the time of registration, the sensitive, accurate, and precise method may be replaced by a limit test method. This method will ensure safety and product quality, but will not have the accuracy and precision of the development method. There are many advantages to replacing quantitative methods with limit test methods, most notably speed, efficiency, and cost.

At the conclusion of development, when methods are locked, it is becoming more common for the analysts to ask the quality unit for input and comment on the proposed method package before final validation. Such a method "test drive" can avoid problems with the formal transfer taking place later. This process has been termed an analytical method evaluation ring test (AMERT) by Crowther and associates and was presented and described elsewhere.[6,7] In brief, this process allows the quality unit to make comments and suggestions to their R&D colleagues before the final validation of the method. It is also becoming more common for the quality unit to approve method validation protocols. This provides the quality unit with an opportunity to ensure that the method meets the performance expectations to enable routine use in the quality environment. In some cases, covalidation by the R&D and Quality Units (discussed below) replaces formal method transfer. Thus, the frequently seen "throw the method over the fence" syndrome is avoided.

When the clinical development plan is proceeding smoothly, the transfer of analytical methods typically takes place during the latter part of clinical phase III. At this point, the requirements for analytical technical transfer are outlined by R&D and operations are suitably executed. The details for the transfer process are discussed later in this chapter. After successfully meeting appropriate acceptance criteria, the receiving laboratory is now considered to be qualified and able to generate reportable data as defined by current Good Manufacturing Practices (cGMPs). In most companies, further transfers taking place after the approval of a product are driven and administered by the quality unit. Included may be transfers to a CRO or to multiple manufacturing and testing sites to support global filing and marketing strategy for the new drug product.

III. TYPES OF METHOD TRANSFER

Method transfer is loosely defined as a process that qualifies a laboratory to use a test procedure or analytical method. According to this definition, any and all means of having a laboratory qualified would meet the criteria for transfer. The most common variations of method transfer are comparative testing, covalidation between two laboratories or sites, complete or partial method validation or revalidation, and the omission of formal transfer processes, sometimes termed the transfer waiver.

A. Comparative Testing

Comparative testing is the most common form of transfer in the pharmaceutical industry. It involves two or more laboratories or sites executing a preapproved protocol that details the criteria by which the receiving laboratory is deemed to be qualified to use the method(s) being transferred. The resulting data are analyzed statistically and compared to the acceptance criteria. Such criteria include defined ranges of linearity, precision, and accuracy. Comparative testing is also used in other scenarios during development and after approval. These include alliance partners, CROs, and other internal development groups, for example, the movement of a project between functional areas or development sites.

B. Covalidation Between Two Laboratories

An alternative to comparative testing is to involve the receiving laboratory in the validation of the method(s). By definition, a laboratory or site that performs validation experiments is qualified to use that method for its intended purpose. To perform such a transfer, it is necessary to identify which validation parameters are to be generated or challenged by each partner. A reasonable approach is to involve the receiving laboratory in the interlaboratory qualification, thereby generating a matrix of data that summarizes the effect of testing site, analyst, date of analysis, and instrumentation. By presenting the validation results versus the predefined acceptance criteria in the method validation report and fully describing the experimental design for the validation exercise, it is possible to have this document stand as proof of the analytical transfer.

C. Method Validation and/or Revalidation

Before the performance of method transfer activities involving protocols and acceptance criteria, it was customary for a receiving laboratory to repeat some or all of the validation experiments. This laboratory was thereby deemed to be qualified when the results were demonstrated to meet the acceptance criteria described in the validation protocol. The choice of validation parameter(s) depends highly on the type of method being transferred. For example, content uniformity assays to determine consistency of product potency depend heavily on the method and system precision. As a second example, a determination of trace impurities in an API could not be reproduced between two sites if their instruments did not yield similar limits of detection and/or quantitation. A detailed discussion on the rational choice of validation parameters that would need to be repeated by the receiving laboratory is beyond the scope of this chapter. The reader is referred to the method validation chapters by Wood (in-process methods) and Clarke and Bretnall (API and drug product methods) for additional information on this subject.

D. Transfer Waiver

Certain situations might warrant the omission of conventional transfer qualification experiments. To proceed without some manner of laboratory comparison between the two sites, it is critical to document the reasons for making such a decision. For line extensions involving dosage forms that are routinely tested by the receiving laboratory, it may be possible to waive the transfer for the following reasons:

- The receiving laboratory is already testing the product and is thoroughly familiar with the procedure(s).
- The new dosage form possesses either a comparable composition and/or concentration of API relative to the existing product.
- The analytical method(s) are the same as or very similar to those that are already in use.
- The method validation package encompasses the new methods.
- The new methods involve changes that do not substantially alter the ability to use the method (e.g., changes in sample preparation procedures or changes in calculation formulas).
- The method(s) are included in one or more pharmacopeial compendia.

The transfer strategy for each method should be considered on a case-by-case basis, keeping in mind that the object of the transfer is to provide confidence that the receiving laboratory can achieve equivalent results at their site, using their equipment, materials, and analysts. In many cases, most of the transfer variations described above are available, but in some instances there may be only a single option. For example, it is not practical for an in-process sample (i.e., a reaction completion sample during manufacture of API or in-process blend sample for a solid dosage formulation) to be used in comparative studies. Therefore, the critical performance attributes of in-process methods are generally revalidated at the receiving site. Likewise, most microbiological methods (e.g., sterility) are either compendial or need to be revalidated (perhaps using fewer batches) to complete the method transfer process.

IV. PRETRANSFER ACTIVITIES

A. Establishing the Transfer Timeline

A rough timeline for the evolution of an analytical method with respect to the major milestones within the development life cycle is presented in Figure 2. Movement of methods from so-called "first in human" or "proof of principle" groups to full development organizations must take place early in the genesis of a product. Such transfers can be considered complex because early projects have less well-defined analytical methods and will therefore be more difficult to transfer according to defined acceptance criteria. Many large pharmaceutical companies use a team structure to develop products for market. Within such teams, it is typical to include individuals who represent the quality unit at the appropriate time in the project history. These individuals are charged with defining the acceptance process for the analytical methodology and the overall technical transfer strategy for analytical methods, including the associated documentation.

The timeline for formal transfer to the release laboratory for commercial drug substance/drug product most often has the process taking place in the middle to late clinical phase III. This assumes that the quality unit has provided input via sample communication or the AMERT has taken place. The timing for the technical transfer must be flexible, depending on the specific product. It must take into account the need to have methods in place in the quality unit before regulatory approval or product launch. For example, Kassal *et al.*[8] documented transfer of an in-process control method for bulk steroid APIs. In this work, the authors successfully transferred an overpressured layer chromatographic (OPLC) method from a quality control unit to an in-process laboratory. They used comparative testing as the basis for the transfer. One must know that the receiving laboratory has met all criteria to be qualified to perform the methods and associated testing before the arrival of actual samples. If process characterization or production of registration batches for pivotal stability studies will be done by the quality unit, the transfer must take place earlier than phase III.

The timing of analytical method transfers to support commercialization of products in global markets should be identified by the regulatory liaison. The requirements for product registration vary by country. Some countries will accept

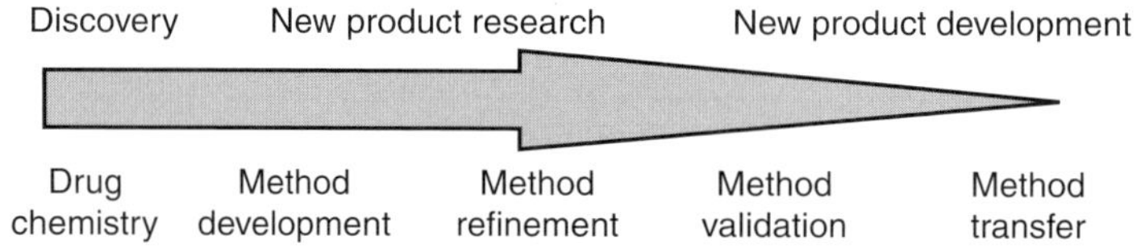

FIGURE 2 **Representative method development, validation, and transfer timeline.**

the Certificate of Analysis (COA) generated by the innovator's manufacturing site; however, other local requirements can be quite stringent. For example, the European Union (EU) requires that release testing be performed within an EU country. Other countries worldwide follow similar regulations. Frequently, this means that product manufactured and released in the United States will be retested at a facility (either associated with the innovator or at a CRO) in Europe. It is important to ascertain whether these transfers should be completed prior to registration or need only be completed prior to marketing approval. Some countries will require a local COA be included in the regulatory submission; local requirements may also require full method validation at the site performing release and/or stability testing.

The timeline and essential tasks of a typical method transfer are described in Figure 3. This timeline does not include the Risk Assessment, which could take place as much as 1 year prior to the required transfer completion date to provide time for any additional method development activities. A typical transfer may take approximately 6 months from the start of transfer activities to the final approval of the report; however, transfer activities for countries that require import permits for supplies may need to be initiated as much as 6 months prior to the expected testing dates. Likewise, some countries will require marketing valuation of supplies to enable excise collection on the imported materials. Close coordination with the commercialization group is required as this valuation may need to occur

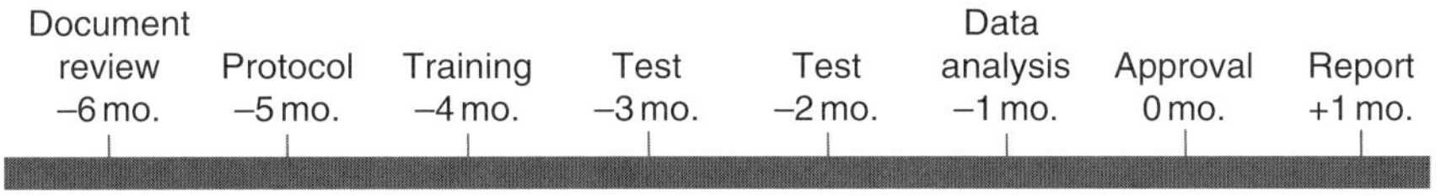

Document review
- General transfer timeline is established.
- Includes review of procedures, validation reports, method development reports, and specifications.
- Step-by-step review of methods is very helpful to ensure procedures are completely understood.
- Receiving lab may request changes to adapt the procedure for use in the QC environment.

Protocol
- Procedures, acceptance criteria, and procedures for investigating failures are established.
- Protocol is reviewed and signed off prior to execution.

Training
- Training usually occurs at the receiving site, but may occur at the expert lab.
- May entail preliminary testing prior to protocol sign-off to enable troubleshooting outside the formal transfer space.

Test
- Perform testing in accordance with the protocol.
- Requires a commitment of resources by the receiving lab.
- Report results quickly to enable timely troubleshooting if results do not meet the acceptance criteria.

Data analysis
- Review results and apply statistical tools as described in the protocol.

Approval
- All of the acceptance criteria are met and the lab is deemed suitable to apply the methods.
- Formal documentation required, a "Qualification memo"may by issued *in lieu* of a formal report.

Report
- Final report describing procedures, acceptance criteria, results, and conclusions.
- Deviations from the protocol are described, as well as any investigations for results that did not meet the acceptance criteria.
- Signed by expert, receiving and quality units, as described by local procedures.

FIGURE 3 Representative method transfer timeline.

over a year in advance of the expected regulatory approval date. Lastly, the purchase, installation (along with any capital improvements, renovations, etc.), and qualification of specialty equipment required by new methodologies introduced into the facility should be accounted for in the timeline.

B. Risk Assessment

ICH Q9, Quality Risk Management,[9] describes the process of assessing risk as it relates to pharmaceutical development and ultimately the quality of the resulting pharmaceutical products. The guidance describes *risk assessment* as the identification of hazards and the analysis and evaluation of the risks associated with exposure to those hazards. The guidance describes methodologies that can be applied to assess the probability of occurrence, the severity of harm, and the ability to detect the presence of the hazard. The principles described in ICH Q9 can also be applied to the technology transfer process to enable ultimate success. While the authors are not recommending the use of any specific risk management methods described in ICH Q9, it is worthwhile to perform a risk assessment to better understand the potential sources for failure and their impact to a successful transfer (and subsequently, product launch), and to develop a risk mitigation strategy to reduce the likelihood of problems during the method transfer. Technology transfers can go awry for a variety of reasons; many of them are not technical in nature. Some transfers are rendered difficult by differences in language and culture. Differences in site laboratory procedures or environmental conditions can cause unexpected differences in the results. Training and education may pose a challenge, particularly if the method(s) use new technology or require special skills and, of course, differences in materials or instrumentation between sites may lead to method performance issues. Even slight changes (e.g., different manufacturer of membrane filters for sample preparation) can lead to differences between lab results. The factors that may affect the successful transfer of high performance liquid chromatography (HPLC) methods is thoroughly discussed in a paper by Kirshbaum,[10] while the effect of instrumentation on performance of chromatographic methods is detailed by Bidlingmeyer.[11]

Once a receiving site has been identified, the expert laboratory should perform a risk assessment to identify variables that may pose a problem during the transfer. The following questions are examples of items that the expert laboratory should consider prior to initiating the method transfer process:

- What are the expert laboratory's experiences transferring methods to this receiving laboratory?
- Are there language or cultural differences that could hinder the transfer?
- What are the environmental conditions at the receiving laboratory, and is their potential to impact method performance understood?
- Do the personnel at the receiving site have the necessary training/education to successfully execute the method?
- Are there any differences in materials and instrumentation between the expert and receiving laboratories and can any of these differences adversely impact method performance?

Once the risks of the transfer are understood, a risk mitigation strategy can be developed. For example, on-site visits are very beneficial to developing relationships required for a successful transfer and to assess the laboratory conditions firsthand. Method training is usually essential and can occur at either the expert laboratory or the receiving site. Lastly, method evaluation or validation by the expert laboratory on the instrument vendor platforms intended for use by the recipient laboratory can greatly reduce the method transfer risk.

Evaluation of risk will also enable the development of a realistic transfer timeline. Extra time should be provided to transfers considered to be high risk, for example, when there is no organizational experience with the receiving laboratory or if the technology is being utilized for the first time.

V. REQUIREMENTS AND ELEMENTS OF ANALYTICAL TECHNOLOGY TRANSFER

The overall method transfer process is summarized in Figure 3. In many companies this process is governed by a hierarchical set of controls or standardized procedures. These documents describing the process include, in order of hierarchy:

A. Standard operating procedure
B. Master plan
C. Transfer protocol
D. Analytical testing methods

Each of these documents is discussed below.

A. Standard Operating Procedure

Analytical method transfers are generally governed by a standard operating procedure (SOP). The SOP describes the responsibilities of the expert laboratory, receiving laboratory, and associated quality organization(s). The SOP should provide requirements for materials used in the transfer (e.g., how many batches should be tested, how to select batches for testing), personnel training requirements, and documentation requirements. Documentation requirements may include a master plan and transfer protocol(s). The SOP will describe the content of these documents as well as procedures for approval, storage, and change control of formal documents such as the transfer protocol and the transfer report.

B. Master Plan

The importance of master plans in the technology transfer process is described by MacLennon Troll.[12] While the master plan is specifically described for the technology transfer process as a whole, it is also applicable to the transfer of analytical methods and its use is becoming more prevalent within the industry. The technology transfer master plan describes the transfer strategy for all of the testing methods described in the specifications for the test article(s) of interest. The master plan is an auditable document that will help the health authorities during the preapproval inspection (PAI) to understand the overall transfer process and make it transparent to a reviewer. A master plan is particularly helpful when testing methods are performed by different laboratories and/or sites. For example, some tests may be outsourced, or performed by different laboratories within the same site. In some companies, chromatographic, physical, and biological testing are all performed by different groups or sites. Each of these different groups may receive transfer protocols for the test methods applicable to their areas. The master plan ties all of these protocols together, providing structure to the overall method transfer strategy. In addition, the master plan may provide justification as to why test methods described in the specifications are not transferred. A master plan is not always required, particularly when the transfer of a complete set of methods can be described within a single transfer protocol.

C. Transfer Protocol

The protocol will describe the scope of the transfer with respect to what laboratories and analysts are affected. In some cases, direct analyst-to-analyst transfer might be necessary because of method complexity or the use of new or unfamiliar equipment. In the case of automated methods, transfer from specific robots or workstations in R&D to comparable systems within operations would need to be performed.[13] The transfer protocol describes the responsibilities and requirements for the method transfer. Typically the transfer protocol contains the following elements:

i. Responsibilities of both the transferring (expert) and receiving laboratories.
ii. Description of the materials to be used during the transfer, such as reference standards, samples for comparative testing, impurity markers, and test organisms for microbiological methods.
iii. Description of method performance verification tests.
iv. Acceptance criteria for method performance verification tests.
v. Procedure(s) for investigation of results that do not meet the acceptance criteria, if this is not already described in the SOP governing method transfer.

Additional details pertaining to each of these elements of the transfer protocol are described further in the following sections.

i. Responsibilities of both the transferring and receiving laboratories

There are three important roles in every technology transfer: the transferring (expert) laboratory, the receiving laboratory, and the transfer facilitator. The responsibility of the transfer facilitator is to lead the transfer process. The facilitator understands the timelines and requirements of the transfer and is usually very familiar with the transfer process itself. The facilitator can be from either the expert or receiving laboratory or an external organization. For example, if development and validation of analytical methods to be transferred to a manufacturing site occurred at a CRO, the expert laboratory would be the CRO, the receiving laboratory would be the quality unit, and the facilitator might be the analytical lead from the R&D organization. The transfer protocol should clearly define the roles of each group in the transfer process.

ii. Description of the materials to be used during the transfer, such as reference standards, samples for comparative testing, impurity markers, and test organisms for microbiological methods

The materials selected for use in comparative studies are critical to the success of method transfer. The selected samples should be well qualified and representative of the quality of the batches expected to be manufactured at the site. For example, to successfully transfer an impurity method, it is important that the samples contain the relevant impurities. The use of degraded samples or samples spiked with impurities (i.e., an impurity cocktail) is often helpful in the transfer of impurity methods. When transferring dissolution or assay methods, batches that represent the range of results should be selected. Batches representing the edges of failure for critical quality attribute(s) should be considered, to appropriately stress the methods during transfer. Such samples provide confidence that the methods perform as intended in the hands of the receiving lab.

Selection of previously released materials for transfer activities should be avoided as the result of not meeting the predetermined acceptance criteria could be an out-of-specification (OOS) investigation. The identity and lot numbers of specific batches should be given. If possible, COA for all samples, including reference standards, should be supplied.

iii. Description of method performance verification tests

The experimental design of comparative testing is critical to a successful method transfer. The ISPE Good Practice Guide for Technology Transfer[3] describes the experimental design and acceptance criteria for many analytical methods, including those not commonly found in the literature, such as cleaning verification, dose delivery methods for inhalation compounds, and particle size. As indicated in the Good Practice Guide, sampling should be statistically based so that sample variability does not contribute significantly to differences in results between the laboratories. Sample size should also be critically evaluated to ensure that statistically sound data analysis can be performed, as described later in this chapter.

Method performance verification tests should be well described by the protocol. Frequent problems during method transfer arise because of misunderstanding the testing procedures. The protocol should describe procedures that are not described by the testing method clearly and unambiguously. For example, if the transfer includes a verification of linearity, it is not sufficient to simply provide the concentration range of the linearity study. The procedure should clearly state the number of points to be tested, the concentrations to be prepared, and the procedure to prepare those concentrations. Higher levels of detail are encouraged.

For comparative testing, the number of lots, replicates, and injections (in the case of HPLC) should be expressly presented. For dissolution transfers, the number of individual dosage forms that will be tested should be stipulated. It is necessary to spell out such details so that small differences in the everyday analytical philosophies do not perturb the transfer process.

The procedure should also describe how results should be presented. Sometimes it is necessary to express results to more significant figures or to report results that are less than the reporting limit to enable a more thorough and robust statistical assessment.

The local procedures may be different between the two sites. The expert laboratory/facilitator should understand the differences between the two local site procedures. For example, the development laboratory may calculate sample results versus an average of the standard responses throughout the test run, while the receiving laboratory may only use the standards that bracket the specific sample. In general, the reporting instructions provided in the protocol should be consistent with the local procedures of the receiving laboratory, since the goal is not just a successful method transfer, but to ensure appropriate method results are achieved routinely after the method transfer using local site practices. However, when local procedures are different, the protocol requirements should take precedence. In general, the use of local procedures is not common or recommended and it is best for both expert and receiving laboratories to use common and agreed procedures to perform the transfer.

iv. Acceptance criteria for method performance verification tests

The transfer protocol must include suitable acceptance criteria relevant to the tests and specific dosage forms. Selection of the acceptance criteria is a critical element of the method transfer. Criteria that are too tight could lead to rejection of acceptable results, thereby causing delays in the transfer process. On the other hand, if criteria are too loose, the receiving laboratory may pass the transfer but be unable to appropriately test the actual product(s), leading to poor decision making with respect to approval of marketed product. Of the two risks, the latter is more critical since, approving a laboratory that is truly not qualified could affect marketed product safety and/or efficacy and ultimately patient safety. This could lead to recalls and severe quality and compliance issues.

Acceptance criteria should take into account method performance attributes and the intended use of the methods. For example, in some instances it may be critical that the method precision and sensitivity (i.e., for impurities) are similar to that obtained by the method development laboratory. In such cases, the samples selected for transfer purposes, the statistical tools applied to demonstrate equivalence and the acceptance criteria should be selected carefully to ensure that the method performance is properly evaluated. On the other hand, in some instances the capabilities (e.g., sensitivity) of the development method may exceed the method performance requirements for commercial release testing. For example, while a gas chromatographic (GC) method may be validated to have sensitivity down to 0.002% for a number of residual solvents monitored during development and, if the specifications are set only on total residual solvents with a limit of 0.5%, it may not be necessary to demonstrate sensitivity for individual solvents to 0.002% to qualify the QC laboratory for routine use. The acceptance criteria should be considered on a case-by-case basis for each method for each product and must be established in advance of the formal testing.

Even when the greatest care is taken to establish acceptance criteria, it is not possible to anticipate all of the potential results that can be obtained during transfer. Sometimes it is necessary to investigate results even when they meet the acceptance criteria. This is particularly true when results are achieved near the point of failure. It is important to remember that the purpose of the method transfer is not just to meet preestablished acceptance criteria, but to ensure the method performs appropriately in the hands of the receiving lab.

v. Procedure(s) for investigation of results that do not meet the acceptance criteria, if this is not described in the SOP

Results that do not meet the acceptance criteria are generally considered OOS and, as such, must be investigated as governed by site SOPs covering this subject unless explicitly addressed in the transfer protocol. Frequently, there are differences between the expert and receiving laboratories investigation procedures, thus the process for investigating these results should be established prior to execution of the protocol. In most cases such transfer failures are not due to nonrobust methods, but are manifestations of either differences between the laboratories and/or a lack of detailed understanding on the part of the receiving laboratory. Such situations where root cause can be identified should lead to a revision and perhaps a greater amount of detail in the write-up of the method.

D. Analytical Testing Methods

The subject methods and method validation reports must be made available to the receiving lab. One should include any idiosyncrasies present in the method. A difficult method is not necessarily nonrugged. Rather, the developer must stipulate very carefully the steps that need to be followed and any pitfalls that might occur. This can even be the case with compendial methods, which do not necessarily need to be formally transferred. Because of the general nature of descriptions of compendial methods, receiving laboratories often run into difficulty executing them. For example, the United States Pharmacopeia (USP) monograph for ethinyl estradiol states in the sample preparation for the drug substance assay, "Transfer about 25 mg of ethinyl estradiol, accurately weighed, to a 2.5 mL volumetric flask, add mobile phase to volume, and mix."[14] The general monograph does not state that one needs to exercise care when preparing standards of this nature because of the effect of humidity or temperature, which could skew the assay results. In all method procedures, step-by-step directions should be given, including "tips and tricks," safety considerations, and clear formulas and calculations.

As in any scientific document, a rationale for the parameters chosen and their effect on the overall success of the transfer should be described. This includes an explanation for the system suitability parameters that have been established for the method. System suitability can be a powerful tool for troubleshooting method discrepancies in addition to being a test of the conformance of the system to a set of parameters before analysis.[15]

VI. PROTOCOL EXECUTION

Once the protocol is approved (e.g., signed) by all relevant parties, the procedures may be executed. As the transfer will most likely be subject to a health authority inspection, cGMPs should be followed. Instruments should be well maintained, qualified, and calibrated as appropriate. Analysts should be trained in the techniques that are the subject of the method transfer and this training documented. Communication between the laboratories should be frequent and transparent. Results should be presented as rapidly as possible to enable troubleshooting in the event the results do not meet the acceptance criteria. The personnel executing the transfer should be the same as those trained in the methods. If the transfer is considered high risk, an on-site presence by the method expert(s) is recommended.

VII. ANALYSIS OF RESULTS/STATISTICAL PACKAGES

The comparative testing type of technical transfer generates results that need to be compared using one or more statistical tools. Although data can be evaluated subjectively, the use of statistics will build objectivity into the data analysis and allow unbiased comparison of the data sets. Because comparative testing involves destructive testing of individual samples, a successful transfer involves proving mathematically that the data sets are equivalent. The complexity of the data treatment is directly related to that of the method transfer.

A comprehensive review of statistical tools for data analysis is beyond the scope of this review; however, a brief summary of the tools described in the literature is pertinent. Historically, many transfers have been governed by the statistics of the standard Student's *t*-test.[16] The *t*-test is useful for determining the statistical difference between like data sets and is well documented. Vial and several French coworkers[17] documented the considerable statistical rigor for transfer of HPLC methods using the *t*-test. Their work was especially relevant as they discussed certified materials, reanalysis, and other important factors. Recent articles challenge the application of the Student's *t*-test to analytical transfer method results, because of the potential risk of passing results that may not be truly equivalent.[4,18,19] Other approaches to establishing similarity between data sets are the equivalence test[18,20] and the application of total error.[20,21] The equivalence test evaluates the accuracy and precision of the method separately, establishing upper and lower confidence limits based on laboratory means and standard deviations. Schepers and Watzig[20] applied the equivalence test and *t*-test approach to simulated results generated in accordance with the experimental design described by the ISPE guidelines for technology transfer. Their statistical evaluation indicated a higher probability of passing a bad transfer using the *t*-test compared to the equivalence test. The total error method, described by Rozet *et al.*[21] and Dewe *et al.*,[22] combines the estimates of systematic and random errors into a single decision criterion. While providing a robust method for analysis, this approach can be complicated and should include the assistance of statisticians. de Fontenay[19] describes yet another statistical tool which takes into account the method precision and the tolerance interval of the specification. Based on the results of thousands of simulations, de Fontenay demonstrates the robustness of this statistical tool to make decisions regarding the similarity of data sets.

There are two risks associated with analytical method transfer decisions: (1) falsely failing a laboratory (creating a risk to the manufacturer) and (2) falsely passing a laboratory (creating a risk to the patient). Kringle *et al.* provides an excellent assessment of these risks and how statistical tools can be used to control this risk.[18] The authors do not endorse any particular statistical tool, but recommend that the advantages and disadvantages of the experimental design, acceptance criteria, and statistical tools selected for use be understood, so that when results at the edge of failure are achieved, appropriate actions can be taken to mitigate both of these risks. All statistical approaches utilized for the analysis of data obtained during the transfer should be clearly delineated in the protocol.

VIII. DOCUMENTATION OF RESULTS

The old adage "if it isn't written, it isn't done" certainly applies to analytical transfer. The expectation of the health authorities is that a final report will be issued documenting the analytical method transfer process and associated results. Two types of records are subject to PAI: Primary records that demonstrate safety, purity, and efficacy of the drug (e.g., batch records, test data) and supporting documentation such as equipment verification records, change control and development and validation reports that demonstrate the cGMP compliance status of the facility.[12] The method transfer report is categorized under the latter set of documentation.

The results of the comparative testing experiments should be documented in the method transfer report. Considerable attention must be paid to details of all aspects, observations, and results of the experiments. In some cases, the results for the receiving laboratory may not meet the established acceptance criteria. As described earlier, such situations should be addressed by a policy agreed upon prior to the start of the method transfer. An investigation into all results that fall outside of the acceptance criteria should be performed and summarized. In addition, deviations from the protocol should also be captured with appropriate justification or explanation. Conditions for method approval should be clearly stated in the conclusion. For example, if residual solvent method was transferred to a vendor of a starting material, and validation or comparative testing was only performed for a subset of the volatiles described by the method (i.e., the volatiles used in that vendor's process), the final report should clearly indicate which solvents the vendor is qualified to test. If the vendor changes the solvents used in the process, during the change control process the report could be consulted to determine if any additional work is required.

IX. ANALYST CERTIFICATION AND TRAINING

During the performance of method transfer activities, the training component must not be taken lightly. If the personnel in the receiving laboratory do not speak the native language of personnel in the developing laboratory, translation of the method procedures may be necessary. It is also important to ensure that proper training of the analysts who will perform the method procedures has taken place. In some respects, the actual process of comparative testing might be considered to be biased as both the transferring and receiving laboratories will be paying close attention to the testing procedures. A better measure of the ruggedness of the method(s) would be a test of the methods once they were in routine use within the quality unit; however, this is impractical for documentation of the success of the transfer.

The actual training of the analysts at the receiving end of the methods is an area that has received considerable attention from regulatory authorities and internal quality assurance auditors. There are several options to certify analysts to perform the required test procedures. One possibility is to qualify individual analysts on each method, thereby creating a matrix of methods and analysts. Such an approach can become cumbersome as it would involve substantial documentation with respect to training records for individual analysts. Compliance liability could also result if an analyst who was not "certified" produced reportable GMP data.

An alternative to an individual method certificate is to qualify analytical chemists by technique, not method. This is the approach most often used by larger companies. Training records must include documentation that the individual is indeed competent to perform routine tests using specific transfers. In these cases, the supervisor or another qualified analyst can certify that the individual analyst is competent to perform the test procedure in question. Johnson and Johnson has taken this approach one step further and has instituted the Laboratory Analyst and Training Certification Program (LATCP), in which all analysts developing or executing methods for the release and/or stability of products for human consumption, either in a clinical study or after approval, attend a comprehensive 2-week course. The LATCP involves both theoretical classroom instruction and laboratory experimentation. At the conclusion of this course, students must pass an examination to become certified. This course and its associated subject material were described by Miller.[23] In addition, the entire course content has been published as a textbook by Miller and Crowther.[24] With the increasing emphasis of the authorities on training and analyst competency, there is no doubt that such programs will proliferate.

X. TRANSFER OF TECHNICAL OWNERSHIP

At the heart of the technical transfer process is documentation. As stated earlier in this chapter, not only is documentation necessary but also attention to detail while writing results, observations, and conclusions related to method transfer. The R&D component of the drug development process depicted in Figure 1 encompasses several years, during which time the knowledge base surrounding a compound increases steadily. At the junction between R&D and operations, it is important to ensure not only that the analytical methods are successfully used by the quality unit but also that the applicable knowledge and data are transferred or are readily accessible by the receiving laboratory. Although the quality unit might not require excessive detail regarding all scientific research performed during the R&D period, some information transfer is necessary. This is becoming more critical in the wake of shifting development toward the Quality by Design paradigm (discussed in Chapter 5). Rather than inundate the quality unit with a large number of lengthy reports, several concise documents that meet the specific needs of an operations-based analytical unit are preferred. In most companies, the most important reports are the method development/validation reports and analytical development reports.

A. Method Development Report

The method development report may be a part of the method validation report. This report should provide a concise summary of the development of the key analytical methods used to assure the identity, quality, purity, potency, and composition of the test article. For HPLC methods, the rationale for the column and conditions chosen should be documented. For discriminating dissolution methods, the choice of media, type of apparatus, and detection method should be given. In all cases, a description or tabular list of unsuccessful approaches should be included. The reason for including such information is to assist the quality unit should it be necessary to redevelop or refine the method after approval because of changes in the process, formulation, or other variables. If and when this does occur, having this information at the disposal of the quality unit will help avoid "redeveloping the wheel" with respect to the analytical methods. A good method development/validation report will include sample chromatograms for approaches that have been tried. For a more detailed discussion of method development and validation, the reader is referred to several other chapters in this text dealing with these subjects.

B. Analytical Development Report

The analytical development report is a documented summary and logic flow of all essential analytical information acquired during the R&D phase of the project. In many cases, this report serves a variety of purposes: it is a repository of all important analytical information pertaining to the project and is therefore a source document for the quality unit, and it is also used in the preparation of the site inspection at the time of product approval. All information contained in this report must be traceable to raw data and other reports, if applicable. Table 1 contains a summary of all information that should be included in the analytical development report. The format of the report is not critical, but the information must be present. For further details on the various sections and chapters in this report, the reader is referred to other chapters in this text that describe the details therein. One alternate approach to authoring a large report is to create a process and system for storing and accessing development knowledge on an intranet site or shared drive. At Bristol-Myers Squibb, the analytical knowledge is stored in an Analytical Knowledge Book, a series of spreadsheets in an Excel workbook, either containing primary information or hyperlinks to electronic documents stored within the Bristol-Myers Squibb (BMS) intranet. In addition to links to method reports (development, validation, etc.), specifications and solid-state characterization, the Knowledge Book will also contain structures, origin and fate of impurities, batch history, and solid-state

TABLE I Suggested Content of the Analytical Development Report

Section	Content
1. Overview of project	Statement of analytical testing Pertinent background information on test article (summary of synthetic scheme for substance, composition/manufacturing process for dosage form)
2. Summary	Summary of drug chemistry of substance Discussion of potential impurities and degradation products Degradation mechanisms and relevance Preformulation data summary
3. Method history	Development history for key methods Discussion of method changes, additions, deletions Rationale for key methods Reference method development/validation reports
4. Specifications	Rationale for proposed marketed product specifications History of specification setting, tightening, etc. Tables of key batches clearly illustrating justification for proposed specification
5. Stability	Summary of stability behavior, trends, etc. for substance and/or product Reference substance or dosage form stability reports Brief discussion of "probe stability" studies as development pharmaceutics
6. Tips and tricks	Guidance for quality unit laboratories Summary of key ingredients of each method Safety issues/precautions Use of compendial methods
7. References	List of important reports/sections "Partner" reports (chemical development, pharmaceutics, etc.)

characterization. Access to the Knowledge Book is provided to the Quality Control laboratory at the beginning of the transfer process.

The analytical development report is one of the most valuable documents written by the R&D unit during the development process. If properly composed and completed, it can serve as a powerful document for many years after product approval.

C. Transfer File

Many companies have developed the practice of composing what is termed the "transfer file" as a means of ensuring that all key documents and relevant information are imparted to operations or the receiving laboratory. This file is merely a collection of important reports. For analytical methods, such documents include the method development and validation reports, impurity profiling report, stability reports and tables, and specification archive. The power of such an approach is that it ensures that all information is conveyed to the receiving laboratory. This strategy is useful if operations will be relied on to continue the development process. Examples include development for new formulation strengths for "brand support" or life cycle management or the support of phase IV clinical trials.

XI. SUMMARY

This chapter highlights the important aspects of the analytical transfer processes as they relate to process, compliance, analytical data, and documentation. Types of method transfers and the timeline of transfer activities are discussed. The risk assessment prior to initiation of transfer activities is also described. The chapter describes content and utility of the transfer protocol and final report, as well as documents that govern analytical method transfers (i.e., SOPs and master plan). The importance of selecting appropriate method transfer acceptance criteria and use of statistical methods to evaluate results are described. The significance of the inclusion of an adequate level of detail in the methods, protocol(s), and other documents cannot be overly stressed. Last of all, the process for transfer of technical ownership of the analytical methods is discussed. Other chapters in this text should be consulted for elaboration on the various important facets of technical transfer, including method development, method validation, documentation, and stability.

REFERENCES

1. McGonigle, E. Assay validation and inter-laboratory transfer. In *Development and Validation of Analytical Methods*, (C. M. Riley and Z. W. Rosanske, eds.), Elsevier, New York, pp. 3–13, 1996.
2. Johnson, Y. An FDA Perspective on Method Transfer of Chromatographic Analytical Methods in the Pharmaceutical Industry. The Chromatography Forum of Delaware, October Meeting, 2004.
3. ISPE Good Practice Guide: Technology Transfer. www.ispe.org, 2003.
4. Minois-Offroy, F., Appriou, Y., Brousset, V., Chapuzet, E., de Fontenay, G., Dewé, W., Dumas, E., Ellie, C., Galiay, M., Lefebvre, N., Mottu, P., Quint, M. P. and Schoeffter, F. Transfert des méthodes analytiques: méthodologie, Rapport d'une Commission SFSTP. *STP Pharma Pratiques* 12(6):337–343, 2002.
5. Scypinski, S., Roberts, D., Oates, M. and Este, J. Pharmaceutical research and manufacturers association acceptable analytical practice for analytical method transfer. *Pharm. Technol.* 2613:84–88, 2002.
6. Crowther, J. B. International method transfers. *Presented at the 1998 Eastern Analytical Symposium and Exposition*, 1998. Abstract 490.
7. Crowther, J. B., Jimidag, M. L., Niemeijer, N. and Salomons, R. Qualification of laboratory instrumentation, validation, and transfer of analytical methods. In *Analytical Chemistry in a GMP Environment*, (J. M. Miller and J. B. Crowther, eds.), Wiley, New York, pp. 453–456, 2000.
8. Kassal, A., Szecsi, A., Koppany, A., Vegh, Z. and Ferenczi-Fodor, K. Analytical transfer of an OPLC purity test and its application for in-process monitoring. *J. Planar Chromatogr.* 13:30–32, 2000.
9. *ICH Q9*. Quality Risk Management. www.ich.org, 2005.
10. Kirschbaum, J. J. Interlaboratory transfer of HPLC methods: Problems and solutions. *J. Pharm. Biomed. Anal.* 7:813–833, 1989.
11. Bidlingmeyer, B. A. Practical HPLC Methodology and Applications. Wiley, New York, 1992.
12. MacLennan Troll, E. Technology transfer and scale-up. In *Preparing for FDA Preapproval Inspections*, (M. D. Hynes, ed.), Marcel Dekker, New York, pp. 215–231, 1998.
13. Howard, B. Experiences and expectations for transferring automated pharmaceutical analysis methods in a global pharmaceutical company. *Proceedings of the 1999 International Symposium on Laboratory Automation and Robotics*, ISLAR, Hopkinton, MA, 1999. (CD-ROM version).
14. Ethinyl estradiol. *USP 32 NF 27.* U.S. Pharmacopeial Convention, Rockville, MD, pp. 2322–2323, 2009.
15. Crowther, J. B., Salomons, E. and Callaghan, C. Analytical method development for assay and impurity determination in drug substances and drug products. In *Analytical Chemistry in a GMP Environment*, (J. M. Miller and J. B. Crowther, eds.), Wiley, New York, pp. 350–351, 2000.
16. Meieg, P. C. and Zund, R. E. Statistical Methods in Analytical Chemistry. Wiley, New York, 1993.

17. Vial, J., Jardy, A., Anger, P., Brun, A. and Menet, J. Methodology for transfer of liquid chromatographic methods based on statistical considerations. *J. Chromatogr. A* 815:173–182, 1998.
18. Kringle, R., Khan-Malek, R., Snikeris, F., Munden, P., Agut, C. and Bauer, M. A unified approach for design and analysis of transfer studies for analytical methods. *Drug Inf. J.* 35:1271–1288, 2001.
19. de Fontenay, G. Analytical method transfer: New descriptive approach for acceptance criteria definition. *J. Pharm. Biomed. Anal.* 46:104–112, 2008.
20. Schepers, U. and Watzig, H. Application of the equivalence test according to a concept for analytical method transfers from the International Society for Pharmaceutical Engineering (ISPE). *J. Pharm. Biomed. Anal.* 39:310–314, 2005.
21. Rozet, E., Mertens, B., Dewe, W., Govaerts, B., Boulanger, B., Chiap, P., Streel, B., Crommen, J. and Hubert, P. The transfer of a LC-UV method for the determination of fenofibrate and fenofibric acid in Lidoses: Use of total error as decision criterion. *J. Pharm. Biomed. Anal.* 42:64–70, 2006.
22. Dewe, W., Govaerts, B., Boulanger, B., Rozet, E., Chiap, P. and Hubert, P. Using total error as decision criterion in analytical method transfer. *Chemometr. Intell. Lab. Syst.* 85:262–268, 2007.
23. Miller, J. M. The training of pharmaceutical analysts. *Am. Lab.* 32(20):13–19, 2000.
24. Miller, J. M. and Crowther, J. B. Analytical Chemistry in a GMP Environment. Wiley, New York, 2000.

14 PHARMACEUTICAL ANALYSIS DOCUMENTATION

HITESH CHOKSHI, KATHLEEN SCHOSTACK, AND RICHARD STEINBACH

Hoffmann-La Roche, Inc., Nutley, New Jersey, USA

I. SCOPE

This chapter describes analytical documentation needs during the life cycle of a pharmaceutical product—from initial candidate screening and selection, through

EIHs/IND, to NDA and postapproval marketed product support. Most of the information is based on US FDA and ICH guidelines. However, EU technical requirements are generally similar, though specific IMPD/CTA requirements may vary from country-to-country. Also, the technical requirements for Japan are similar based on the ICH guidelines.

II. INTRODUCTION

Pharmaceutical analytical data are the foundation and backbone for design, development, and marketing of new drugs. These data are acquired and reported in various ways to demonstrate identity, quality, purity, and stability of an investigational or marketed drug product. The data also serve to bridge changes that occur in synthesis, formulation, and manufacture of a pharmaceutical product from initial concept, through development and postapproval modifications.

The chapter begins with an overview of the life cycle of a new drug and the associated analytical data needed at different stages in order to support development and marketing. These data are captured in various reports as a combination of:

- regulatory documentation (IND, NDA, BLA) for submission to health authorities,
- compliance (GLP and GMP) reports for inspection, and
- internal research and development reports.

Many new guidelines concerning the Chemistry, Manufacturing, and Controls (CMC) requirements for pharmaceuticals have been drafted and issued over the last few years. Most concern regulatory requirements for IND and NDA submissions. There are also a number of GMP-related guidelines. Consistent with this trend, the following sections emphasize regulatory documentation with a less detailed discussion of compliance documents. The importance of internal research and development reports is also discussed. Internal reports are becoming even more important for capturing the history of a product as requirements for regulatory documentation are reduced. All three types of documents are necessary to capture the history of a product and assure the integrity of the data.

This chapter is not intended as a cookbook. It does not provide step-by-step directions on how to prepare analytical documentation for a perfect IND, NDA, or GLP/GMP inspection. However, the information presented on the following pages will aid in identifying the critical documents (and their contents) necessary to attain these objectives. A list of related regulatory guidance and policy documents (FDA and ICH) is provided in Appendix I to assist the reader in locating more detailed information on specific topics. Special attention over the recent years has increased toward the application of the Common Technical Document (CTD) format. In addition, analytical scientists have an increasingly important role as a result of the application of Quality by Design (QbD) and Design Space (Q8(R1)) concepts. The analytical scientist will be called upon as important contributors to the quality team as increasing emphasis is placed on key areas of the CTD through recent ICH guidances such as Q8 (Pharmaceutical Development) and Q9 (Quality Risk Management). The ICH Q8 (Pharmaceutical Development) incorporates elements of risk and QbD throughout the life cycle of the product.

Analytical scientists and their data help to ensure a sound documentation of the risks and quality of a product. ICH Q9 (Quality Risk Management) offers a systematic approach to quality risk management and complements other existing quality practices that are to be maintained throughout the product life cycle.

Note: A glossary is provided at the end of the chapter with definitions for acronyms used in the text.

III. PHARMACEUTICAL ANALYSIS DURING PRODUCT LIFE CYCLE

The life cycle of a new drug is illustrated in Figure 1. As shown, the process consists of sequential phases and milestones. The phases are defined by the stages of preclinical and clinical development leading to market launch. In practice, the development phases are rarely discrete so that only the decision points clearly mark progress.

The activities within analytical research and development progress with the different phases of drug development. The timelines for technical development activities depend on the therapeutic indication and associated clinical and toxicology programs. New drugs with the promise of fulfilling an unmet medical need may undertake a highly accelerated clinical program. The technical activities will then be compressed and overlapped. For chronic use drugs, carcinogenicity studies and clinical studies may be on critical path providing more time to optimize the technical development.

As shown in Figure 2, analytical documentation is comprised of a variety of analytical reports. The major types of analytical documentation are frequently repeated and/or updated throughout development and during postapproval changes (Table 1). The analytical organization must have scientifically sound and compliant practices and well-defined processes to ensure that these critical reports are prepared in a consistent, timely manner. Because of the repetition during development and product-to-product, the analytical group may benefit from developing a template or guideline for each type of report in order to (1) ensure consistency and completeness and (2) keep up with today's highly accelerated pace for development and production changes. Some of the new documentation practices requiring special attention are those concerning genotoxic impurities testing and control strategy, CTD format, as well as QbD concepts.

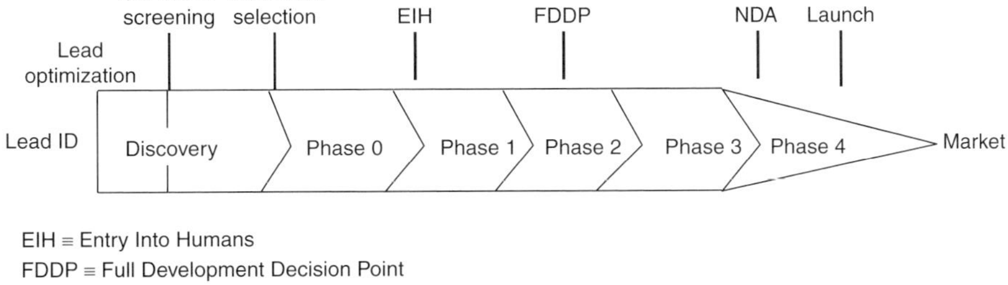

FIGURE 1 **Life cycle for new drug development. EIH, entry into humans; FDDP, full development decision point.**

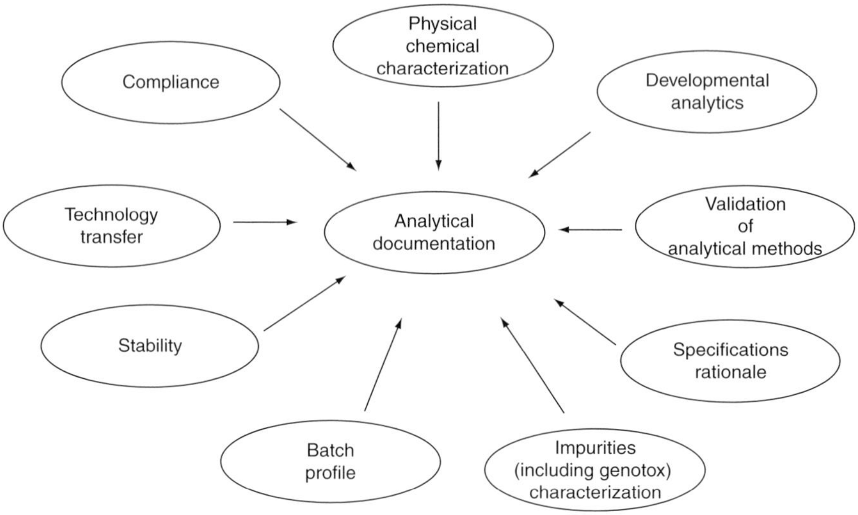

FIGURE 2 **Major types of analytical documents.**

Critical activities for each development phase or milestone in a new product's life cycle are summarized below. The key analytical support, activities, and associated documentation are described for each phase. The activities are generally applicable but may vary considerably for a particular program depending on the nature of the new chemical/biological entity and the therapeutic indication. The activities are also applicable to line-extension products or generics, but will be abbreviated because of the significant body of data available for the original/innovator product.

A. Discovery: Candidate Screening

The primary goal during candidate screening is to rank order a large number of compounds based on efficacy in animal models, biopharmaceutical properties (such as solubility and permeability in caco-2 cells, maximum absorbable dose, potential of food effect, etc.), preliminary pharmacokinetic profile in animals, and potential for metabolic and toxicity liabilities. A non-GLP Ames test may be performed to screen for potential mutagenicity.

Early analytical activities focus on becoming familiar with the chemistry, physical properties, and stability of the new APIs. The purity of the test material(s) and preliminary solid-state and solution stability should be established for candidates prior to use in the Ames test. Candidates are also screened with respect to potential technical issues for further development. Purity and stability testing are performed using a combination of relatively simple chromatographic methods (i.e., HPLC, TLC, GC). A basic solubility profile is developed. Preliminary solid-state characterization is performed using DSC, TGA, and XRD. Early selection of a pharmaceutically acceptable chemical form (where applicable) becomes a key activity to ensure optimal bioavailability (BA), stability, and manufacturability. Biopharmaceutical properties such as potential of food effect, particle size effect, etc., of the proposed clinical candidates are assessed by *in vitro* and *in silico* methods.

TABLE 1 Analytical Documentation: Key Product-Specific Documents During Development

Documentation Type	Phase 0	Phase 1 (IND)	Phase 2	Phase 3	NDA
Physical–chemical characterizations	• Analytical and preformulation report • Ideally, rationale for selection of the solid-state (polymorph) and chemical (salt) forms	• Updated characterization reports linking toxicology and clinical materials	• Updated reports to bridge evolution of synthesis	• Updated reports to bridge evolution of synthesis • Data to bridge supplies	• See attached Appendix II
Analytical methods	• Validated analytical methods for API and toxicology formulation; preliminary analytical methods for clinical formulation	• Updated reports for API and clinical formulation	• Updated reports for API, toxicology, and clinical formulations	• Updated reports for API, toxicology, and clinical formulations • Rigorous validation data; application of QbD concepts for analytical methods	• See attached Appendix II
Batch profiles	• Impurity profile for API and dosage forms used for toxicology supplies	• Updated batch profiles for API, toxicology and clinical formulation; with particular attention to genotoxic impurities	• Updated batch profiles for API, toxicology, and clinical formulation • Updated genotoxic impurities strategy	• Updated batch profiles for API, toxicology, and clinical formulations • Data to bridge supplies	• See attached Appendix II • Proposed product shelf-life
Stability	• Short-term (3 months) stability data on API and toxicology formulation	• Stability reports for API, toxicology and clinical formulation (at least one month accelerated data) • Proposed storage conditions for API and clinical formulations	• Updated stability data • Updated retest period	• Updated stability data • Stability plan for registration batches • Updated retest period	• See attached Appendix II

Continued

TABLE 1 *(Continued)*

Documentation Type	Phase 0	Phase 1 (IND)	Phase 2	Phase 3	NDA
Specifications	• Proposed specifications for API and toxicology formulations	• Proposed specifications for API and clinical formulation	• Updated specifications reflecting changes in synthesis and stability profile	• Updated specifications reflecting changes in synthesis and stability profile • Regulatory strategy for planned or anticipated postapproval changes	• See attached Appendix II
Reference standard characterization	• Characterization and COA	• Characterization and COA	• Characterization and COA	• Characterization and COA • NDA reference standard	• See attached Appendix II
Technology Transfer	• If applicable, for example, CRO testing	• If applicable, for example, CRO testing, other site	• If applicable, for example, CRO testing, other site	• If applicable, for example, CRO testing, other site	• If applicable, for example CRO testing, other site
Certificate of analysis or analytical profile (report)	• API and toxicology supplies	• API, toxicology, and clinical supplies	• API, toxicology, and clinical supplies	• API, toxicology, and clinical supplies	• API, toxicology, clinical, registration, validation supplies

Note: Analytics for biological pharmaceutical products follow similar principles but utilize methods for *in vitro* or *in vivo* (animal model) bioactivity, electrophoretic and chromatographic methods, amino acid analysis, peptide mapping, sequencing, etc. for characterization.

Analytical documentation during the discovery phase consists of:

- Biopharmaceutical classification based on solubility and *in vitro* permeability data
- Internal methods—qualified by partial assessment of validation parameters
- First impurity profile, including potential of genotoxic liability for synthetic by-products or the degradation products
- Short-term solid-state and solution stability
- Stability data for formulations used in efficacy and pharmacokinetic studies in animals

B. Clinical Candidate Selection

Usually, one or two compounds are selected from the candidate screening and evaluated further in at least one rodent and one nonrodent species for gross toxicity (i.e., 1–2 weeks dosing).

The first real technical challenge is development of a toxicology formulation, which will provide adequate exposure in animals—generally several times higher than the expected dosing per body weight in humans. The decision to formally begin EIH enabling studies for the selected new chemical entity depends on a favorable safety profile from these short-term animal studies. The analytical group supports the toxicology program while expanding its knowledge of the compound.

Additional preformulation and physicochemical characterization of the candidate compound are performed and stress stability studies may be initiated. Ideally, the optimal solid-state (polymorphic) and chemical (salt) form of the molecule are identified as part of clinical candidate selection. Selection of the most stable and bioavailable chemical and physical forms will expedite subsequent development. The methods for testing the drug substance are refined and additional methods may be developed, including tests for physical, chemical, and chiral purity. Analytical characterization becomes extremely critical where significant formulation intervention is required such as amorphous formulations for BCS 2/4 compounds, inhalation products, or highly concentrated formulations for monoclonal antibodies.

Analytical methods for assay of the toxicology formulations and cleaning validation are developed and validated in preparation for the first GLP studies. Release and stability testing of the toxicology test articles are performed to support the suitability of the materials through their anticipated period of use. Typically, short-term accelerated stability studies are performed on the toxicology batches for at least 3 months in order to cover the time from date of manufacture through the last dose.

Analytical documentation at clinical candidate selection milestone may consist of:

- Preliminary analytical and preformulation profiles
- Rationale for selection of solid-state and chemical form

- Revised and new internal methods—qualified
- Impurity profiles for new batches, comparison to previous (Ames) batch(es)
- Rationale and quality control for genotoxic impurities in preparation for EIH
- Analytical profiles or certificates of analysis for the API and toxicology test articles
- Short-term stability of toxicology test articles

C. Phase 0: Preclinical Development/Preentry into Humans

Following candidate selection, planning for EIH becomes the focal point of the team's efforts with emphasis on development, manufacture, and release of clinical materials. A formal product development team may be formed and a development plan is prepared to define the target profile for the successful candidate and criteria to be used to assess progress toward that profile.

During phase 0, the analytical research and development group continues to learn about the chemistry and stability of the new compound and supports ongoing toxicology studies. Analytical support for toxicology studies includes GLP studies with additional species and/or longer exposure (i.e., at least 2–4 weeks dosing) as well as the GLP Ames, safety pharmacology, genotoxicity, and pilot teratology studies. Analytical scientists collaborate with the toxicologists and GLP Study Directors to plan for GLP testing of the toxicological supplies. Additionally, the formulations group begins to develop a clinical formulation suitable for EIH, requiring extensive analytical support for excipient compatibility studies and stability studies of prototype formulations. The clinical formulation for EIH is selected based on satisfactory stability data and, preferably, a BA study in a suitable animal model. The animal BA study is performed to assist in predicting BA in humans. The analytical scientist collaborates with the biopharmaceutics scientists such as drug metabolism pharmacokinetics (DMPK), clinical pharmacology, formulation, and modeling and simulation begin identifying potential predictive *in vitro* models for assisting in design and development of dosage form for EIH and beyond.

Analytical methods continue to be improved while evaluating changes in synthesis or quality of the API. The preferred methods of analysis are identified. The analytical methods are qualified using selected validation criteria and data are compiled for establishing IND specifications for clinical materials.

Analytical documentation during phase 0 to enabling EIH generally consists of:

- Updated analytical and preformulation report of the API—summary of physical and chemical properties as well as excipient compatibility
- Salt and polymorph selection report
- Reference standard characterization, including for active drug, and where feasible for impurities, and degradation products
- New and revised internal methods (for cleaning validation; release and stability testing of the API and toxicology formulations)
- Analytical methods validation summaries/reports—qualification of methods for GLP toxicology studies

- Impurity profiles for new batches and comparison to previous toxicology batches
- Rationale and quality controls for genotoxic impurities
- Certificates of analysis—API
- Analytical profiles or certificates of analysis for toxicology batches
- Analytical testing data for in-life toxicological study samples (homogeneity, periodic testing, and end-of-study)—the testing of samples from the animal dosing site required as per GLPs
- Ongoing stability of toxicology test articles or stability studies of new toxicology formulations
- Characterization of novel and functional excipients, if applicable
- Pre- and prototype-formulation studies for clinical formulations

D. Phase I: Entry into Humans/Opening IND

Phase 1 clinical trials include introduction of the investigational drug into humans. The studies may be conducted in patients or normal subjects, depending on the therapeutic area. They are designed to determine the side effects associated with increasing doses and, if possible, to gain early evidence of effectiveness. Phase 1 studies also include research into drug metabolism, pharmacokinetics, and mechanism of action in humans. Total phase 1 exposure is generally in the range of 20–80 subjects.

An IND application must be submitted to the FDA prior to the first clinical trial in the United States. In Europe, IMPD/CTA submission is required, and the content and expectations may vary country-to-country. For IND/IMPD EIH studies, the project team must plan for the manufacture and release of clinical trial materials as well as for the IND/IMPD documentation. A regulatory submission may not be required in certain ex-US/EU countries if the study is conducted in volunteers. For a study in volunteers not requiring an IND/IMPD, documentation needs are reduced but all other activities remain the same.

Within analytical R&D, specifications and methods for clinical trial materials are approved and results from short-term stability evaluations are reported. Phase dependent method validation is completed for critical validation parameters for those methods used in release and stability testing of the clinical materials. Impurities and degradation products are tabulated for the toxicology and clinical batches to assure the safety of the materials for human use. Significant attention is given to the potential genotoxic impurities, and strategy for testing and control should be included in the IND.

The primary analytical mission at this stage is to ensure that adequate documentation for identity, quality, consistency, and stability of the clinical materials is available to bridge the drug development as safely as possible from animals to humans.

IND-enabling analytical documentation includes:

- Characterization of the reference standard for purity, impurities, physical–chemical properties related to BA
- Rationale for selection of polymorphic and salt form, if applicable
- Batch profile—toxicology and clinical batches

- Supporting information for known impurities, degradation pathways, and products
- Rationale and quality control for genotoxic impurities
- Rationale for the IND specifications (API and drug product)
- Specifications (test, limits, and methods) for API and drug product(s), including placebo (release and retest/stability, if different)
- Methods qualification/validation for methods included in IND specifications
- Stability data
  - Available data from stressed studies of API.
  - Available data from accelerated (generally at least 1-month) and long-term studies of clinical formulation(s)
- Functional excipient characterization and control strategy, if applicable
- Certificates of analysis for the API and drug product to be used in the clinic

E. Phase 2: Proof of Concept/Safety Studies

Phase 2 trials are the early controlled clinical studies conducted to evaluate the dose ranging and effectiveness of the drug for a particular indication in patients with the disease and to determine common short-term side effects and risks associated with the drug. General exposure is usually less than several hundred subjects.

Many of the phase 2 activities are simply a continuation of studies initiated during phase 1. Previously, this was a relatively quiet period in the development life cycle. More recently, the respite is short-lived, if at all. The accelerated pace of development has compressed all aspects of drug development and the pace rarely slows down. There is no slowing of activities if FDA designates a compound for fast track development. Phase 2 regulatory activities often include an end of phase 2 (EOP2) CMC strategy meeting with FDA to discuss and obtain concurrence on plans for registration activities and development toward NDA. Analytical testing and documentation become critical for bridging changes during development in the API's properties (e.g., impurities and degradation products, physical form, particle size distribution, surface energy, etc.) and dosage forms (e.g., *in vitro* predictive models, such as dissolution or other physical characterization data such as particle size distribution, physical form, etc.). Analytical characterization data to be generated at the interface of the API and drug product in preparation of the phase 3 and market formulation should be well thought out and planned in advance. The existence of ICH guidances Q1–Q6 (listed in Appendices I and II) are specifically aimed at analytics.

Analytical support is needed for development and optimization of the proposed registration process and potential commercial formulations. Methods may need to be revised in line with changes to the synthesis or formulation. Analytical methods are further optimized for analytical purposes. Changes to the procedures may necessitate (re)validation. Also, characterization of the API continues. Evaluation and control of stereochemistry (for stereoisomeric compounds) and/or potential solid-state forms (e.g., hydrates, solvates, or polymorphs) remain critical concerns. Release and stability testing continues for long-term GLP studies, including carcinogenicity studies, if required, and for ongoing clinical studies.

Analytical documentation during phase 2 includes IND amendments and internal development reports:

- Identification of new impurities or degradation products
- Revised specifications and methods of analysis (including analytical improvements as well as new synthetic routes/processes and new formulations)
- Stability updates/reports
- Certificates of analysis for API batches and clinical supplies

F. Full Development Decision Point

Results of the phase 2 clinical trials form the basis for a firm's decision to proceed with full development. The full development decision is critical because it results in commitment of significant financial resources and capital investment for the larger, longer term clinical trials required for marketing approval. The additional money and resources required for initiation of CMC registration activities are equally critical. This involves the manufacture of registration batches of the API and drug product for NDA/BLA submission. Major capital expenditures are often required at this time for the manufacturing sites to produce the required registration batches.

G. Phase 3: Pivotal Efficacy Trial(s); Pre-NDA Planning/Development

Phase 3 studies are performed after evidence of efficacy of the drug has been obtained. The expanded clinical trials are intended to gather additional data regarding safety and effectiveness, as well as determine the risk–benefit relationship of the drug. Phase 3 studies may range from several hundred to several thousand patients. Often a bioequivalence (BE) or BA study is included to demonstrate comparability of the proposed market formulation with respect to the clinical formulation(s) used in the previous safety and efficacy studies. Plans for NDA preparation are initiated in anticipation of favorable clinical results from the phase 3 trials.

Analytical R&D is a central participant in the various phase 3 activities. Testing and release of clinical trial materials (API and drug product) continue, including supplies of the proposed market formulation for a BA/BE study versus the previous clinical trial formulations. Also, extended physical and chemical characterizations are performed for the registration batches (API and drug product) to demonstrate technical equivalence of the proposed market product and pivotal clinical formulation(s). Long-term stability studies of previous clinical formulations continue. Accelerated and long-term stability studies of the registration batches are initiated.

Analytical characterization of the APIs and dosage forms during definition of the market formulation and registration enabling scale-up studies have become an integral part of the QbD and PAT initiatives. Analytical scientists collect a large amount of characterization data at the interface of the API and dosage form as well as for the establishment of CQAs, CPPs, and control strategy. The QbD concept is a systematic and scientific approach to product and manufacturing process design and development that utilize multivariate data collection and modeling to assist identification of CQAs, CPPs, design, and control space. This wealth of analytical data is important for completing the QOS for the NDA/BLA.

Typically, a pre-NDA meeting is held with FDA to ensure the submission is complete and meets all regulatory requirements. Ideally the pre-NDA meeting should be conducted approximately 6 months prior to the planned NDA submission. At this time the sponsor should gain concurrence on any areas of concern regarding the content of the NDA. The firm should also provide an update on progress against issues discussed at the EOP2 meeting—closure or additional work to be done. Analytical data are presented to link materials used in toxicology studies, human dose ranging, pivotal, and BE studies. The most critical links from the safety and equivalency perspectives concern the data for purity/impurities, stability, and *in vitro* product performance (e.g., dissolution). It may be necessary to obtain agreement that the available data are adequate and no further *in vivo* comparison is needed. Occasionally a firm will propose to submit less than 12-month stability data in the original NDA, with an update provided during the review period, in order to move up the filing date and accelerate the approval date.

Draft documents for the NDA are needed during phase 3 both for the pre-NDA meeting and for compilation and team review of a draft of the CMC section.

Documents required for pre-NDA include:

- Pre-NDA meeting briefing package
- Draft NDA documents:
  - Drug substance:
    - Nomenclature, formulas, and proof-of-structure
    - Physical–chemical characterizations
    - Stability—stress, accelerated, long term; storage, retest/expiry
    - Comprehensive summary table for batch profiles
    - Specifications—methods, tests, limits, rationale, validation
    - Identification of CQAs, CPPs, and control strategy
  - Drug product:
    - Stability—accelerated, long term (stress studies generally not performed for drug product)
    - Comprehensive summary table for batch profiles
    - Specifications—methods, tests, limits, rationale, validation
    - Identification of CQAs, CPPs, and control strategy
    - Batch tracking—drug substance and drug product

An expanded table of contents utilizing the CTD format for the NDA is provided in Appendix II.

Note: Biological pharmaceuticals require extensive biochemical and bioactivity comparability studies for materials used for toxicology, clinical and proposed market supplies.

H. NDA—Submission, Approval, and Launch

The final development milestones for a new pharmaceutical product are the submission and approval of the NDA and product launch. During this stage, the team must finalize all NDA studies (preclinical, clinical, and technical) and study reports. The documents are compiled and the original NDA is submitted to the health authorities for review. Some studies continue following NDA submission. Updated reports may be submitted during the review period or the reports should be prepared and kept within the company in anticipation of questions. Prior agreement may be needed from the health authorities to permit submission of amendments during the review period.

In addition to the NDA documentation, the technical areas must prepare for PAI and validation for launch. Special development reports may be written to provide a comprehensive summary of the development history in a given technical area for PAI. Other documents (e.g., investigation reports, certificates of analysis) are simply assembled for easy retrieval. Protocols and final reports are prepared for validation and technology transfer. Thus, the required documentation at NDA filing includes all documents explicitly listed in the NDA table of contents as well as additional compliance and research reports to support investigations and decisions.

During the review period, any of the CMC sections may be revised in response to comments from the review chemist or field inspector. Responses may include postapproval commitments for additional work or data updates. For a globally marketed product, comments will be received from individual health authorities over a period of time—often extending for years. Changes in response to comments from the health authority for one country may necessitate overall changes if a firm wishes to harmonize internal requirements and keep a single dossier. As a result, the approval and launch period has a very hectic pace, tapering as additional approvals are received but continuing over an extended period of time.

Analytical documentation for NDA and PAI consists of:

- Finalization of all documents for the CMC section of the original NDA submission (listed above and in Appendix II).
- Compliance documents for PAI.
- Revision/update of NDA documents in response to comments by regulatory authorities.
- Change control, if needed.
- Certificates of analysis for corresponding samples of API, drug product samples, reference standards to permit the FDA laboratories to run duplicate analysis to verify analytical validation of the proposed test methods.

I. Phase 4: Marketing

Often even before the first marketing approval, firms begin planning extension and alternate supply for market. New indications, line-extensions, and/or postapproval supply changes are considered and possibly initiated before approval. Line extensions may include new formulations for other routes of administration or more convenient dosing regimens. Development of pediatric formulations in pursuit of additional exclusivity in the United States has been a major area of activity. Similarly, a geriatric indication is becoming more important. Clinical trials for new indications do not require support from the technical team. However, those which involve new formulations require extensive CMC support. These changes may result in an sNDA but often involve a new NDA for a new dosage form.

Analytical documentation for an sNDA or a new NDA is the same as that for the drug product of the first NDA. Additionally, comparative data may be needed to illustrate the similarities or differences versus the reference marketed product.

Analytical documentation for sNDA or NDA for line-extension consists of:

- Drug product:
  - Stability—accelerated, long term (stress studies generally not performed for drug product), statistical analysis
  - Comprehensive summary table for batch profiles
  - Specifications—methods, tests, limits, rationale, validation

The postapproval period is also very active for changes in the manufacturing sites, processes, or controls. All of these changes should be managed globally via the firm's change control system. All changes require supportive analytical data, generally associated with demonstrating equivalence with the approved material so that analytical documentation for postapproval changes is specific to the type of change.

IV. REGULATORY DOCUMENTS

Although the phases in the life cycle of a new drug are sequential, the process for preparation of regulatory documents for IND and NDA submission is not linear. Most of the development activities overlap so that the associated documentation is highly interdependent. The biggest efforts related to regulatory documentation are concentrated around the opening IND and the NDA. Planning and preparation for these and other critical regulatory submissions are necessary to meet development timelines. NDA planning is especially critical to ensure on time completion of a quality dossier. In all cases, authors must collaborate closely with other contributors to ensure completeness, consistency, and accuracy of the regulatory documentation.

Throughout drug development, the analytical group works on multiple tasks simultaneously (e.g., method development or optimization and validation, release and stability testing). Most of the tasks are interrelated so that changes in one study or report may necessitate changes within others. The group needs to ensure integration and consistency across related documents—within and external to analytics. Regulatory updates are prepared periodically or following changes to the API or clinical formulation.

Although some documents are required only for NDA submission, the major types of analytical reports for regulatory submission are very similar for IND and NDA. The differences in these reports for IND or NDA submission primarily concern the amount of data (e.g., number of batches, duration of stability studies) and level of detail. Also, the requirements for amendment of the information are different. The major types of analytical documents common to IND and NDA submissions are listed below:

- Drug substance:
  - Physical-chemical characterizations
  - Reference standard characterization
  - Specifications and analytical methods
  - Certificates of analysis
  - Methods validation
  - Stability (stress, accelerated, and long-term conditions)
- Drug product:
  - Specifications and analytical methods
  - Certificates of analysis
  - Methods validation
  - Stability (accelerated and long-term conditions).

The analytical concerns and activities in support of IND and NDA submissions are described in the following sections. Specific requirements on content and format for US regulatory documents are contained in the Code of Federal Registry and numerous FDA issued guidances. Certain global requirements are captured in the ICH Quality guidelines. Because global development and marketing are becoming the norm for many companies, a modular approach for preparing a global CMC NDA is also described. A list of the most frequently used guidelines and a generic table of contents for a US NDA are provided in Appendices I and II of this chapter.

A. IND Documentation

Under current US regulations, any human use of an investigational drug product first requires submission of an IND to the FDA prior to conducting clinical trials in humans. The sponsor provides short-term animal toxicity testing to predict whether initial human studies will be acceptably safe and, if possible, the drug's expected therapeutic activity. Once an IND is filed with FDA, the sponsor must wait 30 days prior to initiation of clinical studies. During the 30-day waiting period, FDA reviews the submission to ensure that human subjects will not be subjected to unreasonable risks. Testing in humans may begin if no comments or questions are received from FDA within the 30-day review period. Clinical testing proceeds progressively through phases 1–3 where each phase is more extensive than the previous phase with respect to the number of subjects and duration of treatment.

It is required that an IND or IMPD is submitted for FDA/EU country-specific review and authorization prior to the initiation of the first clinical study for EIHs. As more firms undertake global development, EIH studies in healthy volunteers may be performed ex-US/EU. The US IND or EU CTA must still be filed prior to first use of the investigational drug product in the US/EU.

1. Enabling Preclinical Studies

Throughout drug development, preclinical animal studies are used to predict and assure the safety of human clinical trials. The preclinical program proceeds ahead of the clinical program in order to provide the necessary safety/

toxicity information in a timely manner. The sponsor must relate the drug product being proposed for use in clinical studies to material used in the preceding animal toxicology studies that support the safety of the proposed human study. This is often done for the IND using a batch-tracking table. The table lists the toxicology and clinical batches manufactured to date with a summary of the analytical data related to safety and BA (i.e., purity, impurities, and certain physical and chemical characteristics). Additionally, chemistry or manufacturing differences between the toxicology and clinical materials should be described and potential impact of the differences on the safety profile should be discussed.

2. CMC IND Information

The CMC information for an IMPD includes summaries of information related to the quality, manufacture and control of the Investigational Medicinal Product, data from nonclinical studies and from its clinical use. An overall risk–benefit assessment, critical analyses of the nonclinical and clinical data in relation to the potential risks and benefits of the proposed study have to be part of the IMPD. It is in principle similar to the US IND and provides the basis for conducting clinical trails.

The US IND must include sufficient CMC information to ensure the proper quality, identity, purity, and strength of the investigational drug for the planned clinical trial. The type and extent of information submitted to the IND depend on the clinical phase of the investigation with respect to the size of the study and duration of treatment, the dosage form and the amount of information available. Considerations in the CMC data to be submitted concern the nature and source of the API and the clinical dosage form. The methods of preparation of the new API and dosage form are expected to be modified as the clinical investigation progresses. Updates and revisions of the CMC section during clinical trials (e.g., manufacturing process, formulation, and specifications) should be submitted to the IND in an information amendment and routine stability data updates may be submitted at the time of the annual report.

Following enactment of the PDUFA, the FDA issued a guidance on content and format of INDs for phase 1 studies (issue date November 1995). More recently, a FDA guidance on content and format for CMC information required for INDs for phases 2 and 3 studies was issued (May 2003). These guidances restate current regulations, emphasizing the progressive nature of CMC information, and clarify the data to be included in the IND based on the phase of development. The recommendations in the IND guidances focus on information for evaluation of the safety of the subjects in the proposed study. Recommendations are also provided regarding supplementary data and information for the phase 3 pivotal study that corroborate the quality and safety criteria. The IND guidances are intended to expedite drug development and entry into the marketplace while ensuring that FDA has sufficient data to assess the quality of the clinical trial materials and the safety of the proposed clinical studies from the CMC perspective.

The emphasis in the IND for phase 1 studies is on patient/subject safety. The CMC emphasis is placed on the identification and quality control of the new API.

As clinical development progresses, additional information should be submitted for new processes and larger scale manufacturing procedures used to produce materials for the phases 2 and 3 clinical trials. Information for phase 1 studies may be submitted in a summary report with brief descriptions. The summaries are expanded and data are added as development and clinical trials progress. For example, established specifications and method validation need not be submitted for phase 1 but a detailed listing of all tests with a complete description of the corresponding procedures with validation information is expected by phase 3.

Required analytical information for IND includes the following items:

- Description, physical and chemical characteristics of the API
- Evidence of the proposed structure
- Test methods and "proposed acceptable limits" for the clinical trials materials
- Analytical data and certificates of analysis
- Stability study and test methods used to monitor stability of the API including preliminary stability data for representative material

3. IND Meetings

In the United States, the FDA provides firms with opportunities to discuss technical issues directly with the review division. These meetings, scheduled at key development milestones, are critical to the effective development of a new drug product and to the timely submission and approval of the product for market. These meetings provide the sponsor with the opportunity to discuss with and also update FDA on the development progress and to obtain feedback regarding proposed changes that may impact the NDA application. The FDA guidance "IND Meetings for Human Drugs and Biologics Chemistry, Manufacturing, and Controls Information" (May 2001) formalizes procedures for CMC IND meetings including pre-IND, end-of-phase 2 (EOP2), and pre-NDA (or pre-BLA for biological products). Another important FDA guidance is "Formal Meetings between the FDA and Sponsors or Applicants" (May 2009). These meetings can be used to address questions and scientific issues that arise during the course of the technical development. The discussions are intended to aid in resolution of problems and to facilitate evaluation of new drugs. Procedures for IND meetings are described in these guidances.

The pre-IND meeting is intended to discuss potential safety issues related to the investigational drug and to identify and avoid potential clinical hold issues. Reasons for a clinical hold are identification of a safety concern or insufficient data to make an evaluation of safety. Safety concerns related to CMC arise when the physical–chemical characteristics or the impurity profile of the clinical trial material is insufficiently defined so that the potential health hazard cannot be assessed or the product cannot remain stable through the proposed clinical study. Pre-IND meetings are most useful for biotechnological drug products, novel dosage forms, or drug-device combinations.

The EOP2 meeting provides a firm with an opportunity to discuss their plans and protocols for phase 3, NDA and market. A CMC-specific EOP2 meeting is held principally as a CMC strategy discussion. It is the most critical IND meeting for planning the CMC portion of the NDA. The goal of the meeting is to identify potential problems which may delay marketing approval at a sufficiently early stage so as to avoid or minimize their impact on approvability and ensure that meaningful data will be generated during phase 3 studies. Typical discussion points concern reduced stability protocols for registration batches; bridging strategies for clinical, registration and commercial materials; proposed NDA specifications and other product-specific development problems. The outcome of a successful EOP2 meeting is a clear understanding of the CMC data needed for an approvable NDA.

The pre-NDA/BLA meeting is intended primarily to identify and resolve potential filability or approvability issues. Other goals are familiarization of FDA reviewers with the product to facilitate review of the NDA, update on progress or changes since the EOP2 meeting, and to discuss site readiness for inspection. Often, the firm will propose a reduction in the amount of stability data at filing, with a commitment to update during review, in order to move up the filing date. The meeting is critical to ensure that the CMC review team is in agreement with the format and organization of the information to be presented in the upcoming NDA/BLA submission.

B. NDA Documentation

NDA preparation requires extensive planning and coordination among authors to ensure consistency and completeness of the dossier. Analytical data are the common thread to all CMC documents and are highly intertwined with other reports in the CMC section. Documents related to biobatches also need to be integrated between the CMC and clinical sections of the NDA. The value of cross-functional review of documents cannot be overemphasized. It is highly recommended that a review be conducted of the complete CMC section by all functional CMC team members and an independent technical reviewer in order to ensure that there is consistency and continuity of presented data.

The ICH requirement for 12-month stability data in the NDA submission has two major effects on the technical planning and NDA:

(1) The physical registration work (i.e., manufacture of registration batches) is often completed early in phase 3. In principle, the related documents can be written shortly thereafter. In practice, most documents are not written until shortly before filing.
(2) The QOS, stability reports and drug product specifications will frequently define the time critical path for NDA filing. These documents may be prototyped based on data from clinical batches and available data from registration batches. They cannot be finalized until the last stability data points are reported.

The NDA must contain information that establishes the linkage between dosage forms used in clinical trials and that intended as commercial product. The firm must demonstrate that changes in API or drug product during scale-up for market has not affected the safety or BA/BE of the dosage form. The linkage can be made using *in vivo* and/or *in vitro* data (e.g., impurity profiles, dissolution profiles). By the time the NDA is compiled, data should have been assembled to support the safety and BA of the proposed market product relative to the clinical materials from the primary efficacy study(ies). Additionally, data for the registration batches should demonstrate the consistency of the proposed commercial process and the quality and stability of the API and drug product to be marketed.

Specifically, the CMC section of the NDA should have data to support the following statements.

- For the API:

The synthesis (i.e., chemistry and process) is understood and the quality of API is well characterized and controlled. Appropriate controls have been established for the raw materials, starting materials, and in-process tests to ensure that the process can consistently produce API of high specific quality. Appropriate release and retest specifications (methods and limits) have been established based on data from development lots (clinical and registration) to assure the identity, quality, and purity of the material. The impurity profile for registration batches is equivalent or superior to batches used in toxicology or clinical studies—all significant impurities have been qualified at the specified levels in preclinical or clinical studies. Physical characteristics are controlled, if needed, to guarantee manufacturability and BA for the dosage form. The quality of the API is maintained under the recommended storage conditions through the proposed retest period.

- For drug product:

The proposed market formulation is bioequivalent to investigational materials used in the pivotal clinical trial(s). The drug product manufacturing process has been thoroughly studied and is understood and controlled. Appropriate process parameters have been assessed and adequate in-process controls are in place to ensure that the process can reproducibly produce drug product of high specific quality. Appropriate specifications (release and stability, if applicable) have been established for the excipients and finished drug product. The stability profile of the proposed market formulation is equivalent or superior to that used in toxicology or clinical studies. Degradation products have been qualified at the specified limits in preclinical or clinical studies. Stability data on registration batches support the proposed expiration date.

The analytical group produces all of the CMC documents which support the above development work. The support is mostly derived from bridging analytical data which demonstrate technical equivalence between the proposed market formulation and the clinical formulation(s). Technical data from analytical comparability studies or *in vivo* BE studies can be used to demonstrate equivalence in API synthesis, drug product formulation, packaging, site and/or process changes.

BA (and related toxicity) issues are addressed in the physical–chemical characterizations of the API, the solid-state form of the API (rationale for selection and comparative physical–chemical characteristics), particle size controls and/or dissolution data for solid oral and suspension dosage forms. Other safety issues concern purity and impurities of the API and drug product. Within the NDA, the batch-tracking summary table compares the impurity profile during development up to and including material produced by the registration process demonstrating consistent and improved quality. It is finally assured by the proposed quality control measures. In making this case, the discussion of the quality controls should include explicit remarks regarding the choice of routine tests and limits as well as the rationale for the controls and qualification of specified impurities.

The stability of the proposed market product is preferably demonstrated first by satisfactory results for the primary registration batches under accelerated and long-term storage conditions. Alternately (or additionally), the stability profiles for the primary registration batches can be compared to those for supportive development batches where longer real-time data may demonstrate stability over prolonged storage. Stability assurance is also demonstrated by a good understanding of the degradation pathways and products determined during stress stability studies.

C. Global CMC NDA

Through the efforts of the ICH, agreement has been achieved in both format and structure with regards to the presentation of technical information to support a product registration of a new pharmaceutical product for human use. From these efforts a CTD harmonizing the CMC regulatory requirements for global development and marketing has been achieved. The CTD provides a harmonized structure and format for the new product application. The CTD was agreed upon in November 2000 in San Diego, US. The CTD is divided into four separate sections. The four sections address the application organization (M4), the quality section (M4Q), the safety section (M4S), and the efficacy section (M4E) of the harmonized application. The agreed upon implementation date for the CTD, in the ICH regions, was July 2003.

The common format, which standardizes the presentation and location of data, should facilitate the preparation and health authority review. Ultimately, the aim is to attempt to reduce the time and resources used to compile and review marketing applications through the use of common standards and formats. In the United States various guidances have been drafted, issued, and withdrawn for consistency with the ICH guidelines. In August 2001, the FDA issued a guidance for industry entitle "M4Q: The CTD-Quality". In this guidance, the format for the quality section of the CTD is described. The Code of Federal Registry, Title 21 Part 314–Applications for FDA Approval to Market a New Drug or an Antibiotic Drug, section 314.50 speaks to the content and less to the format of an application and therefore should also allow consistency with CTD.

A pre-NDA meeting should always include a discussion and agreement on the presentation of CMC information. The specific areas of analytical focus are within the Module 2.3—Quality Overall Summary and Module 3—Quality. The analytical scientists have a major area of contribution within the following specific sections of these modules:

- Drug substance:
  - S.3 Characterization
  - S.4 Control of Drug Substance
  - S.5 Reference Standards
  - S.7 Stability

- Drug product:
  - P.2 Pharmaceutical Development
  - P.4 Control of Excipients
  - P.5 Control of Drug Product
  - P.6 Reference Standards
  - P.8 Stability
  - 3.2R Regional Information
  - Methods Validation Package

The CMC section for a NDA consists of a complete set of modules (2.3 and 3) where each individual module is global in its contents. Some additional country-specific documents (e.g., Environmental Assessment and Method Validation package for United States and QOS for EU and United States) still need to be prepared but many of the other technical reports may be suitable for both dossiers by simply reordering. The final EU Marketing Application (MA) or US NDA will consist of a specific arrangement of the modules with the appropriate (country-specific) headings. This modular approach reduces the redundancy and effort in preparing the documents.

V. COMPLIANCE DOCUMENTS

Compliance documents refer to those reports required by GLPs and GMPs and/or utilized during the course of inspection by a health authority. Compliance documents are needed in addition to the regulatory and research documents to demonstrate integrity of the data. A good foundation for the compliance documents begins with well thought out and established quality and compliance systems and practices within an organization. A team, consisting of representatives from R&D, QC, and QA, develops these systems. It should be one of the top priorities of management and the QA unit(s) to develop and monitor these systems to comply with the CGMP and GLP expectations.

Both general (GMP) inspections and product-specific PAIs focus on documents related to the quality systems and the product-specific controls and integrity of the analytical data. The main difference between the two types of inspections is the emphasis: General GMP inspections begin with audit of the underlying quality systems and SOPs, evaluating selected product-specific data as representative of compliance to those systems. PAIs initially concern the product-specific data for a pending application and then evaluate some aspects of the quality systems to gauge overall compliance with GMPs.

GLP inspections, in addition to the GLP quality systems, focus on the study specific analytical GLP Toxicology Study Reports describing in-life dose confirmation and stability of the test article, and any deviations during the study.

Typical analytical compliance documents include:

General quality system documents

- SOPs
- Investigation procedures—required for OOS results and product failures
  - Preliminary laboratory investigations or work sheets-required for authorizing reanalysis
  - Tracking system

- Training records
- Equipment maintenance and qualification
- General marketed stability protocol
  - Sampling of batches
  - Testing in accord with protocol
- Methods validation
- Validation of stability chambers
- 21 CFR Part 11 validation of data acquisition system and LIMS
- Sample tracking and storage
- Reference standard control (active and impurities)
- GLP study reports
- CRO audits and qualification

Product-specific documents for PAI

- Analytical development reports—drug substance and drug product
- Investigation reports—required for suspect and OOS results and product failures
- Certificates of analysis for all toxicology and clinical supplies
- Batch profiles—bridging preclinical, clinical, registration, and validation batches, if available
- Reference standard characterization
- Stability reports and updates—drug substance and drug product (registration, clinical, and toxicology)
- Specifications—IND and NDA
  - Method validation
  - Rationale
- Impurities specifications (including genotoxic)—rationale and methods
- Cleaning validation

Technology transfer document

Technology Transfer documents are one of the most reviewed types of documents during an inspection. Responsibilities for analytical support of a new drug are often transferred to other sites or other departments. Typically, there will be at least one transfer between the analytical research and development group at the development site(s) to the quality control unit at the launch site(s). Other technology transfers may arise because of the firm's functional organization or changes in supply strategy. The development history including change control and knowledge must accompany the transfer in responsibilities.

Analytical development report

It is extremely beneficial to have comprehensive analytical development reports describing the evolution of the analytical methods for a new drug substance or drug product. Separate reports are prepared for the drug substance and product. The reports trace the chronological history of the IND specifications up to the proposed NDA specifications, itemizing changes in tests, specifications and methods associated with major revisions. Additionally, the development of key methods used to assure identity, potency, purity/impurities, and performance (e.g., particle size or dissolution) should be discussed in detail. Summaries for the critical methods should explain the rationale for the selected method, reasons for changes made during development, and the equivalency or superiority of the final, optimized methods. Ultimately, the development reports should collect the pivotal analytical data which demonstrate that the quality of the drug substance or product and the associated controls have been established logically and rationally.

The contents of the analytical development report should specifically address the following items:

- Choice of routine tests for determining identity, purity, critical physical characteristics, or performance of the drug substance or product
- Additional/informational tests used during development
- Rationale/supporting data for exclusion of tests from routine QC for commercial products
- Basis for specification limits (including batch history, qualification of specified impurities with particular control strategy for genotoxic impurities)
- Rationale and development of analytical methods
- Validation of proposed methods (including consideration of QbD concepts)

Additionally, two key issues to be discussed in detail are:

- Impurity profile of the drug substance or drug product
- Dissolution profile for the dosage form (if applicable)

Both the impurity and dissolution profiles must be shown to be consistent throughout development, or additional preclinical or clinical studies may be needed to bridge the differences. The impurity profiles must be carefully monitored and controlled batch-to-batch and especially following changes to the process or formulation. Significant changes in the impurity profile may affect the acceptability of prior preclinical and clinical data. When significant changes occur in the dissolution profile, a BE study may be needed to demonstrate equivalence of the dosage forms. Analytical scientists have made advances in identifying predictive *in vitro* and *in silico* models to bridge formulations and assist in identifying control strategy. Documentation describing this learning is usually captured in research reports.

Analytical development reports are not part of the NDA. They are primarily internal documents. The analytical development report serves both as a key reference document during inspection by health authorities and to facilitate technology transfer between departments and/or sites. It is important to capture the QbD concepts for NDA analytical methods. Following marketing approval, they may also provide a retrospective history of the methods for use by the QC laboratory. In order to maximize the utility of the development report, it should provide a thorough summary of the data assembled during the development, including results of successful and unsuccessful experiments leading to the final controls. Emphasis should be placed on *critical* evaluations and conclusions for changes made during development. All data in the report should be fully referenced. References should include previous research reports, journal articles, patents, laboratory notebooks, etc.

QA review of analytical documents

QA's role in review of analytical documents (like all critical compliance documents) is to ensure that data are accurate, complete, and meet all applicable CGMP and GLP regulations. The extent of QA review, audit, and/or approval of analytical documentation depends on the company philosophy and organization. Regardless of a firm's policy for QA oversight, early involvement by QA is becoming more routine because of increased awareness of compliance requirements in the industry, and tighter regulatory standards for both clinical and commercial materials.

Key documents that explicitly require review and approval of the QA unit are as follows:

- Certificates of analysis
- Investigation reports
- SOPs
- Analytical change control
- Technology transfer protocols and reports
- Software (e.g., data acquisition or LIMS) validation protocols and reports

Other documents, which are reviewed or audited, though not necessarily approved by QA, include:

- Specifications and analytical methods
- Methods validation protocols and reports
- Stability reports
- Shelf-life assignment
- Characterization reports
- Preliminary investigation worksheets/laboratory investigations
- Instrument qualification and validation documents
- Laboratory notebooks

A close partnership with the QA unit assists with establishment of quality systems, which are consistent and compliant across departments or sites. QA oversight of the documentation facilitates responses to questions during inspection by health authorities and helps to assure a favorable outcome.

A. Preapproval Inspection Teams

PAI teams may be formed up to 18 months before the targeted NDA filing date. The team is comprised of cross-functional representatives including R&D, QA, DRA, and manufacturing to assess the risks specific to the quality and approvability of a new product. The PAI team assists the project team to develop plans for approvability of a safe and quality product for the market place.

QA personnel with the help of analytical members of the PAI team audit the following product-specific documents:

- CMC NDA documentation and supporting data (preferably, prior to filing to ensure that all data and reports to be submitted in the NDA are accurate and complete)

- Lab notebooks and certificates of analysis for reference standards, clinical and registration batches, etc.
- Updated stability reports
- Technology transfer reports for drug substance and drug product
- Specifications and analytical validation (including raw materials, packaging components)
- Available analytical data
- Laboratory investigations and investigation reports
- Cleaning validation

The above checklist is not all-inclusive and a product-specific checklist should be compiled by the PAI team to assist in preparation for a PAI.

VI. RESEARCH DOCUMENTS

Research reports are a firm's internal mechanism for recording critical events in a product's history. An established mechanism for recording and archiving significant information is usually extremely useful in light of inevitable turnover of people and projects during the discovery and development phases. Other key advantages of a good documentation system for research findings (positive or negative) are information sharing between projects and for helping in decision making or planning for next steps.

Internal summary reports and formal proposals for management are often assembled in preparation for development milestones (e.g., candidate selection, full development decision).

Research reports for critical data are written for products as they progress in development. However, reports for failing studies or discontinued programs are often not prepared and these can be equally useful since history tends to repeat itself. Because of the time and resources required to prepare such reports, a conscious decision should be made if the value added warrants the effort.

Research reports are also emphasized in early development before regulatory documentation is required. However, recent changes with respect to reductions in the documents and extent of data that must be provided in IND submissions are increasing the importance of internal documentation through early clinical development (i.e., phases 1 and 2). Specifically, FDA guidances on IND CMC requirements have relaxed the amount of data required for inclusion in the IND. Summaries are often allowed *in lieu* of detailed data. However, from both a compliance and historical perspective, it is critical to collect these data in a report for future reference. Thus, the documentation requirements overall have not been reduced but have been removed from the regulatory submissions and added to the in-house archives.

Typical research documents include:

- Salt and polymorph screening
- Preformulation reports
- Analytical profiles

- Special technology
- Evaluations of experimental prototype formulations
- Milestones reports and management proposal documents for positive or negative decisions
- Research reports containing critical data needed for future regulatory needs

A. Analytical Document Creation and Electronic Submission

Analytical scientists have welcomed ICH M4 Q—Common Technical Documentation (CTD) as a globally accepted format to provide information and scientific data in drug application submission. Recently, FDA and EU regulatory agencies have encouraged and allowed applicants for the electronic submission of the common technical documentations using ICH eCTD specification.

Computers and information technology have penetrated almost every aspect of pharmaceutical industry over the past few decades. Today, it is common and almost a requirement for any analytical instrument to connect to a computer or to generate computer readable data. An analytical laboratory often uses a LIMS to manage its data and results, and a computer documentation system to manage its protocols, reports, compliance documents, and regulatory documents. A recent trend in the industry is the implementation of ELN to replace conventional paper laboratory notebooks. ELN promises to link laboratory instruments and LIMS together to improve the ability to acquire, manipulate, share, store, and search data, and to serve as a scientific knowledge repository that can be easily accessed. ELN also has the potential to generate various analytical documents using the stored data and results directly for submission purpose. This progress gives hope that one day, a seamless electronic process can be developed from analytical data generation to analytical documentation creation to NDA submission.

VII. SUMMARY

Pharmaceutical analytical documentation ensures the identity, purity, stability, and consistency of drug substance and dosage forms used during investigational studies and for commercial market in accord with the governing regulatory guidances and policies. Prior to marketing approval, analytical research and development personnel support product development and optimization activities. The data and documentation evolve in parallel with the clinical development of the product. Safety is the primary concern throughout the development phases and the analytical data provide the necessary assurance of safety based on BA, purity, and stability. These analytical data become a critical part of the NDA for marketing authorization. Following marketing approval, quality control personnel provide the data to assure consistent quality and stability for the marketed product and to support the inevitable changes, which occur in every product's life cycle.

There are three major types of analytical documentation prepared to support investigational and marketed pharmaceutical products:

Regulatory—documents required for IND, NDA/BLA, and postapproval supplements.

Compliance—including GMP documents (e.g., validation reports, investigation reports, annual product reviews), internal reports (e.g., technology transfer reports), and special development reports (summarizing history of product or chronology of critical issues).

Research—usually prepared to record or transfer product history (e.g., analytical profiles), additional informational studies, non-GMP batch issues and investigations.

All three types of documents are necessary to provide a complete history of a product. The regulatory requirements are well established by FDA guidances and ICH guidelines. Similarly, basic compliance requirements are also described in regulations and guidelines. However, a firm's internal policies and procedures establish the requirements for research reports. Because of recent changes in reporting requirements for regulatory submissions, the research documents are becoming increasingly important to record the history and decisions within the life cycle of the product.

VIII. CONCLUSIONS

The principles for success in the pharmaceutical industry include:

- collaboration and interaction
- focus on quality
- balance of immediate needs and ultimate goals
- scientifically sound and compliant documentation
- sound quality systems for CGMP and GLP

These principles are especially critical to the analytical support group, either analytical research and development or quality control.

APPENDIX I. REFERENCES FOR CMC GUIDANCES

The following list contains relevant CMC guidances, draft and final, related to the regulatory and compliance documents required by FDA for US submissions and by ICH for international dossiers. Although some of the listed guidances are in draft form, they represent FDA's current thinking and are useful references. These guidances can all be found online and are located as part of the US Food and Drug Administration's Drug homepage, specifically, Guidance, Compliance, and Regulatory Information; Guidances (Drugs): http://www.fda.gov/Drugs/GuidanceComplianceRegulatoryInformation/Guidances/default.htm.

IND

Content and Format of Investigational New Drug Applications (INDs) for Phase 1 Studies of Drugs, Including Well-Characterized, Therapeutic, Biotechnology-derived Products, November 1995

INDs for Phases 2 and 3 Studies Chemistry, Manufacturing, and Controls Information, May 2003

IND Meetings for Human Drugs and Biologics Chemistry, Manufacturing, and Controls Information, May 2001

Investigators and Reviewers: Exploratory IND Studies, January 2006

cGMP for Phase 1 Investigational Drugs, July 2008

PET Drug Products—CGMP, September 2005 (DRAFT)

NDA

Analytical Procedures and Methods Validation, August 2000 (DRAFT)

Guideline for Submitting Supporting Documentation in Drug Applications for the Manufacture of Drug Substances, February 1987

Guideline for Submitting Supporting Documentation in Drug Applications for the Manufacture of Drug Products, February 1987

Guideline for Submitting Samples and Analytical Data for Methods Validation, February 1987

FDA's Policy Statement for the Development of New Stereoisomeric Drugs, 5/1/92

Reviewer Guidance Validation of Chromatographic Methods, November 1994

Metered Dose Inhaler (MDI) and Dry Powder Inhaler (DPI) Drug Products Chemistry, Manufacturing, and Controls Documentation, October 1998 (Draft)

Nasal Spray and Inhalation Solution, Suspension, and Spray Drug Products, May 1999 (Draft)

Postapproval

Format and Content of the CMC Section of an Annual Report, September 1994

Immediate Release Solid Oral Dosage Forms: Scale-Up and Postapproval Changes: Chemistry, Manufacturing, and Controls, *In Vitro* Dissolution Testing and *In Vivo* Bioequivalence Documentation, November 1995

Nonsterile Semisolid Dosage Forms: Scale-Up and Postapproval Changes: Chemistry, Manufacturing, and Controls; *In Vitro* Release Testing and *In Vivo* Bioequivalence Documentation, May 1997

Changes to an Approved Application for Specified Biotechnology and Specified Synthetic Biological Products, July 1997

SUPAC-MR: Modified Release Solid Oral Dosage Forms Scale-Up and Postapproval Changes: Chemistry, Manufacturing, and Controls; *In Vitro* Dissolution Testing and *In Vivo* Bioequivalence Documentation, September 1997

Changes to an Approved NDA or ANDA, April 2004

Compliance

Guideline on General Principles of Process Validation, May 1987

Manufacturing, Processing, or Holding Active Pharmaceutical Ingredients, March 1998 (Draft)

Investigating Out of Specifications (OOS) Test Results for Pharmaceutical Production, September 1998

Analytical Document Creation and Electronic Submission

1. FDA Guidance for Industry Part 11, Electronic Records; Electronic Signatures—Scope and Application, August 2003
2. FDA Guidance for Industry Providing Regulatory Submissions in Electronic Format—Human Pharmaceutical Product Applications and Related Submissions Using the eCTD Specifications, June 2008
3. The Common Technical Document For the Registration of Pharmaceuticals for Human Use: Quality—M4q(R1) Quality Overall Summary of Module 2 Module 3: Quality
4. ICH M2 EWG Electronic Common Technical Document Specification

ICH

Q1A: Stability Testing of New Drug Substances and Products, November 2003

Q1B: Photostability Testing of New Drug Substances and New Drug Products, November 1996

Q1C: Stability Testing for New Dosage Forms, May 1997

Q1D Bracketing and Matrixing Designs for Stability Testing of New Drug Substances and Products, January 2003

Q2A: Text on Validation of Analytical Procedures, March 1995

Q2B: Validation of Analytical Procedures: Methodology, May 1997

Q3A: Impurities in New Drug Substances, June 2008

Q3B: Impurities in New Drug Products, August 2006

Q3C: Impurities: Residual Solvents

Q5A: Viral Safety Evaluation of Biotechnology Products Derived from Cell Lines of Human or Animal Origin

Q5B: Quality of Biotechnological Products: Analysis of the Expression Construct in Cells Used for Production of r-DNA Derived Protein Products

Q5C: Quality of Biotechnological Products: Stability Testing of Biotechnological/Biological Products

Q5D: Derivation and Characterization of Cell Substrates Used for Production of Biotechnological/Biological Products

Q6A: Specifications: Test Procedures and Acceptance Criteria for New Drug Substances and New Drug Products: Chemical Substances

Q6B: Specifications: Test Procedures and Acceptance Criteria for Biotechnological/Biological Products

Q7A: Good Manufacturing Practice Guide for Active Pharmaceutical Ingredients (draft)

Q8(R1): Pharmaceutical Development

Q9: Quality Risk Management

Q10: Pharmaceutical Quality System

APPENDIX II. MODULE 3 CTD TABLE OF CONTENTS—QUALITY SECTION OF US NDA SUBMISSION

3.2.S		Drug Substance	Comments—Analytical Focus
	S.1	General Information	Contributing author to overview
	S.1.1	Nomenclature	
	S.1.2	Structural	
	S.1.3	General Properties	A list of physicochemical and other relevant properties of the drug substance.
	S.2	Manufacture	
	S.2.1	Manufacturer(s)	
	S.2.2	Description of Manufacturing Process and Process Controls	
	S.2.3	Controls of Materials	Information related to the quality control and analytical procedures of the materials (i.e., starting material, solvents, reagents, raw materials, etc.) to be used should be provided as appropriate.
	S.2.4	Control of Critical Steps and Intermediates	Tests and acceptance criteria of the manufacturing process.
	S.2.5	Process Validation and/or Evaluation	
	S.2.6	Manufacturing Process Development	
	S.3	Characterization	
	S.3.1	Elucidation of Structure and other Characteristics	Analytical proof of structure (and stereochemistry) including: Spectroscopic data Optical rotation X-ray diffraction Thermal analysis Elemental analysis Chemical (purity) data
	S.3.2	Impurities	Characterization of potential and observed impurities (by- and degradation products)
	S.4	Control of Drug Substance	
	S.4.1	Specification	Tests, limits and methods used for release and stability testing of finished drug substance

S.4.2	Analytical Procedures	Analytical procedures used for the drug substance.
S.4.3	Validation of Analytical Procedures	
S.4.4	Batch Analyses	Release results and impurity profiles for: Registration batches Supportive batches (toxicology, clinical and development)
S.4.5	Justification of Specifications	Choice of routine tests Development of methods Justification for proposed limits and omitted tests Qualification of specified impurities
S.5	Reference Standards or Materials	
S.6	Container Closure	Quality controls for packaging components and materials. Identity of materials of the primary packaging components and their specifications. Noncompendial methods (with validation) should be provided where appropriate.
S.7	Stability	
S.7.1	Stability Summary	Summary of primary and supportive data including forced degradation and stress studies. Potential and observed degradation pathway(s) and product(s).
S.7.2	Postapproval Stability Protocol and Stability Commitment	
S.7.3	Stability Data	Information on the analytical methods used and validation of the procedures. Description of stability indicating methods. Comment on differences (if any) versus NDA specifications and methods.

3.2.P	Drug Product	Comments—Analytical Focus
P.1	Description of the Drug Product	
P.2	Development Pharmaceutics	
P.2.1	Components of the Drug Product	
P.2.1.1	Drug Substance	Compatibility of the drug substance with the excipients. Key influencing characteristics which may influence the drug substance to be discussed.
P.2.2	Drug Product	
P.2.2.1	Formulation Development	Results of comparative *in vitro* studies (e.g., dissolution) discussed
P.2.2.2	Overages	
P.2.2.3	Physicochemical and Biological Properties	Influence of parameters such as pH, ionic strength, dissolution, redispersion, reconstitution, particle size distribution, aggregation, polymorphism, biological potency, or activity.
P.2.3	Manufacturing Process Development	
P.2.4	Container Closure System(s)	Full description
P.2.5	Microbiological Attributes	
P.2.6	Compatibility	Compatibility with reconstitution diluents, sorption on injection vessel
P.3	Manufacture	
P.3.1	Manufacturer(s)	
P.3.2	Batch Formula	
P.3.3	Description of Manufacturing Process and Process Controls	
P.3.4	Controls of Critical Steps and Intermediates	
P.3.5	Process Validation and/or Evaluation	

P.4	Control of Excipients	
P.4.1	Specifications	Reference for compendial items Additional requirements for compendial items, if any Specifications and methods for noncompendial items
P.4.2	Analytical Procedures	Identify use of noncompendial procedure
P.4.3	Validation of Analytical Procedures	Validation of noncompendial procedure
P.4.4	Justification of Specifications	
P.4.5	Excipients of Human or Animal Origin	
P.4.6	Novel Excipients	
P.5	Control of Drug Product	
P.5.1	Specifications	Tests, limits, and methods used for release and stability testing of drug product
P.5.2	Analytical Procedures	Description of analytical test procedures provided
P.5.3	Validation of Analytical Procedures	Analytical validation information for the analytical test procedures
P.5.4	Batch Analyses	Results, impurity profiles and dissolution profiles, if applicable, for: Registration batches Supportive batches (clinical and development)
P.5.5	Characterization of Impurities	Information on the characterization of impurities if not previously provided in 3.2. S.3.2, Impurities
P.5.6	Justification of Specifications	
P.6	Reference Standards or Materials	
P.7	Container Closure System	Bulk and market packages: Quality controls for packaging components and materials Extractables/compatibility data, if needed

	P.8	Stability	
	P.8.1	Stability Summary and Conclusion	
	P.8.2	Postapproval Stability Protocol and Stability Commitment	
	P.8.3	Stability Data	Information on the analytical procedures used to generate the data and validation of these procedures if different than the procedures used for release testing. Comment on differences (if any) versus NDA specifications and methods
3.2.	A	Appendices	
	A.1	Facilities and Equipment	
	A.2	Adventitious Agents Safety Evaluation	
	A.3	Novel Excipients	
3.2.	R	Regional information	Executed batch record, methods validation package (e.g., list of reference standards and exhibit samples of drug substance and product to be provided to FDA laboratory) and comparability protocols
3.3		Literature References	
		Rationale for Specifications	Choice of routine tests Development of methods Justification for proposed limits and omitted tests Qualification of specified impurities

GLOSSARY

API Active pharmaceutical ingredient
BA Bioavailability
BCS Biopharmaceutical classification system
BE Bioequivalence
BLA Biological license application

CGMP Current good manufacturing practices
CMC Chemistry, manufacturing, and controls
COA Certificate of analysis
CPP Critical processing parameters
CQA Critical quality attributes
CRO Contract Research Organizations
CTA Clinical trial authorization
CTD Common technical document
DMPK Drug metabolism and pharmacokinetics
DRA Drug Regulatory Affairs
DSC Differential scanning calorimetry
EIH Entry into human
ELN Electronic laboratory notebook
EOP2 End of phase 2
EU European Union
FDA Food and Drug Administration
FDDP Full development decision point
GC Gas chromatography
GLP Good laboratory practices
GMP Good manufacturing practices
HPLC High-performance liquid chromatography
ICH International Conference on Harmonization
IMPD Investigational Medicinal Product Dossier
IND Investigational new drug
LIMS Laboratory information management system
NDA New drug application
OOS Out of specification
PAI Preapproval inspection
PAT Process analytical technology
PDUFA Prescription drug user fee act
QA Quality assurance
QbD Quality by design
QC Quality control
QOS Quality overall summary
R&D Research and development
sNDA Supplemental NDA
TGA Thermogravimetry analysis
TLC Thin layer chromatography
XRD X-ray diffraction

INDEX

Note: Page numbers followed by f and t indicate figures and tables, respectively.

C

D

G

H

I

J

K

L

N

O

S

T

U

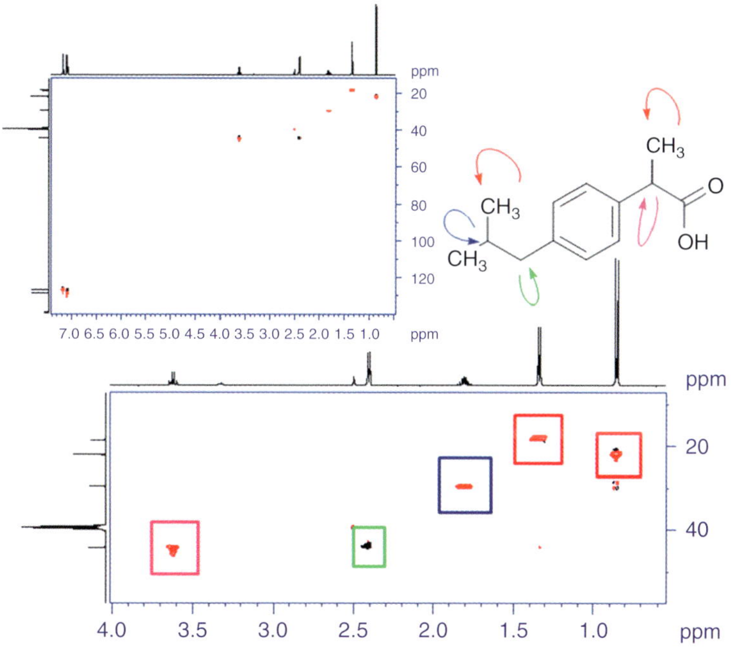

K. M. ALSANTE et al., FIGURE 32 Multiplicity edited HSQC spectrum of ibuprofen. The boxes are color-matched to the arrows indicating the ${}^{1}J_{CH}$ heteronuclear scalar coupling.

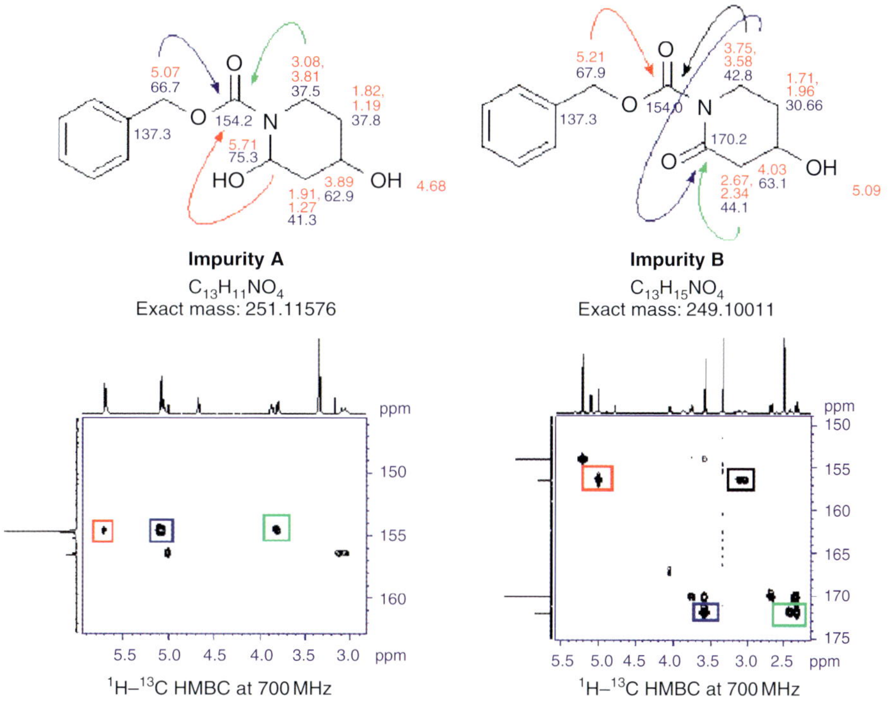

K. M. ALSANTE et al., FIGURE 48 The proposed structures and chemical shift assignments for the two elucidated impurities. The arrows on the structures indicate the key HMBC correlations observed for these molecules. Below are two expansions of the HMBC data for each of the impurities, respectively, from left to right. The colors are meant to quickly indicate the actual correlation in the 2D NMR plot of the HMBC data.

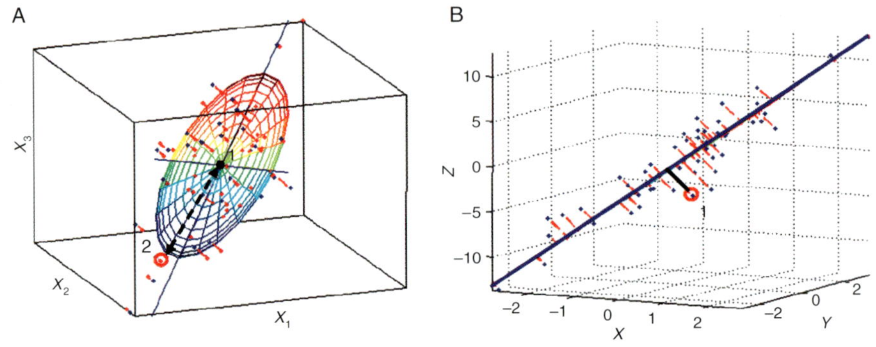

R. W. BONDI AND J. K. DRENNEN, FIGURE 4 **(A) Representation of Hotelling's T^2. Point 1 represents the center of the model plane while point 2 represents an individual sample. Hotelling's T^2 for the sample represented by point 2 is given by its distance to the center of the model, within the model's plane, which is represented by the broken line. (B) Representation of Q residual and SPE. The Q residual for the sample marked as point 1 is defined as the squared perpendicular distance from point 1 to the plane of the model (blue solid line) which is represented by the solid black line. The SPE would be determined by summing the Q residual for all samples shown in the figure.**

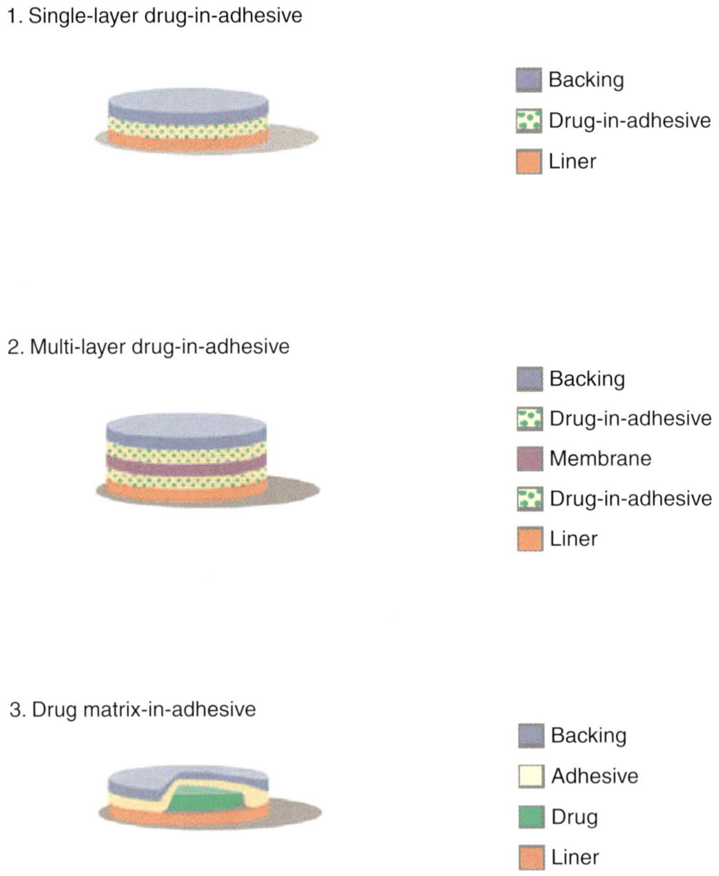

J. ETSE, FIGURE 2 **Schematic of different types of transdermal systems. Reprinted from Ref. 29.**

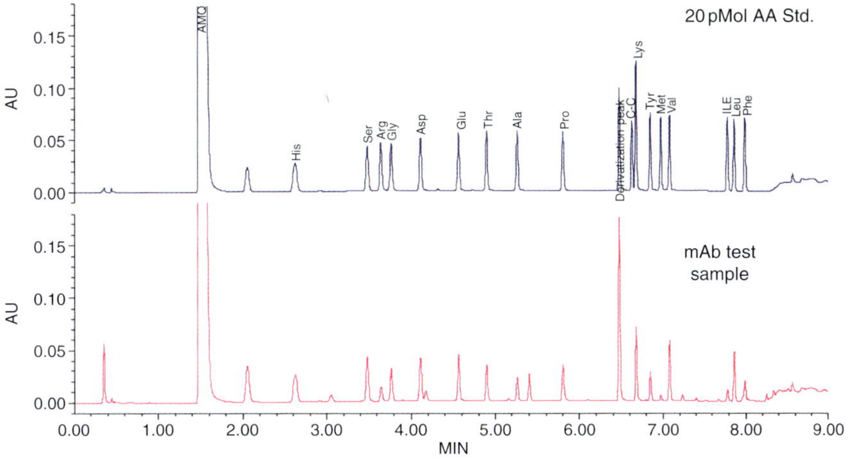

Y. LUO et al., FIGURE 9 **Amino acid analysis of a monoclonal antibody (mAb) using liquid-phase acid hydrolysis (6 M HCl), followed by AccQ-tag labeling and UHPLC separation. The top panel shows a chromatogram of amino acid standards.**

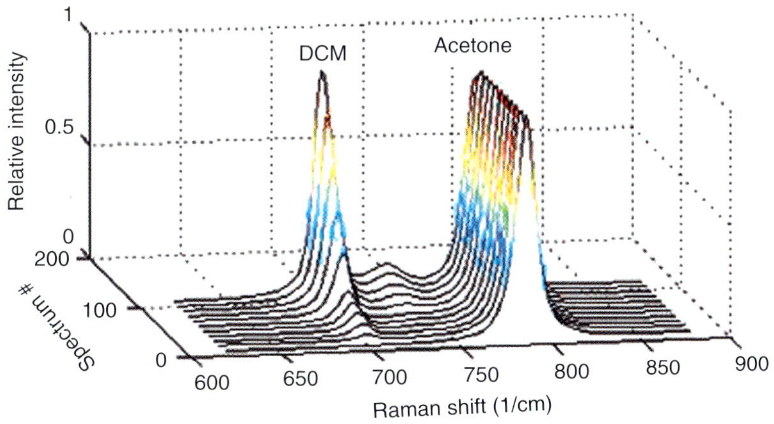

C. WOOD, FIGURE 9 **Waterfall off-line plot Raman spectra of DCM from 0% to 30% v/v standards in acetone.**

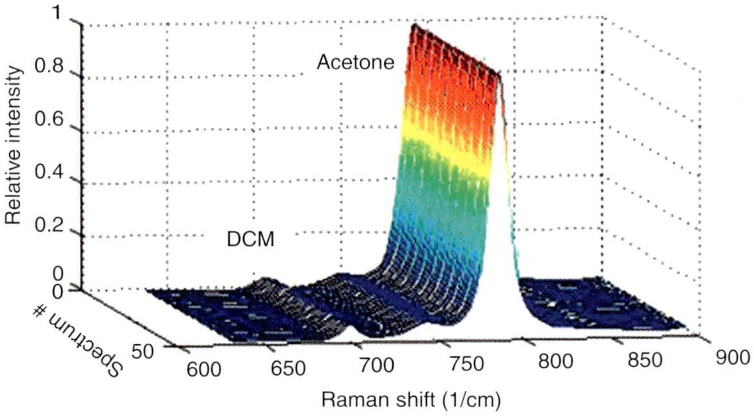

C. WOOD, FIGURE 10 **Waterfall plot off-line Raman spectra of reaction plot samples after solvent exchange.**